PERIODIC TABLE OF THE ELEM

The relative atomic weights are based on the assigned relative atomic mass of $^{12}C = 12$ (inside rear cover). In several cases the atomic weight is rounded from four or five significant figures.

Color key:
- gray symbol denotes gaseous elements
- tint, liquid element
- red, solid element

	IB	IIB	IIIA	IVA	VA	VIA	VIIA	VIIIA
								2 $1s^2$ He 4.0026
			5 (He) $2s^2 2p^1$ B 10.81	6 (He) $2s^2 2p^2$ C 12.011	7 (He) $2s^2 2p^3$ N 14.007	8 (He) $2s^2 2p^4$ O 15.999	9 (He) $2s^2 2p^5$ F 18.998	10 (He) $2s^2 2p^6$ Ne 20.179
			13 (Ne) $3s^2 3p^1$ Al 26.98	14 (Ne) $3s^2 3p^2$ Si 28.09	15 (Ne) $3s^2 3p^3$ P 30.974	16 (Ne) $3s^2 3p^4$ S 32.06	17 (Ne) $3s^2 3p^5$ Cl 35.453	18 (Ne) $3s^2 3p^6$ Ar 39.948
28 (Ar) $3d^8 4s^2$ Ni 58.71	29 (Ar) $3d^{10} 4s^1$ Cu 63.54	30 (Ar) $3d^{10} 4s^2$ Zn 65.37	31 (Ar) $3d^{10} 4s^2 4p^1$ Ga 69.72	32 (Ar) $3d^{10} 4s^2 4p^2$ Ge 72.59	33 (Ar) $3d^{10} 4s^2 4p^3$ As 74.92	34 (Ar) $3d^{10} 4s^2 4p^4$ Se 78.96	35 (Ar) $3d^{10} 4s^2 4p^5$ Br 79.904	36 (Ar) $3d^{10} 4s^2 4p^6$ Kr 83.80
46 (Kr) $4d^{10}$ Pd 106.4	47 (Kr) $4d^{10} 5s^1$ Ag 107.87	48 (Kr) $4d^{10} 5s^2$ Cd 112.40	49 (Kr) $4d^{10} 5s^2 5p^1$ In 114.82	50 (Kr) $4d^{10} 5s^2 5p^2$ Sn 118.69	51 (Kr) $4d^{10} 5s^2 5p^3$ Sb 121.75	52 (Kr) $4d^{10} 5s^2 5p^4$ Te 127.60	53 (Kr) $4d^{10} 5s^2 5p^5$ I 126.90	54 (Kr) $4d^{10} 5s^2 5p^6$ Xe 131.30
78 (Xe) $4f^{14} 5d^9 6s^1$ Pt 195.09	79 (Xe) $4f^{14} 5d^{10} 6s^1$ Au 196.97	80 (Xe) $4f^{14} 5d^{10} 6s^2$ Hg 200.59	81 (Xe) $4f^{14} 5d^{10} 6s^2 6p^1$ Tl 204.37	82 (Xe) $4f^{14} 5d^{10} 6s^2 6p^2$ Pb 207.19	83 (Xe) $4f^{14} 5d^{10} 6s^2 6p^3$ Bi 208.98	84 (Xe) $4f^{14} 5d^{10} 6s^2 6p^4$ Po (209)[a]	85 (Xe) $4f^{14} 5d^{10} 6s^2 6p^5$ At (210)	86 (Xe) $4f^{14} 5d^{10} 6s^2 6p^6$ Rn (222)

Russian scientists who reported the discovery of the element 104; Rf (rutherfordium) by an American for 104.

[c]American scientists proposed the name hahnium, Ha, for element 105.

64 (Xe) $4f^7 5d^1 6s^2$ Gd 157.25	65 (Xe) $4f^9 6s^2$ Tb 158.93	66 (Xe) $4f^{10} 6s^2$ Dy 162.50	67 (Xe) $4f^{11} 6s^2$ Ho 164.93	68 (Xe) $4f^{12} 6s^2$ Er 167.26	69 (Xe) $4f^{13} 6s^2$ Tm 168.93	70 (Xe) $4f^{14} 6s^2$ Yb 173.04	71 (Xe) $4f^{14} 5d^1 6s^2$ Lu 174.97
96 (Rn) $5f^7 6d^1 7s^2$ Cm (245)	97 (Rn) $5f^8 6d^1 7s^2$ Bk (247)	98 (Rn) $5f^{10} 7s^2$ Cf (249)	99 (Rn) $5f^{11} 7s^2$ Es (254)	100 (Rn) $5f^{12} 7s^2$ Fm (255)	101 (Rn) $5f^{13} 7s^2$ Md (256)	102 (Rn) $5f^{14} 7s^2$ No (254)	103 (Rn) $5f^{14} 6d^1 7s^2$ Lr (257)

COLLEGE PHYSICAL SCIENCE

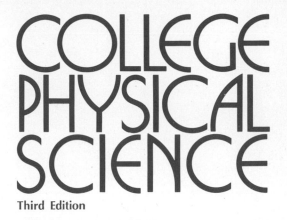

COLLEGE PHYSICAL SCIENCE

Third Edition

VADEN W. MILES
Professor Emeritus
Wayne State University

HENRY O. HOOPER
University of Maine

EMIL J. KACZOR
Wayne State University

WILLARD H. PARSONS
Wayne State University

HARPER & ROW, PUBLISHERS
New York, Evanston, San Francisco, London

Sponsoring Editor: John A. Woods
Project Editor: Lois Wernick
Designer: June Negrycz
Production Supervisor: Stefania J. Taflinska

College
Physical
Science, Third Edition

Library of Congress Cataloging in Publication Data
Miles, Vaden Willis, 1911-
College physical science.
Includes bibliographies.
1. Science I. Title.
Q160.2.M54 1974 500.2 73-13200
ISBN 0-06-044443-6

CONTENTS

PREFACE

College Physical Science was written for a cultural or liberal education in the physical sciences of nonscience majors in liberal arts, business, education, journalism, law, mass communications, social work, and music. In order to understand the physical world, the nonscientist needs a course *in* science, not a course *about* science. The book includes a selection of subject matter from the fields of physics, including astronomy and meteorology, chemistry, and geology.

The essential concepts of physical science are presented simply and clearly, often in their historical context. Emphasis is placed on an understanding of the fundamental principles of science and on procedures and methods in science rather than on memorization of isolated information. Facts are important, but they are of less permanent use and value to the student than the generalizations and theories developed from the facts. This book is designed to encourage a depth of learning that comes from the student's own reasoning and creative thinking. Each subject is approached as if the student were a participant rather than a spectator.

The presentation begins with astronomy—early man observed the heavens and tried to explain what he observed. Astronomy is especially intriguing to students, and the elementary concepts are relatively nonmathematical. The sequence of topics thereafter follows naturally from this subject and is arranged to emphasize the unity of the physical sciences.

One of the characteristic aspects of science is change, the very nature of science providing an inherent procedure for self-alteration. Because this is so, an account that omitted historical development entirely would not be an adequate presentation. Thus, several important concepts are presented in a historical matrix, but the history of the development of ideas is never substituted for the ideas themselves.

We have tried to achieve a level of presentation appropriate to college freshmen and sophomores. Among the nonscience majors in most colleges and universities, less than 20 percent have had high school physics, about one-half to two-thirds have had high school chemistry, and very few have studied astronomy, meteorology, or geology. In our presentation we have assumed no particular preparation in high school science and no preparation in mathematics beyond elementary high school algebra. The mathematical discussions and examples were designed to supplement the basic exposition. A mathematical appendix is provided for the student who needs a quick review, or study for the first time, of fundamental processes, chiefly in arithmetic and algebra.

An effort has been made to discuss physical principles in such a way that the mathematical examples could be omitted if an instructor so desired. To this end, numerous nonmathematical questions have been added in this edition so that students can test their understanding of concepts and principles without the use of mathematical manipulations.

Several additional aids to learning appear throughout the book. Chapters involving problems contain both examples solved in detail within the chapter and additional problems at the end. For the latter, answers are given at the end of the odd-numbered problems. Even-numbered problems are usually similar to the preceding problem. Other aids appearing at the ends of chapters include summaries, lists of important words and terms, questions and exercises, projects and laboratory-type experiences (of particular value in those courses without a laboratory), and annotated lists of supplementary readings. An *Instructor's Manual* includes answers to the questions, exercises, and problems. To provide additional aid to students, a *Study Guide* has been prepared to supplement this text. It includes numerous problems, questions with solutions, and answers.

As previously noted, the ends of some chapters provide laboratory types of experiences to supplement the many lecture–demonstration experiments included in the book. It is hoped, however, that a weekly or biweekly laboratory visit, especially designed for the student in physical science, will accompany the use of this book so that the student will have the opportunity to be a scientist for a day and will thus come to appreciate that progress in physical science is based on observation and experimentation. If at all possible, visits to planetariums and/or museums should be arranged, and for the study of geology, there is no fully satisfactory substitute for an organized field trip.

The authors have, for a number of years, been active in developing, teaching, and administrating a one-year course in physical science for the nonscience major. We have applied the findings from these years of experience to the preparation of this book. Many rough spots have been smoothed out in the present edition by using the suggestions of several generations of university and college students and their instructors, who have used the previous editions of the book. The two previous editions of *College Physical Science* have passed the test of time with rigorous evaluations by scientists, educators, and students, who have used these editions in several hundred universities and colleges, both four-year and community colleges.

We owe special thanks to the various chairmen, nationwide, of the science departments and of the departments of physics and astronomy, chemistry, and geology in large institutions who have encouraged the course. As a result, physical science courses have been fostered, improved, and made available to the two-thirds of our college population, who are nonscience majors. In preparation for writing this edition of *College Physical Science,* one author spent several months traveling through continental United States, visiting many colleges and universities of all types in order to study their physical science courses for the nonscience major. At these institutions, students, professors, department chairmen, deans, and presidents were interviewed extensively. An earlier edition of *College Physical Science* was in use at many of these colleges and universities. The author concluded that its success was the result of widespread satisfaction with its organization and content, and the fact that students found it easy to read and understand. The general organization and much of the content of the first two editions have therefore been retained. Like those editions, this book begins with a judicious selection of material from the fields of physics, including astronomy and some meteorology, continues with chemistry, and concludes with physical geology.

New sections covering recent developments in physical science that are important for the education of the citizen as we approach the twenty-first century have been added to this edition. In the selection and treatment of new topics, the authors have been guided by the suggestions made by many instructors of physical science across the United States. The latest available information has been added to update the subjects that had been previously treated in the second edition.

The most apparent change in the section devoted to astronomy and physics in this edition is a rearrangement of the order in which the material is presented. The first three chapters are devoted to early astronomical observations, the motions of the earth, and a discussion of the solar system. Chapters 4 through 10 are devoted to various areas of physics: mechanics, gravitation, fluids, heat and meteorology, wave motion, electricity, and magnetism. In Chapter 11, stellar astronomy and the origin of the universe are examined, employing the principles of physics and astronomy developed in the earlier chapters. Chapters 12 and 13 provide a natural transition from physics and astronomy into chemistry. Here, in an essentially historical development, the concepts of the atom and its nucleus and some aspects of modern atomic theory are developed and applied to the building up of the periodic table. Chapters 14 through 18 are devoted to topics in chemistry. These chapters on chemistry have been completely rewritten and reorganized to provide an orderly presentation of content from the field of chemistry.

Chapter 19 is a new chapter devoted to energy and its sources available to man on earth. An attempt to put the energy crisis in perspective is made in this chapter. Prospects for the use of solar and geothermal energy are discussed; nuclear fusion and fission processes are also discussed in some detail in this regard.

The last section of the book deals with earth science and geology. Chapter 20 has been expanded to include a description of the total earth including the lithosphere and atmosphere, earth magnetism, geologic processes, and environmental problems associated with geology. In Chapter 21 the materials of the earth are discussed in terms of the chemical combinations found on earth to form crystals, glasses, minerals, and rocks. The remaining chapters (excluding Chapter 28) have not been changed greatly in the overall coverage of various aspects of geology except for an expansion of work on sea-floor spreading, plate tectonics, and mountain building found in Chapter 27. Chapter 28 is completely new and is a discussion of the moon and its geology. The explorations of the moon by the Apollo astronauts are described, and the results of the scientific analysis of the explorations and the samples of lunar material returned to earth are discussed. The questions of age, evolution, and origin of the moon are discussed in light of the new data obtained from man's exploration of the moon's surface.

A book such as this could not have been completed without help from many individuals. To those persons, too numerous to mention, who contributed to our ideas and to the preparation of the manuscript, the authors are gratefully appreciative. Many of our colleagues in the departments of physics and astronomy, chemistry, and geology have through the years offered valuable suggestions. Acknowledgement of permission to use copyrighted material and photographs has been made at appropriate points throughout the book. Finally, we should like to thank Dr. Glenn T. Seaborg for his statement immediately preceding the opening page of the text.

We hope that the use of this book will motivate the student and also give him the background necessary to maintain, throughout his life, a continuing acquaintance with the progress and accomplishments of science.

V. W. M.
H. O. H.
E. J. K.
W. H. P.

COLLEGE PHYSICAL SCIENCE

There simply must be a greater degree of scientific literacy among the general public. The problems that are posed by science enter into the political framework—into the framework of the whole society in many ways. Our economic future has become geared to science. The entire population in a democracy, if a democracy is going to survive, has to learn more about science—even some of the fundamental principles of science.

Glenn T. Seaborg, Former Chairman

UNITED STATES ATOMIC ENERGY COMMISSION

THE SOLAR SYSTEM

Many modern ideas concerning the universe, notably those originating in the Copernican revolution, began not with new facts but from looking at old facts in a new way.

G. J. WHITROW, 1959

Early man must have sensed that the sun was closer to him than most of the thousands of heavenly bodies that he was able to see. He could have reached this conclusion from observing the sun's greater size and the tremendous amount of light and heat it radiated. He must have noticed, too, that the moon was large and varied in appearance from a very thin crescent to a fully lighted disk. Further, he must have seen that the moon rose later each night by about 50 minutes. Early man may also have reasoned that the moon was closer to him than the stars.

How do we know which of the myriads of brightly shining objects that we see on a clear night are relatively close to us?

In this chapter we shall discuss the constellations; the sun, the moon, and the planets in terms of things that any good theory of the universe must explain and how these things are explained by an earth-centered concept of the universe and by a sun-centered concept; and finally the discovery of planets, asteroids, satellites, comets, and meteors.

Early observations

On a clear night go out and look up at the sky. You will observe what the ancient astronomers saw. If you were to make nightly observations of the positions and changes in the positions of the jewellike objects in the sky, you would "rediscover" the work of the ancient astronomers. You could classify and attempt to order your observations as was done in ancient times. The physical sciences began in the science of astronomy. Therefore it appears appropriate to begin a study of the physical sciences by examining some observations of the motions of the "objects" in "our" sky and outline how men attempted to develop models with which to explain the natural motions of the numerous objects in the sky.

THE CONSTELLATIONS

The ancients noticed that all of the jewellike objects as they appeared to cross the sky from east to west during the night kept their same relative positions. Most of these objects are *stars,* which we now define as those objects in the sky that can be seen because they give off their own light. The shape of one group of stars suggested a lion (Leo); of another, a large bear (Ursa Major); of still another, a small bear (Ursa Minor); and similarly, the shapes of other groups of stars suggested other familiar objects. A group of stars in a limited region of the sky set off by arbitrary boundary lines is called a *constellation.*

Astronomers now classify the stars into 88 constellations. Figures 1.1, 1.2, and 1.3 show some of the more important of these groupings.

The student should become familiar with some of the better-known groups such as the Big Dipper (a part of the large constellation Ursa Major); the Little Dipper (a part of Ursa Minor) with the North Star, or Polaris, at the end of its handle; the W-shaped figure in Cassiopeia; the principal stars of Orion; Sirius and neighboring stars in Canis Major; the somewhat reddish star Aldebaran in Taurus and the nearby group called the Pleiades; and a few others. The Dippers and Cassiopeia are visible throughout the year for those living in the mid-northern latitudes.

It should be noted that if we were to group stars into modern constellations, we would no doubt arrive at a much different set of groupings of the same stars; that is, the same stars would create very different patterns for us because our perspective and outlook is much more "modern." An example of a possible "new constellation" is given in Fig. 1.4. The important observation is that the relative positions of the stars in a constellation have remained practically unchanged as observed by the unaided eye since the ancients classi-

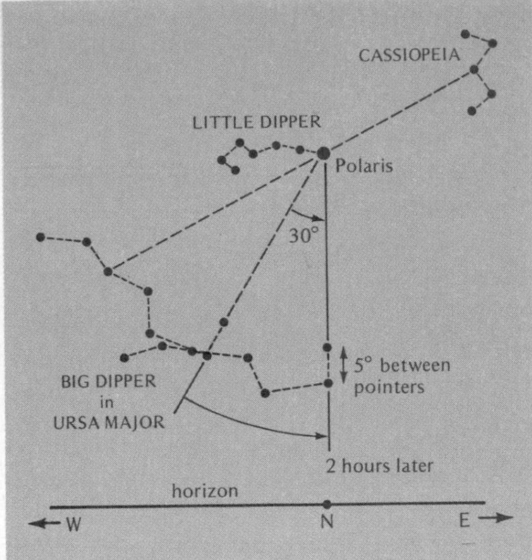

Figure 1.1
North circumpolar constellations. The Big Dipper, a part of the constellation Ursa Major; the North Star, or Polaris; the W of Cassiopeia; and the Little Dipper, a part of the constellation Ursa Minor, as seen from the middle northern latitudes on September 1 at 11 P.M. or October 1 at 9 P.M. or November 1 at 7 P.M. The position of the Big Dipper 2 hours later is shown in color; the latter positions of the Little Dipper and of Cassiopeia are not shown.

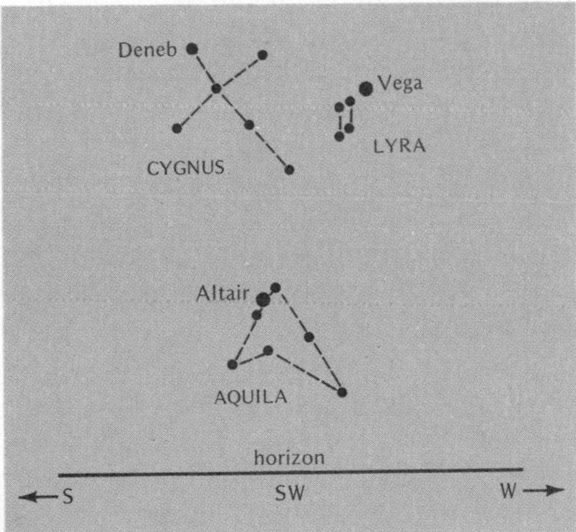

Figure 1.2
Summer constellations. Bright stars—Deneb in Cygnus, the swan (Northern Cross); Vega in Lyra, the harp; and Altair in Aquila, the eagle—as seen from middle northern latitudes on September 1 at 11 P.M. or October 1 at 9 P.M. or November 1 at 7 P.M.

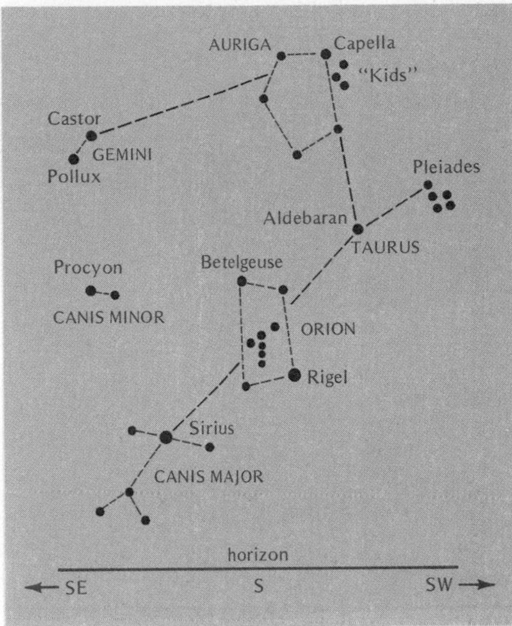

Figure 1.3
Winter constellations. Orion, including the bright stars Betelgeuse and Rigel; Canis Major, including Sirius, which is the brightest star in the heavens; Taurus, including the bright star Aldebaran and the well-known group known as the Pleiades; Canis Minor, with the bright star Procyon; Gemini, including the twin stars Pollux and Castor; and Auriga, including Capella, the goat, with the "kids," as seen from middle northern latitudes on January 1 at 11 P.M. or February 1 at 9 P.M. or March 1 at 7 P.M. A sword hangs from the belt of Orion, the hunter.

fied the stars up until the present time. The fact that the star groupings remain unchanged was noticed by the ancients and led to the expression "the fixed stars." Apparently the stars do not move relative to one another or else they are at such great distances from the earth that changes in their positions are not noticeable. The nearest star beyond our sun is Alpha Centauri and it is more than 20 million million miles away. Alpha Centauri is so far away that its light, traveling slightly more than 186,000 mi/sec, takes over 4 years to reach us. The sun, our

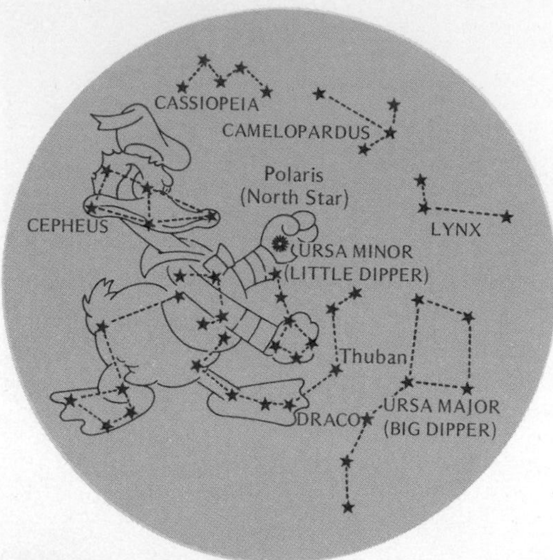

Figure 1.4
Donald Duck, a new constellation.

own daytime star, is, of course, extremely important to us so we shall examine in some detail the motion of the sun relative to the "fixed stars."

THE PLANETS

While classifying groups of stars into constellations, the early peoples noted that there were five bright objects that moved from one group of stars to another. This movement is detected only after the positions of the stars are recorded each night at a given time for several days or weeks. These five bright objects were called *planets,* from the Greek word *planētēs* meaning "a wanderer," to distinguish them from the fixed stars. This motion of the planets among the stars consists mainly in a slow eastward drift. This means that after a period of a few weeks a planet's position relative to the background of fixed stars, some of which are grouped into constellations, will be found east of the original position of sighting. Occasionally, however, a planet will appear to slow in its eastward motion relative to the fixed stars, stop, and then move westward among the stars for a short time, stop, and resume its longer, slow eastward motion.

This motion of a planet is illustrated for the planet Venus in Fig. 1.5 where the actual positions of Venus among the stars are shown from March 1, 1972, to October 1, 1972. The position of Venus was determined each evening over this period and "mapped" with respect to the fixed stars in Fig. 1.5. Near May 1, Venus is in the constellation Taurus; near June 1, in the constellation Gemini. Venus moves rapidly eastward from March to May when the planet appears to reduce its speed until about June 1. Then Venus appears to reverse its direction and move westward (retrograde motion) until on July 1, Venus returns to the constellation Taurus. The planet then resumes its slow eastward drift from July 1 through October 1. This reversal of the direction in which the planet is moving is called *retrograde motion*.

Planets also differ from true stars in that they do not seem to twinkle so much. We know now that this difference between stars and planets can be accounted for by the fact that planets (like our moon) are observed by reflected light only. Galileo observed that planets differ from stars in other ways (p.13).

Many of the early philosophers correctly interpreted the apparent motion of the five planets visible to the unaided eye as due to their relative proximity to the earth and appreciable motion. They were named after the Roman gods: Mercury, Venus, Mars, Jupiter, and Saturn.

The great interest to early peoples in the motions of the moon, sun, and the planets is expressed by the names given the days of the week; Sunday the Sun's day, Monday the Moon's day, and Saturday Saturn's day. From the Romance languages it can be seen that the other days of the week are similarly named for the remaining four planets.

OTHER MOTIONS

There are a number of observations, some very obvious to us, concerning the motions

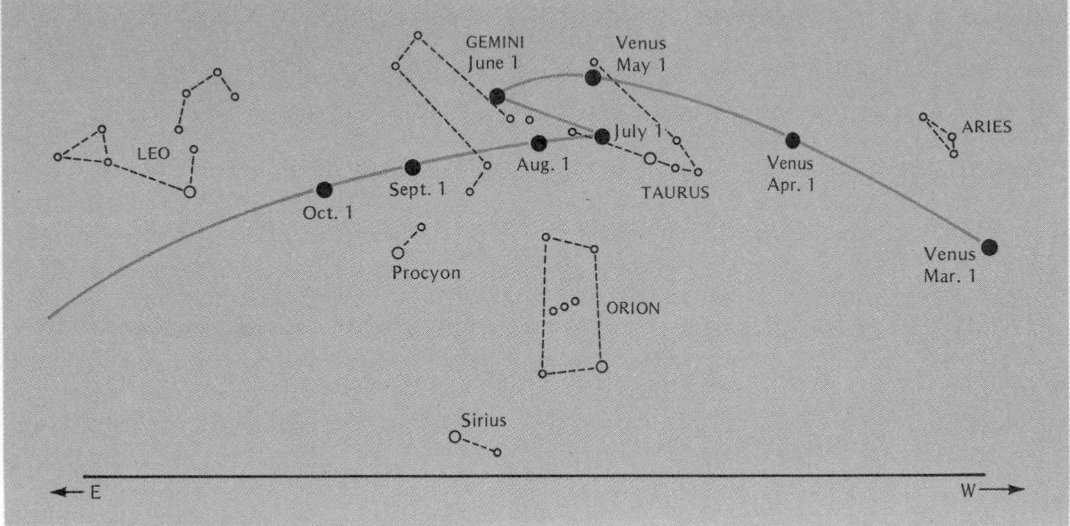

Figure 1.5

Apparent path of Venus for most of 1972. Retrograde (westward) motion is shown from June 1 to July 1, 1972. The open black circles represent stars.

of the sun, moon, planets, and stars which can be made with the unaided eye. We shall describe some of these because they are all observations that must be explained or accounted for in the construction of a theory or model of the universe.

The sun, the moon, the planets, and the stars all rise daily (or nightly) in the eastern sky and set in the western sky. If, for example, one watches the stars for several hours on a given evening, he will note that the positions of the stars change. For observers in the middle northern latitudes those stars and constellations that are fairly close to Polaris, the North Star, seem to circle counterclockwise each night around a point in the sky near Polaris. This phenomenon is illustrated in Fig. 1.1 in which the position of the Big Dipper is shown at two different times separated by 2 hours on the same evening. Figure 1.6 shows the northern circumpolar stars in a time-exposed photograph. In this photograph the star trails are about 45° of arc resulting from a time exposure of about 3 hours. This time can be determined from the fact that in 24 hours there would be an apparent rotation of 360° which results in 15° each hour. If you observed the positions of those stars that are more directly overhead during a given night, you would notice that they rise in the east and set in the west in a manner similar to the nightly motion of the planets and the moon as well as the daily motion of the sun.

In addition to the observations that the sun, moon, and planets seem to rise in the east and set in the west once daily (a 24-hour period), one observes that these objects all move through the same rather restricted region of the sky. The paths of the sun, moon, and planets in the sky change within the region over a full year.

A less obvious observation, but still one that you can make, concerns the position of the sun (and planets) among the fixed stars during the course of a year resulting in a so-called yearly motion of the sun among the stars. If you were to get up each morning just before daybreak and see which constellation is on the eastern horizon just before the sun rises, you would notice that there is a continuous change. In fact during the 12 months of the year, you would find the sun rises into a continually changing background of stars.

Figure 1.6
Star trails, or diurnal arcs, of the northern circumpolar stars. The bright arc near the center of motion is that of Polaris, which is about 1° from the pole. The north celestial pole is established by the common center of these diurnal arcs. The Carillon Tower of Wellesley College is in the foreground. (John C. Duncan.)

The ancients divided this background into 12 equal parts and called each part a sign of the zodiac. These signs were named for the 12 constellations in which the sun resided about 2000 years ago when the zodiac was first defined. At that time the sun was in the constellation Aries in the month of March; that is, one would observe the constellation Aries on the eastern horizon just before sunrise in March. Similar observations can be made for each planet. One finds that the planets reside within different constellations of the zodiac at different times of the year. It is important to note that the sun and the planets are always in that rather narrow band of the sky in which the constellations of the zodiac reside. From an early time the positions of the planets in the zodiac were related to astrological superstitions. It is a bit ironic that the science of astronomy owes much to the ancient astrologers for keeping accurate records of the positions of the planets over many years.

In addition to the daily and yearly motions of the sun and the planets with respect to the fixed stars, the ancients observed, as we can, the fact that the moon rose in the east and set in the west, changed its apparent shape during each month, was not visible during certain times of the month, and in addition moved rapidly eastward (about thirteen times more rapidly than the sun) among the stars when observed throughout the year.

Theories of the Universe

RELATION OF SUN, MOON, AND PLANETS

The actual relation of the earth, the sun, the moon, and the planets to one another greatly interested the Greek philosophers. Most of them followed the lead of Aristotle in assuming a fixed earth. About 125 B.C. Hipparchus set up a hypothesis, which Ptolemy, the Greco-Egyptian astronomer, geographer, and mathematician, elaborated and publicized about A.D. 140, to account for the motions of heavenly bodies. Both men assumed a fixed earth. There are certain things that any good theory of the universe must explain, and Ptolemy had to account for (1) the fast apparent eastward motion of the moon among the stars (about 13°/day); (2) the slower apparent eastward motion of the sun among the stars (about 1°/day); (3) the slow, differing apparent eastward motion of each planet among the stars with occasional short westward, or retrograde, motion; and (4) the apparent daily rising in the east and setting in

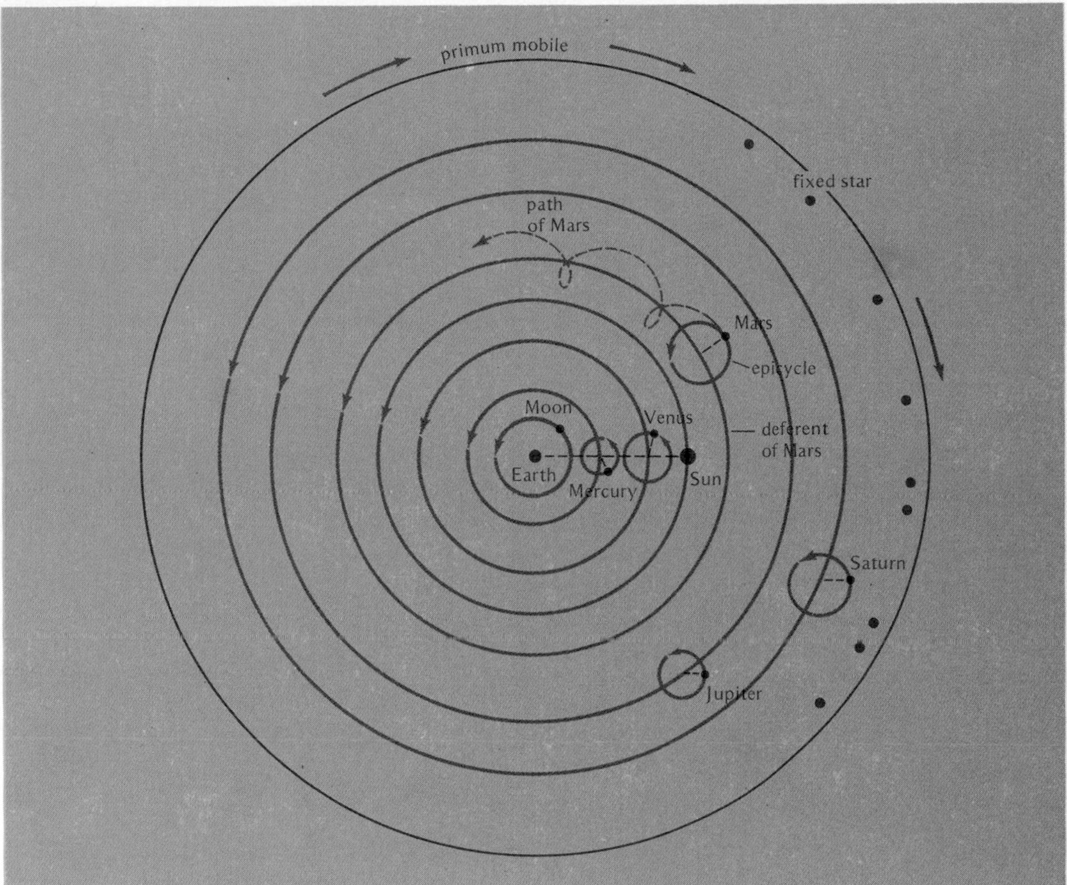

Figure 1.7
Ptolemy's geocentric concept of the universe.

the west of the sun, the moon, the planets, and the stars; and (5) later it was realized that a good theory also had to predict accurately future positions of the heavenly bodies.

THE PTOLEMAIC HYPOTHESIS

Figure 1.7 shows Ptolemy's *geocentric,* or earth-centered, concept of the universe. From the relative eastward speed of the moon, the sun, and the planets among the stars, Ptolemy in about A.D. 140 postulated that the moon moved in an orbit (path) nearest the earth, Mercury in an orbit next farthest out, then Venus, then the sun, and then Mars, Jupiter, and Saturn. Ptolemy accounted for the long eastward motion of these heavenly bodies with respect to the stars by postulating counterclockwise motion along concentric circles called *deferents.* He ingeniously explained the loop formed by a planet when it moves westward for a short time (Fig. 1.5) by assuming that the planets move counterclockwise on smaller circles called *epicycles* as the centers of the epicycles move along the deferents (Figs. 1.7 and 1.8). Thus by a proper choice of size and position of deferents and epicycles, Ptolemy

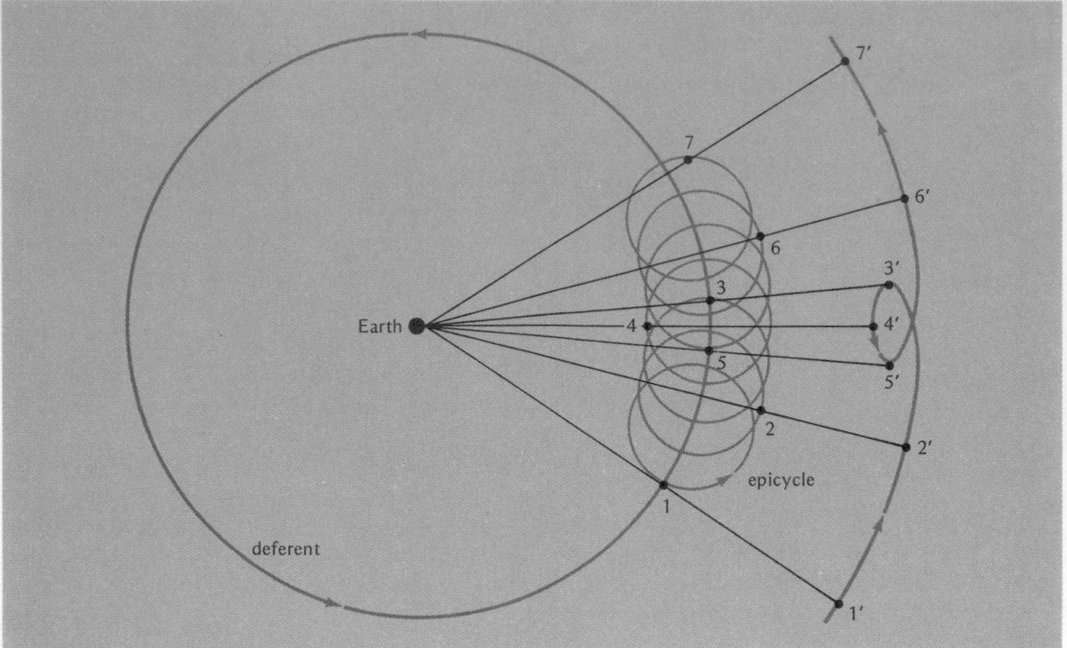

Figure 1.8

Ptolemy's explanation of the looped path of a planet. As both the epicycle as a whole and the planet on the epicycle move forward, the planet makes its usual eastward motion 1′2′3′. Then the backward motion of the planet on the epicycle more than offsets the forward motion of the epicycle as a whole, resulting in the short retrograde (westward) motion 3′4′5′. Then the two effects are again in the same direction, producing the long eastward motion 5′6′7′.

was able to account for the observed apparent paths of the moon, the sun, and the planets, as indicated in items (1), (2), and (3) above.

To account for the apparent daily rising in the east and setting in the west of the sun, the moon, the planets, and the stars, item (4), Ptolemy incorporated into his model a set of concentric crystal spheres which had been proposed earlier by Aristotle. The *fixed stars* were located on the outermost crystal sphere, called the *primum mobile*. The circular orbits of the sun and the moon and the deferents of each planet lay on the surface of individual crystal spheres. All the crystal spheres and the *primum mobile* made one clockwise rotation daily (Fig. 1.7).

It should be pointed out that all of the motions of the planets, sun, and moon were assumed to be circular by Ptolemy, and these circles were all in one plane, the plane of the page, as shown in Fig. 1.7. This arose from the observation that all of the planets, the sun, and the moon move through essentially the same region of the sky, that region in which the constellations of the zodiac reside. The planets, sun, and moon hence are observed to move essentially in the same plane. However, the primum mobile in Ptolemy's model is a sphere, for there are stars in any direction we look above the earth.

The Ptolemaic hypothesis illustrates very well the use of the *scientific method of reasoning*. Many careful observations were made of the positions of the planets, the moon, and

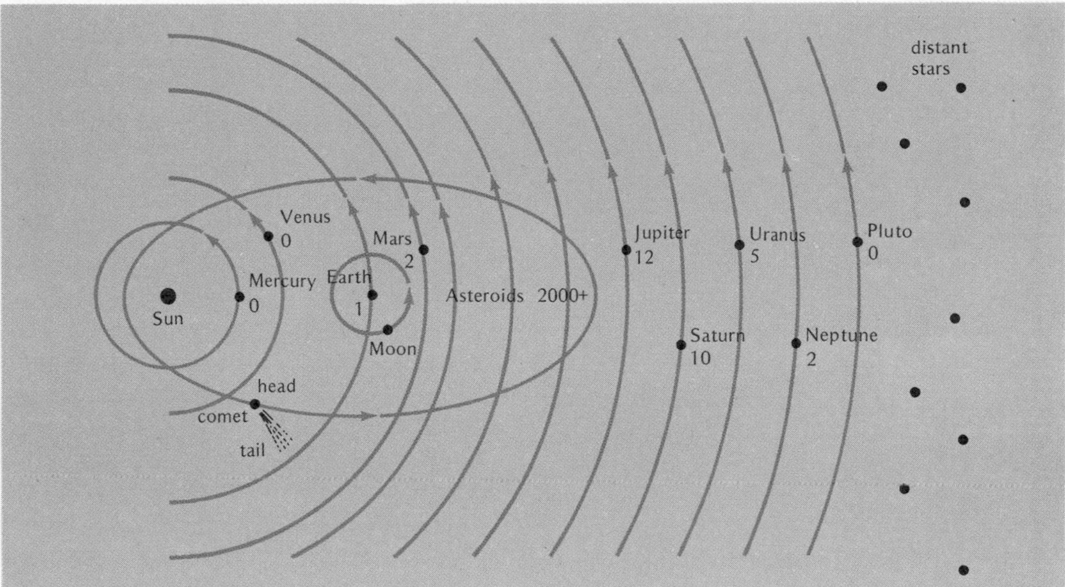

Figure 1.9
Modified Copernican (present) heliocentric theory of the universe. The number beside each planet represents the number of moons.

the sun, and a suggested explanation was worked out and modified until all the known facts were explained. Then the hypothesis was used to predict future positions of the moon, the sun, and the planets, and for a few centuries there appeared to be little discrepancy between observed and computed positions. Any good theory of the universe, however, must accurately predict future positions of heavenly bodies for very long periods. The Ptolemaic hypothesis is therefore a good illustration of the fact that a hypothesis that has been scientifically developed and that has been satisfactorily used for many years may still not be the correct explanation.

HYPOTHESIS VERSUS THEORY

The student might at this point be confused as to the distinction between a hypothesis and a theory. A hypothesis is a tentative explanation based on experimental observations, whereas a theory implies a much greater range of evidence and degree of acceptance. Because the Ptolemaic model of the universe was so widely accepted and agreed so well with observations for more than 14 centuries, it could clearly be called a *theory*.

THE COPERNICAN HYPOTHESIS

After several centuries, the positions of the planets as observed with more precise instruments were found to differ considerably from the positions computed from the Ptolemaic hypothesis. Scientists started working on a new hypothesis that would better explain the motions of heavenly bodies.

Nicolaus Copernicus (1473–1543), a Polish astronomer educated in Poland and Italy, developed the ideas held many centuries earlier by a few Greek philosophers that the sun was the center about which our group

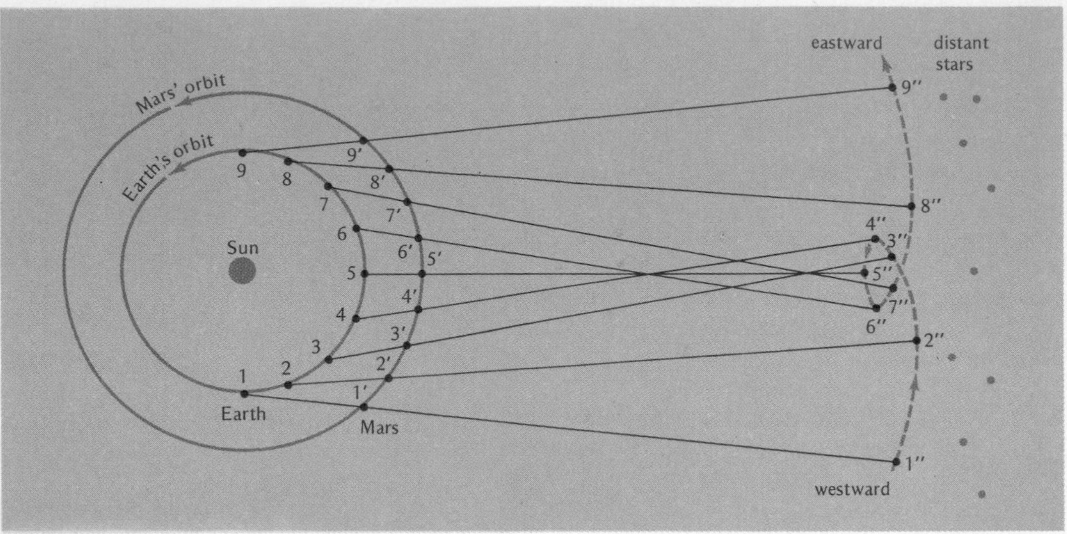

Figure 1.10

Direct and retrograde motion of Mars according to the Copernican theory. Because the period of revolution of Mars is 2 years, it follows that for every 45° of revolution of earth, Mars revolves $22\frac{1}{2}°$. Note that the lines joining the earth and Mars (1 to 1", 2 to 2", etc.) swing counterclockwise from 1 to 1" to 4 to 4", giving rise to an apparent direct (eastward) motion; then clockwise 4 to 4" to 6 to 6", resulting in an apparent retrograde (westward) motion; and then counterclockwise 6 to 6" to 9 to 9" and continuing to give a long direct (eastward) motion. This accounts for the apparent loop path of planets, here shown as a long direct eastward motion 1" to 4" followed by a short apparent westward (retrograde) motion 4" to 6" and a continuing long direct eastward motion 6" to 9".

of heavenly bodies moved and that the earth was one of several planets that revolved around the sun in circular orbits. This is known as the *heliocentric,* or sun-centered, theory and was suggested by the Greek astronomer Aristarchus as early as the third century before Christ. Copernicus also assumed that the earth made one rotation on its axis daily and that the moon revolved around the earth in a little less than a month.

Figure 1.9 shows the general arrangement of Copernicus's heliocentric theory of the universe. We have modified it to include elliptical orbits—we show the complete orbit of a typical comet—and also planets and asteroids (or planetoids) discovered later, with the number of moons given for each planet (below the planet's name).

Like Ptolemy, Copernicus had to explain the same five things that any good theory of the universe must account for. The fast apparent eastward motion of the moon among the stars of 13° each day was due to the counterclockwise motion of the moon around the earth and was computed by dividing 360° by 27.3 days, the time for the moon to revolve around the earth with respect to the stars. The slower apparent eastward motion of the sun among the stars of about 1° a day was due to the counterclockwise revolution of the earth around the sun and was computed by dividing the 360° of one revolution of the earth by the approximately 365 days necessary for that revolution. The general eastward motion of planets, including the occasional retrograde motion (Fig. 1.5), was due to the combined motion of the planet and the earth, as shown in Fig.

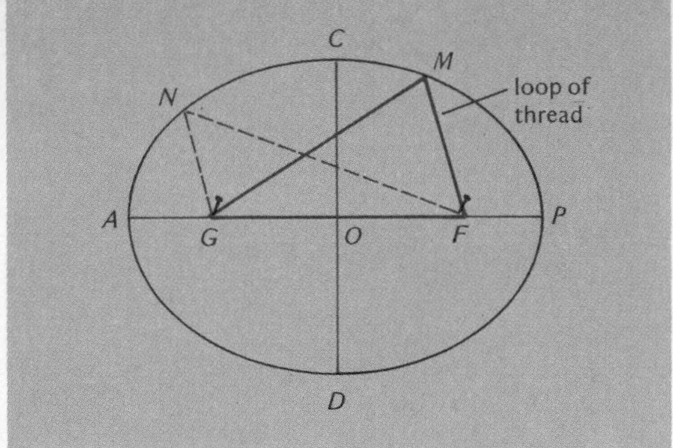

Figure 1.11
Construction of an ellipse. With pins placed at foci *F* and *G*, a loop of thread that is just big enough to cause a pencil point to fall at *M* is used. The loop will then guide the pencil successively through points *C, N, A, D, P,* and *M*, forming the ellipse.

1.10. The daily rising and setting of the sun, the moon, the planets, and the distant stars was due to the daily rotation of the earth.

Copernicus's assumption of circular orbits for the planets and the moon resulted in about the same error between computed and observed positions as given by Ptolemy's theory. Neither the Copernican nor the Ptolemaic hypothesis had been completely accurate in predicting the future positions of heavenly bodies. This made it difficult to decide between the two hypotheses, and the need was felt for more careful and accurate observations of the planets.

TYCHO BRAHE'S OBSERVATIONS

The Danish astronomer Tycho Brahe (1546–1601), using the best equipment of his time, made careful observations of the planets on every clear night for more than two decades. Even though the telescope was unavailable to Tycho Brahe, his data were more than 20 times more precise than those of Copernicus.

ELLIPTICAL ORBITS

Johannes Kepler (1571–1630), the German astronomer, had the mathematical training that Tycho Brahe lacked and the vision to set up various hypotheses concerning the motions of the planets. He carefully checked each hypothesis against Tycho's great volume of accurate observational data and discarded one hypothesis after another when they did not fit the observed data. His progress became more rapid when he discarded combinations of circles and tried elliptical orbits.

All points on an ellipse are so located that the sum of their distances from two fixed points, called *foci*, remains constant. Such a curve can be constructed with the aid of two pins, a loop of thread, and a pencil, as shown in Fig. 1.11.

Ellipses vary in shape from nearly a circle, which is an ellipse where both foci coincide at the center of the circle, to very elongated shapes. The orbits of planets are not far from being circles. For example, Kepler found that the orbit of Mars as measured by Tycho Brahe differed by only about ⅛° from being a circle. Kepler had such respect for the accuracy of Tycho's observations that he could not ignore this ⅛° and had to resort to the use of an elliptical motion to fit the observed path of Mars. The orbits of comets as shown in Fig. 1.9 are usually far from circular.

KEPLER'S THREE LAWS OF PLANETARY MOTION

Kepler finally summarized his work in three statements that we now refer to as *Kepler's three laws of planetary motion*.

Note that we refer to Kepler's statements as "laws" of nature. Kepler's statements when he first proposed them would be considered a hypothesis or perhaps a theory. However, after years of use and without the discovery of any contradictory phenomena, a hypothesis may sometimes be dignified by the term "law" or "principle." Even so there is rarely a "law" in science that has not had to be refined or altered in time due to more and better observations.

Kepler's three laws of planetary motion are

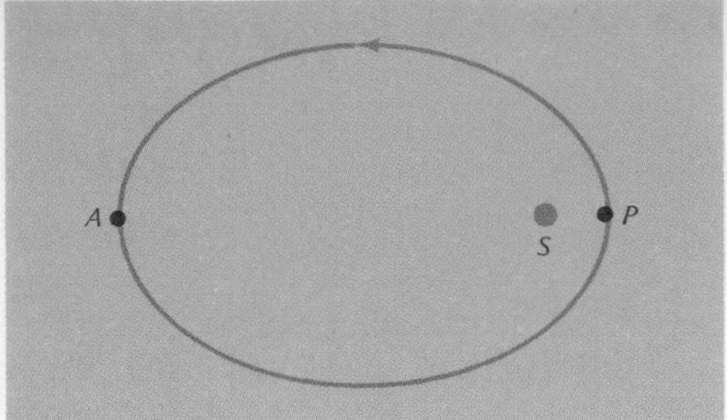

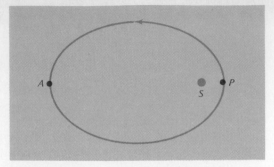

Figure 1.12
Elliptical orbit of a planet, with the sun (S) at one focus, to illustrate Kepler's first law of planetary motion.

1. *Each planet moves in an elliptical orbit with the sun at one focus.*

Figure 1.12 shows the sun at one focus. The *perihelion* is the point *P* in the planet's orbit nearest the sun. The *aphelion* is the point *A* in the planet's orbit farthest from the sun.

2. *The line joining any planet and the sun sweeps over equal areas in equal intervals of time.*

This is known as the law of areas. In Fig. 1.13 if the time for a planet to move from *P* to *Q* equals the time for it to move from *A* to *B*, the area *PSQ* equals the area *ASB*.

The earth's orbit is an ellipse having a major axis (the distance *AP* in Fig. 1.12) of about 186 million miles. When the earth is at perihelion it is 91.5 million miles from the sun, and when it is at aphelion it is 94.5 million miles from the sun. Note that the "average distance" of the earth from the sun is about 93 million miles and that the earth–sun separation never differs from this average by more than about 1.6 percent

$$\frac{94.5 - 93.0}{93.0} \times 100 = 1.6\%$$

The orbit is rather close to being circular. Figure 1.13 shows these relations. Careful measurements of the sun's angular diameter at different days in the year show the earth to be nearest the sun about January 3 and farthest from the sun about July 3. Note that this observation would seem to contradict observed temperature changes of those of us who live in the middle northern latitudes

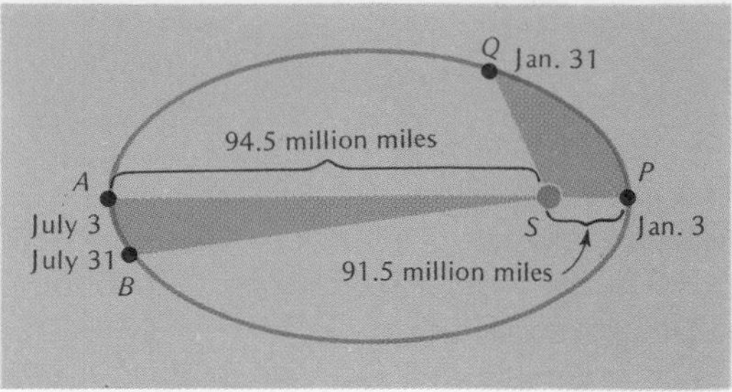

Figure 1.13
The earth in its orbit. The drawing, not done to scale, represents the earth in its orbit around the sun at one focus, with perihelion at *P* and aphelion at *A*.

where it is much colder in January than it is in July when the earth is further away from the sun. This problem will be discussed in Chap. 2.

According to Kepler's second law of motion, the area swept over by the sun–earth line from January 3 to 31 must equal the area swept over during the equal time interval from July 3 to 31; that is, area *PSQ* = area *ASB*, as shown in Fig. 1.13. Because *SP* is shorter than *SA*, in order for the two areas to be equal, it follows that arc *PQ* must be longer than arc *AB*. The longer arc *PQ* is swept over by the planet in the same length of time as the shorter arc *AB;* hence the earth must travel faster over *PQ* than over *AB*. Thus it is seen that the earth moves fastest through perihelion (January 3) and slowest through aphelion (July 3). The actual orbital speed of the earth varies from about 18.2 mi/sec on July 3 to about 18.9 mi/sec on January 3 with an average speed of about 18.6 mi/sec (about 67,000 mi/hour), a rather high speed.

3. *The squares of the times of revolution of any two planets about the sun are directly proportional to the cubes of their average distances from the sun.*

This is called the *harmonic law* and required an additional 10 years of patient work on Kepler's part. It may be written in the form of a proportion (see App. A):

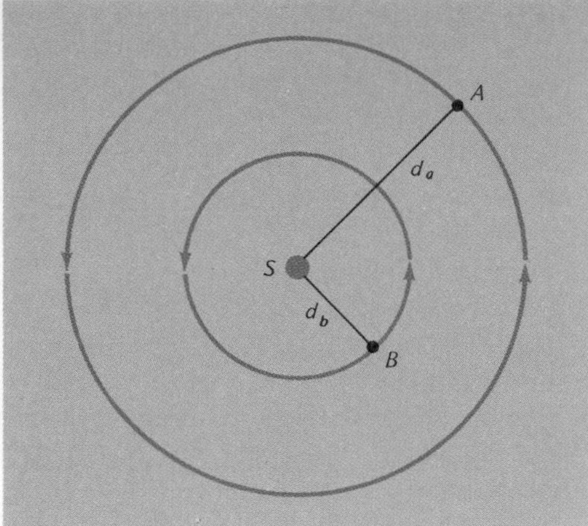

Figure 1.14
Kepler's third law of planetary motion.

$$\frac{T_a^2}{T_b^2} = \frac{d_a^3}{d_b^3} \qquad (1.1)$$

where d_a is the average distance of planet *A* from sun, and T_a is the time of revolution (period) of planet *A* around sun; d_b and T_b have similar meanings for planet *B*.

In Fig. 1.14 planet *A* is at an average distance d_a from the sun and revolves around the sun once in T_a years. Planet *B* is at an average distance d_b from the sun and revolves around the sun in T_b years. If planet *B* is the earth, d_b is the average distance of the earth from the sun, which is called 1 *astronomical unit* (AU) which equals 93 million miles, and T_b is the time of revolution of the earth around the sun, which is 1 year. This makes the proportion

$$\frac{T_a^2}{1^2} = \frac{d_a^3}{1^3} \quad \text{or} \quad T_a^2 = d_a^3 \qquad (1.2)$$

when T_a is expressed in years and d_a in astronomical units. Hence, an alternate and convenient way of stating Kepler's third law is

The square of the time of revolution of a planet about the sun (expressed in years) is equal to the cube of its average distance from the sun (expressed in astronomical units).

Thus knowing that the average distance of Jupiter from the sun is about 5 AU, we can compute its time of revolution about the sun:

$$T_j^2 = d_j^3 = 5^3 = 5 \times 5 \times 5 = 125$$

$$T_j = \sqrt{125} = 11 + \text{years}$$

Using Kepler's laws, astronomers found that predicted positions of the planets agreed very closely with observed positions, so that the Copernican hypothesis of a stationary sun and a moving earth received important support.

Kepler's three laws apply to the orbits of planets, asteroids, and those comets whose orbits are elliptical.

Kepler's laws also apply to any system consisting of a planet with satellites moving about it. For example, the planet Jupiter has 12 moons which move in elliptical orbits about Jupiter as a common focus for each of these orbits. The motions of the moon and man-made satellites about the earth are all elliptical with the earth as a common focus; these motions obey Kepler's laws. However, to apply Kepler's law we must make use of Eq. 1.1 where T_a and T_b refer to the periods of two earth satellites whereas d_a and d_b would refer to the corresponding average distances of the satellites *A* and *B* from the earth; that is, Kepler's law written in the form shown in Eq. 1.2 is written specifically for the system of the sun and its satellites (the planets).

GALILEO'S CONTRIBUTIONS

A few years later Galileo Galilei (1564–1642), the Italian astronomer and physicist and the first to use an astronomical telescope as a scientific instrument (1609), observed that Venus went through phases similar to those of the moon, as Copernicus had predicted from his heliocentric theory. Galileo also observed four moons, or satellites, revolving around Jupiter, and this gave support to the idea that the moon moves around the earth. He noticed that when viewed through his small telescope, the planets appeared as small disks, whereas the stars simply became brighter points of light. (At his first telescope observation the student should note this difference between planets and stars.) This magnification of the size of the planets and not of the stars gave added proof that the planets are relatively near the earth.

ACCEPTANCE OF COPERNICUS–KEPLER HYPOTHESIS

Definite evidence that the earth rotates on its axis and revolves around the sun has been developed. The Foucault pendulum evidence of the earth's rotation and the parallax-of-stars evidence of the earth's revolution are given in Chap. 2. Thus the Copernicus–Kepler theory of the solar system is now firmly established and accepted.

We have followed the development of man's conception of the relation of the sun, the moon, the five naked-eye planets (those that can be seen with the unaided eye), and the stars to one another. Let us now consider the other bodies that we now know to be part of the solar system.

Later discoveries

URANUS, NEPTUNE, AND PLUTO

In 1781 Sir William Herschel, while searching the sky with a new telescope larger than any previously used, discovered an object that he at first thought to be a comet but which on further observations proved to be another planet. The name *Uranus* was finally given to this planet.

From observing the slight deviations of Uranus from its computed orbit, mathematicians came to believe that another planet existed farther out in space and they computed its probable position. In February of 1846, Galle, in Germany, found the new planet, Neptune, within 1° of its computed position.

Similarly, in 1905 Percival Lowell decided that slight deviations from computed positions of the outer planets were due to the existence of still another planet. Failing to find it in his lifetime, he left sufficient funds to establish and maintain an observatory (the Lowell Observatory at Flagstaff, Arizona), with the stipulation that a continued search by photography be made for the predicted planet. Years later, in 1930, a staff member of the observatory, C. W. Tombaugh, in making a routine development of plates taken the previous night, discovered the image of the planet Pluto.

The slight deviations of planets from their computed positions due to the attraction of other planets and other bodies in the solar system indicate that Kepler's laws are only close approximations to what is true. As we improve our instruments and techniques for collecting and interpreting data, we may achieve closer and closer approximations to the truth. The work of natural scientists is characterized by an endless quest in the physical world for closer approximations to complete accuracy.

ASTEROIDS

As early as 1801 Piazzi, an Italian astronomer, discovered with his telescope a body less than 500 miles in diameter and having an orbit around the sun. Within a few years several similar but still smaller bodies were found. Because all these were much smaller than the planets and had orbits between those of Mars and Jupiter, it was decided to call them *asteroids* or *planetoids* to distinguish them from the nine major planets. Since 1891 photography has been employed in the search for asteroids, and more than 2000 have been discovered for which orbits have been computed. The method is quite simple. Photographic equipment, in place of an eyepiece, is attached to a telescope. By means of clockwork the telescope is made to turn just enough to offset the effect of the earth's rotation. Thus stars are photographed as small circular dots (larger dots for brighter stars). A nearby object, such as an asteroid,

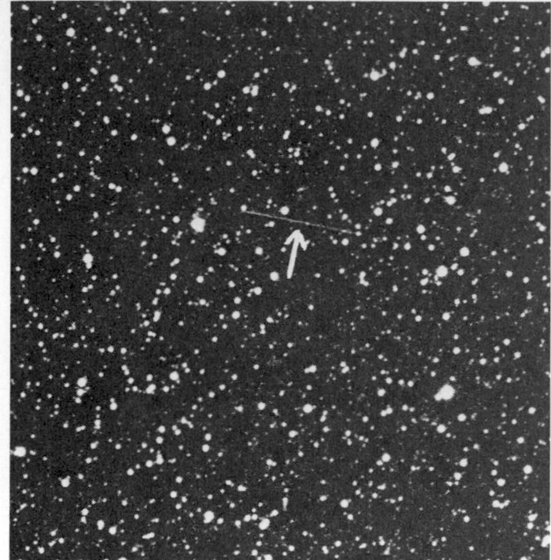

Figure 1.15
Trail of the asteroid Icarus, photographed June 26, 1949. Icarus was last closest to the earth on June 14, 1968, when it was about 4 million miles away. (Photograph courtesy of the Hale Observatories.)

because of its own motion appears as a bright short straight line or trail. This is shown in Fig. 1.15. The Minor Planet Center at Cincinnati Observatory issues official announcements of newly named asteroids.

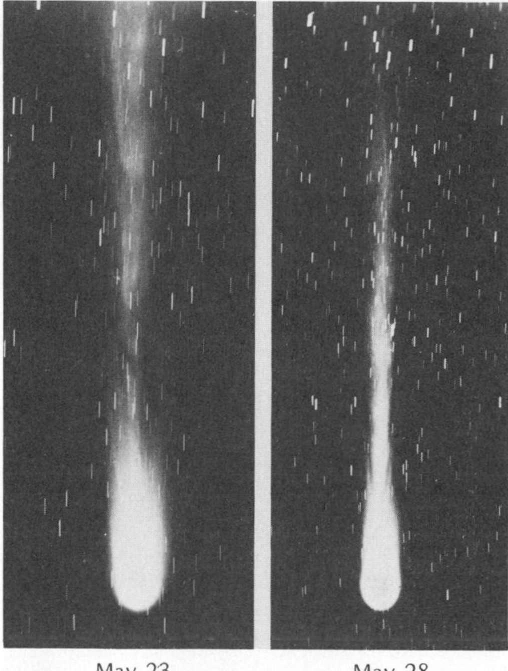

Figure 1.16
Halley's comet, photographed May 23 and 28, 1910. Note the change in its apparent size within 5 days and the star images visible through the comet. (Photograph courtesy of the Hale Observatories.)

ADDITIONAL SATELLITES

Since Galileo's discovery of four satellites, or moons, of Jupiter, other satellites of Jupiter and satellites of other planets have been found. With the aid of photography astronomers have identified 32 satellites, the latest four being the fifth satellite of Uranus in 1948, the second satellite of Neptune in 1949, both by Kuiper at the MacDonald Observatory in Texas, the twelfth satellite of Jupiter in 1951 by Nicholson, and the tenth satellite (Janus) of Saturn in 1966 in France by Dollfus.

COMETS

Each year with the aid of a telescope several large, bizarre, fast-moving objects called comets are observed. Most of these comets have been seen in previous years, since most of them, at least, follow elongated elliptical orbits. Many have perihelion points well within the orbit of Mercury and aphelion points near the orbit of Jupiter. Several have aphelion points beyond Neptune's orbit, but all of them that travel along elongated elliptical orbits are members of the solar system.

Figure 1.16 is a photograph of Halley's comet, which revolves around the sun in a period of about 75 or 76 years and was last seen from the earth in 1910 as the comet passed near its perihelion. The student may look forward to seeing it in 1985 or 1986. Note the very long tail extending from its head, or coma, in a direction away from the sun. Figure 1.17 shows the orbit of Halley's comet.

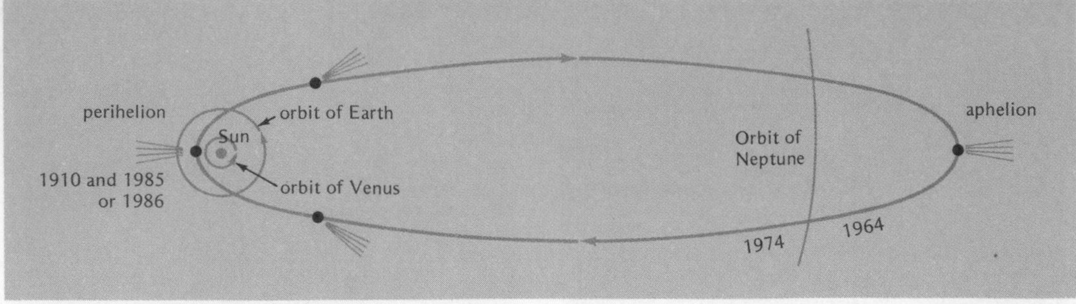

Figure 1.17
Orbit of Halley's comet.

Figure 1.18
Ahnighito meteorite, composed of $34\frac{1}{2}$ tons of iron, nickel, and various other elements. (Courtesy of the American Museum-Hayden Planetarium.)

METEORS AND METEORITES

In addition to the planets and asteroids that are revolving around the sun, millions of very small bodies sometimes called *planetesimals* and weighing a fraction of an ounce to hundreds of tons are likewise moving around the sun throughout the space occupied by the solar system. Occasionally the earth in its orbital motion comes so close to one of these bodies that the body is attracted toward the earth by the earth's large gravitational force. As the object comes hurtling toward the earth's surface, it moves through the extensive layer of gases that we call atmosphere. The friction of this motion produces heat that causes the outside of the small fast-moving body to become incandescent, and we see a *meteor,* or shooting star. Thus a *meteor* is a relatively small body that has become incandescent because of the heat of friction in passing through the earth's atmosphere. Photographs of faint stars, taken by time exposures, frequently show *meteor trails* across the plate.

Showers of swift, bright meteors can be observed with the naked eye each year about August 10 to 12, when the atmosphere of the earth encounters the Perseid meteors. These meteors are distributed fairly equally in a ring that encircles the sun in the manner of the orbit of a comet. A similar shower was the intense rain of Leonid meteors over the western United States on November 17, 1966. The meteors were seen at a rate of around 150,000 per hour for about 20 min. An hourly rate of 50 is normal. The best showers are seen high in the northeast for a few hours after midnight. The streaks of light appear to originate from a common point or constellation (Perseus, Leo, etc.) hence the description "Perseids" or "Leonids."

Most meteors are so small that during their brief passage through the earth's atmosphere they vaporize completely, owing to the intense heat of friction. Occasionally, however, a meteor that is so large it does not completely vaporize actually reaches the earth's surface and is then called a *meteorite.* Figure 1.18 shows the huge (34-ton) Cape York or Ahnighito meteorite, which Peary brought from Greenland to the American Museum of Natural History in New York City. In an authenticated case in 1954 an

Figure 1.19
Aerial view of meteor crater near Winslow, Arizona. (Courtesy of John A. Farrell.)

Alabama housewife was injured by a glancing blow from a small meteorite that fell through the roof of her home. Figure 1.19 shows a large meteor crater made by a meteorite that struck the earth many centuries ago. The crater is 4200 ft across, and its bottom is nearly 600 ft below the edge of its rim. The crater rim rises 120 ft above the surrounding plain.

SUMMARY

Early attempts to account for the relation of the sun, the moon, the naked-eye planets, and the stars to one another resulted first in Ptolemy's geocentric, or earth-centered, hypothesis.

Centuries later Copernicus published his heliocentric, or sun-centered, hypothesis. This was modified by Kepler, who, from a long study of the 20 years of observations of planet positions by Tycho Brahe, suggested elliptical orbits and stated three laws of planetary motion, to give the presently accepted hypothesis.

Galileo's telescopic observations of the phases of Venus and the four brightest moons of Jupiter further supported the Copernicus–Kepler theory.

The solar system as now known consists of

the sun at the center
9 planets (including the earth), large bodies that revolve around the sun
2000+ asteroids, smaller bodies that revolve around the sun, with orbits mostly between those of Mars and Jupiter
32 natural satellites, bodies that revolve around planets
several hundred comets, bodies of huge volume and small mass that travel in elliptical orbits about the sun
millions of meteors, small particles that plunge into the earth's atmosphere daily, and a few larger objects, called meteorites, that reach the earth's surface.

The orbit of the earth is an ellipse with the sun at one focus. The earth at perihelion is 91.5 million miles from the sun and at aphelion 94.5 million miles, the average distance being 93 million miles (1 AU). As expected from the law of areas, the earth moves fastest in its orbit as it passes through perihelion (January 3) and slowest through aphelion (July 3).

Important words and terms

constellations	ellipse
planet	Kepler's three laws
star	perihelion
Polaris	aphelion
the fixed stars	meteor
retrograde motion	meteorite
zodiac	comet
Ptolemaic or geocentric model	period
	asteroid
Copernican or heliocentric model	planetoid
	hypothesis
deferent	theory
epicycles	law
primum mobile	principle

Questions and exercises

1. The W of the constellation Cassiopeia is assumed to be highest in the sky at 11 P.M. tonight for an observer in the United States. At what time tonight will the center star of the open part of the W be directly to the left of Polaris? Explain your reasoning.
2. The bright star Betelgeuse, of the constella-

tion Orion, is seen in the southeast by an observer in the United States at 11 P.M. on November 21. Observed from the same location, where will Betelgeuse be at 11 P.M. on February 21? Explain your reasoning.

3. Write down the third-grade student's mnemonic for the known planets in order of increasing distance from the sun: *My Very Educated Mother Just Served Us Nine Pies*. Using this aid, list the planets. Where would you place the asteroid belt?
4. Draw a fully labeled diagram (names, arrows, etc.) to show the modified (present) Copernican theory of the solar system and universe. Include the planets discovered later, the orbits of all planets, the orbit of our moon, the orbit of a satellite of Jupiter, the orbits of three typical asteroids, the orbit of a typical comet with the comet shown receding from the sun, and the location of a few stars (x). Show the location of a meteor and a meteorite in a separate enlarged diagram properly labeled.
5. Distinguish between a planet, a moon, and a star.
6. What is meant by retrograde motion? How is retrograde motion explained using the heliocentric model? Is retrograde motion "real" or "apparent"?
7. Describe the motion of the stars during a single night. Do the stars directly overhead have the same nightly motion as do the stars in the Big Dipper? Explain.
8. Describe the apparent motion of the sun during the year amongst the stars.
9. How did Ptolemy account for the daily rising and setting of the sun, the moon, the planets, and the stars?
10. (a) Distinguish between a hypothesis and a theory. (b) Arrange these terms in order of increasing weight of evidence supporting them: hypothesis, law (principle), theory.
11. State four things that any good theory of the solar system and the universe must explain, and immediately after each of the four tell (a) how the Ptolemaic theory explained it and (b) how the Copernican theory explained it. What additional thing was found to be true for the modified (present) Copernican theory, but not for the Ptolemaic theory?
12. State Kepler's three laws of planetary motion, and illustrate each of the first two with a labeled diagram.
13. Whereas the Copernican theory used ______ orbits, Kepler modified it, using ______ orbits.
14. (a) Draw a figure of the earth and the elliptical path of an artificial satellite about it; indicate the two foci. (b) Draw a figure of the earth and the corresponding elliptical path of a mail rocket fired to land some 2000 to 3000 miles away; use dotted lines for the extension of the elliptical path within the earth and indicate its two foci.
15. Give Galileo's observations (with an astronomical telescope) that gave support to the Copernican theory as modified by Kepler, and accompany each observation with an indication of how it supported the theory.
16. State several ways in which planets differ from stars.
17. Distinguish between a comet, a meteor, an asteroid, and a satellite.
18. In the text it was mentioned that the earth is moving at about 18 mi/sec or 67,000 mi/hour as it orbits (revolves) about the sun. Why do we not notice this tremendous speed in our daily lives?

Problems

1. If the period of revolution (T) of an asteroid (time to go around the sun once) is 8 years, what is its average distance (d) from the sun?

Answer: 4 AU.

2. If the period of revolution of an asteroid is 5 years, what is its average distance (to the nearest whole number) from the sun?
3. If a heavenly body that is in orbit between the orbits of Jupiter and Saturn is at an average distance from the sun of 6 AU, what is its period of revolution (to the nearest whole number)?

Answer: 15 years.

4. If the average distance of an asteroid from the sun is 4 AU, what is its period of revolution?
5. The time for the return of Halley's comet is variously predicted as 75 or 76 years. Recently computer analysis was used in an effort to tie in this discrepancy, noted for the last 2000 years, to the existence of a tenth planet beyond Pluto and at a distance of 6 billion miles from the sun. (a) Calculate its distance from the sun in astronomical units (AU) to the nearest whole number. (b) Find the period of revolution of this tenth planet, assuming that the distance given is the average distance.

Answer: (a) 65 AU; (b) 524 years.

6. For an earth satellite to be fixed permanently overhead of a spot on the earth's equator, its period of revolution must match that of the earth's rotation. The astronauts' capsules 100 mi above the earth's surface (assume 4000 miles from the center of the earth) made a complete orbit in about 90 min. At what height above the earth must a transmitting TV satellite be placed in orbit to match the earth's speed of rotation?

Projects

1. North circumpolar constellations

a. On a clear night (give date and time of your observation) draw large dots on a diagram to show the stars of the Big Dipper, Polaris, and the main stars of the W of Cassiopeia as you see them. Represent the horizon and label north and the direction toward east and toward west.

b. Observe the same stars 2 or 3 or 4 hours later (give time of second observation), and represent them on the same diagram by means of small circles.

c. Study the relation between the positions of the stars as shown in the diagrams drawn for (a) and (b), and state your explanation of this relation.

d. Given that the angular distance between the pointers of the dipper is 5°, indicate on your diagram the angular distance (1) across the top of the open bowl of the Big Dipper, (2) from Polaris to the middle star of Cassiopeia, and (3) across the top of the W of Cassiopeia. Example: if 5° is 0.5 in., a distance of 1 in. would be 10° (see Fig. 1.1).

e. (Optional) If you have a camera with a time-exposure attachment, make a photograph of the north circumpolar stars, using a time exposure of 2 or 3 or 4 hours. To do this, place your camera facing north and tilted up from the horizon at an angle equal to your latitude, and protect the lens opening from the light from nearby street lights, house lights, etc. When you have developed your film, measure in degrees the lengths of the arcs of the various star trails. Are they the number of degrees to be expected for the time of exposure? Explain.

2. Winter constellations

a. On a clear night (give date and time of your observation) draw dots and lines on a diagram labeled to show the constellations Canis Major, Orion, Pleiades, and Gemini. Label by name major stars indicated by large dots. Represent the horizon and label south and the direction toward southeast and toward southwest.

b. Observe the stars of Orion 2 or 3 hours later and represent them on the same diagram by means of small circles.

c. Study the relation between the positions of the stars in constellation Orion as shown in the diagrams drawn for (a) and (b), and state your explanation of this relation.

3. Construction of ellipses

a. Construct an ellipse with the aid of paper, two pins, a loop of thread, and a pencil.

b. Construct a second ellipse that is much more elongated. Discuss the relationship between the shape of an ellipse, the distance of separation of the foci, and a circle.

Supplementary readings

Textbooks

Books written for use in introductory courses in astronomy are an excellent source of additional information for Chaps. 1, 2, 3, 4, 5, and 11.

Abell, George, *Exploration of the Universe,* Holt, Rinehart and Winston, New York (1969).

Baker, Robert H., *Astronomy,* Van Nostrand-Reinhold, Princeton, N. J. (1964).

Baker, Robert H., and Laurence W. Fredrick, *An Introduction to Astronomy,* Van Nostrand-Reinhold, New York (1971).

Bartky, Walter, *Highlights of Astronomy,* University of Chicago Press, Chicago (1935). (Paperback.)

Degani, Meir H., *Astronomy Made Simple,* Doubleday, Garden City, New York (1963). (Paperback.)

Duncan, John C., *Astronomy,* Harper & Row, New York (1955).

Ebbighausen, E. G., *Astronomy,* Charles E. Merrill, Columbus, Ohio (1971).

Hesse, Walter H., *Astronomy: A Brief Introduction,* Addison-Wesley, Reading, Mass. (1967). (Paperback.)

Huffer, Charles M., Frederick E. Trinklein, and Mark Bunge, *An Introduction to Astronomy,* Holt, Rinehart and Winston, New York (1967).

Inglis, Stuart J., *Planets, Stars, and Galaxies,* Wiley, New York (1972).

Krogdahl, Wasley S., *The Astronomical Universe,* Macmillan, New York (1962).

McLaughlin, Dean B., *Introduction to Astronomy,* Houghton Mifflin, Boston (1961).

Mehlin, Theodore G., *Astronomy,* Wiley, New York (1959).

Motz, Lloyd, and Anneta Duveen, *Essentials of Astronomy,* Columbia University Press, Irvington-on-Hudson, New York (1972).

Nicolson, I., *Astronomy,* Bantam Books, Grosset & Dunlap, New York (1971). (Paperback.)

Page, L. W., *Astronomy, How Man Learned About the Universe,* Addison-Wesley, Reading, Mass. (1969). (Paperback.)

Stokley, James, *Atoms to Galaxies,* Ronald Press, New York (1961).

Struve, Otto, Beverly Lynds, and Helen Pillans, *Elementary Astronomy,* Oxford University Press, Fair Lawn, N. J. (1959).

Wyatt, Stanley P., *Principles of Astronomy,* Allyn & Bacon, Boston (1971).

The Project Physics Course, Holt, Rinehart and Winston, New York (1970). [In addition to the text, The Project Physics Course includes a series of articles and essays in Readers. These Readers are excellent.]

Other books

Cohen, I. Bernard, *The Birth of a New Physics,* Anchor Books, Doubleday, Garden City, N. Y. (1960). (Paperback.) [An eminent historian of science explains the Ptolemaic and Copernican systems; also the contributions of Galileo, Kepler, and Newton. An excellent list of suggested readings is given.]

Holton, Gerald, and Duane H. D. Roller, *Foundations of Modern Physical Science,* Addison-Wesley, Reading, Mass. (1958). [Chapters 6–10 cover the works of Ptolemy, Copernicus, Kepler, and Galileo in considerable detail. The historical development is well treated. Includes a good list of supplementary readings.]

Koestler, Arthur, *The Watershed: A Biography of Johannes Kepler,* Anchor Books, Doubleday, Garden City, N. Y. (1960). (Paperback.) [Chapter 4 is devoted to Tycho Brahe, Chap. 5 to Kepler and Tycho, Chap. 8 to Kepler and Galileo, and Chaps. 6 and 9 to Kepler's laws of planetary motion.]

Kuhn, Thomas S., *The Copernican Revolution,* Vintage Books, Random House, New York (1959). (Paperback.)

Menzel, Donald H., *A Field Guide to the Stars and Planets,* Houghton Mifflin, Boston (1964). [Includes monthly sky maps, moon maps, and maps of constellations; also chapters on the sun, moon, planets, stars, and nebulae, other bodies of the solar system, and time.]

Nininger, H. H., *Out of the Sky: An Introduction to Meteoritics,* Dover Publications, New York (1952). [Meteors, meteorites, and meteor craters; essential facts and major theories presented in terms easily understood; meteoritics—a connecting link between astronomy and geology.]

Omer, Guy G., Jr., et al., *Physical Science: Men and Concepts,* Heath, Boston (1962). [Excerpts (Chaps. 4–6) from Ptolemy's *Almagest* show his system for explaining the observed celestial motions: the postulates, the geocentric universe, the Ptolemaic planetary system. In Part III we see the attack on the Ptolemaic astronomy and the support of the heliocentric theory by Copernicus, Kepler, and Galileo in passages from their writings. The foreword to Part IV contains part of the sentence read to Galileo by the judges of the Inquisition and his forced recantation of his belief in the heliocentric theory. The bibliography lists several general sources for further reading in the history of science.]

Richardson, Robert S., *Getting Acquainted with Comets,* McGraw-Hill, New York (1967). [Excellent illustrations, unclouded explanations, general readability.]

Rogers, Eric M., *Physics for the Inquiring Mind,* Princeton University Press, Princeton, N. J. (1960). [Part 2, "Astronomy," provides a clear example of the growth and use of theory in science: Greek astronomy, Copernicus, Brahe, Kepler, Galileo.]

Watson, Fletcher G., *Between the Planets,* Harvard University Press, Cambridge, Mass. (1956). [A well-illustrated source book that includes the story of the asteroids, comets, meteors, and meteorites.]

Periodicals

Science News. A weekly summary of current science published by Science Service, Inc., 1719 N St., NW, Washington, D. C. 20036. $10.00 a year. [For the last week of each month it includes maps of the evening sky and directions of how, where, and when to observe interesting things in the skies during the following months. Available in almost all libraries.]

Sky and Telescope. A monthly magazine published by Sky Publishing Corporation, 49 Bay State Road, Cambridge, Mass. 02138. $8 a year. [Informative articles that are well illustrated; news items; observing material, including monthly sky maps; and telescope-making notes. The January issue contains a graphic time table of the heavens for the year.]

The American Ephemeris and Nautical Almanac. An annual publication (book) of astronomical data for the calendar year. Available more than a year in advance. For sale by Superintendent of Documents, U.S. Government Printing Office, Washington, D. C. 20402. $6.50. [For astronomers, instructors, and advanced students.]

The Observer's Handbook. An annual publication of the Royal Astronomical Society of Canada, 252 College St., Toronto 130, Ontario, Canada. $2.50, including exchange and mailing. [An extremely useful source of data for the amateur and the professional. Includes such things as symbols and abbreviations; the constellations; miscellaneous astronomical data; principal elements of the solar system; solar, sidereal, and ephemeris time; times of rising and setting of the sun and the moon; the planets; the sky and astronomical phenomena month by month; eclipses; stars; star map for observing the evening sky.]

Articles

Gee, B., "400 Years: Johannes Kepler," *The Physics Teacher,* **9,** no. 9, 510 (December, 1971).

Wilson, C., "How Did Kepler Discover His First Two Laws?" *Scientific American,* **226,** no. 3, 93 (March, 1972). [A discussion of whether Kepler found that a planet's motion fit an ellipse or did the ellipse come first.]

OUR PLANET, THE EARTH

I keep six honest serving men
(They taught me all I knew);
Their names are What and Why and When
And How and Where and Who.

RUDYARD KIPLING, 1902

Now that we have had a bird's-eye view of the solar system, let us consider in greater detail the planet in which we are most interested—the planet Earth. Today practically everyone accepts the idea that the earth is approximately spherical, but as recently as 500 years ago all but a few people thought the earth was a huge, flat disk. It is no wonder that Columbus's sailors became frightened and wished to turn back as they sailed on and on, and no land was sighted.

A few of the early philosophers, however, had the right concept of the earth. Pythagoras, Eudoxus, Aristotle, Aristarchus of Samos, and Archytes of Sicily, all of whom lived a few hundred years before Christ, believed the earth was a globe. Eratosthenes of Alexandria (273–192 B.C.) not only taught that the earth was spherical but carried out some measurements in Egypt that gave the size of the earth fairly accurately.

In this chapter we shall discuss the shape and size of the earth, its motions, its orbit, the cause of seasons on the earth and on other planets, and the measurement of time on the earth. Additional consideration of the earth appears in the last few chapters.

Shape and size

EVIDENCES THAT THE EARTH IS APPROXIMATELY A SPHERE

Thinking persons do not like to accept statements without evidence. Accordingly, throughout this book we will offer supporting evidence whenever possible. For example, how do we know the earth is approximately spherical—what evidence is there?

It is perhaps surprising to learn that Ptolemy concluded that the earth was a sphere. Ptolemy believed the stars were a great distance from the earth which was tiny in comparison to the distance to the stars. He concluded this from the observation that the shapes of the constellations are the same no matter from what point on earth they are viewed. This would happen only if the stars were very far away.

If the earth were flat, then anyone on the earth who looked directly overhead at night would see the same constellations no matter where he happened to be on the earth. Because a different star pattern is seen at different places on earth, Ptolemy concluded that the earth curved uniformly north and south; hence, it must be spherical in shape. The idea of a spherical shape was very pleasing to Greek philosophers because they held a high regard for symmetry and the sphere was, of course, the "most perfect" shape for an object.

The earth's surface is convex rather than plane or concave. Figure 2.1, a photograph of the earth taken from a space vehicle, shows the sphericity of the earth and the convexity of its surface.

SIZE

The exact shape and size of the earth have been determined traditionally as follows: Two stations *A* and *B* in Fig. 2.2 are set up 1° apart. The distance between these two stations is then accurately measured by a method called *triangulation*. This has been done by the United States Coast and Geodetic Survey for several parts of our country, and similar organizations in various parts of the world have also carefully measured the distance between two stations that are 1° apart. If the earth were exactly spherical, each of the many arcs subtending 1° would have the same length. Actually the arc lengths for 1° intervals vary considerably at different parts of the earth; from 68.71 miles near the equator to 69.41 miles near the poles.

Figure 2.1
Curvature of the earth. The earth was photographed from about 98,000 nautical miles away during *Apollo 11* translunar coast toward the moon. Most of Africa and portions of Europe and Asia can be seen in this photograph. (NASA.)

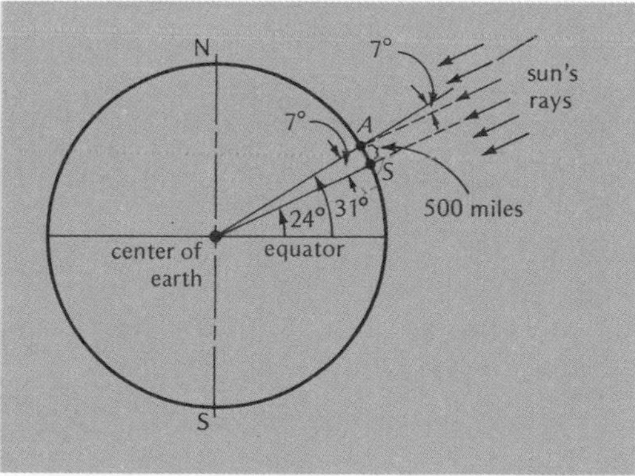

Figure 2.3
Eratosthenes' method of the determination of the circumference of the earth. Point *A* represents Alexandria and point *S* represents Syene.

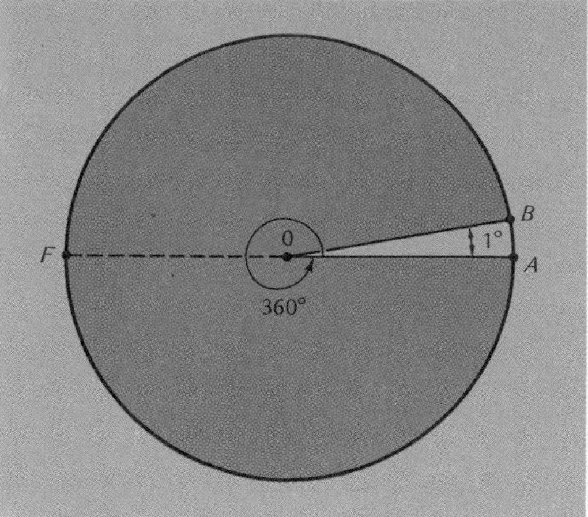

Figure 2.2
Circumference of the earth. By measurement, arc *AB* subtended by 1° of central angle, exaggerated here, is approximately 69 miles. The circumference of the earth equals 360° times 69 miles per degree, equals about 25,000 miles.

COMPUTATION OF THE EARTH'S DIAMETER

Eratosthenes appears to be the first astronomer to have determined the circumference of the earth. He used the distance of 500 miles between Syene (now Aswan) and Alexandria which are located on a north–south line in Egypt. He obtained the number of degrees of angular separation between these two towns from the observation that at noon on June 21 the sun, directly overhead at Syene, was shining directly down a deep vertical well whereas at the same time in Alexandria the sun cast a shadow of a vertical pole of about 7°. From Fig. 2.3 we can see that these two towns are separated by 7° of arc. Because there are 360° in a circle, this 7° of arc that corresponds to a distance of 500 miles means that the total distance around the circle, or circumference, would be $360°/7° \times 500$ miles; this results in a circumference (C) of the earth of approximately $C = 25{,}700$ miles. This value of the circumference of the earth is in good agreement with that obtained using the approximate value of 69 mi/deg. By measurement, $1° = 69$ miles (along the surface). Hence in Fig. 2.2, circumference *ABFA* $= C = 360° \times 69$ mi/deg $= 24{,}840$ miles. A corrected figure is $C = 24{,}901.93$ miles,

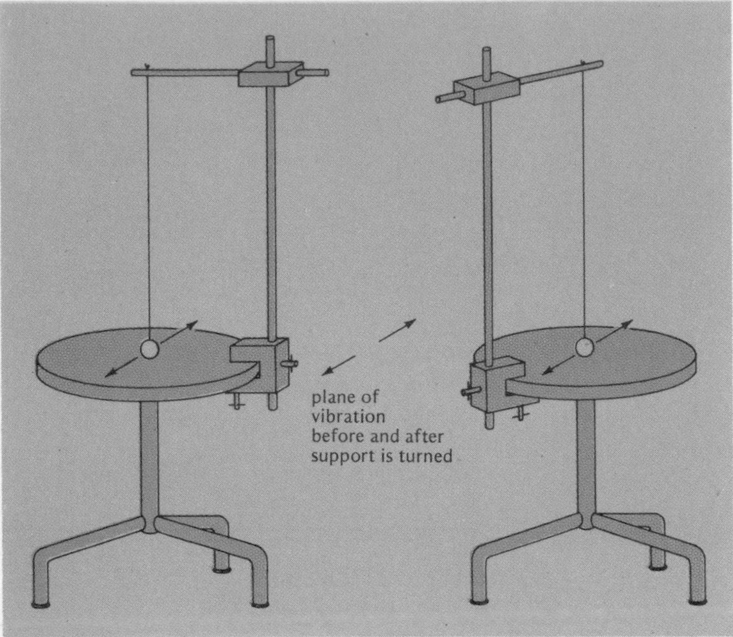

Figure 2.4
Apparatus to demonstrate that a pendulum continues in the same plane of vibration.

based on recent information from earth satellites. For any circle, circumference = $\pi \times$ diameter = πD. Hence using the value of 3.14 for π, we can determine the diameter as follows: $C = 3.14D$, or

$$D = \frac{C}{3.14} = \frac{24{,}902 \text{ miles}}{3.14} = 7926 \text{ miles}$$

EXACT SHAPE

Using careful measurements of the length of 1° over various parts of the earth and in various directions, we can determine the exact shape of the earth. It is found to be an oblate spheroid having a polar diameter of 7900 miles and an equatorial diameter of 7926 miles. A grapefruit pressed firmly with the hands above the stem side and below is similar to an oblate spheroid. It is flattened at the top and bottom and bulges outward around its sides.

Motions and orbit

The two main motions of the earth consist of a daily rotation around its axis and its annual revolution around the sun. Rotation occurs about an internal axis (in this case, the earth's polar axis); revolution occurs about an external axis (in this case, an axis within the sun and through the common center of gravity of the sun and the earth). In addition, the earth, with the rest of the solar system, is plunging through space in the direction of the constellation Hercules at about 12 mi/sec. It also wobbles slightly on its axis, like a top.

ROTATION

As mentioned in Chap. 1, according to the modified Copernican theory the daily rising and setting of the sun, the moon, the stars, and the planets is due to the rotation of the earth on its axis every 24 hours. One proof of the rotation of the earth on an axis will be presented—the Foucault pendulum proof.

As a background for this proof, the important law known as *Newton's first law of motion* is needed.

Every object continues in its state of rest, or of uniform motion in a straight line, unless some force causes that state to change.

For example, a book will remain at rest on a table top unless someone exerts a force on it—that is, gives it a push or a pull. A marble will roll along a smooth horizontal table top for a long distance before coming to rest. According to the law, should not the marble continue in the same direction without slowing up at all? Of course, the point is that there is a force acting on the rolling marble—the retarding force of friction. Making the surface smoother and smoother would result in increasingly less retarding force of friction and increasingly less decrease of speed. It seems reasonable to conclude that if we could completely do away with the retarding force of friction, which we cannot do, the marble would continue moving in the same direction at the same rate. This law will be discussed further in Chap. 4.

A short pendulum (3 or 4 ft long), if set swinging (usually called vibrating or oscillating), will come to rest within 5 min because of the effect of friction as it moves through the air. In 1851 Foucault, a French physicist, set up a pendulum having a heavy lead sphere suspended from the inside of the Pantheon in Paris by a wire 200 ft long. A swivel mount with very little friction was employed to attach the wire to the ceiling. With such a heavy and long pendulum the effect of friction was relatively small, and the pendulum vibrated for more than 24 hours.

When a pendulum is set in vibration, the force of gravity (downward) makes it stop at each extremity and swing back toward the center. The vibration continues in the same plane relative to the fixed stars, because there is no force to make the plane rotate. Figure 2.4 shows how to demonstrate that the orientation of the plane of vibration will remain unchanged relative to the room even if the support is turned. The pendulum is set in vibration, and then the apparatus supporting the pendulum is rotated around a vertical axis through several degrees (care must be taken not to tip the apparatus away from the vertical); the plane of vibration is not affected.

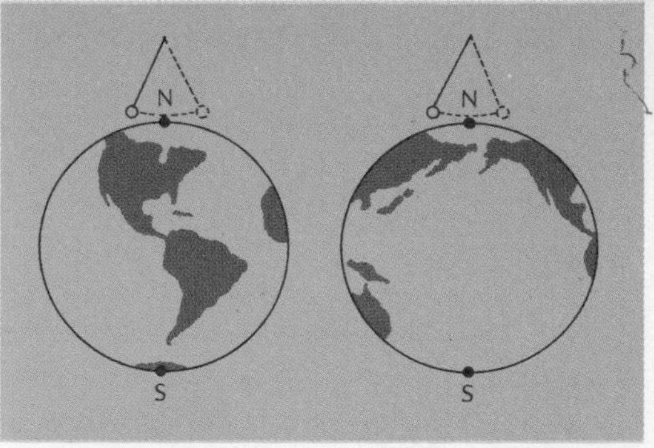

Figure 2.5
Foucault pendulum above the North Pole. As viewed at the North Pole a pendulum set in motion will continue to swing in the direction of the same "fixed stars" in space while the earth rotates under the pendulum.

When a Foucault pendulum is set swinging as the earth turns under it (see Fig. 2.5), the pendulum bob tends to keep swinging back and forth in the same plane (direction in space)—that is, to keep swinging in the general direction of the same stars. Of course, to a person on the earth and in the same room with the pendulum, it appears that the pendulum is changing its direction of swing. Actually, because no forces are acting to change the direction of swing of the pendulum, it is the earth that is rotating underneath the pendulum.

If one holds a short vibrating pendulum over the north-pole end of the axis of a rotating globe representing the earth, it can be seen that one complete rotation of 360° occurs in 24 hours. Hence a Foucault pendulum at a pole of the earth will apparently rotate 360°/24 hours = 15°/hour. Suppose a Foucault pendulum at the earth's equator is vibrating over a certain meridian

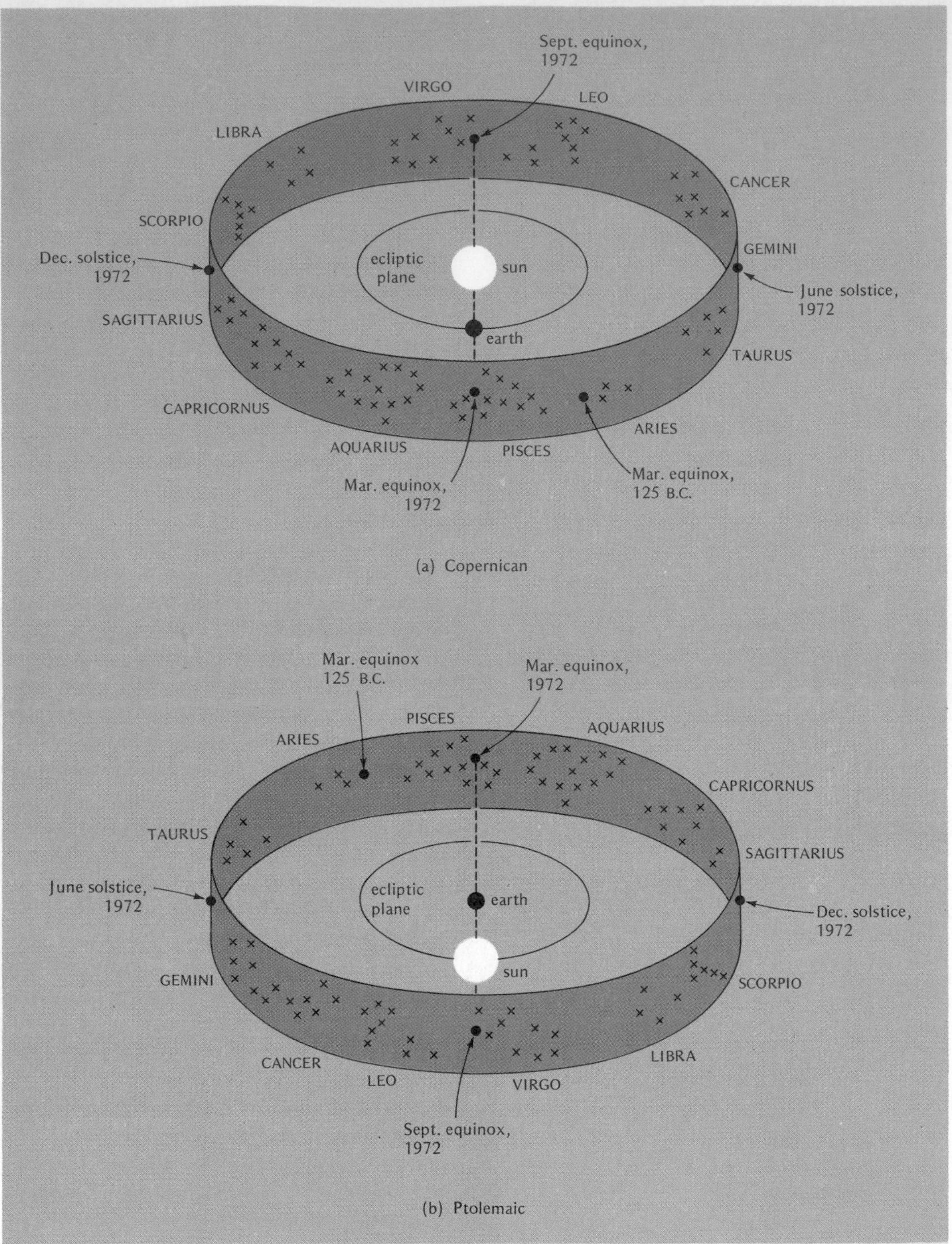

Figure 2.6
Apparent motion of the sun through the band of constellations which makes up the zodiac. In either the Copernican (*A*) or the Ptolemaic (*B*) models the sun appears to observers on the earth to move through the background stars of the constellations of the zodiac. This motion repeats itself once a year.

(a circle on the earth passing through the two poles). A little later the support and pendulum will have moved the same amount eastward, but the pendulum will still be vibrating over the starting line. Thus at the equator the apparent rotation is 0°/hour. At any intermediate latitude, the amount of apparent rotation of the plane of vibration can be shown to be 15°/hour × sine of the latitude. For a place of latitude 42° this becomes approximately 15°/hour × 0.7 = 10.5°/hour. Hundreds of Foucault pendulum experiments have been performed, and some museums and planetaria feature the Foucault pendulum. The United Nations building in New York City has an impressive one.

Because the earth rotates west to east once a day (counterclockwise as seen from above the North Pole), the sun seems to rise in the eastern sky each morning and to set in the western sky each evening, thus giving us day and night. The moon, the planets, and the stars, likewise, seem to have a similar daily motion (see Fig. 1.6, showing star trails).

REVOLUTION

Suppose one observes a bright star or a certain easily recognizable group of stars that is due south at a convenient time in the evening on a certain date. If this same star or group of stars is observed at the same time of evening one month later, it will be seen that the stars have appeared to move about 30° westward from their first position. Because the time of evening is the same, the sun is the same number of degrees west of the meridian in both cases. Hence by observation at a given time of night the stars are found apparently to move westward with respect to the sun by 30° each month, or the sun to move eastward among the stars by 30° each month.

What is actually happening? These observations were described briefly in Chap. 1 where we described how the sun, on rising, appears throughout the year to move into a different constellation within the band of constellations called the *zodiac*. These observations can be explained equally well using either the geocentric (earth-centered) model of Ptolemy or the heliocentric (sun-centered) model of Copernicus. By examining the two diagrams in Fig. 2.6 it is clear that regardless of whether the sun revolves about the earth or the earth revolves about the sun, to an observer on the earth the sun appears to move eastward among the stars making up the constellations of the zodiac. Hence this observation does not enable us to determine which object revolves about the other.

It was not until 1838 that it was shown that the earth does revolve about the sun producing the apparent relative motion of the sun and the stars. In 1838 Bessel with the aid of a telescope succeeded in observing parallax of stars.

Parallax is the apparent motion of a near object with respect to distant objects due to the motion of the observer. To illustrate this, the reader should hold a pencil at arm's length, close one eye, and note where the top of the pencil appears against some distant object on a wall of the room. Without moving the pencil, he should then move his head sideways and note that the top of the pencil seems to move as seen against the wall. With the aid of this knowledge of parallax, the parallax-of-stars evidence that the earth revolves around the sun can be understood.

Photographs of the same region of the sky taken 6 months apart show an apparent forth and back motion of one star with respect to neighboring groups of stars. In Fig. 2.7 there is a simple illustration of stellar parallax. In Fig. 2.7 the earth is shown at two points in its orbit about the sun. These points correspond to a time interval of 6 months. Note that the relative position of the two stars is slightly different when the stars are viewed from the two positions of the earth in space. It should be pointed out that this diagram is highly out of proportion,

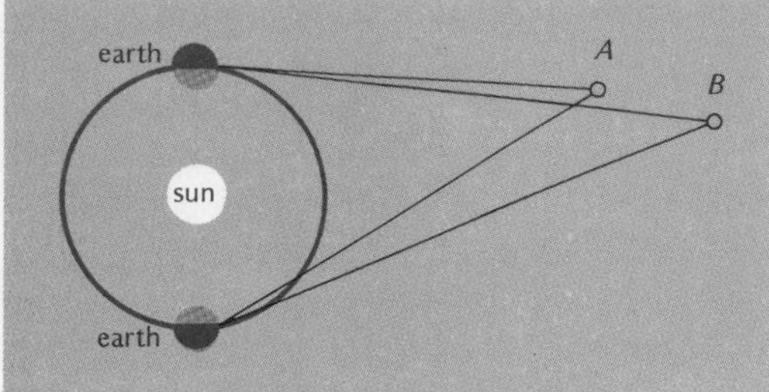

Figure 2.7
Stellar parallax; the Copernican model. The relative position of stars *A* and *B* appears to change when observed over a 6-month interval as the earth rotates about the sun. If the earth and the sun were interchanged resulting in the geocentric model, the earth would be fixed and no parallax would be expected.

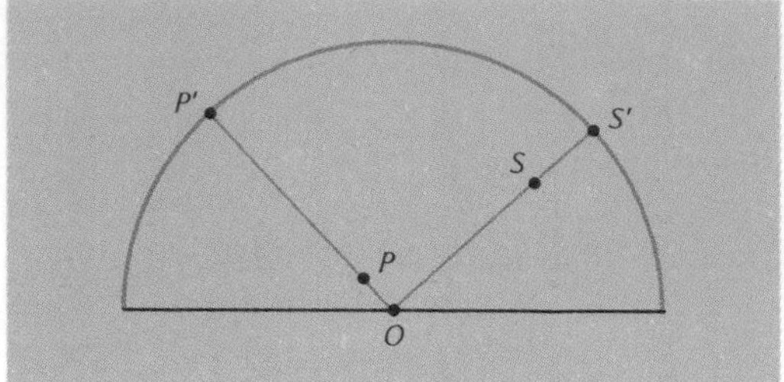

Figure 2.8
Projections on the celestial sphere. To an observer at *O*, a star *S* is projected on the celestial sphere at *S'* and the planet *P* is projected at *P'*.

for the stars should be at a much, much greater distance. This results in the fact that stellar parallax is extremely small and is impossible to observe without a telescope, a fact which bothered many scientists at the time Copernicus proposed that the earth revolved about the sun. These men reasoned that the revolution of the earth about the sun would produce stellar parallax; yet it was not observed. Stellar parallax did exist but was not observable until the telescope had been sufficiently perfected. Note that in the geocentric model there would be no stellar parallax because the earth would be in the center of Fig. 2.7 with the sun revolving about the earth—that is, the sun and earth would be interchanged. The observation of stellar parallax provides us with direct evidence that the earth does, in fact, revolve about the sun.

ORBIT

The evidence just given that the earth is moving around the sun is in line with Kepler's first law of planetary motion, given in Chap. 1, which states that each planet moves in an elliptical orbit with the sun at one focus. Kepler's second and third laws of planetary motion also apply to the earth's orbit, because the earth is one of the planets.

Understanding the seasons

THE CELESTIAL SPHERE

Has the reader not had the experience of gazing at the stars at different times on a clear night and feeling that they seemed to be located on the inside of a huge bowl that was rotating gradually from east to west? Scientists long ago discarded the crystal sphere of the early Greek astronomers inside which all stars were considered to be located and to which orbits of planets were attached and suspended in space as the globe rotated on an axis once a day. There is a distinct advantage, however, in being able to describe easily the apparent location and motion of celestial bodies from a geocentric viewpoint even though these apparent motions are properly attributed to the rotation and revolution of the earth. For this reason we postulate a *celestial sphere,* whose inner surface, on which all heavenly bodies are considered to be projected, is farther away than the farthest star. It should be noted that even today the terminology and procedures of navigation have their basis in a geocentric model. Figure 2.8 shows how

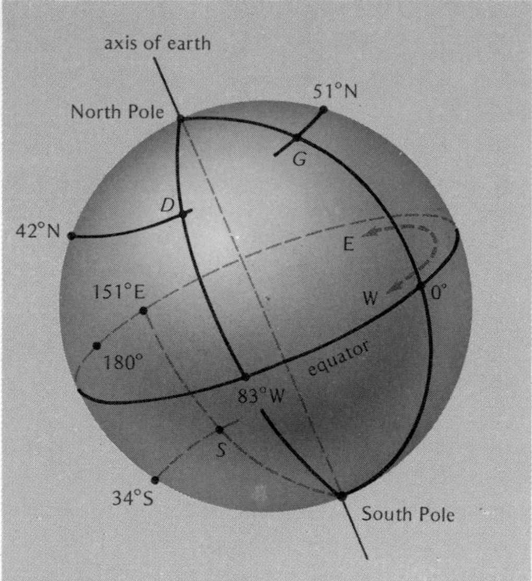

Figure 2.9
Longitude and latitude, showing some locations on the earth. Thus *G* (Greenwich, England) is longitude 0°, latitude 51°N; *D* (Detroit, Michigan) is 83°W, 42°N; and *S* (Sydney, Australia) is 151°E, 34°S.

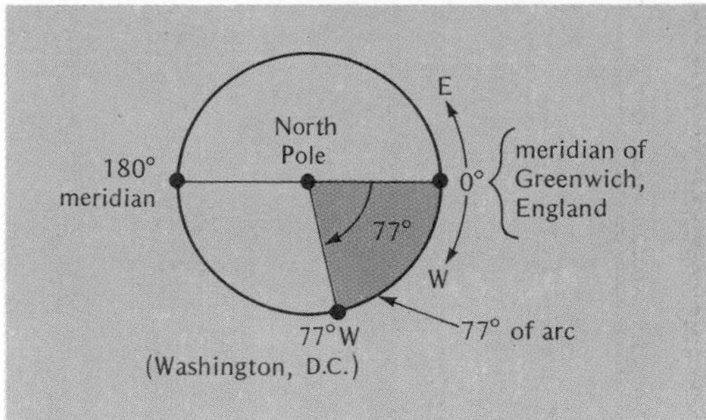

Figure 2.10
Longitude. Diagram showing how longitude of Washington, D. C., is defined relative to the prime (0°) meridian.

this projection is applied in the case of the celestial sphere.

FUNDAMENTAL POINTS AND CIRCLES FOR THE EARTH

The meanings of certain terms used with the earth will be considered here. Figures 2.9–2.11 will assist in an understanding of these terms and of how the location of a place on the earth is described.

The *earth's poles* are the two points at which the earth's axis of rotation pierces the surface of the earth.

A *great circle* of a sphere is a circle on the surface of a sphere with its center at the center of the sphere; for example, a plane passed through the earth's center in any direction would cut the surface in a great circle.

The *earth's equator* is the great circle on the earth midway between the earth's poles.

An *earth meridian* is a great circle on the earth that passes through the earth's poles. The *meridian of a place* is the half of a great circle on the earth that extends from the North Pole through the place to the South Pole.

The *longitude* of a place is the angle subtended by the arc of the earth's equator measured in degrees east or west from the reference or prime (0°) meridian of Greenwich, England, to the meridian through the place (Fig. 2.10). The values of longitude range from 0° at the Greenwich meridian to 180°. Note that meridians are labeled by the number of degrees they are located east or west from the Greenwich or prime meridian. Greenwich, England, was chosen for the location of the zero meridian because the Royal Observatory was located there and England "ruled" the seas at the time the longitude and latitude reference system was established.

The *latitude* of a place on the surface of the earth is the angle subtended by the arc measured along the meridian through the place from the earth's equator north or south to the place (Fig. 2.11). Latitude is measured in degrees north or south of the equator. A circle drawn around the earth that is everywhere perpendicular to the meridians that it crosses is called a *parallel* of latitude. Lines of constant latitude on the earth's surface are small circles, not great circles, located in planes parallel to the equator. They are called *parallels,* meaning parallels of lati-

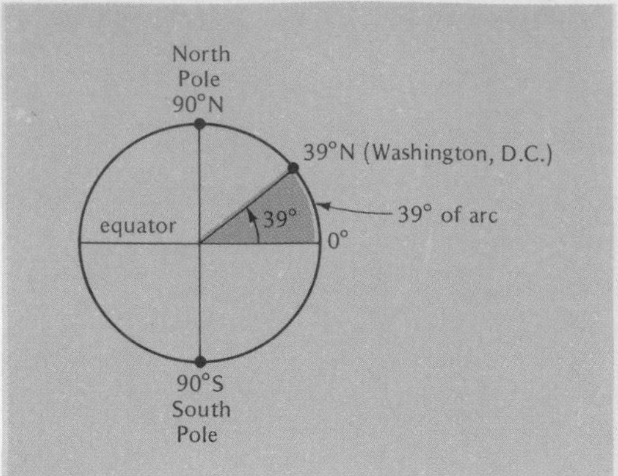

Figure 2.11
Latitude. Diagram showing how latitude of Washington, D. C., is defined relative to the equator.

tude. An illustration of the location of points on the earth's surface by specifying the latitude and longitude of the corresponding meridian and parallel passing through the point is given in Fig. 2.9.

THE ECLIPTIC

The earth revolves about the sun once each year, and the projection from the earth of the path of the sun against the distant background of stars is called the *ecliptic*. We observe that once a year the sun seems to make one complete orbit among the stars. The ecliptic (Fig. 2.12) is then the apparent yearly path of the sun among the stars. Because we cannot see stars in the daytime, other than the sun, we deduce which stars lie behind and beyond the sun by observing at night the stars visible in the opposite direction. We postulate a celestial sphere with an inner surface, on which all heavenly bodies are considered to be projected, that is farther away than the farthest star. The north pole of the celestial sphere is directly over the North Pole of the earth; that is, if the axis on which the earth spins (rotates) each day were extended from the North Pole of the earth, it would pass through the north pole of the celestial sphere. The North Star, Polaris, when projected onto the celestial sphere is within $\frac{1}{2}°$ of, and is said to be at, the north pole of the celestial sphere. The celestial equator lies in the same plane as the equator of the earth. When extended outward, the plane defined by the earth's revolution about the sun becomes the plane of the ecliptic. The plane of the earth's orbit lies in the plane of the ecliptic. Because the earth's axis is not perpendicular to the plane of the earth's orbit and hence is not perpendicular to the plane of the ecliptic, the ecliptic and the celestial equator do not coincide. The axis of the earth tilts away from the perpendicular to the plane of the earth's orbit by $23\frac{1}{2}°$ (Fig. 2.12). The earth is shown revolving about the sun, and the celestial sphere is pictured with the sun at its center. If we visualize the celestial sphere as having a much greater radius than that of the earth's orbit, we can put Fig. 2.12 into a better perspective; that is, in Fig. 2.12 the sun, the earth, and the earth's orbit are drawn much too large in relationship to the size of the celestial sphere.

The zodiac is a belt of sky about 16° wide extending about 8° on each side of the ecliptic. Besides the sun, the moon and the principal planets are always seen within this belt. In Fig. 2.13 is shown an edge-on view of the planes in which the planets move. When projected, these planes would all cross the celestial sphere in that region defined by the zodiac. Spaced along it are the 12 constellations of the zodiac: Aries, the ram; Taurus, the bull; Gemini, the twins; Cancer, the crab; Leo, the lion; Virgo, the virgin; Libra, the scales; Scorpius, the scorpion; Sagittarius, the archer; Capricornus, the goat; Aquarius, the water bearer; Pisces, the fishes. The sun appears to enter a different constellation of the zodiac each month.

EQUINOXES AND SOLSTICES

From the earth in its orbit about the sun, a projection of the sun would trace the ecliptic on the celestial sphere. The two points of intersection of the celestial equator and the ecliptic are called the *equinoxes* (Fig. 2.12). The *vernal equinox* is the intersection of the ecliptic and the celestial equator

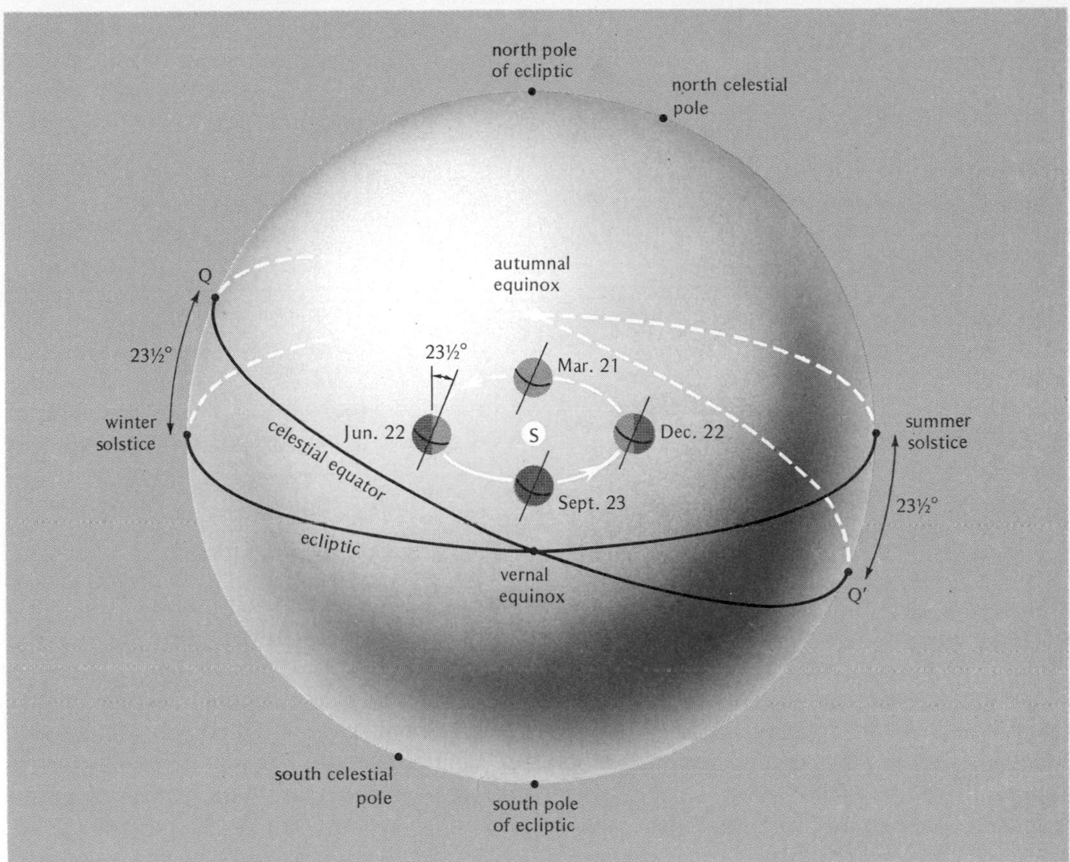

Figure 2.12
Relation of the ecliptic, the celestial equator, the equinoxes, and the solstices. From the earth in its orbit, a projection of the sun (S) would trace the ecliptic on the celestial sphere. Note that this figure is in a sense a mixture of the geocentric and heliocentric model because the celestial sphere should be centered on the earth rather than on the sun as indicated here.

where the sun apparently crosses the celestial equator going northward and is the apparent position of the sun as viewed from the earth on or about March 21. The *autumnal equinox* is the intersection of the ecliptic and the celestial equator where the sun apparently crosses the celestial equator going southward and is the apparent position of the sun as viewed from the earth on or about September 23. The *summer solstice* (Fig. 2.12) is the apparent position of the sun as viewed from the earth on or about June 22, when the sun is farthest north of the celestial equator. The *winter solstice* is the apparent position of the sun as viewed from the earth on or about December 22, when the sun is farthest south of the celestial equator.

Figure 2.12 shows the earth in four positions corresponding to the beginning of the four seasons as it revolves in its orbit

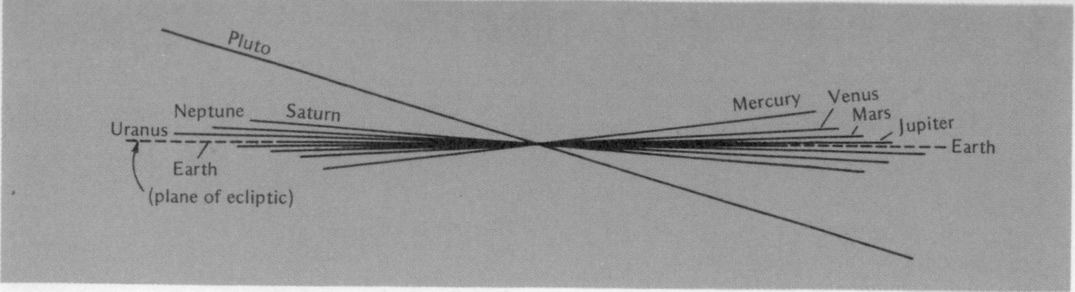

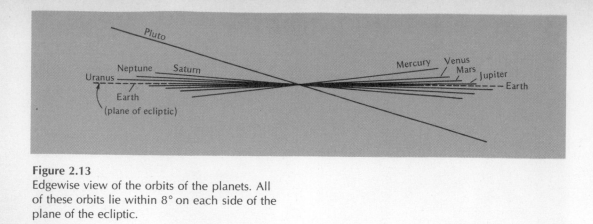

Figure 2.13
Edgewise view of the orbits of the planets. All of these orbits lie within 8° on each side of the plane of the ecliptic.

around the sun. When the earth is at its June 22 position, the sun appears on the celestial sphere at the summer solstice (looking from the earth towards and past the sun); when the earth is at its September 23 position, the sun appears at the autumnal equinox; when the earth is at its December 22 position, the sun appears at the winter solstice; and when the earth is at its March 21 position, the sun appears at the vernal equinox. Thus the apparent yearly path of the sun through the summer solstice, the autumnal equinox, the winter solstice, the vernal equinox, and back to the summer solstice is the ecliptic. The vernal equinox is now in the constellation Pisces, the fishes.

Seasons on planets

CAUSES

There are two principal causes of seasons: (1) variation in distance of a planet from the sun; (2) tilting of the planet's axis from the perpendicular to the plane of the planet's orbit (the earth and Mars are so tilted) together with the revolution of the planet around the sun.

At first one might expect that seasons would be determined primarily by variations in the distance between the sun and a planet. If this were the case, then the earth would experience its warmest season in January when it is closest (about 91.5 million miles) to the sun. However, on the earth the tilt of the earth's axis (2) has a predominant effect.

SEASONS ON THE EARTH

Figure 2.14 shows that because of the tilting of the earth's axis, the sun's rays shine beyond the North Pole on June 22, giving the Northern Hemisphere more heat and light on that date; on December 22 the rays shine beyond the South Pole, resulting in summer for the Southern Hemisphere. In between, at the times of the equinoxes, the sun's rays reach both poles, giving us autumn and spring.

The tilting of the earth's axis results in more hours of sunshine and in more direct rays from the sun on various parts of the earth at various times of the year, thus causing seasons. At the times of the equinoxes day and night are equal everywhere because light rays extend to the poles (Figs. 2.12 and 2.14); there are 12 hours of sunshine.

An additional reason why the tilt of the earth's axis with respect to the sun helps cause seasons is that, for a uniform depth of atmosphere above the earth, oblique rays from the sun must pass through a greater distance of atmosphere (particles of gases, dust, water vapor, etc.) than do perpendicular rays in order to reach a given latitude on the earth. Therefore the shorter the distance through the atmosphere (more nearly perpendicular the rays), the less likely are the

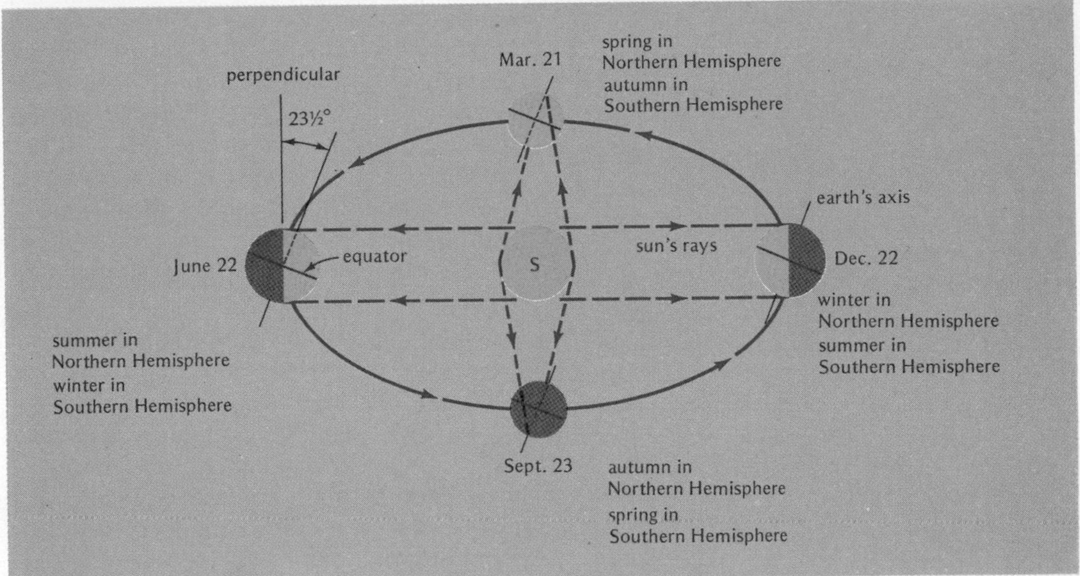

Figure 2.14
Main cause of seasons on the earth. The earth's positions with respect to the sun (S) at the beginning of the four seasons. Note that the earth's axis always points in the same direction with respect to the fixed stars in space—that is, to the north celestial pole. The axis is tilted $23\frac{1}{2}°$ away from the perpendicular to the plane of the earth's orbit.

sun's rays to be reflected backward and the more likely are the rays to be transmitted to the earth. Hence a greater amount of energy from the sun comes to the earth at middle northern latitudes about June 22, when the oblique rays from the sun are more nearly perpendicular to the earth, than about December 22, when the rays slant far more and must pass through a greater distance of atmosphere.

For a latitude of 45°N, the effect of the longer hours of sunshine and the more nearly perpendicular rays on June 22 results in that latitude receiving about four times as much energy from the sun on June 22 as on December 22. An examination of Fig. 2.15 will show how the amount of energy per unit area varies due to the changing angle the sun's rays make at a given position on the earth's surface.

As mentioned in Chap. 1, the earth's orbit is quite circular, the perihelion distance being 91.5 million miles, the aphelion 94.5 million miles. Because the intensity of radiation varies inversely as the square of the distance from the source (the sun), calculations show that the earth as a whole receives nearly 7 percent more heat at perihelion than at aphelion. For a given place on the earth, the effect due to the tilting of the axis is so much greater (300 percent more at 45°N latitude on June 22 than on December 22) than the variation-in-distance factor, that the latter is almost negligible.

Because the earth is at perihelion on January 3, when it is winter in the Northern

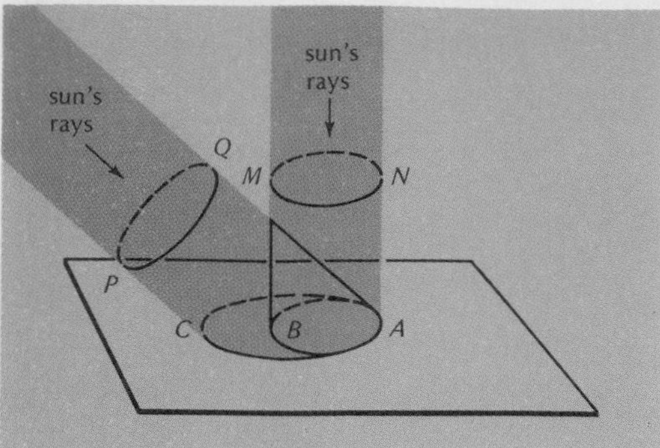

Figure 2.15
One reason why the tilt of the earth's axis helps to cause the seasons. A certain cross section (circle MN) of the sun's rays falls perpendicularly on the earth's surface (circle AB = circle MN). If a cross section of the sun's rays of the same size (circle PQ = circle MN) strikes the earth obliquely, the same amount of energy is spread over a greater, elliptical-shaped area AC, giving less energy per unit area.

Hemisphere and summer in the Southern Hemisphere, the variation-in-distance factor slightly accentuates the seasons in the Southern Hemisphere; that is, the summer is slightly warmer and the winter slightly colder than in the Northern Hemisphere. This effect, however, is quite largely offset by the presence of considerably more water in the Southern Hemisphere, because water has a much higher specific heat (Chap. 8) than the materials making up the land. This means that the same amount of energy falling on water and land will cause a smaller rise in the temperature of the water.

Another equalizing factor is the length of seasons. From March 21 to September 23 (spring and summer for the Northern Hemisphere) there are 186 days as compared with 179 days for the same seasons in the Southern Hemisphere (September 23 to March 21). This is true because the earth moves more slowly near aphelion (July 3) than near perihelion (January 3), as discussed earlier.

LAG OF SEASONS

Although the greatest amount of radiation (energy given off that travels through space) is received from the sun on June 22 in the Northern Hemisphere, the hottest days of the summer occur several weeks later. With increasing daily amounts of heat from the sun, the temperatures rise to June 22. On June 23 less heat from the sun is received than on June 22, but this lesser amount of heat still is more than the heat radiated from the earth on June 23, so that the temperature continues to rise. This also happens on June 24, June 25, and so on, until finally, in the latter part of July or early in August, the earth radiates as much heat as it receives during 24 hours. From then on, the average daily temperatures start to fall. Likewise the coldest days are not near December 22, but because the earth continues to radiate more heat than it receives for several weeks, the coldest weather is in late January or early February.

The balance of heat on hand determines the average daily temperature. As with one's bank balance, the quantity increases as the deposits exceed the withdrawals. In terms of day and night the average daily temperature increases when the earth stores more energy during the day than it loses at night. Conversely, the average daily temperature decreases when the amount of energy stored during the day is less than the amount lost at night. As indicated, a cumulative increase of energy occurs in the Northern Hemisphere until late July or early August, when the warmest weather usually occurs, not at the time of the summer solstice. A cumulative decrease occurs until late January or early February.

SEASONS ON OTHER PLANETS

Because of its proximity to the sun, we cannot detect definite markings on the planet Mercury. Hence we do not have information regarding the amount of tilt of the axis from the perpendicular to the orbit. We do know, however, that Mercury has pronounced seasons, owing to the variation-in-distance factor.

Definite markings are easily observable on Mars with the aid of a telescope, so that its period of rotation and the tilt of its axis can be determined. The axis of Mars is tilted with respect to the perpendicular to the plane of the planet's orbit by 24°, which is nearly the same as for the earth. Hence Mars has seasons similar to ours and for the same reason. The orbit of Mars, however, is much more elliptical than that of the earth, so that the variation-in-distance factor is appreciable. Mars receives nearly half again as much heat and light at perihelion as at aphelion, and this results in considerably warmer summers and colder winters in its southern hemisphere. This is borne out by the observed fact that the southern polar cap varies in size with the seasons considerably more than the northern polar cap.

Measurement of time

Civilized man attempts to keep a record of events—when they happened and where they happened. We need some method of measuring the interval between events: a definite unit of time that is unchanged and may be used to describe the instant of occurrence of events in the past, to keep track of present-day happenings, and to plan for the occurrence of future events. Any periodically recurring phenomenon whose recurrence can be counted may be used as a measure of time, and the interval between two successive recurrences is a unit of time. A clock, for example, is any mechanism that counts the recurrences. Among such mechanisms are astronomical clocks, atomic clocks, quartz-crystal clocks, and radioactive clocks.

ASTRONOMICAL CLOCKS

Astronomical clocks have, without competition until recently, been the foundation of the measurement of time. Why? They are continuous—they never stop running. The unit for the fundamental quantity time is the second.

How did we arrive at our standard of time—the offically accepted definition of the second (the ephemeris second), which we shall state below?

We commonly use the time of rotation of the earth on its axis as a basis for measuring time. This is chosen because it is one of the most constant intervals with which we are acquainted and because, by observing the stars and the sun in their apparent motion, we can easily and exactly measure this interval. The time of rotation of the earth on its axis can be measured by the use of a meridian telescope in conjunction with a pendulum clock, which, so used, may be thought of as an astronomical clock.

There are problems associated with the use of the day—that is, the time for the earth to make one complete rotation—to define our fundamental unit of time, the second. For example, the days are not all of the same length during a year and when we define the day, we need to establish a reference point from which we can determine when one complete rotation of the earth has taken place.

SIDEREAL VS. SOLAR DAY

The day has been defined in two ways. The length of these "days" depends on the reference with respect to which one complete rotation of the earth is measured. A *true* (or apparent) *solar day* is the time of one rotation of the earth on its axis with respect to the center of the actual sun. This is the interval between two apparent successive passages of the center of the sun across the observer's celestial meridian. In order to specify a time that is part way through a day, it is necessary to use a definite starting point such as midnight or noon and to use some smaller units such as hours, minutes, and seconds. The true solar day starts from midnight.

The table of time units with which everyone is familiar is

1 day = 24 hours
1 hour = 60 min
1 min = 60 sec

From this it follows that there are $60 \times 60 \times 24 = 86{,}400$ sec in a day.

The *sidereal day* is defined as the time of one rotation of the earth on its axis with respect to a distant star. This time interval is determined by measuring the time between two successive crossings of the same meridian by a star. In Fig. 2.16 the difference between the sidereal and solar day is illustrated. Sidereal and solar days differ because the earth turns through 360° for a sidereal day but through approximately 361° for a solar day. This, of course, is due to the fact that the earth is revolving in its orbit about the sun while at the same time rotating on its axis. Sidereal days are practically all of the same length; the length of the solar day varies with the earth's position in its elliptical orbit around the sun. However, because the sun figures so much more in our lives than the stars, we use the solar day as our time base.

MEAN SOLAR DAY AND TIME

Because the earth moves in an elliptical orbit with varying speed, the true solar day varies slightly from day to day. This would make it difficult to make a timepiece that kept correct time with 1 day equal to $24^h0^m0^s$, the next $24^h0^m1.2^s$, the next $24^h0^m2.3^s$, and at a later time of year $23^h59^m53.2^s$, the next $23^h59^m52.9^s$, and so on. Certainly if such a timepiece could be made, the large number of complicated gears needed would greatly increase its size. Instead of wearing it on one's wrist, it might have to be hauled on a car trailer. It is thus evident that for purposes of keeping track of time, we must use a day of unvarying length.

The average of all the true solar days of a year is called *mean solar day*. It would be very convenient if we had a body that moved through the sky in such a way as to give us mean solar days instead of the true solar days given by the actual sun.

The *mean solar day* may now be redefined as the time of one rotation of the earth with respect to the mean sun, and a *mean solar second* is 1/86,400 of a mean solar day.

UNITS OF TIME: THE YEAR AND THE SECOND

In order to measure long periods of time, it is convenient to use a larger unit. Because there is an advantage in considering seasons in connection with the passage of time, the year, which is the time of revolution of the earth around the sun, is chosen.

A *sidereal year* is the time of revolution of the earth around the sun with respect to a given star, and its length is $365^d6^h9^m10^s$. The days are mean solar days.

A *tropical year* is the time of revolution of the earth around the sun with respect to the vernal equinox, its length being $365^d5^h48^m46^s$. The days are also mean solar days. The tropical year gets shorter all the time; it is now 0.3 sec shorter than in 1900. A unit of time (a fixed or invariant interval of time) based on the tropical year would have to be defined as a particular fraction of a specified tropical year. Since the basis of ephemeris time in principle is the annual orbital motion of the earth about the sun, as seen in the equivalent motion of the apparent sun about the earth, a second defined in terms of a tropical year would be an ephemeris second.

The most refined astronomical unit—the ephemeris second—was defined by the International Astronomical Union (IAU), and the definition was approved by the General Conference of Weights and Measures.

A large unit of time, a certain tropical year, now determines the second. As defined by the IAU, the *ephemeris second* is

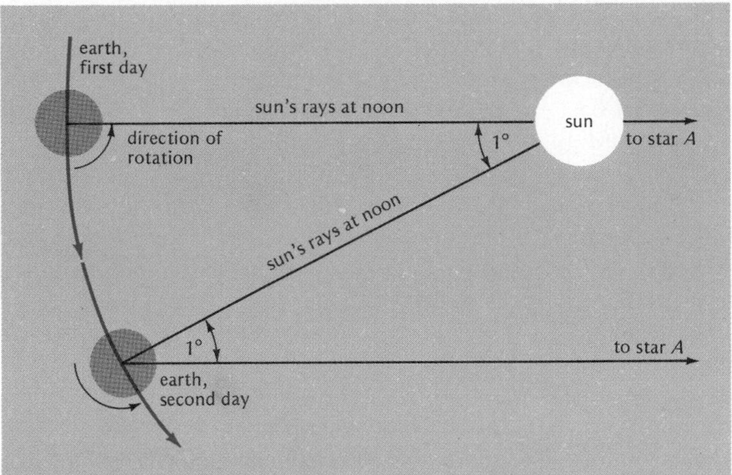

Figure 2.16
Sidereal versus solar day. These two days differ by about 4 min because the earth moves in its orbit while it rotates. The solar day is one rotation of the earth with respect to the sun's rays. The sidereal day is one rotation of the earth with respect to a distant star.

the fraction 1/31,556,925.9747 of the tropical year beginning 1900.0. The length of the tropical year is decreased about $\frac{1}{2}$ sec every century. This variation being known, any particular year can be related to the tropical year starting 1900.0 (an invariable unit) with a precision of 1 in 10 trillion. For most practical purposes, however, the *second* will continue to be considered as the fraction 1/86,400 of a mean solar day. It is seen that 60 sec/min × 60 min/hour × 24 hours/day = 86,400 sec/day.

ATOMIC CLOCKS

As astronomers were refining techniques used to define the second, atomic-frequency standards were being developed. It had long been known that atoms and molecules emitted or absorbed periodic electrical oscillations at sharply defined frequencies (spectral lines), but there were no suitable tools for using frequencies as a basis for a frequency or time-measuring device. (Concepts relating frequency, wavelength, and spectral line are developed in Chap. 9.) In 1934 low frequencies (spectral lines in the microwave range) were first observed. About 1945 Nobel laureate I. I. Rabi suggested that some spectral lines could be used as frequency standards. In 1948 the world's first atomic clock was built at the National Bureau of Standards, based on a definite frequency—an absorption line of ammonia.

In 1958 measurements showed the frequency of cesium (an element) to be 9,192,631,770 ± 20 cycles per ephemeris second, and in 1960 cesium atomic-beam devices were adopted as the U.S. frequency standard and as the international frequency standard. In 1965 a third cesium-beam device provided the U.S. standard. The General Conference of Weights and Measures assigned the frequency mentioned above to cesium and designated its use for the practical measurement of time. Thus the first step was taken toward changing the definition of the unit of time from an astronomical to an atomic basis. The basis of atomic time is an atomic standard of frequency.

Research and refinements continue. The master clock in the U.S. Naval Observatory has available two atomic "watchdogs" to keep it accurate to within 1 sec every 300,000 years. Twin atomic hydrogen masers can be used to provide super-accurate intervals of time for continuously resetting the clock.

The frequency of the hydrogen atom's transition is pinpointed at an incredible 1,420,405,751,694 cycles/sec. Further refinements in the measurement of time are being made by our National Bureau of Standards (NBS) as NBS relates its atomic time scales to astronomical time scales.

Time signals are broadcast by the U.S. Naval Observatory. Time signals and standard frequencies are broadcast to users by radio from NBS Radio Standard Laboratory, Fort Collins, Colorado. Time and frequency are reciprocals (Chap. 9). These time intervals (seconds) and frequencies are calibrated directly to the atomic standard of time. The laboratory has two short-wave (high-frequency) stations and two long-wave (low-frequency) stations.

The high-frequency (short-wavelength) signals with wavelengths of about 30 m travel around the earth by reflections between the earth's surface and various layers of the mirrorlike ionosphere.

This reflective quality of the ionosphere varies with its ever-changing state, thus causing the time of travel of radio signals between two points to change continually, with the signal entirely lost in outer space at times. The low-frequency (long-wavelength) signals with wavelengths of about 10 miles follow the curvature of the earth, moving in a giant duct formed by the ionosphere above and the earth below.

The two high-frequency stations have by far the largest audience, including radio and television stations, electric power companies, radio amateurs, many businesses, and the general public. The two low-frequency stations offer greater accuracy for research laboratories, instrument manufacturers, and installations involved in our space programs, where incredible accuracy is necessary.

QUARTZ-CRYSTAL CLOCKS

Small intervals of time are measured accurately by quartz-crystal clocks. The broadcast frequencies of radio and television stations are regulated by quartz-crystal oscillators. The Federal Communications Commission has assigned a broadcast frequency to each station. It is not practical for NBS to operate cesium-frequency standards for long periods of time because of possible power failures and need for maintenance. Thus it is necessary to use some secondary standard and to calibrate this frequently in terms of the atomic standard. Quartz-crystal oscillators serve NBS well as secondary standards. They are simple, reliable, and operate continuously.

In summary, (1) our basis of timekeeping has moved from the macroscopic to the microscopic; (2) the cesium atom serves as the master control for oscillators which then serve as the continuously running "pendulum" of the atomic clock, and the total units counted by this composite clock form an atomic time scale; (3) atomic time standards permit our technology to accomplish things we could not do before, and to do them relatively economically.

RADIOACTIVE CLOCKS

For dating such things as fossils and geological deposits, which involve extremely long intervals of time, a radioactive clock, based on radioactive isotopes (see Chap. 12), is used, and one method of estimating the age

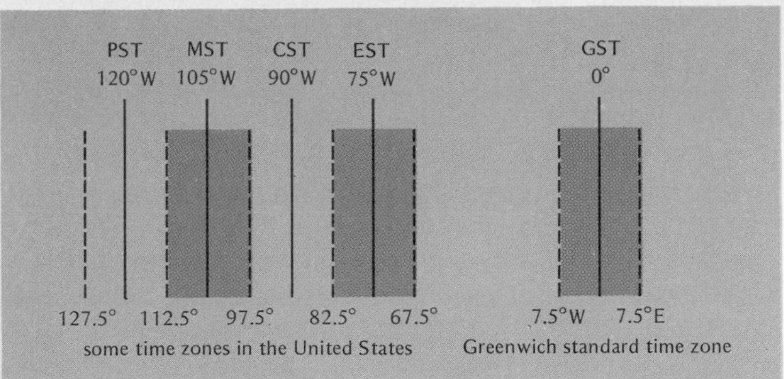

Figure 2.17
The longitude of the standard, or central, meridian of each time zone is shown at the top. The 0° meridian passes through Greenwich, England. In the general geographical plan for zones, the width of each zone is 15°, or 7.5° to the right and to the left of the standard meridian.

of the earth is based on the gradual change of uranium to lead (see Chap. 12). When radioactive carbon from an old material is being studied for dating (determination of age), we may speak of a "carbon clock."

The student may wonder how all the new methods will affect his life. The answer is, probably not much. The astronomical clock, involving the time of rotation of the earth, or more recently the revolution of the earth based on observations of the moon—ephemeris time—will undoubtedly continue to be used in measuring civilian time.

RELATION BETWEEN TIME AND LONGITUDE

Because the sun rises in the east, people living east of us see the sun before we do. If it rises for them at 6 A.M. (as it would on March 21 or September 23), by the time we see it rising (6 A.M. for us) their clock will read a later time—7 A.M. or 8 A.M. or some later time. Because this is an effect due to the rotation of the earth, the change in time due to a difference in longitude will amount to 24 hours for 360°, which is 24 hours divided by 360° = 1 hour/15°, or 4 min/deg; hence the following relation: For every 15° east in longitude, the time of day is 1 hour later; for every 15° west in longitude, the time of day is 1 hour earlier.

STANDARD TIME

The mean solar time for the meridian of a place is *local mean solar time* (LMST). Consider a place that is 1° west of you on the earth's surface (about 60 miles at 30°N and 53 miles at 40°N). When the local mean solar time by your meridian is 10 A.M., the local mean solar time at the place 1° westward is 4 min earlier, or 9:56 A.M. If each community used the time of its own earth meridian, what confusion there would be. To avoid this difficulty, standard time zones are used. Until about 1880 every community used its own local mean solar time. Increasingly, groups of communities used the same time, and in 1884 the standard time zones we now use were set up on a worldwide basis.

Greenwich standard time (GST) is the mean solar time of the 0° meridian and is used as civil time (clock time) by most places within 7.5° of this meridian (Fig. 2.17, right). England has adopted continental time, which is 1 hour later than GST, as her clock time. Greenwich mean solar time is also called *universal time* (UT); data are usually given in astronomical tables. The 0° meridian

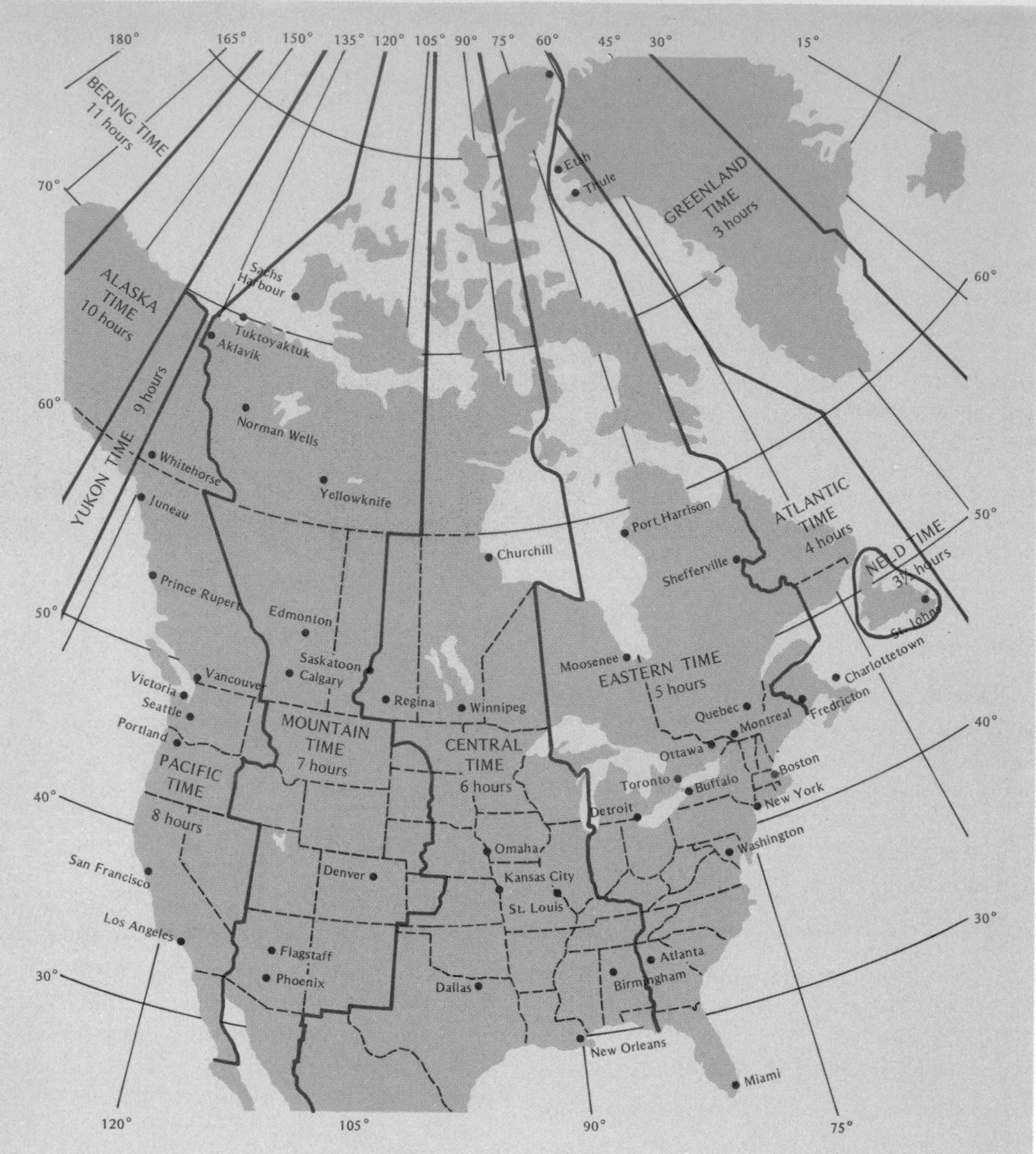

Figure 2.18
Map of standard time zones of the United States, showing the irregularity of the zone lines. Greenwich standard time is later than the standard time of a particular time zone by the number of hours shown for the zone. (From the *Observer's Handbook* (1972), 12–13, Royal Astronomical Society of Canada.)

passes through the Royal Observatory, founded in 1675 in Greenwich, England, a suburb of London. Similarly for the first 48 states of the United States we have eastern standard time, EST (mean solar time of 75°W meridian); central standard time, CST (90°W); mountain standard time, MST (105°W); and Pacific standard time, PST (120°W), as shown in Figs. 2.17 and 2.18.

These standard times are used by most

places within 7.5° of the standard or central meridian of the time zone. The meridian 75°W longitude is near Philadelphia, 90°W is near Chicago, 105°W is near Denver, and 120°W is near Los Angeles. The Atlantic standard time zone (60°W) includes Puerto Rico and the Virgin Islands. Time zones for Alaska include Pacific time (120°W), Yukon time (135°W), official Alaskan time (150°W), and Bering time (165°W). Hawaiian time is also 150°W.

As Fig. 2.18 shows, there are many exceptions to the general geographical setup of the time zone in which standard times are used by places within 7.5° of the standard meridian. Communities may adopt the time of an adjacent zone for their civil time if they so desire. Detroit, Michigan, longitude 83°W, and the lower peninsula of Michigan, for example, have adopted eastern standard time in preference to central standard time, which a strict geographical division would require.

DAYLIGHT SAVING TIME

To take advantage of the maximum amount of sunlight during normal waking hours, most states have daylight saving time during the summer months. It extends from 2 A.M. of the last Sunday in April to 2 A.M. of the last Sunday in October. Daylight saving time is standard time (zone time) plus 1 hour. For example, when it is 6 A.M. EST in New York City, eastern daylight saving time (EDST) would be 7 A.M..

CHANGE-OF-DATE LINE

When you travel westward across the United States, each time you move into a different time zone the time on your watch would be 1 hour fast—that is, 1 hour ahead of the standard time in this new (western) time zone you have entered. This leads to rather interesting timetables for air travel. For example, a flight leaving Detroit at 5:00 P.M. EST is scheduled to arrive in Chicago at 5:10 P.M. (CST time). Actually the air trip is 1 hour and 10 min of elapsed time and not 10 min. If you were to continue to travel westward all the way around the earth, you would lose 24 hours or 1 day.

To understand why a day is lost in traveling around the earth in a westward direction (or a day is gained traveling around the earth in an eastward direction), we can take a make believe trip by airplane. Assume that we start from the 75°W meridian, which passes through Philadelphia in the eastern standard time zone, at true solar noon on October 1, and fly westward at a speed equal to the apparent westward speed of the sun which is 700 mi/hour at latitude 42°N. At this speed we shall appear to move with the sun relative to the earth; that is, the sun will appear to remain stationary with respect to our airplane. At the end of an hour we are about 700 miles west, about over Yankton, South Dakota. We observe that the sun is still on the meridian; it is still true solar noon, October 1. After 2 hours more we are starting over the Pacific Ocean, but the sun tells us that it is still noon, October 1. We cross the Pacific, pass over China and other parts of Asia, over Europe, cross the Atlantic, and come into our original airport after passing over our northeastern states. Our friends are there to meet us after this epoch-making trip, and we look at the sun and say, "Well, it's still noon, October 1." But our friends argue that it is noon, October 2, because they have had three meals and a good night's sleep since we left.

It thus becomes evident that somewhere on our journey we must stop calling it noon, October 1, and start calling it noon, October 2; in other words, we need a change-of-date line. Should we let this line pass through some large city? In that case millions of people just east of the line would have October 1 where as millions of others just west of the line would have October 2. What confusion! In order to have as little inconvenience as possible, the change-of-date line—international date line (IDL)—

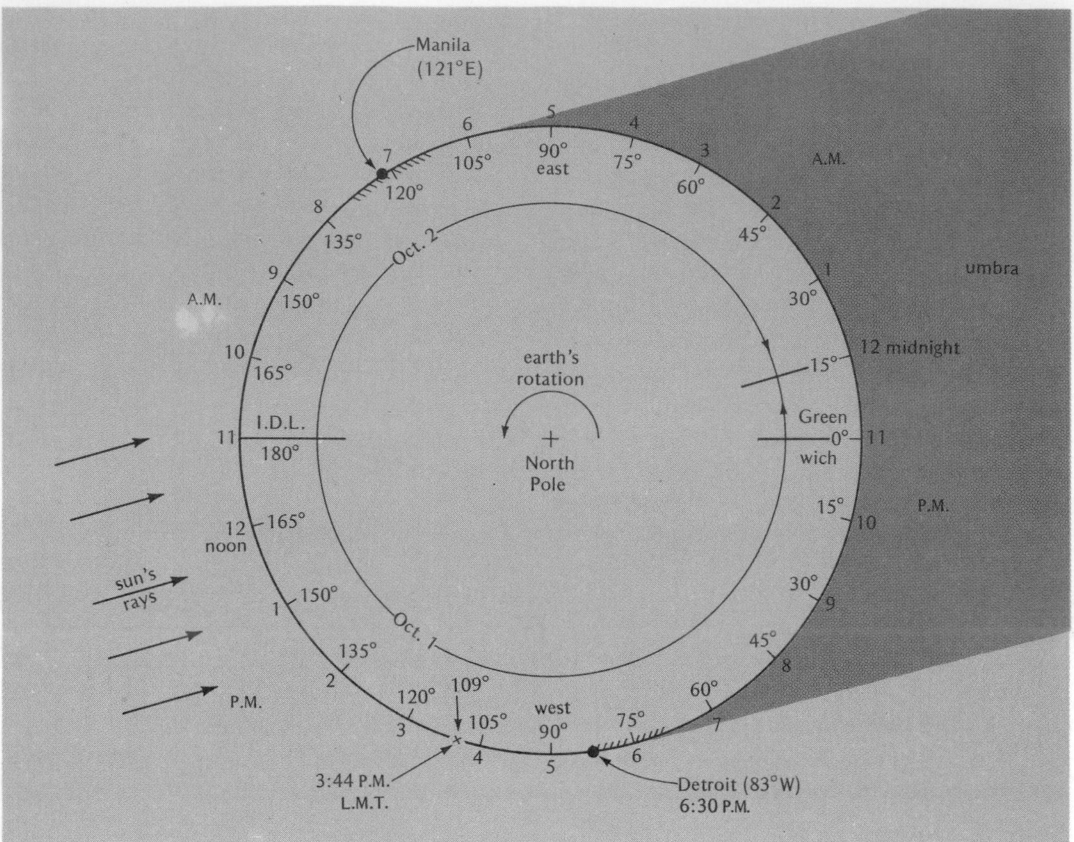

Figure 2.19
Time diagram. The figure represents a view of the earth looking down on the North Pole. Here are illustrated the differences in clock time around the world. On one-half of the earth, it is October 1 while it is October 2 on the other half. The change in date occurs at the International Date Line (IDL) which is the 180° meridian.

coincides with the 180° meridian, which passes through the middle of the Pacific Ocean, except where it deviates to avoid land areas such as the tip of Siberia and groups of islands such as the Fijian group. In crossing the international date line going westward, as from San Francisco to Tokyo, the date and day are set ahead one full day, as from Monday, June 14, to Tuesday, June 15. In crossing the international date line going eastward, the date and day are set back one full day.

In Fig. 2.19 the time and date at various places about the globe are presented when it is 6 P.M. on the 75°W meridian on October 1. We see that when it is 6 P.M., October 1, at 75°W, it is 1 P.M., October 1, at 150°W; noon, October 1, at 165°W; and 11 A.M., October 1, at 180°. The instant after crossing the 180° meridian it is 11 A.M., October 2, as indicated in the diagram. Then it is 10 A.M., October 2, at 165°E; 9 A.M., October 2, at 150°E; 8 A.M., October 2, at 135°E; and finally 7 A.M., October 2, at 120°E.

NAVIGATION AND MILITARY TIME

In navigation it is customary to use standard time of the zone in which the craft is located, but this is expressed from 0 to 24 hours. Four figures are always used: The left two denote the number of hours since midnight and the right two are the number of minutes past the hour. Thus 1:23 A.M. is 0123, 8:02 A.M. is 0802, and 3:12 P.M. is 1512. Midnight of one day may be written 2400 of that day or 0000 of the next day. The military ordinarily use this type of time designation.

THE CALENDAR YEAR

Let us consider the question, "Which of the two kinds of year, sidereal or tropical (p. 38), should be used as our calendar year?" It will be most convenient to use that year which keeps the seasons always at the same time of year. In the heavily populated middle northern latitudes, most people wish always to have winter during December, January, and February rather than have it gradually slip backward through the months, finally occurring in June, July, and August. The tropical year depends on the equinoxes and so do the seasons, so that the tropical year is the seasonal year and should be used as our calendar year.

Because the tropical year consists of 365 days 5 hours 48 min 46 sec, it is not easy to set up a calendar year that will contain a whole number of days and yet will have the required average length. The first good attempt at establishing a calendar year approximating the tropical year was made in the time of Julius Caesar. The astronomer Sosigenes of Alexandria suggested 3 years of 365 days each, followed by a year of 366 days. This gives an average year of $365\frac{1}{4}$ days, or $365^{dh}\,0^{m}0^{6}$, which is within $11^{m}\,14^{s}$ of the tropical year. This year was known as the Julian year, and the Julian leap-year rule was that every year divisible by 4 without a remainder was a leap year. A leap year contains 366 days, the extra day being February 29.

At first thought, the Julian year might seem to be a sufficiently good approximation, but a little calculation shows that a few centuries throw the time of seasons off by several days. The Julian year, first introduced in 45 B.C., when the beginning of the year was moved from March 1 to January 1, was adopted by the Council of Nicaea and put into general use in A.D. 325. By 1582 the time of the vernal equinox had become March 11, that is, 10 days earlier. In 1583 Pope Gregory XIII put into effect for "Catholic countries" an amendment to the leap-year rule suggested by the astronomer Clavius. The revised rule is given below. England and the American Colonies finally made the change in 1752, but Russia waited until 1918. By 1752 the calendar was 11 days ahead of the seasons. An act of the English Parliament then decreed that the day following September 2, 1752, should be called September 14, 1752, thus bringing the vernal equinox back to March 21 and getting the calendar back in step with the seasons. Perhaps you have heard the argument about George Washington's birthday. It is now celebrated on February 22, but some argue that it was February 11. Because he was born in 1732, before we adopted the Gregorian calendar, he was born on February 11 by the Julian calendar (old style), which is February 22 by our present Gregorian calendar (new style).

The revised leap-year rule, called the *Gregorian leap-year rule,* is that every year divisible by four without a remainder is a leap year, except that century years are leap years only if divisible by 400 without a remainder. Only one out of every four successive century years is evenly divisible by 400 (for example, of the century years 1800, 1900, 2000, 2100, only 2000 is a leap year), so that there are 3 days less in 400 years using the Gregorian calendar. Three days is $3 \times 24 \times 60$ min. Hence the average year is shortened by $(3 \times 24 \times 60)/400 = 10.8$ min or 10^m48^s. Subtracting this from the average Julian year gives

average Julian year	$365^d6^h\ 0^m\ 0^s$
less	10^m48^s
average Gregorian year	$365^d5^h49^m12^s$

Thus the average Gregorian year is within 26 sec of the tropical year, so that it will be $(24 \times 60 \times 60)/26 = 3323$ years before the Gregorian year gets even 1 day out of step with the seasons.

THE DAY, MONTH, AND YEAR

We have examined various time standards in this chapter. What is the relationship between the motions that give rise to the three time intervals based on astronomical measurements—that is, the day, the month, and the year? These three time intervals are based on three separate motions: the spinning of the earth on its axis, the day; the time for the moon to revolve about the earth, a 28-day month; and the time for the earth to revolve about the sun, the year. Many attempts have been made to force these natural time intervals to be whole-number multiples of one another. All such attempts have failed. Why should these three motions be related to one another? They appear to be independent motions. Perhaps when we have discovered exactly how the solar system was formed, we will be able to relate these motions to one another.

SUMMARY

The earth is approximately a sphere. It is an oblate spheroid having a polar diameter of 7900 miles and an equatorial diameter of 7926 miles.

There are two main motions of the earth: rotation about its internal axis and revolution about an external axis through the sun. Evidence used in support of these motions includes the Foucault pendulum proof that the earth rotates on its axis and the parallax-of-stars proof that the earth revolves around the sun. The earth rotates on its axis once daily, and it revolves once in its orbit around the sun in 1 year.

Because the planet Earth obeys Kepler's three laws of planetary motion, a portion of the summary of Chap. 1 is here repeated: The orbit of the earth is an ellipse with the sun at one focus. The earth at perihelion is 91.5 million miles from the sun and at aphelion 94.5 million miles, the average distance being 93 million miles (1 AU). The earth moves fastest in its orbit as it passes through perihelion (January 3) and slowest through aphelion (July 3).

An understanding of the meaning of certain fundamental points and circles for the earth and of similar points and circles for the celestial sphere is helpful in developing concepts of the seasons, of the cause of seasons on planets, and of the daily paths of the sun.

The bases for measuring time, the fundamental unit of time, the relation between time and longitude, and the calendar year have evolved from our knowledge of the motions of the earth. With the use of the atomic clock, our basis of time-keeping has moved from the macroscopic to the microscopic.

Important words and terms

rotation	latitude
revolution	ecliptic
Foucault pendulum	equinox
vibration	solstice
parallax	mean solar day
satellite	sidereal day
celestial sphere	tropical year
great circle	sidereal year
equator	atomic clock
meridian	change-of-date line
longitude	

Questions and exercises

1. How did Ptolemy reason that the stars are a very great distance from the earth?
2. State evidence based on observations of the

stars that the earth is not flat. List other evidence that indicates that the earth is not flat.

3. Does a degree of latitude contain the same number of miles everywhere on earth? Explain.
4. Using a chair, a yardstick, and your body, what could you do (a) to demonstrate the difference between rotation and revolution; (b) to demonstrate that a body (your body) can rotate and revolve at the same time?
5. Is a "revolving door" a rotating or a revolving door? Why?
6. Using the terms "rotate" and "revolve" properly, explain how the geocentric and heliocentric theories accounted for day and night. Indicate the "stationary" object in each case.
7. Give fully, with the aid of diagrams, the Foucault pendulum proof that the earth rotates on its axis.
8. Give fully, with the aid of diagrams, the parallax-of-stars proof that the earth revolves around the sun.
9. How is the ecliptic defined? How is the earth's axis of rotation oriented with respect to the plane of the ecliptic?
10. What is an equinox? A solstice? How are they related to the yearly motion of the earth?
11. Suppose that the earth took half a year to complete one revolution about the sun. Explain what effect this change would have on our seasons.
12. Suppose the earth's axis were always to tilt toward the sun. Explain what effect this change from reality would have on the seasons in both the Northern Hemisphere and the Southern Hemisphere.
13. Draw and label a diagram to show the main cause of seasons on the earth, showing the sun and the earth in four positions corresponding to the beginnings of the four seasons. Label each position with the proper date and seasons for each hemisphere. Right now the season in Cape Town, South Africa, is ___________.
14. Give reasons why the tilt of the earth's axis with respect to the sun helps to cause seasons.
15. Explain why for most of the United States the coldest weather is usually not near the first day of winter (December 22) but in late January or February.
16. Explain why for most of the United States the warmest weather is usually not near the first day of summer (June 22) but in late July or early August.
17. If the Rose Bowl football game begins in Pasadena, California, at 2 P.M., January 1, at what hour would it begin for television viewers in Denver, Colorado (MST); in Chicago, Illinois (CST); in Washington, D. C. (EST)?
18. Encircle the leap year(s) in the following list: 1900 1967 1968 1970 2000
19. Use *The World Almanac* to find the latitude and longitude of the approximate location of your college; of your home.

Problems

1. (a) Are the longitude and latitude of each city approximately as given (use an earth globe): Greenwich, England (0°, 51°N); Detroit, Michigan (83°W, 42°N); Istanbul, Turkey (29°E, 41°N); São Paulo, Brazil (47°W, 23°S); Adelaide, Australia (138°E, 35°S)? (b) Using an earth globe, list the longitude and latitude of the following places: Paris, France; Auckland, New Zealand; Cape Town, South Africa; New York, New York; Fairbanks, Alaska.
2. A certain event occurred in Port Moresby, New Guinea (longitude 147°E, latitude 10°S), at 8 P.M., Friday, October 20, Port Moresby standard clock time. For that instant, using a time diagram (shown): (a) Compute Greenwich standard time, day, and date. (b) Express this time as used in navigation by the United States Navy and airlines. (c) Compute standard clock time, day, and date in Washington, D. C. (77°W, 38°N). (d) Compute Washington local mean solar time. (e) What would Washington daylight saving time be? (f) Compute the longitude of city X (latitude 58°N), which has a local mean solar time of 3:08 A.M. for the occurrence of the event. (g) What is the present (October 20) season in Port Moresby?
3. A radio commentator is speaking in Tokyo, Japan (longitude 140°E, latitude 36°N), at 7:00 A.M., Wednesday, February 25, by Tokyo standard clock time. For that instant, using a time diagram (shown): (a) Compute Greenwich standard time, day, and date. (b) Express this time as in navigation. (c) Compute standard clock time, day, and date at your college. (d) Compute local mean solar time, day, and date at your college. An earth globe may be used to determine the longitude and latitude of your

college. (e) People in a certain city hear the broadcast at 3:12 P.M. by their local mean solar time. Compute the longitude of the city. (f) What would Tokyo daylight saving time be? (g) What is the present (February 25) season in Tokyo?

Supplementary readings

Textbooks

See Chap. 1, "Supplementary Readings."

Other books

McGraw-Hill Encyclopedia of Science and Technology, McGraw-Hill, New York (1960), Vols. 1–15. [A reference work that includes articles on all major fields of science. Many articles are by those who made the original discoveries. May be used with any chapter in this textbook.]

Stumpff, Karl, *Planet Earth,* Ann Arbor Paperbacks, University of Michigan Press, Ann Arbor, Mich. (1959). [The shape of the earth, time and the earth's rotation, latitude and longitude, the seasons, and the air around the earth are the subjects of separate chapters, all interesting and readable.]

Articles

Richardson, John M., and James F. Brockman, "Atomic Standards of Frequency and Time," *The Physics Teacher,* **4,** no. 6, 247–256 (September, 1966). [A summary from the National Bureau of Standards.]

Smylie, D. E., and L. Mansinka, "The Rotation of the Earth," *Scientific American,* **225,** no. 6, 80 (December, 1971). [The subtle wobbles of the earth's axis seem to be caused by earthquakes.]

Thomson, M. M., "Sundials," *The Physics Teacher,* **10,** no. 3, 117 (March, 1972).

Reprints from *Scientific American*

W. H. Freeman and Company, 660 Market St., San Francisco, Calif. 94104, has reprints of articles exactly as they appeared in *Scientific American*—complete text and illustrations, full color—available in any quantity and combination at 20 cents each. Write for a descriptive folder and order form, or use the order number.

225 Lyons, Harold, "Atomic Clocks" (February, 1957). [The "pendulums" that regulate them are the vibrating parts of atoms or molecules. So steady are these oscillations that atomic clocks keep better time than the spinning earth itself.]

THE SUN, THE MOON, AND THE PLANETS

Human imagination should not be limited so long as the established laws of physics are not violated.

J. ALLEN HYNEK, 1967

In the first two chapters we examined observations made by the early astronomers with particular attention given to the motions of the planets, the moon, the sun, and the stars. In the present chapter we shall examine the sun, the moon, and the planets and their satellites. Factual information concerning the structure and features of the sun, the moon, and the planets will be presented, and the planets will be compared with respect to the earth. Much of the material in this chapter is rather qualitative and further details on many of the observations presented here will be discussed in later chapters.

In addition to considering the sun, the moon, eclipses, and the planets, this chapter will briefly discuss artificial moons, or satellites, and the possibility of life on other planets.

The sun

The sun is important to us for three reasons:

1. The sun is our main source of light and heat. Most of our energy comes directly from the sun in the form of radiation. A considerable portion of the energy we use comes indirectly from the sun to us through the use of fuels. The energy that we obtain from the combustion of the common fuels —coal, oil, and natural gas—is stored-up energy that plants originally received from the sun in past ages.

2. The sun is by far the most massive body of the solar system and hence controls the motions of the other bodies of the system (Chaps. 1 and 2). The sun is so massive relative to the planets and other smaller objects such as asteroids that more than 99 percent of the mass of the solar system resides in the sun itself.

3. The sun is the nearest of the stars, and by studying the sun with telescopes and other instruments we can indirectly learn a great deal about stars in general even though they may be millions of light-years away.

One *light-year* is the distance light travels in 1 year at its velocity of slightly more than 186,000 mi/sec. The light-year is a unit of distance used in astronomy where distances in miles are so large they become cumbersome to handle. In Chap. 1 it was noted that the nearest star to the sun, Alpha Centauri, is more than 20 million million miles away. It takes light about 4 years to travel from Alpha Centauri to us, so we say that Alpha Centauri is 4 light-years from us. The sun is relatively close to us; in fact, it takes light only 8.3 min to reach us from the sun. Therefore the sun is 8.3 light-minutes, or only a fraction of 1 light-year, away from the earth.

GENERAL INFORMATION

The sun is a huge ball of gas, with the linear diameter of its "apparent disk" being 864,000 miles. The polar diameter is shorter than the equatorial diameter by 43 miles. The diameter of the sun is obtained from the fact that the sun is about 93 million miles from the earth, and the angle subtended by the diameter of the sun is about $\frac{1}{2}°$ as shown in Fig. 3.1.

The mass of the sun is about 300,000 times that of the earth. The diameter of the sun is roughly 100 times that of the earth resulting in the volume of the sun being about 1 million times the volume of the earth. (Remember that the volume of a sphere depends on the cube of the radius.)

It is convenient to define a quantity that results from dividing the mass of an object by its volume as the density of the object. The density of a substance is often used to identify the substance. For example, we might have two pieces of granite rock, one very much larger than the other. Because both of these are made of the same substance, their density (mass divided by volume) will be the same even though their masses and volumes are unequal.

Density will be discussed in greater detail in Chap. 7. However, it is introduced here in order that a comparison of the planets, earth,

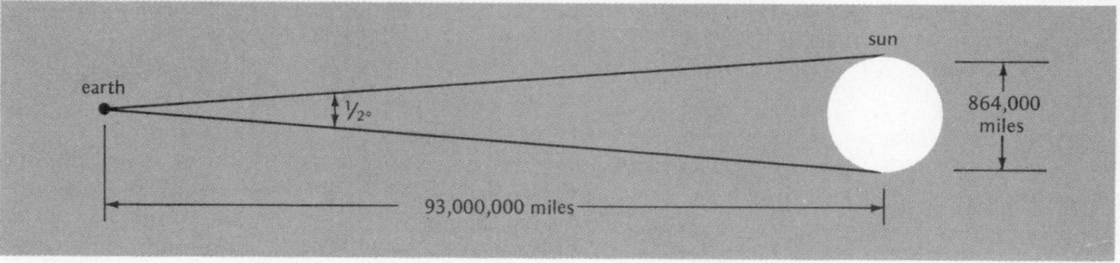

Figure 3.1
Linear and angular diameter of the sun.

sun, and moon may be facilitated. The density of the earth as a whole is about $5\frac{1}{2}$ times the density of water, which is usually taken as the density reference substance for liquids and solids. The density of iron is about eight times that of water, whereas the density of the atmosphere (air) at the earth's surface is about 1/1000 of the density of water.

Using the facts concerning the sun's mass M_S and volume V_S relative to the earth's mass M_E and volume V_E, the density of the sun D_S can be calculated in terms of the earth's density D_E.

$$D_S = \frac{M_S}{V_S} = \frac{300{,}000 M_E}{1{,}000{,}000 V_E} = \frac{0.3 M_E}{V_E}$$

or $D_S = 0.3 D_E$. This small value of density would indicate that a great deal of the sun is gaseous.

The temperature at the apparent surface, or photosphere, of the sun is approximately 6000°C (11,000°F), and the estimated temperature near the center of the sun is 25,000,000°C (45,000,000°F).

Spectral analysis (Chap. 9) shows that more than 60 of the approximately 90 natural elements found on the earth are present in the sun. Most of the elements not yet found in the sun are those of higher atomic weights (see inside front or back cover), and no element not yet known on earth is present in the sun.

Observations of sunspots (see Fig. 3.7) show the sun to be rotating, with the period of rotation increasing from about 25 days at its equator to about 27.5 days at 40° latitude. Although sunspots do not exist at latitudes above 40°, the Doppler effect (Chap. 9) in connection with solar spectral lines indicates that the rotation period continues to increase at higher latitudes to a maximum of 34 days near the poles. This variation in the rotation period of the sun at different latitudes is only one of many solar phenomena not yet well understood. It is, however, rather obvious that the sun could not be a rigid solid object but must be gaseous if it is to rotate with different rates at its poles and at its equator.

LAYERS AND FEATURES

The layers of the sun are the photosphere, the chromosphere, and the corona. The main features of the sun are sunspots, faculae, flares, granules, spicules, and prominences. Although the sun is gaseous throughout, there is a definite visible surface called the *photosphere* (Fig. 3.2). The photosphere is not of uniform brightness but has a mottled appearance (Fig. 3.3) owing to the presence of the relatively small bright areas (600 to 2000 miles in diameter) called *granules*. The granules, which come and go with a mean life of only a few minutes, give the appearance of whitecaps on a choppy lake (Fig. 3.3). The layer that forms the photosphere is probably about 220 miles thick.

The photosphere is surrounded by a layer of gases extending outward for about 9000

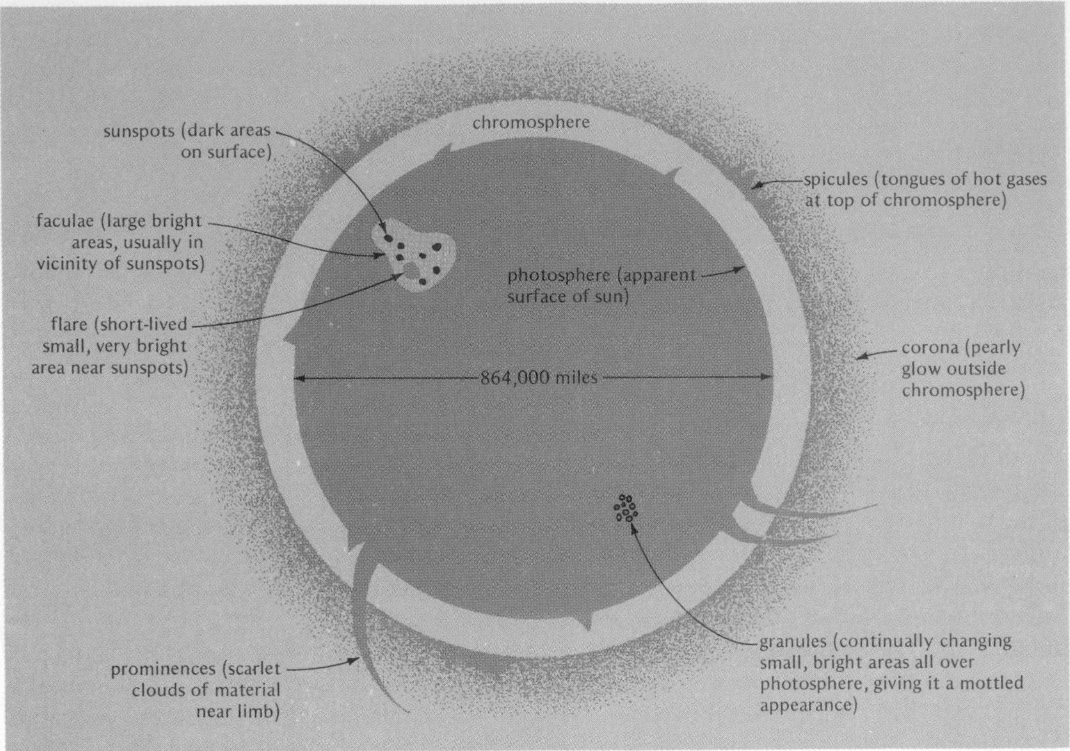

Figure 3.2
Layers and features of the sun.

miles. This layer is called the *chromosphere* (Fig. 3.2) because of its reddish color due to the presence of a large amount of hydrogen. The inner 400 or 500 miles is dense enough to cause the dark absorption lines (color plate facing p. 208, bar 2), called the *Fraunhofer lines,* which appear on photographs of the sun's spectrum, and is sometimes referred to as the reversing layer. Rising from the outer part of the chromosphere are small tongues of hot gases called *spicules*.

Beyond the chromosphere and visible only during a total eclipse, or by means of artificial occultations with a coronagraph, is the pearly glow called the *corona* (Fig. 3.4). At times the corona may be seen out to a distance greater than the sun's diameter, and its appearance changes greatly. During minimum sunspot activity, the corona takes the form shown in Fig. 3.4, top—short brushlike tufts near the poles of the sun and long streamers extending outward from the equator. The other extreme is the form observed at times of maximum sunspot activity, when the corona appears quite uniform, as shown in Fig. 3.4, bottom. The corona is more elliptical at times of minimum sunspot activity and more circular or uniform at times of maximum spot activity.

Earlier spectrographic research on the sun disclosed a green spectral line in the spectrum of the corona, which, because it was not recognized as due to any earth elements, was called the *coronium line*. Further research by Edlén and others, however, indicated that this spectral line and many others subsequently discovered are due to highly ionized atoms of iron, calcium, nickel, and

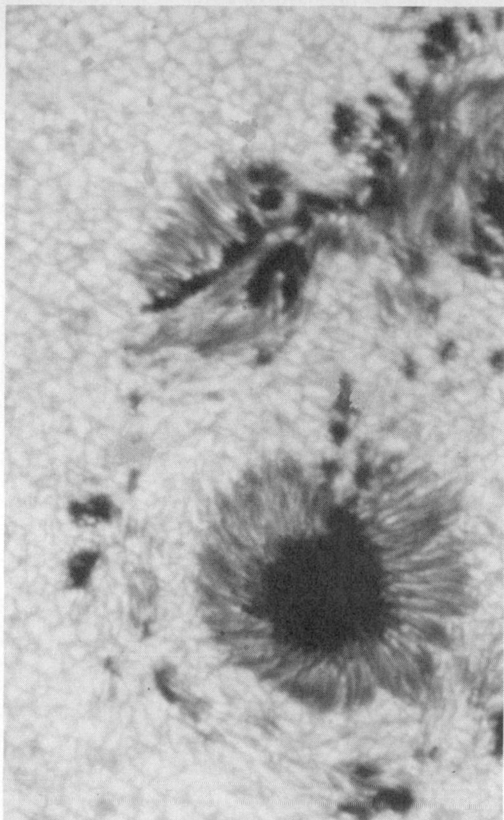

Figure 3.3
An active sunspot. Photographed with a balloon telescope at an altitude of 80,000 ft over Minnesota. The lack of atmosphere above permitted a picture with minimum distortion of light. The cooler dark umbra is depressed, with filaments of gases moving outward and upward to the photosphere of lighter color and higher temperature. (M. Schwarzschild; Project Stratoscope of Princeton University, sponsored by Office of Naval Research, National Science Foundation, and National Aeronautics and Space Administration.)

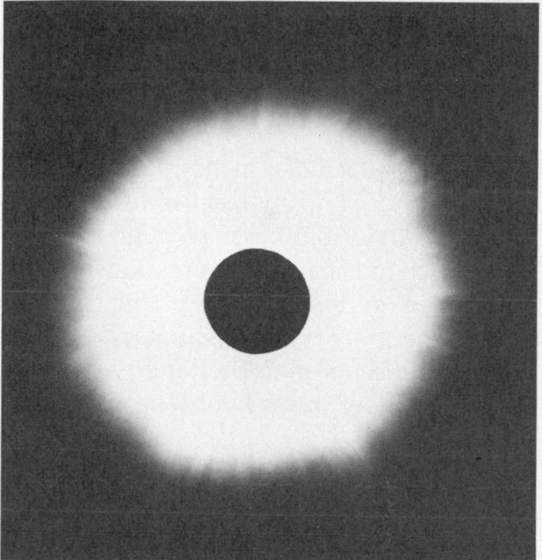

Figure 3.4
Top, shape of corona at time of minimum sunspot activity—total eclipse of June 8, 1918; bottom, shape of corona at time of maximum sunspot activity—total eclipse of May 20, 1947, photographed by Van Biesbroeck. (Top photograph courtesy of the Hale Observatories; bottom photograph courtesy of the Yerkes Observatory.)

argon stripped of a large number of their orbiting or planetary electrons. Theory leads us to believe that such highly ionized atoms can be produced only at temperatures of the order of 1,000,000°C (1,800,000°F). The breadth of the coronal emission lines also indicates a temperature of about 1,000,000°C. A temperature of a million or more degrees is also required to account for the recently discovered "radio noise," or high-frequency radio emission (wavelength band from 1 to 10 m), from the sun.

How to account for this very high temperature of the corona is another of the unexplained problems of the sun. Does the heat causing these high coronal temperatures come (1) from outside the sun (by accretion

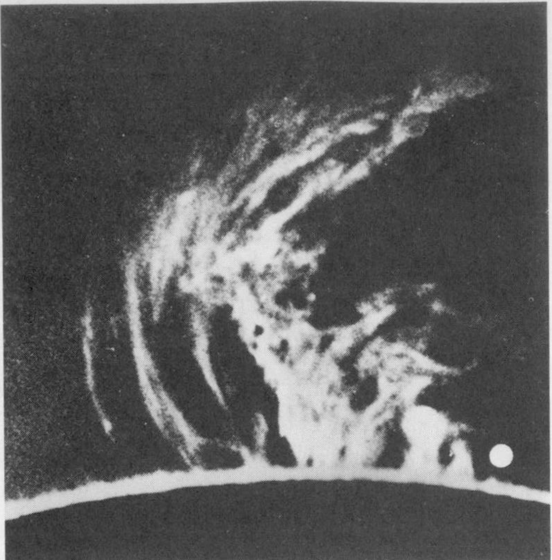

Figure 3.5
Typical eruptive prominence, 140,000 miles high. The white circle shows the comparative size of the earth. (Photograph courtesy of the Hale Observatories.)

Figure 3.6
Typical loop prominence. The two photographs were taken 3 min apart. (Courtesy of the McMath-Hulbert Observatory of the University of Michigan.)

of in-falling interstellar matter); (2) from electric currents set up by electric fields; or (3) from the interior of the sun through the granules, which may cause compressional waves, giving rise to the chromospheric spicules whose energy may be dissipated into heat energy in the corona? The latter view is favored by Kuiper in his book *The Sun*.

PROMINENCES

Larger than spicules are the fantastic *prominences,* which may be seen along the limb of the chromosphere during total solar eclipses or by means of artificial blocking off of the light from the photosphere. Some prominences are eruptive, surging up from the chromosphere; others apparently result from "condensation" of material that may then pour down into the photosphere or float above it like clouds. Photographs and motion pictures taken at the McMath-Hulbert Observatory (Michigan), the Mount Wilson Observatory (California), the High Altitude Observatory (Colorado), and at other observatories show occasional, very rapid changes in prominences, and in some cases material is apparently "blown off" into space. Figure 3.5 shows a typical eruptive prominence and Fig. 3.6 shows a loop prominence.

When prominences that have been observed along the limb, or edge, of the sun are carried away from the limb by the sun's rotation, they appear on the photosphere as long narrow filaments, and they usually appear dark because they are usually cooler than the surrounding photosphere.

SUNSPOTS

Of even greater interest than prominences are the large dark areas, from 500 to 90,000

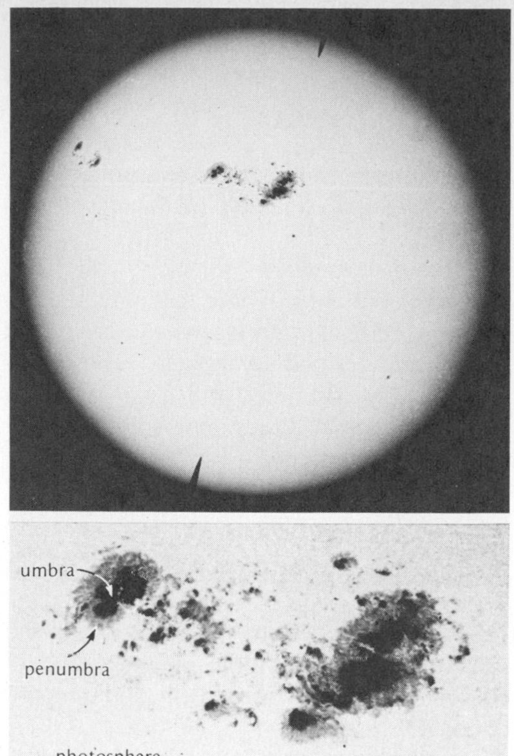

Figure 3.7
Large sunspot group: the whole solar disk and an enlargement of one sunspot group. (Photograph courtesy of the Hale Observatories.)

miles across, called *sunspots,* which frequently appear on the photosphere between latitudes 10° and 40° both north and south of the sun's equator. These differ entirely from the dark filaments previously mentioned. A typical sunspot consists of a relatively black central portion, the umbra, surrounded by an irregular less dark area called the *penumbra* (Fig. 3.7; also Fig. 3.3). Some astronomers who have done research on the sun suggest that a sunspot is a solar storm, like an earth tornado on a huge scale. There is evidence of a rotary motion of the material. In fact, without such a rotation it would be very difficult to account for the huge magnetic fields that are found to accompany sunspots. Quite likely the main part of the vortex, or rotating mass, is below the part we are able to see. Although large single spots do occur, more frequently sunspots occur in pairs and very often many smaller spots accompany each large pair. Figure 3.7 shows a large spot group.

The fact that sunspots appear darker than the surrounding photosphere suggests that they are at a lower temperature. Spectroscopic evidence supports this view and indicates temperatures of about 4000°C (7232°F) for sunspots as compared with about 6000°C (10,832°F) for the photosphere. There seems to be some evidence that the umbra of a sunspot is depressed a few hundred miles, with the penumbra sloping upward to the general surface of the photosphere (Figs. 3.3 and 3.7).

SUNSPOT CYCLES

Sunspots occur in cycles. At the beginning of a cycle a few spots appear at latitudes 35° to 40°. During the next few years the spots increase in number and appear in lower and lower latitudes. Maximum numbers of spots occur about 5 to 6 years after the start of the cycle, at which time the spots are centered around a latitude of 18°. During the next 5 to 6 years the spot zone drifts nearer the sun's equator, and the number of spots decreases. The cycle finally ends with a few spots within 5° of the equator, at which time a few spots of a new cycle again appear at latitudes of 35 to 40°. This periodicity, although it averages slightly over 11 years, has actually varied from 8 to 16 years.

FLARES

Occasionally, intensely bright areas called *flares* (Fig. 3.8) develop in the chromosphere in the neighborhood of sunspots. Estimates of the temperature of solar flares have been placed as high as 2,000,000°C. Flares are smaller and brighter than faculae (Fig. 3.2) and last only a few minutes to a few hours. Flares emit energy, including that of radio frequencies, and seem to affect the ionosphere of the earth, the earth's magnetic field, and the auroras. Chapman suggests that a solar flare is an intensively active region from

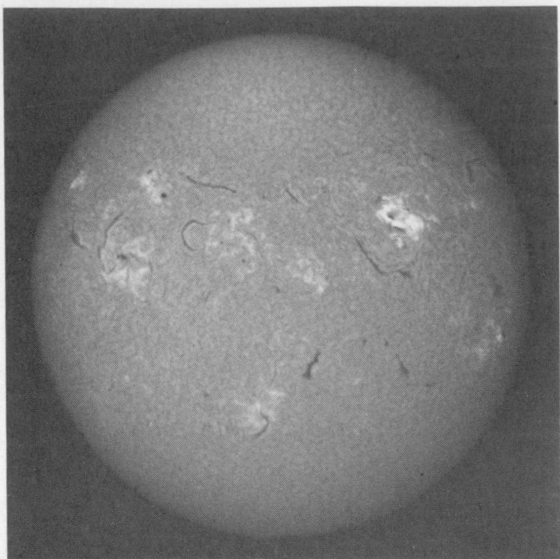

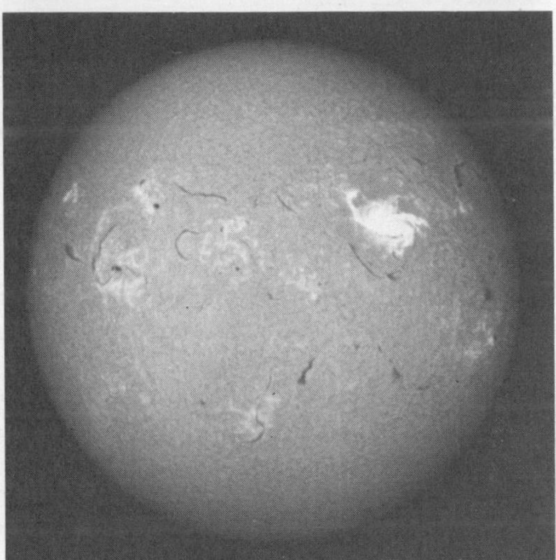

Figure 3.8
Great solar flare of July 16, 1959. The two photographs were taken 44 min apart. (Courtesy of the McMath-Hulbert Observatory of the University of Michigan.)

which a stream of ions and electrons is expelled into space. When the earth, moving in its orbit, encounters this stream of ions and electrons, changes in the ionosphere and the earth's magnetic field follow. These changes may, in turn, produce some interference with radio and television reception and cause auroras. Maximum interference occurs at times of maximum sunspot activity. Figure 3.8 shows a large flare that developed suddenly.

In the disturbed area of faculae and flares, the bright blotches, where calcium is glowing more brightly than elsewhere, are called *plages* (pronounced "plahzhes"). They seem to be closely associated with faculae, which apparently lie at lower atmospheric levels than the bright plages.

PHENOMENA RELATED BY THE 11-YEAR SUNSPOT CYCLE

The following phenomena occurring on the sun and in the earth's atmosphere all vary according to the 11-year cycle and are undoubtedly related:

1. Maximum number of sunspots about every 11 years.
2. Greater percentage of large spots at maximum spot periods.
3. At maximum spot periods, the spots tend to center at about 18° latitude.
4. The corona is more elliptical at time of minimum sunspot activity, more circular or uniform at time of maximum spot activity (Fig. 3.4).
5. More prominences and flares at maximum spot periods (Figs. 3.5 and 3.8).
6. More frequent and more extensive displays of northern lights at maximum spot periods.
7. More magnetic storms (variation in earth's magnetic field) at maximum spot periods, resulting in sudden variations in directions of compass needles and interference with telephone and telegraph communication.
8. More interference with high-frequency radio transmission within the earth's atmosphere at maximum spot periods.
9. A larger percentage of energy received from the sun is in the ultraviolet region at times of maximum spot periods.

THE SUN AS A STAR

Perhaps the most important observation or "feeling" that arises from the many observations of the sun is that the sun is an extremely active, violent, dynamic object and that ac-

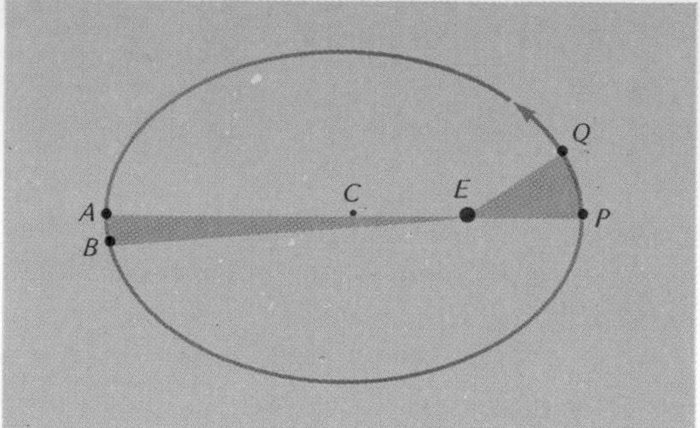

Figure 3.9
Orbit of the moon (*PQABP*). Apogee is at *A* and perigee at *P*.

tivity on the sun has pronounced effects on the earth. In Chap. 11 a further discussion of the activity on the sun will be given when stars are discussed. The sun is the only star that is close enough to the earth to allow us to obtain much information concerning the activity in a star.

The moon

Having discussed the brightest and most important object in the sky—the sun—let us now consider the second brightest heavenly body, our near (astronomically speaking) neighbor the moon. The moon is the only natural satellite of the earth. In comparison with the others, of the 32 natural satellites of the solar system, the moon is fifth in size, ranking behind the three largest satellites of Jupiter and the largest satellite of Saturn.

ORBIT

The moon moves in an elliptical path (Fig. 3.9) around the earth at one focus. The lunar (moon's) orbit is more elongated than that of the earth around the sun. The point *P* in the lunar orbit (Fig. 3.9), where the moon is nearest the earth, is called *perigee,* and the point *A* of the orbit farthest from the earth is *apogee.* Figure 3.9 shows the orbit of the moon with respect to the earth (not to scale) and indicates that the motion of the moon along the orbit is in accord with Kepler's law of areas (p. 12).

The perigee distance *PE* of the moon from the earth is about 222,000 miles, and the apogee distance *AE* about 252,000 miles, with a mean distance of about 239,000 miles. World travelers who have been "around the earth" 10 times have traveled far enough to have reached the moon. The time or period of revolution of the moon about the earth with respect to the stars is $27\frac{1}{3}$ days (sidereal period), but with respect to the sun the period is $29\frac{1}{2}$ days (synodic period).

GENERAL INFORMATION

The moon is a solid, nearly spherical body 2160 miles in diameter. The varying distance of the moon from the earth gives an angular diameter varying from about $29\frac{1}{3}'$ (read "$29\frac{1}{3}$ minutes"; 1 min of angle $= 1°/60$) at apogee to about $33\frac{1}{2}'$ at perigee, compared with the sun's angular diameter of about 32′. This variation of the moon's apparent size causes some solar eclipses to be total and others annular, as shown in Figs. 3.18 and 3.20. The mass of the moon is about 1/80 that of the earth. Because its diameter is roughly $\frac{1}{4}$ that of the earth, its volume is about $(\frac{1}{4})^3 = \frac{1}{64}$ that of the earth. Hence the moon's density ($D = m/V$) is

$$\frac{1/80}{1/64} = \frac{64}{80} = 0.8$$

that of the earth. Using more accurate values of mass and volume gives a value of 0.6. The surface gravity on the moon is one-sixth that of the earth, so that a person weighing 150 lb on the earth would weigh only 25 lb on the moon. (Gravity will be discussed in detail in Chap. 5.)

Careful observations of markings on the moon show that the same markings are always seen at about the same place on the lunar disk. This observed and rather remarkable fact that the moon always keeps the same face toward the earth means that when the moon has revolved halfway around the earth, it must also have rotated one-half about an axis through itself; that is, the moon's rotation period must equal its revolution period. Actually, we can see more than 50 percent of the moon's surface, because certain conditions enable us to see first around one limb, or edge, of the moon and then around the opposite limb. It is as if the moon were oscillated back and forth a few degrees; the term "librations" is used to describe the effect. First, there is the libration of longitude due to the fact that the moon does not move uniformly in its orbit (law of areas). Second is the libration of latitude. The fact that the moon's axis is tilted about 6° from the perpendicular to the plane of its orbit allows us to see alternately beyond the north pole of the moon and beyond the south pole. The third principal libration is the diurnal, or daily, libration. Because our position changes, owing to the rotation of the earth, we see a slightly different 50 percent of the moon near moonset than just after moonrise. Because of the librations, we are able to see 59 percent of the moon's surface at one time or another; that is, 41 percent of the lunar surface is never seen from the earth. Our first knowledge of the appearance of this 41 percent of the moon's surface came as a result of the pictures transmitted back to earth by the Soviet Lunik III as it orbited around the far side of the moon in 1959.

LACK OF ATMOSPHERE

Unlike the earth, the moon has no appreciable amount of atmosphere. How is this known? There is no twilight zone. The division between lighted and dark areas, called the *terminator,* is very distinct. No haze or evidence of storms is ever seen on the moon, although such effects are seen on Mars. Since men have visited the moon, measurements were made to determine if any atmosphere existed. As expected, none was found.

The absence of an atmosphere on the moon is to be expected because of its low surface gravity. If the moon had had an atmosphere in its earlier history, the gas molecules would have escaped molecule by molecule, because the velocities of molecules at temperatures known on the moon often exceed the "velocity of escape"—that is, the velocity necessary for gaseous molecules to leave the moon entirely against its attraction (about 1 mi/sec for the moon as compared with about 7 mi/sec for the earth).

SURFACE TEMPERATURES

Although the moon is the same average distance from the sun as is the earth and hence receives the same energy per unit area as does the earth, the temperature of the moon fluctuates much more than that of the earth. Measurements indicate a variation from about 125°C (257°F) at lunar midday to about −100°C (−148°F) at lunar midnight. The surface temperature of the moon can be as low as −150°C (−238°F). There are two main causes for such a wide variation in temperature. Because the rotation period of the moon with respect to the sun is $29\frac{1}{2}$ days, the sun shines on a given area of the moon $14\frac{3}{4}$ days at a time followed by $14\frac{3}{4}$ days of "night," causing hotter days and colder nights. Also the atmosphere surrounding the earth has a blanketing effect. The atmosphere reduces the radiation reaching the earth's surface when the sun is shining and also reduces the radiation from the earth into space during the night hours. Because the moon has no appreciable atmosphere, there is no such blanketing effect to tend to equalize day and night temperatures.

SURFACE FEATURES

Huge dark areas seen on the face of the moon (Fig. 3.10) make it possible for people with good imaginations to see a "man in the moon." These dark areas are called *maria,* meaning seas, because they were earlier erroneously thought to be bodies of water. Numerous mountains, including ranges like the Apennines, may be seen on the moon with the aid of even a small telescope.

Figure 3.10
Telescopic view of the moon. (Lick Observatory photograph.)

A characteristic feature of the lunar surface is the craters. Figure 3.10 shows the craters Tycho, Kepler, Copernicus, Archimedes, and Plato; Fig. 3.11 is an enlarged photograph of the crater Copernicus. The more than 30,000 craters vary in size from diameters of a fraction of a mile, seen only with the largest telescopes, to 150 miles as in the case of Clavius and Grimaldi. In a crater the fact that the volume of material piled up in the walls above the general ground level seems to be equal in volume to the depression part of the crater lends support to the theory that the craters were formed by impact of large meteorites. The peaks, often found within craters, are

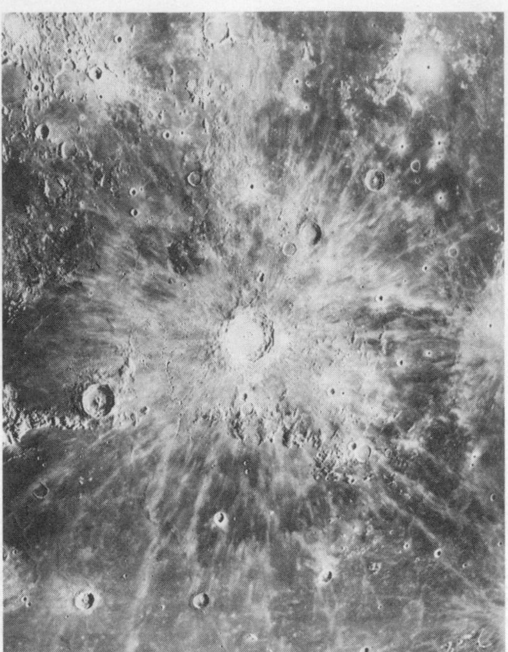

Figure 3.11
The crater Copernicus (center of photograph). The crater is about 56 miles in diameter and is surrounded by white streaks or rays. (Photograph courtesy of the Hale Observatories.)

thought to be a "splashback" affect that was caused when the moon material was made molten owing to the heat of impact. The lack of more large craters on the earth (like Meteor Crater near Winslow, Arizona —see Fig. 1.17) is explained by the effect of our atmosphere in causing obliteration by weathering of all but the most recent craters. Another unusual feature of the moon's surface is the white streaks or rays extending out for hundreds of miles from certain craters, for example, Tycho and Copernicus. These are easily seen in Figs. 3.10 and 3.11. Rills are cracks in the moon's surface as evidenced by the shadows cast by the walls of the cracks. Rays, on the other hand, do not cast shadows, and are much longer than rills. There are rills up to $\frac{1}{2}$ mile wide and 70 miles long and rays up to 5 to 10 miles wide and 1500 miles long. Further discussion of the geology of the moon and the results of recent moon exploration by the astronauts can be found in Chap. 28.

PHASES

Nearly everyone has noticed the varying appearance of the moon in the sky. It ranges from a thin bright crescent, which appears just following the *new moon,* through a half illuminated disk called the *first quarter,* seen about a week later, to a fully illuminated disk (*full moon*) after another week. It then diminishes to a half-illuminated disk (*last quarter*) at the end of the third week and returns to new moon after 29.5 days. The cause of the moon's phases is easily understood if we keep in mind that we see the moon only by reflected sunlight.

The time interval between successive phases = 29.5 days for one revolution/4 phases = about 7.4 days.

In Fig. 3.12 the dark half of the moon and dark half of the earth are shaded. Looking from the earth, the dark half of the moon is seen at new moon, M_1. At a new moon the side of the moon facing the earth is said to be dark. However, light from stars and the earth may cause the moon to be barely visible from the earth. Also the moon is usually a little above or below the plane of the ecliptic so that at new moon we see a thin bright crescent. A new moon is said to be *in conjunction* with the sun. Two bodies are in conjunction if they are in the same general direction from the earth. The new moon rises near the east at sunrise and sets near the west at sunset.

Because the moon revolves around the earth, with respect to the sun in 29.5 days, the moon apparently moves eastward in the sky with respect to the sun (counterclockwise in Fig. 3.12) about 360°/29.5 days, or about 12°/day; this means 13°/day with respect to the stars, since the earth also moves eastward by 1°/day. One day after "new" (M_1), the moon is 12° east of the sun at sun-

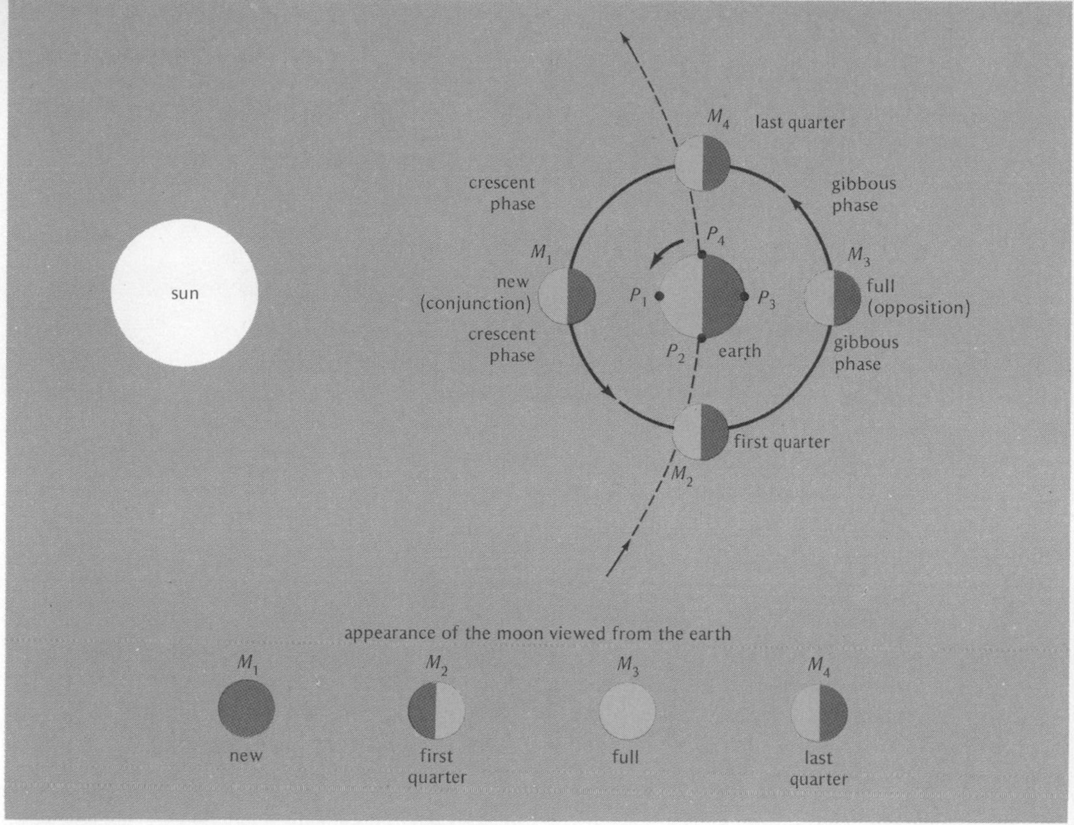

Figure 3.12
Phases of the moon. The dark half of the moon and the earth are shaded.

set and is thus visible low near the western horizon as a slightly larger crescent. At sunset the next night, the moon is about 24° east of the sun and is a larger crescent higher in the sky, somewhat south of west, at sunset. This continues until at about a week after "new," actually 29.5/4 days later, the moon is at position M_2. This first quarter moon M_2 is seen at sunset as a half-illuminated disk on the observer's celestial meridian in the direction of south. If as viewed from the earth the right half of the moon's disk is illuminated, it is a first-quarter moon. Six hours later this first-quarter moon will set near the west. Next morning at sunrise, the first-quarter moon, being 6 hours behind the sun in rising, will be below the horizon on the observer's celestial meridian north. At noon, the first-quarter moon will be seen rising near the east.

A week later the moon is at M_3, having dropped another 6 hours behind the sun in its daily motion. The moon is now opposite in direction to the sun, observed from the earth, and hence is said to be in *opposition*. The disk is now seen fully illuminated and is called the *full moon*. Being opposite the sun, the full moon rises near the east at sunset, is on the meridian south at midnight, and sets near the west at sunrise.

At the end of the third week after "new," the moon is at M_4. It is now 18 hours behind the sun, or 6 hours ahead of the sun. The left half of the moon is now seen to be illuminated, and the phase is called *last quarter*. The last-quarter moon will rise near the east 18 hours after the sun, or 6 hours before it, or about midnight, and will be on the meridian south about 6 A.M., etc.

When the lunar disk is less than half-illuminated, as it is from M_4 through M_1 to M_2, the moon is in *crescent* phase; when the

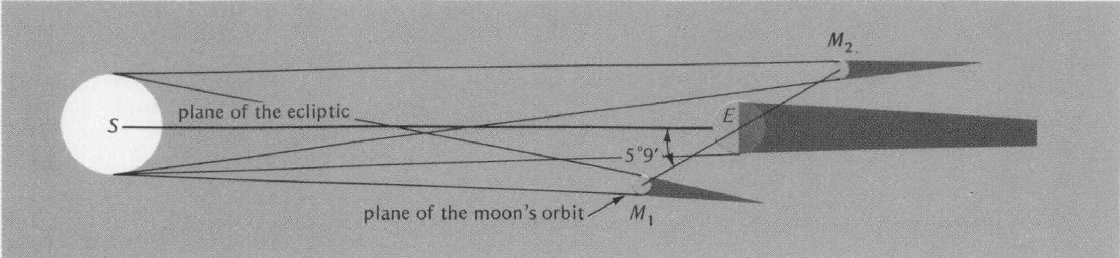

Figure 3.13
Why eclipses do not occur at every new moon and every full moon. The orbital planes of the earth (SE) and the moon (M_1EM_2) are shown edge-on.

lunar disk is more than half-illuminated, as it is from M_2 through M_3 to M_4, it is *gibbous* in phase.

In summary, observing Fig. 3.12, it may be reasoned that (1) a new moon M_1 rises near sunrise (for a person at P_4), is on the observer's celestial meridian south near noon (for the same person now at P_1), and sets near sunset (for the same person now at P_2); (2) a first-quarter moon M_2 rises near noon (for a person at P_1), is on the meridian south near sunset (P_2), and sets near midnight (P_3); (3) a full moon M_3 rises near sunset (for a person at P_2), is on the meridian south near midnight (P_3), and sets near sunrise (P_4); (4) a last-quarter moon M_4 rises near midnight (for a person at P_3) is on the meridian south near sunrise (P_4), and sets near noon (P_1).

Eclipses

For thousands of years man has been interested in eclipses of the moon and the sun. He has progressed from an early superstitious fear that the sun or moon was being eaten by the gods to a present ability to understand the causes of eclipses and to predict accurately, many centuries in advance, the times of their occurrence (as well as to reconstruct for centuries earlier the precise moment of an eclipse, to establish an important historical date).

GENERAL CAUSE

It can be seen from Fig. 3.12 that if the plane of the moon's orbit coincided with the plane of the earth's orbit, at each new moon the moon would be exactly between the earth and the sun, causing an eclipse of the sun, or solar eclipse, and at each full moon the moon would pass through the shadow of the earth, causing an eclipse of the moon, or lunar eclipse. Eclipses, of course, do not happen this often, and the reason for this is brought out by Fig. 3.13. The plane of the moon's orbit makes an angle of 5°9′ with the plane of the ecliptic, which coincides with the plane of the earth's orbit. Thus a new moon at M_1 in Fig. 3.13 will not obscure the sun—hence no solar eclipse—and the full moon at M_2 will pass above the earth's shadow—hence no lunar eclipse.

For an eclipse of the moon to occur, the moon must cross up or down through the plane of the ecliptic at the same time the moon is on or near the earth–sun line. The moon will pass through the earth's shadow (lunar eclipse) if a full moon occurs within a few days of the exact coincidence, and the

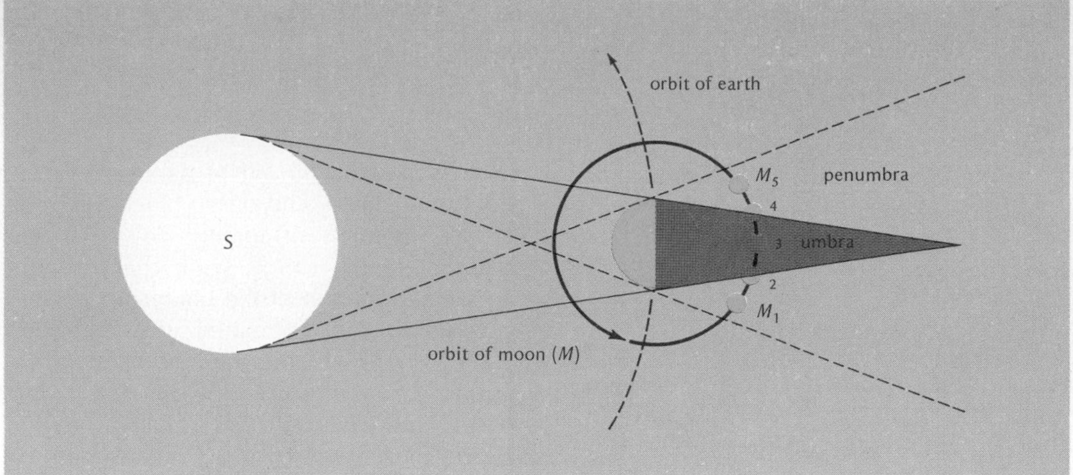

Figure 3.14
Cause of lunar eclipses. The moon (M) is shown at five positions: M_1 and M_5, no eclipse; M_2 and M_4, partial lunar eclipse; M_3, total lunar eclipse (see Fig. 3.15).

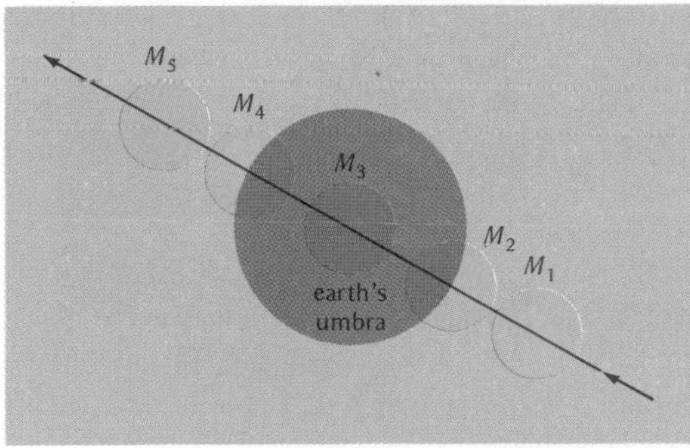

Figure 3.15
Observation of a lunar eclipse. An eclipse occurs as the moon passes into, through, and out of the earth's umbra.

moon's shadow will fall on the earth (solar eclipse) if a new moon occurs within a few days of the exact coincidence. Careful calculations show that there must be at least two solar eclipses per year, and there may be as many as five. Similarly, a year may pass without any lunar eclipses, but as many as three may occur. Conditions cannot exist in the same calendar year, however, for both the maximum number of solar eclipses (five) and the maximum number of lunar eclipses (three). The greatest number of eclipses possible in a given calendar year is seven—five solar and two lunar or four solar and three lunar.

LUNAR ECLIPSES

A lunar eclipse is caused when the moon passes into the earth's shadow, as shown in Fig. 3.14. The length of the umbra, or dark part, of the earth's shadow averages 859,000 miles, and the diameter of the shadow where the moon crosses it is about three lunar diameters. Surrounding the cone of the umbra is the penumbra, which receives light from only part of the sun. At M_1 and M_5 there is no eclipse because there is full illumination; M_2 and M_4 represent partial lunar eclipse with the moon halfway in the umbra; M_3 is total lunar eclipse because all of the moon is within the dark umbra. Figure 3.14 shows the cause of lunar eclipses, whereas Fig. 3.15 represents what the observer sees; the numbering of the moon's positions is the same in both diagrams. One of the necessary conditions for a lunar eclipse to occur is that the moon must be in the full phase (the moon must be on the opposite side of the earth away from the sun).

At times the path of the moon just cuts

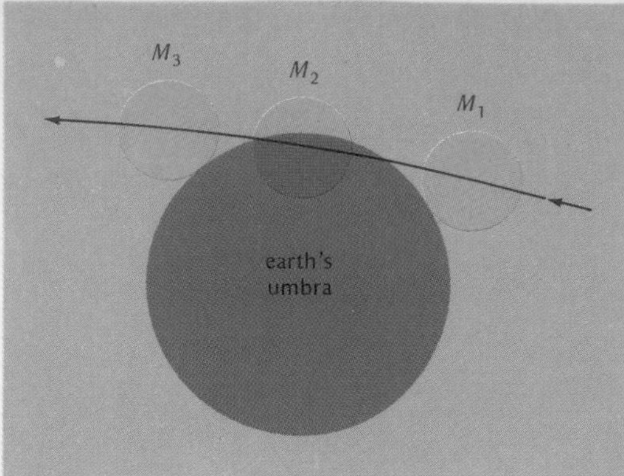

Figure 3.16
Only partial lunar eclipse.

across the top or bottom of the earth's shadow without getting completely within the umbra. Figure 3.16 is a diagram of such a partial lunar eclipse. In this case, the partial eclipse starts at M_1, there is maximum amount of eclipse at M_2, and the partial eclipse is over at M_3.

TOTAL AND PARTIAL SOLAR ECLIPSES

A solar eclipse is caused when the moon gets between the earth and the sun and causes the shadow of the moon to fall upon the earth (Fig. 3.17). One of the essential conditions for a solar eclipse to occur is that the moon must be in the new phase.

The average length of the umbra, the moon's shadow, is about 232,000 miles, which is between the moon's minimum distance from the earth, about 222,000 miles, and its maximum distance, about 252,000 miles. If a solar eclipse occurs when the moon is near perigee the umbra will fall on the earth (*bc* in Fig. 3.17). In this case observers in the region *bc*—a circular area that is drawn out, owing to the rotation of the earth—will experience a total solar eclipse. The width of the path of totality under the most favorable conditions is about 167 miles, and the maximum duration of totality at a given place is $7\frac{1}{2}$ min.

A total solar eclipse is among nature's greatest spectacles (Fig. 3.18); one can be seen from any given geographical location on the average of once in 360 years. To avoid damage to one's eyes, a solar eclipse is viewed with the aid of darkened film—double thickness and dense black. Right up to the moment of totality only the thin crescent of the sun is seen; then, even a very small portion of the photosphere gives a blinding light compared with that from the corona. Suddenly the inner part of the corona bursts into view often with a few bright pinpoints of light, called *Baily's beads,* let through from the photosphere by the irregularities of the moon's limb. Sometimes one large valley of the moon lets through light just before the final thin bright crescent of the photosphere disappears, giving the beautiful diamond-ring effect. Baily's beads or the diamond-ring effect lasts for only a few seconds, and then the full glory of the magnificent pearly white corona is seen. On some occasions, especially during minimum sunspot periods, the coronal "wings" extend out for several diameters. When these long streamers are seen, there is usually a shorter brushlike appearance of the corona near the poles of the sun. At maximum spot periods, the corona extends out almost uniformly in all directions. Frequently one or more of the prominences can be seen, looking like tongues of flame shooting out from the black moon. After a short time—never more than $7\frac{1}{2}$ min—events happen in the reverse order, with Baily's beads appearing, a fading out of the corona, and then the thin bright crescent of the photosphere, which gradually enlarges as the moon continues to move eastward off the face of the sun.

On either side of the region of totality (*a* to *b* and *c* to *d,* Fig. 3.17) observers will see part of the sun blocked off by the moon—that is, a partial solar eclipse. In Fig. 3.17 an observer at *x*, looking past the moon, would see only the top part of the sun obscured—about a 20 percent eclipse in this case. The width of the moon's penumbra at the distance of the earth is nearly 4000 miles, so that a partial eclipse of the sun may be observed from a large portion of the earth's surface.

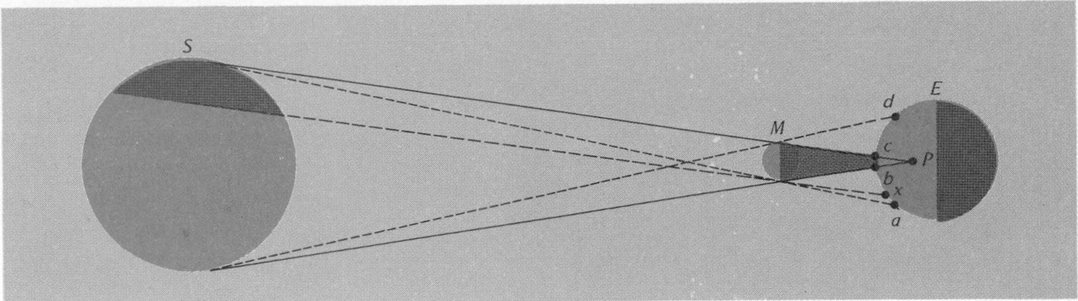

Figure 3.17
Total and partial solar eclipses.

Figure 3.18
Solar eclipse of June 30, 1954. The sun rose in partial eclipse. Photographed at 5-min intervals with a Speed Graphic camera and 12-in. telephoto lens. The partial phases were taken through a number 4 filter, 1/200 sec at *f*/16. The exposure showing the corona was 1/50 sec at *f*/5.6. (Courtesy of the Minneapolis Star.)

The time of duration of the entire solar eclipse from start of partial through totality to end of partial may be up to 4 hours.

ANNULAR AND PARTIAL SOLAR ECLIPSES

Because the length of the umbra of the moon's shadow is less than the average distance of the earth from the moon, more often than not the umbra does not reach the earth. In this case, as shown in Fig. 3.19, the rays forming the umbral cone cross before reaching the earth and give an area *fg* on the earth's surface from which an observer may see the black moon surrounded by a bright ring of sun—an annular eclipse. Note that tangents drawn from within the region *fg*, as from position *y*, and touching the top and bottom of the moon indicate that the top and bottom of the sun may be seen past the moon. When the tangents for three dimensions are considered, the result is a bright ring of sun remaining visible during the eclipse, as shown in Fig. 3.20.

Figure 3.19 also shows that associated with an annular eclipse is a partial eclipse (region *e* to *f* and *g* to *h*), just like the partial eclipse that accompanies a total solar eclipse.

ONLY PARTIAL SOLAR ECLIPSES

About 35 percent of all solar eclipses are only partial. These occur, as shown in Fig. 3.21, when the earth passes through a part of the moon's penumbra but entirely misses

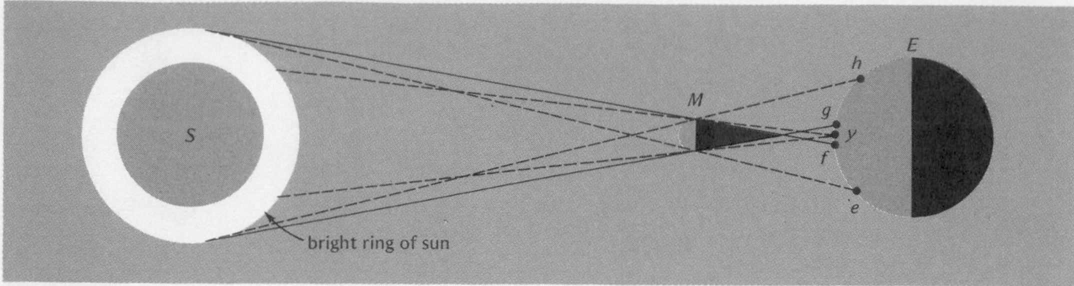

Figure 3.19
Annular and partial solar eclipses.

Figure 3.20
Annular solar eclipse of April 8, 1959. (Ronald W. Boggis, Perth, Australia.)

the umbra. In Fig. 3.21, the earth *E* is as far into the moon's penumbra as it will get, and a partial eclipse is seen over the region *jk*, the amount of eclipse in this case varying from 0 at *j* to about 50 percent at *k*.

Artificial moons

Exploration of space is proceeding so rapidly that only certain early successes are recounted here, with some information of significance. On October 4, 1957, the USSR put into orbit the first man-made satellite, which weighed 184 lb and was called Sputnik I. The perigee distance was about 140 miles above the earth's surface or about 4090 miles from the center of the earth, which is at the focus of the elliptical orbit; the apogee distance was 560 miles. The time of revolution was 1 hour 36 min. Sputnik I spiraled back into the earth's lower atmosphere and disintegrated.

The USSR put up a second artificial satellite on November 3, 1957. The instrument section, weighing 1120 lb, carried a live dog, which finally died in orbit. Sputnik II is no longer in orbit.

The United States placed its first satellite, the 31-lb Explorer, into orbit on January 31, 1958. It was used particularly for exploring the various electrically charged layers of the earth's ionosphere.

On March 17, 1958, the United States orbited the first of its Vanguard series, Vanguard I, which because of its great distance

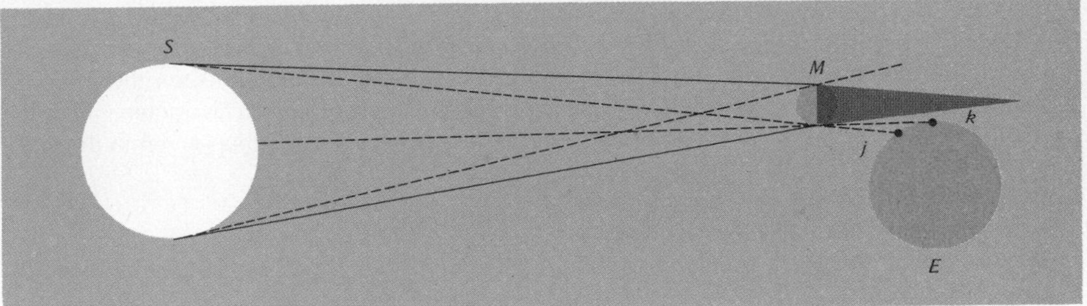

Figure 3.21
Only partial solar eclipse. The earth passes below the umbra in this case.

from the earth's surface—apogee 2460 miles and perigee a little over 400 miles—will not for several hundred years dip down into the denser atmosphere (from 120 miles down) which has caused other satellites to spiral back to the earth's surface. Vanguard I used a solar battery to send back signals. The Vanguards have been used to study the radiation belts that surround the earth (like the dangerous Van Allen radiation belts), to study the exact shape of the earth, and to determine accurately the density of the higher layers of the earth's atmosphere.

The Echo balloon satellites and their successors were used in developing long-distance communications, especially for radio and television.

The Tiros series and their successors represent weather satellites used to make long-range weather forecasts and to improve short-range forecasts by sending back photographs showing cloud coverage over large areas of the earth's surface.

The transits of artificial satellites provide navigational aids that can lead to a determination of one's position on the earth's surface to within 0.1 mile, compared with a 0.5-mile determination by conventional celestial navigation and radio methods.

The Midas satellites and their successors measure infrared radiation from the earth's surface to a sensitiveness that will allow the detection of the firing of a ballistic missile from anywhere on earth.

The Soviets hit the moon with Lunik II (September 12–13, 1959), and Lunik III (October 4, 1959) was used to photograph the side of the moon away from the earth. The Soviet Lunik I (January 2, 1959) had missed the moon and became the first object caused by man to go into orbit around the sun. American Pioneers also orbit the sun and have sent back valuable data from far out in space regarding both solar radiation and cosmic radiation.

On February 12, 1961, the Soviets sent a space probe toward Venus, which, after passing near Venus, went into orbit around the sun. The United States extended this study of Venus with its Mariner series.

Both the United States and Soviet scientists recovered animals and human astronauts that had been sent aloft in earth satellites in preparation for their sending the first man farther out into space. Suborbital flights were made by the Americans Shepard (May 5, 1961) and Grissom (July 21, 1961). The first orbital flight was made by Major Gagarin of the USSR on April 12, 1961; the second by Major Titov, also of the USSR, on August 6, 1961; and the third by Lieutenant Colonel John H. Glenn, Jr., of the United States, on February 20, 1962.

On April 19, 1971, the Soviet Union

placed the scientific station Salyut 1 in orbit. The station consisted of four sections: a compartment containing an airlock and docking attachment for manned space crafts, a biological laboratory, a compartment containing scientific instruments, and a compartment containing a propulsion system. On April 23, 1971, Soyuz 10 was launched and ultimately docked to Salyut 1. Soyuz 10 returned to earth after having orbited with Salyut 1, but no cosmonauts had entered Salyut 1 in space. However, on June 6, Soyuz 11 was launched and docked with Salyut 1. The cosmonauts spent several days carrying out experiments while their space craft was rigidly attached to the space station Salyut 1. The first manned space station had been established. Unfortunately the Soyuz 11 cosmonauts were killed, apparently due to a faulty seal in the space craft which resulted in rapid decompression during landing. It seems clear that permanent manned earth satellites will be established in the near future.

The successful trips to the moon by American astronauts and resulting exploration of the moon are discussed in Chap. 28 because the most apparent scientific work carried out on the moon has been related to geology and the history of the moon.

Manned space vehicles and improved scientific instrumentation continue to add valuable information to man's knowledge of our solar system. We now look forward to further exploration of the moon, information from manned space stations, and trips to Mars and Venus.

An effort should be made to keep outer space from becoming a "junk yard." It now contains some debris (no longer useful satellites, boosters, etc.).

Planets of the solar system

In order to give as much material as possible in a minimum space and to make comparisons possible, considerable information about planets is given in Table 3.1. Facts about the earth are included in the table in order to better understand the other planets by comparison. Some characteristic facts about each planet are then given in the text.

MERCURY

Mercury, the smallest of the planets—diameter 3000 miles—is also nearest the sun, and hence it receives energy from the sun at the greatest rate and travels fastest in its orbit. Mercury has a rotation period of 58.6 days. It turns once on its axis in about two-thirds of the time it takes to make one revolution around the sun. The sunlit half of Mercury becomes extremely hot, probably varying from about 570°F at aphelion to about 770°F at perihelion, a considerable seasonal effect. On the other hand, the half away from the sun is probably not too much above absolute zero, perhaps about −400°F, as there is almost no atmosphere to carry heat to the cold side by convection currents. Thus Mercury has at the same time both the highest and lowest temperatures of any planet in the solar system. On certain areas of Mercury, temperatures would not differ too much from our own, but the lack of any appreciable atmosphere rules out any of the kinds of plant or animal life with which we are familiar.

Mercury is difficult to see in the sky because of its nearness to the sun—never more than 28° east or west of it—but may be seen just after sunset low in the western sky when farthest east of the sun or just before sunrise low in the eastern sky when it is farthest west of the sun.

The plane of Mercury's orbit is inclined to the plane of the earth's orbit by 7°. Hence at inferior conjunction, between earth and sun, the planet is usually north or south of the sun. Occasionally, however, it is exactly between the earth and the sun. The planet then appears as a small black dot on the face of the sun and is said to be "in transit."

Table 3.1
Planetary data

	Mercury	Venus	Earth	Mars	Jupiter	Saturn	Uranus	Neptune	Pluto
Distance from Sun, mean (AU)	0.39	0.72	1.00	1.52	5.20	9.54	19.2	30.0	39.4
Equatorial diameter (miles)	3012	7529	7926	4201	88,800	75,300	29,300	30,600	3900(?)
Rotation period	58.6^d	243^d	24^h	24^h43^m	9^h56^m	10^h14^m	10^h48^m	15^h48^m	6^d9^h
Revolution period sidereal	88^d	225^d	$365\frac{1}{4}^d$	1.9^y	11.9^y	29.5^y	84^y	164^y	247^y
Mass, compared with Earth	0.06	0.82	1	0.11	317.9	95.1	14.5	17.3	0.18(?)
Surface gravity compared with Earth	0.39	0.91	1	0.38	2.31	0.88	0.99	1.1	0.44(?)
Number of natural satellites	0	0	1	2	12	10	5	2	0
Density mean (g/cm^3)	5.5	5.27	5.52	3.95	1.33	0.69	1.7	1.6	4.86(?)
Albedo[a]	0.056	0.76	0.36	0.16	0.73	0.76	0.93	0.62	0.14(?)

[a]Albedo = ratio of light reflected to light received.

VENUS

Venus is often referred to as earth's sister planet. It is somewhat smaller than earth—diameter 7600 miles—and only seven-tenths as far from the sun.

Venus's orbit is nearly circular, and only Venus of all the planets rotates clockwise, as seen from above a planet's north pole. The planet may be as far as 48° east or west of the sun. This means that, when 48° east of the sun, it is an evening star, setting about 3 hours after sunset. At this time it dominates the western sky, being brighter than any other celestial object except the sun and the moon.

Venus is covered with a dense cloud cover which made it extremely difficult for astronomers to determine much information concerning conditions on the planet. For example, no markings are observed on Venus because of the cloud cover, which made it impossible to determine visually if Venus was rotating.

Radar studies by the National Aeronautics and Space Administration provide evidence that Venus rotates once every 243 days, while orbiting the sun in 225 days.

Information from space probes sent up from earth by both the Soviets and the United States have yielded rather interesting data concerning the conditions on Venus. The atmosphere is primarily carbon dioxide, some nitrogen, and a little water vapor but no oxygen. The clouds extend about 35 miles above the surface of Venus. These clouds form a blanket and produce a greenhouse effect. The infrared (heat) radiation from the sun readily penetrates this carbon dioxide layer to reach the surface of Venus. The radiation returned from Venus is longer wavelength infrared radiation that will not pass through the atmosphere. This process creates an extremely high temperature recently measured to be 890°F by the Soviet's Venera 7 space probe. It might be mentioned that some scientists and environmentalists fear that continued pollution of our atmosphere may lead to a similar greenhouse effect on the earth. This process will be discussed further in Chap. 8.

Telescopic views of Venus vary from a thin crescent at inferior conjunction—Venus between earth and sun—to a full disk at superior conjunction—sun between Venus and earth and Venus farthest from the earth. Galileo (1564–1642) was the first to observe that Venus goes through similar phases as the moon, as Copernicus had predicted.

MARS

The planet next farther out than earth is the planet Mars. Mars is the second-smallest planet, having a diameter—about 4220 miles—almost one-half that of the earth or twice the diameter of the moon. Mars is nearest the earth when at opposition, opposite the direction of the sun as viewed from the earth. Being thus closest when the fully lighted side is toward us gives a favorable condition for observation. This together with the fact that Mars has a tenuous atmosphere with only occasional clouds in it allows us, with the aid of powerful telescopes, to see considerable surface detail. The principal features are darker blue-gray areas on a red-orange background and white polar caps (Fig. 3.22). Mars is reddish when observed with the unaided eye.

The polar caps show a definite seasonal effect, increasing to a large white area in the Martian winter and practically disappearing in the Martian summer (Fig. 3.22), when the darker markings become more intensified. Seasons, opposite on the two hemispheres, are to be expected, for photographs of the polar cap and other markings show that the Martian axis is tilted with respect to the perpendicular to the planet's orbit by 25°12′, just slightly more than the tilt of the earth's axis. The Martian year, or time of revolution with respect to the stars, is 687 days, or 1.9 years, making the Martian seasons nearly twice as long as ours.

Spectroscopic evidence indicates that the thin Martian atmosphere contains little water vapor but that the carbon dioxide content is comparable to that of the earth's atmosphere. According to Mariner 4 data, Mars has no detectable oxygen.

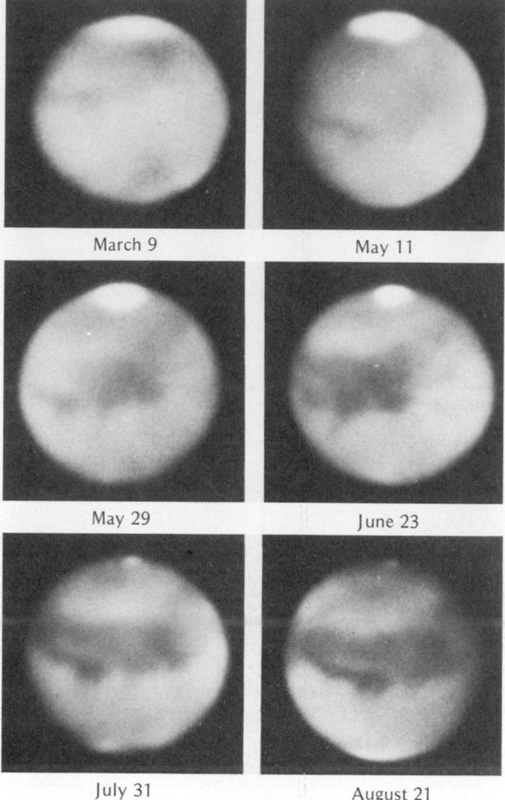

Figure 3.22
Mars. Seasonal changes in the polar cap are shown; as the polar cap decreases in size, the amount of dark area increases. (Lowell Observatory photograph.)

Results of infrared measurements of Mars taken from Mariner 9 indicate that the south polar cap is at a temperature of about −130°C (−202°F) and indicates that the polar caps are primarily frozen carbon dioxide (dry ice) covering water ice on the planet's surface. It should be noted that measurements from the same orbiting satellite indicate the presence of a small amount of water vapor and a larger amount of carbon dioxide gas in the atmosphere above the south polar cap.

Dark areas seen in the photographs in Fig. 3.22 lend credence to the theory that certain low forms of vegetable life like our lichens or mosses may grow there. Temperatures on Mars are favorable for such growth, ranging from about 50°F at Martian noon to about −150°F at Martian midnight, the Martian day being 24^h37^m long.

Of great interest to astronomers are the controversial "canals of Mars." These are networks of fine dark lines first described by the Italian astronomer Schiaparelli in 1877. Lowell and others joined Schiaparelli in attributing the canals to vegetation growing on either side of canals leading water to the desert areas from the melting polar caps. Lowell mapped hundreds of these fine lines, which he saw as great circles on the planet and attributed to intelligent beings on Mars; however, the disagreement as to the detail of the fine markings and the nature of the atmosphere—no detectable oxygen and little water vapor—makes it quite unlikely that any advanced animal life of the type known on earth can exist on Mars.

In November of 1971 Mariner 9 began orbiting Mars. However, because of a planet-wide dust storm, the initial television photographs intended for a mapping experiment yielded little information. The presence of the dust has produced the question of whether or not some of the seasonal variations observed on Mars are caused by dust obscuring to a greater or lesser extent the higher elevations on the Martian surface. Numerous craters have been identified in the Mariner 9 photographs. In addition, a very large trough over 300 miles long and 75 miles wide has been located and is postulated to have been formed by a depression along a line of weakness in the Martian crust. There are numerous branches running into this trough and they resemble a drainage system; however, because of the lack of liquid water on Mars, it is very possible that wind erosion produced these ditches.

Mars has two satellites, one of them, Deimos, behaving like our moon, rising in the east and setting in the west once each Martian day. The other satellite, Phobos, has a revolution period of 7^h39^m. This means that Phobos not only rises in the west and sets in the east but makes three such risings and settings each Martian day. These two moons are rather small, Phobos being about 13 miles in diameter and Deimos being about $7\frac{1}{2}$ miles in diameter.

Mars continues to present unanswered questions for man and there are plans to land on Mars a capsule containing scientific instruments designed to detect the existence of lower forms of plant life.

JUPITER

Jupiter, the largest of the planets—diameter about 89,200 miles—has the shortest day, a little less than 10 hours. This rapid rate of rotation causes a noticeable equatorial bulge.

Visual telescopic observations and telescopic photographs show a succession of bright and dark bands parallel to the equator of Jupiter (Fig. 3.23).

Since the mass of Jupiter is more than 300 times that of the earth, its surface gravity is much greater—about 2.3 times—and it would be expected to "hold" a much more extensive atmosphere. This is indeed the case, and we undoubtedly see only the upper atmosphere of the planet with occasional glimpses of the material of lower layers.

The bands parallel to Jupiter's equator are undoubtedly belts of clouds. Spectroscopic evidence indicates the presence of ammonia and methane gases in large quantities, and a great amount of uncombined hydrogen and helium makes up the bulk of Jupiter's atmosphere.

The density of Jupiter is only 1.33 times that of water, much less than that of the earth which is 5.5 times that of water. In fact, all of the planets farther from the sun than Jupiter have a low density indicating

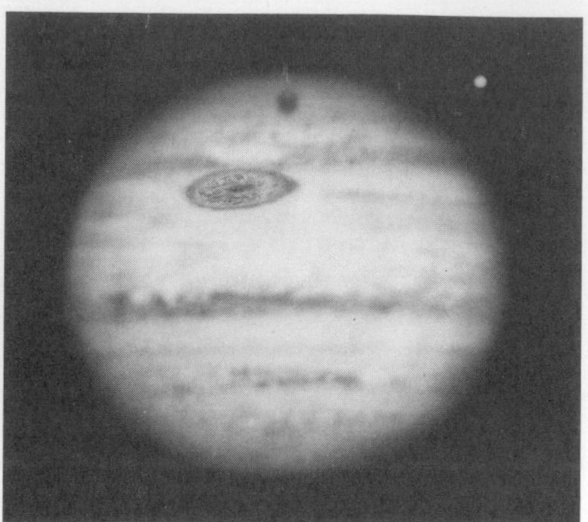

Figure 3.23
Jupiter as photographed with the 200-in. telescope. A satellite (white dot) is above and to the right; its shadow is near the top of the disk. The Great Red Spot is below the shadow. (Photograph courtesy of the Hale Observatories.)

that their volume is predominantly filled with gases. It has been estimated that one might have to penetrate over 20,000 miles into Jupiter before rock is reached.

In addition to the belts of clouds, which change in appearance quite rapidly, there are other markings of a more permanent nature. Chief among these is the Great Red Spot in Jupiter's South Torrid Zone. This huge elliptical spot was first noted in 1831 and has varied in size and appearance ever since, being 30,000 miles long in 1878 and gradually fading but with variations in color since that time. The origin of this spot is still in question. It has been considered to be some kind of eruption from lower levels, and there have been models developed indicating that a violent storm similar to a tropical storm on earth may be responsible for this spot.

Jupiter has 12 known satellites, two of them being larger than the planet Mercury. The brightest four were first discovered by Galileo in 1610 with his earliest telescope, and the twelfth by Nicholson in 1951. Of great interest to amateur astronomers, because observation is possible with the aid of small telescopes, are the phenomena of Jupiter's satellites: the eclipsing of the satellites as they pass through Jupiter's shadow, the transits of the satellites across the face of Jupiter, and the transits of the shadows of the satellites. The outer three satellites seem to revolve in a direction opposite—retrograde motion—to that of the inner nine satellites.

SATURN

Next to Jupiter, Saturn is the largest planet of the solar system, having an equatorial diameter of about 75,000 miles, but its polar diameter is only about 66,000 miles.

Like Jupiter, Saturn has a short rotation period—about 10 hours—and an extensive atmosphere made up of methane and ammonia probably mixed in a large amount of free hydrogen. The surface temperature of −243°F means that more of the ammonia gas is "frozen out" than in the case of Jupiter. Saturn's very low density—only 0.7 that of water—could be accounted for by an atmosphere many thousands of miles thick.

Saturn has had several semipermanent large white spots and has faintly colored belts parallel to its equator (Fig. 3.24).

The most noteworthy feature of Saturn is its series of rings extending, as an apparently flat disk in the plane of Saturn's equator, from about 7000 miles from the surface of the planet to about 37,000 miles from the surface. The rings are seen edge-on every 14.78 years—the next time in 1982. Photographs made with large telescopes show relatively narrow divisions free of material. The thickness of the rings is probably about 10 miles. The displacement of spectral lines of light (Doppler effect, Chap. 9) from the approaching and receding parts of the rings proves that the inner portions of the rings have a greater velocity than the outer portions; the rings, therefore, are not solid disks

Figure 3.24
Saturn and its rings as photographed with the 100-in. telescope. (Photograph courtesy of the Hale Observatories.)

but consist of thousands of individual particles that reflect light. This viewpoint is further supported by the fact that bright stars can be seen through the rings.

Saturn has a family of ten satellites, including Titan, which is nearly twice as massive as our moon and which is known, by spectroscopic evidence, to have an atmosphere. The outermost satellite, Phoebe, moves in retrograde motion.

URANUS

While making a general search of the skies with his telescope, Herschel discovered the planet Uranus in 1781. Uranus is quite similar to Jupiter and Saturn, probably with an atmosphere mainly of hydrogen with some methane and a small amount of ammonia.

The plane of Uranus's equator is tilted 98° to the plane of the ecliptic, giving a retrograde rotation effect. The rotation is rapid, the period being less than 11 hours. Uranus has a diameter about four times that of the earth and has a family of five satellites.

NEPTUNE

As the years passed, Uranus deviated more and more from its predicted orbit. By 1840 the planet was 1.5′ from its predicted position, and Newton's law of universal gravitation, on which the predicted orbit was based, was beginning to be questioned by many leading astronomers. An alternate explanation was put forward that the deviation was caused by an undiscovered planet farther out. Independently Adams in England in 1841 and Leverrier in France in 1845 worked out the location of the suspected stranger. Adams forwarded his calculations to Sir George Airy, Astronomer Royal of England, who did not bother to look for the predicted planet. Leverrier was more fortunate in that the German astronomer Galle looked immediately and found the planet within less than 1′ of the predicted position. This is considered to be one of the greatest triumphs of applied mathematics.

Neptune is about the size of Uranus and is similar to it in most respects. It appears through the telescope as a small green disk, the color probably being due to the methane in its atmosphere. It has two satellites.

PLUTO

Rather slight deviations of both Uranus and Neptune from their predicted orbits led the American astronomer Percival Lowell and others to predict the existence of still another planet. At his death, Lowell left funds for the establishment of the Lowell Observatory at Flagstaff, Arizona. He asked that three main lines of study be followed: a detailed study of the sun (a special solar tower was built), a continuing study of Mars (Lowell believed that the "canals" of Mars indicated the existence of intelligent beings on that planet), and a search in the sky near the ecliptic for another planet. Early in 1930, 14 years after the death of Lowell, Clyde Tombaugh, using a "blink microscope," which compared an earlier photograph of a given sky region with a later one, was thrilled by seeing an object blinking on and off. Photographs were taken for several weeks. Change in the position of the object was observed. It was then concluded that this was indeed the long-sought-for planet. Eventually, sufficient data had been taken to prove not only that the object was a planet but that the approximate period of revolution was about 248 years and the mean distance from the sun was approximately 39.5 AU, the latter being determined from the former by Kepler's third law. All of these data were officially announced in 1930.

Receiving the smallest amount of light from the sun, the new planet has been named for the god of darkness, Pluto. A combination of the first two letters of Pluto suggests the initials of Percival Lowell.

The diameter of Pluto, although difficult to measure accurately, is probably less than one-half that of the earth. No satellites of Pluto have been discovered.

POSSIBILITY OF LIFE ON OTHER PLANETS

In this day of space travel, we should consider the possibility of life on other planets, some of which man may visit in a few decades.

The following conditions must hold in order for plant and animal life, of the kind with which we are familiar, to exist:

1. The planet must have a suitable temperature range. The coldest temperatures must not be much colder and the hottest not much hotter than those reached on the earth. This in turn means that the planet's distance from the sun must not differ too much from that of the earth. Of course, in the case of a planet of a much hotter sun the critical distance for a proper temperature range would be greater.

2. The planet must have a sufficient amount of atmosphere. A planet much smaller than the earth would have such a low "velocity of escape" that a gaseous atmosphere would escape molecule by molecule, thus leaving an "airless" planet, or at least one with an atmosphere too rare to support life as we know it.

3. The planet must have the right kind of atmosphere. In terms of our own carbon dioxide–oxygen cycle, animal life needs oxygen and exhales carbon dioxide; plant life needs carbon dioxide and gives out oxygen. There must be sufficient quantities of these two important gases. Our kind of life could not exist in an atmosphere of "poisonous" gases such as ammonia gas, carbon monoxide, or methane, which are known to exist on certain planets.

4. The planet must have a considerable amount of water on its surface.

Applying these conditions to the other planets of the solar system, we have to conclude that the earth is the only planet on which life as we know it can exist naturally. Mercury has very little atmosphere, and its temperatures are too extreme. Venus and Mars have no, or too little, oxygen and not enough water to support our type of animal life. Perhaps some forms of plant life—lower forms such as lichens and mosses—may exist on these planets. Jupiter, Saturn, and the other outer planets are too cold, and their atmospheres are made up largely of poisonous ammonia and methane gases.

Let us briefly consider the satellites of planets within the solar system. The moon is too small to hold any appreciable atmosphere, so that no water can be present, as it would vaporize and then escape. The satellites of Mars are much too small to have atmosphere. The largest satellites of Jupiter and Saturn may well hold an atmosphere, but at that great distance from the sun, temperatures would be too low for our kind of life.

Thus we reach the conclusion that life as we know it probably does not exist elsewhere in our solar system, with the possible exception of lower forms of plant life on Venus and Mars.

But what about other planets of other solar systems out in space? As can be read in books dealing with cosmology (see Chap. 11 for some other aspects of cosmology), there is no reason to doubt that a large number of the billions of suns out in space have families of planets, and it seems reasonable that one or more planets of a large number of these systems will be of the right size and right distance from their sun to have proper atmospheres and temperatures for the development of life as we know it. Remember, though, that even the nearest of these other possible planetary systems is 20 million million miles from our own solar system. We are not likely to exchange visits with them very soon.

SUMMARY

The sun is important to us because (1) it is our main source of light and heat; (2) it controls the motions of the other bodies of the solar system; and (3) it is the nearest of the stars, and by studying the sun we can indirectly learn about stars in general.

The sun has a diameter of 864,000 miles, is gaseous throughout, is the most massive body in the solar system, and has an estimated temperature near the center of about 45,000,000°F. It contains more than 60 of the natural elements known on the earth.

The layers of the sun are the photosphere, the chromosphere, and the corona. The main surface features are sunspots, faculae, flares, granules, prominences, and spicules. Certain phenomena occurring on the sun and in the earth's atmosphere all vary according to the 11-year sunspot cycle and are undoubtedly related.

The mean distance of the moon from the earth is 239,000 miles, and the period of revolution about the earth is $29\frac{1}{2}$ days with respect to the sun. It has no appreciable atmosphere, and its surface temperature varies from about −150° to +125°C (−238° to 257°F). Its surface features include flat areas (maria), mountain ranges, craters, rills (cracks), and rays.

Data about the planets are summarized in Table 3.1. Pluto is almost 40 times farther from the sun than is the earth and has the longest period of revolution. Mercury is the smallest planet and has the shortest "year." Venus is the brightest of the planets. Jupiter is the largest and has the most moons. Saturn has beautiful rings. Mars has polar caps that are visible with a telescope.

Phases of the moon—new, first quarter, full, last quarter—result from the revolution of the moon around the earth with respect to the sun. The earth has one natural satellite and since 1957 several artificial satellites, which are used in the collection of valuable scientific information and communications transmissions.

A lunar eclipse occurs when the moon passes into the earth's shadow. A solar eclipse occurs when the moon passes between the earth and the sun and causes the shadow of the moon to fall upon the earth.

In order for plant and animal life of the kind with which we are familiar to exist on a planet, it must have (1) a proper temperature, (2) a sufficient amount of atmosphere, (3) the right kind of atmosphere, and (4) a considerable amount of water vapor on its surface. Life as we know it probably does not exist elsewhere in our solar system with the possible exception of lower forms of plant life on Venus and Mars.

There is no reason to doubt that a large number of the billions of suns out in space have planets, and it seems reasonable that one or more is of the right size and distance from a sun to have proper atmospheres and temperatures for the development of life as we know it.

Important words and terms

light-year	sunspot cycle
density	flares
sunspot	perigee
photosphere	apogee
chromosphere	phases of the moon
corona	in conjunction
prominences	solar eclipse
umbra	lunar eclipse
penumbra	partial eclipse

Questions and exercises

1. Explain why the sun is important to us.
2. List several facts of general information about the sun.
3. Discuss the relation of the 11-year sunspot cycle to certain phenomena that occur on earth or in its atmosphere.

4. List several facts of general information about the moon.
5. Imagine that you were with Armstrong when he landed on the moon. Describe some aspects of your trip and the conditions you might expect to have discovered during the explorations.
6. Why do we always see the same face of the moon?
7. Multiple choice: A last quarter moon rises about (a) noon, (b) sunrise, (c) midnight, (d) sunset.
8. Multiple choice: A full moon rises about (a) sunrise, (b) noon, (c) sunset, (d) midnight.
9. Multiple choice: A full moon is seen on the western horizon about (a) sunrise, (b) noon, (c) sunset, (d) midnight.
10. Multiple choice: A first quarter moon would be seen on the eastern horizon about (a) sunrise, (b) noon, (c) sunset, (d) midnight.
11. A person saw the moon on the eastern horizon at sunrise. What was the approximate phase of the moon at that time?
12. From the earth we see phases of the moon. On the moon could we see similar phases of the earth? Explain.
13. Why do eclipses not occur each month?
14. Draw a diagram and label properly to show the cause of lunar eclipse (total, partial, and no eclipse). State the necessary conditions for this type of eclipse.
15. Draw separate diagrams and label each properly to show the cause of (a) total and partial solar eclipse; (b) annular and partial solar eclipse; (c) only partial solar eclipse. On (a) place an observer at *x*, from which he would see about a 30 percent eclipse. State the necessary conditions for each type of eclipse.
16. Give four characteristic facts about each inner planet: Mercury, Venus, Earth, Mars; and about Jupiter and Saturn. Give facts that seem to be important.
17. Few people have seen Mercury. Why?
18. When viewed from the earth why is Venus brighter in its crescent phase than in its full phase?
19. During the different times of the year that it can be viewed from earth, Mars varies greatly in brightness. Why? Explain with a sketch of Mars and earth in orbit around the sun.
20. Discuss the possibility of life as we know it on other planets in our solar system and on planets of other solar systems out in space.

Project

Diameter of the sun

Materials required: One yardstick (or meterstick), two 3- × 5-in. index cards, two thumbtacks, one common pin.

a. Use the common pin to make a pinhole in the center of one of the cards. Make sure that there are no dust or fiber particles in the pinhole. An excellent pinhole can be made by taping a piece of aluminum foil over a larger hole in the card and then punching the pinhole in the foil. Assemble your apparatus with one card at each end of the yardstick.
b. On a clear day point your instrument with the pinhole toward the sun; it should be propped up against a window or held still by some other means. Make two sharp pencil marks to mark the sun's diameter. The image of the sun will not be very bright and you may have to shade the apparatus from all other light.
c. Record the measured diameter of the sun's image. With the measured diameter of the sun's image, the known length of the yardstick, and the information that the earth is 93 million miles from the sun, you can solve for the diameter of the sun.
   (1) Give your equation for solving for the diameter of the sun. (Hint: Use ratios.) Explain, with the aid of diagrams, how you obtained the equation.
   (2) Compute the diameter of the sun using the measurements and the equation.
   (3) Compare the diameter of the sun as computed with the value given in this book.
   (4) Turn in the two index cards as part of your project.

Supplementary readings

Textbooks

See Chap. 1, "Supplementary Readings."

Other books

Bates, D. R. (ed.), *The Earth and Its Atmosphere*, Science Editions, New York (1961). (Paperback.) [Airglow, auroras, and magnetic storms are dealt with in Chaps. 13 and 14.]

Bergamini, D., *The Universe,* Life Nature Library, Time Life Books, New York (1968).

Gamow, George, *The Birth and the Death of the Sun,* Viking Press, New York (1940). [This volume deals in simple language with the story of the stars and the sun.]

Hawkins, Gerald S., *Splendor in the Sky,* Harper & Row, New York (1969). [Of interest for Chap. 3 are the chapters on Venus and Mars; Jupiter and Saturn; and the Moon.]

Kiepenheuer, Karl, *The Sun,* Ann Arbor Paperbacks, University of Michigan Press, Ann Arbor, Mich. (1959). [This is a thorough treatment of the only star whose shape and surface can be observed. There are many excellent photographs and diagrams. Solar eclipses are considered.]

Kuiper, Gerald P., *The Sun,* University of Chicago Press, Chicago (1953). [This is the first of four volumes intended to collect and systematize all significant information available on the sun, planets, satellites, comets, and smaller bodies of the solar system. It is designed as a technical reference book.]

Meeus, J., C. Grosjean, and W. Vanderleen, *Canon of Solar Eclipses,* Oxford University Press, New York (1966). [Utilizing modern computers, this volume shows accurately the paths of visibility for 1450 solar eclipses, some illustrated in the Oppolzer volume cited below. The 58 maps cover eclipses for the period A.D. 1900–2509.]

Menzel, Donald H., *Our Sun,* Harvard University Press, Cambridge, Mass. (1959). [A comprehensive account of the sun as a star.]

Menzel, D. H., *Astronomy,* Random House, New York (1970).

Ohring, G., *Weather on The Planets,* Anchor Books, Doubleday, Garden City, N. Y. (1966). (Paperback.)

Oppolzer, T. von Ritter, *Canon of Eclipses,* Dover, New York (1962). [A monumental work, first published in 1887, showing the paths of visibility for 8000 solar eclipses and 5200 lunar eclipses. The 160 maps cover eclipses from 1207 B.C. to A.D. 2161.]

Pickering, J. S., *1001 Questions Answered about Astronomy,* Dodd, Mead & Company, New York (1958). [Common questions about the universe and its components are answered by a staff member of the Hayden Planetarium, New York City.]

Shklovskii, I. S., and Carl Sagan, *Intelligent Life and the Universe,* Holden-Day, San Francisco (1966). [A stimulating and authoritative survey of modern astronomical developments with technical underbrush avoided. Useful for Chap. 3 are the authors' examination of our own planetary system and their summary of recent ideas on the formation of planetary systems. Twenty-two of 35 chapters are exciting, almost pure astronomy. The last few chapters are adventure and exercise in stretching the imagination.]

Smith, A. G., and T. D. Carr, *Radio Exploration of the Planetary System,* Momentum Books, Van Nostrand, Princeton, N. J. (1964).

Stumpff, Karl, *Planet Earth,* Ann Arbor Paperbacks, University of Michigan Press, Ann Arbor, Mich. (1959). [Chapter 6 is about the earth and the moon; Chap. 7, the sun; and Chap. 11, the earth, the universe, and life.]

Whipple, F. L., *Earth, Moon, and Planets,* McGraw-Hill, New York (1963). [This volume brings the subject to the general reader in a meaningful manner.]

Articles

Becker, John V., "Re-entry from Space," *Scientific American,* **204,** no. 1, 49–57 (January, 1961). [Good tables and discussion of difficulty of landing on various bodies of the solar system.]

Eshleman, V. R., "The Atmospheres of Mars and Venus," *Scientific American,* **220,** no. 3, 78 (March, 1969).

Faller, J. E., and E. J. Wampler, "The Lunar Laser Reflection," *Scientific American,* **224,** no. 3, 38 (March, 1970). [A description of the Apollo 11 experiment by which the distance to the moon from the earth has been measured with an accuracy of 6 in.]

"First Explorers on the Moon," *National Geographic,* **136,** no. 6, 735 (December, 1969). [The story of Apollo 11 told in five articles.]

Hall, A. J., "Apollo 14; The Climb Up Cone Crater," *National Geographic,* **140,** 1, 136 (July, 1971).

Leighton, R. B., "The Surface of Mars," *Scientific American,* **222,** no. 5, 26 (March, 1970). [A discussion of the pictures of Mars taken by Mariner 6 and 7.]

Murray, B. C., "Mars From Mariner 9," *Scientific American,* **228,** no. 1, 48 (January, 1973).

Sagan, C., "Mars A New World to Explore," *National Geographic,* **132,** no. 6, 821 (December, 1967).

"Ten Years of Solar Eclipses," *Sky and Telescope,* **22,** 29–31 (July, 1961). [For the period 1961–1970, tables and maps are given for partial, total, and annular solar eclipses. Information includes date, type, duration, width in miles, path, and where visible. Also, see *The Observer's Handbook* for each year, as listed in "Supplementary Readings" for Chap. 1.]

Weaver, K., "Voyage to The Planets," *National Geographic,* **138,** no. 2, 147 (1970).

Weaver, K. F., "Apollo 15 Explores The Mountains of The Moon," *National Geographic,* **141,** no. 2, 233 (February, 1972).

MOTION AND FORCE

. . . experience is knowledge of individual cases, whereas science is knowledge of universal principles.

ARISTOTLE

A body is said to be in motion if its position with respect to neighboring objects is changing. An automobile may move with respect to the earth's surface, while the earth's surface in turn is moving because of the rotation of the earth on its axis, its revolution around the sun, and the motion of the solar system through space. In this book, however, unless otherwise specified, the earth will be considered as the object of reference and will be thought of as being at rest.

In studying the motion of a body, we will consider its velocity, its speed, (how fast it moves) and its acceleration (how rapidly its velocity changes). We are accustomed to setting a body in motion by exerting a push or a pull on it—that is, by the application of force. A discussion of force in relation to motion will be presented.

Before we can talk intelligently about physical quantities like velocity, speed, acceleration, and force, we must be able to measure the fundamental quantities length, mass, and time. Quantities must be measured in terms of fundamental units; in Chap. 3 we discussed the measurement of time and its fundamental unit, the second. In this chapter we will begin with a discussion of the fundamental units of length and mass, with emphasis on the metric system, followed by consideration of velocity and acceleration, momentum, and Newton's three laws of motion and their applications.

The metric system

For many centuries there were no internationally recognized standards of measurement. France adopted the metric system in 1791. In 1799 the metric system was adopted by several other nations and it is now in worldwide use in scientific circles and in common use in nearly all countries, the notable exceptions being the English-speaking countries. Even in the United States, however, the English standards are defined in terms of metric standards by an act of the United States Congress passed in 1866.

At the present time, Great Britain is well into a period of transition to the metric system. The United States is the only major nation in the world not presently on the metric system or in the process of active transition to the metric system. Agencies of the United States government have been studying the problems and costs involved in making the transition from the English to the metric system. It appears that eventually the United States will adopt the metric system.

Physical concepts or quantities may be classified as fundamental or derived, depending on whether or not they can be expressed in terms of other concepts. In the international system of measurement (the metric system) the fundamental quantities and their fundamental units (given in parentheses) are length (the meter), mass (the kilogram), time (the second), and temperature (degree Kelvin). These units were established independently and arbitrarily.

One can establish his own arbitrary standards to measure length or any other quantity. However, the importance and usefulness of any standard are that everyone agrees to measure with respect to the standard.

METRIC UNITS OF LENGTH

In 1960 the General Conference on Weights and Measures defined the *standard meter* as 1,650,753.73 wavelengths of the orange spectral line of the krypton isotope 86 at 6056 angstroms (Å). This statement will have meaning after your study of Chap. 9, which includes waves and spectra. The length of the meter (39.37 in.) is slightly greater than the English yard.

For all practical purposes, the meter is the distance between two transverse lines on a bar of platinum–iridium when at the temperature of melting ice. It was originally intended to be one ten-millionth of the distance from the earth's equator to one of the

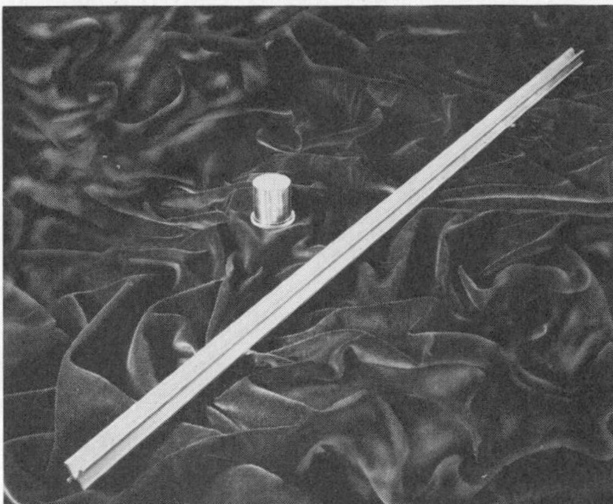

Figure 4.1
Standard kilogram and standard meter bar. They are the United States national standard of mass and the secondary standard of length, and are exact duplicates of the comparable international standards kept at Sèvres, France. (Courtesy of the National Bureau of Standards.)

poles. This standard meter bar is kept at the International Bureau of Weights and Measures, at Sèvres, France, near Paris (see Fig. 4.1). Copies called *substandards* were made, and two of these are located at the National Bureau of Standards, in Washington, D. C.

For measurements of length the smallest unit is the millimeter (mm). The following metric length table shows the relation of the units of length to one another:

10 mm = 1 centimeter (cm)
10 cm = 1 decimeter (dm)[1]
10 dm = 1 meter (m)
1000 m = 1 kilometer (km)

Note the meaning of the prefixes: *milli-* means one-thousandth of, *centi-* means one-hundredth of, and *deci-* means one-tenth of. Thus millimeter means 0.001 m, and centimeter means 0.01 m. The prefix *kilo-* means one thousand times. Thus kilometer means 1000 m. Two additional prefixes, micro- and mega-, are now commonly used, primarily in connection with electrical communication. *Micro-* means one-millionth of, as in microseconds, for a small unit of time. *Mega-* means one million times, as in megahertz (MHz) for the frequency of television and FM radio waves. Atomic bombs and hydrogen bombs are rated in megatons of TNT, referring to their having the energy equivalent of so many million tons of TNT.

The student who knows only the English system of length should observe on a meter stick having the metric system on one side and the English system on the other that (1) 1 meter (m) is slightly longer than a yard (1 m = 39.37 in.); (2) 1 m contains 100 centimeters (cm), the centimeter being about $\frac{1}{2}$ in. in length; (3) 1 cm contains 10 millimeters (mm), hence 1 mm is one one-thousandth of a meter; and (4) 1 in. is about equal to $2\frac{1}{2}$ cm (1 in. = 2.54 cm).

One important advantage of the metric system is that it is a decimal system similar to the monetary system used in the United States. To change from one unit to another in a decimal system one needs only move the decimal point because all units are related by some power of 10. For example, one penny is 0.01 of a dollar and a dime is 0.1 of a dollar and a 10 dollar bill is 10 times a dollar. Note that the United States monetary system is not a perfect decimal system for we have nickels, quarters, half dollars, two dollar bills, and five dollar bills. The handling of measurements in the metric system is illustrated in Example 4.1.

Example 4.1
Express in meters the sum of 1723 mm, 664.9 cm, 13.47 dm, and 1.362 m.

Solution

1723 mm =	1.723 m
664.9 cm =	6.649 m
13.47 dm =	1.347 m
1.362 m =	1.362 m
	11.081 m

This result may be expressed in centimeters by multiplying by 100, that is, by moving the decimal point two places to the right, which gives 1108.1 cm.

The foregoing addition of metric units of

[1] This unit is seldom used.

length illustrates the fact that when quantities are to be added or subtracted, they must be expressed in the same unit. If the student similarly attempts to add $5\frac{1}{4}$ in., $14\frac{1}{2}$ ft, $17\frac{1}{4}$ yd, 62 rods, and $1\frac{3}{4}$ miles, he will appreciate the greater simplicity of the metric system.

The common unit of volume in the metric system is 1 cubic centimeter (cm^3). A larger unit, the liter, which is equal to 1000 cm^3, is also frequently used. From this relation it can be seen that 1 cm^3 is one-thousandth of a liter; it is therefore often called a *milliliter*.

Common units of volume in the English system are cubic inches ($in.^3$), cubic feet (ft^3), and cubic yards (yd^3).

CONCEPT OF MASS

It is within the experience of everyone that a body at rest can be set in motion by a sufficient push or pull exerted upon it. For the time being the word *force* will be taken as equivalent to a push or a pull. A precise definition of force will be given later. It is also common knowledge that two bodies, identical in size and shape, may start off at entirely different rates when given the same push or pull. Consider an iron ball and a wooden ball of the same diameter at rest on a tabletop. If the same push is applied to each, the wooden ball will move faster than the iron ball. The fundamental difference between the two bodies, which is responsible for this difference in their motions, lies in that property called *mass*. Thus the mass of a body is a measure of the resistance it offers to an applied force. The masses of two bodies are equal if each offers the same resistance to equal applied forces. If body *A* offers more resistance to an applied force than body *B*, then body *A* has the greater mass. In this discussion it is assumed that the two objects being pushed experience the same retarding frictional forces.

Mass is a measure of the resistance an object offers to any change in its motion. Not only is it more difficult to start an object with a larger mass, it also requires a greater force to stop an object whose mass is larger. The mass of a body may also be thought of as the quantity of material in the body; under ordinary circumstances that quantity does not change. To obtain a better feeling for the concept of mass, consider two cars, one a Volkswagen the other a standard Cadillac. Suppose we were to put the same engine, the VW engine, in each car. Assuming the engines are capable of exerting the same force on the road, it is obvious that the VW would be easier to start, and once going, it would also be easier to increase its speed. We would say the VW has a smaller mass than the Cadillac. The student might question, "Aren't you just referring to weight?" Mass and weight are related, but they are different. The distinction between mass and weight is extremely important.

CONFUSION OF WEIGHT AND MASS

What is weight? The weight of a body is the force with which the earth attracts the body; for example, when one steps on scales "to weigh himself," the scales register the force with which the earth pulls him downward. We shall discuss in the next chapter how the force the earth exerts on you—that is, your weight—changes slightly depending on whether you are at sea level or on the top of a high mountain. In addition, almost everyone is aware from the flight of our astronauts that his "weight" on the moon would be considerably less than on earth (reduced to about one-sixth, giving a 150-lb man a weight of 25 lb on the moon).

However, if you travel to the moon or to the top of a mountain, you would still contain about the same amount of matter. Your mass remains unchanged but your weight has changed. Mass is a more fundamental quantity whereas weight depends on mass.

Direct measurements of mass are seldom made; instead, weights are compared and the assumption is made that the masses are proportional to the corresponding weights. This assumption has been confirmed by experiments.

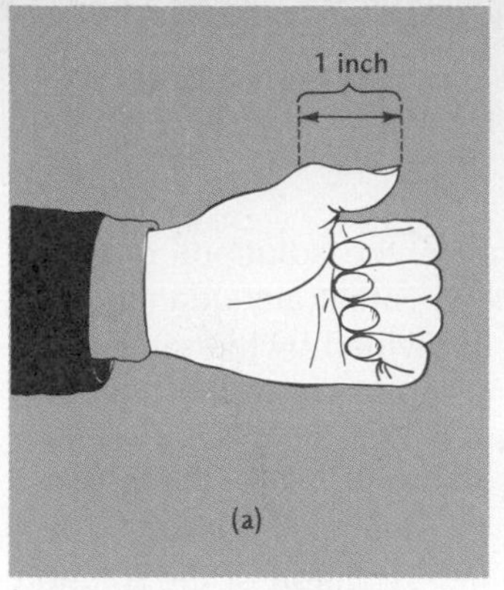

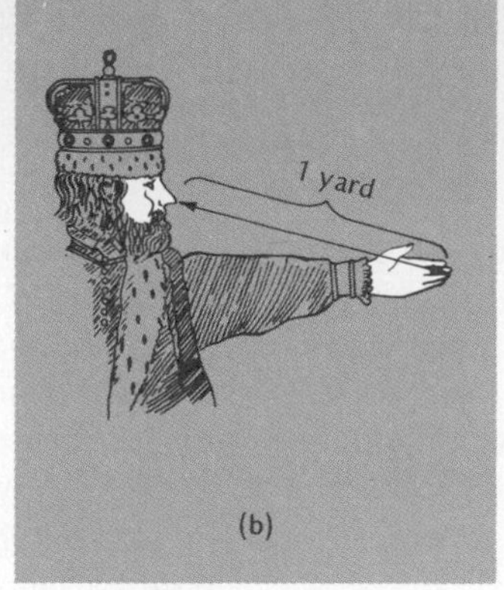

Figure 4.2
Folklore measurements that underlie the English system: (a) the inch, the knuckle of the thumb; (b) the yard, measured from the tip of the nose to the tip of the middle finger on the outstretched arm of King Edgar; (c) the acre, the amount of land plowed by a yoke of oxen in a day. (From "Conversion to the Metric System," Lord Ritchie Calder, *Scientific American,* **223,** no. 1, 17, July, 1970. Copyright © 1970 by Scientific American, Inc. All rights reserved.)

Later in this chapter, in connection with Newton's second law of motion, and then in the next chapter with Newton's law of universal gravitation, we consider further the relation between mass and weight; here our chief purpose is to introduce an awareness that there is a difference between mass and weight.

METRIC UNITS OF MASS

The metric standard of mass is called the kilogram. The standard *kilogram* is the mass of an object of platinum–iridium kept at the International Bureau of Weights and Measures. It is approximately equal to the mass of 1 liter of pure water at 4° centigrade (℃).

For common measurements of mass the smallest unit is the milligram (mg). The following metric mass table shows the relation between mass units:

10 mg = 1 centigram (cg)
10 cg = 1 decigram (dg)
10 dg = 1 gram (g)
1000 g = 1 kilogram (kg)
1000 kg = 1 metric ton

Note the similarity of the metric length and mass tables. Note also that it is customary not to place a period after the abbreviations of units. The centigram and decigram are units that are seldom used.

ENGLISH UNITS OF LENGTH AND CONVERSION OF UNITS

The English standard of length is the *yard,* which was defined by an act of the United States Congress in 1866 as 3600/3937 of the standard meter. You are familiar with other units: inch, foot, rod ($5\frac{1}{2}$ yd), mile. The following table shows the relationship between the length units in the English system:

12 inches (in.) = 1 foot (ft)
3 ft = 1 yard (yd)
5280 ft = 1 mile (mi)

The basis for the units of the English system is illustrated in Fig. 4.2 in which the folklore measurements for the inch, yard, and the acre are pictured.

Because the English system is still the one mainly used in the United States, although the metric system is being used more and more by scientists, it is frequently necessary to change (or convert) from one system to the other. Such conversions are of great importance in this era of increasing international trade and foreign travel. In order to accomplish changes of length units, the following conversion factors are usually employed:

1 in. = 2.54 cm
1 m = 39.37 in.
1 km = 0.621 mile
1 mile = 1.61 km

ADVANTAGES OF THE METRIC SYSTEM

A review of the preceding paragraphs reveals two main advantages of the metric system. The arrangement of units in multiples of 10 enables the student to remember the relations easily and facilitates the change from one unit to another. The similarity of the metric length and mass table assists us to remember each, in contrast to the English system, in which there is no systematic relation between units of the same table and no relation whatsoever between the arrangements of the two tables.

Even though these advantages appear so great, the adoption of the metric system in the United States is prevented in large measure by the habitual use of the older English system and by the high cost of changing over industrial machinery and products manufactured using the English system to those using the metric system.

Example 4.2
A person is 6 ft tall. Express this in centimeters and in meters.

Solution

$$6\ \cancel{\text{ft}} \times \frac{12\ \cancel{\text{in.}}}{1\ \cancel{\text{ft}}} \times \frac{2.54\ \text{cm}}{1\ \cancel{\text{in.}}} = 183\ \text{cm}$$

$$183\ \cancel{\text{cm}} \times \frac{1\ \text{m}}{100\ \cancel{\text{cm}}} = 1.83\ \text{m}$$

A fundamental idea, which is needed in this solution and in the solutions of many problems to follow, is that any quantity may be multiplied and divided by the same thing without changing the value of the quantity; that is, any quantity may be multiplied by a fraction in which the numerator (12 in.) and denominator (1 ft) are equal. The reader will note that in the first equation of the solution the given length (6 ft) is multiplied by two fractions each of which is equal to unity (one). Also, after all possible cancellations of the units involved have been made, the answer is in centimeters. To change the answer from centimeters to meters, the decimal point is moved two places to the left.

HANDLING OF UNITS

When quantities are added or subtracted, they must be expressed in the same unit, and the answer will be in that unit. This is illustrated in the problem of combining various metric lengths, as in Example 4.1.

When quantities are multiplied or divided, the units must be treated in the same manner as the numbers. Example 4.3 illustrates the handling of units in multiplication.

Example 4.3
Compute the area of a rectangular floor that is 20 ft long and 10 ft wide.

Solution
$\text{area} = \text{length} \times \text{width} = 20\ \text{ft} \times 10\ \text{ft} = 200\ \text{ft}^2$

When quantities are divided, if the same unit appears in both the numerator and the denominator, it may be canceled as in Example 4.2. If unlike units appear in the numerator and denominator of an expression, then division is expressed in the result by the inclusion of a fraction containing only the units. This is illustrated in the problems at the end of this chapter. Cancel units in the same way that numbers are canceled, or, in algebra, letters. At this time the student should work Probs. 1 to 7 at the end of this chapter.

Speed, velocity, and acceleration

SPEED AND VELOCITY

Average speed may be defined as the distance traveled divided by the time of travel. This definition is expressed by the following:

$$\text{average speed} = \frac{\text{distance traveled}}{\text{time taken}}$$

Designating average speed by $\bar{v}$ (to be read "v bar"), distance by *d*, and time by *t* gives

$$\bar{v} = \frac{d}{t} \qquad (4.1)$$

This is a fundamental definition expressed in equation form. It can be applied to any type of motion.

The student might wonder why the letter *v* was chosen to represent speed. Speed is the magnitude of the velocity, and the *v* in equation refers to velocity. Speed and velocity are not the same even though they may be numerically identical and expressed in the same units. Speed tells only "how fast"; that is, it gives only the magnitude of the velocity. The velocity of an object tells both the magnitude (the speed) and the direction in which the body is moving. Equation (4.1) defines the velocity of an object if at the same time the direction of the object is specified. Often the words "speed" and "velocity" are used interchangeably. However, because the difference between these words is important in physics, they should be used properly to avoid confusion.

Throughout this book important equations which the student is expected to know, to understand, and to use are numbered in parentheses. Equation (4.1) is read: Average speed equals distance traveled divided by time taken. Learn to read equations in words; the reading will help you develop an understanding of each equation. To say $\bar{v}$ equals *d* over *t* is simply memory work and has little or no meaning. You should write or visualize an equation in abbreviated form [Eq. (4.1), for example], but read it in words that make a sentence, as you write or visualize it. This technique will help with understanding and decrease meaningless memory work. There is too much in this book for you to try to memorize it all—though some students will try, without success. Work to develop understanding of concepts, definitions, and equations.

The student has undoubtedly had experiences with the relation expressed in Eq. (4.1). For example, on an automobile trip, if a distance of 400 miles is covered in 10 hours,

$$\bar{v} = \frac{d}{t} = \frac{400 \text{ miles}}{10 \text{ hours}} = 40 \text{ mi/hour}$$

This result is read, "The average speed equals 40 miles per hour." The average velocity of this car would be 40 mi/hour west, or some other direction. The speedometer on a car's dashboard is correctly named. It indicates only how fast a car is moving (speed); it does not tell direction.

In the automobile trip just mentioned, it is obvious that the car did not maintain a continuous speed of 40 mi/hour. Part of the time the speed was zero, owing to stops for traffic lights, to purchase gasoline and oil, etc., and therefore the car probably traveled at a rate of 50 mi/hour or more much of the time. At a certain time the speedometer may have indicated 52 mi/hour, which means that the speed at that instant, called *instantaneous speed,* was 52 mi/hour.

More properly the *instantaneous speed* could be defined as the distance traveled during an infinitesimally short time ("an instant of time") divided by this instant of time. The magnitude of the instantaneous velocity—that is, the instantaneous speed—is defined as

$$v_i = \frac{\Delta d}{\Delta t}$$

where Δd is the very small change in distance traveled along a straight line occurring in the corresponding small time interval Δt. The Δ is read "delta" and signifies an infinitesimally small change.

Because velocity is obtained by dividing a distance by a time, any distance unit divided by any time unit gives a proper speed or velocity unit. Some of the common units of speed and velocity are miles per hour, feet per second, miles per second, kilometers per hour, centimeters per second, and kilometers per second. These may be abbreviated mi/hour, ft/sec, mi/sec, km/hour, cm/sec, and km/sec.

It is frequently advantageous to know the relation between miles per hour and feet per second. For example, 15 mi/hour can be changed as follows:

$$15\,\frac{\cancel{\text{miles}}}{\cancel{\text{hour}}} \times \frac{5280\text{ ft}}{1\,\cancel{\text{mile}}} \times \frac{1\,\cancel{\text{hour}}}{60\,\cancel{\text{min}}} \times \frac{1\,\cancel{\text{min}}}{60\text{ sec}} = 22\,\frac{\text{ft}}{\text{sec}}$$

Carrying out all possible cancellations gives a result of 22 ft/sec. Thus

15 mi/hour = 22 ft/sec
30 mi/hour = 44 ft/sec
45 mi/hour = 66 ft/sec
60 mi/hour = 88 ft/sec

These conversions are often very useful when you attempt to visualize the distance covered during a single second. For example, 30 mi/hour is equivalent to 44 ft/sec which can be visualized as about 2½ car lengths per second. Note that an ordinary full-size American automobile is about 18 ft long. The student may obtain some perspective to help in his understanding by relating speeds and distances to his everyday experiences.

Often the unit of a derived quantity is given a special name. Thus in navigation a speed of 1 nautical mile (6080.2 ft) per hour is called a *knot*.

From the relation $\bar{v} = d/t$, it is possible to obtain an expression for distance d in terms of average velocity $\bar{v}$ and time t. Because an expression for d is desired, we must have only d on the right side of the equation (instead of d/t). This may be obtained by multiplying the right side of the equation by t. If one side of an equation is multiplied by a certain quantity, however, the other side must be multiplied by the same quantity. This gives

$$\bar{v}t = \frac{d}{\cancel{t}}\cancel{t} \qquad \text{or} \qquad d = \bar{v}t$$

If an automobile is traveling with an average speed of 40 mi/hour, the distance traveled in 4 hours will be

$$d = \bar{v}t = 40\text{ mi/hour} \times 4\text{ hours} = 160\text{ miles}$$

UNIFORM MOTION

Uniform motion is motion in which the velocity is constant; that is, equal distances are traveled in equal times along a straight line. In this kind of motion the velocity is unchanged, therefore, there is no need to distinguish between average and instantaneous velocity for they are numerically equal at all times during *uniform motion.* For uniform motion only, Eq. (4.1) for the magnitude of the velocity becomes

$$v = \frac{d}{t}$$

from which follows

$$d = vt \quad \text{and} \quad t = \frac{d}{v}$$

When the speed is constant, the bar over v is not necessary.

This relationship is useful in assisting a driver to visualize what a particular speed means. For example, if you are traveling 45 mi/hour (66 ft/sec) along a road, the length of road traveled during a 5-sec time interval would be 330 ft.

$$d = vt = 66\text{ ft/}\cancel{\text{sec}} \times 5\,\cancel{\text{sec}} = 330\text{ ft}$$

In terms of lengths of cars this is about 18 car lengths, which means that at 45 mi/hour you could pass 18 parked cars in 5 sec.

The student will now find it helpful to do Probs. 8 to 17 at the end of this chapter.

ACCELERATION

Again referring to experience with an automobile, there is a pedal that when pushed results in an increase in speed of the car. This is well named the "accelerator."

Average acceleration is the change in *velocity* divided by the time to make the change. Thus

$$\text{average acceleration} = \frac{\text{change in velocity}}{\text{time}}$$

$$\bar{a} = \frac{v_f - v_b}{t} \tag{4.2}$$

where

$\bar{a}$ = average acceleration
v_f = velocity at end of time t (final velocity)
v_b = velocity at beginning of time t (beginning velocity)

For the time being we shall be concerned with motion along a straight line so that changes in velocity are due only to changes in speed. The values of v_f and v_b are therefore the final and beginning speeds, respectively.

Example 4.4
While an automobile is in second gear, its velocity along a straight road is increased from 6 mi/hour to 21 mi/hour in 3 sec. Compute the average acceleration.

Solution

$$\bar{a} = \frac{v_f - v_b}{t} = \frac{21\ \text{mi/hour} - 6\ \text{mi/hour}}{3\ \text{sec}}$$

$$= \frac{15\ \text{mi/hour}}{3\ \text{sec}} = 5\ \text{mi/hour/sec}$$

This result is read, "5 miles per hour per second" and means that on the average the velocity has increased by 5 mi/hour for each second.

The student will better grasp the full meaning of an average acceleration of 5 mi/hour/sec if we consider that in the above example the acceleration were uniform—that is, constant. We can then consider what happens each second.

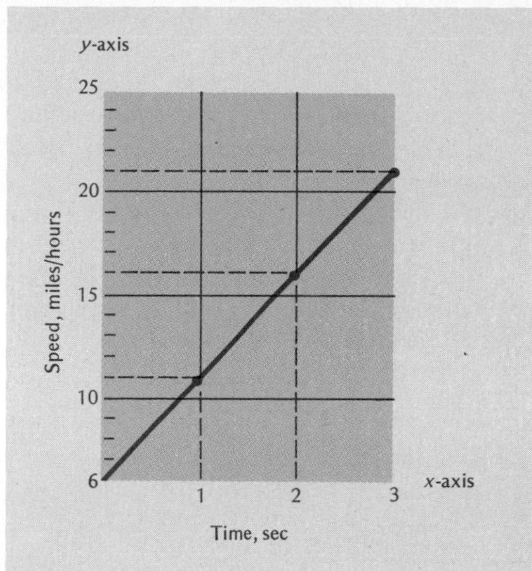

Figure 4.3
Graph relating speed and time. The acceleration, the change in velocity per unit time is 5 mi/hour/sec along a straight line.

Time	*Speed*
0 sec	6 mi/hour
1 sec	11 mi/hour
2 sec	16 mi/hour
3 sec	21 mi/hour

This tabular outline clearly shows an increase in speed of 5 mi/hour each second. This is the equivalent of an average acceleration of 5 mi/hour/sec. The speeds listed are the magnitudes of the instantaneous velocities at the end of the time interval.

Using the time-speed data from the tabular outline and plotting the speed at the end of each time interval as a point in a graph grid, one observes (see Fig. 4.3) that the line extending from the origin (starting point, lower left) through the plotted points is straight. The line is straight because the speed increases at a uniform rate (5 mi/hour/sec), as shown in the tabular outline. (See App. A for suggestions on techniques of graphing and the interpretation of graphs.)

A change in velocity along a straight line may be due to an increase in speed, as in the example above, or it may be due to a decrease in speed, which results in a negative acceleration, sometimes called "deceleration."

Example 4.5

A motorist traveling with a speed of 30 mi/hour applies the brakes and brings his car to a stop in 5 sec. Compute the average acceleration.

Solution

$$\bar{a} = \frac{v_f - v_b}{t} = \frac{0 - 30 \text{ mi/hour}}{5 \text{ sec}}$$

$$= \frac{-30 \text{ mi/hour}}{5 \text{ sec}} = -6 \text{ mi/hour/sec}$$

This result means that the average change in speed, for each second of time, is a decrease of 6 mi/hour, the decrease, or deceleration, being indicated by the negative sign (−).

Again, a tabular outline will help to show a change (decrease) in speed of 6 mi/hour each second, if the stopping process is constant so the acceleration is constant;

Time	*Speed*
0 sec	30 mi/hour
1 sec	24 mi/hour
2 sec	18 mi/hour
3 sec	12 mi/hour
4 sec	6 mi/hour
5 sec	0 mi/hour

Too many drivers follow the car ahead too closely. The distance at which it is safe to follow another car depends on a number of conditions. It is a greater distance than most drivers realize. Here is a simple rule of thumb followed by many professional drivers: A driver with an average reaction time should stay at least 20 ft (about one car length) behind the car he is following for each 10 mi/hour of his speed.

The acceleration at any instant, called *instantaneous acceleration* a_i, is

$$a_i = \frac{\Delta v}{\Delta t}$$

where Δv is the very small change in velocity occurring in corresponding small time Δt. The Δ is read delta. On an automobile there is no device for measuring acceleration corresponding to the speedometer, which measures instantaneous speed, but on many airplanes there is such an instrument, called an "accelerometer," which gives the acceleration at any instant (the instantaneous acceleration).

In any acceleration unit, one time unit is involved in the change of velocity and a second time unit is used in expressing the time during which the change of velocity takes place. Usually the two time units are the same, as in centimeters per second per second (cm/sec/sec) and feet per second per second (ft/sec/sec), in which case, treating the units as we do numbers, we may write them "cm/sec^2" and "ft/sec^2." Any velocity unit divided by any time unit, however, is a proper acceleration unit; for example, some common acceleration units are miles per hour per second (mi/hour/sec) and kilometers per hour per second (km/hour/sec).

In the preceding discussions examples of uniform or constant acceleration have been given. In these cases the average acceleration and the instantaneous acceleration both have the same value. There are several important motions to be discussed later in which the acceleration is constant. In many real cases, however, such as the stopping or starting of a car, there is no reason to believe that the acceleration would be uniform. The average value of acceleration may then differ from the instantaneous acceleration.

Also in the preceding discussion only motion along a straight line has been considered. However, acceleration is defined as a change in *velocity* over a time interval. So far we have considered only changes in the magnitude of the velocity—that is, changes in the speed. If, however, the *direction* of the velocity were to change (for example, when a car moves around a curve), there would be an acceleration even if the speed were to remain constant; that is, changes in velocity may be due to either changes in speed or in the direction of the motion or to changes in both. An example of this type of acceleration is given later in this chapter when "uniform circular motion" is considered. A few simple problems involving acceleration will be found at the end of this chapter (Probs. 18 to 21).

Newton's three laws of motion

FIRST LAW

Probably the best understanding of force and its relation to motion is to be obtained from Newton's statements, in his great scientific work *Principia* (1687), that are now called in his honor, "Newton's three laws of motion." His first law of motion may be stated as follows:

Every body continues in its state of rest or of uniform motion in a straight line unless some force acts on the body to cause it to change that state.

Force is that which tends to change the motion of a body. We commonly think of force as a push or a pull on a body. Newton's first law also includes the concept expressed by the term *inertia,* which is the tendency of a body to persist in its state of rest or of uniform motion in a straight line. A body possesses inertia in direct proportion to its mass. This is in accord with the student's experience that it requires a much greater force to set an automobile in motion than to produce a similar motion of a child's wagon.

A few simple examples will make the concept of inertia easy to understand. Most people have stood in crowded buses or in subways. When a bus starts forward, the tendency of a standing passenger to remain where he is results in his seeming to fall toward the rear of the vehicle. Actually the rear of the vehicle moves toward the passenger. He tends to be left behind on account of his inertia, for the motor exerts a forward force on the bus but not on him. He must quickly find a force to accelerate him along with the bus and commonly does this by grasping a seat or a strap which can pull on him to overcome his inertia. Similarly, when the bus stops suddenly, the tendency of the passenger to continue in his forward motion causes him to tend to fall toward the front of the bus. Because of his inertia he must quickly find a force to stop his forward motion, perhaps by grasping the top of a seat.

A more desperate situation occurs when an automobile collides head on with a large tree. The car quickly stops due to the force that the large tree exerts on the car. However, the inertia of a person in the car carries him forward until a force stops him. All too often that force is provided by the windshield, the dashboard, or the steering wheel. The use of a seat belt and a shoulder harness is a means of providing the necessary restraining force. The use of an air bag provides an even greater chance for reducing injuries, for the force on the person would be more uniformly distributed over the body. Safety standards for automobiles have their basis in Newton's laws of motion. It is rather sad that these laws written down about 300 years ago were not applied to auto safety until almost 100 years after the invention of the car. It might be noted that Newton's first law applies equally as well to the tree with which the car collided. The tree was at rest and tends to remain at rest; that is, a large tree (one with a large mass) has a large inertia.

Other examples of inertia occur in other everyday situations. When coal is shoveled into a furnace, the shovel and coal are both set in motion. A sudden stopping of the shovel in front of the furnace door allows the coal to continue its forward motion into the furnace. The shoveling of dirt, sand, gravel, or snow provides a similar experience. Perhaps the student would enjoy performing a simple experiment. A glass filled with water is placed on a sheet of paper. A quick jerk will remove the paper without spilling any water. This is possible because of the inertia of the glass of water. In this example the glass at rest tends to stay at rest unless a force is exerted on the glass. The only force that can exist is that of the paper on the glass, a friction force. Now, if the paper is removed quickly, the frictional force is very small. It might be said that there is not sufficient time for the force you exert on the paper to be transferred to the glass. On the other hand if you pull the paper slowly, then the glass is pulled along with the paper. The student should be prepared to give several other examples of inertia.

It should be noted that Newton in his first law put two seemingly different states of motion into the same category. He states that an object at rest or in a state of uniform motion will remain so unless a net force acts on the body. The state of rest is specified by $v = 0$ and $a = 0$; the state of uniform motion is specified by v = constant. The common condition here is that the acceleration is zero.

NEWTON'S SECOND LAW

As indicated above, Newton's first law of motion tells what force does. Frequently we wish to measure the amount of force involved. The earlier discussion of the concept of mass indicates a relation between force and mass. Moreover, a body can be accelerated positively (velocity increasing) only as long as a net force acts on it.

It is within the experience of many students that when unequal forces are applied to two chairs of equal mass in order to slide them across a smooth floor, the greater force produces greater acceleration. If, however, the forces on the two chairs are equal and the masses of the two chairs are unequal, the chair of greater mass receives smaller acceleration. Thus, acceleration bears an inverse relation to the mass (the greater the mass, the less the acceleration; and vice versa).

The foregoing relations are more specifically and quantitatively stated in one form of Newton's second law of motion.

Whenever an unbalanced force is applied to a body that is free to move, the force is equal to the product of the mass of the body and the acceleration given to the body by the force.

The equation form of the statement of Newton's second law is

$$\text{force} = \text{mass} \times \text{acceleration}$$

Using the usual symbols f for force, m for mass, and a for acceleration, we have the important equation

$$f_{\text{net}} = ma \qquad (4.3)$$

Thus another way of stating the second law is

An unbalanced force acting on a body produces an acceleration which is directly proportional to the force and in the same direction as the force.

An unbalanced force is a net force (f_{net}). Mass in the equation is a constant of proportionality that measures inertia.

Because units are treated in the same manner as numbers, it follows that a force unit may be obtained by multiplying a mass unit by an acceleration unit. Thus in the metric system if the mass is 1 g and the acceleration is 1 cm/sec^2, or if the mass is 1 kg and the acceleration is 1 m/sec^2, the force will be

$$f = ma = 1 \text{ g} \times 1 \text{ cm/sec}^2 = 1 \text{ g-cm/sec}^2 \text{ or } 1 \text{ dyne}$$
$$f = ma = 1 \text{ kg} \times 1 \text{ m/sec}^2 = 1 \text{ kg-m/sec}^2 \text{ or } 1 \text{ newton}$$

The unit of force 1 g-cm/sec^2 is called 1 *dyne*, which is the force that can give a mass of 1 g an acceleration of 1 cm/sec each second. The unit of force 1 kg/m-sec^2 is called 1 newton. Thus, by definition, the *newton* (abbreviated N) is the force that can give a mass of 1 kg an acceleration of 1 m/sec^2. One newton = 10^5 dynes. The use of mks units has increased rapidly and seems destined to secure universal adoption, particularly in scientific and electrical-engineering circles. The cgs system continues to be used; for example, in connection with pressure of the atmosphere (Chap. 7).

The reading of equations as a learning technique was suggested earlier. Definitions can often be read from an equation. Consider the equation that measures force,

$$\begin{aligned} f &= m \times a \\ 1 \text{ dyne} &= 1 \text{ g} \times 1 \text{ cm/sec/sec} \\ &= 1 \text{ g-cm/sec}^2 \\ 1 \text{ newton} &= 1 \text{ kg} \times 1 \text{ m/sec/sec} \\ &= 1 \text{ kg-m/sec}^2 \end{aligned}$$

The definition of the dyne can be read from the first two equations; that is,

One dyne is the force that can give a mass of one gram an acceleration of one centimeter per second per second.

Likewise, the definition of a newton can be read from $f = ma$:

One newton is the force that can give a mass of one kilogram an acceleration of one meter per second per second.

In the solution of problems, the units should be of a consistent scheme.

Example 4.6

Compute the unbalanced force necessary to give a body having a mass of 40 g a constant acceleration of 10 cm/sec².

Solution

$$\begin{aligned} f_{net} &= ma \\ &= 40 \text{ g} \times 10 \text{ cm/sec}^2 \\ &= 400 \text{ g-cm/sec}^2 \\ &= 400 \text{ dynes} \end{aligned}$$

Example 4.7

Compute the unbalanced force necessary to give a body having a mass of 60 kg a constant acceleration of 5 m/sec².

Solution

$$\begin{aligned} f_{net} &= ma \\ &= 60 \text{ kg} \times 5 \text{ m/sec}^2 \\ &= 300 \text{ kg-m/sec}^2 \\ &= 300 \text{ newtons} \end{aligned}$$

Example 4.8

Compute the acceleration of a body having a mass of 200 g when acted upon by a net force of 60,000 dynes.

Solution

Using $f_{net} = ma$ and dividing both sides by m, it follows that

$$\begin{aligned} a &= \frac{f_{net}}{m} = \frac{60{,}000 \text{ dynes}}{200 \text{ g}} = \frac{60{,}000 \text{ g-cm/sec}^2}{200 \text{ g}} \\ &= 300 \text{ cm/sec}^2 \end{aligned}$$

Similarly, the mass of a body can be computed (Probs. 29 and 30) if the force acting on it and its acceleration are given.

Throughout this discussion we have carefully noted that the force f in Newton's second law is the net, or unbalanced, force acting on the object. For example, if two men each apply a force in the same direction to a box that sits on a very smooth surface, the acceleration on the box is determined by the sum of the two forces. Another practical example would be a man pushing on a large truck. Suppose that no matter how hard the man pushes, the truck does not move. How is that possible? A force on an object is supposed to give the object an acceleration. The problem is that we must consider the net force on the truck. There is a large frictional force acting on the truck, and unless the man pushes on it with a force greater than the frictional resisting force between the ground and the truck wheels, the truck will not move. It is the net or unbalanced force that gives the object an acceleration. Almost every object at rest has some forces acting on it. However, if the object is at rest or in uniform motion, there is *no net* force acting on it.

Problems 25 to 33, involving $f_{net} = ma$ in its various forms, are included at the end of this chapter.

MASS AND INERTIA

Inertia has been defined, in connection with Newton's first law of motion, as the tendency of a body to persist in its state of rest or of uniform motion in a straight line; *inertia* may also be defined as the tendency of a body to resist acceleration. From Newton's second law of motion, $f_{net} = ma$ [Eq. (4.3)], it may be stated that the acceleration a is proportional to the net force applied f_{net}. This means that in a given system of units the ratio of the magnitude of the net force to the magnitude of the acceleration is always the same—constant—for a given body. Regardless of where the body is located in the universe, the ratio should be the same for all sorts of net forces in all sorts of motion. In symbols the ratio would be

$$\frac{f_{net}}{a} = \text{constant}$$

The constant is called the mass m of the body being considered. Thus we see that

$$\frac{f_{net}}{a} = m$$

which is known as Newton's second law of motion [Eq. (4.3)].

MASS AND WEIGHT

The foregoing definition of mass implies that unit force applied to unit mass produces unit acceleration. In order to satisfy this condition, consistent units are required. Unfortunately several different systems of consistent units for $f_{net} = ma$ are in fairly common use. In order to proceed by a short path from our statement of Newton's second law of motion to simple applications that will encourage a feeling for its meaning, we sidestep, in a sense, the question of units by substituting for mass m in the equation $f_{net} = ma$ the equivalent ratio of the weight of the body being considered to the acceleration of gravity g.

Measurements show that any freely falling body near the earth's surface has an acceleration of approximately 980 cm/sec², 9.8 m/sec², or approximately 32 ft/sec², if friction is neglected. These values are customarily represented by g. Experiments have verified that the acceleration g has the same value for all bodies at any given location on the earth.

Consider the case of the earth attracting a body that is free to fall. In the equation $f_{net} = ma$, the f becomes the weight w of the body, and the acceleration becomes the acceleration of a freely falling body, represented by g. Thus

$$f_{net} = ma$$

becomes

$$w = mg \tag{4.4}$$

and

$$m = \frac{w}{g}$$

The ratio w/g can be used interchangeably with the mass symbol m in Newton's second law,

$$f_{net} = ma$$

or

$$f_{net} = \frac{w}{g}\, a \tag{4.5}$$

Equation (4.5) can be useful for nonscientists because it is valid for any choice of units, provided only that the same unit is used for the forces f and w and the same unit for the accelerations a and g. The weight of a body on earth is the force with which the earth attracts the body; that is, the force exerted on a body by gravity is its weight w. If when you step on scales you are pulled downward by gravity (next chapter) with a force of 120 lb, the scales will register that force and you are said to weigh 120 lb.

Mass and weight will be discussed again in Chap. 5, where we study Newton's law of universal gravitation. Step by step at appropriate places in your study, the difference between mass and weight will become clearer.

BRITISH GRAVITATIONAL SYSTEM OF UNITS

In this system the unit of force is the *pound-force,* defined as the weight of a certain standard object (a platinum cylinder known as the British Imperial standard pound) at a location where the acceleration of gravity is the standard value 32.1725 ft/sec², the mean observed acceleration of gravity at latitude 45° and sea level.

The unit of mass in this system is the mass of a body that experiences an acceleration of 1 ft/sec² when acted on by a standard pound-force. This unit of mass has been called a *slug* to indicate a measure of inertia, or "sluggishness."

Equations (4.3) and (4.4) with proper units for this system would be

$$f\ (\text{lb-force}) = m\ (\text{slugs}) \times a\ (\text{ft/sec}^2)$$

$$m\ (\text{slugs}) = \frac{w\ (\text{lb})}{g\ (\text{ft/sec}^2)}$$

The usefulness of knowing the mass of an object in slugs, when we normally deal with weights rather than masses, becomes apparent in applications of Newton's second law of motion, $f_{net} = ma$. To obtain the mass in slugs, the weight in pounds is divided by the constant g (acceleration due to gravity rounded to 32 ft/sec²).

Example 4.9
An automobile weighing 3200 lb is pushed with a uniform net force of 200 lb. Neglecting friction, compute the acceleration of the car.

Solution

$$m\ (\text{slugs}) = \frac{w\ (\text{lb})}{g\ (\text{ft/sec}^2)}$$

$$m = \frac{3200\ \text{lb}}{32\ \text{ft/sec}^2} = 100\ \text{slugs}$$

$$f_{net} = ma$$

$$a = \frac{f_{net}}{m} = \frac{200\ \text{lb}}{100\ \text{slugs}} = 2\ \text{ft/sec}^2$$

NEWTON'S THIRD LAW

The rowing of a boat consists of action: successive backward pushes of the oars against the water, the reaction to which can produce forward motion of the boat with a positive acceleration. In track, a runner leans forward and pushes against the ground with his feet. The ground pushes forward on the runner, giving him a positive acceleration. If the ground is soft, the reality of the backward force exerted by the runner can be observed, because in each footprint the soil is observed to have been pressed backward.

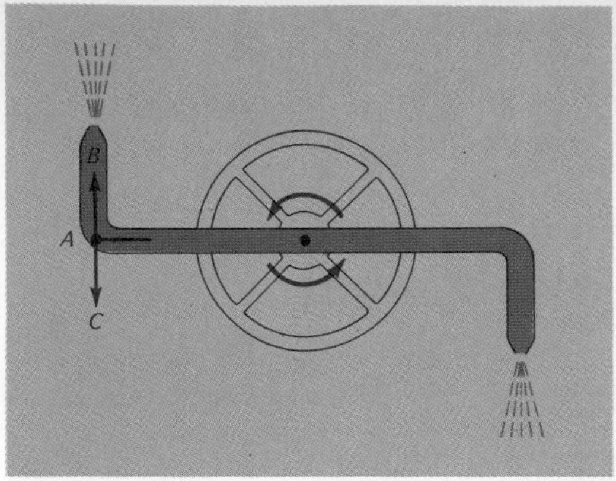

Figure 4.4
Typical rotary lawn sprinkler, as seen from above.

The idea of forces always occurring in pairs is the basis for Newton's third law of motion, which perhaps may be best stated as:

When one body exerts a force on a second body, the second body exerts an equal and opposite force on the first body.

Among the common illustrations of Newton's third law are a hammer hitting a nail, the rotary lawn sprinkler, a jet airplane, and rockets.

A carpenter swinging a hammer may cause the hammer to exert a force on a nail to drive the nail into a piece of wood. The nail exerts an equal and opposite force on the hammer that stops the motion of the hammer.

In a rotary lawn sprinkler (Fig. 4.4), the wall of the sprinkler at the bend *A* exerts a deflecting force (represented by arrow *AB*) to send the water forward and out the opening *B*; however, the equal and opposite reacting force of the water (represented by arrow *AC*) pushes backward against the wall, causing the sprinkler to rotate counterclockwise and spread the water.

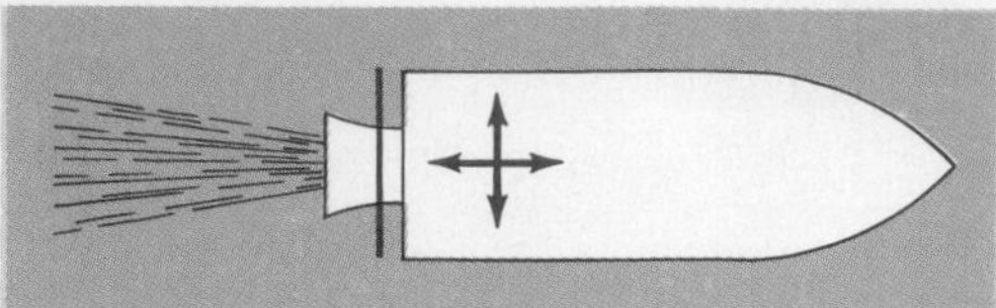

Figure 4.5
Operation of a rocket.

In a jet airplane, expanding hot gases from an exploding mixture of fuels such as alcohol and oxygen, or low-grade gasoline and oxygen, are forced out at very high speed from the rear of the jet engine, or engines, of the plane. The reaction force exerted by the gases on the plane pushes the plane forward. The principle of a rocket can be demonstrated by blowing up a rubber balloon and releasing it. As the balloon contracts, it forces the air out, the reaction force of the escaping air on the balloon causing the balloon to dart forward.

The essential difference between a jet plane and a rocket is that in the case of the jet plane the oxygen necessary for combustion is obtained by allowing outside air, which is about 20 percent oxygen, to enter the combustion chamber. Thus a jet plane can operate only if it is flying through an appreciable atmosphere. In the case of the rocket, the oxygen, in the form of liquid oxygen or some type of solid fuel, is contained within the rocket, so that a rocket could operate even in a vacuum.

The National Aeronautics and Space Administration operates rocket launching pads, such as those at Cape Kennedy, from which space vehicles are launched with several rockets attached. The rockets are jettisoned in space after they have served their function. Space capsules and probes have retrorockets for use in directional guidance and to slow the vehicles for landing—on the earth or on the moon, for example.

The operation of a rocket is illustrated in Fig. 4.5. If the rear compartment were closed, there would be equal forces in all directions against the containing walls, owing to the gases formed. With an opening at the rear, there is no surface there to be pushed against, thus there is a net push forward. The net force to the right in Fig. 4.5 sends the rocket to the right. This is in accordance with Newton's second law of motion. From $f_{\text{net}} = ma$, the acceleration of the rocket is directly proportional to the net force forward. Both Newton's second and third laws of motion are used in explaining the operation of a rocket.

It is easy to make a model rocket that will take off. Small cylinders, about the size of a thumb, of compressed carbon dioxide gas used to make carbonated water and a mechanism to puncture them can be purchased inexpensively at drug stores. Attach the cylinder firmly to the underside of a light board cut to resemble a rocket. Suspend the rocket on a strong guide wire by using wheels or screw-eyes. Puncture one end of the cylinder and observe the rocket as it takes off (Fig. 4.6).

Because of Newton's third law of motion, the soldier and the hunter must learn how to hold a gun as it is being fired in order to avoid bruising the shoulder. The student should be able to think of many other examples of this law, because for every acting force there is an equal and opposite reacting force. Note that the "acting" force and the opposite "reacting" force each act on different objects.

As a further example of Newton's law consider the weight force acting on a body, that is, the force with which the earth attracts the body. The Newton's third law reaction force is the force of attraction the body exerts on the earth. Clearly these forces are oppositely directed and act on different objects, the body and the earth, and according to Newton's third law these forces must be equal in magnitude.

To summarize Newton's three laws in terms of force, the first law tells about inertia and the consequences (rest or constant motion) when there is no net force; the second law tells how force is measured and how an

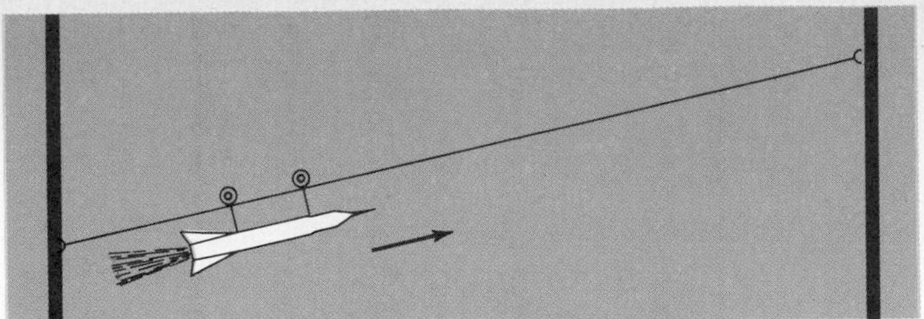

Figure 4.6
Experimental demonstration of a model rocket.

unbalanced (net) force changes the state of motion of an object; and the third law states that forces always occur in pairs, the two forces of a pair being equal and opposite and always acting on different objects.

MOMENTUM

A quantity that is very useful in considering the interactions between two objects and the resulting changes in the motions of the objects is the momentum of those bodies. The *momentum* of a body is the product of the mass of that body and its velocity. This definition is expressed by the equation

$$\text{momentum} = \text{mass} \times \text{velocity} \qquad (4.6)$$
$$M = mv$$

Clearly, since velocity has a magnitude and a direction, the momentum must also have a magnitude and a direction. The direction of the momentum will be that of the velocity, and the magnitude of the momentum will be the mass of the object times the speed of the object.

Example 4.10
Compute the momentum of a car which weighs 3200 lb and is moving with a speed of 30 mi/hour.

Solution

$$v = 30 \text{ mi/hour}$$

and

$$m = \frac{w}{g} = \frac{3200 \text{ lb}}{32 \text{ ft/sec}^2} = 100 \text{ slugs}$$

$$M = mv = 100 \text{ slugs} \times 30 \text{ mi/hour}$$
$$= 3000 \text{ slug-mi/hour}$$

As indicated in Example 4.10, any mass unit multiplied by any velocity unit gives a proper momentum unit. Some of the common units of momentum are gram-centimeters per second, kilogram-meters per second, and kilogram-kilometers per hour. Problems 23 to 26 at the end of this chapter are designed to give the student familiarity with momentum and its units.

CONSERVATION OF MOMENTUM

The fact that an acting force and the reacting force referred to in Newton's third law are equal in magnitude and are oppositely directed may be expressed by the equation

$$f_1 = -f_2$$

where f_1 is the magnitude of the force acting on a body of mass m_1 exerted by a body of mass m_2, and f_2 is the magnitude of the force acting on a body of mass m_2 exerted by body of mass m_1. The minus sign is used to indicate that these two forces are oppositely directed. These forces, of course, act along the same direction, so one of the two directions can be specified as a positive quantity and the other by a negative sign. Because, $f = ma$, in general, $f_1 = -f_2$ may be written

$$m_1a_1 = -m_2a_2$$

If initially the two bodies under consideration were at rest with respect to each other ($v_b = 0$), then after a time t body 1 will have a speed v_1 and body 2 will have a speed v_2. Recalling the definition of acceleration

$$a = \frac{v_f - v_b}{t}$$

then

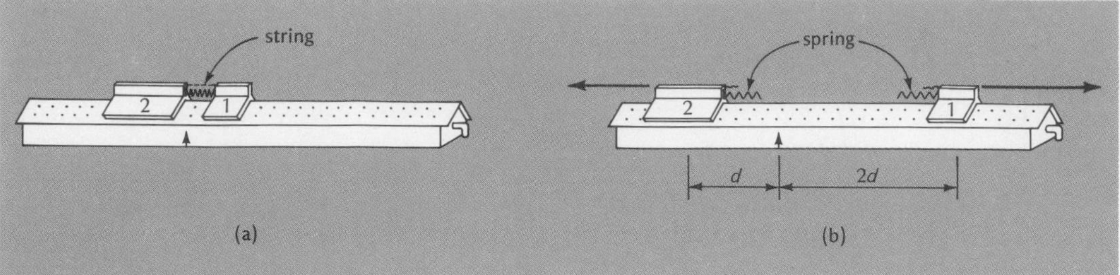

Figure 4.7
Conservation of momentum; two carts on an air track. In (a) two carts, 1 and 2 of mass $m_1 = \frac{1}{2}m_2$ are coupled together with a string compressing the springs on each cart. The carts ride on a cushion of air which comes through small holes in the air track. The air originates from a blower attached to one end of the air track. After the string is cut (b), the two carts separate due to the Newton's third-law pair of forces. Because the momentum of the two carts initially in (a) is zero, the sum of the momenta will also remain zero in (b). This means that the momentum of cart 1 will be equal to and opposite that of cart 2 after the expansion of the spring (b). Note that because $m_1 = \frac{1}{2}m_2$, cart 2 moves with one-half the speed of cart 1 in (b). Cart 1 will travel twice as far as cart 2 during the same time.

$$a_1 = \frac{v_1 - 0}{t} = \frac{v_1}{t}$$

and

$$a_2 = \frac{v_2 - 0}{t} = \frac{v_2}{t}$$

and

$$m_1 a_1 = -m_2 a_2$$

may be written

$$\frac{m_1 v_1}{t} = -\frac{m_2 v_2}{t}$$

An example of such a situation is illustrated in Fig. 4.7 where two frictionless carts on a level air track are initially at rest coupled together by a string with a compressed spring between them. Air is forced through small holes on the air track providing a cushion of air on which the carts slide with little friction. The two carts 1 and 2 are pushing on one another (via the spring) with equal and opposite forces. When the string is carefully cut, the two carts separate with each cart gaining speed due to the acceleration. If we return to Newton's third law, we recall that two objects always push upon each other with the same magnitude of force; thus when one ceases to push, the other must cease to push back. This requires that the times over which the two forces act must be the same. This allows for the cancellation of the t from both sides of this equation or

$$m_1 v_1 = -m_2 v_2 \tag{4.7}$$

This equation states that the final momentum of two objects that were initially at rest (that is, their initial momentum was zero) is equal and oppositely directed. Equation (4.7) because of the minus sign indicates that body 1 moves in the opposite direction to body 2. This result also states that the total change in the momentum of body 1 equals the total change in the momentum of body 2 or that the total momentum of a given system (two bodies in this example) remains unchanged unless some outside force is applied to the system.

The meaning of Eq. (4.7) is illustrated in the firing of a cannon. The equation tells us that if a certain momentum is imparted to the projectile by the cannon, an equal momentum in the opposite direction is imparted to the cannon by the projectile. If the mass of the cannon is 1000 times that of the projectile, the velocity of the projectile will be 1000 times the velocity of recoil of the cannon.

The foregoing example illustrates the principle of conservation of momentum.

The total momentum of a given system of bodies remains unchanged unless some outside force is applied to the system.

On offense a football player depends on his momentum, resulting from mass and velocity, to gain that extra yard after impact with the defensive line or tackler.

If an empty 5-ton truck going 20 mi/hour were in a head-on collision with a 1.5-ton car going 70 mi/hour, the car would have the greater momentum, 105 ton-mi/hour to the truck's 100 ton mi/hour. The car would therefore push the truck backward.

It can readily be seen why many automobile insurance companies provide a form on which information about a collision between two cars is to be reported. Information such as (1) make, year, and model of car to provide knowledge of mass, (2) direction of travel and estimated speed, and (3) direction and length of skid marks is valuable to a lawyer should a damage claim reach a court. Knowledge of mass, velocity, acceleration, and momentum can be used in proving or disproving claims involving several thousands of dollars of liability insurance.

The student will benefit now by working Probs. 38 to 42, given at the end of this chapter.

Friction

Because in uniform motion the velocity is constant, it follows that the acceleration must be 0. Hence, applying $f_{net} = ma$, $f_{net} = m \times 0 = 0$. This means that no force is necessary to keep a body moving if it is already in motion. In practice, however, we find that if a book is caused to slide along a table top by giving it an initial push, it will slow up and soon come to rest. This is attributed to a resisting force, called "friction," that acts on the moving book.

CAUSE OF FRICTION

Friction is the resisting force offered to the passage of one surface over another. It is easy to understand the cause of friction if it is realized that an ordinary smooth surface such as a table top, when viewed through a microscope, will appear full of elevations and depressions. When a wooden block is pulled across the table top, the elevations and depressions of the two surfaces tend to interlock and cause friction.

METHODS OF REDUCING FRICTION

It is evident that friction can be reduced by making surfaces smoother and by lubricating surfaces with oil, grease, soap, graphite, or some other lubricant. The lubricant tends to fill in the depressions. If one surface is at rest over another there is more "settling" of elevations into depressions than is possible when one surface is moving over another; thus "starting friction" is greater than "moving friction." When one surface rolls over another, there is usually less friction than when one surface slides over the other; that is, "rolling friction" is usually less than "sliding friction." Hence roller bearings are used in lawn mowers, bicycle wheels, and many other machines.

UNDESIRABLE FRICTION

We usually think of friction as being undesirable. Machines wear out because of friction. Friction results in our obtaining less useful energy from a machine than we put into it; the difference is converted into nonuseful heat energy.

The friction of air against moving automobiles and airplanes greatly reduces their speed and makes the cost of operating at high speeds with a great deal of friction much more than at lower speeds with less friction. The frictional force of the air increases about as the square of the velocity. Doubling the velocity, for example, would make the retarding frictional force four times greater than at the original velocity.

USEFUL FRICTION

In many cases friction is not only desirable but necessary. If it were not for friction, we could not walk, automobiles would not run, our houses would fall apart, brakes would not operate, etc.

USE OF NET FORCE IN $f_{net} = ma$

Suppose that a force of 20,000 dynes is required to keep a block of wood of mass 400 g moving uniformly along a horizontal table top. Because "no force" should be necessary to keep the block moving uniformly, the entire force of 20,000 dynes is used in overcoming friction; that is, the force of friction is 20,000 dynes. Calling the applied force positive (+20,000 dynes) and the retarding force of friction negative (−20,000 dynes), the *net* force is zero, as would be expected from $f_{net} = ma$. If the applied force in the above problem were 60,000 dynes, and the retarding frictional force remained 20,000 dynes, the net force would be 40,000 dynes, and the acceleration given the object would be

$$a = \frac{f_{net}}{m} = \frac{40{,}000 \text{ dynes}}{400 \text{ g}} = \frac{40{,}0\not{0}\not{0} \text{ g-cm/sec}^2}{40\not{0} \text{ g}}$$

$$= 100 \text{ cm/sec}^2$$

Note that in $f_{net} = ma$ net force must be used and a net force of zero means uniform motion for a body already moving or that the body is at rest.

Uniformly accelerated motion

Again consider the relation $f_{net} = ma$. If the net force is constant, then because the mass m of a body does not usually change, the acceleration a must be constant. Hence if a constant force is applied to a body free to move, it will move with constant acceleration—that is, with uniformly accelerated motion.

In *uniformly accelerated motion* the velocity remains along the same straight line whereas the magnitude of the velocity, the speed, changes by the same amount each second; that is, the acceleration is constant but not zero. This differs from uniform motion in which the velocity is constant (but not zero, therefore the acceleration is zero).

FREELY FALLING BODIES

Any object near the surface of the earth is attracted toward the center of the earth with a force that is practically constant. Hence, in accordance with Newton's second law, if air friction is neglected, freely falling bodies move with uniformly accelerated motion.

Careful experiments have been performed that show that the *acceleration due to gravity,* as the acceleration of freely falling bodies is called, is about 32 ft/sec^2 or about 980 cm/sec^2. The exact value depends on the latitude and the elevation above sea level.

Until the time of Galileo (1564–1642), most people followed the lead of Aristotle in believing that heavy objects fall faster than light objects. They reasoned that if the mass of one object is greater than that of another, the force of attraction of the earth on the greater mass is correspondingly greater, and hence the object of greater mass will be given a greater acceleration; that is, it will fall faster. Can the student spot the fallacy in this reasoning?

Galileo, often called the father of experimental science, believed in trying things out. If a dense ball such as one made of steel or lead and a short piece of chalk are held as

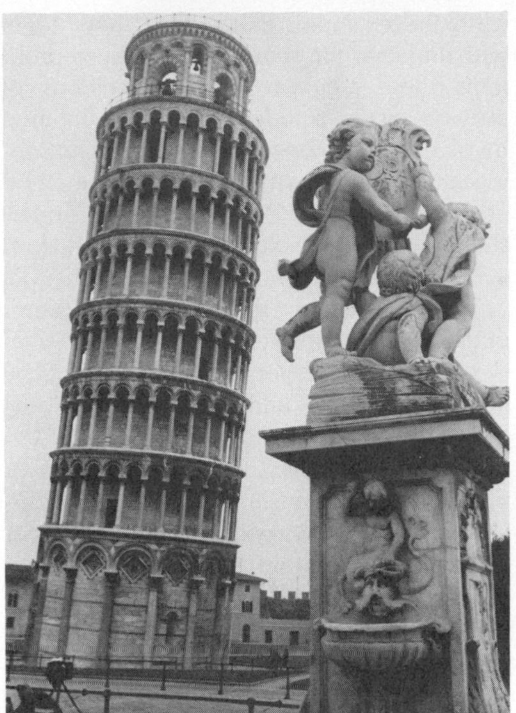

Figure 4.8
Leaning Tower of Pisa. From this tower Galileo reputedly performed some of his experiments on falling bodies. (Courtesy of the Italian Government Travel Office.)

high above the floor as possible and dropped simultaneously, they will seem to reach the floor at the same time. A better test would be to drop them from a higher level. Galileo is reputed to have had his students drop objects from the upper floors of the Leaning Tower of Pisa (Fig. 4.8). No difference in time could be observed, and he correctly concluded that all objects fall with the same acceleration, if air friction is neglected.

Going back to Aristotle's argument, although it is true that the force on the heavier objects is greater than on the lighter objects, there is a correspondingly greater mass to be moved. For example, in $f = ma$ or $a = f/m$, if both f and m are doubled, a remains the same. If the weight were halved by cutting the object in two, f and m would also be halved and a would remain unchanged. If, however, a short piece of chalk and a flat piece of notebook paper side by side are dropped from above a table, the chalk will reach the table top first. The paper, with a large surface area in proportion to its mass, is slowed up by friction in moving through the air much more than the chalk, which has a small surface area in proportion to its mass. This effect of surface area may be demonstrated by wadding the paper into a tight ball and repeating the experiment. The chalk and the paper ball now fall with the same acceleration, and one can observe that they reach the table top at the same time.

The problem with the visualization of free fall is to neglect the effect of the air. This can be demonstrated by placing a penny and a feather in a tube from which the air can be removed using a pump. In the evacuated tube the penny and the feather both fall with the same acceleration. This experiment was performed by an astronaut in 1971 while on the surface of the moon where there is no atmosphere, so dropping any object on the moon is truly free fall.

FINAL VELOCITY AND DISTANCE IN UNIFORMLY ACCELERATED MOTION

To obtain an expression for the velocity of a body moving with uniformly accelerated motion, start with Eq. 4.2, which defines average acceleration. Because the acceleration is constant in uniformly accelerated motion, the acceleration is the same as the average acceleration. In Eq. 4.2 the bar over the a can be removed to yield

$$a = \frac{v_f - v_b}{t}$$

where v_f is the final velocity and v_b is the beginning velocity. Multiply both sides of the equation by t to get

$$v_f - v_b = at$$

By transposing v_b to the other side of the equal sign and, in accordance with mathematical procedure, changing the sign of v_b from minus to plus, we get

$$v_f = v_b + at \tag{4.8}$$

As you look at Eq. 4.8 you should read "Final velocity equals beginning velocity plus acceleration times time," instead of "v sub f equals v sub b plus a times t."

An expression for distance traveled in time t by a body moving with uniformly accelerated motion follows from the fundamental relation distance = average velocity × time,

$$d = \bar{v}t$$

where $\bar{v}$ = average velocity. The average of two quantities is one-half their sum. Hence in uniformly accelerated motion,

$$\bar{v} = \frac{v_b + v_f}{2}$$

Because $v_f = v_b + at$, the substitution of $v_b + at$ for v_f in the equation for average velocity would give

$$\bar{v} = \frac{v_b + v_b + at}{2} = \frac{2v_b + at}{2}$$

$$= v_b + \frac{at}{2}$$

Hence

$$d = \bar{v}t = \left(v_b + \frac{at}{2}\right)t$$

$$= v_bt + \frac{at^2}{2}$$

$$= v_bt + \tfrac{1}{2}at^2$$

Eliminating from the foregoing series of equations the central portions showing the substitutions, the important equation is

$$d = v_bt + \tfrac{1}{2}at^2 \tag{4.9}$$

Equation 4.9 is read "Distance equals beginning velocity times time plus one-half the acceleration times time squared." If the beginning velocity is 0, $v_bt = 0$, leaving $d = \frac{1}{2}at^2$.

In the examples that follow, the student will find that for some parts of these problems there are two different methods of arriving at a solution. The first method involves a slow development using reasoning and understanding. The second method (indicated by "or") involves an understanding of the development of Eqs. 4.8 and 4.9, followed by substitution of numerical values with units in the equation and computation of the answer. The second method is considerably faster, but its use may be more mechanical and thus beget less real understanding.

In whichever system is used (English or metric), the student should be able to determine the only units possible for answers for velocity, acceleration, distance, and time. In applying Eqs. 4.8 and 4.9 to problems, the student will speed his work by reading both number and unit as he makes the substitution but recording only the number. The answer is, of course, numerical with units.

Simple problems on uniformly accelerated motion may be solved by applying the meaning of acceleration to get final velocity and using the relation distance = average velocity × time, $d = \bar{v}t$ (Eq. 4.1).

The student should study the results (see Table 4.1) of an analysis of free fall, an example of uniformly accelerated motion, where $a = g = 32$ ft/sec², the acceleration due to gravity at the earth's surface.

Certain conclusions can be stated from an analysis of Table 4.1. It is observed in column v_f that the velocity at the end of each successive second is 32 ft/sec greater than at the end of the preceding second. Moreover, because of the acceleration of a freely falling body, the distance d fallen by the end of each successive second is increasingly greater, as shown in the next to the last column, and the distance fallen during each successive second is 32 ft greater than the distance for the preceding second, as shown in the last column.

Table 4.1
Freely falling body starting from rest

t	v_f	$\bar{v} = \frac{v_b + v_f}{2}$	$d = \bar{v}t$ or $d = \frac{1}{2}at^2$	
time, sec	v_f at end of the second, ft/sec	$\bar{v}$ during the elapsed time ($v_b = 0$), ft/sec	d fallen by end of the second, ft	d fallen during the second, ft
1	32	16	16	16
2	64	32	64	48
3	96	48	144	80
4	128	64	256	112
5	—	—	—	—
6	—	—	—	—

Example 4.11
How far does a body fall during the first second after it is dropped?

Solution
The beginning velocity is 0, and the final velocity at the end of 1 sec is 32 ft/sec. Hence the average velocity for the first second of fall $\bar{v}_1$ is

$$v_1 = \frac{v_b + v_f}{2} = \frac{0 + 32 \text{ ft/sec}}{2} = 16 \text{ ft/sec}$$

$$d_1 = \bar{v}t = 16 \text{ ft/}\cancel{\text{sec}} \times 1 \cancel{\text{sec}} = 16 \text{ ft}$$

or

$$d_1 = v_bt + \tfrac{1}{2}at^2$$

$$= 0 + \tfrac{1}{2} \times 32 \times (1)^2 = 16 \text{ ft}$$

Example 4.12
An object is dropped from a stationary helicopter that is 1000 ft above the earth's surface. Neglecting the effect of friction, compute in English units (a) the velocity of the object 3 sec after release; (b) the distance fallen during the 3 sec; (c) the distance fallen during the third second.

Solution
(a) Because the acceleration is 32 ft/sec², the body is falling faster each second by 32 ft/sec. Thus starting from rest, the velocity at the end of 1 sec will be 32 ft/sec; at the end of 2 sec, 32 ft/sec + 32 ft/sec = 64 ft/sec; and at the end of 3 sec, 32 ft/sec + 32 ft/sec + 32 ft/sec = 96 ft/sec

or

$$v_f = v_b + at$$

$$v_3 = 0 + 32 \times 3 = 96 \text{ ft/sec}$$

(b) The average velocity for the 3-sec period is

$$\bar{v}_3 = \frac{v_b + v_f}{2} = \frac{0 + 96 \text{ ft/sec}}{2} = 48 \text{ ft/sec}$$

$$d_3 = \bar{v}t = 48 \text{ ft/}\cancel{\text{sec}} \times 3 \cancel{\text{sec}} = 144 \text{ ft}$$

or

$$d = v_bt + \tfrac{1}{2}at^2$$

$$d_3 = 0 + \tfrac{1}{2} \times 32 \times (3)^2$$

$$= 0 + 16 \times 9 = 144 \text{ ft}$$

(c) Because the velocity at the beginning of the third second is 64 ft/sec and at the end of the third second is 96 ft/sec,

$$\bar{v}_{3\text{rd}} = \frac{64 \text{ ft/sec} + 96 \text{ ft/sec}}{2} = \frac{160 \text{ ft/sec}}{2}$$

$$= 80 \text{ ft/sec}$$

$$d_{3\text{rd}} = \bar{v}_{3\text{rd}}t = 80 \text{ ft/}\cancel{\text{sec}} \times 1 \cancel{\text{sec}} = 80 \text{ ft}$$

This answer of 80 ft could also be obtained by computing separately the distance fallen in 3 sec (144 ft) and the distance fallen in 2 sec,

$$d_2 = \bar{v}_2 t = \frac{0 + 64 \text{ ft/sec}}{2} \times 2 \text{ sec} = 64 \text{ ft}$$

and subtracting the latter from the former,

$$d_{3\text{rd}} = d_3 - d_2 = 144 \text{ ft} - 64 \text{ ft} = 80 \text{ ft}$$

Example 4.13

A person standing on a bridge high over the Colorado River throws a stone vertically downward with an initial velocity of 500 cm/sec. Neglecting friction, compute in metric units (a) the velocity of the stone 4 sec after release; (b) the distance fallen by the stone during the 4 sec; (c) the distance fallen during the fourth second.

Solution

(a) $v_4 = 500 \text{ cm/sec} + 980 \text{ cm/sec} + 980 \text{ cm/sec} + 980 \text{ cm/sec} + 980 \text{ cm/sec}$

$= 4420 \text{ cm/sec}$

or

$$v_f = v_b + at$$

$$v_4 = 500 + 980 \times 4 = 4420 \text{ cm/sec}$$

(b)

$$d_4 = \bar{v}t = \frac{500 \text{ cm/sec} + 4420 \text{ cm/sec}}{2} \times 4 \text{ sec} = 2460 \text{ cm/}\cancel{\text{sec}} \times 4\ \cancel{\text{sec}} = 9840 \text{ cm}$$

or

$$d = v_b t + \tfrac{1}{2}at^2$$

$$d_4 = 500 \times 4 + \tfrac{1}{2} \times 980 \times (4)^2 = 2000 + 490 \times 16 = 2000 + 7840 = 9840 \text{ cm}$$

(c) $d_{4\text{th}} = \bar{v}_{4\text{th}} \times t$

$$= \frac{3440 \text{ cm/sec} + 4420 \text{ cm/sec}}{2} \times 1 \text{ sec} = \frac{7860 \text{ cm/}\cancel{\text{sec}}}{2} \times 1\ \cancel{\text{sec}} = 3930 \text{ cm}$$

UNIFORMLY ACCELERATED MOTION IN GENERAL

Although freely falling bodies represent the most common examples of uniformly accelerated motion, there are many cases in which the acceleration is less than 32 ft/sec² (980 cm/sec²) and some in which it is greater.

Example 4.14

An automobile starting from rest accelerates uniformly along a highway at the rate of 5 ft/sec². Compute (a) its velocity at the end of 5 sec; (b) the distance traveled in these 5 sec.

Solution

(a) $v_5 = 0 + 5 \text{ ft/sec} + 5 \text{ ft/sec} + 5 \text{ ft/sec} + 5 \text{ ft/sec} + 5 \text{ ft/sec}$

$= 25 \text{ ft/sec}$

or

$$v_5 = v_b + at = 0 + 5 \times 5 = 25 \text{ ft/sec}$$

(b) $\bar{v}_5 = \dfrac{v_b + v_f}{2} = \dfrac{0 + 25 \text{ ft/sec}}{2} = 12.5 \text{ ft/sec}$

$$d_5 = \bar{v}_5 t = 12.5 \text{ ft/}\cancel{\text{sec}} \times 5\ \cancel{\text{sec}} = 62.5 \text{ ft}$$

or

$$d_5 = v_b t + \tfrac{1}{2}at^2 = 0 + \tfrac{1}{2} \times 5 \times 25 = 62.5 \text{ ft}$$

Example 4.15

An object slides down an incline with a uniform acceleration of 400 cm/sec². It passes a mark on the incline with a velocity of 200 cm/sec. Compute in metric units (a) its velocity 3 sec after passing the mark; (b) the distance it slides during these 3 sec; (c) the distance it slides during the third second.

Solution

(a) $v_3 = 200 \text{ cm/sec} + 400 \text{ cm/sec} + 400 \text{ cm/sec} + 400 \text{ cm/sec}$

$= 1400 \text{ cm/sec}$

or

$$v_f = v_b + at$$

$$v_3 = 200 + 400 \times 3$$

$$= 1400 \text{ cm/sec}$$

(b) $\bar{v}_3 = \dfrac{200 \text{ cm/sec} + 1400 \text{ cm/sec}}{2}$

$$= 800 \text{ cm/sec}$$

$$d_3 = \bar{v}_3 t = 800 \text{ cm/}\cancel{\text{sec}} \times 3 \cancel{\text{sec}} = 2400 \text{ cm}$$

or

$$d_3 = v_b t + \tfrac{1}{2}at^2$$

$$= 200 \times 3 + \tfrac{1}{2} \times 400 \times (3)^2$$

$$= 600 + 200 \times 9 = 2400 \text{ cm}$$

(c) $d_{3\text{rd}} = v_{3\text{rd}}t =$ (student should complete as in Example 4.12c)

NEGATIVE ACCELERATION

If a constant retarding force is applied to a moving body, the body will slow up with a constant acceleration. In this case the acceleration is considered negative (sometimes called "deceleration").

Example 4.16
An automobile is traveling at a rate of 30 mi/hour (44 ft/sec) when the brakes are suddenly applied. Assume that the braking force is constant and results in a negative uniform acceleration of 6 ft/sec². Compute in English units (a) the velocity of the car 4 sec after the brakes are applied; (b) the distance the car travels in the 4 sec.

Solution

(a) $v_4 = 44 \text{ ft/sec} - 6 \text{ ft/sec} - 6 \text{ ft/sec} - 6 \text{ ft/sec} - 6 \text{ ft/sec}$

$$= 20 \text{ ft/sec}$$

or

$$v_f = v_b + at$$

$$v_4 = 44 + (-6) \times 4$$

$$= 44 - 24 = 20 \text{ ft/sec}$$

(b) $\bar{v}_4 = \dfrac{44 \text{ ft/sec} + 20 \text{ ft/sec}}{2} = 32 \text{ ft/sec}$

$$d_4 = \bar{v}_4 t = 32 \text{ ft/}\cancel{\text{sec}} \times 4 \cancel{\text{sec}} = 128 \text{ ft}$$

or

$$d = v_b t + \tfrac{1}{2}at^2$$

$$d_4 = 44 \times 4 + \tfrac{1}{2}(-6) \times 4^2$$

$$= 176 - 48 = 128 \text{ ft}$$

The student should now check his understanding of uniformly accelerated motion by trying to solve Probs. 43 to 50, at the end of this chapter.

Terminal velocity

EFFECT OF FRICTION ON MOVING BODIES

In the earlier discussion on falling bodies, the effect of air friction was neglected. Actually after a few seconds, when the velocity of the falling body has become quite high, air friction is very great. In fact, if an object the size and shape of a human body is dropped, it will after less than 5.5 sec be moving with uniform motion (terminal velocity of about 176 ft/sec) rather than uniformly accelerated motion. At the beginning the acceleration is 32 ft/sec². Soon, because of the increasing retarding force of air friction, which makes the net downward force less and less, the acceleration will drop to 31 ft/sec², then to 30 ft/sec², later to 20 ft/sec², to 10 ft/sec², to 1 ft/sec², and finally the acceleration becomes 0, which means that the body is moving with uniform velocity. This final velocity attained is called "terminal velocity." Raindrops have fallen a long enough time as they near the earth's surface to attain a terminal velocity. In the bailing out of men and equipment from an airplane or the ejection of a pilot from a jet plane, the parachute is used to increase friction and thus the lower terminal velocity is attained more rapidly. Parachutes along with "retro" rockets are also used to slow down space capsules as they reenter the earth's atmosphere, so they will not disintegrate. Sky diving, a rather recent popular sport, is made possible by the fact that a terminal velocity is obtained and the speed of descent of the diver toward earth does not continue to increase as he falls.

TERMINAL VELOCITY OF AUTOMOBILES

The student may have had experience with a similar change from uniformly accelerated motion to uniform motion in the case of an automobile. Suppose on a straight country highway with no traffic in sight an auto salesman demonstrates how fast his car will go. As the forward force due to the engine is increased by supplying more gasoline, the car for a time increases its velocity quite uniformly (uniformly accelerated motion) from 20 to 30 to 40 mi/hour. As the velocity increases, however, the retarding force of air friction increases at a much greater rate than the velocity, so that as the net forward force increases, it does so at a much slower rate (diminishing acceleration). The car increases in velocity from 70 to 75 mi/hour quite slowly; the change from 75 to 80 mi/hour takes still longer. Finally, the speedometer needle may slowly creep up to 85 mi/hour and stay there. The car has reached its highest possible velocity (terminal velocity) and from then on moves with uniform motion until allowed to decelerate. At the top velocity, the acceleration is zero. Does this mean that the forward force of the engine is 0? No. In fact, the forward engine force is at a maximum. The retarding force of friction, however, has become equal to the forward force of the engine, giving a net force of zero.

Vectors

VECTOR AND SCALAR QUANTITIES

A *vector quantity* is one that has both magnitude and direction. Displacement (a given distance in a specified direction), velocity, momentum, acceleration, and force (including weight) are vector quantities. A *scalar quantity* is one that has magnitude but not direction. Area, volume, and mass are scalar quantities because they are expressed only in terms of a certain amount, or magnitude, and no direction is involved.

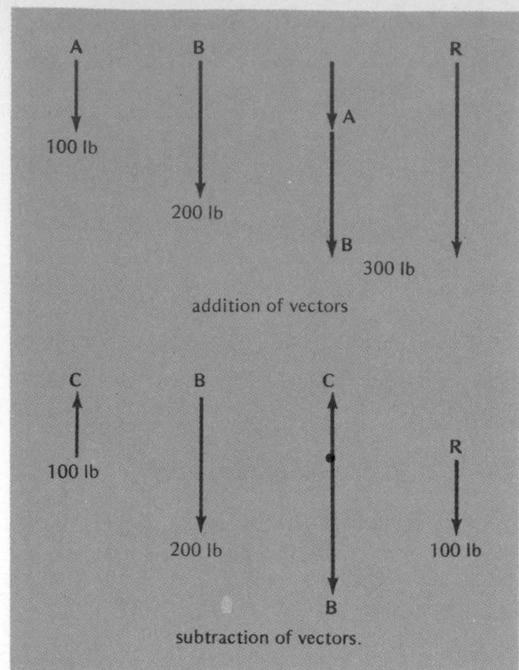

Figure 4.9
Vectors. A vector quantity has both magnitude and direction. Vectors may be added or subtracted to obtain a resultant **R.** The resultant of two or more vectors is the single vector which acting alone will produce the same effect as the two or more vectors acting independently at the same time.

REPRESENTATION AND COMPOSITION OF VECTOR QUANTITIES

A vector quantity may be represented by a vector line, or vector, the length of the line representing the magnitude and an arrow the direction. The combining of vectors is called "composition" of vector quantities.

Vectors in the same direction are added together arithmetically; those in opposite directions are subtracted. Thus, in Fig.

4.9 top, a downward force of 100 lb (vector **A**) and a downward force of 200 lb (vector **B**) could be represented by a resultant net downward force of 300 lb (vector **R**). If the scale adopted is 1 cm = 100 lb, then **A** = 1 cm, **B** = 2 cm, and **R** = 3 cm. In this vector addition, **A** + **B** = **R** and 100 lb + 200 lb = 300 lb, the resultant net downward force.

In vector addition when two or more vectors are parallel to one another in the same direction, their action is equivalent to that of a single vector (called "resultant") in the same direction whose magnitude is the sum of the separate vectors.

In Fig. 4.9 bottom, an upward force of 100 lb (vector **C**) and a downward force of 200 lb (vector **B**) could be represented by a resultant net downward force of 100 lb (vector **R**). In this vector subtraction, **B** − **C** = **R** and 200 lb − 100 lb = 100 lb, the resultant net downward force.

In vector subtraction when two vectors are parallel to one another but in opposite directions, their action is equivalent to that of a single vector (called "resultant") whose magnitude is the difference between the magnitudes of the separate vectors and whose direction is that of the larger separate vector.

It should be clear that the vector nature of a quantity such as force is very important because to use Newton's second law $f_{net} = ma$ we must add up *all* the forces on the mass to determine its acceleration. This addition must be a vector addition in which we account for the directions of the forces.

When vectors are not along the same line, vector addition becomes more complicated. One analysis of a problem in which there are two nonparallel vectors will be given as an illustration.

Example 4.17

A ship steams 8 miles east and then 6 miles northeast. Represent these two displacements by vectors and determine the distance and direction of the ship from the starting point.

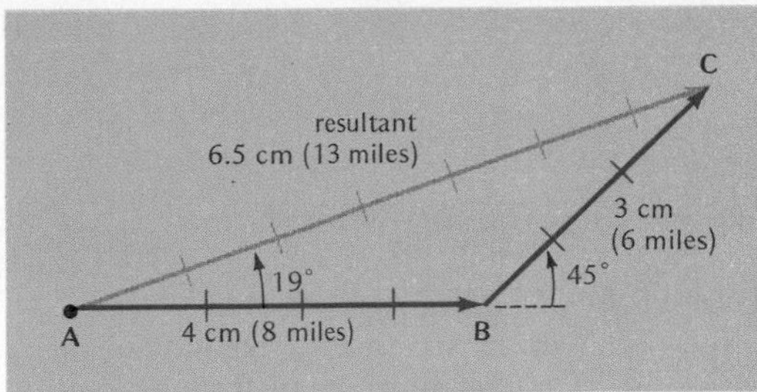

Figure 4.10
Combination of two vectors by the triangle method. **AC** is the resultant of vectors **AB** and **BC.** Scale: 1 cm = 2 miles.

Solution

(Refer to Fig. 4.10) Choose a convenient scale, one that is neither too large nor too small. A scale using centimeters is usually more convenient than one employing inches. In Fig. 4.10 the scale is 1 cm = 2 miles. Choosing *A* as the starting point, lay off the line **AB** horizontally to the right (east) and 4 cm (8 miles) in length. The second displacement is then represented by the line **BC** drawn from *B* in a direction 45° north of east and 3 cm (6 miles) long. The line drawn from *A* to *C* represents the resultant displacement. Measure the length **AC.** It will be found to be 6.5 cm, which is 13 miles on the scale chosen. A measurement of the angle *BAC* by means of a protractor gives 19° for its value. Thus the resultant displacement is found to be 13 miles in a direction 19° north of east. As in Fig. 4.10, two vectors may be combined, or added, by laying off the first vector **AB** from its starting point *A* to its terminus *B*, and then the second vector **BC** from *B* to the terminus *C*, and finally joining the starting point *A* of the first vector to the terminus *C* of the second vector to obtain the resultant vector **AC.** The starting point is often called the tail; the terminus, the head.

ADDITION OF ANY TWO VECTORS NOT IN A STRAIGHT LINE

If two vectors are not in the same straight line (Fig. 4.11), place the tail of the second one at the head of the first one with the second one parallel to its original direction. The resultant vector is drawn as a straight

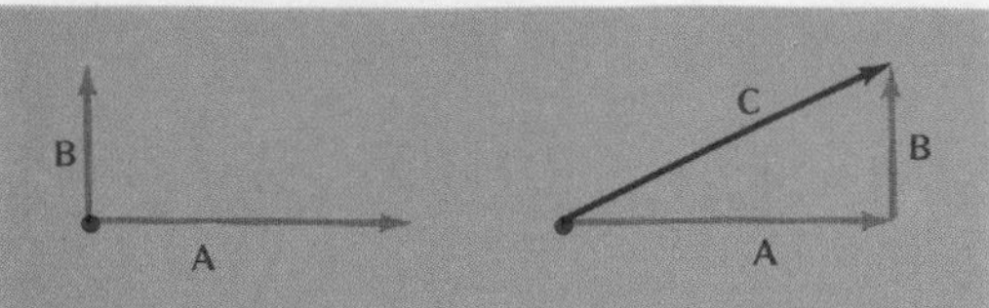

Figure 4.11
Vector addition of two vectors. To add vectors **A** and **B,** place the tail of vector **B** at the head of vector **A,** parallel to its position and in the same direction. Vector **C** is the sum, and the single equivalent, of the combination of vectors **A** and **B.**

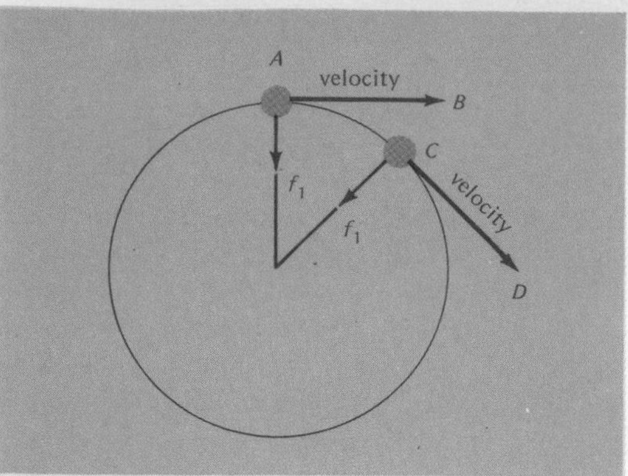

Figure 4.12
Uniform circular motion, change of velocity. When the ball is at **A** it is moving in the direction **AB** at a velocity represented by the length of the line **AB.** When the ball is at **C** it is moving in the direction **CD.** Line **CD** is of the same length as **AB,** indicating that the magnitude of the new velocity is the same. However, the velocity is changed in direction. The string provides the inward centripetal force f_1 which acts on the ball.

line with tail of resultant at tail of the first vector and head of resultant at head of the second vector. This method is applicable whether the vectors are or are not at right angles to one another.

Uniform circular motion

Consider an object moving in circular motion and with a constant speed. Even though the magnitude of a velocity is unchanged, a change in the direction of motion means that there is a change in velocity, hence an acceleration. An example of this kind of change in velocity occurs when a ball tied to a string is caused to move in a circle so that the motion of the ball along the circumference is uniform. Figure 4.12 represents such motion. Because the direction of the velocity changes, the ball must have an acceleration. From Newton's second law $f_{net} = ma$, a net force must act on the ball to provide the acceleration. When one thus causes a ball to revolve, he can feel that he is exerting a force on the string pulling the ball inward from its original path. *Centripetal force* is the force acting toward the center of a circle when a body moves uniformly around a circle (f_1 in Fig. 4.13). Because the force is toward the center of the circle, so is the acceleration, which is called "centripetal" acceleration.

By Newton's third law of motion, if the string exerts a force f_1 inward on the ball, the ball must exert a force f_2 outward on the string. This equal and opposite reaction to centripetal force is called *centrifugal force.* If one force disappears, so must the other. If the string breaks, both centripetal force f_1 and centrifugal force f_2 disappear, and the ball (by Newton's first law of motion) will then move along the tangent to the circle at the point it was when the string broke. This is shown in Fig. 4.13.

It can be shown that the inward centripetal acceleration $= v^2/r$, where v is the uniform speed along the circumference of the circle and r is the radius of the circle. Hence from $f = ma$, the centripetal force acting on the ball (and hence the equal and opposite centrifugal force) is

$$\text{centripetal force} = \frac{mv^2}{r} \qquad (4.10)$$

Note that to keep an object moving in a circular motion with a constant speed, some real inward force (the centripetal force) with a magnitude of mv^2/r must exist. A net force is required to keep an object moving in a circle even with a uniform speed.

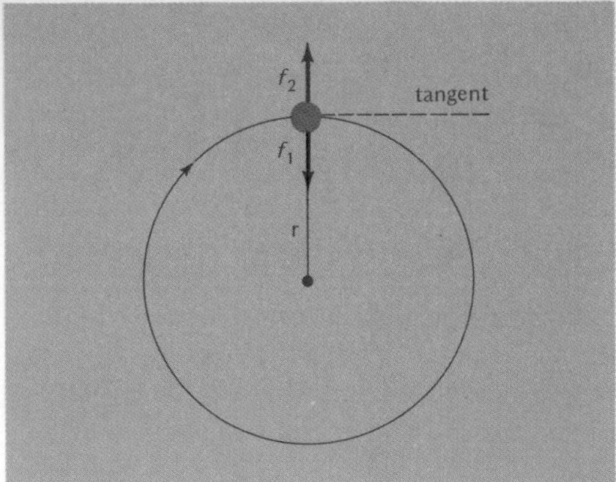

Figure 4.13
Uniform circular motion, centripetal and centrifugal forces. Force f_1 is the centripetal force of the string on the ball. Force f_2 is the equal and opposite reacting force of the ball on the string, called *centrifugal force*. If the string should break when the ball is at the position shown, both of these forces would disappear and the ball would move along the tangent.

APPLICATIONS OF CENTRIPETAL AND CENTRIFUGAL FORCE

An interesting application of centripetal force is the banking of highways. Frictional force alone will not keep a car on the road if it is traveling too fast around a curve. Hence an additional inward centripetal force is needed. By banking the highway it is possible to increase this inward or centripetal force of the bank on the tires rather than to rely on only the frictional force between the tires and the road to keep the car from skidding.

Note that the centripetal force mv^2/r varies directly as the square of the velocity v and inversely as the radius r of the curve. If cars are to negotiate turns safely at high speeds, the radius of the curve should be as large as possible, and the curve should be banked as a further aid.

There are numerous practical illustrations of centripetal force such as a cream separator and the drying cycle in an automatic clothes dryer where the clothes are kept rotating by the inward push of the inside of the drum. The water on the other hand slips through the small holes in the drum.

The slight bulge of the earth near its equator is attributed to the rotation of the earth on its axis. Other planets also show this bulge due to the rotation about an axis through their center.

Almost everyone has been aware of being "pushed" against the side of the car while riding in a car that is negotiating a sharp turn. Actually what is happening is that you were originally moving in a certain direction and then the car turned. To keep you turning with the car an inward force must be exerted on you. Hence you will slide across the seat until you hit the inside of the car door or the person next to you; the person or the door will then push on you providing the necessary centripetal force.

The student can no doubt think of other examples.

SUMMARY

The discussion in this chapter has carried us far from our starting point—that a body is in motion if its position with respect to neighboring objects is changing. In any system the basic quantities of measurement are length, mass, time, and temperature. There are two systems of units for these quantities: the English and the metric. For length and mass the metric system has two main advantages: arrangement of units in multiples of 10, and similarity of length and mass tables.

Average speed is the distance traveled divided by the time of travel. Average acceleration is the change in velocity divided by the time to make the change. In uniform motion the velocity is constant, and equal distances are traveled in equal times along a straight line. In uniformly accelerated motion the velocity changes by the same amount each second along a straight line; the acceleration is constant but not zero.

Momentum is the product of the mass of a body and its velocity. The principle of conservation of momentum is as follows:

The total momentum of a given system of bodies added in a vector manner remains unchanged unless some outside force is applied to the system.

Force is that which tends to change the motion of a body. The weight of a body is the force with which the earth attracts it. Inertia is the tendency of a body to persist in its state of rest or of uniform motion in a straight line. Newton's three laws of motion help in the understandi 3 of force and its relation to motion. The laws may be stated as follows:

First law: *Every body continues in its state of rest or of uniform motion in a straight line unless some force acts on the body to cause it to change that state.*

Second law: *Whenever a net force is applied to a body that is free to move, the force is equal to the product of the mass of the body and the acceleration given to the body by the force.*

Third law: *When one body exerts a force on a second body, the second body exerts an equal and opposite force on the first body.*

Newton's second law ($f_{net} = ma$) may be regarded as defining force and telling how force is measured. Writing the law as $a = f_{net}/m$ shows that acceleration is the ratio of the force to the mass. The first law may be regarded as a special case of the second law: When the force is zero, the acceleration is zero. The third law reveals that forces always occur in pairs, the two forces of a pair being equal and opposite and not acting on the same object. Among the common illustrations of the third law are jet airplane engines and rockets.

Friction may be useful or undesirable. Frictional forces always act to oppose the motion of moving bodies yet we require a frictional force between the road and tires to drive an automobile.

Vectors in a straight line in the same direction are added together arithmetically to get the vector sum or resultant; if in opposite directions they are subtracted arithmetically to get the vector sum or resultant. If two vectors are not in the same straight line, place the tail of the second one at the head of the first one with the second one parallel to its original direction; the resultant vector is drawn as a straight line with tail of resultant at tail of the first vector and head of resultant at head of the second vector. The resultant vector is the sum of and the equivalent of the other two.

The problems at the end of this chapter should have been solved at the points indicated. Solution of the problems should have helped the student develop understanding of the fundamental concepts of motion, force, momentum, and acceleration.

Important words and terms

metric system
British system
meter
gram
mass
inertia
weight
speed
velocity
average velocity
instantaneous velocity
uniform motion
acceleration
Newton's three laws of motion
net force
newton (unit)
dyne
momentum
conservation of momentum
friction
uniformly accelerated motion
free fall
terminal velocity
vectors
scalar
uniform circular motion
centripetal acceleration
centripetal force
centrifugal force

Questions and exercises

1. Write a brief essay on the concepts of weight and mass.
2. What are the main advantages of the metric system? What are some specific examples of

the high cost of changing to the metric system in the United States?

3. Why are two time units necessary in any quantitative statement about acceleration?
4. A car hits a telephone pole at 45 mi/hour and comes to a complete stop in 0.02 sec after the front end crumples. The driver and occupant in front have no seat belts. (a) With what velocity do they move forward while the car is coming to a stop? (b) By what principle do we arrive at this conclusion? (c) With regard to speed of impact, would there be any difference if the occupants were sitting in their seats and the steering wheel and windshield propelled toward the stationary occupants at 45 mi/hour?
5. Why do football coaches usually want fast halfbacks for broken-field running and heavy fullbacks for line plunges? Explain in terms of Newton's laws and momentum.
6. Assume that you are sitting in the middle of a frozen lake and the ice is so smooth it offers no frictional resistance to any kind of motion. How could you get off the lake?
7. Is it necessary that the rocket pad from which satellite-bearing rockets take off be used in explaining how rockets take off in accord with Newton's third law of motion? Discuss.
8. State Newton's first and second laws and indicate how they complement each other.
9. Distinguish between a vector and a scalar. Which of the following are vectors and which are scalars: velocity, area, volume, weight, mass, speed, acceleration, force.
10. A ball is whirled at a uniform rate from the end of a string. Its __________ is constant, but the __________ is constantly changing; therefore the __________ is also changing and there is a constant __________ towards the center. (Fill in with the following words: direction, speed, velocity, acceleration.)

Problems

1. Compute the number of centimeters in 12 in.

Answer: 30.48 cm.

2. The average reading at sea level of a mercurial barometer is 76 cm. Express this height in inches.
3. In the Olympic games the 100-m dash is one of the principal events. Express this distance in yards.

Answer: 109.4 yd.

4. The diameter of the earth is approximately 7900 mi. Express this in kilometers.
5. The reading on a sign post in France is "Paris 20 km." Compute this distance in miles.
6. Compute the number of seconds in a year of 365 days, handling the units as in the solution to Example 3.2.

Answer: 31,536,000 sec.

7. Compute your own height in centimeters and change it to meters.
8. An automobile is driven from Detroit, Michigan, to Buffalo, New York, a distance of 280 miles, in 5 hours. Calculate the average speed.

Answer: 56.0 mi/hour.

9. A cyclist travels from Paris, France, to a certain youth hostel, a distance of 150 km from Paris, in 9.5 hours. Compute his average speed.
10. On February 20, 1962, American astronaut John Glenn orbited the earth three times, a distance of 75,000 miles (approximately), in 4 hours 56 min. Compute the average speed of his Mercury capsule in mi/hour.

Answer: 15,203 mi/hour.

11. The record for the 100-yd dash was 9.1 sec. Compute the average speed of the runner in ft/sec and in mi/hour.
12. Mark Spitz set the 1972 Olympic record for the 100-m swim (butterfly style) with his time of 54.27 sec. Compute his average speed in mi/hour.
13. Jim Ryun set the world record for the mile in 1967 with a time of 3 min 51.1 sec. Compute his average speed in mi/hour.
14. Calais, France, is 298 km north of Paris. Compute the time required for a train to travel this distance if its average speed is 80 km/hour.

Answer: 3 hours 43.5 min.

15. In 1935 Jesse Owens ran 220 yd at an average rate of 10.83 yd/sec. Compute the time.
16. Light travels from the sun to the earth in 8.33 min. The speed of light is approximately 186,000 mi/sec. Compute the distance of the earth from the sun.

Answer: 93,000,000 miles.

17. The time interval between seeing a flash of lightning and hearing the accompanying thunder is often used in computing the distance of the flash from the observer. The speed of sound may be taken as 1100 ft/sec, and the speed of light is so great (186,000

mi/sec) that the time of light travel may be neglected. Compute the distance of a flash of lightning when the sound of thunder is heard 5 sec after the flash is seen.

Velocity and acceleration

18. An automobile with an automatic transmission accelerates from rest to 25 mi/hour in 6 sec. Compute the average acceleration.

Answer: 4.17 mi/hour/sec.

19. A cake of ice slides down an inclined chute. When part way down the chute, its velocity is observed to be 80 cm/sec; 4 sec later its velocity has become 500 cm/sec. Compute the average acceleration.

Answer: 105 cm/sec^2.

20. A marble starting from rest rolls down an incline with an average acceleration of 40 cm/sec^2. Compute the velocity of the marble 6 sec after starting.

Answer: 240 cm/sec, downward.

21. In order to become airborne, a certain plane must attain a velocity of 90 mi/hour (132 ft/sec). If its average acceleration is 5 ft/sec^2, compute the time required for the take-off run.

Answer: 26.4 sec.

Newton's second law

22. Show that 1 newton is equal to 10^5 dynes.
23. When the acceleration a is known, it is possible to associate the dyne and the newton with a given mass. (a) Accordingly, on earth what mass in grams would cause a spring balance to register 1 dyne? (b) Similarly, what mass in grams would cause a spring balance to register 1 newton?

Answer: (a) approximately 0.001 g (1 mg); (b) approximately 100 g.

24. Knowing your weight, compute your mass in kilograms.
25. Calculate the net force required to give a body of mass 400 g an acceleration of 80 cm/sec^2.

Answer: 32,000 dynes.

26. A steel ball has a mass of 620 g. Compute the net force necessary to give it an acceleration of 300 cm/sec^2.
27. A net force of 80,000 dynes is applied to a body having a mass of 250 g. Compute the resulting acceleration.

Answer: 320 cm/sec^2.

28. A body that has a mass of 5 kg is acted upon by a net force of 45 newtons. Compute the resulting acceleration.
29. A body is given an acceleration of 980 cm/sec^2 by a net force of 294,000 dynes. Compute the mass of the body.

Answer: 300 g.

30. Compute the mass of a body that is given an acceleration of 8 m/sec^2 by a net force of 168 newtons.

Answer: 21 kg.

31. A force of 10 newtons is applied by a man to push a box. A resisting frictional force of 5 newtons opposes this motion. (a) What is the net force causing the box to move? (b) If the mass of the box is 2.5 kg, compute the resulting acceleration.
32. (a) A car *weighs* 4800 lb. What net force is needed to give it an acceleration of 2 ft/sec^2? (b) Another car has a *mass* of 120 slugs. What net force is needed to give it an acceleration of 2 ft/sec^2?
33. In Question 4 assume that the 96-lb torso of the driver contacts the steering wheel for an average duration time of 0.002 sec before the car comes to a complete stop. (a) With what average acceleration does his body meet the steering wheel? (b) What is the average force of the steering wheel against the driver in tons during this 0.002-sec interval?

Answer: (a) 3.3×10^4 ft/sec^2; (b) 49.5 tons.

Momentum

34. A certain airplane has a weight of 12,000 lb and is moving with a speed of 300 mi/hour. Compute the momentum of the airplane.

Answer: 112,000 slug-mi/hour.

35. A stone having a mass of 200 g is thrown and at a certain instant has a speed of 1500 cm/sec. Compute its momentum.
36. A bowling ball has a weight of 16 lb and a momentum of 7.5 slug-ft/sec. Compute its speed.

Answer: 15 ft/sec.

37. A rifle bullet has a speed of 700 m/sec. If its momentum is 105,000 g-m/sec, compute the mass of the bullet.

Answer: 150 g.

Conservation of momentum

38. Car A (300 g) and car B (1500 g) are tied together with a compressed spring between

them on an air track as in Fig. 4.7. One second after the string is cut, car *A* is 600 cm from its starting point. At what distance is car *B* from the starting point?

39. A cannon has a weight of 12,000 lb. If a projectile having a weight of 80 lb is fired with a muzzle velocity of 1500 ft/sec, compute the velocity of recoil of the cannon.

Answer: 10 ft/sec.

40. A 20-mm cannon fires a projectile having a mass of 800 g with a muzzle velocity of 750 ft/sec. If the mass of the cannon is 80 kg, compute the velocity of recoil of the cannon in feet per second.
41. A motorboat with contents has a weight of 2000 lb, and a rowboat with contents has a weight of 600 lb. In order to transfer a passenger, a rope is thrown from one stationary boat to the other. If pulling on the rope results in a velocity of the rowboat of 3 ft/sec, compute the speed of the motorboat.

Answer: 0.9 ft/sec.

42. A bullet having a mass of 20 g is fired into a block of wood that has a mass of 1000 g. As a result the block and the embedded bullet start moving with a speed of 19.7 cm/sec. Compute the speed of the bullet as it strikes the block. (Such knowledge of ballistics is valuable in the field of crime detection.)

Uniformly accelerated motion

43. A steel ball starting from rest rolls down an incline with a uniform acceleration of 50 cm/sec². Compute in metric units (a) the speed of the ball 5 sec after it starts rolling; (b) the distance rolled during these 5 sec; (c) the distance rolled during the fifth second.

Answer: (a) 250 cm/sec; (b) 625 cm; (c) 225 cm.

44. An automobile starts from rest and moves with a uniform acceleration of 5 ft/sec². Compute (a) the speed at the end of 10 sec; (b) the distance traveled during these 10 sec; (c) the distance during the tenth second.
45. A ball is dropped from an upper window of a skyscraper and is observed to hit the ground 6 sec later. Neglecting friction, compute in English units (a) the speed at impact; (b) the height of the window above the ground.

Answer: (a) 192 ft/sec; (b) 576 ft.

46. An object dropped from a stationary balloon strikes the ground 8 sec later. Neglecting friction, compute in metric units (a) the speed at impact; (b) the height of the balloon.
47. The driver of an automobile that has a speed of 30 mi/hour (44 ft/sec) "steps on the gas," causing a uniform acceleration of 6 ft/sec²: (a) Compute the speed of the automobile 4 sec after applying the foot to the accelerator ("gas pedal"); (b) compute the distance the car moves during this 4 sec.

Answer: (a) 68 ft/sec; (b) 224 ft.

48. A block slides down an incline (neglect friction) with a uniform acceleration of 10 ft/sec². When first observed, it has a speed of 12 ft/sec. Compute (a) the speed of the block 3 sec after the first observation; (b) the distance the block slides during these 3 sec.
49. How long will it take for a body starting from rest to fall 144 ft if friction is neglected?

Answer: 3 sec.

50. Neglecting friction, compute the time for a body to fall 7840 cm starting from rest.

Answer: 4 sec.

Vectors

51. A man rows a boat so that its velocity in still water would be 4 mi/hour. He heads the boat directly across a stream in which the velocity of the water is 3 mi/hour. Find the resultant velocity (both magnitude and direction) of the boat.

Answer: 5 mi/hour at an angle of 53° with the bank.

52. A balloon rises with a vertical velocity of 15 mi/hour. A west wind (wind blowing from the west) is blowing with a velocity of 10 mi/hour. Obtain the resultant velocity (both magnitude and direction).
53. An airplane is headed east with an airspeed (velocity through the air relative to the ground) of 200 mi/hour. If a wind from the northwest is blowing with a velocity of 40 mi/hour (velocity of the air over ground), find the resultant velocity of the airplane relative to the ground.

Answer: 230 mi/hour in a direction 7° south of east.

Supplementary readings

Books

Andrade, E. N. da C., *Sir Isaac Newton: His Life and Work,* Anchor Books, Doubleday, Garden City, N. Y. (1958). (Paperback.) [A personal portrait of one of the great geniuses of history.]

Holton, Gerald, and Duane H. D. Roller, *Foundations of Modern Physical Science,* Addison-Wesley, Reading, Mass. (1958). [In Chap. 4 Newton's three laws of motion are discussed, as are weight and mass, including inertial and gravitational mass.]

Omer, Guy G., Jr., et al., *Physical Science: Men and Concepts,* D. C. Heath, Boston (1962). [Excerpts from original writings by Galileo and by Newton. In Galileo's *Dialogues Concerning Two New Sciences* (Chaps. 11–14) some see the beginning of the modern scientific age. Galileo considers falling bodies and uniform and accelerated motion. Chapters 15–18 include an excellent biographical vignette of Newton before presenting material from his *Principia* on the basic laws of motion, the law of gravitation, and how he used these principles to explain the motions within the solar system.]

Rogers, Eric M., *Physics for the Inquiring Mind,* Princeton University Press, Princeton, N. J. (1960). [Chapter 1 is a particularly useful discussion of accelerated motion, including inductive and deductive treatment and the algebra of motion.]

Taylor, Lloyd William, *Physics: The Pioneer Science,* Houghton Mifflin, Boston (1941); or, a paperback, L. W. Taylor and F. G. Tucker, *Physics: The Pioneer Science,* vols. 1 and 2, Dover Publications, New York (1962). [In Chaps. 9–11 Taylor clarifies the confusion that exists about weight and mass, related mass and acceleration, and thoroughly considers Newton's three laws of motion.]

The Project Physics Course, Holt, Rinehart and Winston, New York (1970).

Articles

Freilich, F. G., "A Modern Visit to Galileo," *The Physics Teacher,* **8,** no. 2, 63 (February, 1970).

Ritchie-Calder, Lord, "Conversion to the Metric System," *Scientific American,* **223,** no. 1, 17 (July, 1970).

LAW OF UNIVERSAL GRAVITATION

$$f_G = G\frac{m_1 m_2}{d^2}$$

There is no mathematics — whether algebra, geometry, or the calculus — to justify this bold step. One can say of it only that it is one of those triumphs that humble ordinary men in the presence of genius.

I. BERNARD COHEN, 1960

Figure 5.1
Sir Isaac Newton (1642–1727), English mathematician and philosopher. Newton contributed notably to mathematics and to the study of light, and invented a reflecting telescope. He is best known for his formulation of the law of universal gravitation and the laws of motion. (Detail from painting, courtesy of the National Portrait Gallery, London.)

As mentioned in Chap. 4, Sir Isaac Newton in his *Principia* (1687) formulated the relations between force and motion. Further consideration of these relations and recognition of the fact that the moon was held in its orbit by the same kind of force that caused objects near the earth to fall to its surface resulted in his statement in the third book of the *Principia* of a much more general law—the law of universal gravitation.

Newton's law of universal gravitation

SIR ISAAC NEWTON

A portrait (Fig. 5.1) and a word—genius—are appropriate here to help convey to you a feeling for the man whose mind formulated (1) the nature of the relations between force and motion—the laws of motion—and (2) the law of universal gravitation. As you study this latter concept and learn of its impact on man's thinking you will realize that Newton's genius produced one of the greatest concepts of all time. Newton's law of *universal gravitation* may be stated as follows:

Any two mass particles (or bodies composed of particles) in the universe attract each other with a force that is directly proportional to the product of the masses of the two particles (or bodies) and inversely proportional to the square of the distance between them.

Homogeneous bodies (same texture throughout) that are spherical may be treated as if their entire masses were concentrated at their centers (see Fig. 5.2). In this case the distance between them becomes the distance between their centers. The error introduced by treating ordinary bodies of various shapes as if their masses were concentrated at the center is small, so that in this textbook the distance used will always be the *distance between the centers of the bodies.*

The general method used in arriving at an equation from the statement of a law will now be explained and applied to the above law. If two quantities are directly proportional, they are written in the numerators on opposite sides of an equation. If two quantities are inversely proportional, one is written in the numerator of one side of an equation and the other in the denominator of the other side of the equation. Also included in an equation is a proportionality constant, whose value depends on the units used.

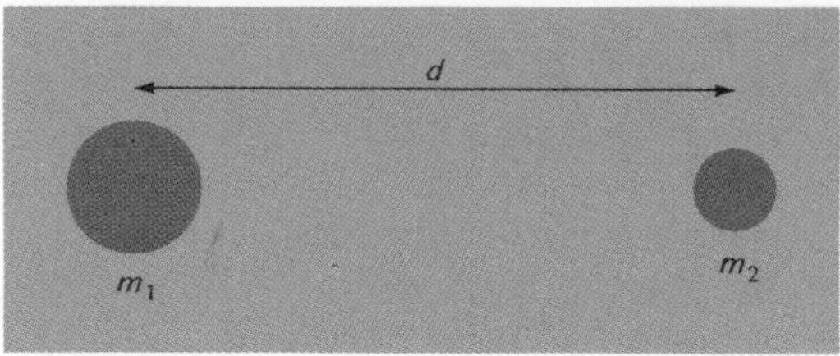

Figure 5.2
Meaning of Newton's law of universal gravitation. For two bodies of masses m_1 and m_2 with a distance d between their centers, $f_G = G(m_1m_2/d^2)$, where f_G is the force of attraction between the two bodies and G is a constant.

Thus, from the statement of Newton's law of universal gravitation, if the force of gravity f_G is placed on the left side of the equation, the product of the masses m_1m_2 will be in the numerator of the right side, and the square of the distance between the centers of mass d^2 will be in the denominator of the right side of the equation. A constant, represented by G, is introduced to change the proportionality to an equation. Thus the equation

$$f_G = G\frac{m_1m_2}{d^2} \tag{5.1}$$

It is evident from Eq. (5.1) that

$f_G \propto m_1$ (read "force is proportional to mass one")

and

$f_G \propto m_2$

and

$f_G \propto 1/d^2$ (read "gravitational force is inversely proportional to distance squared")

Hence the gravitational force of attraction f_G between two bodies of masses m_1 and m_2 is increased when either m_1 or m_2 is increased and decreased when either m_1 or m_2 is decreased. Moreover, the force f_G is decreased when the distance d between the centers of m_1 and m_2 is increased and increased when d is decreased.

The G in Eq. 5.1 is called the *constant of universal gravitation.* Its value depends on the units used for the quantities of the equation, and it has the same value for any pair of bodies in the universe if the same units are used. If m_1 and m_2 are in grams and d is in centimeters, f_G will be in dynes (g-cm/sec²) if the constant G has the value 0.0000000667 cm³/g-sec², usually written 6.67×10^{-8} cm³/g-sec² (see below). In the mks system of units in which m_1 and m_2 are in kilograms and d is in meters, f_G will be in newtons if the constant G has the value 6.67×10^{-11} m³/kg-sec². Note that it is convenient to write G using powers of 10 notation as described in App. A. Students unfamiliar with powers of 10 notation should examine App. A carefully.

DETERMINATION OF THE VALUE FOR G

Newton obtained his law of universal gravitation by examining the data that described the motions of the planets. In fact, he was able to derive Kepler's second law (Eq. 1.1) from his law of universal gravitation with the aid of his other three laws just discussed in Chap. 4. However, the value for G could not be obtained from the data on the motions of the planets in the solar system, primarily because there had been no independent determination of the masses of the planets. An experimentally determined value for G was not obtained until about 100 years after Newton proposed his law of universal gravitation. In 1797 Henry Cavendish, an English scientist, using a sensitive torsion balance determined G by measuring the force between known masses in his laboratory.

A method used a few years ago at the Massachusetts Institute of Technology in determining the value of G is easily understood. Two spherical bodies of equal mass m_1 were hung from an equal-arm beam balance (Fig. 5.3), giving equilibrium. A huge lead sphere of mass m_2 was rolled in from the next room and placed directly under one of the small

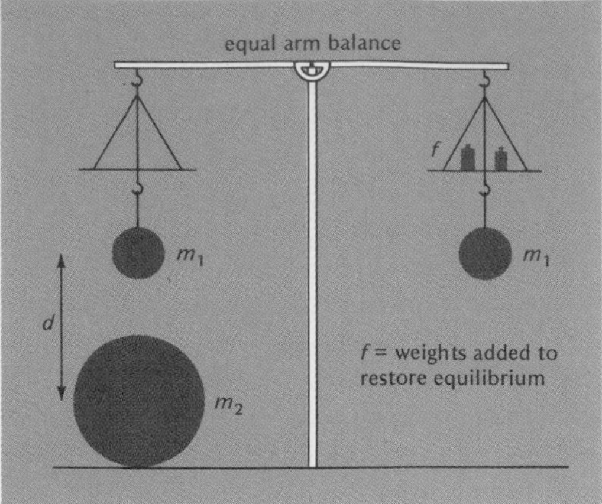

Figure 5.3
One method used to determine the gravitational constant *G*.

spheres m_1. The mutual attraction of the large sphere and the small sphere directly above it caused the left side of the balance to go down, but equilibrium was then regained by adding a few milligrams to the right-hand scale pan. Thus the force of attraction f_G was obtained and the values of m_1 and m_2 and d were easily determined by weighings and measurements. Hence in the equation $f_G = G(m_1m_2/d^2)$, all the quantities except G were known and G was computed.

The force of attraction between two ordinary objects, such as this textbook and a blackboard eraser, is so small that it cannot be measured easily. However, when the attraction is between a large body, such as the earth, and another body, the force becomes quite large. When the earth is one of the two bodies, the force of attraction is called the *force of gravity* or simply *gravity*.

As defined in Chap. 4, the *weight* of a body on earth is a measure of the force with which the earth attracts the body.

DIFFERENCE BETWEEN WEIGHT AND MASS

The relationship that links weight with mass was developed in Chap. 4 in a discussion of force, weight, and mass. However, the distinction between these concepts is so often not well understood following only an initial introduction that further discussion is in order. We shall then consider some problems, similar to those encountered by our National Aeronautics and Space Administration in space programs involving the moon, earth satellites, and planets, whose solutions involve Newton's law of universal gravitation.

The difference between weight and mass can perhaps be better understood if we consider the discrepant results obtained upon transporting a block of iron, a spring scale, an equal-arm balance, and a set of weights to a different location—the moon in this case. On the earth a 6-lb block of iron will register six on the spring scale and will also be counterbalanced by 6 lb of weights on an equal-arm balance. On the moon the spring scale will register only one; but with the beam balance, the block of iron will still require in the opposite pan, in order to counterbalance it, the weights that total 6 lb on earth.

The spring scale registers weight—the magnitude of the force pulling down on the spring in the scale. The tension properties of the spring have not changed by virtue of its being on the moon, for the spring retains its property of being extendible a certain length for a given force regardless of where it is used. Thus the smaller extension of the spring is a true indication that on the moon there is less pull on the spring by the block of iron—that is, that the iron block weighs less on the moon than on the earth.

The equal-arm balance, however, permits one to compare the gravitational forces (not masses) acting on the contents of each of the pans and thereby to compare quantitatively the masses of matter in the two pans. The comparison is possible because of a simple, directly proportional relationship between the magnitude of the gravitational force and the quantity of mass. In both locations—on earth, on moon—the equal-arm balance indicates that the two forces on the pans are equal, from which fact we infer that the masses in the two pans are equal. This so-called weighing on an equal-arm balance measures mass and thus might be called "massing" rather than "weighing."

If we transported our equipment out to free space far from any large bodies such as the earth or moon, our spring balance would register zero for the block of iron, and no forces would be acting on the pans of the beam balance either. Consequently we would be unable to use the balance to compare the force on the block of iron with the force on the standard weights we brought along. We would thus have to devise some other means to compare masses. For example, we might measure the relative accelerations given different masses by the same force and then make computations using the equation $m_1a_1 = m_2a_2$ (p. 95), or $m_1 = m_2 \times a_2/a_1$, where m_2 is a standard mass.

Mass is an *intrinsic* property of matter itself. *Weight* is a *relative* property: as we saw in Chap. 4, it expresses a relationship, the magnitude of the force of attraction, between the object weighed and the body (earth, moon, etc.) on which the weighing is done. The relationship that links mass with weight was developed in connection with Eqs. 4.4 and 4.5.

DETERMINATION OF g

Let us reexamine the meaning of the weight force defined in Eq. 4.4; that is, weight $= mg$. The weight of a body is the force with which the body is attracted to the earth. From Newton's second law, the weight is the product of the mass m of the body times the acceleration due to gravity g near the earth's surface. However, Newton's law of universal gravitation says that the force of attraction between the body of mass m and the earth of mass M_E is given by

$$f_G = \frac{GmM_E}{d^2} = \text{weight}$$

If the earth is considered to be a sphere, then the separation d between the center of the earth and the center of any ordinary object on the earth's surface will be essentially the radius of the earth R_E. The earth's pull on the body or weight of the body will be

$$\text{weight} = \frac{GmM_E}{R_E^2}$$

The weight of an object was originally defined as mg in Eq. 4.4. These two expressions must be equal,

$$mg = \frac{GmM_E}{R_E^2}$$

If both sides of this equation are divided by m, then the acceleration due to gravity g on the earth's surface is given by

$$g = \frac{GM_E}{R_E^2} \qquad (5.2)$$

It is clear from Eq. 5.2 that g would be a constant everywhere on the surface of the earth if the earth were a perfect sphere, for then R_E would be constant; M_E can be considered a constant and G is a universal constant. The fact that the mass of the body does not appear in Eq. 5.2 indicates that g is the same for all objects; that is, all objects have the same acceleration due to gravity near the surface of the earth. This expression for g also indicates that the value of g will be smaller as a body is moved to the top of a mountain because R_E will now be increased.

The above analysis could be repeated on the moon to define g_M the acceleration due to gravity on the moon. In that case the mass of the moon would replace the mass M_E of the earth and the radius of the moon replaces the radius R_E of the earth. Making substitutions for M_E and R_E, the value of g_M can be easily computed. The result of such a calculation indicates that g_M is about $\frac{1}{6}g$ on the earth.

PROBLEMS ON NEWTON'S LAW OF UNIVERSAL GRAVITATION

The equation $f_G = G(m_1m_2/d^2)$ states that for a given distance between the centers of bodies (G also remains the same), the force

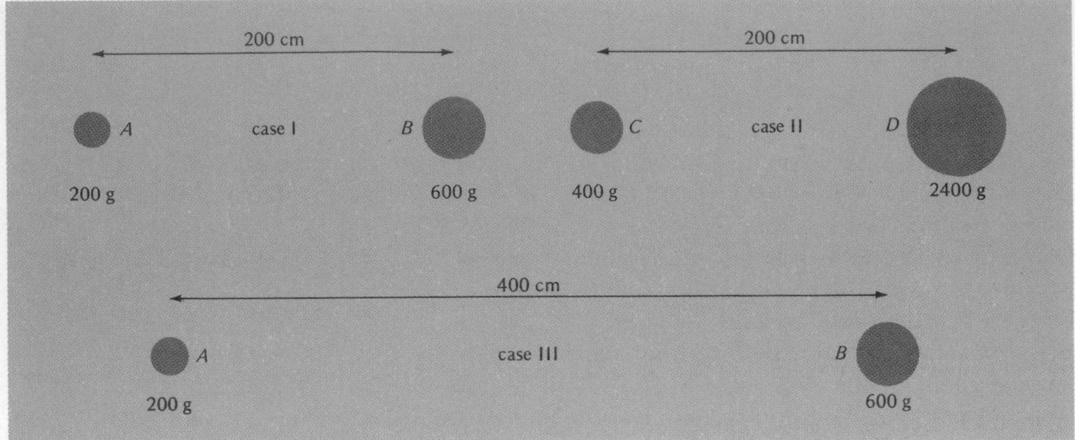

Figure 5.4
Variation of masses and distances apart affecting the force of attraction between bodies.

is directly proportional to the product of the masses m_1m_2. This means that if the product m_1m_2 is twice as much, the force f_G is twice as much; if the product m_1m_2 is 10 times as much, f_G is 10 times as much; etc. The equation $f_G = G(m_1m_2/d^2)$ also tells us that for the masses remaining the same (G also remains the same), f_G is proportional to $1/d^2$. This means that, if d is doubled, f_G is $\frac{1}{2^2}$, or one-fourth as much; if d is tripled, f_G is $\frac{1}{3^2}$, or one-ninth as much; if d is four times as great, f_G is $\frac{1}{4^2}$ or one-sixteenth as much; if d is one-half as great, f_G is $1/(\frac{1}{2})^2$, or 2^2, or four times as great, etc. Seldom does a problem arise that involves simple substitution in the equation $f_G = G(m_1m_2/d^2)$. Usually the variation of the force due to a change in both masses and distance is required.

The meaning of this important law can best be made clearer by the solution of several problems involving the law. Example 5.1, which follows, may appear to be complex, though the mathematics used is simple proportion; however, once the student has gained from this problem an understanding of the variation in force due to change in both masses and distance, a simplified method, as in Example 5.4, may be used.

Example 5.1

Using Fig. 5.4, compare the force of attraction (a) between bodies C and D of masses 400 g and 2400 g, respectively, with a distance of 200 cm between their centers (case II), with the force between bodies A and B of case I, having the same distance between their centers; (b) between bodies A and B of case III with the force between bodies A and B of case I; (c) between bodies C and D of case II with the force between bodies A and B of case III.

Solution

(a) Because the force of attraction is directly proportional to the product of the masses and the distances are the same, in cases II and I

$f_{\text{II}} \propto$ product of masses in II

$f_{\text{I}} \propto$ product of masses in I

Therefore

$$\frac{f_{\text{II}}}{f_{\text{I}}} = \frac{\text{product of masses in II}}{\text{product of masses in I}}$$

or

$$f_{\text{II}} = \frac{\text{product of masses in II}}{\text{product of masses in I}} f_{\text{I}}$$

$$= \frac{400 \times 2400}{200 \times 600} f_{\text{I}} = 8f_{\text{I}}$$

(b) Because the force of attraction varies inversely as the square of the distance, and the distance in III is twice the distance in I, but the masses are the same, in cases III and I

$$\frac{f_{III}}{1} \propto \frac{1}{(\text{distance in III})^2}$$

$$f_I \propto \frac{1}{(\text{distance in I})^2}$$

$$\frac{1}{f_I} \propto \frac{(\text{distance in I})^2}{1}$$

Therefore

$$\frac{f_{III}}{f_I} = \frac{(\text{distance in I})^2}{(\text{distance in III})^2}$$

or

$$f_{III} = \frac{(\text{distance in I})^2}{(\text{distance in III})^2} f_I = \frac{200^2}{400^2} f_I$$

$$= \left(\frac{200}{400}\right)^2 f_I = \frac{1}{2^2} f_I = \frac{1}{4} f_I$$

(c) Considering masses only, in cases II and III

$f_{II} \propto$ product of masses in II

$f_{III} \propto$ product of masses in III

Therefore

$$\frac{f_{II}}{f_{III}} = \frac{\text{product of masses in II}}{\text{product of masses in III}}$$

or

$$f_{II} = \frac{400 \times 2400}{200 \times 600} f_{III} = 8 f_{III}$$

Considering distances only, in cases II and III

$$\frac{f_{II}}{1} \propto \frac{1}{(\text{distance in II})^2}$$

$$f_{III} \propto \frac{1}{(\text{distance in III})^2}$$

$$\frac{1}{f_{III}} \propto \frac{(\text{distance in III})^2}{1}$$

Therefore

$$\frac{f_{II}}{f_{III}} = \frac{(\text{distance in III})^2}{(\text{distance in II})^2}$$

or

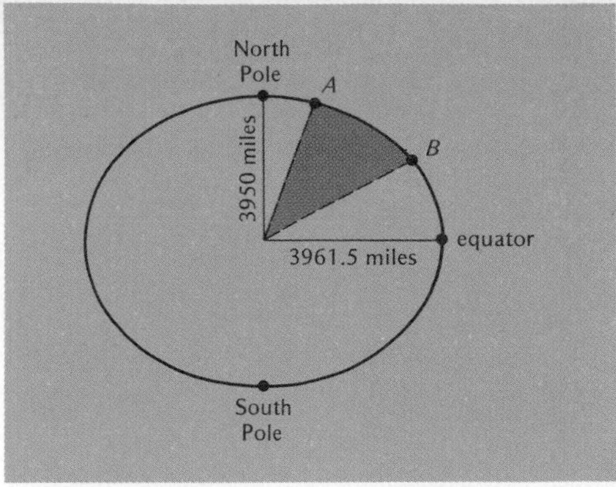

Figure 5.5
Variation of weight with latitude.

$$f_{II} = \frac{400^2}{200^2} f_{III} = 4 f_{III}$$

Considering both masses and distances,

$$f_{II} = (8 \times 4) f_{III} = 32 f_{III}$$

Or, considering both masses and distances, one could compute f_{II}, then f_{III}, and finally the relation between f_{II} and f_{III} as follows:

$$f_{II} = G \frac{400 \times 2400}{(200)^2} = 24\ G$$

$$f_{III} = G \frac{200 \times 600}{(400)^2} = \frac{3}{4} G$$

$$\frac{f_{II}}{f_{III}} = \frac{24\ G}{\frac{3}{4} G} = 24 \times \frac{4}{3} = 8 \times 4$$

Solving for f_{II},

$$f_{II} = (8 \times 4) f_{III} = 32 f_{III}$$

Example 5.2
Does a given body weigh the same near the earth's equator as near the earth's poles or does it weigh more or less near the poles (Fig. 5.5)? Explain.

Solution
The earth has the shape of an oblate spheroid because of the effect of its rotation. As indicated in Fig. 5.5, the equatorial radius of the earth (3961.5 miles) exceeds its polar radius (3950 miles) by 11.5 miles. A given mass will weigh less in position *B* than at *A* (*A* being of greater latitude), because the value of *d* in the equation $f = G(m_1 m_2/d^2)$ is greater at *B* than at *A* and the value of m_1, m_2, and *G* remain the same at both positions. This difference, however, is not very great, and, unless

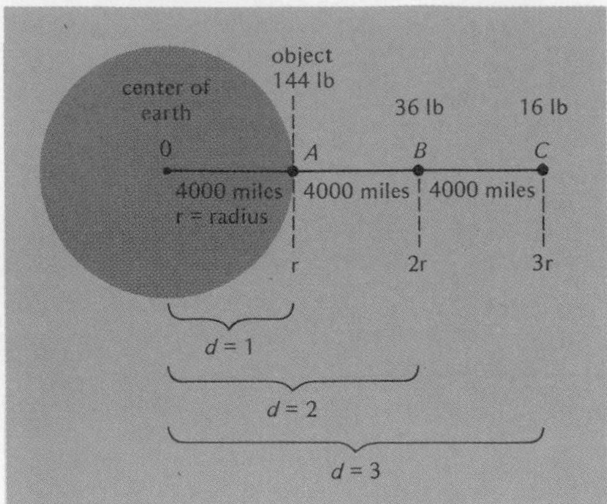

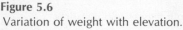

Figure 5.6
Variation of weight with elevation.

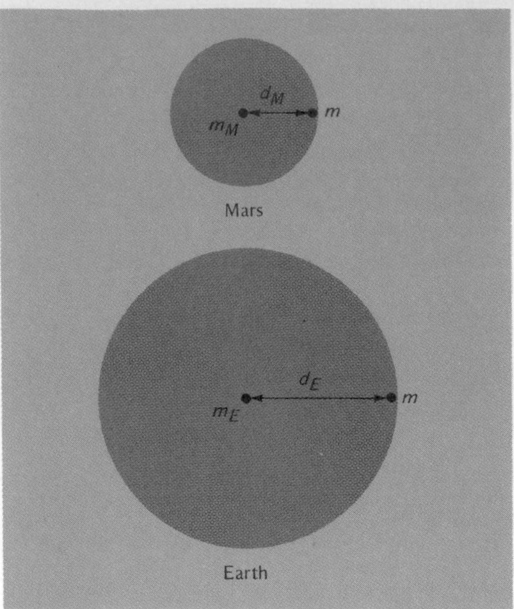

Figure 5.7
Object on surface of Mars and on surface of the Earth.

otherwise stated, it will be considered in working out problems that the earth is a sphere so that a 1-lb weight will weigh 1 lb anywhere around the earth.

The difference just discussed, though not great, may in some situations be significant. The effect of latitude on athletic records is appreciable, and a correction factor should be applied to take into account differences in the force of gravity at the various latitudes where the records were established if the latitudes are such that differences in the distance to the center of the earth are considerable. For example, the Olympic Games have been held in Melbourne, Australia, near the equator, and in Helsinki, Finland, near an earth pole. (See Heiskanen, under "Supplementary Readings.")

Example 5.3

Using the approximate value of 4000 miles for the earth's radius, compute the weight of a 160-lb man as shown in Fig. 5.6, (a) at the earth's surface *A*; (b) at point *B*, 4000 miles above the surface; (c) at point *C*, 8000 miles above the surface.

Solution

(a) By the usual definition, the weight of a 160-lb man means that the pull of the gravity on the man at the surface of the earth is 160 lb. The mass of the man is determined by dividing the weight of the man by the acceleration due to gravity g,

$$m = \frac{160 \text{ lb}}{32 \text{ ft/sec}^2} = 5 \text{ slugs}$$

(b) Because the force of attraction on the man varies inversely as the square of the distance, the man at *B*, which is twice as far from the center of the earth as is *A*, "weighs" $1/2^2 = \frac{1}{4}$ as much or $\frac{1}{4} \times 160 = 40$ lb. The man still has a mass of 5 slugs.

(c) Point C is three times as far from the center of the earth so that the force on the man of C will be reduced by $1/3^2 = \frac{1}{9}$. He "weighs"

$$\tfrac{1}{9} \times 160 = 17.8 \text{ lb}$$

The attraction of other planets and heavenly bodies for objects on their surfaces (called their "surface gravities"), as compared with the earth's attraction for objects on its surface, can be computed with the aid of the law of universal gravitation.

Example 5.4

The planet Mars has a mass about one-tenth that of the earth and a diameter about one-half that of the earth (see Fig. 5.7). Compute (a) the surface gravity for Mars; (b) the weight of an object on Mars if it weighs 120 lb on the earth.

Solution

(a) This is similar to Example 5.1c and may be solved in the same manner. Designate by f_M the force of attraction of Mars for the object m on its surface and by f_E the force of attraction of earth for the same object on its surface. Considering masses only,

$$f_M = \tfrac{1}{10} f_E$$

Considering distances only,

$$f_M \propto \frac{1}{(d_M)^2} \qquad \text{and} \qquad f_E \propto \frac{1}{(d_E)^2}$$

where d_M is the radius of the moon and d_E is the radius of the earth. If we take the ratio of these two forces

$$\frac{f_M}{f_E} = \frac{1/d_M^2}{1/d_E^2} = \frac{d_E^2}{d_M^2} \qquad \text{or} \qquad f_M = \frac{d_E^2}{d_M^2} f_E$$

Because $d_M = \frac{1}{2} d_E$ then f_M will equal

$$f_M = \frac{d_E^2}{\frac{1}{4} d_E^2} f_E = \frac{1}{\frac{1}{4}} f_E = 4 f_E$$

Considering both masses and distances,

$$f_M = (\tfrac{1}{10} \times 4)\, f_E = 0.4 f_E$$

(b) Because $f_M = 0.4 f_E$, the surface gravity on Mars = 0.4 the surface gravity on earth. From the meaning of "weight on a planet," a 120-lb object on earth would "weigh" 0.4 × 120 lb = 48 lb on Mars.

In this day when interplanetary travel is becoming a reality (for example, probes from earth to Venus and Mars), it is of interest to compute what a person weighing 100 lb on earth would weigh on other planets. Such computations can readily be made in the manner just shown for Mars and also by using the information on surface gravity given in Table 3.1.

Newton's law of universal gravitation and the solar system

So far in this chapter the discussion of Newton's law of universal gravitation has been limited to "static" considerations. Now let us consider the sun and the planet earth and apply Newton's law of universal gravitation. Universal gravitation states that the earth and the sun are attracted to one another by the force

$$f_G = G\frac{M_E M_S}{d_{SE}^2}$$

where M_E is the mass of the earth, M_S is the mass of the sun, and d_{SE} is the distance between the center of the sun and the earth. Therefore the earth and the sun should accelerate toward one another and collide. Something must be wrong? The resolution of this problem lies in the fact that the earth revolves with high speed about the sun in essentially a circle. We recall from the preceding chapter that to keep a body moving in a circular orbit, such as a ball on a string, a centripetal force is necessary. In the case of the earth and the sun, this centripetal force arises from the mutual gravitational attraction between the sun and the earth.

This gravitational force f_G must provide the centripetal force

$$\frac{mv^2}{r} = \frac{M_E v^2}{d_{SE}}$$

where m is the mass of the orbiting body the earth, and r is the radius of the orbit that is equal to d_{SE}, the distance between the center of the earth and the center of the sun (see Fig. 5.8). Equating the gravitational attractive force and the centripetal force yields

$$\frac{M_E v^2}{d_{SE}} = G\frac{M_E M_S}{d_{SE}^2}$$

Dividing both sides of the equation by M_E and multiplying by d_{SE}, an expression for the speed of the earth in its yearly orbit about the sun can be obtained:

$$v^2 = \frac{GM_S}{d_{SE}}$$

This relationship can be adapted equally as well to any of the planets of the sun; only the distance d_{SE} need be changed for the various planets. This relationship indicates that the planets closer to the sun (smallest value of d) will have the highest speed v and hence will have the shortest period. This expression

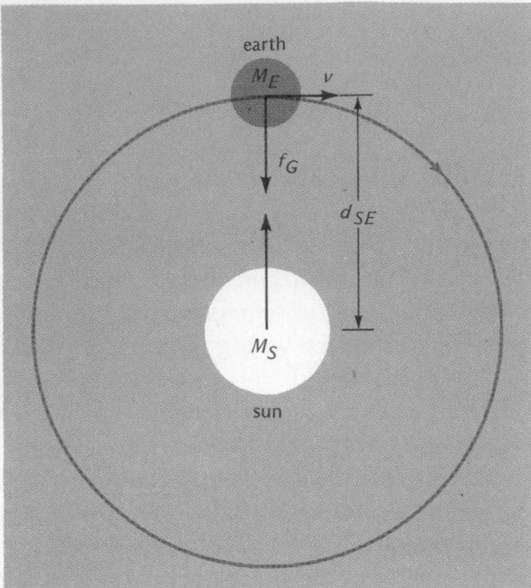

Figure 5.8
The earth in its orbit about the sun. The gravitational attractive force (f_G) of the sun on the earth provides the centripetal force required to keep the earth accelerating in its orbit.

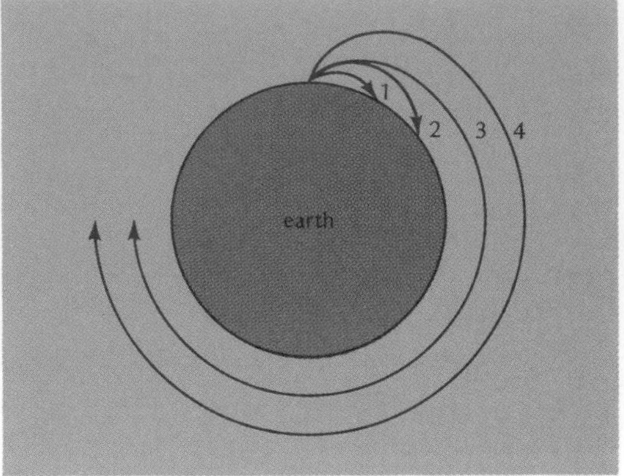

Figure 5.9
Paths of rockets and satellites. A rocket given sufficiently high speed in the correct direction will go into a stable circular orbit indicated by paths 3 or 4. Rockets following path 1 or 2 return to earth. A satellite in orbit 3 must move more rapidly than a satellite in orbit 4.

can be easily converted into Kepler's second law.

ORBITS OF ARTIFICIAL SATELLITES

The preceding analysis is very similar to that required to compute how fast a rocket must move to place a satellite into a certain orbit about the earth. When a satellite is in a stable orbit about the earth, the gravitational attraction force f_G between the earth and the satellite must provide the necessary centripetal force to keep the satellite moving in its circular orbit. Real orbits are usually elliptical but it is simpler to consider only circular orbits. This problem can be examined with the help of Fig. 5.9. The projectiles following paths 1 and 2 did not have sufficiently high speed in the proper direction to go into a stable orbit. However, the rockets following paths 3 and 4 were given exactly the proper speed in the required direction, so the gravitational attraction of rocket by the earth was equivalent to the centripetal acceleration required for the particular orbit. The satellite following path 3 would have to move faster than the satellite following path 4 because the satellite following path 3 is closer to the earth. It is of interest to note that Newton showed us how to determine all the necessary information required to put a satellite into orbit; however, man was not able to develop sufficiently powerful rockets with which to launch satellites until more than 200 years after Newton had recorded his laws of motion.

THE TIDES

The tides had been correlated with the motion of the moon long before Newton was born; however, Newton was the first to show the specific connection between the tides and the motion of the moon. The ocean tides are caused by the gravitational pull of the moon, and to a lesser extent by the gravita-

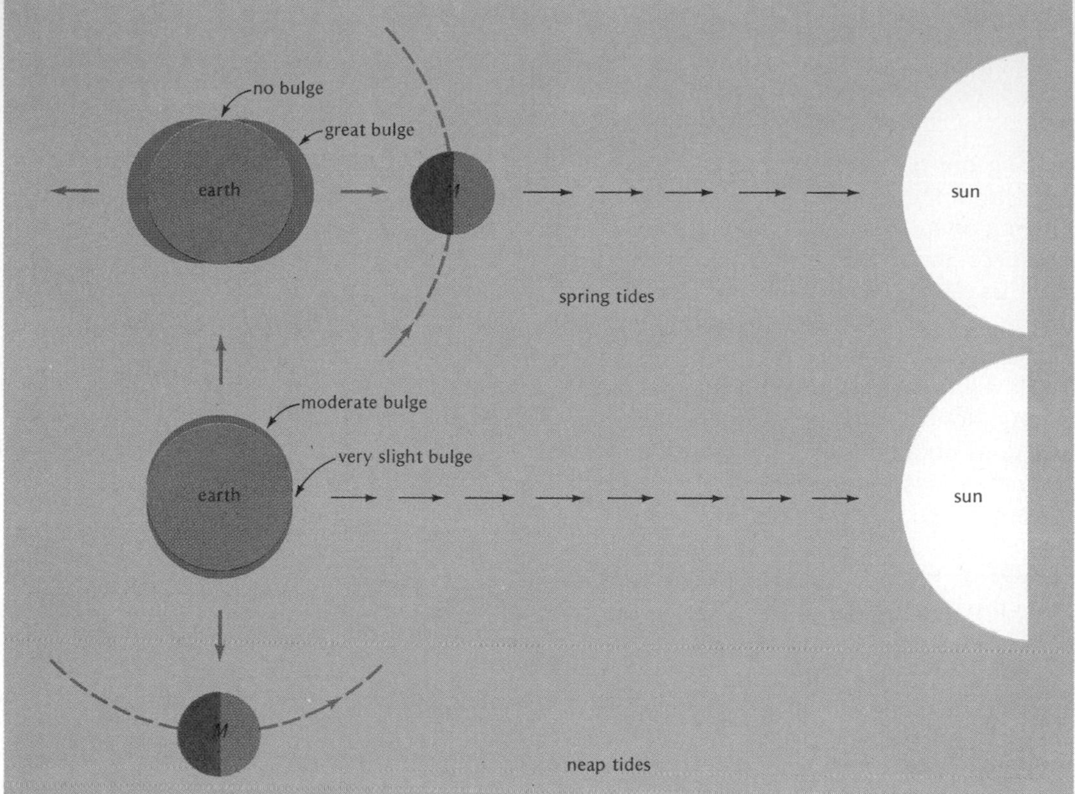

Figure 5.10

The tides. Top, spring tides. When the earth, the moon and the sun are in line (at the time of full moon and of new moon), the pull of the sun (represented by light arrows) reinforces the pull of the moon (heavy arrows), and the highest high tides and lowest low tides occur. Bottom, neap tides. When the moon is in a position at 90° to the sun's position relative to the earth (at half moon), the pull of the sun and moon work against each other, so that high tides are of only medium amount and low tides are not as low. Relative sizes of the earth, the moon, and the sun, and the distances apart, are not drawn to scale, and tidal bulges are greatly exaggerated.

tional pull of the sun acting on the earth. Although these tidal effects are manifest on all parts of the earth (both the solid and liquid), they are most noticeable in the oceans.

If the motion of the moon, relative to the earth, is examined carefully, it is found that the moon does not revolve about the center of the earth but that both the moon and the earth actually revolve about a "center of gravity" which lies about 1000 miles below the surface of the earth. The centripetal force for the earth and the centripetal force for the moon necessary to keep the earth and the moon revolving about this common center of gravity are provided by the mutual gravitational attraction between the moon and the earth.

Consider the earth to be completely covered by one large ocean. The portion of this ocean nearest the moon is pulled toward the moon with a greater force than average and a "bulge" of water will occur on the side of the earth facing the moon. The portion of the ocean on the far side of the earth from the moon will experience a less than average gravitational force required to keep the earth moving in its orbit. This portion of ocean will therefore bulge out as shown in Fig. 5.10 because of the reduced centripetal force. While this motion is occurring and the water is bulging out on the near and far sides of the earth from the moon (Fig. 5.10), the earth is rotating on its axis once every 24 hours. This produces the effect of the bulges or high tides traveling about the earth each day so that at any one place on the earth there is a sequence of high tide, low tide, high tide, and low tide each separated by approximately 6 hours (one-quarter of 24 hours). The high tide does not occur exactly below the moon but rather lags by about 6 hours behind the motion of the moon due to inertia and friction. The position of land masses greatly interferes with the motion of the tide causing great tidal complexities and a lag in the arrival time of the high-tidal bulge.

When the sun and the moon pull together —that is, when earth, moon, and sun are in a line—unusually high tides result: the so-called *spring tides*. On the other hand, when the sun and the moon are at right angles to each other relative to the earth's position, the respective pulls of the sun and moon do not reinforce each other and high tides are lower than normal; these are called *neap tides*. Each type occurs twice in the lunar month. These relationships are shown in Fig. 5.10.

High tides occur twice a day on most coast lines, or to be exact, every 12 hours and 20 to 25 min. The variation from high to low tide may be only 1 or 2 ft, as is the case in many mid-Pacific Islands. At the other extreme, the tidal variation is as much as 70 ft in the Bay of Fundy. Differences of this kind in the height of tides depend on the configuration of the land surface and the ocean bottom offshore. Tidal currents can move into and out of narrow bays, such as the Bay of Fundy, at velocities up to 15 mi/hour. Such currents have an important erosional effect, deepening channels and keeping narrow bays and inlets free of sediment. The effects of these tidal currents in deeper water is not fully known.

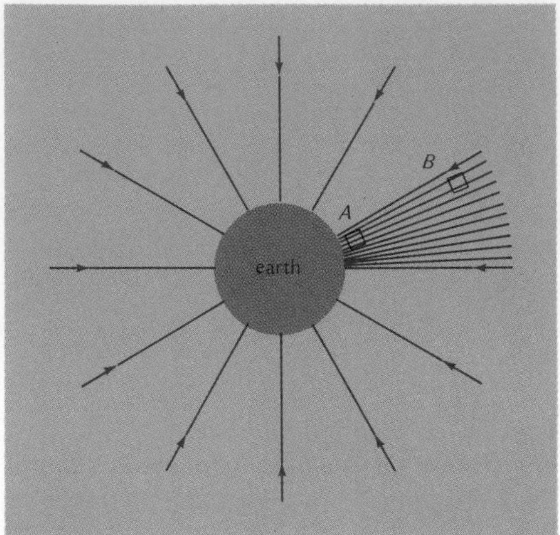

Figure 5.11
Strength of the earth's gravitational field. The strength of the field (number of lines per unit area) is greater at *A* than at *B*.

THE EARTH'S GRAVITATIONAL FIELD

The space surrounding the earth in which a mass is affected by the force of the earth's attraction is called the *earth's gravitational field*. Theoretically it extends throughout all space. The strength of the earth's gravitational field at a given point is the force per unit mass exerted by the earth on a body at that point and is directed toward the center of the earth. This can be represented by the value of the acceleration due to gravity at each point in space about the earth—that is, by $g = GM_E/d^2$, where d is the distance from the center of the earth to the point in space.

The earth's gravitational field may be represented by imaginary lines in space. These lines must be in the direction of the earth's center and are usually considered to terminate at the earth's surface, as in Fig. 5.11. The strength of the earth's gravitational field at a given point is indicated by the proximity of the lines in that region. At region *A* near the earth's surface, where the field is strongest, the lines are more closely spaced than at region *B*, where the field is weaker. Electrostatic fields and magnetic fields (Chap. 10) are treated in a similar manner.

It is worth noting in conclusion that gravitation remains one of the greatest riddles of science. Today scientists are attempting to examine gravity by both experimental and theoretical studies. We know how gravity affects objects, but we have little understanding of where gravity originates. Einstein worked on the problem of gravity and was unable to produce a complete description or theory of gravitation. However, his work on general relativity provides the basis for the modern studies of gravity.

SUMMARY

The *weight* of a body on earth is a measure of the force with which the earth attracts the body. When the body is near the surface of the earth, this force is just sufficient to give the body an acceleration of about 980 cm/sec^2, or about 32 ft/sec^2, which is called the *acceleration* of gravity and is designated by the symbol g. The force of attraction and hence the weight of a body varies slightly with latitude and with elevation above the center of the earth.

Kepler's laws of planetary motion and Newton's general laws of motion served as background for the development of Newton's law of universal gravitation, which states that any two bodies having mass attract each other with a force whose magnitude is given by the equation

$$f_G = G\frac{m_1 m_2}{d^2}$$

Newton assumed that all bodies in the universe attract each other with forces governed by this equation.

The gravitational attractive force between each planet and the sun provides the necessary centripetal force to keep the planet in its orbit about the sun. Likewise, the gravitational attractive force between the moon and the earth provides the centripetal force necessary to keep the moon in its orbit. The gravitational pull of the moon on the earth is the force primarily responsible for the tides of the oceans on the earth.

The space surrounding the earth in which a mass is affected by the force of the earth's attraction is called the *earth's gravitational field.*

Important words and terms

law of universal gravitation
constant of universal gravitation, G
weight
g
conditions required to keep a satellite in orbit
spring tides
neap tides
gravitational field

Questions and exercises

1. State Newton's law of universal gravitation (a) in words; (b) as an equation, using symbols. Indicate the meaning of each symbol.
2. Draw a diagram properly labeled to illustrate the method that has been used to determine the value of G. Describe this method. In addition to the method described in this chapter, some students will wish to study the method used by Henry Cavendish (1731–1810). Some of the supplementary readings listed at the end of this chapter and some encyclopedias show this method.
3. Once a value has been determined for G it can be used to determine the mass of the earth. How would you compute the "weight of the earth"? What does "weight of the earth" mean? Supplementary readings such as those by Holton and Roller and by Rogers (end of this chapter) may be consulted (see also Prob. 6 below).
4. What is the meaning of gravity?
5. (a) What is meant by the weight of a body? (b) What is your weight in pounds? (c) What is your mass?
6. Would a 16-lb metal ball used in the shot-put event in the Olympic games weigh the same in Melbourne, Australia, as in Helsinki, Finland? more in Melbourne than in Helsinki? or more in Helsinki than in Melbourne? Explain your answer.
7. (a) As an airplane takes off and its altitude increases, does the weight of a passenger increase, decrease, or does he weigh the same? Explain. (b) Does the mass of the passenger change? If so, how?
8. (a) As an airplane descends for a landing, does the weight of a passenger increase, de-

crease, or remain the same? Explain your answer. (b) How does the mass of the passenger change?

9. Why does the moon cause water level to rise on both sides of the earth?
10. The moon is being constantly accelerated toward the earth but remains in orbit about the earth, that is, it does not "fall" to the earth. Explain.
11. Write a paper of several hundred words on the life and achievements of Sir Isaac Newton. Point out the importance and significance of his law of universal gravitation.
12. Suppose you are in a space ship in outer space. (a) How far would you have to be from the earth before you would be out of the earth's gravitational field? (b) While you are on your way to the moon, you would arrive at a point where you could say that you are under the influence of the moon's gravity? (c) What does this mean in terms of the gravitational fields of the moon and the earth?
13. Suppose that you are in a space ship way out in outer space and, in fact, so far away from all other objects in the universe that you do not "feel" the pull of gravity from any object. Suppose you then went outside the space ship for a "walk" and became detached from the ship. What would happen to you? Are you separated from the ship forever?

Problems

1. (a) Calculate the acceleration given to a body that has a mass of 4 kg by a net force of 2 newtons. (b) What is the weight of the 4-kg mass?
2. A net force of 49,000 dynes is applied to an object resulting in an acceleration of 70 cm/sec^2. (a) Compute the mass of the object. (b) What is the weight of the object?
3. Compute how many times greater the force of attraction between a 600-g mass and a 2000-g mass with a distance of 40 cm between their centers is than the force of attraction between masses of 300 g and 400 g that have a distance of 20 cm between their centers.

Answer: 2.5.

4. (a) Compute the surface gravity on the planet Jupiter, which has a diameter of 88,000 miles compared with 8000 miles for the earth, and a mass equal to three hundred times that of the earth; (b) compute the "weight" on Jupiter of a person who weighs 100 lb on earth. (Hint: Compare the radius of Jupiter with radius of earth.)
5. The gravitational force between Jupiter and one of its moons, when the two are separated by a distance *d*, has a value *F*. For the following changes in mass, distance of separation, or both, how will the gravitational force change, expressed as a fraction or multiple of *F*? The factors not mentioned remain unchanged. (a) the moon's mass is doubled; (b) the moon's mass is halved; (c) Jupiter's mass is doubled; (d) Jupiter's mass is halved; (e) the distance between the two is doubled; (f) the distance between the two is tripled; (g) the distance between the two is halved; (h) both masses are doubled; (i) both masses are doubled and the distance halved; (j) the mass of the moon is doubled and the distance is halved; (k) the mass of the moon is halved and the distance is doubled.
6. How can the mass of the earth be calculated by using Newton's law of universal gravitation? (Hints: What is the numerical value of the gravitational constant? What is the force of the earth on a 1-g mass, and would this be a mutual force of attraction? What is the radius of the earth in cm; the radius is 6400 km?) Compute the mass of the earth in grams by using the foregoing information in Newton's law of universal gravitation.

Supplementary readings

Textbooks

See Chap. 1, "Supplementary Readings."

Other books

Andrade, E. N. da C., *Sir Isaac Newton: His Life and Work*, Anchor Books, Doubleday, Garden City, N. Y. (1958). (Paperback.) [A personal portrait of one of the great geniuses of history.]

Cohen, I. Bernard, *The Birth of a New Physics*, Anchor Books, Doubleday, Garden City, N. Y. (1960). (Paperback.) [After the student has studied the first four chapters of this textbook, he can read Cohen's concluding chapter to see the grand design in Newton's laws of motion and in his law of universal gravitation. There is an excellent guide to further reading about Copernicus, Galileo, Kepler, and Newton.]

Gamow, George, *Gravity*, Anchor Books, Doubleday, Garden City, N. Y. (1962). (Paperback.)

[A writer with the technical qualifications and the ability to simplify complex ideas discusses the nature of gravity and explains the ideas of three men—Galileo, Newton, and Einstein—before indicating the unsolved problems of gravity.]

Holton, Gerald, and Duane H. D. Roller, *Foundations of Modern Physical Science,* Addison-Wesley, Reading, Mass. (1958). [Chapters 11 and 12 treat the theory of universal gravitation and show some consequences of Newton's work.]

Omer, Guy G., Jr., et al., *Physical Science: Men and Concepts,* Heath and Company, Boston (1962). [In Chaps. 19–22 excerpts from Newton's *Principia* introduce the reader to (1) the inverse square law of gravitation; (2) a series of propositions in which Newton explained the solar system in terms of his laws of motion and the inverse square law; and (3), in summary, the steps, as given in the *Principia,* by which Newton arrived at the law of gravitation.]

Rogers, Eric M., *Physics for the Inquiring Mind,* Princeton University Press, Princeton, N. J. (1960). [Chapter 22 deals with Newton's laws of motion; Chap. 23, with universal gravitation.]

Sutton, Richard M., *The Physics of Space,* Holt, Rinehart and Winston, New York (1965). (Paperback.) [Chapter 10 explains weightlessness, a concept that is fairly common in this age of space capsules.]

Article

Heiskanen, Weikko A., "The Earth's Gravity," *Scientific American,* **193,** no. 3, 164–174 (September, 1955). [Tells how gravity is measured. Shows an appreciable effect of latitude on athletic records, and gives the correction required to account for differences in the force of gravity at the various latitudes where the records were established.]

WORK, ENERGY, AND POWER

One machine can do the work of 50 ordinary men. No machine can do the work of one extraordinary man.

ELBERT HUBBARD, 1859–1915

"Work" is a word of many meanings. Most students consider that they work in preparing an assignment. A teacher may consider he is working while conducting classes. In the physical sciences, however, work has a definite and restricted meaning, as will appear in this chapter.

The consideration of work will be followed by a study of energy—kinds of energy, conservation of energy, transformations of energy. Knowledge of work and energy will then be used in developing the concept of power.

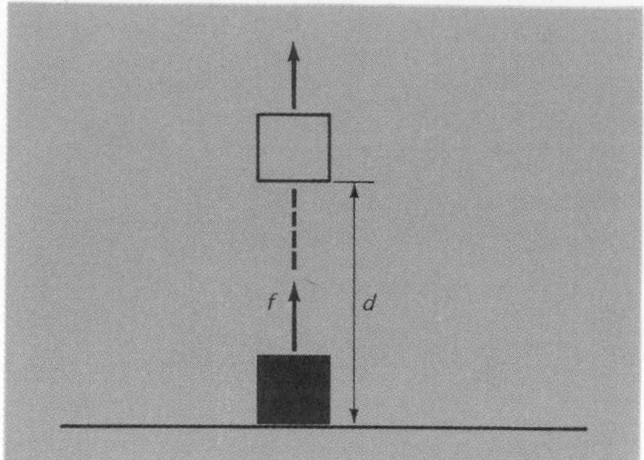

Figure 6.1
Measurement of work when both force and distance are vertical. Work is measured by the product of the vertical force *f* and the vertical distance *d;* thus $W = fd$.

Work

Work is the product of a force acting and the distance through which the force acts, the force and the distance being in the same direction. In Fig. 6.1 the vertical force *f* acts through the vertical distance *d,*

work = force × distance

or

$$W = fd \qquad (6.1)$$

Example 6.1 shows how work is computed when both the force and the distance are vertical.

Consider the case of an object being pulled horizontally along a tabletop by a horizontal force whose magnitude may be registered by a spring balance, as in Fig. 6.2. Example 6.2 shows how work is computed when both the force and the distance are horizontal.

In general, the force and distance may be in any direction. In cases in which the force and distance are not in the same direction, as with a sled being pulled by means of a rope, the work may be computed, but this involves the idea of "components" and the use of trigonometric relations that are not dealt with in this book.

WORK UNITS

Because $W = fd$, any force unit multiplied by any distance unit, if both units are in the same system, gives a proper work unit, as will be seen in the following examples.

Example 6.1
Compute the work done in lifting an object weighing 20 lb through a vertical distance of 3 ft at constant speed.

Solution
To lift a weight of 20 lb at constant speed, a vertical force of 20 lb is required. Because the force and distance are in the same direction,

$$W = fd = 20 \text{ lb} \times 3 \text{ ft}$$
$$= 60 \text{ foot-pounds (ft-lb)}$$

Calling the unit "foot-pound" instead of "pound-foot" is a matter of custom. A *foot-pound* is the work done by a force of 1 lb acting through a distance of 1 ft, the force and distance being in the same direction. This reversing of the units is not done in the metric system, as is shown in the following.

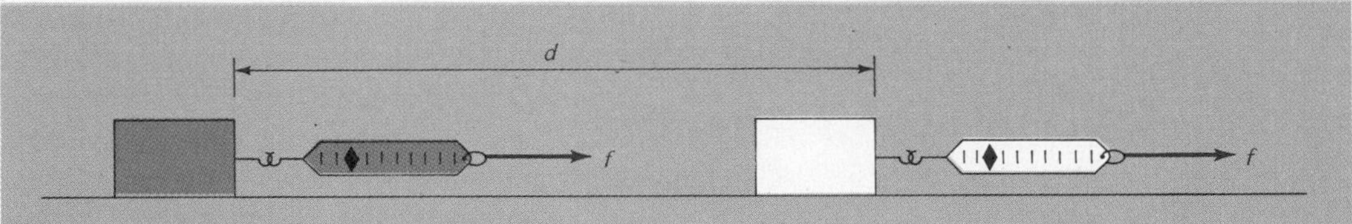

Figure 6.2
Measurement of work when both force and distance are horizontal. The magnitude of constant force f is shown by the indicator on the scale $W = fd$.

Example 6.2
Calculate the work done when a force of 200,000 dynes moves an object a distance of 300 cm in the direction of the force.

Solution
Using cgs units,

$$W = fd = 200{,}000 \text{ dynes} \times 300 \text{ cm}$$
$$= 60{,}000{,}000 \text{ dyne-cm}$$
$$= 60{,}000{,}000 \text{ ergs}$$

Solution
Using mks units, in the mks system, forces are expressed in newtons (100,000 or 10^5 dynes = 1 newton) and lengths are expressed in meters (100 cm = 1 m).

$$f = 200{,}000 \text{ dynes} \times \frac{1 \text{ newton}}{100{,}000 \text{ dynes}} = 2 \text{ newtons}$$

$$d = 300 \text{ cm} \times \frac{1 \text{ m}}{100 \text{ cm}} = 3 \text{ m}$$

$$W = fd = 2 \text{ newtons} \times 3 \text{ m} = 6 \text{ newton-m} = 6 \text{ joules}$$

Note that the special name "erg" is given to the dyne-centimeter.

$$W \quad = f \quad \times d$$
$$1 \text{ erg} = 1 \text{ dyne} \times 1 \text{ cm}$$

Thus an *erg* is the work done by a force of 1 dyne acting through a distance of 1 cm, the force and distance being in the same direction. An erg is such a small unit of work that we more frequently express work in a larger unit, called a *joule,* which equals 10 million, or 10^7, ergs. Thus in Example 6.2 the answer of 60,000,000 ergs is better expressed as 6 joules. One joule is 1 newton-m.

Energy

Nearly everyone recognizes that water at the top of a falls is capable of doing work. The spring of a clock when wound has the ability to do the work necessary to keep the clock running for at least the next 24 hours. A charged automobile battery can be used to operate car lights and to do the work in starting the engine by means of the electric self-starter. The ability to do work, possessed by all these bodies, is called *energy*. We will now discuss the different kinds of energy and how one kind can be transformed into another kind for our benefit.

KINDS OF ENERGY

Usually the forms of energy are said to be mechanical energy; heat, or thermal, energy; electrical energy; chemical energy; nuclear energy; and radiant energy, including light energy, radiant energy from the sun, and alpha, beta, and gamma radiation from radioactive materials. *Radiant energy* is energy that travels through space. Each form of energy is the major subject of one or more chapters in this book.

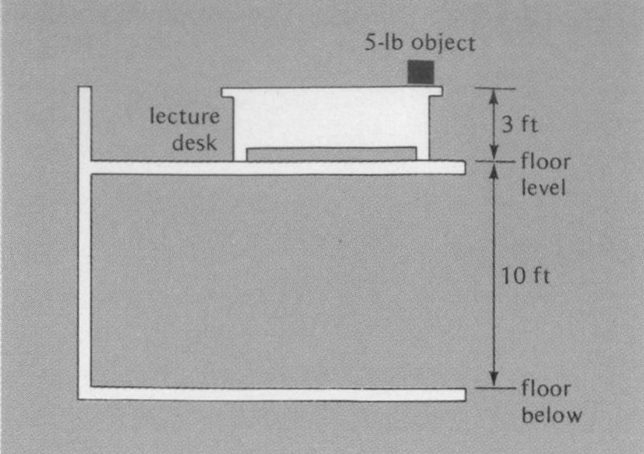

Figure 6.3
Potential energy due to position.

Mechanical energy is expressed in work units since energy is possessed by a body because work has been done on it. There are two main kinds of mechanical energy: potential and kinetic.

POTENTIAL ENERGY

Potential energy is energy due to position or state of stress. Consider an object weighing 5 lb that is on the floor 3 ft below the top of a lecture table (Fig. 6.3). The work required to lift this object from the floor to the top of the table would be $W = fd = 5 \text{ lb} \times 3 \text{ ft} = 15$ ft-lb. In its elevated position the object has the ability to do work. If allowed to fall, it might be made to hit a paddle wheel, which in turning would wind up a weight, thus accomplishing work. This may be visualized if one remembers that water falling over a dam may be made to turn a large water wheel, which in turn operates the machinery in a mill.

In the above analysis the force necessary to raise the object is equivalent to its weight *mg;* hence the work to lift an object at constant speed to a height *h* could be expressed

$$W = fd = mg \times h$$

where $f = mg$; and *d*, the distance moved, is represented by the vertical height *h*. It is common to say that the gravitational potential energy of an object is *mgh*. This energy is stored in the object due to the position of the object.

Potential energy of position depends on the location of the starting level, or plane of reference. The 5-lb object on the lecture desk 3 ft above the floor has 15 ft-lb of potential energy with respect to the floor. If there is another floor 10 ft below, however, the potential energy of the object on the lecture desk with respect to the lower floor is equal to the work done in raising the object from the lower floor to the lecture desk. In this case, $W = 5 \text{ lb} \times 13 \text{ ft} = 65$ ft-lb. If a plane of reference is not specifically mentioned in a problem, it is usually assumed to be the nearest floor below, if inside, or the level of the ground, if outside, or the lowest point in its subsequent motion.

If a spring is compressed or elongated, it is said to be in a state of stress. In returning to its normal condition, it may be made to push a weight or pull an object through a distance, thus doing work. A toy dart gun utilizes a compressed spring. A slingshot sends a stone some distance as the rubber returns from its stretched state of stress to its normal unstretched condition.

Potential energy due to position is sometimes called "gravitational" potential energy, and potential energy due to state of stress is sometimes referred to as "elastic" potential energy.

KINETIC ENERGY

Kinetic energy is the energy possessed by a body because of its motion. Nearly everyone has seen a ball carrier in football hit the defensive line with a force great enough to push the line back several feet. This indicates that work is done by a moving object as it expends energy in slowing down. A bullet shot into a freely suspended block of wood will cause the block to move. Winds (moving air) exert forces that cause various objects to move.

Let us consider here a direct method of computing kinetic energy. Suppose an object of mass *m* at rest is acted on by a constant net force f_{net} and caused to move along a smooth horizontal plane; neglect friction. The work done in pushing the mass m a distance d is $W = fd$. In this case $f = ma$ and $d = \frac{1}{2}at^2$. The kinetic energy the object will have because of the work done is, with substitutions,

$$\begin{aligned} KE = W &= f \times d \\ &= ma \times \tfrac{1}{2}at^2 \\ &= \tfrac{1}{2}ma^2t^2 \end{aligned}$$

Therefore

$$KE = \tfrac{1}{2}mv^2 \qquad (6.2)$$

Equation 6.2 follows from the equation preceding it because in uniformly accelerated motion starting from rest, $v = at$ and hence $a^2t^2 = v^2$. It is seen from Eq. 6.2 that

$$KE \propto m$$

and

$$KE \propto v^2$$

A car with twice the mass of a second car would have twice as much kinetic energy as the second car owing to mass alone. However, doubling the speed of a car, as from 40 to 80 mi/hour, would cause the kinetic energy of the car to be four times as much. In automobile accidents serious damage is often done by and to a car traveling at high speed because of the great kinetic energy of the car resulting from its high velocity.

CONSERVATION OF ENERGY

In discussing potential energy due to position, it was pointed out that a 5-lb object lifted from the floor to a tabletop 3 ft above the floor would require an expenditure of 15 ft-lb of work. The object on the tabletop is then capable of doing 15 ft-lb of work if allowed to fall to the floor and is said to have a potential energy of 15 ft-lb relative to the floor. This idea that the energy that a body has owing to position is exactly equal to the work done in giving it that position is more generally expressed in the fundamental and important principle of conservation of energy:

Energy can neither be created nor destroyed, or *the total amount of energy in the universe or in any isolated system is constant.*

A third and very common statement of the *principle of conservation of energy* is

Energy can neither be created nor destroyed; it can be transformed from one form to another with exact equivalence.

As we study changes of energy from one kind to another, we must keep in mind this idea that the total amount of energy does not change.

MEASUREMENT OF POTENTIAL AND KINETIC ENERGY

As indicated above, the potential energy of position is exactly equal to the work done in giving the object its position (above the plane of reference):

potential energy = work = force × distance

or

$$PE = W = f \times d = mg \times h \qquad (6.3)$$

In the case of the 5-lb object (that is, $mg = 5$ lb) on the lecture desk 3 ft above the floor (Fig. 6.3),

$$PE = W = fd = 5 \text{ lb} \times 3 \text{ ft} = 15 \text{ ft-lb}$$

Suppose the object falls from the desk. As it passes a point 1 ft down (2 ft above the floor), its

$$PE = W = fd = 5 \text{ lb} \times 2 \text{ ft} = 10 \text{ ft-lb}$$

In accordance with the law of conservation of energy, the loss of 5 ft-lb of potential energy has resulted in 5 ft-lb of kinetic energy. At a point 1 ft above the floor, the potential energy has decreased to 5 ft-lb and the kinetic energy has increased to 10 ft-lb. Just at the instant before the object strikes the floor, it is moving with maximum velocity. Because it has reached the plane of reference, its PE = 0, and hence its KE = 15 ft-lb. A moment later it has struck and stopped moving.

What has become of its energy? Its potential and kinetic energy are now both zero. Owing to the impact, the little particles called "molecules" making up the floor and the object have been caused to vibrate more rapidly; that is, the floor and object have become a little hotter. Thus the original 15 ft-lb of potential energy was gradually changed to 15 ft-lb of kinetic energy and then on impact suddenly changed to 15 ft-lb of heat energy. A nail gets warm when pounded into a hardwood plank and the brakes of an automobile may get overheated in braking down a long hill—these are other illustrations of the change of kinetic energy into heat energy.

Example 6.3

An object with a mass of 8 lb is at rest on a window sill 200 ft above the ground. Compute its kinetic energy 3 sec after it falls from the window sill.

Solution

Let us obtain the kinetic energy from the law of conservation of energy. At the top,

$$PE = W = fd = 8 \text{ lb} \times 200 \text{ ft} = 1600 \text{ ft-lb}$$

From the discussion of Chap. 4, in 3 sec the object will fall

$$d_3 = \bar{v}_3 t = \frac{0 + 96 \text{ ft/sec}}{2} \times 3 \text{ sec} = 144 \text{ ft}$$

At a point 144 ft down,

$$PE = 8 \text{ lb} \times 56 \text{ ft} = 448 \text{ ft-lb}$$

so that

$$KE = \text{total energy} - PE = 1600 \text{ ft-lb} - 448 \text{ ft-lb}$$

$$= 1152 \text{ ft-lb}$$

TRANSFORMATIONS OF ENERGY

Changes of energy from one kind to another are continually taking place. One of the simplest transformations of energy is the change from potential to kinetic and back to potential that occurs in a rope swing as shown in Fig. 6.4. At either extremity, the potential energy is a maximum and the kinetic energy is zero. When the swing is in its lowest position, the potential energy with respect to a horizontal plane of reference through this lowest position is zero, and the kinetic energy is a maximum. At any intermediate position, there is some potential energy and some kinetic energy. Thus the energy changes back and forth, potential to kinetic to potential, etc. Because of the friction at the support and against the air, mechanical energy is gradually transformed into heat energy. Hence if there is no pumping action, the swing does not go quite as high on the right side as it was when released on the left side. This effect causes a smaller and smaller amplitude of swing until the swing finally comes to rest.

Water at the top of a dam has potential energy of position. As the water falls over the dam, its potential energy decreases, but it has a corresponding increase in kinetic energy, owing to its motion. At the Niagara Falls power plant, some of the water is diverted above the falls and allowed to fall so as to rotate the turbines connected to the armatures of huge generators, thus changing the kinetic energy of the falling water to electrical energy. This electrical energy is often transferred many miles over high-voltage lines. After the voltage has been reduced, the electrical energy is used to operate lights (electrical energy changed to light energy and heat energy), electric stoves (electrical energy to heat energy), and motors (electrical energy back to kinetic energy).

Another interesting series of transformations of energy takes place in the automobile. Chemical energy stored up in the battery is changed to electrical energy, which in turn is changed to light energy and heat energy when the headlights are on and to mechanical energy when the starter button is pushed. The electrical energy from the battery also causes the electric sparks that ignite the explosive mixture of gasoline vapor and air, thus changing the chemical energy stored up in the gasoline to the kinetic energy of the moving car. The explosion also results in a great deal of heat

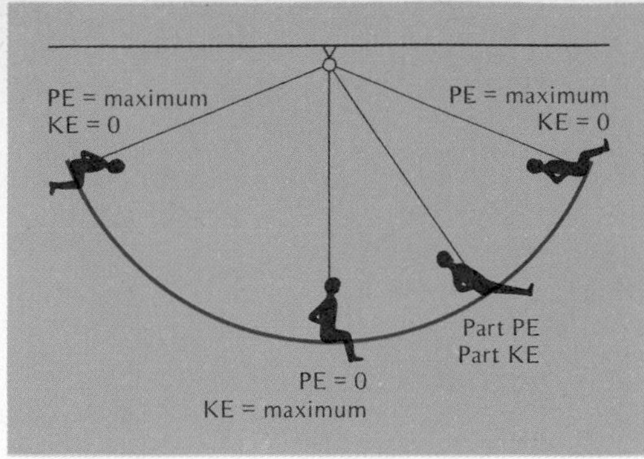

Figure 6.4
Continual transformations of energy in a rope swing.

energy, which is taken care of by the circulating water of the cooling system. When the brakes are applied, the kinetic energy of the moving car is changed to heat energy.

Still another example of transformation of energy occurs as a ball falls from a position of rest.

Example 6.4
A ball with a mass of $\frac{1}{2}$ kg is dropped from rest from a height of 10 m. How much energy does the ball have when it is 10 m, 7.5 m, and 2.5 m above the ground level? What is the form of the energy at each of these heights?

Solution
At the 10-m height above the ground level, the ball is at rest so

$$KE = 0$$

The ball has potential energy,

$$PE = mgh = 0.5 \text{ kg} \times 9.8 \text{ m/sec}^2 \times 10 \text{ m} = 49 \text{ joules}$$

The total energy will be the sum,

$$PE + KE = 49 \text{ joules}$$

From the conservation of energy we know that the total energy will remain constant. Therefore, at any position while the ball is falling, the PE + KE must add up to 49 joules.

At the instant just before the ball hits the ground, h will be zero so PE = 0. The energy will have been converted completely into kinetic energy so KE = 49 joules.

At the intermediate positions part of the energy will be PE and part will be KE. The PE can be computed from PE = mgh at every point. Then, because PE + KE = 49 joules, the KE can be obtained by subtracting the PE from the total energy (49 joules). The table at the bottom of the page shows the results of these calculations.

It should be pointed out that conservation of energy is very useful when we wish to determine the speed of an object. In Example 6.4 potential energy (PE) of the dropping ball at its highest point, a distance h above the ground, is completely transformed into kinetic energy (KE) at the instant just before the ball hits the ground. Therefore we could equate the PE at the top to the KE at the bottom or

$$mgh = \tfrac{1}{2}mv^2$$

where v is the speed of the ball at the instant just before the ball hits the ground after having dropped a distance h. This equation can be simplified and solved for v^2 to yield

$$v^2 = 2gh \qquad \text{or} \qquad v = \sqrt{2gh}$$

Using the data in Example 6.4, the speed of the ball just before it hits the ground can be readily calculated because h = 10 m and g = 9.8 m/sec²

$$v^2 = 2 \times 9.8 \times 10 = 196 \text{ m}^2/\text{sec}^2$$

or

$$v = \sqrt{196 \text{ m}^2/\text{sec}^2} = 14 \text{ m/sec}$$

height above ground level	form of the energy	amount of PE, joules	amount of KE, joules	total energy KE + PE, joules
10.0 m	all PE	49	0	49
7.5 m	$\frac{3}{4}$ PE, $\frac{1}{4}$ KE	36.75	12.25	49
5.0 m	$\frac{1}{2}$ PE, $\frac{1}{2}$ KE	24.5	24.5	49
2.5 m	$\frac{1}{4}$ PE, $\frac{3}{4}$ KE	12.25	36.75	49
0	all KE	0	49	49

This method of computing the speed of a falling object is usually a little easier than that discussed in Chap. 4. In addition, the energy approach outlined above applies equally as well to a swinging object (Fig. 6.4). Here the PE at the highest point h above the lowest point of the swing equals the KE at the lowest point and one again finds that $v^2 = 2gh$.

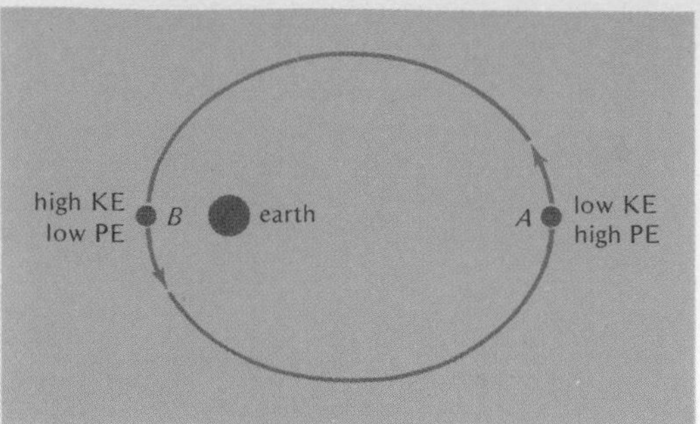

Figure 6.5
Energy transformation in the motion of a satellite in orbit about the earth. The total energy (PE + KE) of the satellite remains the same at all points in its orbit. When the satellite is farthest from the earth at *A*, the PE is at a maximum and, hence, the KE must be a minimum. The satellite has its lowest speed at *A*. When the satellite is closest to the earth at *B*, its PE is at a minimum and its KE is a maximum. Hence, the satellite moves fastest at *B*.

ENERGY OF AN EARTH SATELLITE

When a satellite is in orbit about the earth, the satellite is moving (that is, it has a speed) and hence has kinetic energy. In addition, it has been raised above the earth so that it has gravitational potential energy. From conservation of energy we know that the total energy of the satellite must remain constant. The total energy is the sum of the potential energy and the kinetic energy. Consider for the moment a satellite that is moving in an elliptical orbit about the earth (see Fig. 6.5). Because the total energy will remain constant when the satellite is closest to the earth, its potential energy will be at its lowest value, and therefore the kinetic energy will be at its maximum. When the kinetic energy increases, the speed of the satellite must also increase. Such an analysis predicts that when the satellite is closest to the earth, it is moving at its greatest speed, and when the satellite is farthest from the earth, its speed will be at its lowest value. This is in agreement with the results of Kepler's laws.

It should be noted that the earth in its orbit about the sun moves in an elliptical path, and in January, when the earth is closest to the sun, it has its greatest speed. In July, when the earth is farthest from the sun, the earth is moving at its lowest speed.

EFFICIENCY AND MECHANICAL ADVANTAGE OF MACHINES

Although energy cannot be destroyed, the useful energy obtained from any machine, whether a simple one like a pulley or a jackscrew or one of the large, complex machines used in industry, is always less than the energy supplied to the machine. This is because the friction of moving parts causes a certain amount of the applied energy to be changed to nonuseful heat energy. In this case, "energy in" equals "useful energy out" plus heat energy due to friction.

The *efficiency* of a machine is the ratio of the useful energy out to the total energy in. Thus

$$\text{efficiency} = \frac{\text{useful energy out}}{\text{total energy in}}$$

In most cases the useful energy out is the "work out," W_o, and the total energy in is the "work in," W_i, so that

$$\text{eff} = \frac{W_o}{W_i} \qquad (6.5)$$

Although one cannot gain in energy by the use of a machine, one can gain in force by applying a small force in, f_i, to the machine acting through a large distance in, d_i, enabling the machine to exert a large force out, f_o, acting through a smaller distance out, d_o. The *mechanical advantage* of a machine is the ratio of the force out of the machine to the force put into the machine.

Thus

$$\text{mechanical advantage} = \frac{\text{force out}}{\text{force in}}$$

or

$$MA = \frac{f_o}{f_i} \qquad (6.6)$$

A simple example will show the meaning of efficiency and mechanical advantage.

Example 6.5

Consider a machine that enables a worker to lift an object weighing 500 lb vertically 4 ft by applying a force of 125 lb through a distance of 20 ft. Compute (a) the efficiency of the machine; (b) the mechanical advantage of the machine.

Solution

(a) $\text{eff} = \frac{W_o}{W_i} = \frac{f_o d_o}{f_i d_i} = \frac{500 \text{ lb} \times 4 \text{ ft}}{125 \text{ lb} \times 20 \text{ ft}}$

$= \frac{2000 \text{ ft-lb}}{2500 \text{ ft-lb}}$

$= \frac{4}{5} = 0.8$ or 80 percent

(b) $MA = \frac{f_o}{f_i} = \frac{500 \text{ lb}}{125 \text{ lb}} = 4$

Power and power units

Suppose a man lifts 1 ton of sand vertically 3 ft by shoveling. Neglecting the work the man does in lifting his own weight—part of his body is raised every time he straightens up—the work done in lifting the sand is

$W = fd = 2000 \text{ lb} \times 3 \text{ ft} = 6000 \text{ ft-lb}$

The amount of work is the same whether the man accomplishes the task in 30 min or 10 hours. This matter of time involved, however, is important. To take this factor into account, we use the term *power,* which is the amount of work done, or energy expended, in one unit of time,

$$\text{power} = \frac{\text{work}}{\text{time}}$$

or

$$P = \frac{W}{t} \qquad (6.7)$$

From Eq. 6.7 it follows that any work unit divided by any time unit gives a proper unit for power. The two most common units of power are the watt for the metric system and the horsepower for the English system. A *watt* is the power when work is done, or energy is expended, at the rate of 1 joule/sec.

Example 6.6

Compute (a) the work done (in joules) and (b) the power (in watts) when an object weighing 100 kg is lifted 3 m in 10 sec at constant speed—that is, with no increase in kinetic energy.

Solution

(a) $W = fd = mg \times h = 100 \text{ kg} \times 9.8 \text{ m/sec}^2 \times 3 \text{ m}$
$= 2940 \text{ newton-m} = 2940 \text{ joules}$

(b) $P = \frac{W}{t} = \frac{2940 \text{ joules}}{10 \text{ sec}} = 294 \text{ joules/sec}$

$= 294 \text{ watts}$

When English units are used in $P = W/t$, the power is usually in foot-pounds per second.

It is customary to express the power in horsepower. A *horsepower* is the power when work is done, or energy is expended, at the rate of 550 ft-lb/sec, or 33,000 ft-lb/min. A careful treatment of units shows that 1 horsepower (hp) = 746 watts.

Example 6.7

(a) Calculate the work done in foot-pounds, and (b) express the rate of doing this work in horsepower when an elevator weighing 5000 lb, including load, is lifted 40 ft in 20 sec at constant speed.

Solution

(a) $W = fd = 5000 \text{ lb} \times 40 \text{ ft} = 200{,}000 \text{ ft-lb}$

$$\textbf{(b)}\ P = \frac{W}{t} = \frac{200{,}000 \text{ ft-lb}}{20 \text{ sec}} = 10{,}000 \text{ ft-lb/sec}$$

$$= \frac{10{,}000 \text{ ft-lb/sec}}{(550 \text{ ft-lb/sec})/\text{hp}}$$

$$= 18.2\ \frac{\text{ft-lb/sec} \times \text{hp}}{\text{ft-lb/sec}} = 18.2 \text{ hp}$$

In computing work done, it is important always to keep in mind that the force and distance that are multiplied must be in the same direction. Sometimes the only force applied is the force to overcome friction. These ideas are illustrated in Example 6.8.

Example 6.8

A horse exerts a steady horizontal pull, along the traces, of 80 lb in uniformly moving a wagon. The wagon, weighing 2500 lb including contents, is moved a distance of 0.25 mile in 2.5 min along a horizontal road. Compute (a) the work done by the horse on the wagon, and (b) express this rate of doing work in horsepower. (c) What is the frictional force involved in this problem?

Solution

Because the distance traveled is in a horizontal direction, the force used in computing the work done must be in the horizontal direction. This is the forward force exerted by the horse. The weight, which is a vertical force, does not enter into the computation.

$$\textbf{(a)}\ W = fd = 80 \text{ lb} \times 1320 \text{ ft} = 105{,}600 \text{ ft-lb}$$

$$\textbf{(b)}\ P = \frac{W}{t} = \frac{105{,}600 \text{ ft-lb}}{150 \text{ sec}} = \frac{704 \text{ ft-lb}}{\text{sec}}$$

$$= \frac{704}{550} \text{ hp} = 1.28 \text{ hp}$$

(c) Considering that the motion in this problem is uniform—the additional force to start the wagon is not considered—the only force exerted by the horse is the force to overcome friction. Hence the force of friction is 80 lb.

Had the horse been pulling the wagon up a hill, the force exerted by the horse would have been the sum of two amounts—namely, the force to overcome friction and the force to overcome the component of the weight along the surface of the road.

In practice, the work done in lifting one's body, or a part of it, often exceeds the work done in lifting an object, as may be seen in Example 6.9.

Example 6.9

(a) Compute the work done by a housewife who weighs 140 lb in carrying a basket of wet clothes weighing 25 lb from the basement up the stairs to the ground level, which is 8 ft higher than the basement floor; (b) if the time required to do this is 6 sec, compute the power in horsepower.

Solution

In this case the housewife raises not only the 25 lb of basket and wet clothes but also her own weight of 140 lb.

$$\textbf{(a)}\ W = fd = (25 + 140) \text{ lb} \times 8 \text{ ft}$$

$$= 165 \text{ lb} \times 8 \text{ ft} = 1320 \text{ ft-lb}$$

$$\textbf{(b)}\ P = \frac{W}{t} = \frac{1320 \text{ ft-lb}}{6 \text{ sec}} = 220 \text{ ft-lb/sec}$$

$$= \frac{220}{550} \text{ hp} = 0.4 \text{ hp}$$

SUMMARY

Work is the product of the force acting and the distance through which the force acts, the force and the distance being in the same direction.

Energy is the ability to do work. There are two main kinds of mechanical energy: potential energy and kinetic energy. There are other forms of energy such as heat energy, electrical energy, chemical energy, light energy, other forms of radiant energy, and nuclear energy. Potential energy is energy due to position or state of stress; kinetic energy is the energy possessed by a body because of its motion.

Changes of energy from one kind to another occur in accordance with the principle of conservation of energy: Energy can be neither created nor destroyed; it can be transformed from one form to another with exact equivalence.

Although energy cannot be destroyed, the

useful energy got out of any machine is always less than the energy supplied to the machine; thus the efficiency of a machine is always less than 100 percent.

The rate at which work is done, or energy is expended, is called *power;* it is customary to express power in horsepower.

Important words and terms

work
joule
erg
energy
kinetic energy
potential energy
principle of conservation of energy
efficiency
mechanical advantage
power
watt

Questions and exercises

1. Distinguish between work and energy by defining them.
2. Distinguish between potential energy and kinetic energy by defining them. Give three examples of each of these kinds of energy.
3. Describe the energy interchanges that occur as and after an automobile engine is started; after the headlights are turned on.
4. Give different ways of stating the principle of conservation of energy. Which statement do you prefer? Why?
5. Explain the statement, "Mechanical advantage is a ratio of forces; efficiency is a ratio of energies."
6. Give two examples of processes that involve the conversion of potential energy into kinetic energy.
7. Consider a swinging pendulum. During which part of its swing is (a) the potential energy at a maximum; (b) the potential energy at a minimum; (c) the kinetic energy at a maximum; (d) the kinetic energy at zero; (e) the kinetic energy and potential energy equal.
8. Consider an artificial satellite orbiting the earth in an elliptical orbit such that at one end of its orbit the satellite is 200 miles above the earth and at the other end of the orbit it is 1000 miles from the earth. Apply the principle of conservation of energy to this situation and describe how the speed of the satellite changes in orbit.
9. From what you have learned in this chapter state why scientists do not believe that a perpetual motion machine is possible.
10. In its revolution about the sun why cannot the earth be considered a perpetual motion machine?
11. Three skiers, Dick, Jane, and Sally, were together at the top of a hill. Dick went off a jump, Jane skied down the slope, and Sally rode the ski-lift down. Compare the changes in their gravitational potential energy. Compare the speed at which Dick and Jane would have arrived at the bottom of the hill, assuming that they had "lost" no energy due to frictional processes.

Problems

1. A youth weighing 150 lb runs up a flight of stairs in 3 sec. The vertical distance between floors is 10 ft. Compute (a) the work done against gravity; (b) his rate of doing this work in horsepower.

Answer: (a) 1500 ft-lb; (b) 0.91 hp.

2. A high jumper who weighs 160 lb in clearing the bar at 6 ft 6 in. actually lifts his own weight a vertical distance of 3 ft in a time of 0.6 sec. Compute (a) the work done by the jumper; (b) his rate of doing work in horsepower.
3. An object having a mass of 5 kg is pulled uniformly a distance of 4 m along a horizontal tabletop by a constant force of 2.9 newtons in a time of 8 sec. Compute the work done in (a) newton-meters; (b) joules; and (c) ergs. (d) Compute the power in watts. (e) What is the value of the frictional force in this case?

Answer: (a) 11.6 newton-m; (b) 11.6 joules; (c) 11.6×10^7 ergs; (d) 1.45 watts; (e) 2.9 newtons.

4. An electric motor raises an elevator cage, which with contents weighs 3000 kg, a height of 20 m in 10 sec. Compute the work done (a) in newton-meters; (b) in joules; (c) Compute the rate of doing this work.
5. A hammer having a mass of 2 kg lies on a roof parapet 30 m above the ground. Compute (a) its potential energy; (b) its kinetic energy. A workman accidentally knocks the hammer off the parapet. As it passes a window ledge which is 10 m above the ground,

compute (c) its potential energy; (d) its kinetic energy. Just as it reaches the ground level, then traveling at its maximum velocity, compute (e) its potential energy; (f) its kinetic energy. (g) After striking the ground and coming to rest, what happened to the mechanical energy it previously possessed?

Answer: (a) 60 newton-m; (b) 0; (c) 20 newton-m; (d) 40 newton-m; (e) 0; (f) 60 newton-m; (g) changed to heat energy.

6. The "hammer" of a drop pile driver weighs 4000 lb. When it is stationary in a position 30 ft directly above the pile, which is to be driven into the ground, compute (a) its potential energy with respect to the pile; (b) its kinetic energy. The hammer is released. Considering it to drop freely, compute (c) its location 1 sec after being released; (d) its potential energy at that time; (e) its kinetic energy at that time. Just as it reaches the pile, compute (f) its potential energy; (g) its kinetic energy. (h) Assuming that 30 percent of the energy is transformed into heat energy on impact, compute the work that will be done by the hammer in forcing the pile into the ground.
7. By means of a set of block-and-tackle pulleys a man is able to lift a weight of 600 lb vertically a distance of 8 ft in 2 min by applying a force of 120 lb through a distance of 50 ft. Compute (a) the mechanical advantage of the machine (the set of pulleys); (b) the work accomplished; (c) the work done by the man; (d) the efficiency of the machine; (e) the rate in horsepower at which the man works.

Answer: (a) 5; (b) 4800 ft-lb; (c) 6000 ft-lb; (d) 80 percent; (e) 0.091 hp.

8. A jackscrew is used to raise one corner of a sagging garage. By having a worker apply a force of 50 lb to the end of a jack handle, the jackscrew exerts a force of 5000 lb on the garage. In order to lift that corner of the garage a distance of 6 in., the force of 50 lb applied by the worker must act through a distance of 200 ft (this may be accomplished by the turning action of the jack handle). Compute (a) the mechanical advantage of the machine (the jackscrew); (b) the work accomplished; (c) the work done by the worker; (d) the efficiency of the machine.
9. (a) How much kinetic energy does a 10-g mass moving at a speed of 2 cm/sec have? (b) Suppose the speed of this mass were to increase to 3 cm/sec, 4 cm/sec, and then 8 cm/sec; compare the kinetic energy at each speed. (c) By what factor has the kinetic energy increased?

Answer: (a) 20 ergs; (b) 45 ergs, 80 ergs, 320 ergs; (c) ratio of square of speeds.

10. A mass *m* slides down a frictionless plane that is 10 m long. The top of the incline is 7 m above the horizontal surface on which the incline sits. How much work was required (a) to lift the mass to the top of the incline, (b) to slide the mass up the incline to the top? How much potential energy is transformed into kinetic energy when the mass (c) slides down the incline, (d) falls off the top of the incline and drops vertically to the horizontal platform? (e) What is the speed of the mass as it reaches the horizontal in (c) and (d)?

Project

Run up a flight of stairs as quickly as you can, obtaining your "time of flight" in seconds as accurately as possible.

1. Compute the work done in foot-pounds in lifting your own weight. Show by a diagram what distance you use.
2. Compute in horsepower the rate at which you worked.
3. What is your potential energy at the top of the stairs with reference to the starting floor level?
4. Express your mass in grams and the distance in centimeters, and compute the work done in dyne-centimeters, in ergs, and in joules.
5. Compute your rate of doing work in watts. Check this answer against that in part 2 by the relation 1 hp = 746 watts.

Supplementary readings

Holton, Gerald, and Duane H. D. Roller, *Foundations of Modern Physical Science,* Addison-Wesley, Reading, Mass. (1958). [In Chap. 18 the concept of work and the principle of conservation of energy are discussed.]

McCue, J. J., and Kenneth W. Sherk, *Introduction to Physical Science: The World of Atoms,* Ronald Press, New York (1963). [Work and energy are discussed in Chap. 9.]

Rogers, Eric M., *Physics for the Inquiring Mind,* Princeton University Press, Princeton, N. J. (1960). [A thorough treatment of energy appears in Chaps. 26–29.]

MECHANICS OF FLUIDS

Give me matter and motion, and I will construct the universe.

RENÉ DESCARTES, 1640

The previous chapters have dealt almost entirely with solid objects, their motions, and the effect of forces on them. Matter, however, which may be defined as anything that occupies space and has mass, may exist in three states: solid, liquid, and gaseous. The lecture desk and this book are common examples of solids. Water is the most common liquid, and air is the most familiar gas. Ice is a solid, melted ice or water is a liquid, and water changed to water vapor or steam is a gas. Liquids and gases are often classified together as fluids because they both flow.

Scientists, beginning with the early Greek philosophers, have carefully studied the various phenomena of fluids and as a result have arrived at statements of several important principles, or generalizations, that describe the behavior of fluids. In this chapter we will discuss three of the most important of these principles named in honor of the scientists who did the greatest amount of the experimentation and reasoning leading to these generalizations: Archimedes, Pascal, and Bernoulli. Because much of our understanding of the behavior of matter in its three states depends on some assumptions that we now group together under what we call the "kinetic molecular theory," we will begin the chapter with this topic.

Kinetic molecular theory of matter

For centuries man has wondered to what extent matter can be subdivided. An aluminum wire can be cut into two pieces, and each of the two pieces can be cut into two pieces, and so on. Is there any limit beyond which a minute piece of aluminum cannot be further subdivided without destruction of the aluminum?

For almost 2500 years there has been recurrent speculation about similar questions involving the ultimate character of matter such as whether substances consist of certain extremely small particles that cannot be further subdivided without destroying the identity of the substances. The Greek philosophers decided that there must be a limit beyond which a sample of substance could not be subdivided. They gave the name "atom" (meaning "indivisible," from the Greek word *atomos,* "uncuttable") to each of these tiny invisible and indivisible theoretical particles. The idea that matter is composed of invisible and indivisible atoms appeared as early as the fifth century before Christ in the writings of Leucippus and Democritus (460?–362? B.C.). They supposed that matter must consist of hard, indivisible particles, or atoms, of a common substance but of different sizes and shapes.

Many of the early Greek scientists were not very practical, and little application of their idea was made until about A.D. 1800. The atomistic hypothesis had remained for more than 2000 years a speculative idea that did not suggest new experiments. As explained in the first few pages of Chap. 12, the reasonings of Dalton, Gay-Lussac, Avogadro, and Cannizzaro, mostly in the first half of the nineteenth century, lead us to the conclusion that the atoms of the Greeks are the molecules of our present-day concept. This had led to a *statement of modern kinetic theory* as follows:

All matter, whether in the solid, liquid, or gaseous state, is made up of very small particles, called molecules, that are continually in motion.

If the stopper is removed from a bottle of perfume, the odor can soon be detected several feet away. This happens because the individual small molecules of the perfume vapor (a gas), moving about rapidly and colliding frequently with invisible air molecules, have rapid random (irregular) motion (Fig. 7.1). Note that the molecules move in straight lines between collisions, as would be expected from Newton's first law of motion.

Actual measurements of the speeds of gas

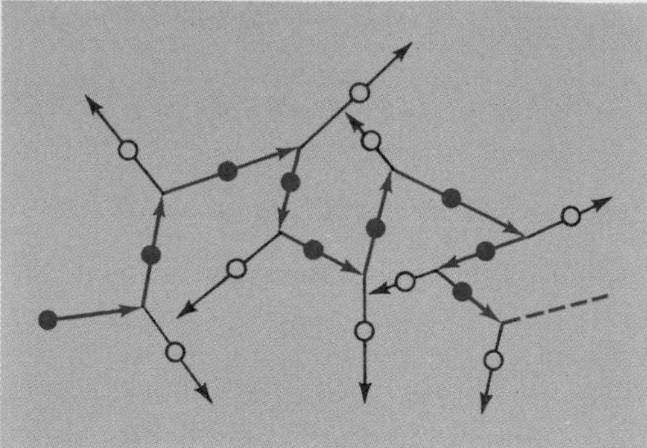

Figure 7.1
Random motion of a molecule due to collisions with other molecules.

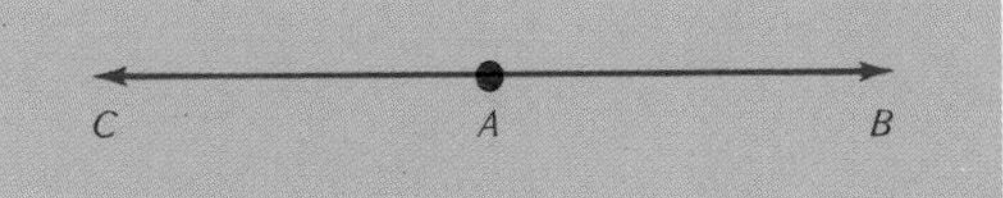

Figure 7.2
Vibratory motion of a molecule. The motion is through a position of rest *A* to *B* to *A* to *C* to *A*, and so on.

molecules in air indicate that the molecules travel at speeds similar to those of rifle bullets. However, because of the frequent collisions of the molecules, the net motion of a molecule through a gas is reduced greatly. The molecule is said to "diffuse" through the gas, and the perfume would be said to undergo *diffusion* through the air.

If colored liquid from a medicine dropper is released under the surface of water in a glass, the color spreads outward gradually. It will be several minutes before the water in the glass has attained a uniform color. This indicates that in liquids molecules move with relatively slow random motion.

This further indicates that in liquids the molecules diffuse due to the random motion and frequent collisions of the molecules with one another. In actual practice convection currents cause a more rapid distribution of color through the liquid than that due to random motion of molecules. *Convection* is the motion of a mass of liquid or gas in the fluid due to variations in density produced by unequal heating throughout the fluid. The rising of hot air is an example of a convection phenomenon. The effect caused by convection should not be confused with the effects due to random molecular motion—that is, diffusion.

Even the molecules of a solid have a very slight random motion. In the case of a solid, however, the motion of the molecules is mainly vibratory, a forth-and-back motion through a central point, as from *A* to *B* to *A* to *C* to *A* to *B*, etc., in Fig. 7.2.

One of the earliest uses of the idea of the kinetic molecular theory was made by Count Rumford to account for the large amount of heat that resulted from the boring of a cannon. He attributed this heating of the cannon, the boring tool, and the borings to increased kinetic energy of the molecules of the metals, owing to the application of mechanical energy in the process of boring. The student will note many applications of the kinetic molecular theory throughout the rest of this textbook, as in the explanations in Chap. 8 of evaporation, fusion, and heat of fusion. Looking back, it will be recalled that the kinetic energy of a falling body "disappeared" on impact and resulted in the heating of the body and the surface at and near which impact occurred; that is, the molecules of the fallen body and also of the impact surface moved faster.

In addition to the abstract idea of invisible moving molecules, we now have indirect experimental evidence of their existence. As early as 1827 the English botanist Brown, using a microscope, noticed that fine grains of pollen suspended in water kept in seemingly perpetual zigzag motion, which became known as Brownian motion. It was later established that this motion was caused by the uneven bombardment of the colloidal particles by the moving molecules of the surrounding water. One may observe Brownian motion by looking into a chamber containing puffs of cigarette smoke illuminated by a beam of strong light (Fig. 7.3). This motion is convincing evidence that very small particles of matter do exist and that they are in continuous random motion.

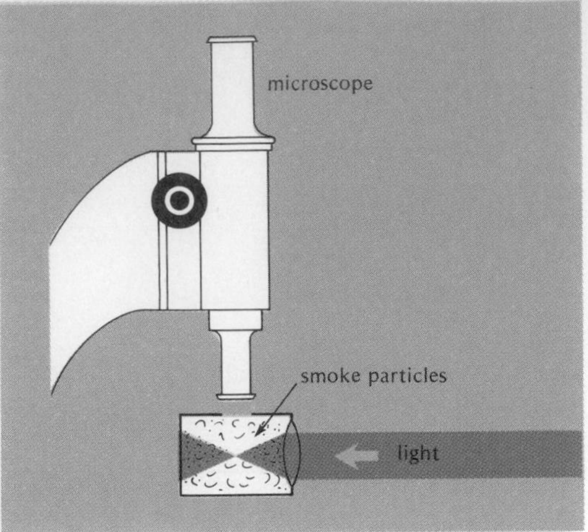

Figure 7.3
Arrangement for observing Brownian motion.

OTHER METHODS OF DISTINGUISHING SOLIDS, LIQUIDS, AND GASES

In addition to distinguishing among the three states of matter by the motion of molecules, we may recognize them by their boundary surfaces. Solids have complete boundary surfaces of their own that will not change unless a force is applied. A chunk of iron or a rock requires a considerable force to change its shape and an even greater force to change its volume. Other solids, such as a blob of putty, change shape under small forces, although large forces are necessary to change their volumes. The boundary surfaces of a liquid are the surfaces of the containing vessel except for the upper surface, which is always horizontal. Liquids offer practically no resistance to a change in shape, as is noted when water is poured from a pitcher into a glass. Liquids, however, are highly incompressible; that is, very large pressures, of the order of thousands of tons per square inch, are required to diminish their volumes appreciably. For all ordinary purposes, liquids are considered to be incompressible. Because of the high velocities of the molecules of a gas, a gas tends to expand indefinitely and hence fills any container into which it is admitted. The boundary surfaces of a gas are the walls of the container. Methods of distinguishing solids, liquids, and gases from one another are summarized in Table 7.1.

DENSITY OF A SUBSTANCE

In a gas, where the molecules are relatively far apart, the actual amount of material, or mass of the substance, in a given volume is small. In a liquid or solid, the molecules are much closer together, resulting in much more material in a given volume. *Density* is the mass per unit volume of a substance,

$$\text{density} = \frac{\text{mass}}{\text{volume}}$$

or

$$D = \frac{m}{V} \tag{7.1}$$

For example, since 10 cm³ of steel weighs 78 g (mass is 78 g),

$$D_{\text{steel}} = \frac{m}{V} = \frac{78\ \text{g}}{10\ \text{cm}^3} = 7.8\ \text{g/cm}^3$$

From the definition of a standard kilogram, it follows that for water at 4°C

$$D_{\text{water}} = \frac{m}{V} = \frac{1000\ \text{g}}{1000\ \text{cm}^3} = 1\ \text{g/cm}^3$$

A cubic foot of water weighs 62.4 lb. To compute the density of water in English units, the mass contained in a cubic foot of water must be computed. Recalling that weight equals mg, 62.4 lb of water is equivalent to a mass,

$$m = \frac{62.4\ \text{lb}}{32.2\ \text{ft/sec}^2} = 1.94\ \text{slugs}$$

The density of water in the English system is

$$D_{\text{water}} = \frac{m}{V} = \frac{1.94\ \text{slugs}}{1\ \text{ft}^3} = 1.94\ \text{slugs/ft}^3$$

It should be noted that in the United States it is common practice for engineers to define a weight density D_{wt} as the weight per unit volume. In the English system 62.4 lb of water occupies 1 ft³, so that the weight density of water is

$$D_{\text{wt}}\ (\text{water}) = \frac{62.4\ \text{lb}}{1\ \text{ft}^3} = 62.4\ \text{lb/ft}^3$$

Table 7.1
Methods of distinguishing solids, liquids, and gases

state of matter	examples	volume and shape	motion of molecules
solid	ice, wood	definite volume and shape	vibratory motion, as in Fig. 7.2
liquid (fluid)	water, gasoline	definite volume and takes shape of container except that upper surface is horizontal	slow random motion, as in Fig. 7.1
gas (fluid)	water vapor, air	tends to expand indefinitely to fill and to take shape of container	quite rapid random motion, as in Fig. 7.1

The weight density equals the product of the mass density and *g*, the acceleration due to gravity.

A liter (1000 cm³) of air under normal atmospheric pressure and at a temperature of 0°C has a mass of 1.293 g. Hence,

$$D_{\text{air}} = \frac{m}{V} = \frac{1.293 \text{ g}}{1000 \text{ cm}^3} = 0.001293 \text{ g/cm}^3$$

The last value indicates the relatively low density of gases, compared with solids and liquids.

SPECIFIC GRAVITY OF A SUBSTANCE

Frequently it is desired to compare the densities of various substances with that of some standard, usually water. The *specific gravity* of a substance is the ratio of the density of the substance to the density of water. Actually, "relative density" would be a more appropriate term than "specific gravity,"

$$\text{specific gravity} = \frac{\text{density of substance}}{\text{density of water}}$$

or

$$\text{sp gr} = \frac{D_s}{D_w} \tag{7.2}$$

For steel,

$$\text{sp gr}_{\text{steel}} = \frac{D_{\text{steel}}}{D_{\text{water}}} = \frac{7.8 \text{ g/cm}^3}{1 \text{ g/cm}^3} = 7.8$$

From the definition, it follows that the specific gravity of a substance tells how many times denser the substance is than water, or it compares the mass of a certain volume of the substance with the mass of an equal volume of water. Because mass and weight at the same place are proportional to one another (the constant of proportionality being the acceleration due to gravity), the specific gravity also compares the weight of a certain volume of a substance with the weight of an equal volume of water when the substance and the water are weighed at the same place. In other words, the weight density of, say, steel would be different on the moon from that measured on earth; however, the specific gravity of steel would be the same on any body in the universe and in free space. We will return to specific gravity after discussing Archimedes' principle later in this chapter.

Table 7.2 shows the relation of density and specific gravity for a few common substances. Notice that in the metric system density and specific gravity are numerically equal but the latter has no units. For steel, mercury, and brass the quantity given is in boldface, and the arrows indicate how the other quantities are obtained. In the case of solids it is the large amount of matter in individual atoms that accounts for the increased density. The basis for this will be discussed in Chap. 12.

Table 7.2
Density and specific gravity

substance	density (metric)		specific gravity		density (English)
water	1 g/cm³		1		62.4 lb/ft³
steel	**7.8 g/cm³**	$\longrightarrow$	$\frac{7.8 \text{ g/cm}^3}{1 \text{ g/cm}^3} = 7.8$	$\longrightarrow$	7.8×62.4 lb/ft³ $= 486.7$ lb/ft³
mercury	13.6 g/cm³	$\longleftarrow$	**13.6**	$\longrightarrow$	848.6 lb/ft³
brass	8.5 g/cm³	$\longleftarrow$	$\frac{530.4 \text{ lb/ft}^3}{62.4 \text{ lb/ft}^3} = 8.5$	$\longleftarrow$	**530.4 lb/ft³**
air[a]	0.001293 g/cm³		0.001293		0.0807 lb/ft³
ice	0.92 g/cm³		0.92		57.4 lb/ft³
granite	2.64–2.76 g/cm³		2.64 – 2.76		165–172 lb/ft³
alcohol	0.81 g/cm³		0.81		50.5 lb/ft³
lead	11.3 g/cm³		11.3		705 lb/ft³
aluminum	2.7 g/cm³		2.7		168 lb/ft³
gold	19.3 g/cm³		19.3		1204 lb/ft³

[a]At standard pressure (76 cm of mercury) and standard temperature (0°C).

Pressure

MEANING OF PRESSURE

Pressure is the force per unit area acting on a surface,

$$\text{pressure} = \frac{\text{force}}{\text{area}}$$

or

$$p = \frac{F}{A} \tag{7.3}$$

Suppose the block on the tabletop in Fig. 7.4 is in the shape of a parallelepiped and has a base 5 in. wide and 6 in. long. If the block weighs 60 lb, a force of 60 lb is exerted against the tabletop. This 60 lb, however, is distributed uniformly over 30 in.² of surface. Hence the pressure or force per unit area on the tabletop is

$$p = \frac{F}{A} = \frac{60 \text{ lb}}{30 \text{ in.}^2} = 2 \text{ lb/in.}^2$$

Pressure is an extremely important quantity, particularly when we wish to express the strength of materials. To illustrate the importance of pressure consider a man weighing 200 lb standing in normal shoes on a floor. Consider that each of his shoes is 1 ft long and about 4 in. ($\frac{1}{3}$ ft) wide. If we imagine that his shoes were rectangular in shape, then each shoe would have an area of $1 \times \frac{1}{3}$ ft $= \frac{1}{3}$ ft². Therefore the man is in contact with the floor over an area of $2 \times \frac{1}{3}$ ft² $= \frac{2}{3}$ ft². The pressure he exerts on the floor is

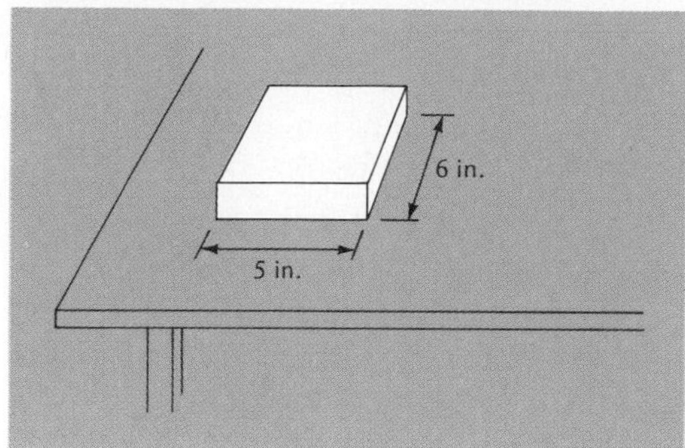

Figure 7.4
Pressure on a tabletop due to a 60-lb block.

$$P = \frac{F}{A} = \frac{200 \text{ lb}}{\frac{2}{3} \text{ ft}^2} = 300 \text{ lb/ft}^2$$

This means that to support the man on his two feet the floor must be strong enough to withstand a pressure of at least 300 lb/ft². However, consider what would happen if the man were to wear a pair of cowboy boots with a square heel about 1 in. on a side. If the man were to stand with all of his weight on one heel, the pressure he would exert on the floor would be

$$p = \frac{F}{A} = \frac{200 \text{ lb}}{\frac{1}{12} \text{ ft} \times \frac{1}{12} \text{ ft}} = \frac{200 \text{ lb}}{1/144 \text{ ft}^2} = 28{,}800 \text{ lb/ft}^2$$

The material on the floor must now withstand this large pressure. In many cases tile floors or carpets will not withstand such a pressure; and such a heel, or a "spike" heel on a woman's shoe, produces such large pressures that the floor covering is damaged. When we want to walk on snow, we put on snowshoes which spread our weight out over a large area, thus reducing the pressure so that we do not sink deeply into the snow. In all of these examples, the force we are dealing with is the man's weight, yet the pressure the man exerts on the floor or snow changes greatly.

When we ice skate, we exert such a high pressure on the ice that we cause the ice to melt under the runners on the skates, and we then skate easily on the layer of water.

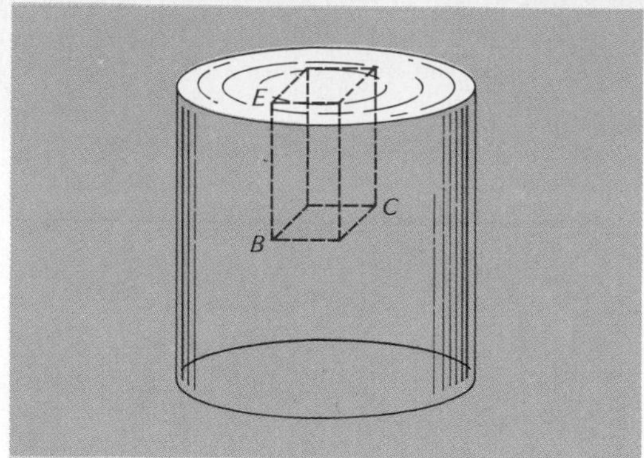

Figure 7.5
Pressure of a fluid on an immersed surface.

PRESSURE OF A FLUID

Figure 7.5 represents a jar filled with a liquid. Consider a horizontal surface *BC* having an area *A*. A force is exerted down on the surface that is equal to the weight of the column of liquid *BE* above the surface. The pressure on the surface is equal to this force divided by the area of the surface. If the area of the surface is made half as great, the weight of the column of liquid directly above the smaller surface will also be half as great, so that the pressure remains the same. Similarly, the area can be made smaller and smaller, with the weight of the liquid above becoming correspondingly smaller, so that the pressure still remains the same. If the area becomes so small that it approaches a point, one speaks of "pressure at a point," which, for a given depth in a given fluid, is the same as the pressure on an area of any finite size.

PRESSURE SAME IN ALL DIRECTIONS

An experiment may be performed, using the apparatus shown in Fig. 7.6, to demonstrate that at a given depth the pressure due to a fluid, water in this case, is the same in all directions (downward, sideways, upward, etc.). If the center of the lower opening of each J-tube is at the same level and if the same amount of mercury is poured into each tube, the upper levels of mercury will be the same in each tube and the lower levels of mercury will be the same in each tube, showing equal pressures for the same depth.

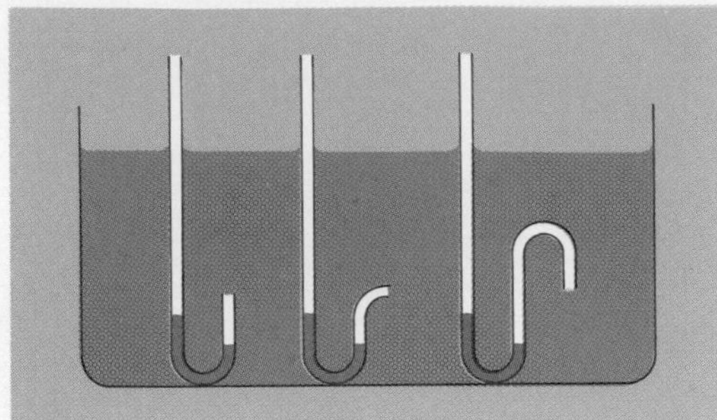

Figure 7.6
Pressure within a fluid is the same in all directions at a given depth.

FORCE AND PRESSURE OF FLUIDS ON IMMERSED SURFACES

If we now refer again to Fig. 7.5, the force down, F, on the horizontal surface BC of area A is the weight of the column of fluid BE directly above the immersed surface BC. Hence

F = weight of column of fluid BE

To compute the weight of the column of fluid BE one could find the mass of this column of fluid and multiply this mass by g, the acceleration due to gravity. To find the mass of the fluid it is convenient to use the density $D = m/V$, from which mass = volume × density. The weight would then equal the volume × density × g.

F = weight of column of fluid BE

= mass of column of fluid $BE \times g$

= volume of column of fluid BE × density of the fluid × g

= area (A) of the column × height (H) of the column × Dg

$= AHDg$

In this expression H is the depth of the horizontal surface BC below the surface of the fluid. The force

$$F = AHDg \qquad (7.4)$$

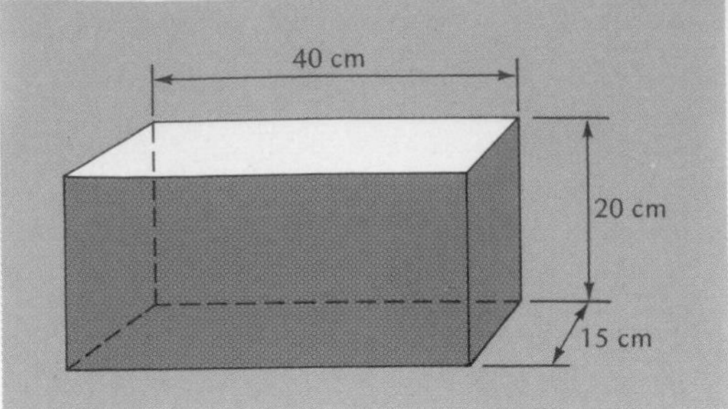

Figure 7.7
Force and pressure on immersed surfaces.

is the force on an horizontal surface of area A at BC.

Whether the immersed surface is horizontal or not, the relation $F = AHD$ holds if H is taken as the depth of the center of area of the immersed surface below the top surface of the fluid. Hence for all immersed surfaces Eq. 7.4 is valid.

Because pressure = force/area, the pressure at a depth H in a fluid of density D is

$$p = \frac{F}{A} = \frac{AHDg}{A} = HDg \qquad (7.5)$$

Note that the force given in Eq. 7.4 is the downward force on a horizontal area A at a distance H below the surface of the fluid of density D. However, the pressure p given in Eq. 7.5 is the pressure at a depth H below the surface and has the same value in all directions at a point in the fluid. Pressure does not have a direction associated with it as is the case for force, which is a vector quantity.

Example 7.1

Consider the tank of Fig. 7.7 to be filled with alcohol of specific gravity = 0.8. Compute the force due to the alcohol on (a) the bottom surface (F_b); (b) the front face (F_f); and (c) one end (F_e). Also compute (d) the pressure on the bottom (p_b); and (e) the pressure at a point on the front face 5 cm below the surface (p_c).

Solution

(a) $F = A \times H \times D \times g$

$F_b = (40\ \text{cm} \times 15\ \text{cm}) \times 20\ \text{cm} \times 0.8\ \text{g/cm}^2 \times 980\ \text{cm/sec}^2 = 9.4 \times 10^6$ dynes

(b) $F = A \times H \times D \times g$

$F_f = (40\ \text{cm} \times 20\ \text{cm}) \times 10\ \text{cm} \times 0.8\ \text{g/cm}^3 \times 980\ \text{cm/sec}^2 = 6.27 \times 10^6$ dynes

(c) $F = A \times H \times D \times g$
$F_e = (15 \text{ cm} \times 20 \text{ cm}) \times 10 \text{ cm} \times 0.8 \text{ g/cm}^3 \times 980 \text{ cm/sec}^2 = 2.35 \times 10^6$ dynes
(d) $p = H \times D \times g$
$p_b = 20 \text{ cm} \times 0.8 \text{ g/cm}^3 \times 980 \text{ cm/sec}^2 = 1.57 \times 10^4 \text{ dynes/cm}^2$
(e) $p = H \times D \times g$
$p_c = 5 \text{ cm} \times 0.8 \text{ g/cm}^3 \times 980 \text{ cm/sec}^2 = 3.92 \times 10^3 \text{ dynes/cm}^2$

The pressure on the bottom could also be computed by using the equation which defines pressure.

$$p_b = \frac{F}{A} = \frac{9.4 \times 10^6 \text{ dynes}}{40 \text{ cm} \times 15 \text{ cm}} = 1.57 \times 10^4 \text{ dynes/cm}^2$$

Example 7.2
Compute (a) the force exerted by water against the vertical face of a dam which is 60 ft long and 40 ft high (water comes just to the top of dam); and (b) the pressure against the dam at a point 15 ft below the surface.

Solution

(a) $F = A \times H \times D \times g$
$F = (60 \text{ ft} \times 40 \text{ ft}) \times 20 \text{ ft} \times 62.4 \text{ lb/ft}^3 = 3.0 \times 10^6$ lb
(b) $p = H \times D \times g$
$p = 15 \text{ ft} \times 62.4 \text{ lb/ft}^3 = 936 \text{ lb/ft}^2$

Note that in this problem, in which the English units were used, the weight density for water 62.4 lb/ft² is equivalent to product of the (mass) density and g.

PRESSURE OF THE ATMOSPHERE

The earth is surrounded by a layer of gas that extends outward from the surface for several hundred miles. There is little air above 250 miles from the earth's surface. Approximately 90 percent of the entire air and 95 percent of the water vapor constituent are within the troposphere, the part within 10 miles of the earth's surface. See Chap. 20 for a more complete discussion of the earth's atmosphere.

Several interesting experiments may be performed to demonstrate that our atmosphere exerts a pressure. An empty can, like that used for duplicating fluid, is fitted with a metal tube in its screw top. The tube is connected by a rubber hose to a vacuum pump. Before the pump starts, the walls of the can are stationary because the air pressure outside the can is the same as that inside. When the pump is started, some air is removed from the inside and the excess pressure of the outside air pushes the walls of the can inward. Compressed air can be used to push the walls outward to their original position. From this and similar experiments discussed below we conclude that the atmosphere exerts pressure, and we derive the generalization that

Figure 7.8
Pressure of air against a rubber diaphragm.

Fluids tend to move from regions of higher pressure toward regions of lower pressure.

In another experiment a vacuum pump is connected to a pump stand on which a jar with an open top is placed (Fig. 7.8). Before the pump starts, a rubber diaphragm is stretched across the top and remains horizontal because the pressure just above on the top is the same as the pressure just below the bottom. When the pump is started, air is removed from below the rubber diaphragm, and the excess pressure of the outside air causes the diaphragm to bulge downward.

A classic experiment on air pressure consists of removing the air from the inside of a pair of Magdeburg hemispheres about 4 in.

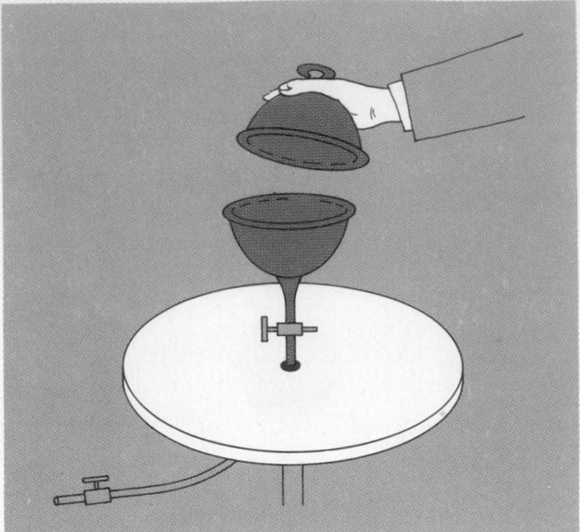

Figure 7.9
Magdeburg hemispheres.

in diameter placed together as in Fig. 7.9. After the air has been exhausted from the hemispheres by vacuum pump, two people standing on the floor and without other support probably will not be able to pull them apart. From the relation $F = pA$, where p is the pressure of the atmosphere (standard value of 14.7 lb/in.^{2} is used) and A in this case is the cross-section area, not the surface area,

$$F = pA = p \times \pi R^2 = 14.7 \text{ lb/in.}^2 \times \frac{22}{7} \times 4 \text{ in.}^2$$
$$= 185 \text{ lb}$$

Of course, a perfect vacuum is not obtained with the pump, so that a force of 180 lb will probably be sufficient to pull the hemispheres apart. These hemispheres are called the "Magdeburg hemispheres" because Otto von Guericke of Magdeburg, Germany, first used them. He constructed a pair nearly 2 ft in diameter, and eight horses pulling on each handle could not separate them. From Newton's third law, this was a waste of eight horses (one handle could be tied to a tree), but it was good showmanship.

Practical applications of air pressure or of pressure reduced by vacuum pumps occur in vacuum cleaners, medicine droppers, milking machines, syphons, lift pumps, force pumps, and ordinary breathing.

Atmospheric pressure may be measured indirectly by measuring the height of a column of mercury that the air will support. This can be done by the use of a mercury barometer, which can be made by completely filling a glass tube, closed at one end and at least 80 cm long, with mercury and then inverting the tube in a dish of mercury, as illustrated in Fig. 7.10. If the barometer is carefully made—the tube full of mercury should be heated to drive out any air bubbles that might cling to the glass walls—the space above the mercury will be almost a vacuum, although there will be some mercury vapor molecules present. What keeps the mercury up in the tube?

To "see" the answer to this question examine Fig. 7.10c where a cross section view of the mercury barometer is shown. The small arrows indicate the force of the atmosphere pushing down on the mercury in the dish. At point A in the tube there is no air pushing down on the mercury in the tube. At point A the weight of the mercury in the tube is pushing downward but the force of the air on the mercury in the dish is transmitted through the mercury and pushes up at point A. Because the system is stable, the push of the atmosphere must just equal the push of the column of mercury; that is, the pressure of the column of mercury is equal to the pressure of the atmosphere.

If the air pressure increases, mercury is pushed higher in the tube; when the air pressure decreases, the mercury column falls correspondingly.

The average height of mercury in a barometer at sea level and latitude 45°N is 76 cm (29.92 in.), so that a *standard atmosphere* is said to be 76 cm of mercury, which in ordinary pressure units is 76 cm of mercury × 13.6 g/cm^3 × 980 cm/sec^2 = 1,013,000 dynes/cm^2. The corresponding values in English units are 29.92 in., or 14.7 lb/in.^{2}; however, the approximate value of 15 lb/in.^{2} is often used for atmospheric pressure.

In the metric system the unit of pressure is 1 bar = 1,000,000 dynes/cm^2. Millibars (thousandths of a bar) are used to express atmospheric pressure; they are abbreviated

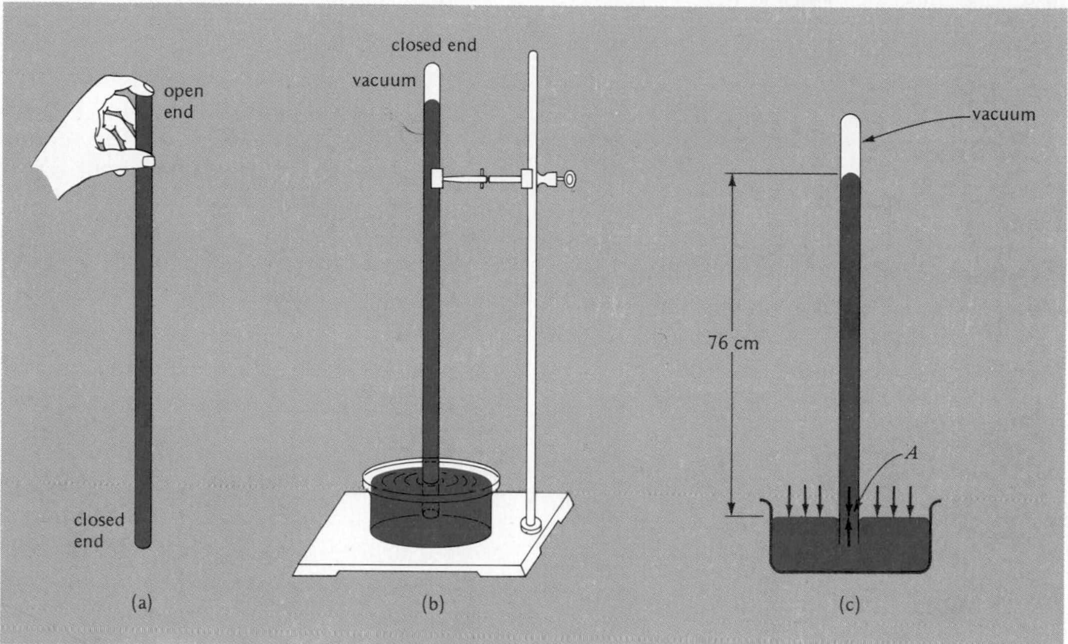

Figure 7.10
Mercury barometer. (a) A glass tube 80 cm long filled with mercury. (b) A mercury barometer. (c) Cross-section view of a mercury barometer.

"mb." The standard atmospheric pressure in the English system of 29.92 in. = 1013 mb in metric system.

Thus atmospheric pressure expressed in millibars has an equivalent pressure (F/A) expressed in dynes per square centimeters. On the weather maps near the end of Chap. 8 an isobar (line of equal barometric pressure) has at one end the pressure in inches (English system) and at the other end in millibars (metric system). Such maps are suitable for use internationally.

The higher a place is above sea level, the less air there is above the place. Near the surface of the earth, there is an approximate reduction of about 0.1 in. (0.25 cm) of mercury for each 100 ft of elevation, and pressure at a given place never varies more than 1 in. (2.54 cm) of mercury from the average value for that place.

The mercury barometer is a rather cumbersome instrument and also presents some hazard because mercury is poisonous. Any spillage of mercury should be cleaned up carefully. However, mercury is the most dense liquid we have available to us and the use of any other liquid in a barometer is impractical. For example, if we were to substitute water for mercury, we would require a tube over 34 ft long, for 1 atmosphere (atm) of pressure will support 34 ft of water. This is readily calculated using Eq. 6.5 by equating 1 atm of pressure (14.7 lb/in.^{2}) to HDg with proper conversion of units

$$14.7 \text{ lb/in.}^2 = HDg$$

or

$$H = \frac{14.7 \text{ lb/in.}^2}{Dg} = \frac{14.7 \text{ lb/in.}^2}{62.4 \text{ lb/ft}^3}$$

$$= \frac{14.7 \text{ lb/in.}^2 \times 144 \text{ in.}^2/\text{ft}^2}{62.4 \text{ lb/ft}^3} = 34 \text{ ft.}$$

This result also tells us that we can not

Figure 7.11
Automatic recording barometer, or barograph. When the atmospheric pressure decreases, a spring pushes the metal bellows up, moving a lever. A second lever, connected to the first, pushes the recording pen down. (Courtesy of the Taylor Instrument Division, Sybron Corporation.)

"pump" water with a typical hand pump higher than 34 ft (that is, we are limited in the depth of a well) by simply creating a vacuum in the top end of a pipe while the bottom end is in the water. It is the atmospheric pressure that pushes the water up a tube and 1 atm of pressure can push water only to a height of 34 ft.

ANEROID BAROMETERS, RECORDING BAROGRAPHS, AND ALTIMETERS

Although a well made mercury barometer is probably the most accurate instrument for measuring air pressure, it is not easily portable. A more practical instrument is the aneroid barometer, which consists of a thin metallic box from which some air has been removed and whose top moves down or up as the air pressure increases or decreases. This motion is registered and magnified by a series of levers and gears that cause a hand to move along a direct-reading scale.

To see easily how pressure is changing with time, the barograph (automatic recording barometer) is used. Figure 7.11 is a photograph of such an instrument, and Fig. 7.12 shows the recorded pressure for one week. This type of instrument is particularly helpful in the prediction of weather, although of course the barometer reading is only one of the many factors considered in weather prediction (see Chap. 8).

Because air pressure falls with increasing elevation, aneroid barometers may be made to read elevation instead of air pressure. The altimeter, located on the instrument panel of airplanes, is just such a device.

Pascal's principle

Figure 7.13 shows a top view of a glass syringe often used for a demonstration in science lectures. As the piston is pushed to the right, the water squirts out in similar streams of equal length from all the horizontal openings, indicating that the pressure applied to the liquid by the piston is transmitted equally to all parts of the liquid. This idea about liquids was first stated by Blaise Pascal (1623–1662), the French mathematician and philosopher, as follows:

Pressure applied to a confined fluid is transmitted unchanged to all parts of the fluid and to the walls of the containing vessel.

The hydraulic press is an important and easily understood example of Pascal's principle. Figure 7.14 shows the idea of a hydraulic press, the valves being omitted for simplicity. The space between the two pistons is filled with some liquid, usually oil. When a force F_1 is applied to the small piston of area A_1, a pressure $p_1 = F_1/A_1$ is exerted against the liquid by the small piston. As Pascal's principle states, this pressure is transmitted through the liquid unchanged, so that the pressure p_2 against the large piston of area A_2 is equal to p_1. Also $p_2 = F_2/A_2$, where F_2 is the force exerted by the liquid against the movable large piston, which in turn acts on the object O. These relations lead to the determination of the mechanical advantage of the hydraulic press as follows:

Pascal's principle $\quad p_2 = p_1$

$$\frac{F_2}{A_2} = \frac{F_1}{A_1}$$

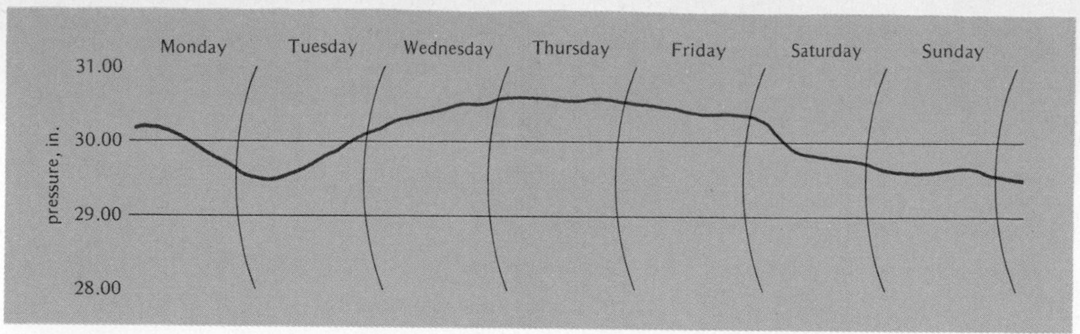

Figure 7.12
Typical barograph record. The chart shows that the atmospheric pressure decreased, then increased, and again decreased.

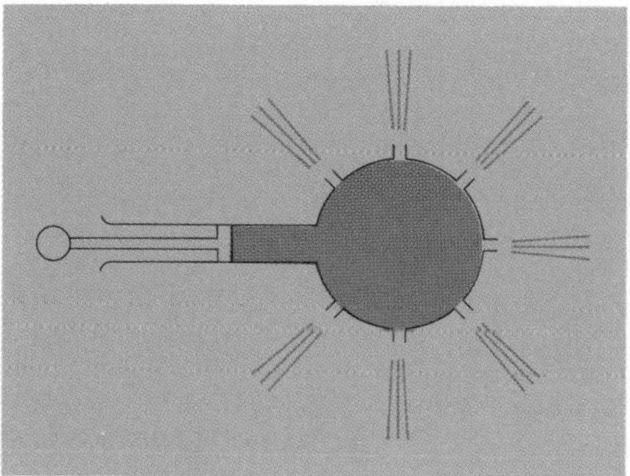

Figure 7.13
Equal pressures at all the horizontal openings of a syringe. Top view of a glass syringe held so that all the openings are in the same horizontal plane.

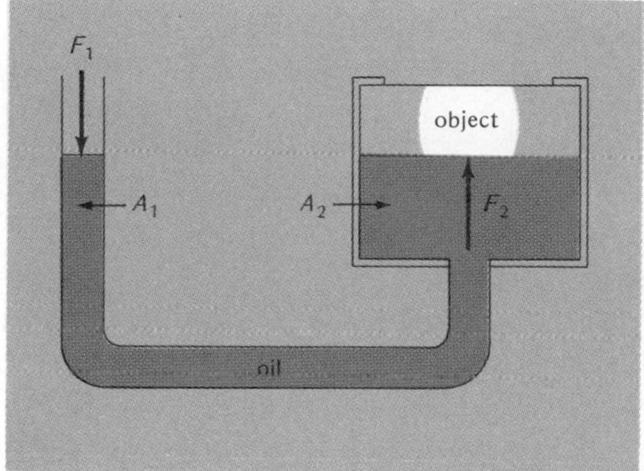

Figure 7.14
Hydraulic press.

or

$$F_2 = F_1 \left(\frac{A_2}{A_1}\right)$$

If A_2 is larger than A_1, then the applied force F_1 will be multiplied by this device; that is, F_2 will be larger than F_1. This can be stated in terms of mechanical advantage. Neglecting friction, the mechanical advantage f_o/f_i (Eq. 6.6) is F_2/F_1 and is also, from the foregoing relation, A_2/A_1 or $\pi R_2^2/\pi R_1^2$ (π times the radius squared is the area of a circle). The mechanical advantage of the hydraulic press is thus seen to be equal to the ratio of the areas of the two pistons, or the ratio of the squares of the radii of the pistons. Hydraulic presses of various types are used in factories, and a barber's chair or dentist's chair is essentially a hydraulic press.

It should be pointed out that in the analysis of Fig. 7.14 we have not included the effect of the atmosphere because the atmosphere exerts the same pressure on area A_1 as it does on A_2. We were required to examine only the increased pressure due to the applied force F_1. In the hydraulic press the "transmitted" pressure is so large as to make the effects of the weight of the fluid negligible, so that any effects of $p = HDg$ can be ignored.

Hydraulic brakes used in automobiles depend on Pascal's principle. The pressure exerted by the foot pedal against the brake fluid is transmitted unchanged to the brake

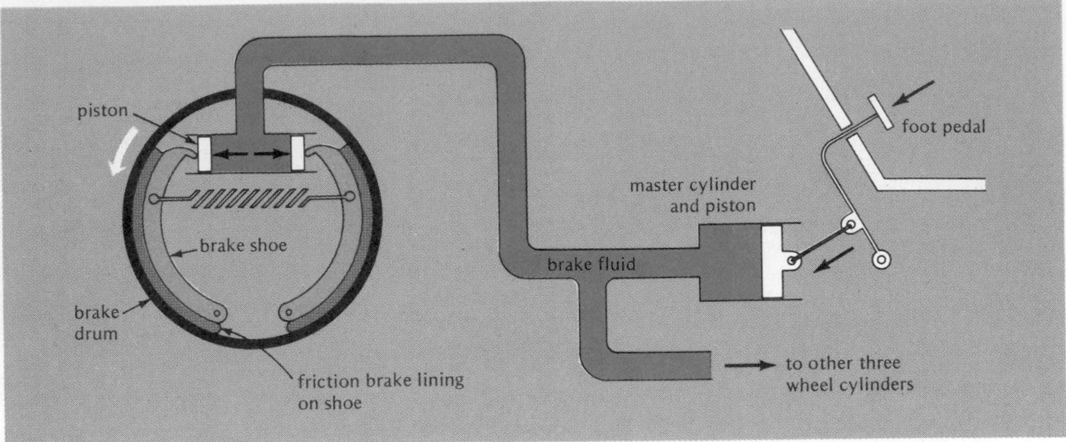

Figure 7.15
Hydraulic brakes.

pistons, which exert a large stopping force against the wheels as the brake bands are pushed against the drum (Fig. 7.15).

Example 7.3
Suppose that in the use of a dentist's chair a force of 30 lb (F_1) is applied by the operator, by means of a hand lever or foot pedal, to a small piston having a cross section of 2 in.^{2} (A_1). The cross section A_2 of the large piston, which supports the chair and occupant, is 20 in.^{2}. Compute (a) the pressure of the small piston against the fluid; (b) the pressure of the fluid against the large piston; (c) the maximum weight F_2, including the large piston, chair, and occupant, that can be raised; (d) the mechanical advantage of the chair.

Solution

(a) $p_1 = \frac{F_1}{A_1} = \frac{30 \text{ lb}}{2 \text{ in.}^2} = 15 \text{ lb/in.}^2$

(b) $p_2 = p_1 = 15 \text{ lb/in.}^2$

The pressure p_2 of the fluid on the large piston must equal the pressure p_1 at the small piston because Pascal's principle states that pressure is transmitted unchanged throughout the fluid.

(c) $F_2 = p_2 A_2 = 15 \text{ lb/in.}^2 \times 20 \text{ in.}^2 = 300 \text{ lb}$

(d) $\text{MA} = \frac{F_2}{F_1} = \frac{300 \text{ lb}}{30 \text{ lb}} = 10$

or

$$\text{MA} = \frac{A_2}{A_1} = \frac{20 \text{ in.}^2}{2 \text{ in.}^2} = 10$$

Pascal's principle is involved in the hydraulic hoist so often used at garages and gasoline stations to raise cars for ease in oil change or in making repairs. Figure 7.16 shows the general arrangement. If the pressure gauge indicates a compressed air pressure of 100 lb/in.^{2}, this is also the pressure against the bottom of the supporting piston. If the supporting piston is 1 ft in diameter, its bottom area is $A = \pi R^2 = 3.14 \times 6^2 = 113$ in.^{2}. Hence the force up against the bottom of the piston is $F = pA = 100 \text{ lb/in.}^2 \times 113 \text{ in.}^2 = 11{,}300$ lb, which is sufficient to support an ordinary passenger car or light truck.

Archimedes' principle

If a cork stopper is held under water and released, it quickly moves upward to the surface. The magnitude of this upward, or buoyant, force was understood by Archimedes (287–212 B.C.), the Greek mathematician and inventor, who stated the important principle now bearing his name.

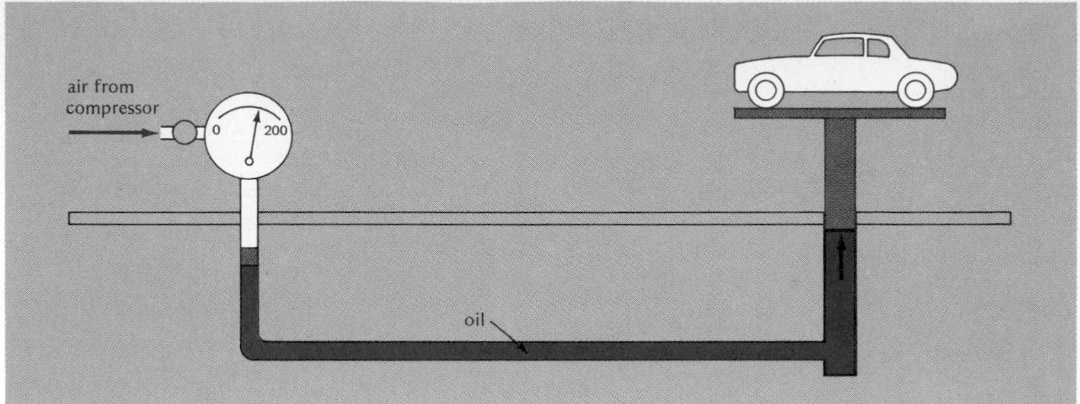

Figure 7.16
Hydraulic hoist.

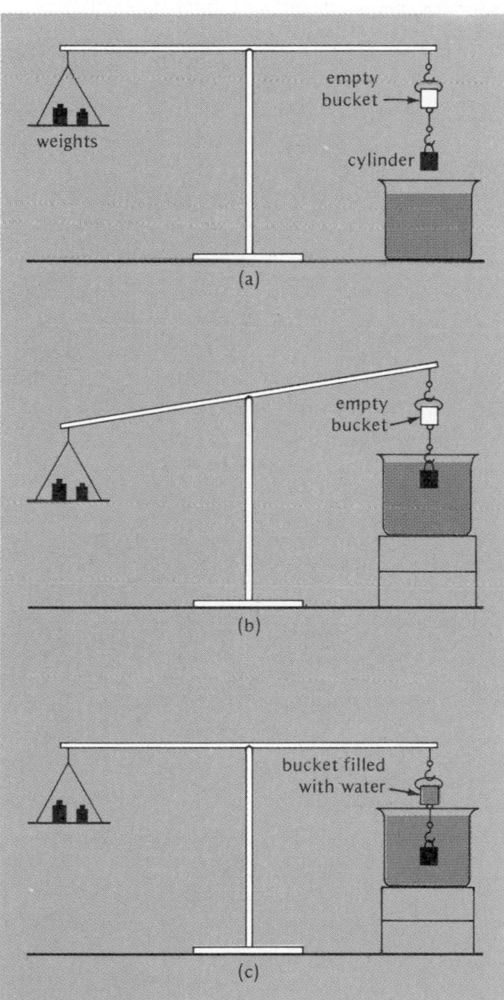

Figure 7.17
Demonstration of Archimedes' principle for immersed bodies; the bucket-and-cylinder experiment.

Any object immersed in, or floating in, a fluid is buoyed up by a force equal to the weight of the fluid displaced.

EXPERIMENTS TO DEMONSTRATE ARCHIMEDES' PRINCIPLE

How can we show that Archimedes' principle applies to submerged, or immersed, bodies? An empty bucket, from which a cylinder that will just fit into the bucket is hung, is brought to equilibrium on an equal-arm balance (Fig. 7.17). The cylinder is then immersed in water (center). The cylinder side goes up, indicating a buoyant force on the immersed cylinder. The bucket is then carefully filled with water (bottom). Equilibrium will be restored, showing that the buoyant force is equal to the weight of the water added to the bucket. Because this is the same amount of water that the cylinder displaces—the cylinder has the same volume as the inside of the bucket—the experiment demonstrates that the buoyant force equals the weight of the water displaced. Thus Archimedes' principle applies to submerged, or immersed, bodies.

Does Archimedes' principle apply to floating bodies?

To illustrate that Archimedes' principle does apply to floating bodies, the apparatus shown in Fig. 7.18 may be employed. Here a metal can with a drain tube is placed on one pan of a platform balance and water is added to a height just above the drain tube. After excess water has run out and the dripping

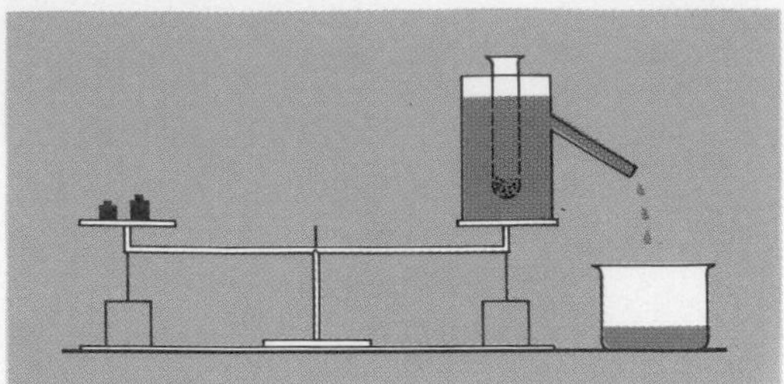

Figure 7.18
Demonstration of Archimedes' principle for floating bodies.

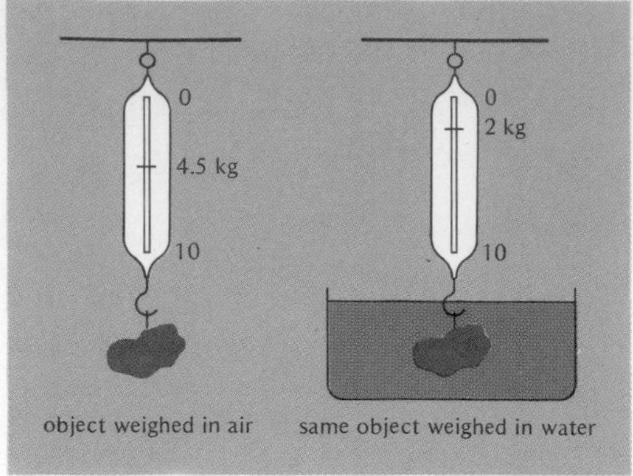

Figure 7.19
In air, the object weighs 4.5 kg × 9.8 m/sec²; in water, 2 kg × 9.8 m/sec².

has stopped, the balance is brought to equilibrium. A test tube with some weight, for example, lead shot or small nails to keep it upright, is floated in the can. The side with the can will go down; but after dripping has again ceased, equilibrium will be restored. This shows that the weight of the test tube with its contents equals the weight of the water displaced. The weight of the test tube with its contents, however, also equals the buoyant force because the tube floats. Therefore the buoyant force equals the weight of the water displaced. Thus Archimedes' principle does apply to floating bodies. A common example is a floating ship.

DETERMINATION OF SPECIFIC GRAVITY

Weigh an object in air and then weigh it immersed in water (Fig. 7.19). Subtract the weight in water from the weight in air; this apparent loss of weight in water is the buoyant force. By Archimedes' principle this buoyant force is equal to the weight of water displaced by the object. This leads to the definition of specific gravity (sp gr; for solid, sp gr_S; for liquid, sp gr_L):

$$\text{sp gr}_S = \frac{D_{\text{object}}}{D_{\text{water}}}$$

$$= \frac{\text{weight of object in air}}{\text{weight of equal volume of water}}$$

$$= \frac{\text{weight of object in air}}{\text{weight of water displaced}}$$

$$\text{sp gr}_S = \frac{\text{weight of object in air}}{\text{loss of weight in water}} \qquad (7.6)$$

Applying this relation to the object in Fig. 7.19,

$$\text{sp gr} = \frac{4.5 \text{ kg} \times 980 \text{ cm/sec}^2}{(4.5 - 2.0) \text{ kg} \times 980 \text{ cm/sec}^2} = \frac{4.5}{2.5} = 1.8$$

Note that the spring balance is calibrated in mass units (kg) and one should convert these units to weight units by multiplying by $g = 980$ cm/sec².

Example 7.4
A stone weighs 5 newtons in air and 375 newtons in water. Compute its specific gravity.

Solution

$$\text{sp gr}_S = \frac{\text{weight in air}}{\text{loss of weight in water}}$$

$$= \frac{500}{500 - 375} = \frac{500}{125} = 4$$

To obtain the specific gravity of a liquid, use

$$\text{sp gr}_L = \frac{D_{\text{liquid}}}{D_{\text{water}}}$$

$$= \frac{\text{weight of liquid displaced when an object is weighed in the liquid}}{\text{weight of liquid displaced when an object is weighed in water}}$$

$$= \frac{\text{loss of weight of solid in the liquid}}{\text{loss of weight of solid in water}} \qquad (7.7)$$

Example 7.5
A buyer of crude oil found that a solid metal weight weighed 3 lb in air, 1.5 lb in crude oil, and 2 lb in water. Compute (a) specific gravity of metal (solid); and (b) of crude oil (liquid). (c) What is the density of the crude oil?

Solution

(a) $\text{sp gr}_S = \dfrac{\text{weight in air}}{\text{loss of weight in water}}$

$$= \frac{3 \text{ lb}}{3 \text{ lb} - 2 \text{ lb}} = 3$$

(b) $\text{sp gr}_L = \dfrac{\text{loss of weight in liquid}}{\text{loss of weight in water}}$

$$= \frac{1.5 \text{ lb}}{3 \text{ lb} - 2 \text{ lb}} = 1.5$$

(c) Because the specific gravity of the crude oil is 1.5, its density is 1.5 g/cm³.

ACTION OF BALLOONS AND SUBMARINES

Consider a balloon whose bag has been filled with 10,000 ft³ of helium gas. The approximate weight density of helium is 0.02 lb/ft³, and the approximate weight density of air near the earth's surface is 0.08 lb/ft³. According to Archimedes' principle, the buoyant force is equal to the weight of the air displaced, which is 10,000 ft³ × 0.08 lb/ft³ = 800 lb. The weight of the contained 10,000 ft³ of helium is 10,000 ft³ × 0.02 lb/ft³ = 200 lb. If the containing bag, ropes, gondola, and contents weigh less than 600 lb, the upward, or buoyant, force exceeds the downward force, or total weight, so the balloon will rise. As the balloon rises, the weight of the displaced air decreases because the density of the displaced air decreases and finally the level is reached where the weight of the displaced air just equals the total weight, and the balloon will stop rising. To prevent too rapid rising of the balloon plus contents from the ground when the balloon is released, ballast in the form of bags of water or sand is frequently carried. The ballast can gradually be released at higher levels to give a slower, more uniform upward motion. To descend, a rip cord that releases the enclosed gas from certain compartments is pulled.

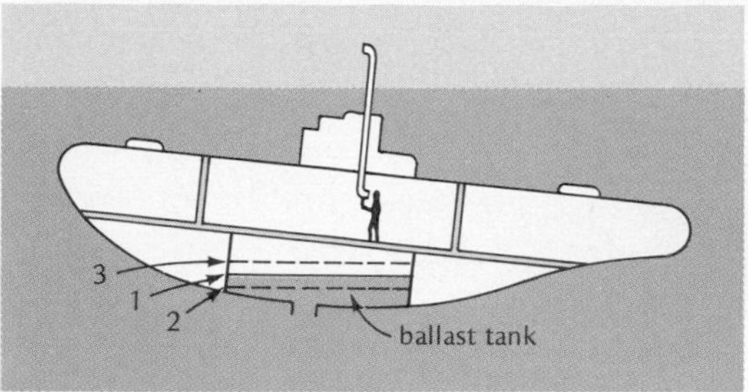

Figure 7.20
Action of a submarine.

Another important application of Archimedes' principle is the submarine (Fig. 7.20). With the submarine at rest under the surface, water is at level 1 in the compartment, or ballast tank, open to sea water; the valves are not shown. In this position the downward force, or total weight—submarine and its contents—just equals the upward, or buoyant, force, which is the weight of the outside water displaced by the submarine. If the water level inside is forced down to level 2 by pumping compressed air into the upper part of the compartment, more outside water is displaced, giving an increased buoyant force, which causes the submarine to rise. To submerge, air is released from the compartment, allowing more water to come in, to some such level as 3. This reduction in water displaced makes the upward, or buoyant, force less than the downward force, or total weight, and the submarine sinks. Of course, when the submarine is moving forward, its up-and-down motion can be controlled by a horizontal rudder.

Bernoulli's principle

Consider the two sets of glass tubes shown in Fig. 7.21. At the top, the water is flowing

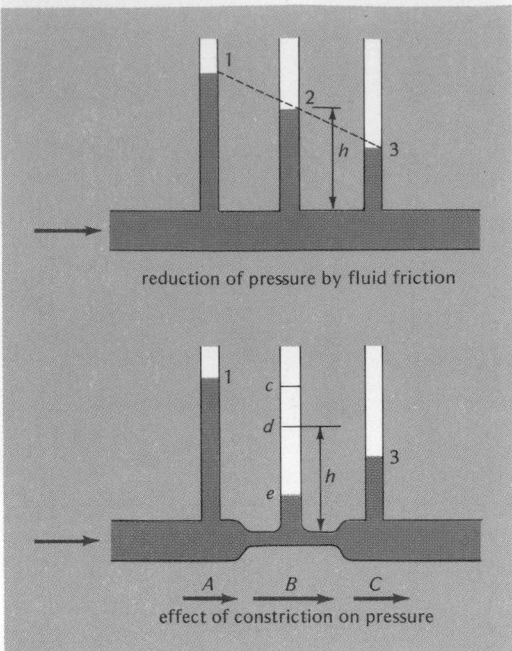

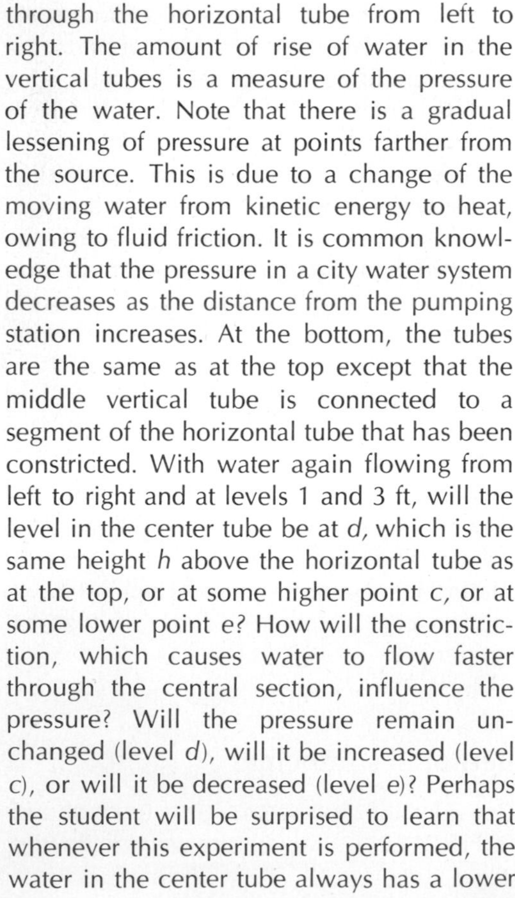

Figure 7.21
Illustration of Bernoulli's principle.

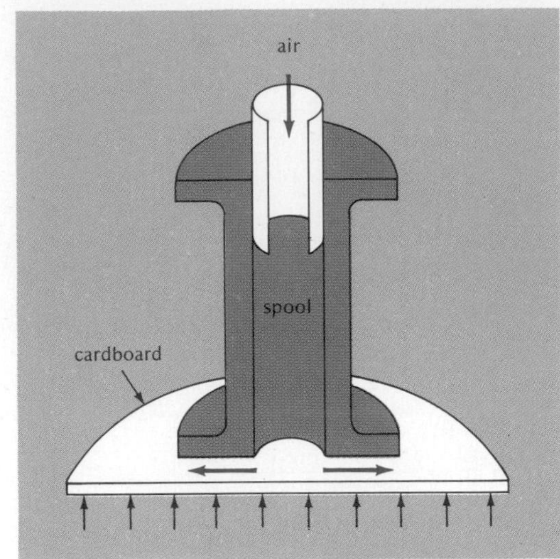

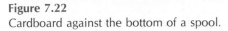

Figure 7.22
Cardboard against the bottom of a spool.

through the horizontal tube from left to right. The amount of rise of water in the vertical tubes is a measure of the pressure of the water. Note that there is a gradual lessening of pressure at points farther from the source. This is due to a change of the moving water from kinetic energy to heat, owing to fluid friction. It is common knowledge that the pressure in a city water system decreases as the distance from the pumping station increases. At the bottom, the tubes are the same as at the top except that the middle vertical tube is connected to a segment of the horizontal tube that has been constricted. With water again flowing from left to right and at levels 1 and 3 ft, will the level in the center tube be at *d*, which is the same height *h* above the horizontal tube as at the top, or at some higher point *c*, or at some lower point *e*? How will the constriction, which causes water to flow faster through the central section, influence the pressure? Will the pressure remain unchanged (level *d*), will it be increased (level *c*), or will it be decreased (level *e*)? Perhaps the student will be surprised to learn that whenever this experiment is performed, the water in the center tube always has a lower level, as at *e*. Thus there is a considerable decrease in pressure at the place of constriction, where the water is flowing fastest.

Daniel Bernoulli (1700–1782), a Swiss scientist, described the above results as follows:

Where the velocity of a stream of fluid is increased, the pressure is decreased, and vice versa.

This is the Bernoulli effect, commonly termed "Bernoulli's principle." In Fig. 7.21 the horizontal arrows represent velocities of the fluid stream. Note that the principle implies that where the velocity of the stream of fluid is greatest (*B*), the pressure of the fluid is least.

There are many interesting applications of Bernoulli's principle, and a discussion of a few of them will enable the student to understand the principle better.

If air is blown through the tube that is attached to the spool (see Fig. 7.22), a large piece of cardboard will be held against the bottom of the spool, in spite of the force of gravity. When no air is blown through the spool, the pressure of the atmosphere down on the cardboard and up on the cardboard

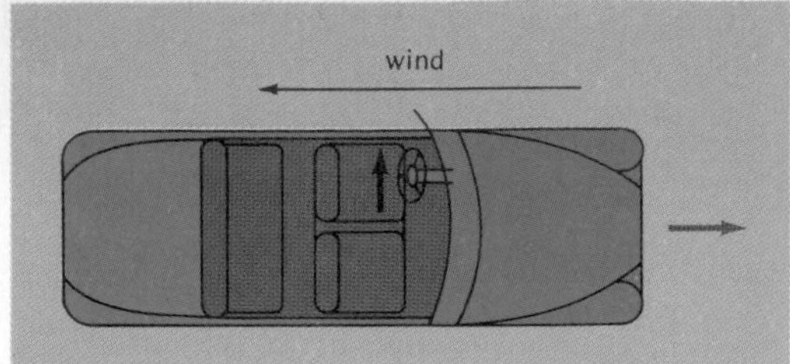

Figure 7.23
Top view of an automobile, showing nondraft ventilation.

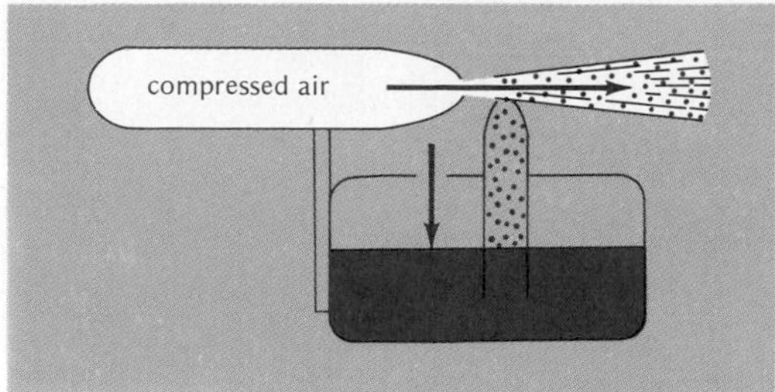

Figure 7.24
Action of a sprayer.

is the same and the force of gravity will cause the cardboard to drop. When air is blown through the spool, it rushes out the thin opening between the bottom of the spool and the cardboard with a high velocity —note the long horizontal velocity arrows. According to Bernoulli's principle, the effect of the high-velocity stream of air above the cardboard is to reduce the pressure of the atmosphere against the top of the cardboard.

Because the pressure of the atmosphere up against the cardboard remains unchanged, this gives a net pressure, and hence a net force, up against the cardboard that is sufficient to overcome the force of gravity on the cardboard. The fact that the cardboard moves about slightly under the spool is proof that although the cardboard seems to be touching the spool, it is really separated from it by the thin layer of air that is rushing out in all directions.

Nondraft ventilation of automobiles (Fig. 7.23) depends on Bernoulli's principle. When the car is not moving, the air pressure inside and outside the opening of the ventilation window is the same. When the car is traveling forward through still air at, say, 30 mi/hour, this is equivalent to a stationary car and a gale of air blowing past the opening at the rate of 30 mi/hour. This high-velocity stream of air reduces the air pressure on the outside, and the excess of the unreduced air pressure on the inside causes air, including any tobacco smoke that may be present, to flow out through the opening. This gives ventilation without the blast of incoming air that occurs when the large windows are open.

An atomizer or sprayer (Fig. 7.24), such as the one used for flies and bugs or for spraying paint, also works on this principle. Before the sprayer is used, there is equal atmospheric pressure on the liquid inside and outside the vertical tube. A high-velocity stream of air is then forced over the top opening of the vertical tube. This reduces the air pressure against the surface within the tube, and the normal air pressure against the liquid outside the tube pushes the liquid into the tube and out the top opening, where the horizontal stream of air sends it forward as a spray. Nitrogen gas is an example of the compressed gases used in spray cans containing liquids such as hair spray, insecticide, shaving cream, "whipped cream," paint, etc.

If the side openings of a Bunsen gas burner (Fig. 7.25) are closed, a smoky yellow flame results; if the side openings are adjusted to admit air, a hot blue flame can be obtained. This is caused by the burning of a mixture of illuminating gas and air. What causes the air to go in? Before the gas is turned on, there is normal air pressure inside and outside the horizontal opening. When the gas is turned on, a high-velocity stream of illuminating gas rushes up through the small opening, below the shaft of the vertical arrow. The effect of this high-velocity stream of gas is to reduce the air pressure on the inside, and the excess outside pressure pushes in air that mixes with the illuminating gas to give the hot blue flame.

The Bernoulli lift on an airplane wing

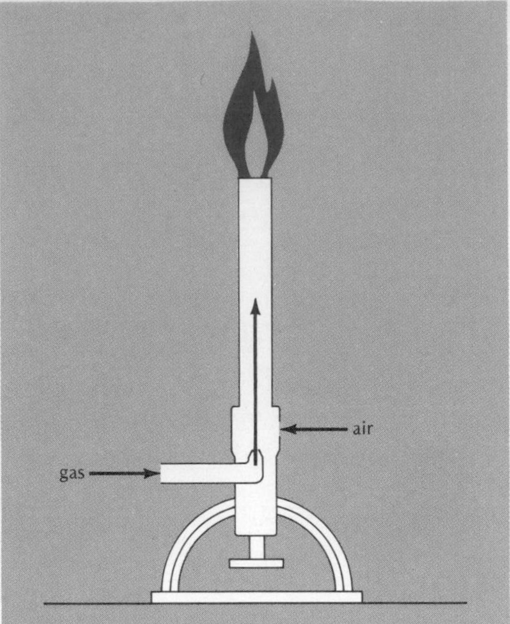

Figure 7.25
Action of a Bunsen burner.

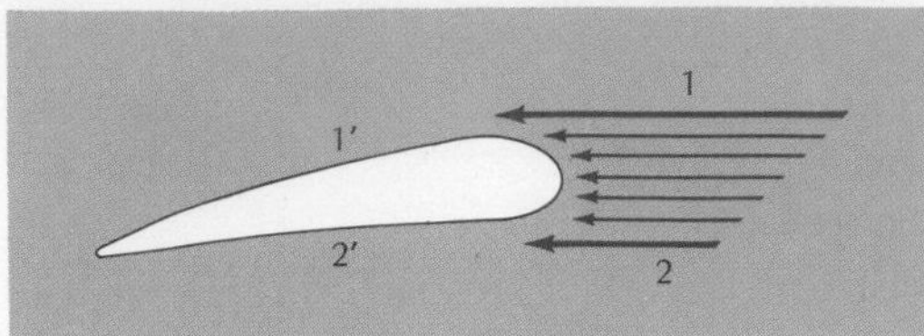

Figure 7.26
The Bernoulli lift on an airplane wing. The airplane is flying from left to right.

(Fig. 7.26) is very important in aviation. The movement of the plane to the right is equivalent to a stationary plane with a gale of wind moving to the left past the wing. The wing is shaped in such a manner as to cause more air to be forced above it than below it. There is a stream of air with a greater velocity above the wing (1) than below it (2). Hence the normal air pressure above the wing (1′) is reduced more than the air pressure at the bottom of the wing (2′). This gives a resultant upward pressure, producing a large lifting force on the wing.

Other applications of Bernoulli's principle occur in the action of the carburetor of an automobile, the aspirator or jet pump, the curving of a baseball, etc.

SUMMARY

Matter, which has been defined as anything that occupies space and has mass, may exist in three states: solid, liquid, or gaseous. The atoms of the Greeks are now considered the molecules of our present-day concept.

A modern statement of the kinetic molecular theory of matter follows:

All matter, whether in the solid, liquid, or gaseous state, is made up of very small particles, called molecules, *that are continually in motion.*

Density is the mass per unit volume of a substance. In a gas the molecules are relatively far apart and the density is small. In a liquid or solid the molecules are much closer together and the density is greater. The specific gravity, or relative density, of a solid or liquid substance is the ratio of the density of the substance to the density of water. Thus the specific gravity tells how many times more dense the substance is than water. Gaseous substances are usually compared with air as a standard, solids and liquids with water.

The kinetic molecular theory of matter served in this chapter as a basis for distinguishing between the three states of matter and also for explaining the cause of pressure of liquids and gases. Pressure is the force per unit area acting on a surface. The air exerts pressure and barometers are used for measuring it.

There are three important principles that describe and illustrate the behavior of fluids.

Pascal's principle:

Pressure applied to a confined fluid is transmitted unchanged to all parts of the fluid and to the walls of the containing vessel.

Archimedes' principle:

Any object immersed in, or floating in, a fluid is buoyed up by a force equal to the weight of the fluid displaced.

Bernoulli's principle:

Where the velocity of a stream of fluid is increased, the pressure is decreased, and vice versa.

There are many applications of these principles.

Important words and terms

liquid
solid
gas
fluid
kinetic molecular theory of matter
atom
convection
Brownian motion
density
specific gravity
pressure
atmospheric pressure
barometer
Pascal's principle
Archimedes' principle
Bernoulli's principle
diffusion

Questions and exercises

1. State the kinetic molecular theory of matter, and explain the pressure of a gas against a surface in terms of this theory.
2. If water instead of mercury were used in a barometer, how high would the column of water be?
3. (a) Explain how to make a simple mercury barometer. (b) What is a barometer supposed to measure? Does it do this directly or indirectly? Discuss briefly.
4. If you fill a metal can full of water, close the can tightly, and then freeze the water in the can, you will find that the can has burst open after the water turned to ice. (a) From this observation, what can you say about the density of ice relative to that of water? (b) Explain how this answer agrees with the observation that ice floats in water.
5. What evidence leads you to believe that atoms are in motion and not stationary?
6. Explain how a drinking straw works.
7. Assume that the water in a lake is at the top of the dam but not flowing through the spillway. How would the pressure at any point—that is, the force on a unit area of the immersed surface of the dam—be affected (be increased, be decreased, remain the same) if the contour of the land under water permitted the water to back up from the dam a short distance (a few miles) or a longer distance (many miles)? That is, does the horizontal distance of water behind a dam affect the pressure on the dam, assuming that the water is only up to the top of the dam? Give reason(s) for your answer.
8. Diagram and explain at least one application, not given in this textbook or by your instructor, of Pascal's principle; of Archimedes' principle; of Bernoulli's principle.
9. With the aid of a properly labeled diagram explain in terms of Bernoulli's principle the action of a Bunsen gas burner; an atomizer or sprayer; nondraft ventilation of an automobile.
10. Do Pascal's principle, Archimedes' principle, and Bernoulli's principle apply to both liquids and gases? Explain your answer and give illustrative examples for each principle.
11. Why do enclosed gases tend to leak rapidly through very small openings?
12. If Brownian motion is a reality, why is it that the bombardment by air molecules of large household objects such as chairs, tables, etc., does not produce random motion of these objects?

Problems

1. Compute the volume of a girl who weighs 125 lb and can just float in water.

Answer: Approx 2 ft^3.

2. Given that the specific gravity of mercury is 13.6, compute the volume in cubic centimeters of 1 kg of mercury.
3. Suppose the living room in your house is 20 ft long, 12 ft wide, and 8 ft tall. If the weight density of air is 0.08 lb/ft^3, how much does the air in the room weigh?
4. A man who weighs 180 lb stands with both feet flat on the floor. The area of each shoe that makes contact with the floor is 30 $in.^2$. Compute (a) the pressure on the floor due to the weight of the man; (b) the pressure on the floor when he balances on one foot.

Answer: (a) 3 $lb/in.^2$; (b) 6 $lb/in.^2$

5. A girl walks in high-heeled shoes in such a manner that periodically her only contact with the floor is one of the heels. If the girl weighs 130 lb and the area of the bottom of one of these sharp heels is 0.05 $in.^2$, compute

the maximum pressure exerted on the floor by the girl in walking.

6. A barber's chair is connected with a large cylinder that has a cross-section area of 15 in.^{2}. The chair is raised by applying a force of 45 lb to a small piston that has a cross-section area of 1.5 in.^{2}. The two pistons are part of a hydraulic-press arrangement with a heavy oil in the chamber between them, the proper valves, etc. Compute (a) the pressure of the small piston against the fluid; (b) the pressure of the fluid against the large piston; (c) the maximum weight, including the weight of the chair, large piston, and occupant, that can be lifted; (d) the mechanical advantage of this "machine."

Answer: (a) 30 lb/in.^{2}; (b) 30 lb/in.^{2}; (c) 450 lb; (d) 10.

7. A passenger automobile is driven on the rack of a hydraulic hoist. The automobile weighs 4000 lb, and the rack, large cylinder, and piston of the hoist weigh 1000 lb. The compressed air is turned on, and a gauge shows that the pressure is 110 lb/in.^{2}. The bottom of the hoist piston has a cross-section area of 90 in.^{2}. (a) Compute the upward force against the hoist piston. (b) If the car is replaced by a truck that, with contents, weighs 5 tons, what will happen when the compressed air is turned on?
8. A rectangular-shaped barge 60 by 30 ft with vertical sides sinks into the water an additional 6 in. when loaded with trucks and cars for a river crossing. Compute the weight of the trucks and cars that the barge ferried.

Answer: 56,160 lb.

9. An object weighs 100 g in air, 88 g in water, and 90 g in alcohol. Compute (a) the density of the object; (b) the density of the alcohol.
10. A fish tank 10 ft long, 8 ft wide, and 6 ft high is full of pure water. Neglecting the weight of the air above the water and using a weight density of water = 62.5 lb/ft^3 (instead of the correct 62.4), compute (a) the force on the bottom of the tank; (b) the pressure on the bottom of the tank; (c) the pressure at a point 2 ft from the top surface of the water; (d) the force on the larger front face.

Project

Locate two containers of fluids in your home or in a nearby store. One is to be a rectangular parallelepiped (such as a home aquarium, can of maple syrup, can of furniture polish, etc.). The other is to be a cylinder (tin can of fruit juice, etc.).

1. For the rectangular parallelepiped (draw a diagram to show dimensions and state what it contains), compute pressure on the bottom surface; force on the bottom surface; force on the largest lateral (side) surface. (Note: Unless the instructor makes suggestions regarding finding the density of the fluid, you are to assume a density of 1.0 gm/cm^3 for water, down to 0.8 gm/cm^3 for liquids containing alcohol, and up to 1.2 for other common liquids. The thinking that accompanies the methods of computation is of greater importance here than the accuracy of the density.)
2. For the cylindrical container (draw diagram to show dimensions and state what it contains), compute pressure on the bottom surface; force on the bottom surface; force on the total lateral surface. (Note: Area of a circle $= \pi D^2/4$, for which D = diameter. Lateral area of a cylinder $= \pi D \times L$, for which D = diameter of base, L = height of cylinder, $\pi = 3.14$.)

Supplementary readings

Conant, James Bryant, *On Understanding Science: An Historical Approach,* Yale University Press, New Haven, Conn. (1947). [In Chap. 2 the work of Galileo, Torricelli, and Boyle is presented.]

McCue, J. J., and Kenneth W. Scherk, *Introduction to Physical Science: The World of Atoms,* The Ronald Press, New York (1963). [In Chap. 14 the pressure on a wall by particles of gas is discussed mathematically and related to the results of experiments on real gases.]

Rogers, Eric M., *Physics for the Inquiring Mind,* Princeton University Press, Princeton, N. J. (1960). [The kinetic molecular theory of gases is discussed in Chaps. 25 and 30. Concepts therein are found in several places in this book, notably Chap. 8.]

Scott, Ewing C., and Frank A. Kanda, *The Nature of Atoms and Molecules: A General Chemistry,* Harper & Row, New York (1962). [The kinetic molecular theory of matter and the gas laws (Chap. 8 of this book) are treated in Chap. 2.]

Article

Tucker, V. A., "The Energetics of Bird Flight," *Scientific American,* **220,** no. 5, 70 (May, 1969).

HEAT AND METEOROLOGY

Heat consists in a minute vibratory motion of the particles of bodies.

ISAAC NEWTON, 1704

Temperature and other fundamental concepts of the branch of science usually called "heat" will be discussed in the early part of this chapter; humidity and the various causes of precipitation will then be studied.

Meteorology, sometimes called "physics of the atmosphere," is the study of the earth's atmosphere with particular reference to weather. Weather data are collected by nearly 500 stations in the United States and are submitted to ESSA, the Environmental Science Service Administration (U.S. Weather Bureau), Washington, D.C. Weather data include the temperature of the air; the pressure of the air; the direction and speed of winds; the amount of water vapor in the air as compared with what the air could contain at the given temperature; the condition of the sky (clear, partly cloudy, etc.); the kind and amount of precipitation; visibility; and the location of high-pressure areas, low-pressure areas, and fronts.

The three basic elements of weather are pressure (of the air); temperature; and humidity, or the amount of water vapor in the air.

Heat

Until the latter part of the eighteenth century, heat was thought of as a weightless fluid called "caloric," more of which entered a body when the body became hotter and some of which left when the body cooled. An understanding of the true nature of heat was reached largely because of careful observations and experiments by two scientists. Benjamin Thompson (Count Rumford, 1753–1814), the British physicist and statesman, noted that a great deal of heat was produced in the boring of a cannon, and Sir Humphrey Davy (1778–1829), the British chemist, carried out an experiment in 1799 in which he caused two pieces of ice to melt by rubbing them together at a temperature below the melting point of ice. Both these phenomena indicated a close relation between heat and work and led to the realization that heat is a form of kinetic energy associated with the motions of the small particles of ordinary matter.

Heat is the *total* kinetic energy of the random motion of the individual molecules of a body. This is easily understood if we consider what happens to the kinetic energy of a moving hammerhead when the hammerhead is stopped by the act of pounding a piece of metal—an iron nail, for example. A clue to the answer is obtained by touching the nail, which is found to be warm. This is easy to understand in terms of the kinetic molecular theory, which was discussed at the beginning of Chap. 7. The blow of the hammer causes the molecules of the metal to vibrate faster, thus giving them a greater amount of kinetic energy. Of course, the molecules of the hammerhead have also been given a greater kinetic energy, for the hammerhead is also warmer than it was before.

It should be stressed that this thermal or heat energy is the total kinetic energy of the random chaotic vibrational motion of the molecules of the hammer and nail. Heat is a form of energy.

MEANING OF TEMPERATURE

When heat energy is absorbed, the temperature of the body usually increases. Heat and temperature are not the same, as will become clear as we discuss these concepts in this chapter. In a simple manner, *temperature* can be thought of as a measure on a scale of the degree of hotness of a body. In terms of kinetic theory, the *temperature* of a body is a measure of the *average* kinetic energy of the random motion of the individual molecules of the body. When we heat an object, the molecules vibrate on the average at a higher speed; however, some molecules vibrate at much higher speeds than others. The vibration of the molecules is chaotic and random in direction. It is the *average* kinetic energy that is the meaningful quantity when dealing with an object that contains so many molecules.

Kinetic theory can easily explain why heat flows from a hotter to a colder object. If two bodies are placed in thermal contact, the collisions of the molecules of the two bodies will result in energy being transferred from the faster-moving molecules of greater energy to the slower ones of lesser energy; that is, heat is always transferred from a hotter body, said to be of higher temperature, to a cooler body, said to be of lower temperature.

TEMPERATURE SCALES

Because temperature is a relative matter, some reference standards must be set up. For the Celsius or centigrade scale, the temperature of melting ice is chosen as 0°C, read "zero degrees centigrade"; and the temperature of boiling water at standard atmospheric pressure—76 cm of mercury—is taken as 100°C. For the Fahrenheit scale, the temperature of melting ice is called 32°F, read "32 degrees Fahrenheit," and the temperature of boiling water at a pressure of 76 cm of mercury is 212°F. Since 1948 the centigrade scale (°C) has officially been the Celsius scale (also °C); however, the word "centigrade" continues to be used.

A thermometer is an instrument used for measuring temperatures. If the bulb of a mercury-in-glass thermometer is placed in a mixture of pure ice and water, the level or mark at which the mercury comes to rest may be labeled 0°C or 32°F (Fig. 8.1). If the thermometer is then placed in the steam above boiling water at standard atmospheric pressure, the mercury will expand more than the glass container and will rise to a higher level or mark, which may then be labeled 100°C or 212°F. The space between the two marks on the Celsius scale may then be divided into 100 equal parts, each of which will represent 1°C. These same equal spaces may be extended above 100°C and below 0°C. Similarly, the space between the two marks on the Fahrenheit scale may be divided into 180 (212 − 32) equal parts, each of which will represent 1°F. These equal spaces may be extended above 212°F and below 32°F. A third scale, called an absolute scale, or Kelvin scale, starts with the coldest possible temperature, which will later be shown to be approximately −273°C, and uses the same size degrees as the Celsius scale. The relation of these three scales to one another is shown in Fig. 8.1. F is a Fahrenheit reading, C the corresponding Celsius or centigrade reading, and K the Kelvin or absolute reading.

The same size of degree is used on the Celsius and Kelvin scales. Starting with −273°C as 0°K, 0°C must be 273°K, and 100°C must be 373°K. Therefore, in general,

$$K = 273° + C \tag{8.1}$$

Thus 20°C expressed on the absolute scale is $K = 273° + 20° = 293°K$.

In most European countries the Celsius scale is commonly used, but in the United States the Fahrenheit scale is most frequently used, although American scientific laboratories are increasingly using the Celsius scale. In either case, it is sometimes necessary to convert from one scale to the other. In Fig. 8.1 it may be seen that corresponding readings, indicated by horizontal lines, are 0°C and 32°F for the freezing point of water and 100°C and 212°F for the boiling point of water. The number of Celsius and Fahrenheit degrees between one pair of corresponding temperatures must be directly proportional to the corresponding numbers of degrees between any other pair of corresponding temperatures. Hence,

$$\frac{C - 0}{F - 32} = \frac{100 - 0}{212 - 32}$$

or

$$\frac{C}{F - 32} = \frac{100}{180}$$

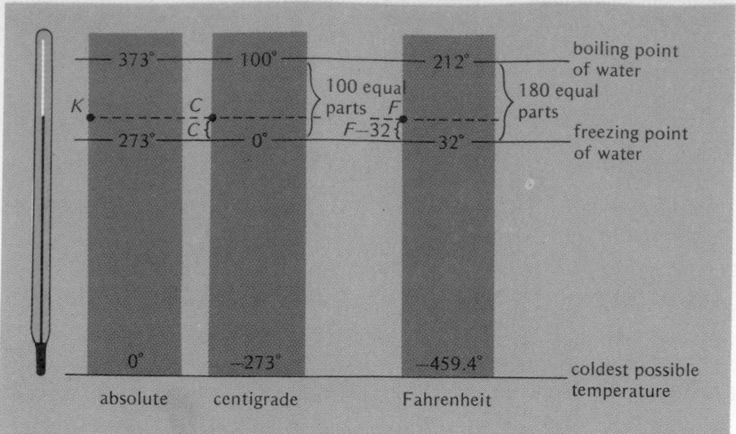

Figure 8.1
Three temperature scales.

which reduces to

$$\frac{C}{F-32}=\frac{5}{9} \qquad (8.2)$$

This fundamental relation may be solved for either C or F; thus

$$C=\frac{5}{9}(F-32)$$

and

$$F=\frac{9}{5}C+32$$

The simplest way to convert from Fahrenheit to Kelvin is first to change to the Celsius scale and then to the Kelvin scale.

Example 8.1
Express 68°F in Celsius and in Kelvin.

Solution

$$\frac{C}{F-32}=\frac{5}{9}$$

$$\frac{C}{68-32}=\frac{5}{9}$$

$$\frac{C}{36}=\frac{5}{9}$$

$$C=\frac{5}{9}\times 36=20°C$$

$$K=273°+C=273°+20°=293°K$$

Example 8.2
Change −12°C to Fahrenheit and Kelvin.

Solution

$$\frac{C}{F-32}=\frac{5}{9}$$

$$\frac{-12}{F-32}=\frac{5}{9}$$

Cross multiplying,

$$5F-160=-12\times 9$$

Transposing and combining,

$$5F=+160-108=52$$

Solving for F,

$$F=\frac{52}{5}=10.4°F$$

or equivalently,

$$F=\frac{9}{5}C+32$$

$$=\frac{9}{5}\times -12+32$$

$$=-21.6+32=10.4°F$$

$$K=273°+C$$

$$=273°+(-12°)=261°K$$

The student can make a rough check on the conversion by noting that on such a diagram as Fig. 8.1, a horizontal line drawn through −12° on the Celsius scale crosses the Fahrenheit scale line between 0 and 32° and crosses the absolute scale line slightly below 273°.

EXPANSION

In most thermometers use is made of the expansion of a substance to indicate increases in temperature. Most substances—solids, liquids, and gases—expand when heated and contract when cooled. Iron railroad rails laid on wooden ties increase in length in hot weather. For this reason, the rails are laid with about 0.25 in. of space between the ends of successive rails to allow for expansion. Allowance must often be made for contraction, too. If a wire clothesline is stretched tightly between two hooks in summer, it may contract enough the following winter to snap into two pieces, or more likely one of the supporting hooks will be pulled out. In other instances contraction may serve a useful purpose. Rivets are heated before being pounded into metal plates. The contraction of the rivets accompanying cooling produces a very tight fit of the metal plates riveted together.

Although we have been speaking of linear expansion, substances actually expand in all directions, producing volume expansion. When concrete pavements are laid, the space between blocks is filled with a tarry substance that is forced up by the expanding blocks in summer. Ordinary mercury-in-glass thermometers are possible because the mercury has a greater coefficient of volume expansion than the glass of the tube that contains the mercury.

HEAT UNITS

Work is required to rub one body back and forth over another. The friction involved results in raising the average kinetic energy of the molecules of the two surfaces of the bodies. Hence the two bodies become hotter and contain more heat than before. According to the law of conservation of energy, the increase in heat energy possessed by the two bodies must exactly equal the amount of work (or mechanical) energy used in producing the increase. It might be expressed in the same units as work—that is, in foot-pounds or joules.

Unfortunately for the student an independent set of heat units was established before it was realized that heat is a form of energy. The heat units were defined in terms of the amount of heat necessary to increase the temperature of a substance. Even though we now recognize that heat is just another form of energy, we continue to use the original units for heat.

The *calorie* (cal) is the amount of heat required to raise the temperature of 1 g of pure water 1°C. This is a rather small unit of heat, so it is customary in expressing energy values of food to use a unit 1000 times as great, often called the *kilocalorie* (kcal) or large calorie. The *kilocalorie* is the amount of heat required to raise the temperature of 1 kg of pure water 1°C. The *British thermal unit* (Btu) is the amount of heat required to raise the temperature of 1 lb of pure water 1°F. Because calories and joules are both units of energy, there must be a definite relation between them, just as there is a definite relation between the two units of volume, gallons and quarts. Careful laboratory experiments in which various amounts of work energy are changed to heat energy give the relation

$$1 \text{ cal} = 4.19 \text{ joules}$$

Similarly,

$$1 \text{ Btu} = 778 \text{ ft-lb}$$

This conversion factor for the equivalence of work energy to heat energy is called the "mechanical equivalent" of heat or "Joule's equivalent," after James Prescott Joule, an Englishman who carried out numerous experiments in which he converted known amounts of mechanical energy into heat energy and determined that 4.19 joules of work will produce 1 cal of heat energy. It should be stressed that the mechanical equivalent of heat is more significant than just being a conversion unit, because it is not possible to convert 1 cal completely into 4.19 joules of work. This process of the conversion of heat to work is discussed more fully under the heading of "thermodynamics" later in this chapter.

SPECIFIC HEAT

Experiment shows that the same amount of heat will raise the temperature of equal masses of different substances by different amounts. This is expressed in terms of *specific heat,* which is the amount of heat required to raise the temperature of unit mass of a substance 1°. The definition of a calorie (heat required to raise the temperature of 1 g of water 1°C) means that the specific heat of water is 1 cal/g-°C. Likewise, in terms of English units, the specific heat of water is 1 Btu/lb-°F. Water has the highest specific heat (1.000 cal/g-°C) of all substances.

The specific heat of aluminum (Al) is $s_{\mathrm{Al}} =$ 0.22 cal/g-°C. This means that 0.22 cal of heat will raise the temperature of 1 g of Al 1°C, say from 20 to 21°C. Suppose it is desired to compute the amount of heat required to raise the temperature of 200 g of aluminum from 5 to 95°C. To raise 1 g of aluminum 1°C requires 0.22 cal. To raise 200 g of aluminum 1°C will require 200 × 0.22 cal. To raise the 200 g of aluminum 90°C (95°C − 5°C) will require ninety times as much heat as for 1°, so that the heat required is

$$0.22 \text{ cal/g-°C} \times 200 \text{ g} \times 90\text{°C} = 3960 \text{ cal}$$

This procedure is equivalent to saying that the heat Q required to raise the temperature of a mass m of a substance of specific heat s from an original temperature of t_1 to a final temperature t_2 is

$$Q = sm(t_2 - t_1) \tag{8.3}$$

read "heat gained or lost equals specific heat times mass times change in temperature."

Example 8.3
Compute the heat required to raise the temperature of a 2-kg aluminum cooking pan from room temperature, 20°C, to 100°C. The specific heat of aluminum is 0.22 cal/g-°C.

Solution
Heat gained = specific heat × mass × change in temperature

$$Q = sm(t_2 - t_1)$$
$$= 0.22 \text{ cal/g-°C} \times 2000 \text{ g} \times 80\text{°C}$$
$$= 35{,}200 \text{ cal} = 35.2 \text{ kcal}$$

The same amount of heat is given out by a body in cooling from a high to a low temperature as is required to raise the body from that same low temperature to the same high temperature.

The heat required to warm up a radiator in the home is furnished by the hot water from the furnace. Heat is then radiated to the room by the radiator, and the returning cooler water is reheated at the furnace and circulates back to the radiator. This is the principle of hot-water heating.

From these concepts of specific heat and temperature, it can be reasoned that cool water in a lake contains far more heat than is contained by a cup of hot coffee at a much higher temperature. The much greater mass of the lake more than offsets the higher temperature of the coffee in the relation $Q = sm(t_2 - t_1)$. The lake contains a greater total kinetic energy (heat) than the cup of coffee, but the coffee has greater "hotness" (temperature) because the average kinetic energy of the molecules in the cup of coffee is higher than that of the molecules in the lake. The student should now see more clearly the difference between heat and temperature. The amount of heat energy in an object depends on the mass of the object, on the material of which it is made, and on the temperature.

Change of state

In the preceding discussion only one result of heat being absorbed by a substance was mentioned—that is, a change in temperature. However, sometimes a substance goes through a *change of state;* for example, when a solid melts to form the liquid state or when a liquid vaporizes to form a gas, a change of state by the substance takes place but *with no accompanying change in temperature*. When a change of state takes place, heat is either absorbed or released but there is no change in temperature.

FUSION

Fusion is the change from the solid to the liquid state. This process, as in the melting of ice or iron, also requires heat. The *heat of fusion* is the heat required to change a unit mass of a solid to a liquid without change in temperature. For ice, the heat of fusion is 80 cal/g or 144 Btu/lb. A change of pressure has very little effect on the temperature of melting. All crystalline substances have definite melting points, and hence there is only one value of the heat of fusion for each crystalline substance.

In ice refrigeration, for every gram of ice that melts, 80 cal of heat is taken from the contents of the refrigerator, thus cooling the food. For this reason it is false economy to wrap ice used in a cooler in newspapers. When the ice is insulated it is kept from melting, but that in turns means that not enough heat is taken from the food, so spoilage may occur.

SUBLIMATION

Sublimation is the direct change from the solid to the gaseous state. Wet clothes hung outdoors when the temperature is below freezing will "freeze stiff" but later will be soft and dry. In this case the water molecules leave the surface of the ice and go into the air as water vapor. Furs and winter clothes are frequently protected from moths by placing balls of "para" (paradichlorobenzene) with them when they are packed away for the summer. In the fall, the para balls will be greatly reduced in size or may have entirely disappeared. However, the persistence of the odor of para proves that para molecules are present in the gaseous state. Because the boiling point of para is higher than any summer temperature, the para balls must have sublimated. Another common substance that undergoes sublimation is carbon dioxide (CO_2) in its solid state, commonly called Dry Ice. At room temperature and under normal atmosphere pressure Dry Ice turns directly to carbon dioxide gas and does not form the liquid state.

EVAPORATION

Evaporation, the change from a liquid to a vapor at the surface of the liquid, and *boiling,* the rapid change from a liquid to a vapor within the liquid, are often included in the more general term *vaporization*. In general, heat is required to change a substance from the liquid to the gaseous state. The *heat of vaporization* is the amount of heat required to change a unit mass of a substance from the liquid to the vapor state without change in temperature. In the case of water, the heat of vaporization is 538.7 cal/g (539 cal/g) at 100°C, 569 cal/g at 50°C, 585 cal/g at 20°C, and 595 cal/g at 0°C. Because the molecules of water of lower temperatures have a lower average kinetic energy than those of water at higher temperatures, it is to be expected that more heat will be required to vaporize the liquid at lower temperatures.

In a steam heating plant, for every gram of steam that condenses in the radiator—condensation takes place near 100°C except in vacuum systems—539 cal of heat is given up to the radiator, which in turn warms the air of the room. Of course, this same amount of heat is supplied by the furnace in changing water to steam in the boiler.

COMPUTATION OF HEAT WHEN BOTH CHANGES IN TEMPERATURE AND CHANGES IN STATE ARE INVOLVED

The earlier material on computing the amount of heat required to raise the temperature of a given mass of a substance through a given temperature range and the present dis-

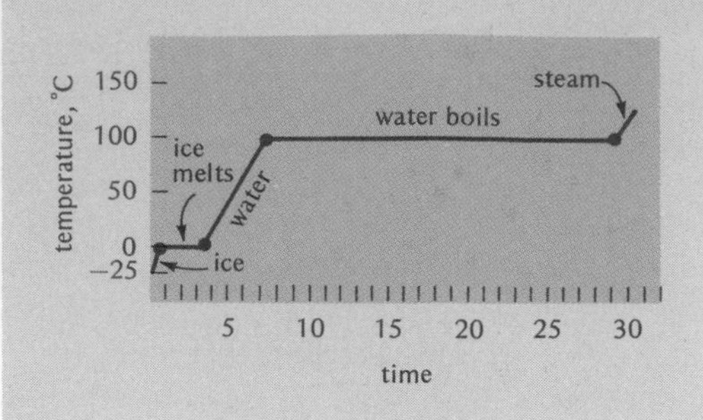

Figure 8.2
Graph of temperature versus time for 10 g of ice initially at −25°C, to which heat is added at a constant rate. The ice first rises in temperature, then turns to water, the water rises in temperature, then the water boils, and finally the water turns to steam.

cussion of heat of fusion and heat of vaporization may be combined to compute the amount of heat required to change a given mass of a solid at one temperature to a gas at a higher temperature when the specific heat, melting point, boiling point, heat of fusion, and heat of vaporization are known.

These latter quantities are, of course, determined experimentally for each substance by making a "heating or cooling curve." Such a curve is shown in Fig. 8.2 in which temperature is plotted against time when a block of ice at −25°C is heated at a constant rate. First, the ice warms and its temperature increases until the total block is at 0°C. Then the block of ice is observed to melt while the temperature remains at 0°C. Only after all the ice has melted will the temperature again begin to rise. At 100°C boiling occurs and again the temperature remains constant until all the water has turned to steam. Then as more heat is absorbed the temperature of the steam increases. In this experiment it is necessary to keep the substance being heated thoroughly mixed. This experiment is easily performed and illustrates that heat is absorbed by a substance and produces one of two effects: The temperature either rises or the substance changes from one state to another. Note on the graph in Fig. 8.2 that from the time required it takes about the same amount of heat to melt the ice as it does to raise the temperature of the water 100°C to the boiling point. However, it takes over five times as much heat to boil the water as it does to raise the temperature of the same quantity of water 100°C.

Example 8.4
Compute the heat necessary to change 100 g of ice at −30°C to steam at 120°C, given $s_{\text{ice}} = 0.5$ cal/g-°C, $s_{\text{water}} = 1$ cal/g-°C, $s_{\text{steam}} = 0.46$ cal/g-°C, heat of fusion of ice $= F = 80$ cal/g, heat of vaporization of water at 100°C $= V = 539$ cal/g. For purposes of simplification, it will be assumed that all the change from the liquid to the gaseous state takes place at the boiling point, although actually some of the water on being heated would evaporate at all temperatures between 0° and 100°C.

Solution
Ice at −30°C to ice at 0°C, where H = heat gained:

$$H = sm(t_2 - t_1)$$
$$= 0.5 \text{ cal/g-°C} \times 100 \text{ g} \times 30\text{°C} = \quad 1500 \text{ cal}$$

Ice at 0°C to water at 0°C:

$$H = Fm = 80 \text{ cal/g} \times 100 \text{ g} = \quad 8000 \text{ cal}$$

Water at 0°C to water at 100°C:

$$H = sm(t_2 - t_1)$$
$$= 1 \text{ cal/g-°C} \times 100 \text{ g} \times 100\text{°C} = \quad 10000 \text{ cal}$$

Water at 100°C to steam at 100°C:

$$H = Vm = 539 \text{ cal/g} \times 100 \text{ g} = \quad 53900 \text{ cal}$$

Steam at 100°C to steam at 120°C:

$$H = sm(t_2 - t_1)$$
$$= 0.46 \text{ cal/g-°C} \times 100 \text{ g} \times 20\text{°C} = \quad 920 \text{ cal}$$

total heat required = 74320 cal

Gas laws

CHARLES'S LAW

Gases, like solids and liquids, tend to expand on heating and contract on cooling. The expansion of gases is complicated, however, by the fact that the volume of a gas depends on its pressure as well as its temperature. Any gas, under constant pressure, contracts by 1/273 of its volume at 0°C for each degree Celsius lowering of temperature. Thus at −1°C, the volume of a gas has diminished by

Table 8.1
Gas volumes at different temperatures with pressure constant (for an ideal gas)

gas volume, cm³ (1)	Celsius temperature (2)	Kelvin temperature (3)	gas volume, cm³ (4)
293	20	293	586
273	0	273	546
272	−1	272	544
271	−2	271	542
73	−200	73	146
3	−270	3	6
1	−272	1	2
?	−273	0	?

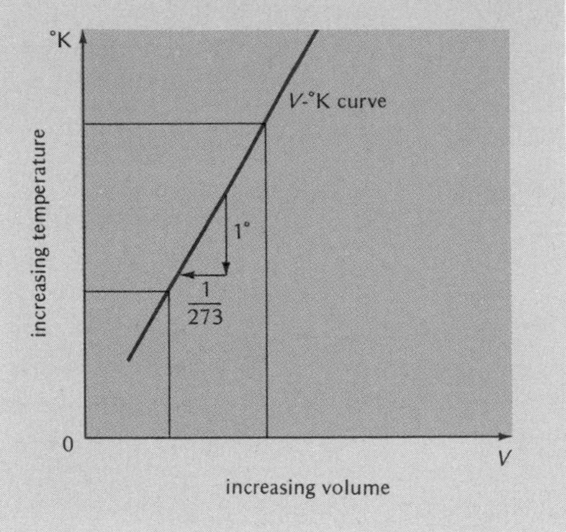

Figure 8.3
Graph of Charles's law, showing relation between the volume V of gas and its absolute temperature (°K) with pressure constant. As the temperature increases, the volume increases. The V vs °K "curve" is a straight line. For each 1°C or 1°K drop in temperature, the volume of gas contracts 1/273 of its volume at 0°C. Not drawn to scale.

1/273, or the volume at −1°C is 272/273 of its volume at 0°C. At −270°C, with no change in pressure, a given mass of an ideal gas whose properties are described in the next paragraph will have contracted by 270/273 of its original volume at 0°C, giving a final volume of 3/273 of its original volume at 0°C. At −272°C, the volume will be only 1/273 of its original volume at 0°C. One must not expect the gas to disappear at −273°C, but as the temperature is lowered more and more, the volume of the ideal gas approaches zero as the temperature approaches −273°C. For this reason, −273°C – actually −273.16°C or −459.6°F – is considered to be the lowest possible theoretical temperature – never to be quite reached experimentally. As mentioned earlier in this chapter, −273°C is taken as the starting point (0°K) on the absolute, or Kelvin, temperature scale, with Kelvin degrees equal in size to Celsius degrees.

The term "ideal gas" was used above because any actual gas would become liquefied long before reaching absolute zero. Experimental efforts to approach the value of absolute zero do not involve gases but use entirely different methods. By means of a demagnetization technique, experimenters have reached a temperature that is less than 0.0001°K from the theoretical absolute zero.

The relation between the volume of an ideal gas and its temperature, under constant pressure, is shown in Table 8.1, which gives the volumes at various temperatures of a gas that occupies 273 cm³ at 0°C. The values of volume in column 1 are directly proportional to the values of the absolute temperatures in column 3. If 546 cm³ (column 4) was the original volume at 0°C, instead of the 273 cm³ of column 1, it also works out that the volume of a given mass of gas, under constant pressure, is directly proportional to the absolute temperature; for example, 546 cm³/146 cm³ = 273°K/73°K. This relation holds for any original volume of the gas and is known as *Charles's law:*

Under constant pressure, the volume of a given mass of gas is directly proportional to the absolute temperature.

Charles's law, stated using symbols, is that if

$P =$ constant, $\quad V \propto T \quad$ in °K

Capital T denotes temperature in degrees Kelvin. This relation was discovered by Charles and later confirmed by Gay-Lussac. (See Fig. 8.3 for a graph of Charles's law.)

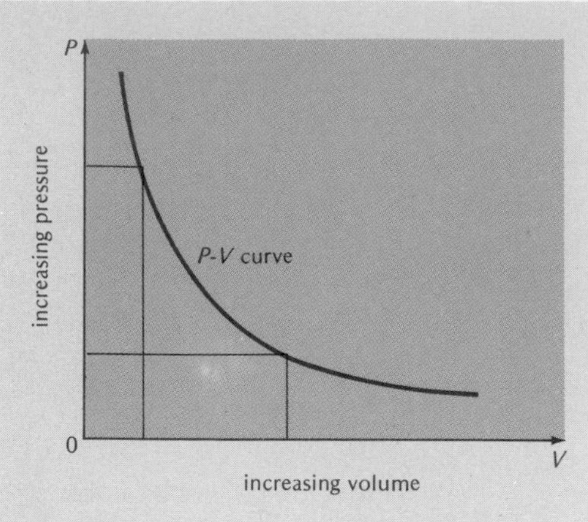

Figure 8.4
Graph of Boyle's law showing relation between the volume V of a gas and its pressure P with temperature constant. As the pressure increases, the volume decreases, and vice versa.

BOYLE'S LAW

In the expansion and contraction of a gas, the volume of the gas depends on its pressure. If the temperature of the gas is held constant, this relation is known as *Boyle's law:*

The volume of a given mass of an enclosed gas varies inversely with the pressure of the gas if the temperature is kept constant.

Boyle's law, using symbols, is that if

$$T = \text{constant}, \qquad P \propto \frac{1}{V}$$

(See Fig. 8.4 for a graph of Boyle's law.)

GAY-LUSSAC'S LAW

There is also a law, formulated by Gay-Lussac, to the effect that

If the volume of an ideal gas is constant, the pressure varies directly as the absolute temperature.

Gay-Lussac's law, using symbols, is that if

$$V = \text{constant}, \qquad P \propto T \quad \text{in °K}$$

(See Fig. 8.5 for a graph of Gay-Lussac's law.)

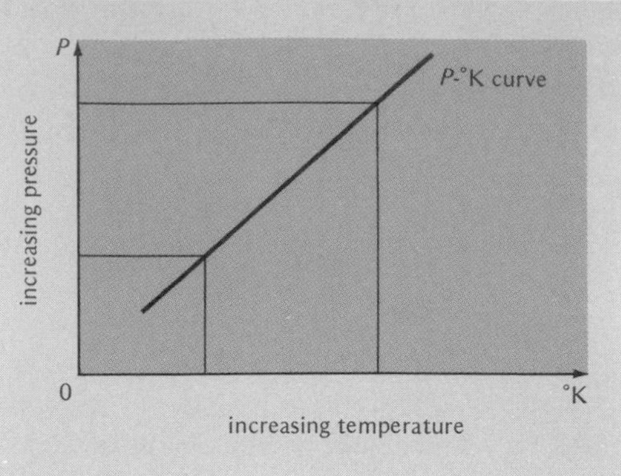

Figure 8.5
Graph of Gay-Lussac's law showing relation between the temperature (°K) of a gas and its pressure P with volume constant. The P vs °K "curve" is a straight line. As the temperature increases, the pressure increases.

GAS LAWS AND KINETIC THEORY

A real understanding of the relation we call Boyle's law may be arrived at by considering the application of the kinetic molecular theory to this relation.

In Fig. 8.6, left, the movable piston is considered to be weightless, so that the pressure down on the enclosed gas is 1 atmosphere (atm) due to the bombardment of the air molecules above the piston. This is just balanced by the pressure of 1 atm due to the bombardment of the gas molecules below the piston. In Fig. 8.6, right, a weight that increases the pressure down on the gas to 2 atm is added. The new volume, at right, is found to be one-half of the original volume at left.

In terms of the kinetic molecular theory, decreasing the volume to one-half means that the same number of molecules is now in one-half of the original space. Hence the time of travel between opposite walls, as top to bottom to top, is one-half as much, so that the time between hits of the same wall by molecules is one-half as much, or the frequency of the bombardment of that wall by the molecules is twice as great. Doubling the number of hits (the frequency of bombardments) doubles the force, and doubles the pressure because $p = F/A$. Thus the pressure is doubled when the volume is halved. This

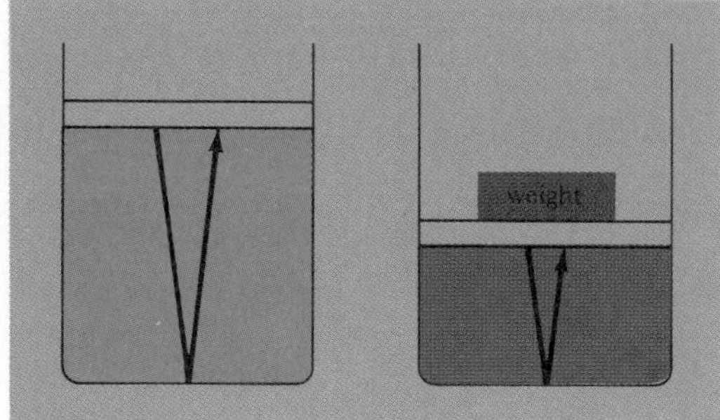

Figure 8.6
Boyle's law in terms of the kinetic molecular theory. At right, where the pressure of the enclosed gas is doubled, owing to the addition of the weight, the volume of the enclosed gas is half of what it was at the left. Conversely, at right, where the volume of the enclosed gas is half of what it was at left, owing to the addition of the weight, the pressure of the enclosed gas is doubled.

is assuming the average velocity of the molecules remains the same, which from the definition of temperature will be the case, since the temperature is to be constant. In the foregoing elementary discussion, the effect of the attraction of the molecules for one another and of the very small volume occupied by the molecules themselves has been neglected.

The kinetic molecular theory of gases can be used in explaining the pressure of a gas on the wall of an inflated rubber balloon. The pressure is due to the infinite number of impacts of molecules on the wall of the balloon. When a balloon is filled with more air, there are more molecules and more impacts and therefore greater pressure and distention of the rubber. According to Charles's law ($V \propto T°K$, if pressure is constant), the expansion of a rubber balloon when the gas in it is heated is due to the fact that the increased temperature produces greater kinetic energy of the molecules of gas. Hence there is increased velocity, more impacts, greater momentum associated with each impact and therefore greater pressure resulting in expansion of the rubber.

THE GENERAL GAS LAW

In experimental situations it is often extremely difficult to maintain a constant pressure, or temperature, or volume. If two or all of these change, one cannot use any of the three gas laws. Is there another mathematical relation of pressure, temperature, and volume?

It has been found experimentally, using an enclosed volume of gas containing a fixed amount of gas, that the product of the pressure P and volume V divided by the absolute temperature T is always equal to the same numerical constant k. For any two cases this could be expressed in the following way:

$$\frac{P_2V_2}{T_2} = \boxed{k \quad k} = \frac{P_1V_1}{T_1}$$

Mathematically, two things equal to the same thing are equal to each other; that is, if $A = C$ and $B = C$, then $A = B$. Therefore, by eliminating k from cases 1 and 2 above, we have the following equation, which is called *the general gas law:*

$$\frac{P_2V_2}{T_2} = \frac{P_1V_1}{T_1} \qquad (8.4)$$

The effect of both a change in pressure and a change in temperature on the volume of a gas is expressed in the above relation. Note also that (1) if the pressure remains constant ($P_2 = P_1$), the equation for the general gas law (Eq. 8.4) reduces to Charles's law, $V_2/V_1 = T_2/T_1$, on cancellation of P_2 and P_1; (2) if the temperature remains constant ($T_2 = T_1$), the equation for the general gas law reduces to Boyle's law, $V_2/V_1 = P_1/P_2$, on cancellation of T_2 and T_1; (3) if the volume remains constant ($V_2 = V_1$), the equation for the general gas law reduces to Gay-Lussac's law, $P_2/P_1 = T_2/T_1$, on cancellation of V_2 and V_1. All pressures must be in the same units, all volumes in the same units, and all temperatures in degrees Kelvin.

Therefore, in the solution of problems involving any of the gas laws, it is relatively simple to start with the equation for the general gas law, cancel any factor that is held constant, make the proper substitution of values for the factors that have changed, and solve the equation for the value of the unknown pressure, volume, or temperature.

The procedure described above is used in Examples 8.5, 8.6, and 8.7, which follow.

Example 8.5
A 500-cm^3 volume of hydrogen gas is at a temperature of 20°C and a pressure of 74 cm of mercury. Compute the volume of this gas if its temperature is increased to 40°C, the pressure remaining 74 cm of mercury.

Solution
Use the equation for the general gas law. Then, because the pressure remains unchanged, cancel P_2 and P_1 and substitute known values of volume and absolute temperatures for the first and second cases, as below:

$$T_2 = 273° + 40° = 313°\text{K}$$

$$T_1 = 273° + 20° = 293°\text{K}$$

$$V_1 = 500 \text{ cm}^3$$

$$\frac{\cancel{P_2}V_2}{T_2} = \frac{\cancel{P_1}V_1}{T_1}$$

$$\frac{V_2}{313°\text{K}} = \frac{500 \text{ cm}^3}{293°\text{K}}$$

$$V_2 = 313°\text{K} \times \frac{500 \text{ cm}^3}{293°\text{K}} = 534 \text{ cm}^3$$

Note that 40°C is double 20°C, but 313°K is not double 293°K.

Example 8.6
A 600-cm^3 volume of nitrogen gas is at a temperature of 25°C and a pressure of 75 cm of mercury. Compute the volume of this gas if its temperature remains at 25°C and the pressure is decreased to 73 cm of mercury.

Solution
Because the temperature is constant, cancel T_2 and T_1 in Eq. (8.4), leaving
$P_2V_2 = P_1V_1$

$$73 \text{ cm } V_2 = 75 \text{ cm} \times 600 \text{ cm}^3$$

$$V_2 = \frac{75 \text{ cm} \times 600 \text{ cm}^3}{73 \text{ cm}} = 616 \text{ cm}^3$$

Example 8.7
In a laboratory 200 cm^3 of oxygen gas is collected at a pressure of 73 cm of mercury and a temperature of 30°C. Compute the volume of this oxygen gas under standard conditions – that is, a pressure of 76 cm and a temperature of 0°C.

Solution

$$\frac{P_2V_2}{T_2} = \frac{P_1V_1}{T_1} \quad \text{or} \quad \frac{76 \text{ cm } V_2}{273°\text{K}} = \frac{73 \text{ cm} \times 200 \text{ cm}^3}{303°\text{K}}$$

$$V_2 = 200 \times \frac{73}{76} \times \frac{273}{303} = 173 \text{ cm}^3$$

Applications of kinetic theory

There are numerous phenomena in nature that can be explained on the basis of kinetic molecular theory. We shall discuss only a few here.

EVAPORATION

According to the kinetic molecular theory, the molecules in a liquid are constantly moving about and frequently colliding with one another. Occasionally some of the faster-moving molecules will break through the surface of the liquid into the space above, thus becoming molecules of vapor (gas), as in Fig. 8.7, top. *Evaporation* is the change from the liquid to the gaseous state due to the escape of faster-moving molecules from the surface of a liquid. If a cover is put on the vessel containing the liquid, there will soon be as many molecules of the vapor returning from the space to the liquid as there are molecules going from the liquid to the space (Fig. 8.7, bottom). When this condition of equilibrium is reached, the space above the liquid is said to be saturated.

The change from the liquid to the vapor state, more generally called *vaporization,* takes place faster as the temperature is raised. Finally, at a certain temperature called the *boiling point,* bubbles of the vapor form within the liquid and rise rapidly to the sur-

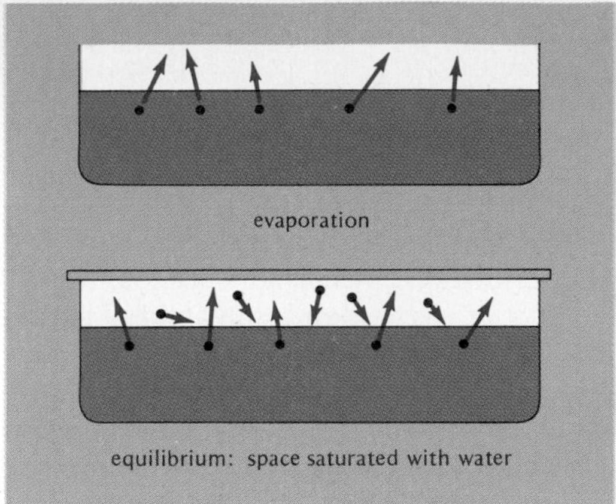

Figure 8.7
Evaporation.

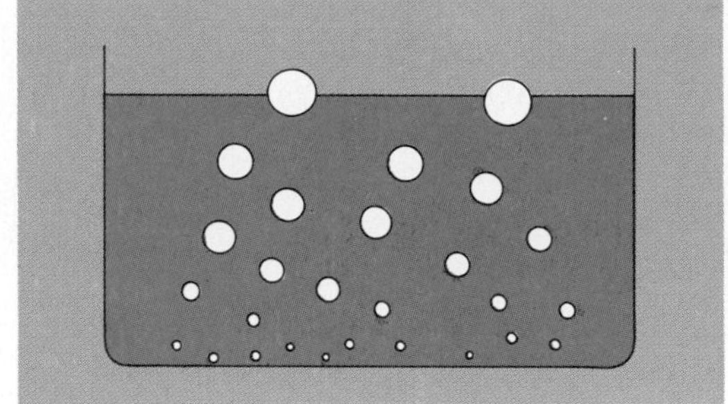

Figure 8.8
Boiling. The changing size of the bubbles is exaggerated somewhat.

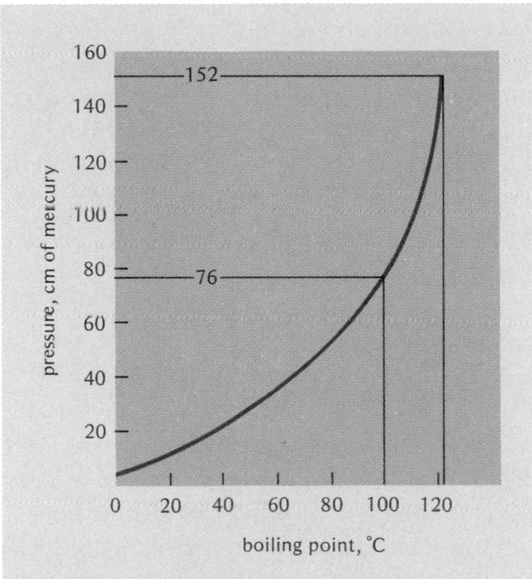

Figure 8.9
Boiling point of water at various pressures.

face; the liquid boils. Note that in accordance with Boyle's law the small bubbles formed at the bottom increase in size as they rise to levels of somewhat lower pressure (Fig. 8.8).

Boiling occurs when the pressure within the bubble of vapor becomes equal to—actually, slightly greater than—the pressure on the surface of the liquid. This vapor pressure within the liquid when boiling occurs is called *vapor tension*. Evidently the higher the pressure on the surface of a liquid, the greater the vapor tension and the higher the temperature of the boiling point. In the case of water, boiling occurs at 100°C at standard atmospheric pressure (76 cm of mercury), but the boiling point is lowered about 0.37°C for each centimeter of mercury lowering of pressure. Thus at 75 cm air pressure water boils at 99.63°C, at 73 cm air pressure it boils at 98.89°C, and at 77 cm air pressure it boils at 100.37°C. This variation of 0.37°C per centimeter mercury change of pressure holds only near 100°C. Figure 8.9 gives the boiling point of water for various pressures against the surface of the water.

This change of the boiling point with change of pressure has many important applications. The increase of boiling point with increased pressure is used in pressure cookers. As Fig. 8.9 indicates, a pressure of 2 atm (152 cm of mercury) gives a boiling point of about 120.5°C or 248.9°F. At this temperature, food cooks much more rapidly. In obtaining sugar from sap, it is necessary to boil off the water. To prevent scorching the sugar crystals, this is usually done under reduced pressure.

When evaporation takes place, it is the faster-moving molecules that leave the surface, so that the average kinetic energy of the remaining molecules is lower; that is, the liquid is cooler. Hence evaporation is a cooling process. A sick person with a high body temperature (103° or 104°F) is often given a sponge bath with alcohol. The rapidly evaporating alcohol takes heat from the body, usually reducing the temperature by 1° or 2°.

THE ATMOSPHERE ON PLANETS

In Chap. 3 a brief description of each planet was given. Referring to data in Table 3.1 and then combining the ideas of kinetic molecular theory with Newton's law of universal gravitation, it is possible to gain some insight into the conditions under which a planet can retain an atmosphere. The force of gravity holds the atmosphere to the planet. The strength of this force at the surface (the "surface gravity") of each planet is given in Table 3.1. According to kinetic molecular theory, the molecules are moving at high speeds in a random manner. If a molecule has a sufficiently high speed, it may escape the gravitational pull of the planet and be lost into space. If the temperature is raised on the planet, then the average speed of the molecules increases giving more molecules sufficient speed to leave the planet. Mercury is the hottest planet, and in addition, due to its small mass, it has a smaller surface gravity than the earth. These two factors apparently account for the fact that Mercury has little or no atmosphere.

The moon has no atmosphere. It has temperatures not too different from those on earth but the surface gravity on the moon is about one-sixth that of the earth; thus the moon has insufficient surface gravity to hold an atmosphere.

Venus, which is similar to the earth in size, has a surface gravity of about 0.8 that of the earth, but it is much hotter on Venus. Most of the atmosphere on Venus is made up of carbon dioxide, and it is questionable whether Venus could hold lighter gases such as helium, hydrogen, or even oxygen or nitrogen in its atmosphere, because at the temperature found on Venus these molecules would have sufficient speed to leave the planet.

Mars has a surface gravity of less than 0.4 that on the earth and similar to that on Mercury. However, Mars is able to hold a small amount of gas to form a thin atmosphere because it is cooler on Mars than on Mercury or on earth.

All of the large planets have sufficient surface gravity and low enough temperatures to hold an atmosphere.

The preceding discussion is a rather sketchy analysis of a difficult problem. It should illustrate to the student that there is a delicate balance between surface gravity, temperature, and the mass of the gas molecules, factors that determine whether or not a planet can hold an atmosphere. Such an analysis may explain why certain gases, particularly the light gases such as helium and hydrogen, do not exist in the atmosphere of certain planets. On the other hand these gases may never have existed on a given planet.

Thermodynamics

Mechanical energy may be transformed into heat energy, for example, when we rub two objects together or when we drill a hole in an object. According to kinetic molecular theory when heat is absorbed by a substance, the individual molecules will move more rapidly—that is, their individual kinetic energies will increase. Often this kinetic energy of the individual molecules is called *internal energy*. The relation between work, heat, and internal energy is of fundamental importance in both physical and natural science. A branch of physics known as *thermodynamics* has developed from the analysis of phenomena involving transformations between thermal and other forms of energy, particularly mechanical energy. Any given situation may be thought of as a system plus its surroundings. Thermodynamics is concerned with macroscopic, or large-scale, properties (pressure, volume, temperature) of the system and its surroundings.

FIRST LAW OF THERMODYNAMICS

The principle of conservation of energy (Chap. 6) may now be extended to include the first law of thermodynamics, which is the application of the principle to thermal processes. We shall assume that all systems possess internal energy. We define an increase in the internal energy of any system as equal to the sum of the amount of heat, measured in mechanical units, that flows into the system from its surroundings and the external work done on the system.

The following equation represents the first law of thermodynamics (where ΔE is the change in internal energy of the system in ergs; QJ the heat absorbed by the system in ergs or joules, with Q the resultant heat in calories and J the mechanical equivalent of heat $= 4.19 \times 10^7$ ergs/cal or 4.19 joules/kcal; and W the work done on the systems in ergs or joules):

increase in internal energy = heat *absorbed by* system + work *done on* system

$$\underset{\text{ergs}}{\Delta E} = \underset{\text{ergs}}{+QJ} + \underset{\text{ergs}}{+W} \tag{8.5}$$

If heat is absorbed by the system, QJ is positive; and if work is done on the system, W is positive. Similarly, if the system cools, QJ is negative; and if the system does work, W is negative, and there is a decrease in internal energy; that is, ΔE will be negative.

The first law of thermodynamics is just an expression of the conservation of energy. However, the first law of thermodynamics expressed by Eq. 8.5 has severe restrictions placed on it in natural processes. These restrictions are expressed by the second law of thermodynamics.

SECOND LAW OF THERMODYNAMICS

It is clear that work can be transformed completely into heat or internal energy. Such a process can be considered as the changing of orderly motion of molecules into random motion of molecules. For example, when you hit a nail with a hammer, the orderly motion of the molecules in the hammer moving in unison toward the nail is converted into random thermal energy of the molecules in both the nail and in the hammer. The nail may also move doing some work. However, eventually all of the orderly mechanical energy has been converted into random kinetic energy, that is, thermal energy. Such a process obeys the first law and is covered by Eq. 8.5. However, consider the reverse process of transforming thermal energy into work. The conversion factor J tells how much heat or thermal energy is equivalent to how much mechanical work, but there is a slight problem. How do you get all of the molecules that have random, disordered motion (thermal or internal energy) to get together and all push in the same direction to produce work or mechanical energy? The problem is one of completely ordering the motion of the molecules. Such processes do not occur naturally. For example, suppose we heat a gas such as steam in a cylinder that has a piston at one end. As the gas expands, the piston will move and do work. It is not possible to convert all of this heat energy into work because in the heating process the thermal energy, or internal energy of the steam, will increase and the temperature of the steam will increase. It is not possible in this process to convert all of the heat into work leaving the temperature and internal energy of the steam unchanged. In such a process the first law of thermodynamics is still obeyed but it is restricted in that the heat is converted into work and some internal energy. The conversion of heat to work can never be 100 percent efficient. This is due to the problem of trying to order the motion of the many molecules that are moving randomly.

These observations are summarized in one of several statements of the second law of thermodynamics:

In any energy transformation some of the energy is transformed into thermal energy which is no longer available for further energy transformation.

Table 8.2
Humidity values

temperature, °C	grams of water vapor to saturate 1 m^3 of air
−5	3.4
0	4.8
5	6.8
10	9.3
15	12.7
20	17.1
25	22.8
30	30.4
35	41.0

This statement of the second law expresses what is observed in nature; that is, in energy transformations eventually all energy will end up as random thermal energy resulting in what has been called the "heat death of the universe."

Another process that satisfies the first law but does not occur in nature is the following: Consider a bucket of water containing molecules that have random kinetic energy, thus giving the water a large amount of internal energy. Why is it that we cannot take the internal energy out of the water, thereby cooling the water, and use this energy to do work. We can do this with a machine, but this process does not happen naturally. A careful examination of other processes in nature leads to a different, but equivalent, expression of the second law of thermodynamics:

It is impossible for a system or machine unaided by an external source of energy to cause heat to flow from a cooler to a warmer region, or from one body to another body that is at higher temperature.

This law is an assumption based on wide and varied experience. For example, it is impossible for a refrigerator unaided by a motor to cause heat to flow from the cooler air and food inside the refrigerator to the warmer air outside. Heat tends to go from a hotter body to a cooler body.

The connection between the two statements of the second law of thermodynamics lies in the fact that natural processes move from more orderly motion of molecules to greater disorder in the motion of molecules—that is, to greater internal energy. It is possible to create an increase in the order of molecules that is equivalent to cooling the substance, but it requires work or energy from outside to increase this order.

Humidity and relative humidity

The constant evaporation of water from such sources as oceans, lakes, and rivers furnishes a huge amount of water vapor to our air. Condensation of this water vapor, in such forms as rain, snow, hail, and dew, returns water vapor from the atmosphere back to the surface of the earth. The amount of water vapor required to saturate the air increases considerably as the temperature is raised.

Table 8.2 gives the number of grams of water vapor that 1 m^3 of air is capable of containing when saturated; the graph of Fig. 8.10 is obtained by plotting these values and may be used for in-between values.

This means that if 1 m^3 of air saturated at 20°C could be squeezed dry like a sponge, 17.1 g of water would come from it. A similar procedure at 0°C would yield only 4.8 g of water. The air of a living room 20 by 12 ft and 8 ft high when saturated with water vapor at 68°F contains a little over 2 lb of water (more than 1 qt).

Relative humidity (RH) is the ratio of the actual amount of water vapor (WV) present in the air at a given temperature to the maximum amount that the air is capable of containing at that temperature:

$$\text{RH} = \frac{\text{amount of WV present}}{\text{amount of WV for saturation}}$$

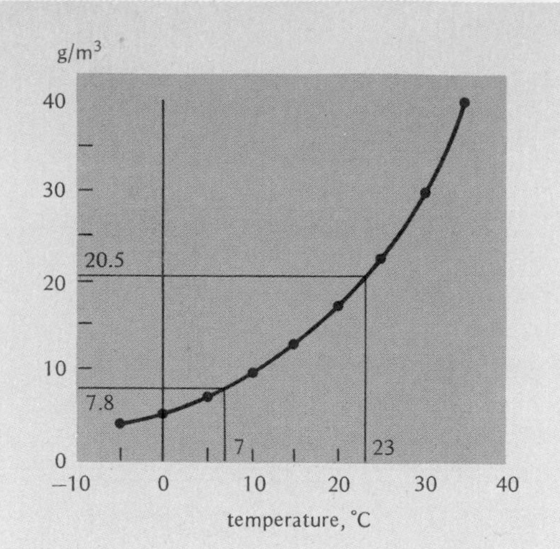

Figure 8.10
Humidity graph. Grams of water vapor to saturate 1 m³ of air at different temperatures.

The actual amount of water vapor in the air can be determined by passing a measured volume of air—a gas meter may be used—through a glass tube containing dry calcium chloride ($CaCl_2$). Because calcium chloride absorbs water vapor, the difference in weight before and after the air is passed through the tube gives the amount of water vapor contained by the air. For example, if the increase in weight is 93 g when 10 m^3 of air at 20°C is passed through the tube, there is 9.3 g/m^3 of air. In this case the relative humidity is

$$\mathrm{RH} = \frac{\text{amount of WV present}}{\text{amount of WV for saturation}} = \frac{9.3}{17.1}$$

$$= 0.54 = 54 \text{ percent}$$

The value of 17.1 used in the denominator is obtained from the humidity table.

If the air mentioned in the above experiment is cooled, the relative humidity will increase, and finally, at 10°C, the 9.3 g/m^3 contained is the amount necessary for saturation. Any further cooling results in the air having more water vapor than it is capable of holding, in which case the excess water vapor condenses out as liquid water (dew) that forms on surfaces present.

The *dew point* (DP) is the temperature at which the air becomes saturated with the amount of water vapor present, and below which excess moisture condenses out as dew. "Room temperature" (RT) is used in Eq. 8.6 instead of "that temperature" because determinations of relative humidity are usually made in a room. Relative humidity (RH) may now be defined,

$$\mathrm{RH} = \frac{\text{amount of WV present at DP}}{\text{amount of WV for saturation at RT}} \qquad (8.6)$$

or

$$\mathrm{RH} = \frac{\text{amount to saturate at DP}}{\text{amount to saturate at RT}}$$

The "condensation" method of measuring dew point and relative humidity is basic and can be used by anyone to determine relative humidity in the home. Any container with a bright surface—a tin can with the paper removed will do—is half-filled with water. Small pieces of ice are added and the water stirred until a film of moisture appears on the surface. This film means that the dew point has been passed and condensation from the air that has been cooled by the cold surface has started. The temperature at which the film just starts to appear, as measured by a thermometer in the water being stirred, may be taken as the dew point. A more accurate determination is obtained by averaging the temperatures of appearance and disappearance of the film. The latter temperature may be obtained by stirring in a small amount of warm water, a little at a time, until the film starts to disappear. The relation in Eq. 8.6 is then used to compute relative humidity. The "amount of water vapor present" is obtained from the humidity table as the amount to saturate at the dew point, and the humidity table is also used to obtain the "amount of water vapor for saturation" at the room temperature.

Example 8.8
What are the dew point and relative humidity for the air of a home when moisture starts to condense on the outside surface of a bright can at 7°C, the room temperature being 23°C?

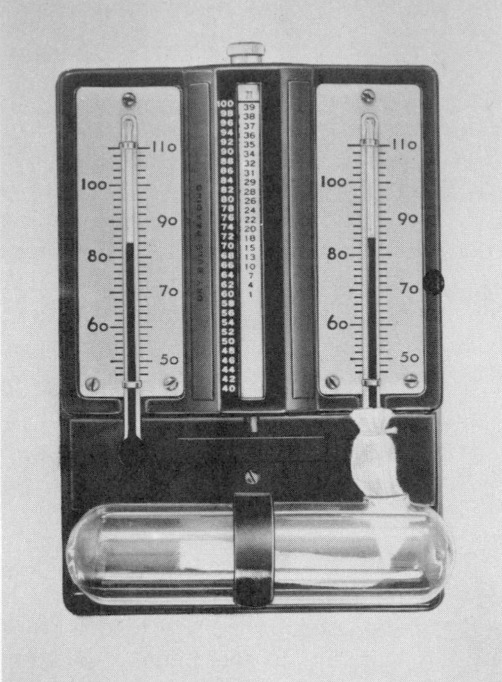

Figure 8.11
Wet and dry bulb hygrometer. There is no water in the reservoir, so both thermometers indicate the same temperature. (Courtesy of the Taylor Instrument Division, Sybron Corporation.)

Solution
Because only the temperature of "appearance of the condensation film" is given, it is assumed that 7°C is the dew point; actually 7° is slightly low. To determine the relative humidity (RH) we use the formula:

$$\text{RH} = \frac{\text{amount of WV present}}{\text{amount of WV for saturation}}$$

$$\text{RH} = \frac{\text{amount to saturate at dew point (7°C)}}{\text{amount to saturate at room temperature (23°C)}}$$

$$= \frac{7.8 \text{ g/m}^3}{20.5 \text{ g/m}^3} = 0.38 = 38 \text{ percent}$$

The student will note that the value of 7.8 as the amount of water vapor to saturate the air at the dew point 7°C is obtained from the curve of Fig. 8.10. Dashed lines are drawn in on that graph to show how this value is obtained. Similarly the value 20.5 is obtained from the curve.

The fact that evaporation is a cooling process leads to a simple method of determining relative humidity. A wet-and-dry-bulb hygrometer (Fig. 8.11) consists of two similar thermometers, one of which is surrounded by a wick kept wet by a container of distilled water. If the surrounding air is saturated, both thermometers will register the same temperature. If the air is not saturated, the cooling produced by evaporation from the wick surrounding the "wet" thermometer causes this thermometer to read lower than the "dry" thermometer. Experiments have been carried out to determine the relative humidity for all combinations of "wet" and "dry" thermometer readings, and these data are assembled in tabular form for easy use. Such a table shows that for a "dry" thermometer reading 20°C and a "wet" thermometer reading of 15°C, the relative humidity is 59 percent.

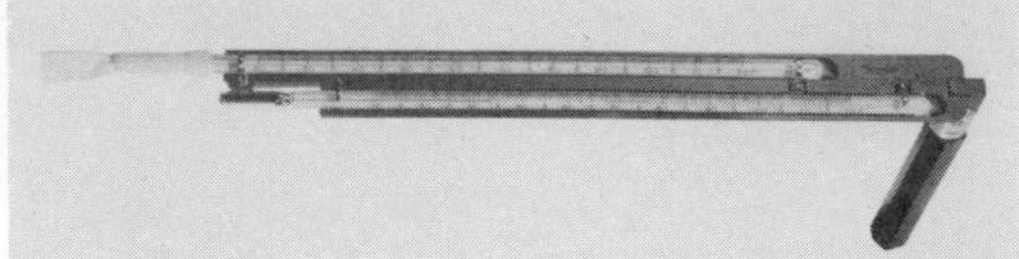

Figure 8.12
Sling psychrometer. (Courtesy of the Taylor Instrument Division, Sybron Corporation.)

It should be noted that the wet-bulb reading is not the dew point. If a wet-and-dry-bulb hygrometer is used, air against the thermometers should be kept in motion by fanning to prevent the air in the immediate vicinity of the wet bulb from becoming more nearly saturated than the air in general.

Another form of hygrometer, which automatically takes care of air motion, is the sling psychrometer (Fig. 8.12). The sling psychrometer is whirled to give the necessary relative motion of air.

Relative humidity is very important in connection with body health and comfort. For best body health and comfort, relative humidity should be kept within the range 35 to 60 percent. For values higher than 60 percent the net evaporation of perspiration from the body is reduced to such an extent that we lose most of the cooling effect due to evaporation. In summer we are then very uncomfortable. We speak of hot, muggy weather and say, "It isn't the heat, it's the humidity." On the other hand, if the air in a home is too dry in winter, we are cooled by too much evaporation and feel chilly even though the thermometer may read 73° to 75°F. We feel as warm at 73°F and relative humidity of 50 percent as at a temperature of 76°F and a humidity of 25 percent. Physicians agree that if the relative humidity in a home is too low, we are more susceptible to colds because of the excess evaporation of moisture from the membranes of the mouth, nose, and throat. On the other hand, too high a relative humidity indoors in winter will result in considerable condensation of moisture on windows with a consequent fogging up, and often sufficient condensation to affect the paint on window sills and to ruin window drapes.

The fact that warm air is capable of containing much more moisture than cold air, as shown by the humidity table, means that air heated in a home in winter will be too dry for best body health and comfort unless water vapor is added to the air by some method. Suppose winter air at 0°C has a relative humidity of 80 percent and is heated by a furnace up to 20°C. If no water vapor is added, the relative humidity is $(0.8 \times 4.8)/17.1 = 3.84/17.1 = 0.22$, or 22 percent. This is as dry as the air over the Sahara Desert.

Various types of humidifiers have been devised to add water vapor to the air. One of the simplest is a water pan attached to the side of a hot-air furnace. In another type, water drips on and runs along a metal trough that extends across the hot-air chamber at the top of the furnace. In winter, several quarts of water are used daily by these humidifiers in bringing the relative humidity up to the desired 40 to 50 percent.

In summer, the reverse difficulty is experienced. Air in homes, particularly in basements, has too high a relative humidity, and moisture then condenses on cold-water pipes. Leather and other materials will mildew. The cold coils of dehumidifiers cool the air sufficiently to condense some of the moisture from the air. Various air conditioners not only cool the air but cause condensation of excess moisture on the cold coils of the units.

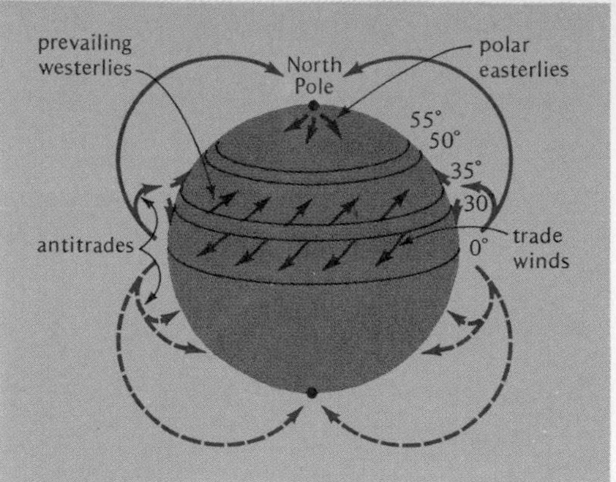

Figure 8.13
Principal winds.

Winds

Winds are huge convection currents of the earth's atmosphere. Figure 8.13 shows the circulation of the principal winds. The air above the earth's equator is heated, on the average, more than the air above any other portion of the earth. The warm air rises above the equator and flows both north and south from the equator. These high-level winds are called the *antitrades*. Most of this air cools and returns to the surface between latitudes 30° and 35°, some flowing back toward the equator as the *trade winds* and some away from the equator as the *prevailing westerlies*.

In order to understand wind directions, it is necessary to keep in mind the fact that the rotation of the earth causes projectiles, including moving molecules of gases that make up winds, to deviate to the right in the Northern Hemisphere and to the left in the Southern Hemisphere. Applying this idea to winds, Fig. 8.13 shows that air of the antitrades returning to the surface at latitude 30°N will flow not directly south toward the equator but, owing to deflection to the right, toward the southwest (from the northeast). Likewise air that flows away

Figure 8.14
Hurricane Gladys. Remarkable photograph taken about 150 miles southwest of Tampa, Florida. Altitude 97 nautical miles. Hurricane winds were at a speed of 80 knots when photograph was taken from *Apollo 7*. (NASA.)

from the equator is deflected to the right, making the predominating direction of the surface winds in the north midlatitudes from the west (or slightly south of west). Some of the air rising at the equator descends near the poles, and the deviation to the right gives us the *polar easterlies* in the Northern Hemisphere.

There is a narrow air stream, called the *jet stream,* located 6 to 8 miles above the earth's surface over the northern midlatitudes and traveling eastward at velocities of up to 250 mi/hour. This stream has northward and southward dips, which, as they move eastward, cause cooling or warming over large sections of the United States. Many weather stations now send up weather balloons and use a system called "radiosonde" for measuring temperatures and relative humidity at various levels of the upper air. By following the flight of the balloons with a small telescope, the direction and speed of the winds at the various levels can also be recorded. Such data have indicated the existence of several more or less parallel jet streams.

This type of upper-air data, including the location of the jet stream, not only has greatly improved the accuracy of day-to-day forecasts but has enabled the U.S. Weather Bureau to issue monthly forecasts that, although necessarily of a general nature, are fairly accurate. Five-day forecasts of probable temperature and precipitation are now given for most areas. These are quite reliable and are used by many organizations whose operations or products are affected by temperature or humidity or precipitation.

Other very helpful aids to weather forecasting are the photographs taken regularly from weather satellites orbiting the earth. These photographs clearly show the cloud formations over the earth. By examining a sequence of photographs taken at different times, the movement of the clouds can be readily seen. An excellent example of a useful weather photograph taken from a satellite is shown in Fig. 8.14.

LOWS AND HIGHS

The effect of sun shining on such areas as the sands of the desert of southern Nevada and southeastern California is to heat the earth, and the air above it, more than the surrounding areas, which are covered with vegetation. The heated air, being less dense than the surrounding cooler air, rises. This causes a lower than normal pressure at the center of the desert area, with gradually increasing pressures outward from the center. A large area of lower than normal pressures is called a *cyclone,* or *low*. Figure 8.15 shows the isobars, or curves of equal pressure, for a typical low. The pressures are given in terms of inches of mercury "corrected to sea level." The *L* indicates the center of the low, or low-pressure area. Remembering that projectiles, including

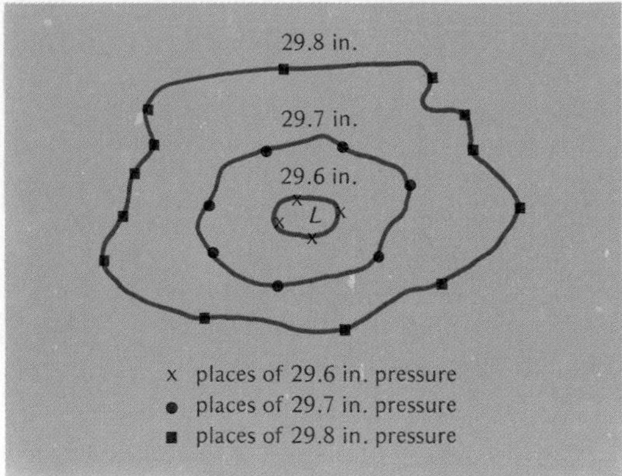

Figure 8.15
Isobars for a typical low, *L*.

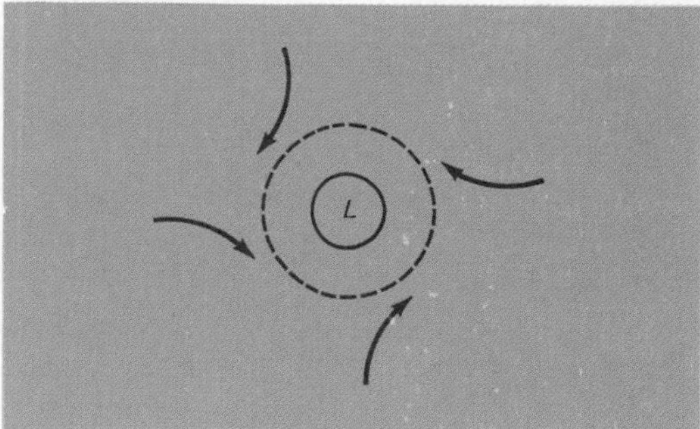

Figure 8.16
Counterclockwise circulation around a low, *L*.

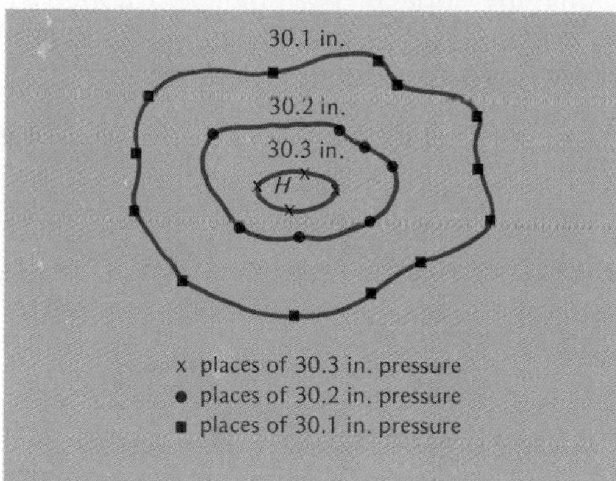

Figure 8.17
Isobars for a typical high, *H*.

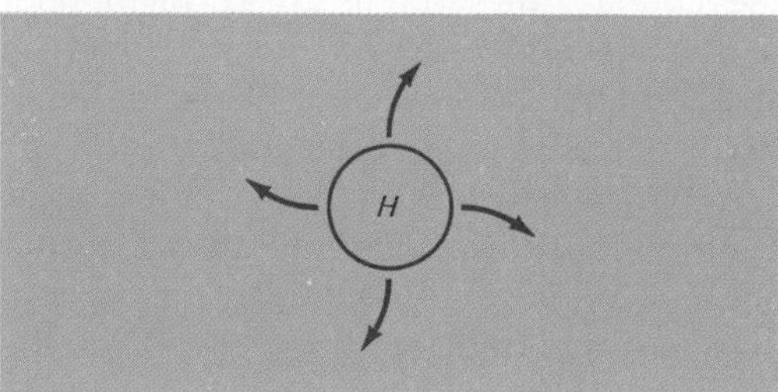

Figure 8.18
Clockwise circulation around a high, *H*.

moving air molecules, are deflected to the right in the Northern Hemisphere, it follows that surface air, which rushes along the surface toward the center of low pressure, becomes a counterclockwise rotation of air, as shown in Fig. 8.16.

Air rising from some heated areas means that air will be returning to the surface at other areas. The effect of a column of air descending over a region is to increase the pressure at the center of this area. Such a large area with pressures higher than normal is called an *anticyclone,* or *high* (Fig. 8.17). The *H* indicates the center of the high, or high-pressure area. As the descending air reaches the surface, it moves outward in all directions. As shown in Fig. 8.18, the "deflection to the right" results in a clockwise circulation of air around a high.

In summary, the movement of air for a low is inward, upward, and counterclockwise, and for a high downward, outward, and clockwise.

LOCATING OF LOWS AND HIGHS

There are more than 500 weather stations in the United States, and additional data are obtained from Canada, Mexico, and ships off the Pacific and Atlantic Coasts. By plotting on a map the barometric readings,

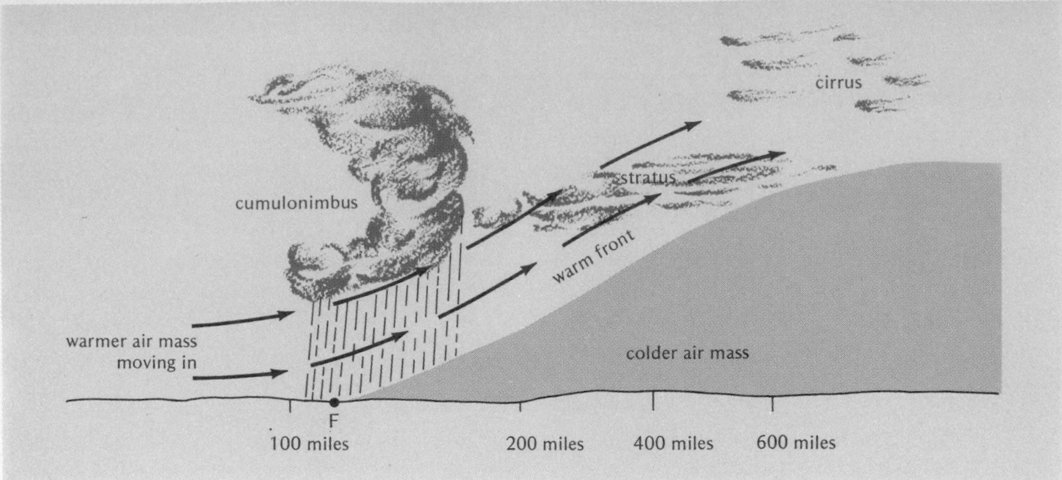

Figure 8.19
Warm front.

corrected to sea level, reported by these stations and then connecting points of equal pressure, the isobars of Figs. 8.15 and 8.17 are obtained, and the positions of the lows and highs are thus known.

WEATHER FRONTS

The flow of air due to circulation around highs and lows brings masses of warm air into contact with masses of cold air. The area of contact of two air masses is referred to as a *weather front*. If warm air moves into a region of cold air, the less dense warm air rises over the more dense cold air, and we have a *warm front*. Figure 8.19 shows a cross section of a warm front, with the surface front at *F* running perpendicular to the cross section. As the diagram shows, cirrus clouds may form as much as 600 miles in advance of a front, owing to condensation of moisture on dust particles resulting from the cooling of the warmer air as it rises above the cold air and expands. This condensation is much greater nearer the front, and towering cumulonimbus clouds, or rain clouds, form, so that rain often falls in advance of the front, frequently as much as 100 to 150 miles, and may occur back of the front, up to 50 or 100 miles.

If cold air moves into a region of warm air, the cold air, being more dense, moves along the surface, pushing the warmer air to higher levels and causing condensation of moisture on dust particles. Figure 8.20 shows a cross section of a cold front. Again, rain may fall both in advance of the front and behind it, as well as along the front.

If the boundary between the mass of warm air and the mass of cold air shows very little motion, we speak of a *stationary front*.

WEATHER MAPS AND FORECASTING

In general the highs and lows and weather fronts move west to east over the United States, although the motion is sometimes northeast or southeast, or even occasionally north or south. The average velocities across the country are usually 20 to 30 mi/hour and generally faster in winter than in summer, but occasionally the highs, lows, and fronts slow up considerably, even becoming stationary for a day or a few days. Figure 8.21 shows simple weather maps for successive days, illustrating the typical eastward progress of weather across the country.

A student with the material of this chapter as a background should readily interpret weather maps carried in daily papers similar

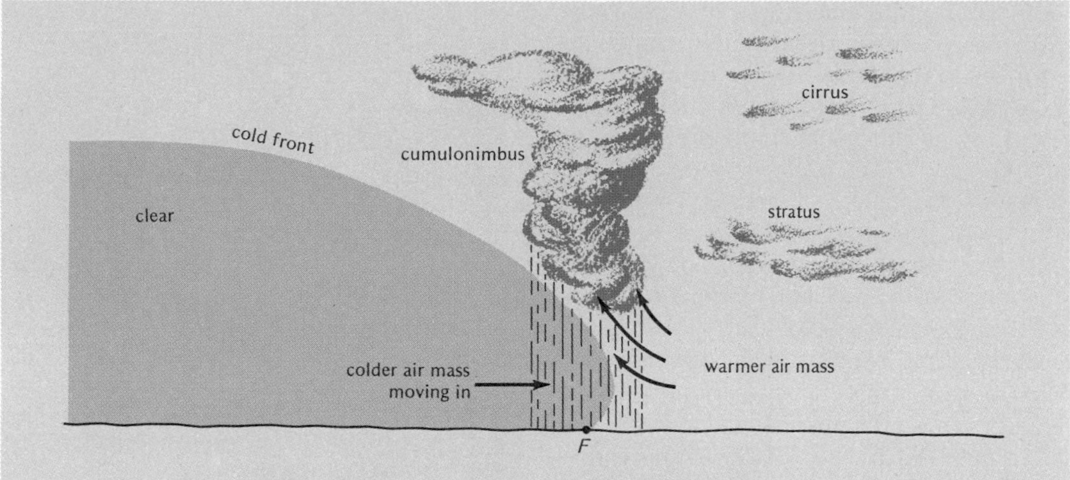

Figure 8.20
Cold front.

to those in Fig. 8.21. Perhaps the student will be intrigued sufficiently to try a weather forecast or two on his own. For this, a knowledge of the kinds of precipitation and their causes will be useful.

Condensation and precipitation

KINDS OF CONDENSATION

On clear nights during spring, summer, and fall, considerable heat is radiated from the earth into space, making the surface of the earth cooler than the atmosphere in contact with it. Hence the air in contact with the earth's surface is cooled. If this cooling carries the temperature of the air below the dew point, moisture from the air condenses as dew on blades of grass, leaves, etc.

If the dew point is below 32°F and the air is cooled below this dew point, the excess moisture in the air condenses in the solid state and frost is formed. Water is not formed as an intermediate step in the formation of frost, so frost cannot be considered frozen dew.

When the air not only at the earth's surface but also for several feet above the surface is cooled below the dew point and dew is formed, excess moisture condenses on minute dust particles in the air, and fog is formed. When the sun shines in the early morning, the air is warmed above the dew point, and the minute water droplets evaporate into the air.

Air at high levels is often cooled, owing to expansion on rising or to coming in contact with colder air. If this cooling carries the temperature below the dew point, the excess moisture condenses on dust particles and a cloud is formed. Thus the main distinction between fog and cloud is a matter of elevation. What is thought of as a cloud enveloping a mountain top is found by a mountain climber to be a fog.

KINDS OF PRECIPITATION

Under certain conditions various kinds of liquid or solid water, H_2O, form and fall to the earth's surface as precipitation. The common forms of precipitation are rain, snow, hail, and sleet.

When rapid condensation of water vapor on dust particles occurs high above the earth's surface, droplets large enough to fall toward the surface are formed. Many of these small droplets probably coalesce (collide and stay together), forming larger droplets that fall to the earth's surface as raindrops. The rapid condensation is usually due to cooling by the expansion of rising air or to cooling of warm humid air by mixing with colder layers of air.

If this rapid condensation takes place below freezing, snow flakes are formed without liquefaction as an intermediate process.

Sometimes, in falling toward the earth's surface, raindrops pass through a layer

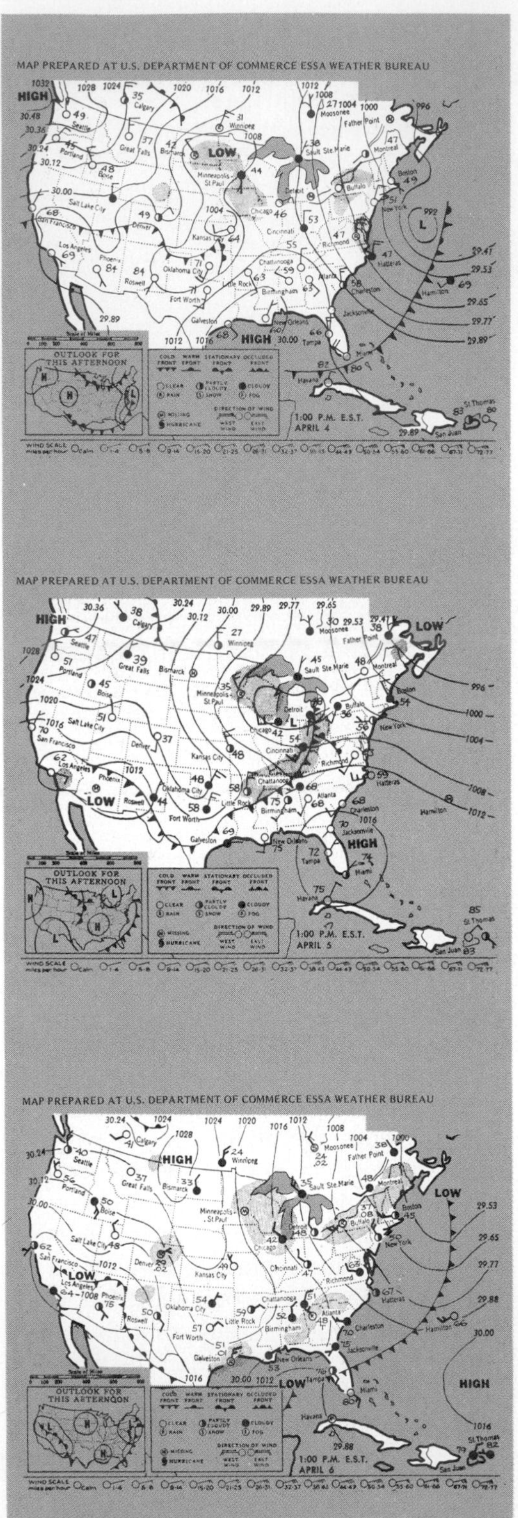

Figure 8.21
Daily weather maps. Note how, in general, the high, lows, weather fronts, and precipitation areas move across the country from west to east on successive days. Figure beside station circle indicates current temperature (°F); a decimal number beneath temperature indicates precipitation in inches during the 6 hours prior to time shown on map. Cold front: A boundary line between cold air and a mass of warmer air, under which the colder air pushes like a wedge, usually advancing southward and eastward. Warm front: A boundary between warm air and a retreating wedge of colder air over which the warm air is forced as it advances, usually northward and eastward. Stationary front: An air mass boundary that shows little or no movement. Occluded front: A line along which warm air has been lifted from the earth's surface by the action of the opposing wedges of cold air. This lifting of the warm air often causes precipitation along the front. Shading on the map indicates areas of precipitation during the 6 hours prior to time shown. Isobars (solid black lines) are lines of equal barometric pressure and form pressure patterns that control air flow. Labels are in millibars and inches. Winds are counterclockwise toward the center of low-pressure systems, and clockwise and outward from high-pressure areas. Pressure systems usually move eastward, averaging 500 miles a day in summer, 700 miles a day in winter.

that is below freezing, and sleet is formed. A theory currently held by many meteorologists is that, in most instances of rain, snow or ice crystals are first formed at elevations where the temperature is below freezing and then melt to rain in falling through warmer air.

The air that rises rapidly in the "chimney" of a cumulonimbus cloud sometimes carries raindrops to such an elevation that they freeze. They may then fall to the earth as hail, or they may fall back into the warm, moist air, receive a coating of moisture, and again be buffeted into the higher, colder regions. This may happen several times and accounts for the concentric layers of ice found in large hailstones, which in cross section exhibit concentric layers like onions.

The average annual rainfall over the entire earth is about 33 in., varying from 0.02 in. in Arica, Chile, to 426 in. in Khasi Hills, India. Because there is equilibrium between total precipitation and evaporation, it follows that the average annual rate of evaporation from the entire earth is also 33 in. To evaporate this much water requires a tremendous amount of energy, from the sun, and just as much energy is emitted near the earth as vapor condenses to water or snow. Some of this energy under certain conditions sets up a very fast rotation of air, producing tornadoes.

SUMMARY

Heat is the total kinetic energy of the random motion of the individual molecules of a body. Temperature is a measure of the average kinetic energy of the random motion of the individual molecules of a body.

A thermometer is an instrument for measuring temperature. In the United States the Fahrenheit scale is most frequently used on thermometers for measuring temperatures though scientific laboratories are increasingly using the Celsius (centigrade) scale. Most thermometers are constructed employing the fact that most substances expand on heating and contract when cooled. The lowest possible temperature is called "absolute zero" and is −273°C or −459°F; this is the zero point on the Kelvin temperature scale.

The calorie, kilocalorie, and British thermal unit are units of heat. Work (or mechanical) energy can be converted into heat energy with a conversion relation (the mechanical equivalent of heat) of 4.19 joules/cal and 778 ft-lb/Btu.

The same amount of heat will raise the temperature of different substances by different amounts. The amounts of heat necessary to raise the temperature of the same amounts of different substances 1° is expressed in terms of specific heat. Water has the highest specific heat of all substances.

When heat is absorbed or given up by a substance, either of two types of processes can occur: The temperature can change or a change of state can occur. Such processes as melting, boiling, sublimation, condensation, and evaporation are examples of changes of state.

The volume of a given mass of an enclosed gas varies inversely with the pressure of the gas if the temperature is kept constant. This is known as Boyle's law, and it can be explained in terms of the kinetic molecular theory. Usually the pressure, temperature, and volume of an enclosed gas change. The effect of the changes is expressed in the general gas law. The kinetic molecular theory can be used in explaining evaporation, boiling, vaporization, fusion, expansion, and such complicated questions as under what conditions can an atmosphere exist on a planet.

Relative humidity is the ratio of the amount of water vapor present in the air to the amount that the air is capable of containing at that temperature. Relative humidity is an important factor in body comfort and health.

Winds are huge convection currents of the earth's atmosphere. Heat and the rotation of the earth influence the movements of large masses of air. In general, masses of air move eastward across the United States as lows and highs. The area of contact between them is a weather front. Changes of weather usually accompany lows, highs, and fronts.

Water vapor can condense as dew or frost, fog or cloud, rain, and snow. The common forms of precipitation are rain, snow, hail, and sleet.

Important words and terms

heat
temperature
thermometer
caloric
British thermal unit
Fahrenheit temperature scale
Celsius temperature scale
Kelvin temperature scale
absolute zero
specific heat
change of state
fusion
sublimation
evaporation
boiling
Boyle's law
Charles's law
ideal gas
ideal gas law
first law of thermodynamics
second law of thermodynamics
internal energy
mechanical equivalent of heat
relative humidity
dew point
condensation
hygrometer
jet stream
trade winds
warm front

Questions and exercises

1. Distinguish between heat and temperature. How are they alike? How are they different?
2. Is −40°F the same temperature represented by −40°C?
3. Explain in terms of molecular motion the expansion of solids, liquids, and gases on heating.
4. Evaporation is said to be a cooling process; that is, the liquid remaining is cooler. Explain in terms of molecular motion.
5. How does perspiration provide the body with a means of cooling itself?
6. A soldier stationed near the Arctic Circle washes his trousers and hangs them out to dry. They freeze quickly, but after a few hours the ice disappears and the trousers are dry. Explain.
7. Ice on the sidewalk disappears even though no one walks on it and the temperature does not rise above the freezing point. Explain.
8. A steam burn is usually more serious than a hot-water burn. Explain.
9. How does the kinetic theory explain the fact that gases exert pressure on all surfaces with which they come into contact?
10. In the Rocky Mountains at, say, 12,000 ft above sea level it is impossible to cook potatoes in an open container no matter how long you boil them. Why? How does a pressure cooker overcome this problem?
11. If a piston is pushed rapidly into a container of gas, what happens to the kinetic energy of the molecules of the gas? What happens to the temperature of the gas?
12. One explains Brownian motion of large smoke particles as being due to the bombardment of the smoke by the randomly moving, very much smaller, air molecules. The air and the smoke must be at the same temperature. What must be true of the relative speeds and the relative kinetic energies of the smoke particles and the air molecules? Discuss.
13. A sample of hydrogen gas is compressed to half its original volume while its temperature is held constant. What happens to the average speed of the hydrogen molecules?
14. On a very hot day in your home you decide to cool off your kitchen by opening the refrigerator door and closing all the kitchen doors and windows. Would this process cool off the kitchen? Explain why or why not.
15. If you put a pinhole in a helium-filled balloon, the helium diffuses throughout the whole room. How does this illustrate the second law of thermodynamics?
16. Describe how the first law of thermodynamics

can be applied to explain (a) the warming of a metal poker in a fire, (b) boiling away of water in an open pan on a stove.

17. The ocean contains a tremendous amount of heat energy. (a) Under what conditions could a ship use this heat to propel it? (b) How would this be done? (c) What would be the change in the temperature of the water? (d) Does this violate the laws of thermodynamics?
18. What determines whether or not a planet will retain its atmosphere?
19. Very often on a household barometer are written the words, "Stormy," "Unsettled," "Fair," etc. How are these words related to the atmospheric pressure? What type of atmospheric pressure is generally associated with fair weather?
20. Why is a cold front characterized by more violent weather change than a warm front?
21. Condensation and precipitation products of water vapor: Using a horizontal line to represent the freezing point of water, complete a diagram by using labeled lines to indicate the proper relation for (a) condensation below and above the freezing point of water; (b) condensation on objects and in air; (c) dew, rain, frost, and snow; (d) rain, sleet, and hail.

Problems

1. Most humans have a normal body temperature of 98.6°F. Convert this temperature (a) to centigrade; (b) to absolute.

Answer: (a) 37.0°C; (b) 310.0°K.

2. Mercury, which is a liquid at ordinary temperatures, freezes at −39°C. Compute the freezing point of mercury on the Fahrenheit scale and on the absolute.
3. Convert room temperature (72°F) and the temperature of a very hot day (100°F), to centigrade and to absolute.

Answer: 22.2°C, 295.2°K; 37.8°C, 310.7°K.

4. It requires 490 calories to melt 10 g of a certain substance at its melting temperature. What is the heat of fusion of this substance?
5. Suppose that you heat 100 g of room-temperature water in a small beaker and 1000 g of room-temperature water in a large beaker until the temperature in each is 60°C. (a) How much heat was put into each beaker of water? (b) Do the water molecules in the large beaker now move as fast as do those in the small one?
6. How many calories are needed to heat 1 g of water from its freezing point to its boiling point?
7. A bathtub contains 100,000 g of water at 25°C. How much water at 60°C must be added to provide a hot bath of 40°C?

Answer: 75,000 g.

8. A cylinder contains 100 cm^3 of air at a total pressure P. What will the total pressure become if (a) the volume is reduced by half to 50 cm^3 while the temperature remains constant? (b) the volume is doubled to 200 cm^3 while the temperature remains constant? (c) the volume is doubled and the absolute temperature is also doubled? (d) the volume is halved and the absolute temperature is doubled?
9. (a) Compute the volume of air, starting at 1 atm of pressure (14.7 lb/in.^{2}), which would be required to fill a 10-ft^3 tank with air to a pressure of 8 atm. (b) What will a pressure gauge, which measures excess of inside pressure over outside pressure, register for this compressed air?

Answer: (a) 80 ft^3; (b) 7 atm or 102.9 lb/in.^{2}.

10. Ten liters of air at 20°C is confined in a vertical cylinder by a piston resting on top of the gas. As the air in the cylinder is heated, the piston rises, thus keeping the pressure constant. Compute the temperature in degrees centigrade to which the enclosed air must be heated to increase the volume to 20 liters.

Answer: 313°C.

11. On a day when the temperature is 19°C and the barometer registers 74 cm of mercury, a balloon is inflated with helium gas to a volume of 100,000 ft^3 (at local atmospheric pressure) before being released from the earth's surface. After the balloon has risen for about 3.5 miles, the new atmospheric pressure is 37 cm of mercury and the temperature is −10°C. Compute to the nearest cubic foot the volume occupied by the helium gas inside the balloon at the higher level.
12. A 200-cm^3 volume of oxygen gas is collected in a vertical graduated tube by displacement

of water. The water level inside and outside the tube is the same, which means that the pressure of the enclosed oxygen gas is the same as the outside atmospheric pressure. In the laboratory a thermometer registers 25°C and the barometer registers 73 cm of mercury. Compute the volume of the oxygen gas under standard conditions (0°C and 76 cm of mercury).

13. Compute the amount of heat required to raise the temperature of a 3-lb electric flatiron from 75° to 275°F, given that the specific heat of the iron is 0.12 Btu/lb-°F.

Answer: 72 Btu.

14. Compute the heat radiated by an aluminum kettle weighing 3000 g in cooling from 99° to 19°C, given that the specific heat of aluminum is 0.22 cal/g°C.

Answer: 52,000 cal.

15. Compute the heat required to change 1000 g of ice from −40°C to steam at 150°C, given: $s_{ice} = 0.5$ cal/g-°C; $s_{steam} = 0.46$ cal/g-°C; heat of fusion for water = 80 cal/g; heat of vaporization for water at 100°C = 539 cal/g. Assume that all the vaporization takes place at 100°C.

Answer: 762,000 cal.

16. In an experiment to determine by the condensation method the relative humidity of the air in a room where the temperature is 22°C, a film of moisture starts to appear on the outside of a bright can at 6°C. A little warm water is added, and the film starts to disappear at 10°C. (a) Determine the dew point; (b) compute the relative humidity.

Projects

1. Relative humidity

In the determination of dew point and relative humidity in your residence or laboratory, first determine the dew point of the air by the condensation method. Note the room temperature at the time.

Using values from a humidity table (Table 8.2), plot a large humidity graph similar to Fig. 8.10. Sheets of graph paper may be obtained at most bookstores. Show on the graph how the number of grams of water vapor to saturate 1 m³ of air was determined for the dew point and for the room temperature.

Compute the relative humidity. Would a person remaining in the room probably be comfortable or uncomfortable?

Make a second, entirely separate determination of dew point and relative humidity for a different room or for the same room at a considerably later time.

Describe the method that is used in your residence to add water vapor to the air in the winter.

2. Weather predictions

a. For one week, record the predicted weather, maximum temperature, minimum temperature, all as given on television, radio, newspapers or telephone, and the observed weather, maximum temperature, and minimum temperature.

If rain, or snow, occurs when not predicted or does not occur when predicted, encircle the prediction and call it a miss. If either the maximum temperature or the minimum temperature differs more than 5° from the predicted values (5° from the middle value, if a range is given; for example, treat 48° through 52° as a middle value 50°) call it a miss and encircle the prediction that was wrong. Divide the number not missed by total number of predictions to get the percentage of accuracy. Predictions of the U.S. Weather Bureau are accurate about 85 percent of the time. Persons tend to remember those times when the predictions proved to be wrong.

b. Follow the weather maps in a daily newspaper for a week. State whether you noticed a usual west to east progress of highs, lows, fronts, and precipitation across the United States.

Supplementary readings

Textbooks

Ashford, Theodore A., *The Physical Sciences: From Atoms to Stars,* Holt, Rinehart and Winston, New York (1967). [In Chap. 7 we find the kinetic molecular theory of gases and its extension (1) to real gases, liquids, solids, and solutions; (2) as a theory of heat; and

(3) by use of Avogadro's hypothesis to obtain the relative weights of molecules.]

Bates, D. R. (ed.), *The Earth and Its Atmosphere,* Science Editions, New York (1961). (Paperback.) [The circulation of the atmosphere and meteorology are treated in Chaps. 9 and 11.]

Battan, Louis J., *The Nature of Violent Storms,* Anchor Books, Doubleday, Garden City, N. Y. (1961). (Paperback.) [Topics include clouds, air motions and the forces that produce them; thunderstorms, tornadoes, hurricanes, and cyclones.]

Battan, L. J., *Harvesting the Clouds,* Anchor Books, Doubleday, Garden City, N. Y. (1969). (Paperback.) [A discussion of recent advances in weather modification.]

Blanchard, D. C., *From Raindrops to Volcanoes,* Anchor Books, Doubleday, Garden City, N. Y. (1967). (Paperback.) [A look at raindrops and their origin and sea surface meteorology.]

Conant, James Bryant, *Robert Boyle's Experiments in Pneumatics,* Harvard University Press, Cambridge, Mass. (1950). (Paperback.) [A case history, documenting the experiments that led to the discovery of Boyle's law.]

Cowling, T. G., *Molecules in Motion,* Harper Torchbooks, Harper & Row, New York (1960). (Paperback.) [An introduction to the kinetic theory of gases. Chapter 1 treats the early days of the atomic theory, including Boyle's law. In Chap. 2 heat and the gas laws are discussed.]

Edinger, J. G., *Watching for the Wind,* Anchor Books, Doubleday, Garden City, N. Y. (1967). (Paperback.) [A discussion of the seen and unseen influences on local weather.]

Fisher, Robert Moore, *How About the Weather?* (rev. ed.) Harper & Row, New York (1958). [A popular book that gives the reader scientific bases for weather and shows him how to follow the weather by interpreting newspaper weather maps as well as by watching the sky and wind. An appendix has good forecasting guides for the layman.]

Holton, Gerald, and Duane H. D. Roller, *Foundations of Modern Physical Science,* Addison-Wesley, Reading, Mass. (1958). [The kinetic theory of matter and heat are brought together in one conceptual scheme in Chap. 25.]

Keenan, Charles W., and Jesse H. Wood, *General College Chemistry,* (4th ed.), Harper & Row, New York (1971). [Kinetic theory and the gas laws are discussed in Chap. 8; calculations involving the gas laws in Chap. 9; the second law of thermodynamics in Chap. 16.]

Kemble, Edwin C., *Physical Science: Its Structure and Development,* vol. 1, Massachusetts Institute of Technology Press, Cambridge, Mass. (1966). [Heat and the conservation of energy are discussed in Chap. 14; the second law of thermodynamics in Chap. 15; the kinetic molecular theory of matter and heat in Chap. 16.]

MacDonald, D. K. C., *Near Zero,* Anchor Books, Doubleday, Garden City, N. Y. (1961). (Paperback.) [The physics of very low (near absolute zero) temperatures.]

McCue, J. J. G., *Introduction to Physical Science: The World of Atoms,* Ronald Press, New York (1963). [Heat and the theory of gases are discussed in Chap. 14.]

Reiter, E. R., *Jet Streams,* Anchor Books, Garden City, N. Y. (1967). (Paperback.)

Roller, Duane, *The Early Development of the Concepts of Temperature and Heat,* Harvard University Press, Cambridge, Mass. (1950). (Paperback.) [A case history of the development of the concepts of temperature and heat.]

Scott, Ewing C., and Frank A. Kanda, *The Nature of Atoms and Molecules: A General Chemistry,* Harper & Row, New York (1962). [The kinetic molecular theory of matter and the gas laws are discussed in Chap. 2.]

Sutton, O. G., *The Challenge of the Atmosphere,* Harper & Row, New York (1961). [Climate and weather; the atmosphere in motion; the physics of clouds and rain, weather-producing systems; hurricanes, tornadoes, and thunderstorms; forecasting.]

Articles

Johnson, A. W., "Weather Satellites II," *Scientific American,* **220,** no. 1, 52 (January, 1969). [A description of the network of weather satellites that report daily on the atmospheric conditions around the earth.]

Kuroiwa, D., "A Life of Snow," *The Physics Teacher,* **7,** no. 1, 13 (January, 1969). [A description of snow crystals.]

Lounasmaa, D. V., "New Methods for Approaching Absolute Zero," *Scientific American,* **221,** no. 6, 26 (December, 1969).

Roberts, W. O., "We're Doing Something About the Weather," *National Geographic,* **141,** no. 4, 518 (April, 1972).

Wheatley, J. C., and H. J. Van Till, "Attaining Low Temperatures," *The Physics Teacher,* **8,** no. 2, 67 (February, 1970). [An excellent review of the meaning of absolute zero and how to experimentally attain ultra low temperatures.]

Zemansky, M. W., "The Use and Misuse of the Word 'Heat' in Physics Teaching," *The Physics Teacher,* **8,** no. 6, 294 (September, 1970).

WAVE MOTION, SOUND, AND LIGHT

The light in the world comes principally from two sources—the sun, and the student's lamp.

BOVÉE, 1842

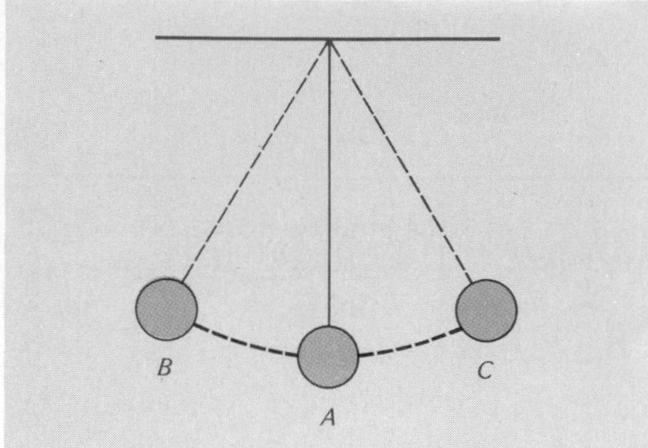

Figure 9.1
Vibrating pendulum.

The concepts of waves and wave motion are involved in explanations of the transmission of sound and of light. Because light, as well as other forms of electromagnetic waves, such as radio and television waves, which will be discussed in the latter part of this chapter, and sound are forms of wave motion, it is possible to treat all these types of waves simultaneously by discussing general properties of waves.

General characteristics of waves

VIBRATION

When a pendulum bob that has been pulled away from its equilibrium position of rest (point *A* of Fig. 9.1) to a point *B* is released, it will start toward the original position of equilibrium. As it approaches the original position of equilibrium, its inertia carries it past that position, and if it were not for friction, it would continue to a point *C*, an equal distance vertically above the position of equilibrium. The pendulum bob would then retrace its path to *B*, and the entire motion would be repeated indefinitely. Because of friction, the bob does not move quite as high above the position of equilibrium on successive swings, so that the motion soon dies out.

The type of motion just described is called a *vibration*. A *complete vibration* consists of the path from one extremity to the other and back again (path *BACAB* of Fig. 9.1). The time required to make one complete vibration is the *period* (T). The *frequency* (f), which is the number of vibrations per second, can always be obtained by the relation $f = 1/T$. Thus if the time required for one vibration is 3 sec, the frequency $f = 1/T = \frac{1}{3}$ vib/sec; if the time for one vibration is $\frac{1}{5}$ sec, the frequency $f = 1/T = 1/\frac{1}{5} = 5$ vib/sec. Of course, if $f = 1/T$, $fT = 1$ and $T = 1/f$. Hence if the frequency is 10 vib/sec, the period will be

$$T = \frac{1}{f} = \frac{1}{10 \text{ vib/sec}} = \frac{1}{10} \text{ sec/vib}$$

The period—that is, the time for one vibration—is $\frac{1}{10}$ sec. Usually the units of the period T are given as seconds. The *amplitude* of vibration is the maximum displacement from the position of rest (*AB* or *AC* in Fig. 9.1).

WAVES AND WAVE MOTION

If a clothesline is fastened at one end and held taut, a jerk upward by the hand holding it will cause a wave form or wave pulse to travel the length of the rope (see Fig. 9.2). If the hand is moved up and down several times, thus causing an up and down vibration of the part of the rope being held, a series of waves will be seen to move along the rope. This leads to the following definition of wave motion:

A wave motion is a propagated disturbance in a deformable medium. The disturbance (wave pulse in Fig. 9.2) travels through the medium or material (the rope in Fig. 9.2). The speed of the wave through the material depends on the nature of the material.

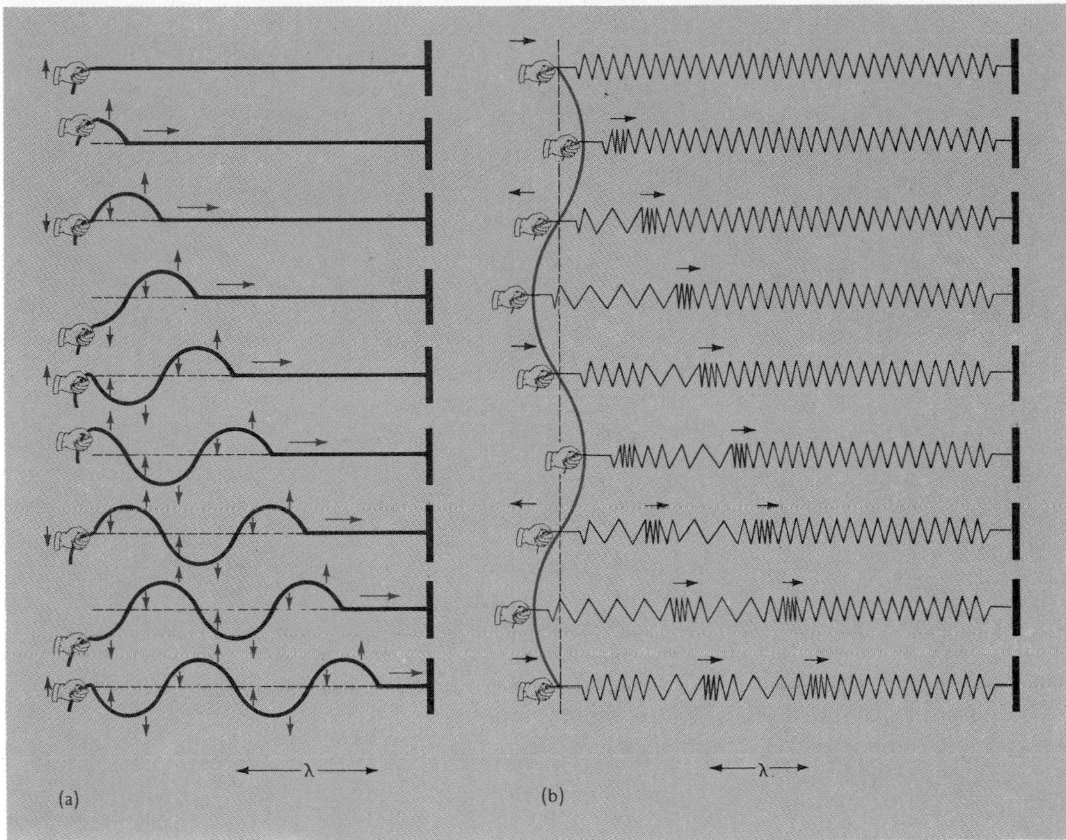

Figure 9.2
Motion of a wave pulse. (a) A transverse wave on a rope. The particles of the medium, the rope, vibrate at right angles to the direction in which the wave (disturbance) is propagated. (b) A horizontal wave on a spring. The particles of the medium, the spring, vibrate back and forth along the same direction in which the wave (disturbance) is propagated. In both examples the wavelength λ is the distance between two adjacent, similar points on the wave.

This definition can also be stated:

A wave motion is the passing on of a wave form from particle to particle of the medium without the individual particles of the medium moving onward; each particle, however, vibrates about a fixed point in the path of the wave motion.

There is one important exception to this definition—namely, electromagnetic waves, which can travel through empty space and require no medium to support them.

TYPES OF WAVE MOTION

There are two kinds of simple wave motion, transverse and longitudinal. In *transverse*

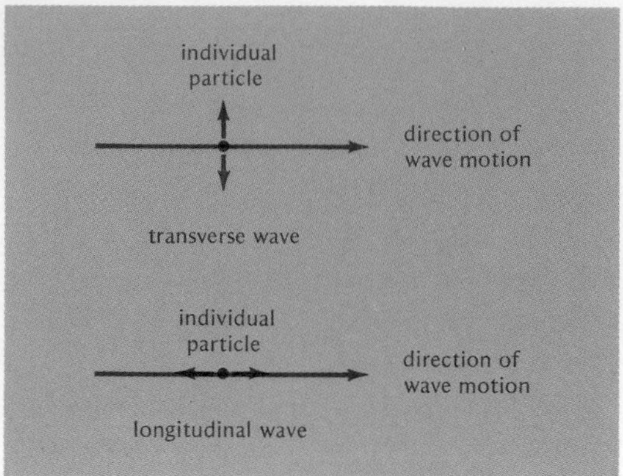

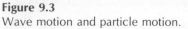

Figure 9.3
Wave motion and particle motion.

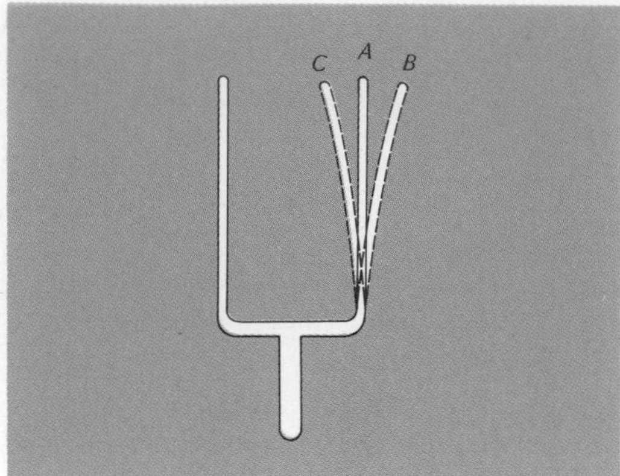

Figure 9.4
Vibrating prongs of a tuning fork.

wave motion, the individual particles of the medium vibrate at right angles to the direction of the wave propagation (Fig. 9.3, top). In the case of the rope the particles of the rope vibrated up and down vertically while the wave form moved forward horizontally. All electromagnetic waves, including light, radio waves, television waves, microwaves, etc., are transverse waves.

If several coils of a stretched spring are pinched together and then released, a wave form—in this case a compression—may be seen to travel along the spring (see Fig. 9.2). As the wave travels onward, the individual particles of the spring move forward and then backward along the direction in which the wave moves. This is an example of the second type of wave motion, or *longitudinal wave motion,* in which the individual particles of the medium vibrate forward and backward parallel to the direction of the wave propagation (Fig. 9.3, bottom). Sound waves propagating through air are perhaps the most common example of a compressional, longitudinal wave motion.

There are numerous other examples of wave motion that are observed in nature. Some of these wave motions are complex and cannot be classified as simple longitudinal or transverse motion. Water waves are an example of a more complex motion. Here the water particles (molecules) rotate in a circular motion as the wave disturbance passes through the water. Sound waves in a solid material such as rock in the earth can be both longitudinal and transverse.

SOUND WAVES IN AIR

When a tuning fork is struck, the prongs swing in and out together. Consider only the right-hand prong in Fig. 9.4. As the prong swings from its equilibrium position *A* to one extremity *B,* it pushes molecules of air ahead of it. Thus the region just ahead of the forward-moving prong temporarily contains many more than the normal number of air molecules; this is called a *compression* or *condensation.* Because there are more than 1 billion billion molecules in a cubic centimeter of air, it is easy to understand that when the forward motion of the prong drives additional millions of molecules into the cubic centimeter just ahead of the prong (to the right of the prong in Fig. 9.4), these molecules will collide with molecules already present, causing them to be pushed forward. Thus additional molecules arrive in a space farther ahead of the forward-moving prong, and these in turn, by collision, push other molecules forward (to the right in Fig. 9.4). In this manner the condensation moves rap-

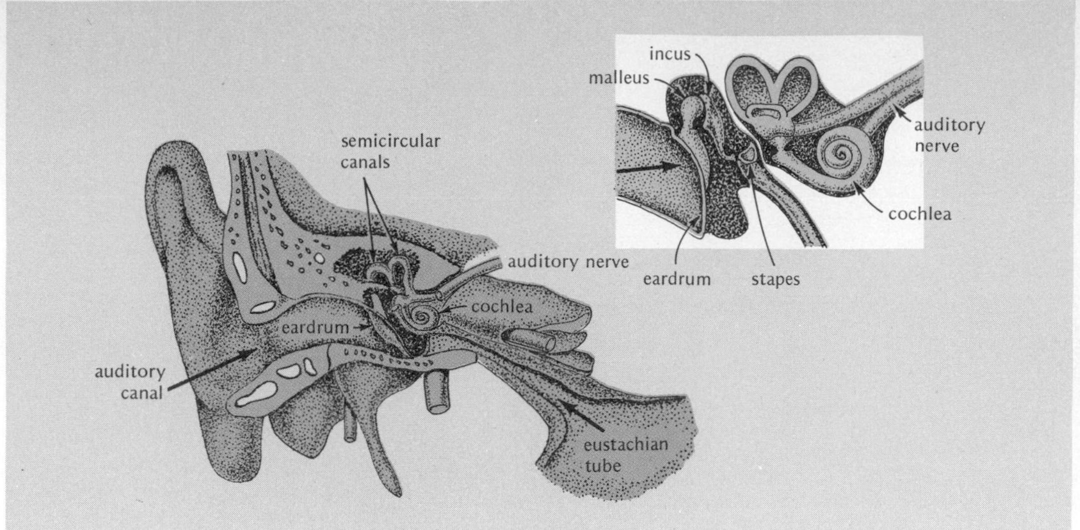

Figure 9.5
The human ear. The small drawing shows how sound waves (arrow) strike the eardrum, moving three small bones—malleus, incus, and stapes—which pass the vibrations to the inner ear. There, cells in the cochlea send sensations of hearing to the auditory nerve.

idly outward from the source of sound, the vibrating prong of the tuning fork, producing a longitudinal wave.

The prong then swings from *B* back to *C*. Immediately to the right of the prong (region *BC*) there are fewer molecules than normal per unit volume. This is a partial vacuum and is called a *rarefaction*. Air molecules rush in from all directions to equalize the pressure. Thus molecules from the region to the right of *B* rush into the volume *BC*, because this region to the right of *C* is a rarefaction. Molecules of air rushing into this new area of reduced pressure cause the rarefaction to move rapidly outward to the right just behind the previously mentioned condensation. As the fork continues to vibrate, a series of condensations and rarefactions proceeds outward through the air with a speed of about 1100 ft/sec.

What happens when these condensations and rarefactions reach a human ear (Fig. 9.5)? When a condensation impinges on the eardrum, the above-normal pressure of the condensation pushes the eardrum in a little. A short time later a rarefaction arrives and the eardrum swings outward. In the middle ear are three little bones—the hammer, the anvil, and the stirrup—which are arranged to act as levers in transmitting and magnifying the vibrations to the fluid of the inner ear. The vibrations thus transmitted stimulate hairlike nerve endings, located in the fluid of the inner ear, that report to the brain, and we have the sensation of sound.

When an earthquake occurs, the slippage of layers causes complex vibrations that travel along the earth's crust and through the interior of the earth for thousands of miles. From the effects on a seismograph, it is learned that earthquake waves consist of both transverse and longitudinal waves. Earthquake waves and the seismograph are considered in Chap. 25.

WAVELENGTH

Wavelength is the distance a wave moves forward during the time required for one complete vibration of an individual particle of the medium. The symbol used for wavelength is lambda (λ). Figure 9.6 (top) represents a transverse periodic wave that is

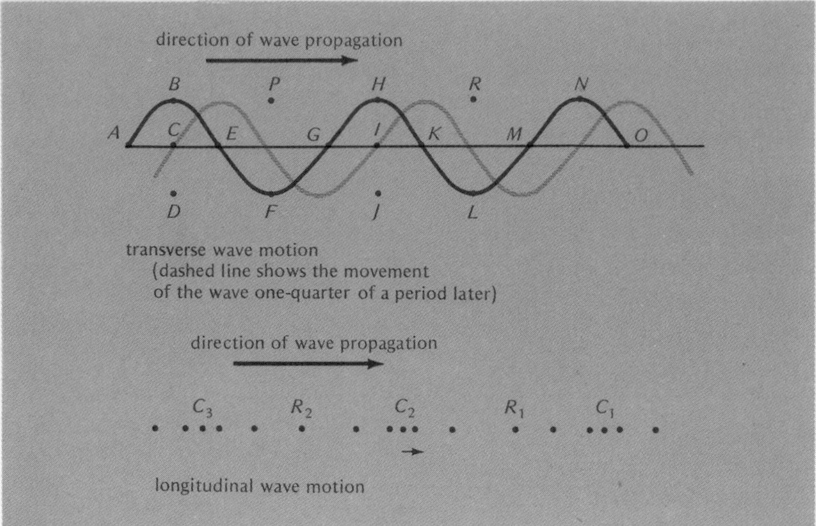

Figure 9.6
Representation of a wave.

moving to the right. During the time it takes for the crest (top point) now at B to move forward through P to crest H, the particle B must have traveled the path $BCDCB$. From the definition of wavelength and the diagram, it follows that a wavelength is the straight-line distance BPH between two successive crests (BH), or FJL between two successive troughs (FL), and also the straight-line distance ($EGIK$; $GIKM$) between two successive "corresponding points" (E and K; G and M); that is, from a particle moving in one direction to the next particle moving in exactly the same direction. The amplitude of the wave is BC.

In Fig. 9.6 (top) the position of the wave after an elapsed time equal to one-fourth the period of the wave motion has passed is shown by the dashed line. As the crest at B moves forward to a point directly above E, the point at E must have moved vertically upward; the motion of the particles vertically results in the wave propagation to the right.

Figure 9.6 (bottom) similarly represents a longitudinal wave with the first condensation having reached C_1, the second C_2, and the third C_3 as the waves move to the right. Note that a particle in the middle of a condensation, as at C_2, is moving forward in the direction of the wave. Its amplitude is represented by the short arrow pointed to the right. Later, as the rarefaction now at R_2 moves through the region now occupied by C_2, the particle mentioned above will be moving to the left, or opposite the direction in which the wave is moving. Then as C_3 comes along, this particle will again move forward. The original definition of wavelength applies to longitudinal as well as to transverse wave motion, and it follows that in longitudinal wave motion the wavelength is the distance between two successive condensations (C_3 C_2 or C_2 C_1) or the distance between two successive rarefactions (R_2 R_1).

SOUND TRANSMISSION

A fundamental difference between longitudinal and transverse waves is that although transverse waves, such as electromagnetic waves, may be transmitted through a vacuum, longitudinal waves require a material medium for transmission. This can be demonstrated in the case of sound by sounding an electric bell that is fastened inside a bell jar. Then when the air is pumped from the bell jar, the sound can no longer be heard although the clapper is seen to move. Sound waves (longitudinal waves) require a material medium (in this case, air) for transmission. The fact that the clapper can still be seen to move shows that transverse waves, light waves in this case, can be transmitted through a vacuum.

SPEED OF SOUND

The speed of sound through air can be measured quite accurately and is found to vary with the temperature of the air. At 0°C, the speed of sound in air is 1088 ft/sec and increases by 2 ft/sec for each 1°C rise in temperature. Thus at 25°C the speed of sound in air is 1088 ft/sec + 25 × 2 ft/sec = 1138 ft/sec.

Sound is transmitted through different media at different rates. The speed of sound is, in general, greatest in solids, less in liquids, and least in gases. The speed of sound through steel at 15°C is about 16,000 ft/sec. Through water at 15°C, sound travels at the rate of 4760 ft/sec, but only 1118 ft/sec in air at the same temperature.

THE WAVE MOTION EQUATION, $v = f\lambda$

All waves move outward from the source with a constant speed, so that the equation that gives speed in the case of uniform motion applies as follows:

$$\text{speed} = \frac{\text{distance}}{\text{time}} \qquad \text{or} \qquad v = \frac{d}{t}$$

We may choose any value of distance for d provided that we use for t the time required for the wave motion to travel the distance d. From the definitions of wavelength and period, it follows that

$$v = \frac{d}{t} = \frac{\text{wavelength}}{\text{period}} = \frac{\lambda}{T}$$

where λ is used for wavelength and T for period. Because frequency $f = 1/T$,

$$v = \frac{\lambda}{T} = \frac{1}{T}\lambda = f\lambda \qquad \text{or} \qquad v = f\lambda \tag{9.1}$$

This wave equation, read "velocity equals frequency times wavelength," applies to all wave motion, both transverse and longitudinal. Examples 9.1 and 9.2 show the use of this relation.

Example 9.1

Compute the wavelength of the sound emitted by a tuning fork of frequency 256 vib/sec when the temperature is 20°C.

Solution

$$\lambda = \frac{v}{f} = \frac{1088 \text{ ft/sec} + 20 \times 2 \text{ ft/sec}}{256 \text{ vib/sec}}$$

$$= \frac{1128 \text{ ft/sec}}{256 \text{ vib/sec}} = 4.41 \text{ ft/vib} \qquad \text{or} \qquad 4.41 \text{ ft}$$

Because the wavelength is by definition the distance a wave travels during one complete vibration, the units of wavelength are simply units of length. In terms of Fig. 9.6 (bottom) this means that waves are proceeding outward from the fork at the rate of 1128 ft/sec. As the prong of the tuning fork swings forward the second time, the first condensation is 4.41 ft ahead of the second condensation, which is just starting out.

Example 9.2

Compute the wavelength of the radio wave sent out by a radio station that broadcasts on an assigned frequency of 1500 kc/sec (kHz), or 1,500,000 vib/sec.

Solution

All electromagnetic waves, including light, radio waves, television waves, X radiations, etc., move with a speed of 186,000 mi/sec, or 3×10^{10} cm/sec. From $v = f\lambda$

$$\lambda = \frac{v}{f} = \frac{30 \times 10^{9} \text{ cm/sec}}{15 \times 10^{5} \text{ vib/sec}} = 2 \times 10^{4} \text{ cm}$$

$$= 2 \times 10^{2} \text{ m}$$

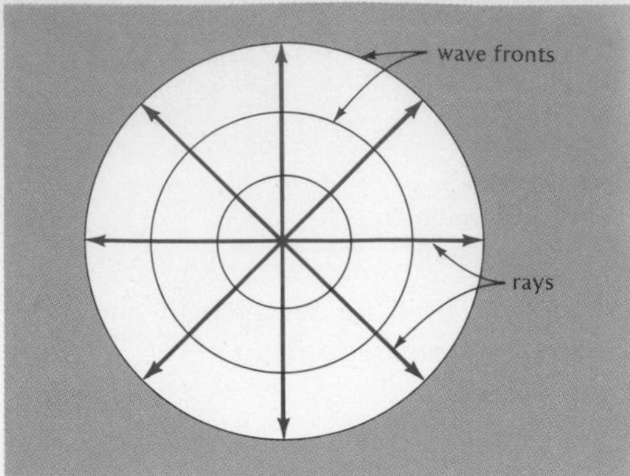

Figure 9.7
Rectilinear propagation.

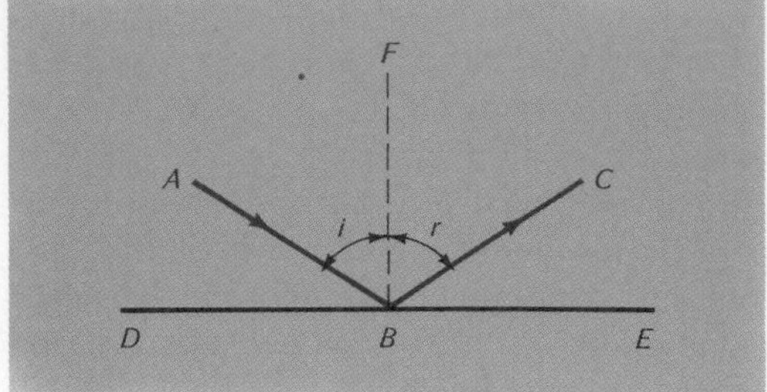

Figure 9.8
Law of reflection.

Properties of waves

There are several properties common to all types of wave phenomena. All types of waves exhibit rectilinear propagation, reflection, refraction, diffraction, interference, and the Doppler effect.

RECTILINEAR PROPAGATION

Rectilinear propagation refers to the fact that waves are propagated along a straight line in all directions from the source. As an illustration, consider a point source of sound at *S* of Fig. 9.7. At the end of a certain time, the sound will have traveled the same distance in all directions, giving rise to a spherical wave front. The arrows along radii of the sphere pictured in Fig. 9.7 indicate the direction in which the wave front is moving. Such arrows are perpendicular to the wave front and are called *rays*.

REFLECTION

Reflection of light is a phenomenon familiar to everyone because of the use of mirrors. A ray of light *AB* striking a reflecting surface *DE* (Fig. 9.8) will be reflected in a direction *BC* such that the angle of reflection *r* between the reflected ray *BC* and the perpendicular *BF* is equal to the angle of incidence *i*, which is between the incident ray *AB* and the perpendicular *BF*. This relation—the angle of reflection equals the angle of incidence for smooth, flat surfaces—is known as the *law of reflection*.

Sound, too, is reflected. The law of reflection also applies to sound. A reflected sound that reaches the ear so much later than the original sound that it is heard as a distinct separate sound is called an *echo*. You may have heard an echo from the far wall of a canyon, from a group of trees, or from the walls of houses on the next street. Reflected sounds that reach our ear soon enough to be heard as a continuation of the original sound produce *reverberation*. Too great a reverberation in auditoriums is detrimental, but a certain amount is helpful in that it reinforces the original sound, making it louder. The science of acoustics deals with obtaining the proper amount of reverberation in auditoriums by the careful use of sound-absorbing materials, such as acoustical tile, drapes, and seat and floor coverings.

Long-wavelength electromagnetic waves or electrical waves, discussed later in this chapter, are reflected by various ionized layers of our atmosphere, and hence signals from powerful radio stations may be heard for hundreds of miles (see Chap. 20, discussion of ionosphere). The shorter electrical waves used for television are not reflected appreciably by these ionized layers of the atmosphere, so that dependable television reception is limited to direct transmission from the television sending towers to the receiving antennas. Even with high television

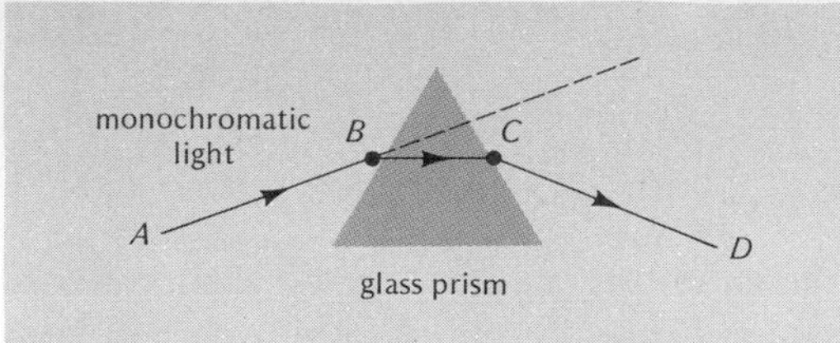

Figure 9.9
Refraction.

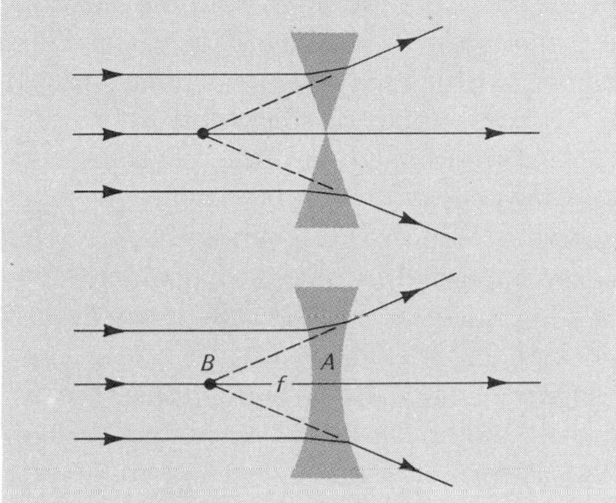

Figure 9.11
Concave lens.

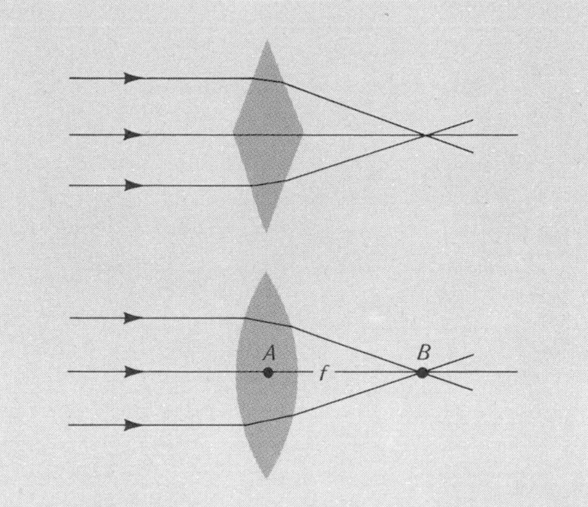

Figure 9.10
Convex lens.

broadcasting towers, often over 1000 ft, good reception is limited to less than 100 miles. Television waves are now received from communication satellites and other orbiting devices which act as "transmitting stations" above the earth.

REFRACTION

Refraction is the change in direction of a wave in passing from one medium into another medium of different density. Figure 9.9 shows the path of a ray of monochromatic light (light of a "single color" or wavelength) through air (*AB*), then through a glass prism (*BC*), and finally through the air (*CD*). Note that the changes are abrupt, taking place at the boundary of the two media, and that the change in direction is toward the base of the prism of the more dense substance, glass.

The most important application of refraction is lenses. If two similar glass prisms are placed base to base, as in Fig. 9.10 (top), the three parallel rays of light shown will, after refraction, all pass through the same point. If one piece of glass the size of the two prisms is ground to the shape shown in Fig. 9.10 (bottom), the result is called a *convex lens*—thicker in the center than at the extremities. The line through *A* toward *B* is called the *principal axis*. The point *B*, through which all rays parallel to the principal axis pass after refraction, is called the *focus*, or *focal point*. The distance *AB* from the center of the lens *A* to the focus *B* is called the *focal length* and is usually designated by *f*.

If two similar glass prisms are placed vertex to vertex, as in Fig. 9.11 (top), the parallel rays shown will, after undergoing refraction, all seem to come from a common point. If one piece of glass about the size of the two prisms is ground to the shape shown in Fig. 9.11 (bottom), the result is called a *concave lens*—thinner in the center than at the extremities. The point *B*, from which all rays parallel to the principal axis seem to come after refraction, is called the *focus*, or the *virtual focus*, and the distance *AB* from the center of the lens *A* to the focus *B* is again called the *focal length* and designated by *f*.

As shown in Fig. 9.12, for a given diam-

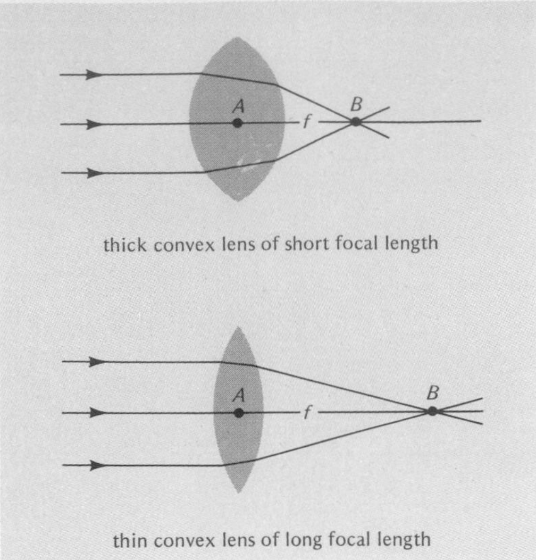

Figure 9.12
Convex lenses of different focal lengths.

eter of lens the thicker the lens, the shorter the focal length *AB;* and the thinner the lens, the longer the focal length.

Lenses are used in telescopes, microscopes, and other optical instruments, but their most important use is in eyeglasses. To understand how lenses are used in aiding vision, it is necessary to understand a little about the action of the normal eye and its defects. Figure 9.13 shows that, in the case of the normal eye, parallel rays from distant objects are refracted by the eye lens and brought to a focus on the retina of the eye. The ciliary muscles should vary the thickness of the lens inside the eye in order to focus the light rays on the retina so that clear vision occurs and eyestrain is avoided. For example, the fact that rays of light from a nearby object tend to focus at a greater distance from the center of the lens than normal is taken care of in the human eye by the ciliary muscles, which cause the eye lens to thicken, thus giving a shorter focal length and causing the image to fall on the retina. This is referred to as the power of accommodation. The ciliary muscles usually do not work well in later years, perhaps owing in part to hardening of the lens, thus causing many middle-aged adults to start wearing eyeglasses or if already using them, to change to bifocal lenses, which allow a different correction for near and distant objects.

One of the common defects of the eye is nearsightedness, or myopia. In this case, as shown in Fig. 9.14, the eye lens is too thick or the eyeball is too long. This type of eye sees nearby objects well, hence the term nearsightedness; but rays from a distant object that enter the eye as parallel rays cross before reaching the retina and hence are not brought to a focus on the retina. This results in a blurred image. In order to cause these rays to focus on the retina, they must be diverged a little before entering the eye. This is accomplished by means of a concave lens in an eyeglass, as shown by the use of broken lines in Fig. 9.14.

Farsightedness, or hypermetropia, the opposite kind of eye defect, which is shown in Fig. 9.15, is due to an eye lens that is too thin in the center or to an eyeball that is too short. In this case, the rays from nearby objects reach the retina before they can be brought to a focus, thus resulting in indistinct vision; vision for distant objects, however, is practically unimpaired. Because the rays should be focused on the retina for clear vision, a converging, or convex, lens must be used. The correcting effect is shown by the broken lines of Fig. 9.15.

A third defect, astigmatism, which often accompanies one of the other defects, results in the eye seeing objects more distinctly in one plane than in other planes. The student should look at the lines of Fig. 9.16. The normal eye sees all lines as equally distinct and sharp. If one line appears darker than the others, he should turn the book 90° and see if the same line is still darker or if the new line that is now in the position previously occupied by the darker line now appears darker. If the latter is true he has astigmatism, which is due to the eye lens not being entirely spherical but of a shape that is the equivalent of a spherical lens combined with a cylindrical lens. To correct for this, the eyeglass lens is ground to compensate for the cylindrical component.

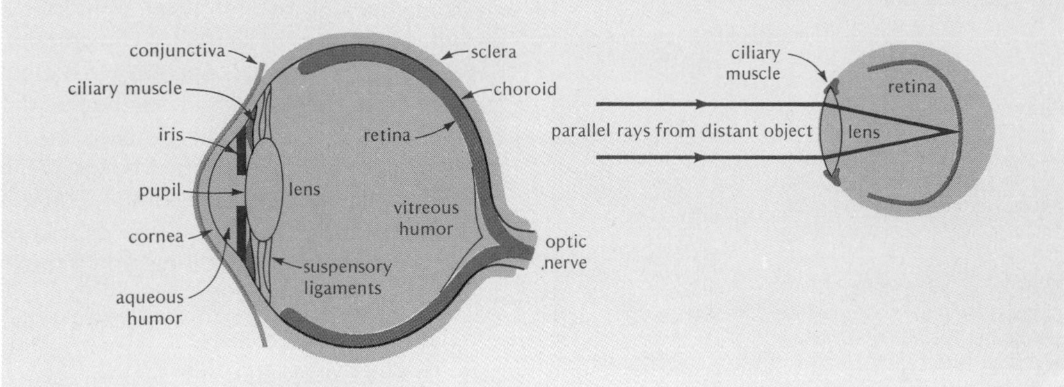

Figure 9.13
The normal eye.

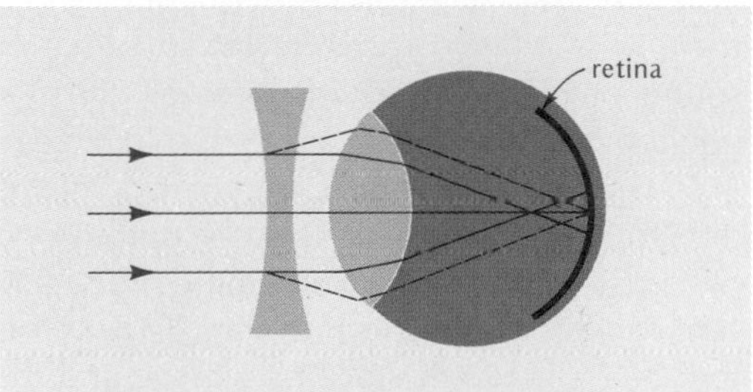

Figure 9.14
Nearsightedness.

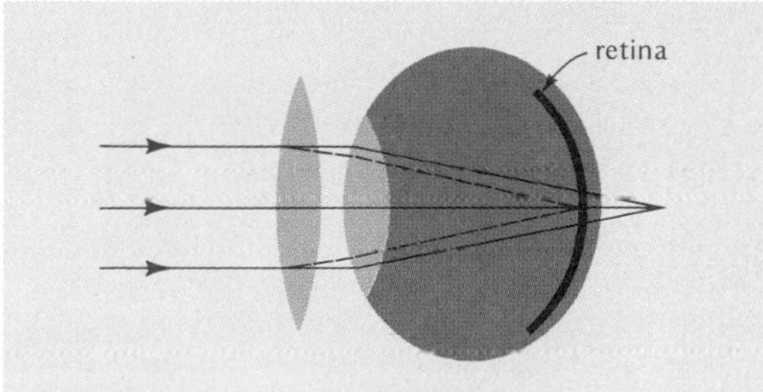

Figure 9.15
Farsightedness.

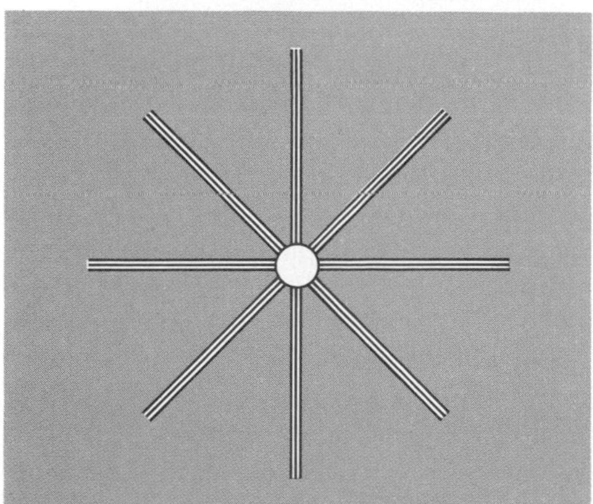

Figure 9.16
Astigmatism. The normal eye sees all lines equally distinctly and sharply.

DIFFRACTION

Diffraction is the bending of waves around objects and the bending of waves when they pass through openings. In Fig. 9.17 water waves in a ripple tank are shown being diffracted through a small opening. Diffraction of sound waves in air is readily observed. There are many everyday instances in which we can observe sound waves being bent, for example, around corners of buildings or through openings such as doorways. Diffraction of light is not as easily observed because of the very short wavelengths of light waves. In order for diffraction to be observable, the opening through which the wave passes should not be too different in size from the wavelength of the wave. Diffraction effects

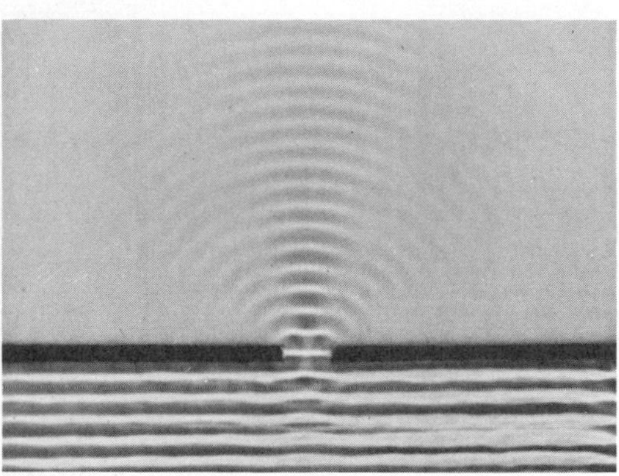

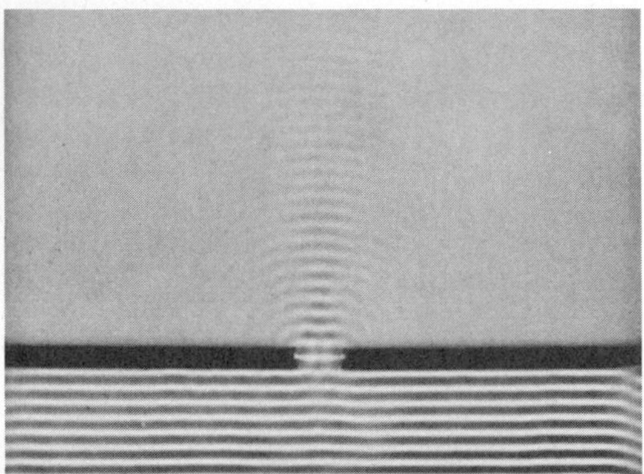

Figure 9.17
Three views of waves passing through the same opening. Note the decrease in bending of the shorter wavelengths. The wavelength is longest in the top picture and shortest in the lowest picture. (From PSSC *Physics*, D. C. Heath and Company, Lexington, Massachusetts, 1965.)

are sometimes observable when light from a distant lamp, for example, a street light, passes through the small openings in a cloth curtain or a fine screen. Usually the diffracted light produces, in addition, an interference pattern as described below. With the use of an inexpensive helium–neon gas laser it is possible to demonstrate to a large group the diffraction of light passing through a very small opening.

THE DOPPLER EFFECT

When there is a relative approach of a source of sound or of light and an observer, the frequency of the sound heard or the light seen is higher than the source frequency; when there is a relative recession of a source of sound or of light and an observer, the frequency of the sound heard or the light seen is lower than the source frequency. This is the *Doppler effect,* which can be understood by considering the waves being emitted and sent forward by the fork in Fig. 9.18. If this fork has a frequency of 280 vib/sec and the velocity of sound is 1120 ft/sec, the waves are 4 ft long (from $v = f\lambda$). Thus the distance from *A* to *B,* successive condensations, is 4 ft, and from *B* to *C* is 4 ft. For an ear located at *A* and stationary relative to the tuning fork, 280 waves would strike the eardrum each second; that is, the true frequency is 280 vib/sec. If during a second, however, the observer moved from *A* to *C,* a distance of 8 ft in the direction of the source, his ear would "run into" two additional waves; the number of waves impinging on the eardrum would then be 282, so that the observed frequency as heard by the ear would be higher —282 vib/sec. Similarly, if the observer were originally at *C* and during a second moved away to *A,* his ear would miss two of the waves that would have reached it, so that the observed frequency would be lower—278 vib/sec.

If a person were standing at the side of a highway, and an automobile went by with its horn sounding, the person would hear a sud-

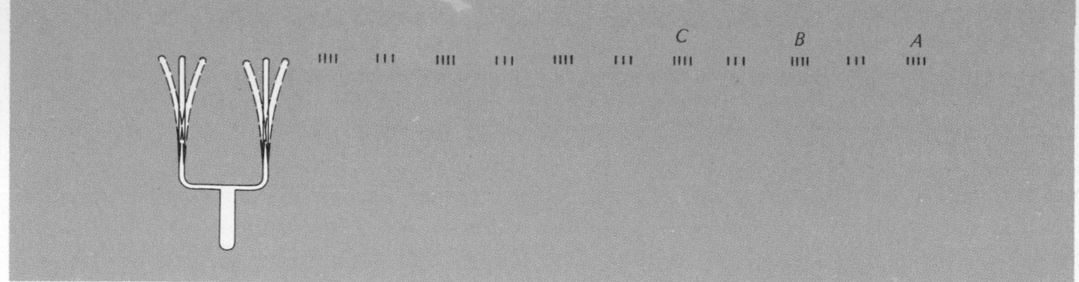

Figure 9.18
The Doppler effect for sound waves. If an observer moves from *A* to *C* in 1 sec, two additional waves strike the ear, thus increasing the frequency heard by 2 vib/sec.

den drop in pitch, or frequency heard, just as the automobile went by. If the auto horn had a frequency of 280 vib/sec, and the automobile were traveling at 60 mi/hour, or 88 ft/sec, the change in pitch as the automobile passed would be from

$$280 \text{ vib/sec} + \frac{88 \text{ ft/sec}}{4 \text{ ft/vib}}$$

or 302 vib/sec, to 280 − 22, or 258 vib/sec. This change in pitch takes place suddenly at the time of passing and should not be confused with the change in loudness. Loudness increases more and more up to the time of passing and then decreases gradually after the instant of passing.

Applications of Doppler's principle to light are given in relation to the Doppler effect for double stars (Fig. 9.25), to the evidence that we live in an expanding universe (Chap. 11), and to the estimation of the age of the universe (Chap. 11).

INTERFERENCE

Interference is the phenomenon of the superposition of two sets of waves passing through the same medium at the same time whereby they may tend to reinforce or cancel one another. Consider two tuning forks A and B with fork A having a frequency of 300 vib/sec and fork B a frequency of 302 vib/sec. An ear at some distance from the two forks may receive condensations from both forks simultaneously, so that the eardrum will be pushed in to a greater degree, registering at the brain as a louder sound. During the next 0.25 sec, the ear will receive 300/4 = 75 complete waves from fork A and 302/4 = 75.5 waves from fork B. This means that at the end of 0.25 sec a condensation from fork A will reach the ear, but a rarefaction from B will arrive. Because a condensation pushes the eardrum inward and a rarefaction tends to allow the eardrum to swing outward, the two arriving together will annul one another, causing minimum sound. Half a second later than the original condition, fork A will have completed 150 vibrations and fork B 151 vibrations, so that condensations from both forks will reach the ear again giving the condition for reinforcement or maximum sound. Again at the end of 0.75 sec, the two waves will oppose each other in their effects on the eardrum, causing minimum sound. At the end of the second or the beginning of the next second, condensations from both forks will reach the eardrum. Thus, in this case, there will be two times of maximum sound and two times of minimum sound each second, or two beats per second. Had the difference in frequency been 3 vib/sec, there would have been three beats per second, etc. *Beats* are the periodic increase and decrease of intensity of sound due to the interference of two sound waves that have similar but not exactly equal frequencies. The number of beats per second is equal to the difference in frequency in the two sets of waves.

Interference can also be demonstrated by sending transverse waves along a rope. If waves are sent along the rope from left to

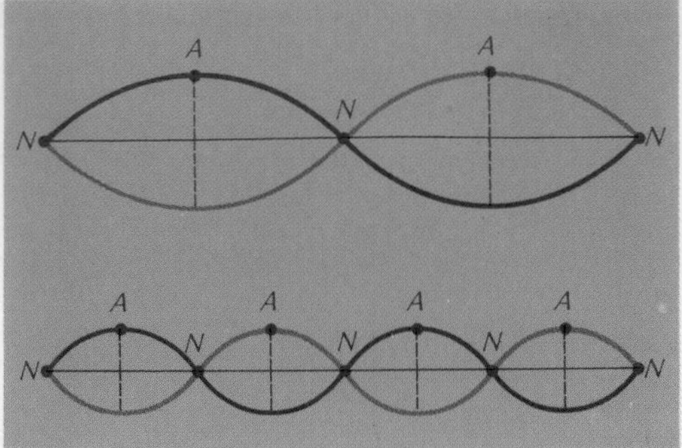

Figure 9.19
Stationary waves.

right, a crest can be seen to travel along the rope and be reflected at the far fixed end as a trough that will travel back along the rope. If a second crest is started along the rope just as the initial crest is being reflected back as a trough, the forward-moving crest and the backward-moving trough will arrive at the center of the rope at the same time, thus annulling one another. A little later, a direct trough and a reflected crest will pass through the center point simultaneously, so that the center point will continue to be a point of minimum motion, with points of maximum motion at the one-quarter and three-quarter points. The effect is to give two loops that appear stationary, as in Fig. 9.19 (top). *Stationary waves* result from the interference of direct and reflected waves. They exist in most musical instruments that are emitting sounds.

If the rope is jerked twice as often, the frequency is doubled, or the wavelength halved, and points of minimum motion occur at the one-quarter, one-half, and three-quarter points, as shown in Fig. 9.19 (bottom). Points of minimum motion, including the fixed ends, are called nodes, marked *N* in Fig. 9.19, and points of maximum motion are called antinodes, marked *A* in Fig. 9.19. Because in Fig. 9.19 (top) one wavelength extends from one end to the other—straight line—the distance between two successive nodes is always half a wavelength, and the distance from a node to the next antinode is one-quarter of a wavelength.

Another example of a standing wave is shown in Fig. 9.20. Here a longitudinal sound wave is generated by a speaker, which, in turn, sets a rubber membrane into vibration, which then produces a sound wave in illuminating gas in a long tube. Along the top of the tube is a series of pinholes. The gas escaping from these holes is lit, producing a row of flames. At certain frequencies a standing wave is set up in the gas tube. At points along the tube where the gas pressure is high, the flame increases in height. At other points between these places of high pressure the flames are lower. The distance between two regions of high flames corresponds to one-half a wavelength just as the distance between the nodes *N* in Fig. 9.19.

If two pebbles are dropped on a quiet water surface at the same time, two sets of waves can be seen. Because they both produce a crest on the surface at the same time, a beautiful interference pattern is set up, as in Fig. 9.21, where point sources have generated circular waves. Where a crest crosses a crest, a bright region is produced; places where troughs meet are dark. Where a crest from one source meets a trough from the other source, the water is practically undisturbed; that is, interference has occurred. These are the gray areas like spokes from the center of a wheel.

Interference also takes place in the case of light waves, although the shortness of the waves makes it more difficult to produce experimentally (for visible violet light, $\lambda_V =$ 0.000038 cm, and for visible red light, $\lambda_R =$ 0.000076 cm). However, by using diffraction gratings (glass or metal) on which 2000 to 20,000 lines/in. are ruled, interference of the light passing between the lines does take place, giving multiple diffraction images of a light source. The colors of thin films, such as an oil film on the surface of water, are due to interference with reinforcement of different colors, or different wavelengths, occurring at slightly different places.

DISPERSION

Another important property, which, however, is found only in transverse waves, is *dispersion*.

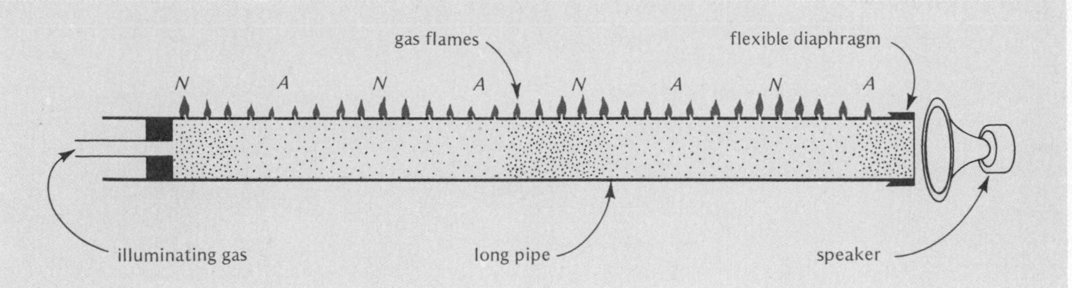

Figure 9.20
Flames show the presence of standing waves in a tube filled with illuminating gas. The *A* and *N* refer to displacement antinodes and nodes, respectively.

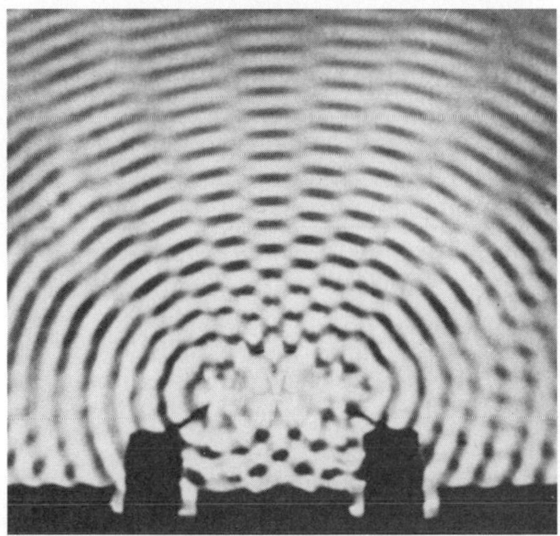

Figure 9.21
Interference of water waves. (From PSSC *Physics*, D. C. Heath and Company, Lexington, Massachusetts, 1965.)

Dispersion occurs because different wavelengths of light travel at different speeds in the same medium, except in a vacuum where all wavelengths of light (corresponding to different colors of light) travel at essentially the same speed. However, in a piece of glass different wavelengths of light travel at different speeds. The most common example of dispersion is the separation of ordinary white light into its component colors, owing to this fact that different colors are refracted differently. Sunlight and light from the common electric lamp are examples of ordinary white light.

As shown in Fig. 9.22, the rays of the various colors that make up the initial beam of white light are refracted by different amounts, the violet end of the spectrum being refracted more than the red end. The spectrum formed is labeled in the figure (as the Roy G. Biv colors; R for red, etc.) to indicate the seven so-called colors of the rainbow, although each of them may be subdivided into many gradations. The seven colors are red, orange, yellow, green, blue, indigo, and violet. Actually the scientist thinks of an infinite number of colors—a separate "color" for each wavelength. (See the color plate, opposite p. 208.)

A complex sound, which is a mixture of many frequencies, cannot be separated into its component frequencies by passing the sound through any device like a prism used with light. Various frequencies, however, can be removed from the complex sound by the use of sound filters, which absorb sounds of certain frequencies.

Dispersion is important because a study of the spectra produced from the light from incandescent bodies, including distant stars, enables us to tell what substances are present in these bodies. An instrument for viewing

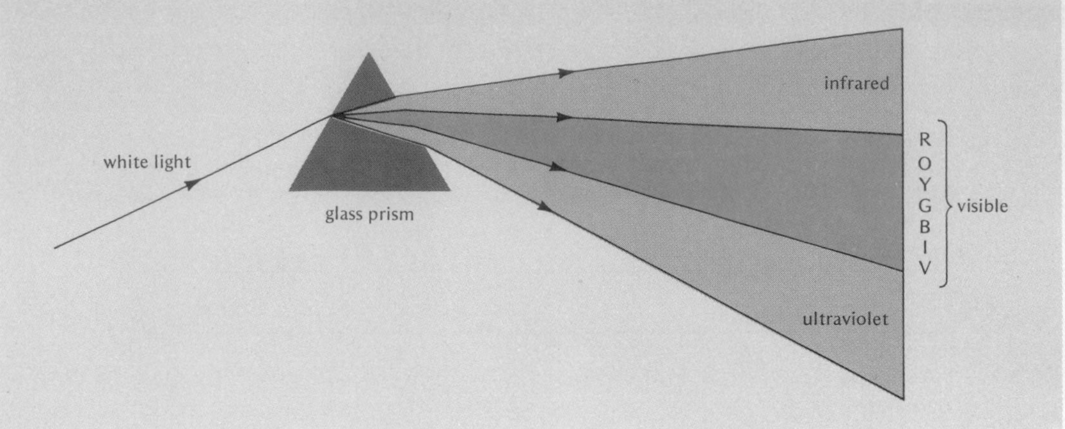

Figure 9.22
Dispersion. A beam of white light is "broken up" by passing through a prism into the various colors forming the visible spectrum, which is flanked by invisible infrared energy on the long-wavelength side and by invisible ultraviolet energy on the short-wavelength side.

the spectrum produced by dispersion is called a *spectroscope* (Fig. 9.23). This may be used visually, or photographs may be taken of the spectrum formed.

SPECTRA

There are three different types of spectra: *continuous, bright-line,* and *absorption* (see color plate). Light from a source that is an incandescent solid, an incandescent liquid, or an incandescent gas under high pressure will give a *continuous spectrum* (bar 1 in color plate)—that is, a spectrum with all colors present from the red to the violet. Light from an ordinary tungsten wire (an incandescent electric lamp) or a carbon arc will also give this type of spectrum.

If a gas or vapor under low pressure is caused to glow by an electric discharge through it, the light from the gas when passed through the prism of the spectroscope gives only certain colors, or a *bright-line spectrum* (color plate, all bars from Na, sodium, down). Salt (sodium chloride) heated in the flame of a Bunsen burner will be vaporized, and the bright lines characteristic of sodium, particularly the two yellow lines near 6000 Å (color plate, sodium bar, Na), can be seen through the spectroscope. The wavelength of 6000 Å is 6000 angstrom units, where 1 Å is equivalent to 10^{-8} cm. Wavelengths of light are often expressed in angstrom units (Å).

Bright lines characteristic of hydrogen are shown in H (hydrogen) of the color plate. The light from a mercury-vapor fluorescent lamp will give an entirely different set of bright lines, as in Hg (mercury) of the color plate.

When light that would give a continuous spectrum on being dispersed by a prism is first passed through a relatively cooler gas, the spectrum that would have been continuous is found to be crossed by dark lines at the same wavelength, or same position in the spectrum, as the bright lines that are emitted by the cooler gas when it is made the source by being heated to incandescence. This spectrum due to absorption by the cooler gas is called a *dark-line* or *absorption spectrum.* Notice in the color plate that two dark lines (Na in bar 2) are in the same position as two bright yellow lines in the spectrum of sodium (Na).

When light from the sun is examined with a spectroscope, the continuous spectrum is found to be crossed with thousands of dark lines, called the *Fraunhofer lines* (not shown in 1 of the color plate but similar to those in 2 of the color plate). Such dark lines in the sun's spectrum are absorption lines caused by the light from the photosphere of the sun, which would give a continuous spectrum,

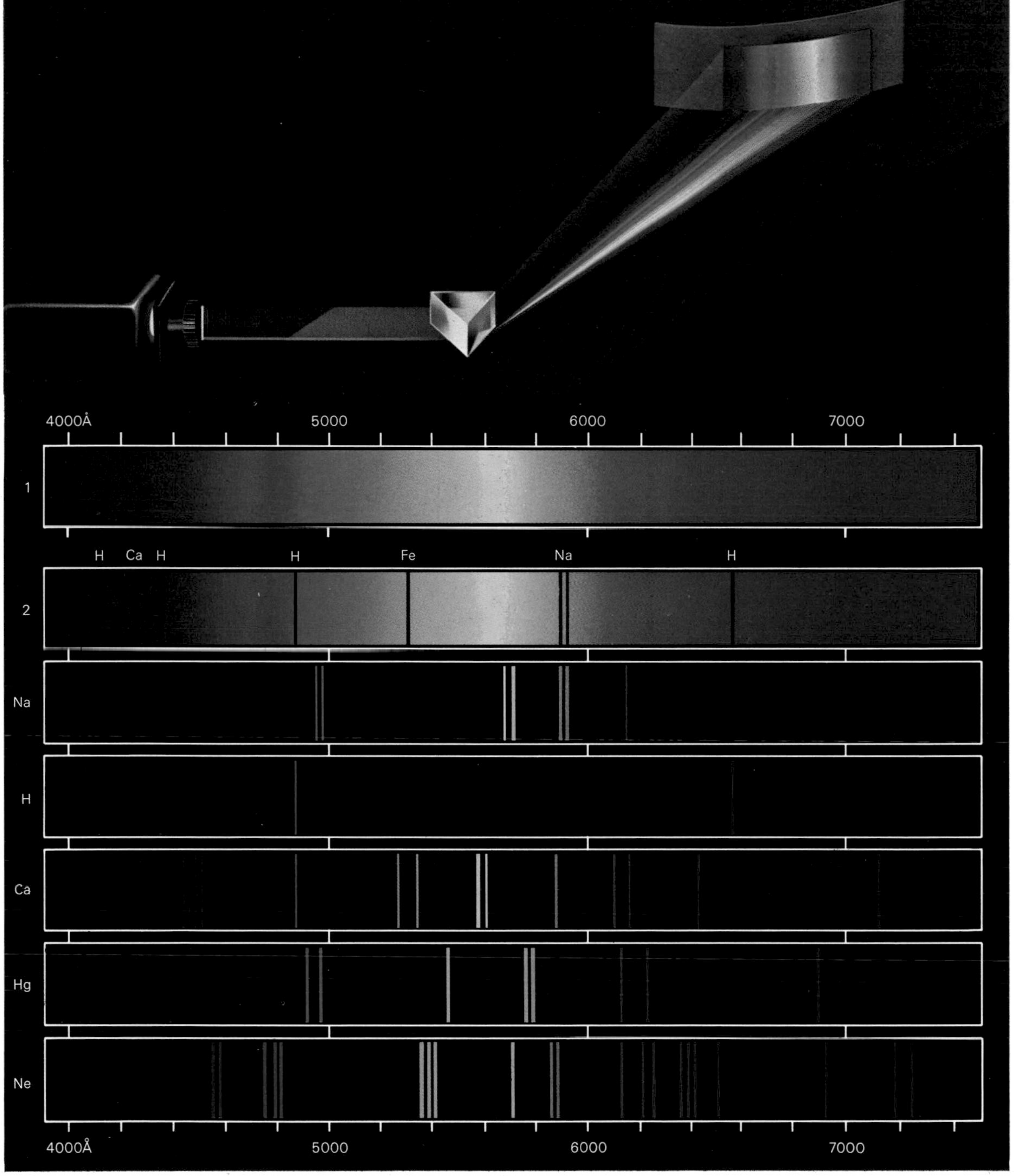

Dispersion, top color plate, is the separation of ordinary white light into its component colors—the visible spectrum. The colors that make up the initial beam are, on passing through a glass prism, refracted by different amounts, red being refracted the least and violet the most.

There are three different types of spectra, bottom color plate: continuous, 1; absorption or dark line, 2; and bright line such as those of sodium (Na), hydrogen (H), calcium (Ca), mercury (Hg), and neon (Ne). (See also Dispersion, p. 206.)

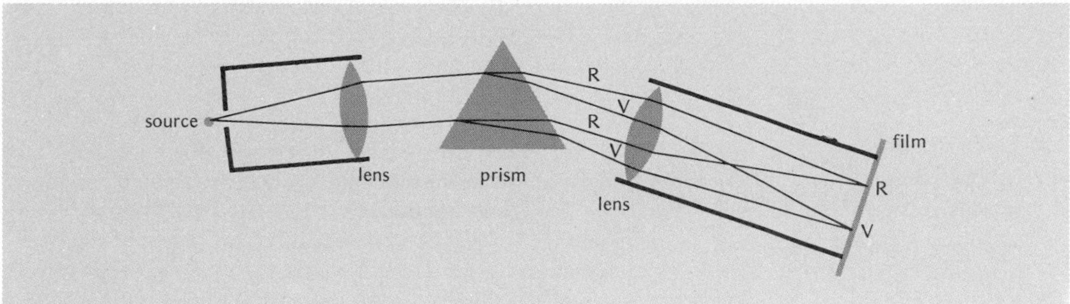

Figure 9.23
The spectroscope.

Figure 9.24
Detection of calcium in the sun.

passing through the outer, relatively cooler layers of the sun's overlying gases and the gases of the earth's atmosphere.

Every gas or vapor will absorb exactly those wavelengths that it is capable of emitting when incandescent.

(The cause of emission lines and absorption lines is given early in Chap. 13 in terms of the Bohr theory of the atom.)

Because the position of the bright lines or dark lines is characteristic of the element causing the lines, a gas may be analyzed by comparing its bright-line or dark-line spectrum with the spectra of known gases. This method, called *spectrum analysis,* has been used to identify more than 60 elements as being present in the sun. This is illustrated in Fig. 9.24, where bright (or dark) lines of calcium, representing those from a photograph, are compared with the dark lines from a photograph of the sun's absorption spectrum. Note that for every calcium line representing those of the laboratory photograph, there is a matching line in the sun's spectrum. Hence we know that calcium is present in the sun. There are methods of telling which of the Fraunhofer lines are due to the sun's "atmosphere" and which to the earth's—Doppler-effect method and others.

In 1868, Janssen noticed certain strong Fraunhofer lines in the sun's absorption spectrum that did not correspond to any element known on earth. Because of its presence in the sun, Lockyer suggested that the new element be named "helium" (from the Greek *helios,* meaning "sun"). In 1895, Ramsay discovered helium on earth as a gas emitted when certain uranium-bearing minerals were heated. Later he showed that helium is present in the earth's atmosphere in minute quantities.

Similarly, we are able by spectrum analysis to tell which elements are mainly predominant in the distant stars, even though they may be thousands of light-years away. Quantitative results as well as qualitative are now obtained for laboratory samples of steel and other products by spectrum analysis.

DOPPLER EFFECT AND THE RED SHIFT

Certain stars that appear to be a single star when observed visually, are known to be double stars, even though thousands of light-years away, because their photographed

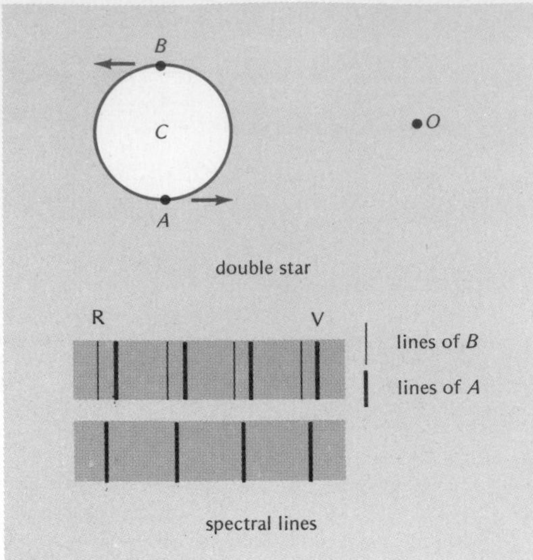

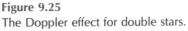

Figure 9.25
The Doppler effect for double stars.

spectral lines are found to be double. In Fig. 9.25 the full spectral lines of star *A*, of the double star *AB*, having a center of mass at *C*, are all shifted to higher frequency—toward the violet (V) end of the spectrum—owing to the approach of star *A* to the distant observer at *O*. The spectral lines of star *B* are all shifted to lower frequency—toward red, R—owing to the receding of star *B* from the observer *O*. The shift toward lower frequencies, toward red, is sometimes referred to as the *red shift*, which is interpreted as an indication of relative recession. The red shift is also used as evidence that we live in an expanding universe (Chap. 11).

The Doppler effect is also used to determine the period of rotation of the sun by carefully measuring the slight shift of the spectral lines from the approaching limb of the sun compared with the lines from the center and from the receding limb.

Sources and differences of sounds

Sounds are always produced by vibrating bodies (solid or fluid), which send out condensations and rarefactions through the air.

There are three general types of sources of sounds:

1. Vibrating strings or bars, which include all stringed instruments, such as the violin and guitar, and tuning forks and the xylophone.
2. Vibrating surfaces, such as drumheads and telephone-receiver diaphragms.
3. Vibrating columns of air, which include all the wind instruments, such as the organ pipe, cornet, and trombone.

Two sounds may differ in loudness, or intensity; pitch, or frequency; and quality. Loudness, pitch, and quality (timbre) are auditory attributes of sound; the corresponding physical terms are intensity, frequency, and quality (wave shape). The differences in the meanings of corresponding terms are explained in the paragraphs that follow.

LOUDNESS

The intensity of sound is measured by the flow of energy through a unit area. The intensity of a sound is directly proportional to the square of the amplitude of vibration of the source of a sound; is directly proportional to the area of the vibrating body; and, for outdoor sounds, is inversely proportional to the square of the distance of the observer from the source. The intensity is also related to frequency, density of the medium, and the speed of the sound in the medium. Intensity and loudness are commonly used interchangeably.

If a tuning fork that has been struck a light blow gives a sound of a certain intensity and is then struck a harder blow that doubles the amplitude of vibration of the fork, the sound emitted will then be 2^2, or 4, times as loud; if the amplitude is tripled, the loudness will be 3^2, or 9, times as great. A tuning fork that has been struck and is held in the hand will give a sound of a certain loudness. If the base of the fork is put against a tabletop, however, the sound will be much louder because part of the tabletop has been set in vibration, thus giving a much larger area of vibration.

When a person speaks outdoors, the energy transmitted by sound waves is spread out over ever-increasing spherical wave fronts. Because the area of a sphere varies directly as the square of the radius ($A = 4\pi R^2$ for a sphere), the same energy is distributed over an area varying directly as the square of the radius. Hence the *intensity,* which is measured by the kinetic energy passing through a unit area, *must vary inversely as the square of the radius,* or *distance out from the source.* A person standing 40 ft from a speaker outdoors hears the sound $(\frac{1}{2})^2$, or $\frac{1}{4}$, as loud as a person 20 ft from the speaker. A person at a distance three times as great would hear the sound $(\frac{1}{3})^2$, or $\frac{1}{9}$, as loud. This inverse-square relation just mentioned does not hold for sounds indoors, because the direct waves are reinforced by waves reflected from the side walls, floors, ceilings, etc.

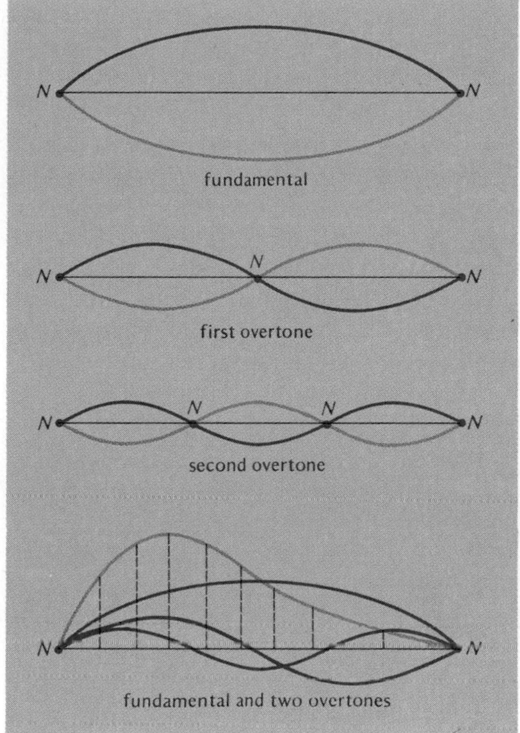

Figure 9.26
Fundamental and overtones.

PITCH

The sound sensation called "pitch" is a psychological or auditory phenomenon that is determined by a physical phenomenon, the frequency of the sound wave that strikes the ear. *Frequency* is the number of vibrations per second. *Pitch* is the sensation produced in the ear of the listener. Tone is a sound sensation having pitch, which thus depends on the number of vibrations per second.

The average human ear can detect sounds over a wide range, extending from a low limit of about 20 to 30 vib/sec to an upper limit of 20,000 to 25,000 vib/sec. The piano scale, based on international pitch (A = 440 vib/sec), extends from 32.7 to 4186 vib/sec. A bass singer usually has a range from about 130 to 350 vib/sec, and a soprano a range from about 250 to 1200 vib/sec.

QUALITY

Imagine yourself blindfolded in a room. One person strikes middle C on the piano and another bows middle C on the violin with the same loudness. Although the two notes have the same frequency and same loudness, you would detect a definite difference. This third way in which two sounds differ is *quality,* which depends on the number, kind, and relative intensity of the overtones that accompany the fundamental. The fundamental is the tone of lowest frequency and is usually more intense than the overtones. The overtones are sounds having frequencies that are integer multiples of the frequency of the fundamental—twice the frequency, three times the frequency, etc. For example, if in Fig. 9.26 the fundamental were 256 vib/sec, the first overtone would be 512 vib/sec; the second, 768.

In nearly all musical sounds, overtones are present. Their presence for a sound emitted by a vibrating wire may be shown as follows: If a wire 100 cm long is under sufficient tension to give a musical note, the wire vibrates mainly as a whole when bowed, as in Fig. 9.26 (top). If while the wire is still emitting sound, it is touched at the center point with the edge of a cork, a fainter note of twice the frequency of the fundamental, called the *oc-*

tave, will be heard. The point of damping being restricted to minimum motion is a node, so that the wavelength is now one-half as long as it was for the fundamental, and the frequency of this first overtone is twice that of the fundamental. Similarly, if the string is bowed and then touched with the cork one-third of the way from one end, the spot touched must become a node. Another node will automatically occur at the two-thirds point, so that the wavelength is now one-third that of the fundamental and the frequency of this second overtone is three times that of the fundamental. In this manner, the presence of as many as nine or ten overtones may be demonstrated. Hence the wire, instead of having a simple vibration as shown in Fig. 9.26 (top) has a complex vibration as indicated in Fig. 9.26 (bottom), the wire simultaneously vibrating in one, two, three, etc., segments. The curve in color in Fig. 9.26 (bottom) results when the fundamental and two overtones are present at the same time.

Complex motion, as in Fig. 9.26, may be shown on an oscilloscope, an instrument that can be used to show the wave forms electronically. With the aid of an oscilloscope, sounds of the same frequency and loudness may be studied, and the shape of the complex wave is a clue to which overtones are present and to their relative intensities.

Loudness and pitch are subjective characteristics. Intensity and frequency can be measured and are therefore objective characteristics. Quality is usually a psychological or auditory attribute of sound and is therefore a subjective characteristic. Quality, however, is an objective characteristic of sound in terms of the fundamental and the number, the kind, and the intensity of the overtones, which, for example, produce complex wave shapes on an oscilloscope.

FREE VIBRATIONS, FORCED VIBRATIONS, AND RESONANCE

When a body that is free to vibrate is given a single blow, it will vibrate with a period that depends on its structure, and the vibrations will gradually decrease in amplitude. Such vibrations are called "free vibrations." Examples are a tuning fork given one blow, a plucked violin string, and a pendulum given a single push.

If a periodic force is applied to a body free to vibrate, "forced vibrations" will result, the frequency of vibration of the body being the same as that of the applied force. A good example of this is seen when the base of a vibrating tuning fork is placed against a tabletop. The tabletop will vibrate with the frequency of the tuning fork placed against it. Similarly, the sounding board of a piano intensifies the sound by vibrating in response to the frequency of whatever piano cord is sounded.

Nearly everyone has had the experience of pushing a youngster on a rope swing. To get maximum results with the least effort, one repeatedly pushes for a short time just as each swing is starting forward. *Resonance* is the special case of a forced vibration in which the frequency of the applied force is the same as the natural frequency of the vibrating body. Thus the pushing of the rope swing is seen to be an example of resonance. Another example of resonance is the sounding of certain piano cords when a person in the room sings or shouts. Also, pictures on a wall or a shelf of dishes in a china closet may vibrate in resonance with vibrating parts of a passing truck or airplane. The natural frequency of the item in the home is the same as that of some vibrating part of the truck or plane. Automobile manufacturers mount the engine on heavy rubber blocks to keep most of the vibrations from being transmitted from the engine along the frame to windows, doors, etc.

The complete electromagnetic spectrum

If a beam of sunlight is passed through a glass prism, dispersion occurs and a visible spectrum extending from red to violet results, as previously shown in Fig. 9.22 and in the color plate.

If a very sensitive thermometer is moved slowly from the visible violet to the visible red, the liquid in the thermometer will rise more and more. Surprisingly, the thermometer will show an even greater response when it is moved into the region beyond the visible red. This demonstrates the existence of energy of wavelengths longer than the visible red; it is called *infrared* radiation. Infrared lamps are now often used in place of a hot-water bag to provide heat energy in the treatment of muscle strains and bruises. Because infrared radiation is absorbed less by water vapor than is the visible and ultraviolet radiation, photographs from airplanes can show features on the ground in spite of haze and light clouds if special films that are sensitive to infrared radiation are used.

If small pieces of photographic film are exposed in successive positions from the visible red to the visible violet, an increased blackening of the film occurs. If film is placed beyond the visible violet, an even greater effect takes place. Thus there is shown to be energy of wavelengths shorter than visible violet, which is called *ultraviolet* radiation.

It is the ultraviolet component of the sun's radiation that causes our summer tan and is instrumental in providing vitamin D in our bodies. Ultraviolet radiation is largely absorbed by water vapor and by glass. In winter the sun is lower in the sky, so that we get less radiation from it. Also the increased cloudiness of winter skies causes more absorption of the ultraviolet portion of sunlight. Finally, in winter we stay indoors more, where window glass keeps the short-wavelength ultraviolet radiation from reaching us.

Sun lamps consisting of special mercury-vapor bulbs or other sources rich in ultraviolet radiation are often used in winter to give a more normal quantity of this short-wavelength radiation, a certain amount of which is conducive to good health. Ultraviolet light is also used in irradiating milk to produce vitamin D.

Of longer wavelength than infrared radiation are electrical waves (Fig. 9.27), which include microwaves adjacent to infrared in wavelength, television and FM radio waves of still longer wavelength, AM radio waves, and very long electrical waves up to several miles in wavelength. Shorter in wavelength than ultraviolet radiation are the rays produced in an X ray tube and used for photographing the bones and other parts of the human body and for the treatment of cancer and some other diseases. Gamma (γ) rays are of even shorter wavelength than X rays. They are emitted by radioactive substances (Chap. 12).

Gamma radiation is used in the treatment of cancer, goiter, and other diseases and in radioactive tracer work.

Cosmic rays coming from outer space are of even shorter wavelength than γ rays.

The difference in the radiations mentioned above is due to a periodic variation of electric and magnetic fields, so that they are grouped together to form the electromagnetic spectrum shown in Fig. 9.27. All these electromagnetic waves are propagated with the speed of light, approximately 186,000 mi/sec, or 3×10^{10} cm/sec in air or vacuum. Waves from 0.3- to 3-cm wavelength can be obtained either from infrared radiation sources or from microwave sources, so that there is a small region of overlapping of infrared and electrical radiations. Likewise, ultraviolet and X rays overlap, and so do X rays and γ rays.

As shown in Fig. 9.27, certain portions of the electromagnetic spectrum devoted to electrical waves are reserved for allocation by the Federal Communications Commission to commercial and educational television and radio stations. Television Channels 2 to 13, which are VHF channels (very high frequencies), were allocated first. Channels 14 to 83, which are UHF channels (ultrahigh frequencies), are available for allocation to new TV stations; UHF channels include most of the so-called educational TV stations. Most commercial TV stations are on VHF channels.

The wavelength band that is reserved for UHF TV stations is the band with wave-

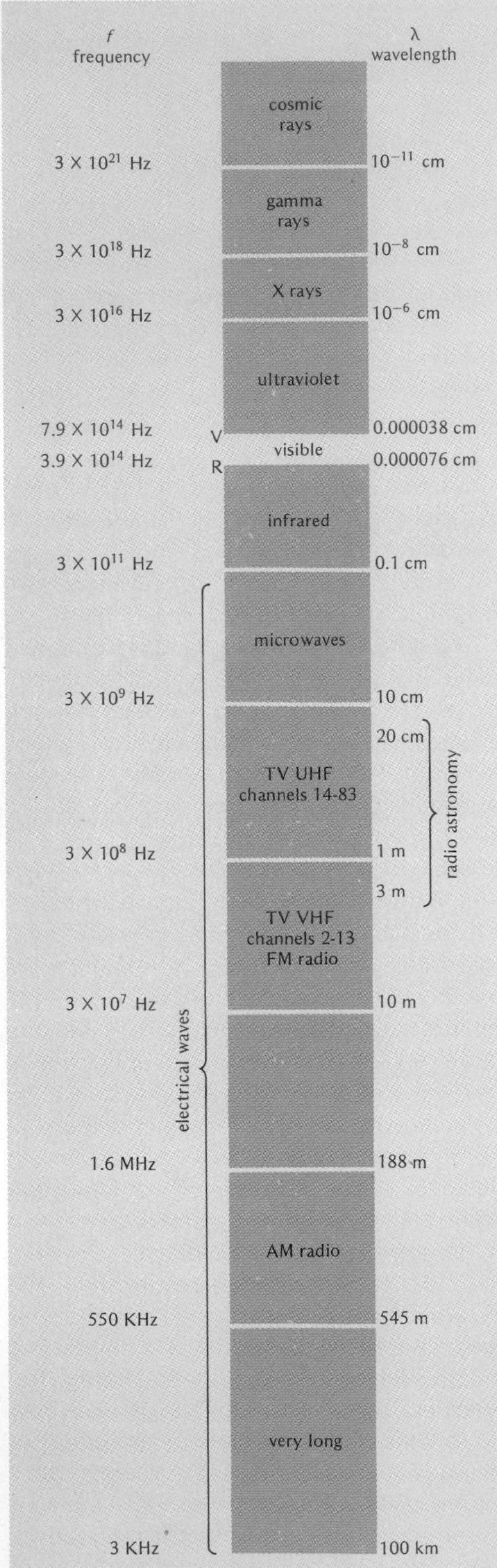

Figure 9.27
The electromagnetic spectrum. On the left-hand side of the diagram appear the frequencies whereas the corresponding wavelengths appear on the right-hand side of the diagram. The frequencies and wavelengths are related by $v = f\lambda$, where $v =$ the speed of light (3×10^8 m/sec) and is the same for all wavelengths of the electromagnetic spectrum.

lengths between 10 cm (0.1 m) and 1 m. The frequencies for these limits are 3000 to 300 megahertz (MHz), respectively. A megahertz, until recently called a "megacycle," is a million cycles, which may be written 10^6 cycles. This UHF band from 0.1 to 1 m ranges from Channel 14, with frequency 475.75 MHz and wavelength 0.63 m, to Channel 83, with frequency 889.76 MHz and wavelength 0.337 m.

The wavelength band that was reserved for the first allocations made to commercial TV stations was the VHF band, with wavelengths between 1 and 10 m. The frequencies for these limits are 300 and 30 MHz or 300,000 and 30,000 kilocycles (kc) or kilohertz (kHz), respectively. The VHF band from 1 to 10 m ranges from Channel 2, with frequency 59.75 MHz and wavelength 5 m, to Channel 13, with frequency 215.75 MHz and wavelength 1.39 m.

The electromagnetic waves used for FM radios (frequency modulation) have wavelengths close to 3 m. This very narrow FM band is between the wavelengths for certain TV channels. The wavelengths used by police cars and taxicabs are about 1 m. The band for standard AM radio (amplitude modulation) has a frequency range from 1600 to 550 kHz with wavelengths from 188 to 545 m, respectively.

The wavelengths given in the preceding paragraphs were computed in the manner shown in Example 9.2 as applied to radio and in Example 9.3, which follows, as applied to television.

Example 9.3
Compute the wavelength in meters of the television wave sent out by a Channel 2 station tele-

casting on an assigned frequency of 60 MHz (60×10^6 cycles).

Solution

$$\lambda = \frac{v}{f} = \frac{3 \times 10^{10} \text{ cm/sec}}{60 \text{ MHz/sec}} = \frac{3 \times 10^{10} \text{ cm/sec}}{60 \times 10^6 \text{ cycles/sec}}$$

$$= \frac{30 \times 10^9 \text{ cm/sec}}{6 \times 10^7 \text{ cycles/sec}} = 5 \times 10^2 \text{ cm/cycle}$$

$$= 500 \text{ cm/cycle} = 5 \text{ m/cycle or } 5 \text{ m}$$

Radio telescopes are now being used to study distant bodies many of which cannot be seen on photographs made using optical telescopes. One radio telescope, for example, is being used to study the incoming waves from distant bodies with frequencies of 100 to 600 MHz and wavelengths of 3 to 0.5 m, respectively. The field of radio astronomy is frequently providing us with additional information about bodies in our universe.

SUMMARY

There are two kinds of wave motion, transverse and longitudinal. Sound waves are longitudinal waves. Transverse waves include water waves; waves on a string; and electromagnetic waves, such as those of light, radio, and television. The spectrum of electromagnetic waves (Fig. 9.27) is arranged from shortest to longest wavelengths as follows: certain cosmic rays, γ rays, X rays, ultraviolet radiation, the visible spectrum, infrared, and electrical waves, including microwaves, waves used in radio astronomy, television waves, and radio waves.

Sound waves require a material medium for transmission; electromagnetic waves do not. Sound is transmitted through different media at different rates. The speed of sound in air is 1088 ft/sec at 0°C and increases 2 ft/sec for each 1°C rise in temperature. The speed of electromagnetic waves is 3×10^{10} cm/sec in air or vacuum, which is about 186,000 mi/sec.

The speed of the different wavelength radiations that make up the electromagnetic spectrum does vary in a material medium such as glass. This is illustrated by dispersion.

The wave equation $v = f\lambda$ applies to all wave motion. The equation can be used to compute the wavelength of sound waves or of radio and television waves.

Six important properties of wave motion, which are common to all types of waves, are rectilinear propagation, reflection, refraction, diffraction, the Doppler effect, and interference. Dispersion is another important property of waves, but it applies only to transverse waves.

Waves travel in straight lines from a point source—rectilinear propagation. For a smooth, flat surface, the angle of incidence equals the angle of reflection; this is the law of reflection.

Refraction is the change in direction of a ray of sound or light in passing from one medium into another medium of different density. A knowledge of refraction is used in correction of the eye defects of nearsightedness and farsightedness. When monochromatic light (light of a single wavelength) is passed through a glass prism, refraction, but not dispersion, occurs. When white light from the sun or an incandescent lamp is passed through a glass prism, refraction occurs at the surfaces and dispersion is produced within the glass. Dispersion is the separation of ordinary white light into its component colors—red, orange, yellow, green, blue, indigo, violet—owing to the fact that different colors, or different wavelengths travel at different speeds in any material (except vacuum) and are refracted differently. An instrument for viewing the spectrum produced by dispersion is called a *spectroscope*. There are three different kinds of spectra: continuous, bright-line, and absorption.

When there is a relative approach of a source of sound or of light and an observer, the frequency of the sound heard or the light seen is higher than the source frequency; when there is a relative recession of a source of sound or of light and an observer, the frequency of the sound heard or the light seen is lower than the source frequency. This is known as the *Doppler effect.*

Interference is the phenomenon of the superposition of two sets of waves whereby they may tend to reinforce or annul one another. Interference applies to both sound waves and light waves.

Diffraction is the bending of waves as they pass the edge of an obstacle or as the waves pass through openings that are of dimension not too different from the wavelength of the wave.

Sounds are produced by vibrating bodies. There are three general types of sources of sound: vibrating strings or bars, vibrating surfaces, and vibrating columns of air. Sounds may differ in loudness, or intensity; pitch, or frequency; and quality. The quality of a sound depends on the overtones that accompany the fundamental, or tone of lowest frequency. Overtones, or frequencies that are integer multiples of the fundamental, are present in nearly all musical sounds.

The electromagnetic spectrum is an orderly arrangement of energy bands from those of short wavelengths to those of long wavelengths. Included are such common things as X rays, the visible spectrum, television, and radio. All of these waves travel through a vacuum at the same speed.

Important words and terms

vibration
frequency
period
wave
transverse wave
longitudinal wave
sound wave
electromagnetic wave
mechanical wave
rarefaction
condensation
wavelength
the wave equation $v = f\lambda$
rectilinear propagation
reflection
refraction
lens
reverberation
focal length
hypermetropia
astigmatism
diffraction
Doppler effect
interference
beats
standing waves
dispersion
spectrum
bright-line spectrum
absorption spectrum
continuous spectrum
red shift
loudness
amplitude
pitch
timbre
resonance
infrared
concave lens
convex lens
farsightedness
nearsightedness
myopia
ultraviolet
nodes
antinodes
electromagnetic spectrum

Questions and exercises

1. Distinguish between transverse and longitudinal waves.
2. State the law of reflection, and draw a fully labeled diagram to illustrate the law for a plane mirror.
3. Draw a fully labeled diagram to illustrate the dispersion of a beam of sunlight by a glass prism. On the diagram place at the appropriate places the names of two common forms of invisible radiant energy from the sun.
4. Multiple choice: Sound waves (a) have their origin in vibrating bodies; (b) require a material medium for transmission; (c) are transverse waves; (d) travel faster in cold night air than in warmer day air. Which are correct?
5. What kind of mechanical waves (a) can propagate in a solid; (b) can propagate in a fluid?
6. When two identical waves of the same frequency travel in opposite directions and interfere to produce a standing wave, what kind of motion does the medium exhibit at (a) the nodes of the standing wave; (b) the antinodes?
7. How does the appearance of the diffraction pattern from a narrow slitlike opening change as the wavelength increases?
8. (a) Distinguish between diffraction and interference. (b) Can there be diffraction without interference? (c) Can there be interference without diffraction?
9. When a wave slows in going from air into glass, what happens to (a) its frequency? (b) its wavelength? (c) its direction?
10. Which of the following has the lowest frequency? the shortest wavelength? (a) γ rays, (b) blue light, (c) infrared rays, (d) ultraviolet rays.
11. The speed of sound in air or any gas is related to temperature of the medium. In terms of molecular motion (kinetic theory) why should such a relationship be reasonable?
12. If you walk past a bell while it is ringing, the pitch will not appear to change. However, if you ride past the same ringing bell in a rapidly moving car, the pitch changes markedly. Explain the changes observed and state why the relative speed is important.

13. Draw and label a diagram, with solid lines showing the eye defect of nearsightedness, and use broken lines to show how to correct the defect. What type of lens should be used in spectacles worn by such a person? Draw the diagram and answer the question for *farsightedness*.
14. Distinguish between continuous spectrum, bright-line spectrum, and absorption spectrum by explaining how they are produced.
15. Explain how man on earth can tell which elements are in the sun or distant stars.
16. If the respective frequencies of two sources producing sound are 49 and 56 vib/sec, what is the beat frequency, that is, number of beats per second?
17. Assign the following to the proper columns, objective or subjective, with correct pairs on the same line: pitch, wave form, loudness, frequency, intensity, quality.
18. Completion: Listener *A* is 10 ft from a sound source and listener *B* is 50 ft away (outdoors). The sound will be ________ as loud for *A* as for *B*.

Problems

1. (a) Compute the wavelength of the sound emitted by a tuning fork of frequency 460 vib/sec when the temperature is 31°C. (b) If you were moving toward this tuning fork with a velocity of 10 ft/sec, what frequency would you hear?

Answer: (a) 2.5 ft; (b) 464 vib/sec.

2. (a) Compute the wavelength of the sound emitted by a G fork (384 vib/sec) when the temperature is 32°C. (b) If you were moving away from this tuning fork with a velocity of 12 ft/sec, what frequency would you hear?
3. Compute in centimeters and in meters the wavelength of the broadcast wave from an AM radio station that broadcasts on an assigned frequency of 800 kHz.

Answer: 37,500 cm; 375 m.

4. Compute in centimeters and in meters the wavelength of the broadcast wave from an FM radio station that broadcasts on an assigned frequency of 102 MHz [1 MHz = 1 million (10^6) cycles].

Answer: 294 cm; 2.94 m.

5. Compute in centimeters and in meters the wavelength of the television visual carrier wave telecast by a station on an assigned frequency of 700 Mhz.
6. Assume that the lowest audible note has a frequency of 16 vib/sec and the highest audible note has a frequency of 20,000 vib/sec. Taking the speed of sound in air as 1100 ft/sec, find the corresponding wavelengths in air. In each case think of some familiar object that has a size comparable with the wavelength in question.
7. Approximately how long does it take a radio signal or light to travel the distances in the following examples? (a) from this book to your eyes; (b) from a radio station 10 miles away to your radio; (c) from New York to Los Angeles; (d) from the sun to the earth.
8. A flash of lightning is seen by an observer 3 sec before the thunder is heard. If the sound traveled at 330 m/sec, at what approximate distance from the observer did the lightning flash occur?

Answer: 990 m.

9. A vibration of 200 cycles/sec produces a wave. (a) What is the frequency of the wave? (b) What is the period of the wave? (c) If the wave speed is 10 m/sec, what is the wavelength?

Project

Law of reflection

Materials required: Small plane mirror, a few common pins, protractor (available at bookstore, dime store, etc.), paper, a small block of wood or a small cardboard box to hold the mirror upright, and a flat surface into which pins may be stuck (cardboard, breadboard, or acoustical tile, for example).

a. Place a sheet of paper on a horizontal surface into which pins may be stuck. Place a plane mirror vertically against a block on the paper and draw a line across the base of the mirror to indicate its position. Would it be better to put the line in front or in back of the mirror? Explain. Place two pins, P_1 and P_2, somewhere in front of the mirror, but do not place both pins on a line perpendicular to nor parallel to the mirror. Place the two pins about 1.5 in. apart.
b. Place a third pin P_3 in line with the *images* of the first two pins. Place a fourth pin P_4 in line with the third pin and the images of the first two. Pins 3 and 4 should be about 1.5 in. apart. Remove block and mirror. Draw two lines, one connecting holes of pins 1 and 2, and one connecting holes of pins 3 and 4. Continue both lines until they intersect. Where do they intersect? Draw a third line perpendicular to the

mirror through the point of intersection. Label locations of pins, using P_1, P_2, P_3, P_4; label the incident ray with arrow and words; and label the reflected ray.

c. Using a protractor, measure the angle between the incident ray and the perpendicular. This is the angle of incidence; mark it *i*. Measure the angle between the reflected ray and the perpendicular. This is the angle of reflection; mark it *r*. Does the angle *i* equal angle *r*?
d. State the law of reflection.
e. Hand in the paper used. Questions may be answered on the same paper in space not used for the experiment, or on the other side.

Supplementary readings

Books

Ashford, Theodore A., *The Physical Sciences: From Atoms to Stars,* Holt, Rinehart and Winston, New York (1967). [In Chap. 5 Ashford considers the properties of light that can be explained by either the wave or the particle theory, then the evidence for the wave nature of light, and finally the particle theory of light.]

Bascom, W., *Waves and Beaches,* Anchor Books, Doubleday, Garden City, N. Y. (1964). (Paperback.) [A general discussion of the dynamics of the ocean surface including the tides.]

Benade, Arthur H., *Horns, Strings, and Harmony,* Anchor Books, Doubleday, Garden City, N. Y. (1960). (Paperback.) [A physicist, flutist, and teacher gives a clear and comprehensive account of both the scientific and the aesthetic nature of music.]

Conant, James Bryant, *Robert Boyle's Experiments in Pneumatics,* Harvard University Press, Cambridge, Mass. (1950). (Paperback.) [A case history: Boyle's experiments on air, including those on air as a medium for transmitting sound (pp. 38–46).]

Greene, Earnest S., *Principles of Physics,* Prentice-Hall, Englewood Cliffs, N. J. (1962). [Chapter 14, "Sound," and Chap. 15, "The Physical Basis of Music," are based on a solid foundation of physical principles and show that music is not solely an art.]

Holton, Gerald, and Duane H. D. Roller, *Foundations of Modern Physical Science,* Addison-Wesley, Reading, Mass. (1958). [Chapter 29 deals with characteristics of any wave motion and electromagnetic waves; Chap. 30, with the particle and wave theories of light.]

Jaffe, Bernard, *Michelson and the Speed of Light,* Anchor Books, Doubleday, Garden City, N. Y. (1960). (Paperback.) [This biographical sketch combines an account of Michelson's dramatic career with sufficient explanation of the properties of light and the theory of relativity to make his scientific achievements understandable. He determined the velocity of light and proved that space had no "ether"—a medium postulated as necessary for the transmission of light waves.]

Kock, W. E., *Sound Waves and Light Waves,* Anchor Books, Doubleday, Garden City, N. Y. (1965). (Paperback.) [A simple description of the fundamentals of wave motion.]

Newhall, B., *Latent Image,* Anchor Books, Doubleday, Garden City, N. Y. (1967). (Paperback.) [The history of the discovery of photography.]

Physical Science Study Committee: *Physics,* D. C. Heath, Boston (1970). [Light is described in terms of waves and particles, Chaps. 1–8 and 16.]

Taylor, Lloyd William, *Physics: The Pioneer Science,* Houghton Mifflin, Boston (1941); or, a paperback, L. W. Taylor, and F. G. Tucker, *Physics: The Pioneer Science,* Vols. 1 and 2, Dover Publications, New York (1962). [The section on sounds deals with the nature and intensity of sound, the acoustics of rooms, and the pitch and quality of musical tone. Chapter 36 treats wavelengths of light; Chap. 37, spectra.]

The Project Physics Course, Holt, Rinehart & Winston, New York (1970). (Paperback.)

Van Heel, A. C. S., and C. H. F. Velzel, *What is Light?,* World University Library, McGraw-Hill, New York (1968). (Paperback.)

Reprints from *Scientific American*

44 Von Békésy, Georg, "The Ear" (August, 1957). [For his work on this topic he was awarded the Nobel prize.]

Articles

Greenewalt, C. H., "How Birds Sing," *Scientific American,* **221,** no. 5, 126 (November, 1969).

Michael, C. R., "Retinal Processing of Visual Images," *Scientific American,* **220,** no. 5, p. 104 (May, 1969).

Stark, L., "Eye Movements and Visual Perception," *Scientific American,* **224,** no. 6, 34 (June, 1971).

Weaver, K. F., "Remote Sensing: New Eyes to See the World," *National Geographic,* **135,** no. 1, 46 (January, 1969).

ELECTRICITY AND MAGNETISM

A great French philosopher . . . said to me once, "I understand everything in the book except what is meant by an electrically charged body."

HENRI POINCARÉ, 1885

Electricity plays a universal role in the life of millions of people throughout the world. As early as 600 B.C., the Greek philosopher Thales of Miletus had knowledge of what we now call static electricity. He observed that when amber was rubbed with silk, the amber would attract light objects, such as small pieces of dry grass, thread, and the pith from the center of stalks of weeds. It was not, however, until about A.D. 1600 that Dr. William Gilbert, physician to Queen Elizabeth and sometimes called the father of the modern science of electricity and magnetism, discovered that the effect could also be produced by rubbing together a great variety of other substances. Gilbert gave the name "electrification," from the Greek word *electron,* meaning amber, to the effect that was produced on various substances by friction. This phenomenon is now called "electricity."

The observations of Thales and Gilbert pertain to that branch of study which is now referred to as electrostatics, and is concerned principally with electric charges at rest. On the other hand, if the electricity, regardless of how it is produced, moves along a substance, such as a wire, it is called *current electricity*. Almost all current electricity is now produced by generators in which coils of copper wire are rotated between magnets.

Static electricity

ELECTRIC CHARGES

Rubbed objects, like the amber just mentioned, are said to be electrically charged or to have electric charges. The student will be able to find many pairs of substances such as rubber and fur, glass and silk, leather and wool (as in shuffling across carpeting), and wood and flannel, that are found to possess electric charges when rubbed together.

The question at once arises of whether there is just one kind of electric charge, or two, and only two kinds, or many different kinds. A few simple experiments will lead one to the correct answer to this question. Rub a glass rod vigorously with a piece of silk cloth and place the rubbed glass rod in a stirrup, or holder, suspended from a bar by a thread, as in Fig. 10.1 (top). Bring a second glass rod that has also been rubbed with silk close to the suspended rubbed glass rod. The suspended charged glass rod will swing away from the second charged glass rod, showing a force of repulsion between these charges of like sign.

Next bring a rubber rod that has been charged by rubbing with fur close to the suspended charged glass rod. The charged glass rod will now swing toward the charged rubber rod, as in Fig. 10.1 (bottom). Thus the kind of charge on the rubber rod is different from that on the glass rod, and there is a force of attraction between these unlike charges. They are distinguished by arbitrarily calling *positive* the kind on glass that has been rubbed with silk, and *negative* the kind on rubber that has been rubbed with fur. It has been found that if a rod of some other substance is rubbed with any other material—such as a wooden meter stick rubbed with wool—the rod will either repel or attract the charged glass rod. Thus there are two, and only two, different kinds of electric charge.

The previous discussion also brings out the fact that there is a *law of attraction and repulsion* for electrostatics:

Unlike charges attract each other, and like charges repel each other.

Quantitative experiments, which measure the amount of force of attraction or repulsion, also establish a *law for electrostatics,* first stated by Coulomb as follows:

Two unlike charges attract each other, and two like charges repel each other, with a force that is directly proportional to the product of the magnitude of the charges and is inversely proportional to the square of the distance between the charges.

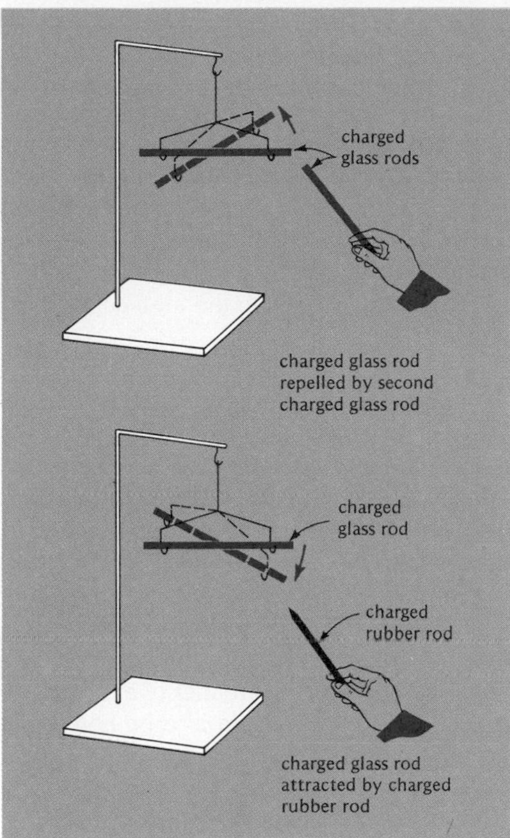

Figure 10.1
Like electric charges repel each other, and unlike attract.

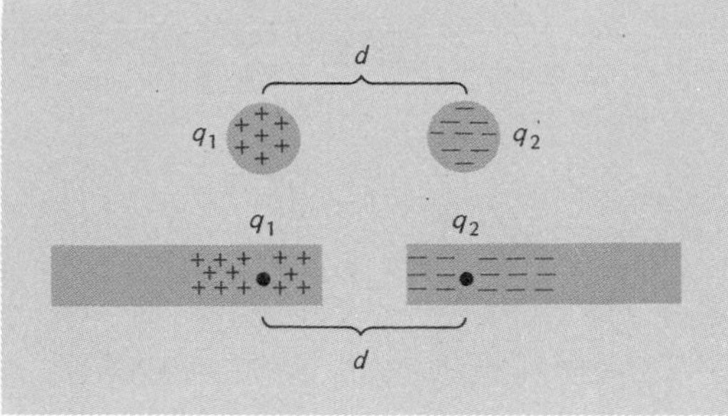

Figure 10.2
The Coulomb force between two sets of electrostatic charges: $F = k(q_1q_2/d^2)$. (a) Two spherical oppositely charged objects. (b) Two charge distributions. Coulomb's law applies to any shape object if the distance between the two charge distributions (d) is large compared to the size of the charged region.

This relation may be expressed as follows:

$$F = k\frac{q_1q_2}{d^2} \tag{10.1}$$

where F is the magnitude of the force between the two sets of charges q_1 and q_2, which are separated by the distance d, and k is a proportionality constant depending on the units used and the medium between the charges. For ordinary cgs units, k is unity for a vacuum and practically unity for air. Figure 10.2 illustrates this relation. Note that the distance d is between the centers of the two sets of charges. Also in this case, the net charges are represented by q_1 and q_2.

The student should note the great similarity between the expressions for the force of attraction between two masses, as expressed by the law of universal gravitation in Eq. (5.1),

$$f = G\frac{m_1m_2}{d^2}$$

and the force of attraction between unlike electrostatic charges, as given in Coulomb's law in Eq. (10.1).

$$F = k\frac{q_1q_2}{d^2}$$

It should be noted that Coulomb's law applies exactly only for pointlike spherical charged objects. However, Coulomb's law is highly accurate if the distance between the two objects, regardless of their shape, is large with respect to the size of the object. This approximation is also true in the case of the gravitational force.

MOVEMENT OF NEGATIVE CHARGES

Only negative charges move. In order to understand what happens when a glass rod is rubbed with silk, a brief discussion of the Bohr theory of the atom will be given at this time. Further discussion follows in Chap. 13.

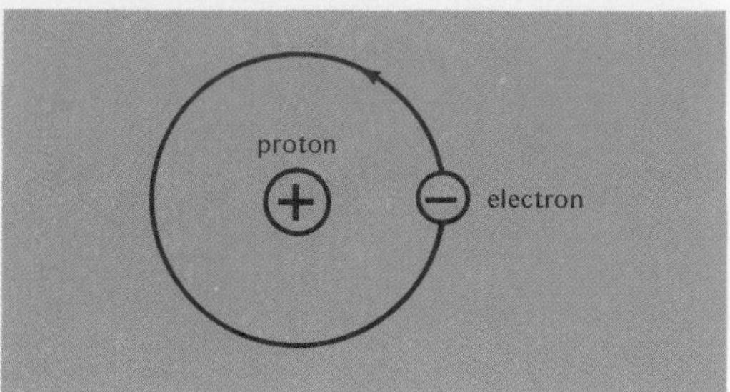

Figure 10.3
The Bohr concept of the hydrogen atom. The mass of the proton is 1836 times greater than the mass of the electron.

First, it should be noted that it has been found experimentally that electric charge occurs in nature in multiples of a smallest charge, a "quantum" of charge. The electron is one of the extremely small particles that carries *this smallest quantity of charge;* it is negatively charged. The proton carries a positive charge of amount equal to that carried on the negative electron.

According to the theory first proposed by Niels Bohr (1885–1962), the Danish physicist, in 1913, a hydrogen atom is made up of a central nucleus with a single positive electric charge, called a *proton,* and a planetary electron with a single negative charge that is considered to revolve around the proton as a planet revolves around the sun. The mass of the proton is 1836 times that of the revolving electron, as indicated in Fig. 10.3.

Atoms of other elements, according to the Bohr theory, have all their positive charges and practically all their mass located in the positive nucleus, with varying numbers of negative electrons revolving around the nuclei. The number of orbiting electrons equals the number of protons in the nucleus of an atom. This leads to matter being electrically neutral unless we do something to it to "charge it electrically."

This concept leads to the important idea that only negative charges move, since the negatively charged parts of the atom are less massive than the positively charged parts and lie nearer the periphery of the atom. Thus when glass is rubbed with silk, some of the negative electrons are rubbed off the glass onto the silk, leaving the glass with a net positive charge and the silk with a net negative charge. When rubber is rubbed with fur, some of the negative electrons are rubbed off the fur onto the rubber, giving the rubber a net negative charge and the fur a net positive charge.

If a glass rod is rubbed with a small piece of Saran wrap, it is very easy to show that the Saran wrap has obtained a charge opposite to that on the glass rod. This is a little more difficult to show when cloth or fur is involved because the charge is distributed over the fibers of the cloth or the individual hairs of the fur.

From the above discussion, it follows that the charging of a substance by rubbing it with another substance is a matter of a redistribution of charges. It should also be noted that an "uncharged," or neutral, body is not a body with no charges. Rather, it is a body with an equal number of well-mixed positive and negative charges, the net external effect being as if no charges were present (Fig. 10.4, top).

CONDUCTORS AND INSULATORS

The planetary electrons are held more firmly by the attraction of positive nuclei in some substances than in others. In some substances, notably the metals—copper, iron, aluminum, etc.—the negative charges move from atom to atom quite readily. *Conductors* are substances along which negative charges, or electrons, move quite freely. Nonconductors, or *insulators,* are substances along which negative charges do not move freely. Some common insulators are glass, hard rubber, plastics, and dry wood. This explains why wires of copper or other metals are used for conducting electricity and why the supports that hold the wires to the poles are

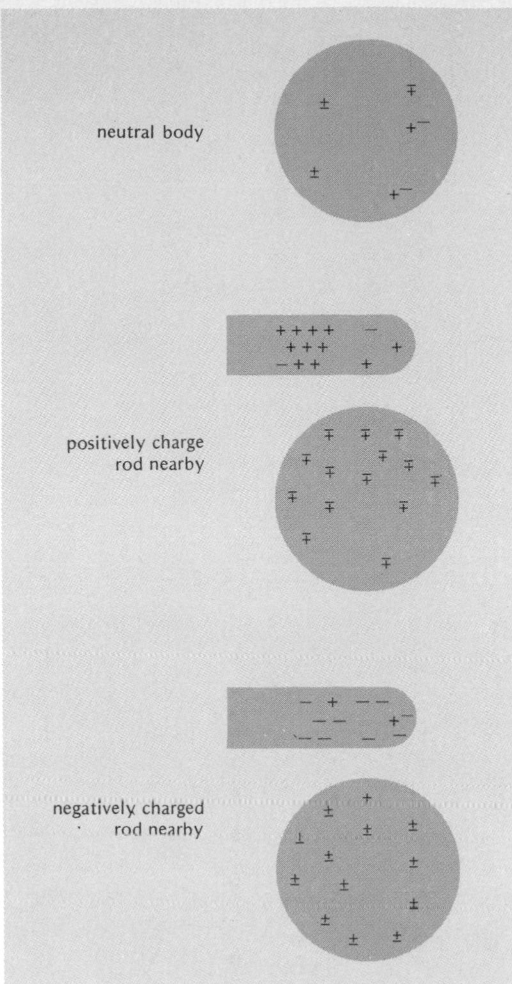

Figure 10.4
A charged object attracts a neutral object. The charges of a neutral body (top) are rearranged in the presence of a positively charged rod (middle) and a negatively charged rod (bottom). Such a rearrangement is called polarization and results in a net attractive force between the neutral object and the charged object.

made of glass, a plastic, or some other insulator.

It also explains why insulators when charged at a particular point, for example, at the end of a rubber rod, remain charged and the other parts of the insulator do not become charged. On the other hand if a metal object that is insulated from its surroundings is charged at any point, the whole metal object becomes charged and holds its charge with little leakage through the insulator.

ATTRACTION OF CHARGED OBJECTS BY NEUTRAL INSULATORS

If a strongly charged rod is brought near a neutral body, a rearrangement of the charges of the neutral body takes place, the body becomes polarized and acts as if it were given a charge opposite to the charge on the rod brought near to it. An explanation of Fig. 10.4 will lead to an understanding of why a charged body, like a charged glass rod, attracts a neutral body, like an uncharged pith ball or a small piece of paper.

For the time being let us consider the atoms of the neutral body to consist of simply a positive (+) and a negative (−) charge. The negative charges are bound to the positive charges by the electrical force. When a positively charged object is brought near to the neutral insulator (the pith ball in Fig. 10.4, middle), the atoms will be reoriented by the electrostatic force, so that the negative part of each atom will move to the side closest to the positively charged rod. The heavier positive nucleus of each atom will then be farther away from the negative rod than the negative part of each atom. In such a state the neutral insulating object has become polarized. Because the Coulomb force $F = k(q_1q_2/d^2)$ is smaller when d is increased, it is clear that the attractive force between the positively charged rod and the nearby negative part of each atom will be larger than the repulsive force between the more remotely placed positive nucleus of each atom and the positively charged rod. The net result is that the polarized neutral object is attracted by the charged positive rod. A similar effect will also occur when a negatively charged rod is placed near a neutral insulator as shown in Fig. 10.4 (bottom). Here the negative part of each atom is repelled away from the negatively charged rod. Now the negative rod and the nearby positive nucleus of each atom will have a larger attractive force than the repulsive force between the negative rod and the farther away negative part of the atom. A net attractive force again exists. In either of the

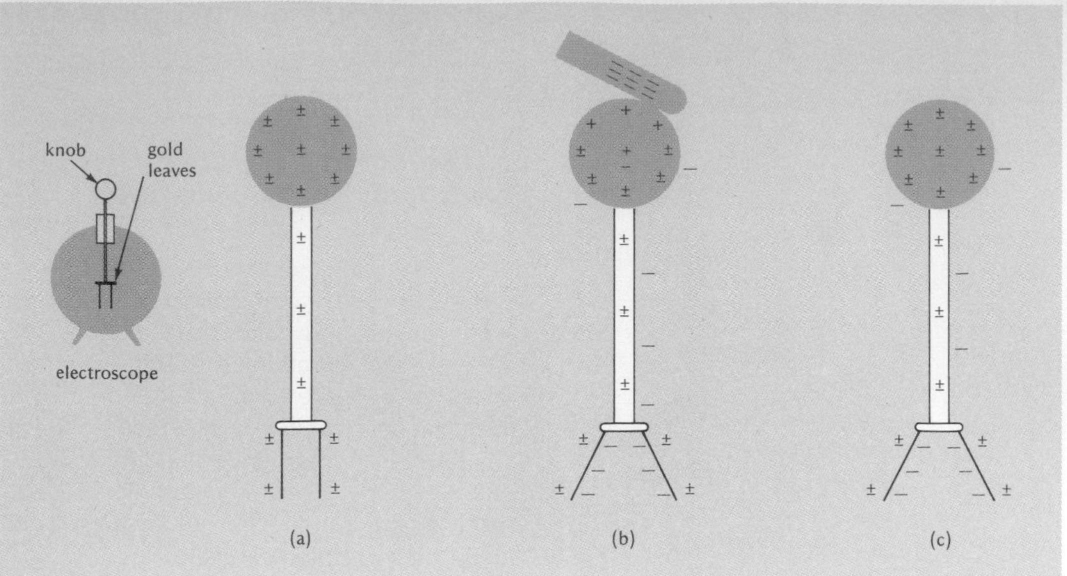

Figure 10.5
Enlarged view of the events in an electroscope charged by contact. In (a) the knob, stem, and leaves have an equal number of positive and negative charges and are neutral as a unit. In (b) charges move off of the negative rubber rod that contacts the knob; some electrons are also repelled from the knob to the leaves that diverge. In (c), after the removal of the negative rod, the electrons regroup themselves toward the knob, leaving a net negative charge over the entire electroscope; the leaves will remain separated. Note that the net charge on the electroscope is the same as that on the charging rod, which is negative in this example.

above cases we say that the charged rod "induced" a charge on the neutral object.

A charged rubber rod (negative) will readily pick up very small pieces of paper. The paper becomes polarized and is attracted to the negative rod; but then as some charges flow off the rod onto the paper, the paper becomes negatively charged and flies off the rod by virtue of the repulsive forces.

It follows that the only sure test of kind of charge is the repulsion test. If a charged glass rod repels a body to be tested, the body must be positively charged. If the body to be tested is attracted by the glass rod, however, it may be negatively charged, or it may be neutral and attracted because of the induced-charge effect. To determine which of these two possibilities is correct, a charged rubber rod should also be used. If the rubber rod repels it, the body is negatively charged, but if the rubber rod also attracts it the body is neutral.

CHARGING OF AN ELECTROSCOPE BY CONTACT

In the preceding paragraph the discussion was primarily concerned with insulators. Suppose we consider how a negatively charged rubber rod affects an insulated metallic conductor. Consider Fig. 10.5a in which a neutral gold leaf electroscope is pictured. An electroscope is an instrument used to detect the presence of excess charge on an object and can also be used to measure quantities of charges. The electroscope usually consists of a metal rod with a knob on the top and a pair of leaves on the lower end. The leaves are usually made of very thin pieces of gold foil. The rod is held by means of an insulator in a stand which encloses the gold leaf portion. If a negatively charged rod touches the metal knob on the electroscope, some of the negative charges will move from the charged rod to the electroscope (Fig. 10.5b and c). Because the electroscope rod is made of a metal, the excess negative charges will distribute over the electroscope rod causing the light movable leaves to separate due to the electrostatic repulsive force resulting from the like charges on the leaves. When the charging rod is removed, the electroscope will remain charged with its leaves separated.

If a positively charged rod is brought up to a neutral electroscope and makes contact with the neutral electroscope, the net result would "look" the same; that is, when the rod is removed, the leaves of the electroscope remain apart. However, in the charging process negative electrons would move from the metal electroscope knob to the positively charged rod, thereby resulting in the electroscope losing negative charge. Thus the electroscope would become positively charged.

CHARGING AN ELECTROSCOPE BY INDUCTION

The result of bringing a negatively charged rod *near* (but neither in contact with nor so close that a spark leaps from the rod to the knob) a neutral electroscope is shown in Fig. 10.6. The negatively charged rod repels the negative electrons on the metal electroscope, forcing the electrons away from the top of the electroscope. The leaves of the electroscope will contain an excess of negative electrons and will separate. The electroscope is still neutral, but the presence of the negatively charged rod has redistributed the charge on the electroscope. If you touch the knob on top of the electroscope with your finger while the negatively charged rod remains near the electroscope (Fig. 10.6b), negative electrons that are free to move in the electroscope will move to your body. You have "grounded" the electroscope, and the electrons have been repelled off the electroscope. If you remove your finger and then remove the charged rod (Fig. 10.6c), the electroscope will remain positively charged, and its leaves will be separated. In charging by induction the electroscope was not contacted by the charging object, and the resulting charge on the electroscope was opposite to that on the charging object. In this process it is the negative electrons that move.

In the charging of an electroscope by induction, it was necessary to ground the electroscope at one step in the process. This means that you connect the electroscope to an object, usually the earth, which is essentially an infinite source or sink of electrons. This is illustrated further in the section in which we discuss lightning.

DISTRIBUTION OF CHARGE

If a metallic sphere is charged negatively by rubbing a charged rubber rod against it, the excess negative charge on the sphere will be uniformly distributed, as in Fig. 10.7 (left). If the charged rubber rod is applied to an egg-shaped conductor, tests show that more of the excess negative charge accumulates near the pointed portions than near the flatter portions, as shown in Fig. 10.7 (right). In other words, charges tend to accumulate on the more pointed portions of a conductor. At sharp points the accumulation of negative charge may be so great that some of the excess charge is repelled into the air, and the charge is said to "leak off the points." If the excess charge is positive, negative charges from the air "leak" to the points.

It is difficult to explain on the basis of Coulomb's law why charges build up on sharp points of a metal object. This phenomenon can be explained, however, in terms of

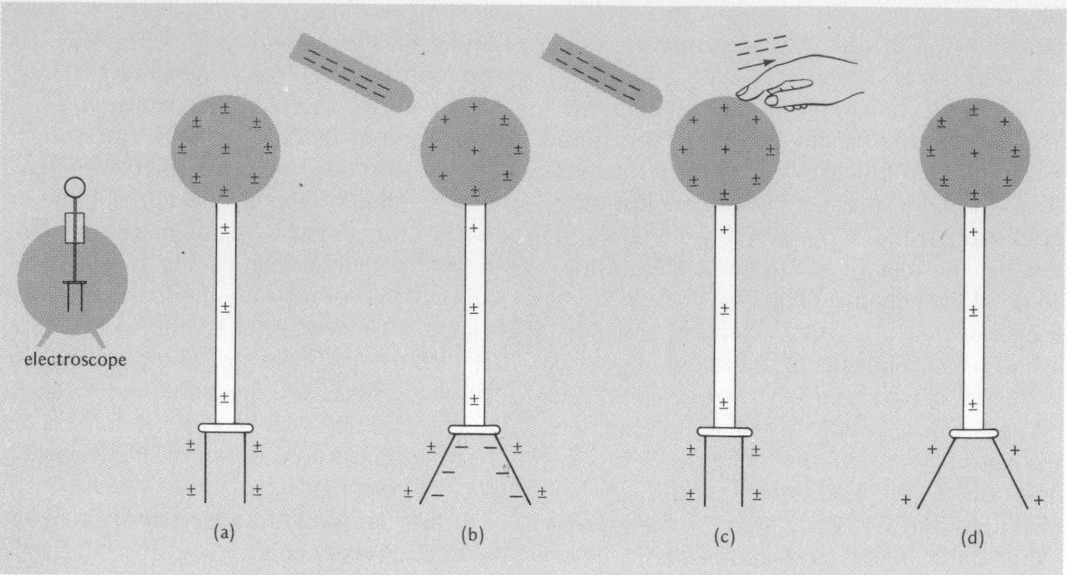

Figure 10.6
Enlarged view of the events in an electroscope charged by induction. In (a) the knob, stem, and leaves of the electroscope have an equal number of positive and negative charges and are neutral as a unit. In (b) the negative rod brought near the knob repels electrons toward the stem and leaves, giving a net negative charge below, as indicated by the divergent leaves. If the knob is grounded by touching the knob as in (c) while the negative rod remains nearby, the surplus electrons on the leaves will flow to earth causing the leaves to collapse; electrons in the knob are still repelled by the nearby rod. After the finger is removed, *followed* by the withdrawal of the rod, the electrons redistribute themselves throughout the electroscope, leaving a net positive charge on the electroscope and the leaves diverged. Note that the net charge on the electroscope is the opposite to that on the charging rod.

a discussion that is slightly beyond the scope of this text.

LIGHTNING AND LIGHTNING RODS

In a thunderstorm the falling of raindrops through the air often causes an accumulation of negative charge on the rising moist air, leaving the falling raindrops with a net positive charge (Fig. 10.8). This process results in the bottom of the cloud becoming negatively charged as shown in Fig. 10.8. If there is a sufficient buildup of opposite charges, a discharge of lightning (a huge electric spark) may take place between the upper positively charged part of the cloud and the lower neg-

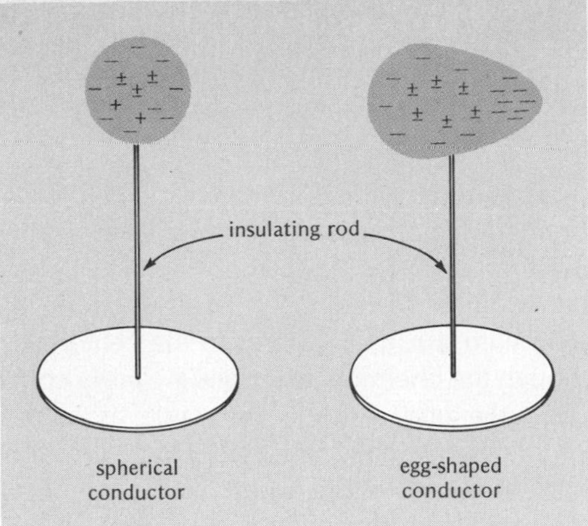

Figure 10.7
Distribution of excess negative charges. Excess negative charges on a spherical conductor are distributed uniformly but on an egg-shaped conductor accumulate mostly near the more pointed part.

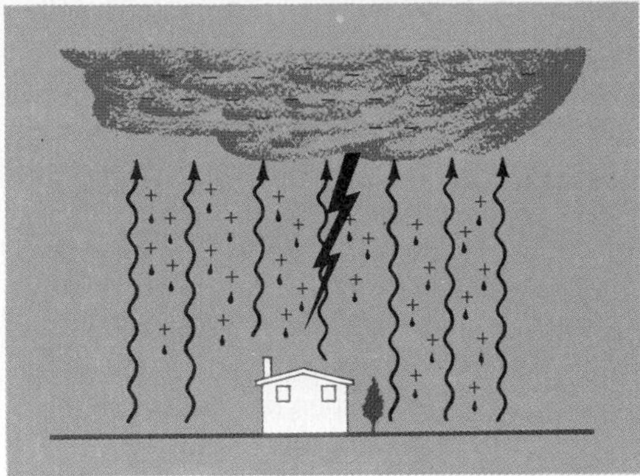

Figure 10.8
Rising moist air becomes negatively charged; falling raindrops, positively charged.

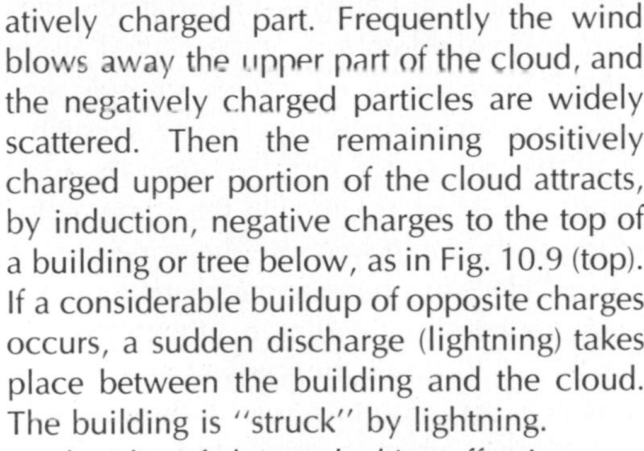

atively charged part. Frequently the wind blows away the upper part of the cloud, and the negatively charged particles are widely scattered. Then the remaining positively charged upper portion of the cloud attracts, by induction, negative charges to the top of a building or tree below, as in Fig. 10.9 (top). If a considerable buildup of opposite charges occurs, a sudden discharge (lightning) takes place between the building and the cloud. The building is "struck" by lightning.

The idea of charges leaking off points may be utilized in protecting buildings from lightning. A lightning rod is fastened to the building and the lower end buried deeply enough to extend into moist earth. When a positively charged cloud moves over the building, as in Fig. 10.9 (bottom), negative charges move up the rod to the top by induction and quietly leak off the upper point toward the positively charged raindrops. As the negative charges reach the raindrops, they neutralize the positive charges on the drops and stop any dangerous buildup of opposite charges, thus preventing lightning.

Sometimes a cloud will become negatively charged and hence induce a positive charge on the earth below. If a discharge or lightning takes place, then the negative charges will move to the earth in the process.

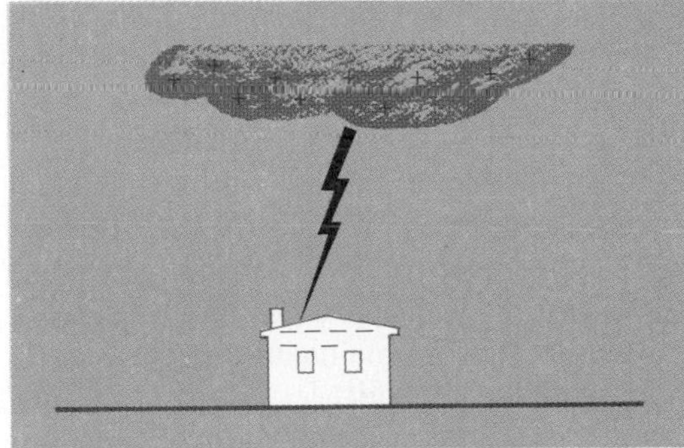

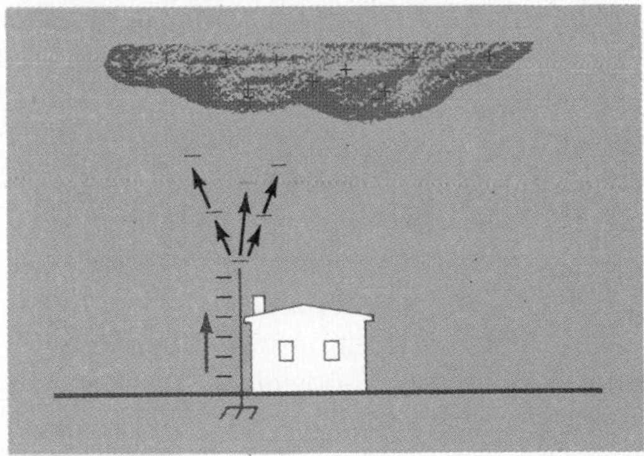

Figure 10.9
Lightning and lightning rods. Top, the induced charge on a building below a positively charged cloud results in lightning; bottom, the discharge from a lightning rod neutralizes the charges of a cloud; $\pm$ indicates neutralized charges.

It is primarily the electrons that move in these processes.

All of the details regarding lightning and the formation of charges on clouds are not clearly understood. However, the general processes described above are correct. It should be clear that it is usually the highest point above the earth, in a region below a charged cloud, to which the electric discharge or lightning will occur. Hence a man standing in a large open field during a thunderstorm is taking a dangerous risk.

Current electricity

If a negative rod is touched to a copper wire connected to an electroscope, two pieces of metal foil on the latter separate because of repulsion of their electrons that came through the wire. *Current electricity* in a wire, then, is a flow of electrons and is called *electron current*. To obtain an electron current large enough to be of any practical use, a continuous supply of electrons at one end of the wire and a lack of electrons, excess positive, at the other end must be maintained. Such a situation is called an *electrical potential difference* in analogy to a gravitational potential difference. Recall that if an object is held above the floor, it will have a greater potential energy with respect to the floor, hence, a gravitational potential difference exists between any position above the floor and the floor. If the object is released, it will fall to the floor. In the electrical case if there is an excess of negative charge at one end of a wire and an excess of positive charge (that is, a deficiency of electrons) at the other end, there is an electrical potential difference and negative electrons will "drop" or move from the negative region to the positive region. To maintain this electrical potential, some source of energy must be supplied.

About 1800, Count Alessandro Volta (1745–1827), the Italian physicist, found that this could be done by the chemical action within a certain type of cell, now called a *voltaic cell* in his honor.

The modern form of the voltaic cell is the *dry cell*. This consists of a moist paste of ammonium chloride and other chemicals contained in a sealed cylindrical electrode of zinc with a rod of carbon in the center. Although the chemical reactions are fairly complex, the end result is the same as in the simple voltaic cell, and a fairly large electron current is obtainable when a wire connects the two electrodes (zinc and carbon).

Atoms of an acid in dilute solution tend to separate into positive ions and negative ions. An *ion* is essentially a charged atom or molecule. Atoms and molecules are normally electrically neutral. However, if an electron is removed by some process, such as the frictional processes described earlier, the atom from which it was removed will become positively charged, and in its charged state it is called an *ion*. Some neutral atoms attract extra electrons rather strongly and often become negatively charged. In such a state they are called *negative ions*. The mass of an ion is approximately the same as the mass of the atom because the added or removed electrons have a much smaller mass; it was pointed out earlier that the proton has a mass 1836 times the mass of an electron. Ions play an important role in the chemical interactions among atoms and molecules; such interactions will be discussed in Chaps. 14–18.

A dilute solution of hydrogen chloride (HCl) will conduct electricity (see Fig. 10.10). The HCl in dilute solution in water tends to separate into hydrogen ions (H^+) and chloride ions (Cl^-). A negative chloride ion is attracted to a positive terminal (of a closed electrical circuit), where it releases an electron; a positive hydrogen ion is attracted to a negative terminal, where it takes on an electron. Thus current electricity may be the movement of ions in a solution. In neon signs high voltage causes neon atoms to break up into positive ions and electrons (Chap. 12). Both positive ions and electrons, by their movement in opposite directions, transfer charges; they constitute an electric current.

The electrical conductivity of a solution or

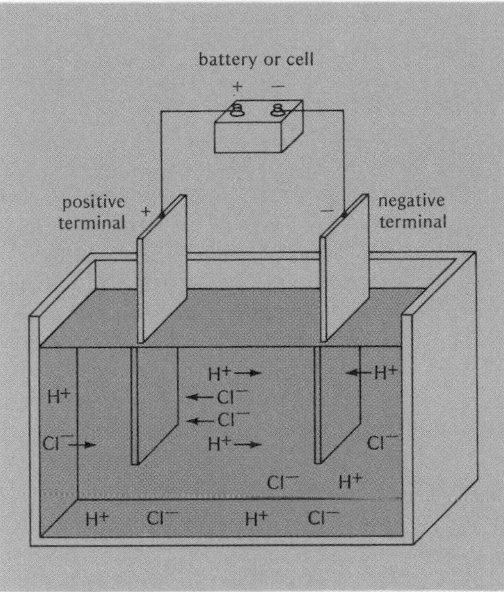

Figure 10.10
Conduction in dilute hydrochloric acid. The hydrochloric acid (HCL) separates into positive hydrogen ions (H) and negative chlorine ions (Cl) in a water solution. If a potential difference, set up by the battery or cell, is attached to two pieces of metal that are inserted in the dilute hydrochloric acid, an electric current will be set up through the acid.

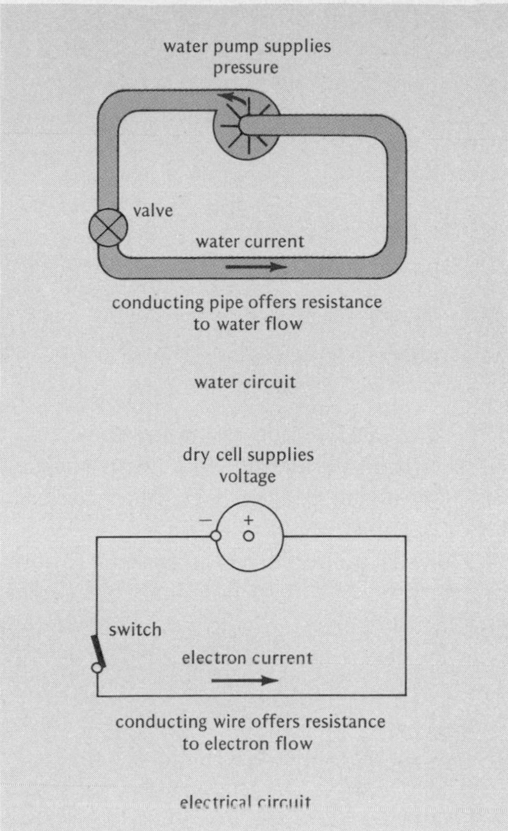

Figure 10.11
Analogy between water and electrical circuits.

of a gas differs from the charge transfer seen when rubber or glass is rubbed. In the latter case only electrons move. Likewise in a wire only electrons move. But in liquid solutions and in ionized gases some of the conductivity is due to the movement of positive and negative ions through the fluid.

WATER CIRCUIT ANALOGY OF AN ELECTRICAL CIRCUIT

The action of a simple electrical circuit can probably be best understood by comparing it with the action of a simple water circuit. Figure 10.11 shows the corresponding parts of the two circuits properly labeled. In the water circuit, the water pump creates water pressure, which tends to move the water. If the valve is open, there is a flow of water, or water current, in spite of the resistance caused by the rough inside surface of the pipe, which impedes the water current to some extent. In the electrical circuit, even with the switch open, the chemical action within the dry cell results in a tendency to cause a flow of electrons. If the switch is closed, this electrical pressure—more often called *voltage*, or electromotive force (emf), or, potential difference—results in a flow of electrons, or electron current, along the wire from the negative electrode to the positive electrode. The conducting wires offer a resistance to the flow of electrons, the amount of resistance being directly proportional to the length of the conductor, inversely proportional to the cross-section area of the conductor, and dependent on the material of which the conductor is made. This relation is expressed by the equation,

$$R = \rho \frac{l}{A} \qquad (10.2)$$

where

$R =$ resistance of conductor

$l =$ length of conductor

$A =$ cross-section area

$\rho =$ constant depending on material of wire and units used (Greek letter rho)

The total resistance in a simple series circuit is the sum of all the individual resistances in the circuit, including the "internal resistance" of the cell.

There is a simple and important relation of the three electrical quantities of current, voltage, and resistance to one another that was discovered by Georg Ohm (1787–1854), the German physicist, in 1826. Ohm's law, with the usual units placed in parentheses following the name of the quantity, is

$$\text{current} = \frac{\text{voltage}}{\text{resistance}}$$

or

$$I \text{ (amperes)} = \frac{V \text{ (volts)}}{R \text{ (ohms)}} \qquad (10.3)$$

where I is the electron current in amperes caused by voltage V expressed in volts through a circuit of resistance R expressed in ohms.

The voltage of a new dry cell is about 1.5 volts, and the voltage of a single cell of an automobile storage battery is about 2 volts. The usual automobile storage battery is a "battery" of three or six cells, each of 2 volts, resulting in 6-volt or 12-volt batteries.

The voltage of the ordinary house circuit, maintained by electric generators or dynamos, is about 118 volts; 120 volts will be used in Examples 10.1 to 10.3 for convenience. Ohm's law needs to be modified only slightly in order to apply to house circuits with alternating current; here the modification is disregarded.

Example 10.1
Compute the electron current through an electric flatiron that has a resistance of 24 ohms when the flatiron is plugged into a house circuit having a voltage of 120 volts.

Solution

$$I = \frac{V}{R} = \frac{120 \text{ volts}}{24 \text{ ohms}} = 5 \text{ amperes (amp)}$$

Example 10.2
In a simple circuit, a dry cell having an emf (voltage) of 1.5 volts and an internal resistance of 0.5 ohms is connected to a coil of wire that has a resistance of 19.5 ohms (this includes the resistance of the connecting wires). Compute the current through the coil.

Solution

$$I = \frac{V}{R} = \frac{1.5 \text{ volts}}{(19.5 + 0.5) \text{ ohms}}$$

$$= \frac{1.5 \text{ volts}}{20 \text{ ohms}} = 0.075 \text{ amp}$$

Example 10.3
Compute the resistance of an electric-heater coil through which 10-amp current flows when the heater is plugged into a house circuit (120 volts).

Solution

$$I = \frac{V}{R} \quad \text{or} \quad IR = V \quad \text{or} \quad R = \frac{V}{I}$$

$$R = \frac{V}{I} = \frac{120 \text{ volts}}{10 \text{ amp}} = 12 \text{ ohms}$$

ELECTRICAL POWER AND ELECTRICAL ENERGY

Electrical power may be expressed in terms of voltage and current, the relation being

$$\text{electrical power (watts)} = \text{voltage (volts)} \times \text{current (amperes)}$$

or

$$P = VI \qquad (10.4)$$

This electrical watt of power is the same watt that is used in the mechanical energy and power relation, where power (watts) = work

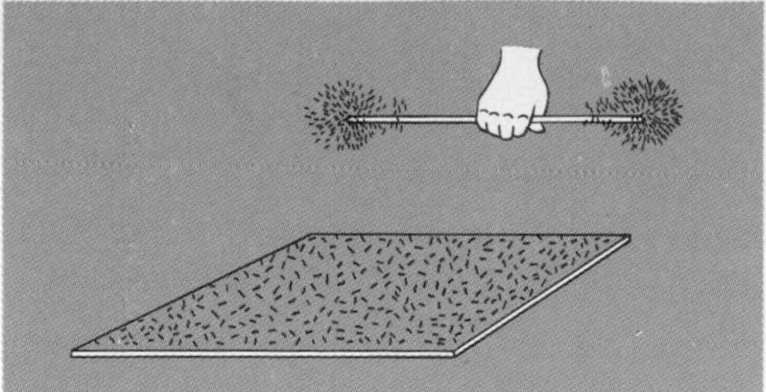

Figure 10.12
Magnet that has been rolled in iron filings.

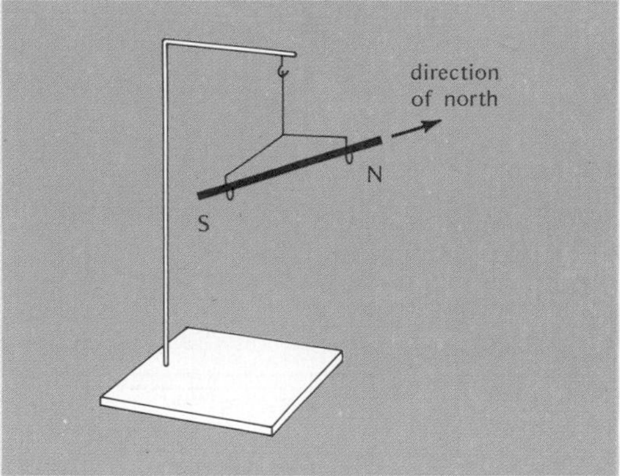

Figure 10.13
Freely suspended magnet comes to rest in a generally north-south direction.

or energy (joules) divided by time (seconds), whence energy = power × time, or joules = watt-seconds. Applying this general relation to electrical energy, the power is usually expressed in kilowatts [1 kilowatt (kW) = 1000 watts] and time in hours (1 hour = 3600 sec). Hence electrical energy = electrical power (kilowatts) × time (hours) can be and usually is expressed in kilowatt-hours (kWh). Also from the above relations a kilowatt-hour of energy is seen to be 1000 × 3600 watt-sec, or 3,600,000 joules of energy. We are billed for so many kilowatt-hours of electrical energy by the electric companies. You should check your understanding of the foregoing by doing Probs. 5, 6, and 7 at the end of this chapter.

MAGNETISM

The tremendous advance in the use of electricity depends to a great extent on the magnetic effect of electron currents. Although it is not considered advisable in this textbook to go into the details of the action of generators, motors, transformers, etc., we will briefly discuss simple magnets and the magnetic fields due to electron currents and the basic principles on which the generation of most of our electrical energy is based.

TWO PROPERTIES OF MAGNETS

Properties that characterize magnets are (1) they attract small bits of iron or steel (Fig. 10.12) and (2) if freely suspended, they will assume a general north-south direction (Fig. 10.13). Few substances have these magnetic properties. Iron and steel, and to a slight degree, nickel and cobalt, are magnetic. In recent years a few alloys that can be made into stronger magnets than steel have been developed. Among these, and of great commercial importance, is Alnico, which is an alloy of aluminum, by itself not magnetic, and nickel and cobalt, both weakly magnetic.

A simple experiment can be performed to show that most substances are nonmagnetic. Hold a strong magnet successively over small plates of glass, copper, brass, cardboard, zinc, aluminum, and iron. The iron plate will be picked up by the magnet, but no effect can be observed on any of the other materials.

When using magnets, care should be taken to keep the magnet away from a watch, for unless the watch is made of nonmagnetic metals, the magnet could damage the watch.

MAGNETIC POLES

When a bar magnet is rolled in a bed of iron filings, as in Fig. 10.12, the filings are found to cling mainly around two spots near the ends of the magnet. *Magnetic poles* are the two points near the ends of a bar magnet where the magnetism seems to be concentrated. When a bar magnet is freely suspended, as in Fig. 10.13, it will oscillate

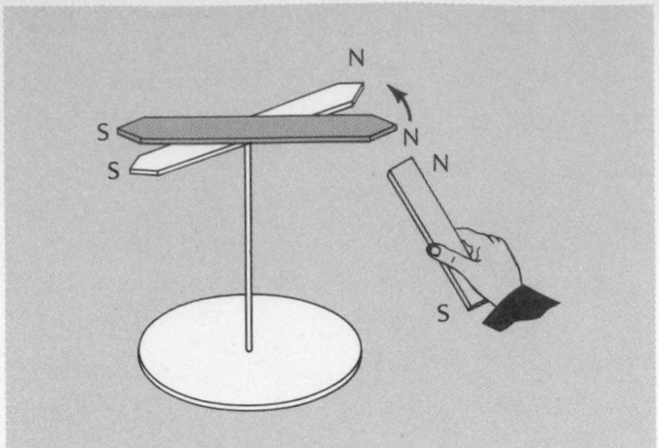

Figure 10.14
Repulsion of like magnetic poles, involving a bar magnet and a compass needle.

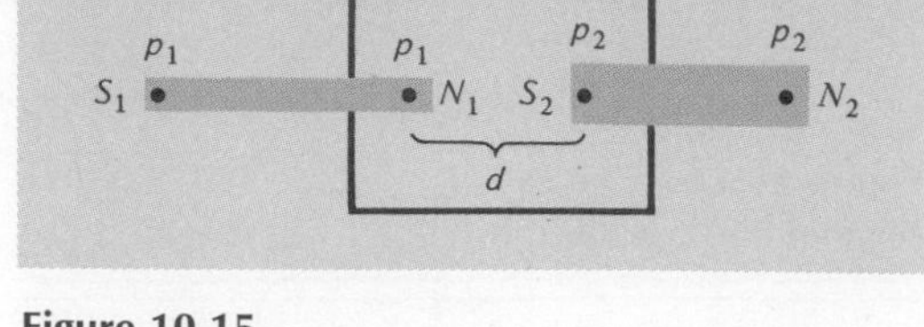

Figure 10.15
The force between magnetic poles: $F = (1/\mu)\ (p_1p_2/d^2)$.

back and forth and finally come to rest in a general north-south direction. The magnetic pole of a freely suspended magnet or compass needle that is then nearest north is called the *north pole,* and the other pole is the *south pole*.

If the north pole of one magnet is brought near the north pole of another magnet that is freely suspended, the latter will move away, as indicated in Fig. 10.14. If the south pole is brought near the north pole of the suspended magnet, the latter will move toward the nearby south pole. If the south pole is brought near the north pole of the suspended magnet, attraction is noted, and if the south pole is brought near the south pole of the suspended magnet, repulsion takes place. The foregoing effects lead to the *law of attraction and repulsion for magnetism:*

Unlike magnetic poles attract each other and like poles repel each other.

Careful measurements of the amount of force of attraction or repulsion between magnetic poles resulted in the relation first stated by Coulomb and now known as *Coulomb's law of magnetism:*

Two unlike magnetic poles attract each other, and two like magnetic poles repel each other, with a force that is directly proportional to the product of their pole strengths, and is inversely proportional to the square of the distance between the poles.

This relation may be put in equation form as follows:

$$F = \frac{1}{\mu}\left(\frac{p_1p_2}{d^2}\right) \tag{10.5}$$

where F is the force between the magnetic poles of strengths p_1 and p_2, respectively, which are at a distance d apart (Fig. 10.15), $1/\mu$ is a constant depending on the medium between the poles, μ being called the *magnetic permeability* of the medium and being exactly unity for a vacuum and practically unity for air. The student will note the similarity of this relation and the previous expression for the force of attraction between masses [Eq. (5.1)] and the force of attraction between unlike charges [Eq. (10.1)].

Actually, isolated magnetic poles are not possible. As shown in Fig. 10.15, the left-hand magnet has two poles of equal strength p_1 and the right-hand magnet has two poles of a different equal strength p_2. In this case there are four Coulomb forces: the attractions between N_1 and S_2 and N_2 and S_1 and the repulsions between N_1 and N_2 and S_1 and S_2. To simplify the problem, we often think of two very long magnets, in which case the attraction between the two near poles N_1 and S_2 is so much greater than the other three forces that the latter are neglected. In Fig. 10.15 the rectangle is drawn to signify that only the effect of poles N_1 and S_2 is to be considered.

Considering only the two nearest poles, Eq. (10.5) tells us that, when the distance between the poles is doubled, the force is $(\frac{1}{2})^2$, or $\frac{1}{4}$, as great; if the distance is tripled, the

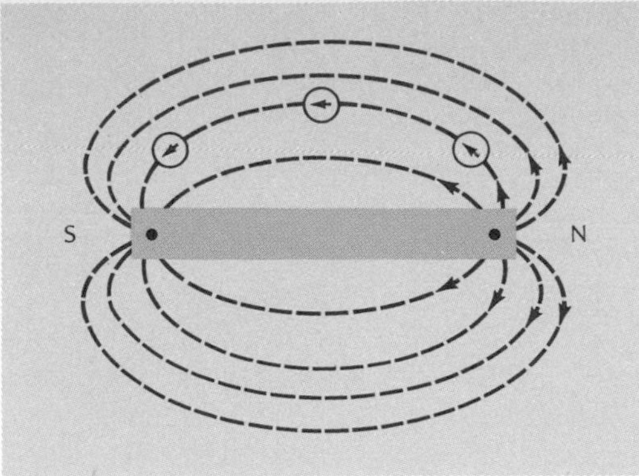

Figure 10.16
Magnetic field around a bar magnet, represented by magnetic lines of induction coming out of the north pole and entering the south pole. The circles represent a small compass.

force is $(\frac{1}{3})^2$, or $\frac{1}{9}$, as great; etc. If the distance *d* remains the same but a magnetic pole that is twice as strong as p_1 is substituted for p_1 and another magnetic pole that is three times as strong as p_2 is substituted for p_2 the force becomes 2×3, or six times as great.

To understand why magnetic poles always come in pairs, it is necessary to ask from what does the magnetism arise? If a magnet is physically cut in pieces, each piece will always have a north pole and a south pole. In fact, if the magnet were to be divided into individual atoms, one would find that the individual atoms have a north and a south pole. Magnetism is a property of the individual atoms. This property will be discussed further at a later point in this chapter.

MAGNETIC FIELDS

The strength of magnetic poles and their effects on one another can be portrayed by the use of imaginary lines of magnetic induction, sometimes called *lines of force,* that are considered to come out of the north pole and to enter the south pole (Fig. 10.16). This idea probably originated from the appearance of iron filings sprinkled on a paper or glass plate above a bar magnet. The space around a magnet that is affected by the magnet is called its *magnetic field.* Thus Fig. 10.16 represents the magnetic field of a bar magnet.

The lower half of the magnetic field in the vicinity of two bar magnets with unlike poles is shown in Fig. 10.17. The student should draw in the upper half, symmetrically. Likewise Fig. 10.18 shows the lower half of the magnetic field in the vicinity of two bar magnets with like poles adjacent. Again the student is urged to draw in the upper half of the magnetic field.

If a small pocket compass is placed in the vicinity of a bar magnet, the needle will point in the direction of the magnetic lines of induction, as shown in Fig. 10.16, where the arrow represents the north-pole end of the compass. This idea may be used to map the magnetic lines of induction of any magnetic field.

MOLECULAR THEORY OF MAGNETISM

Although today our knowledge of magnetism has greatly increased, and we have a correspondingly better, but more complicated, theory to explain magnetism, the older *molecular theory* of magnetism is still often used because of its simplicity. According to this theory, each molecule of a magnetic substance, like iron, is itself a magnet. In an unmagnetized bar of iron these small individual magnets are in random position, as in Fig. 10.19 (top), so that the external effect is nil. When the same bar is magnetized, the individual magnets are arranged with all their north poles in the same direction, as in Fig. 10.19 (bottom). Such an arrangement is called *ferromagnetic order* and in practice not all of the atomic or molecular magnets are aligned in the same direction.

INDUCED MAGNETISM

If one end of a piece of unmagnetized soft iron is slowly brought near, but not touching, the north pole of a large compass needle, attraction results. It would be reasonable to conclude that this end of the soft iron is a south pole, but when the same end of the

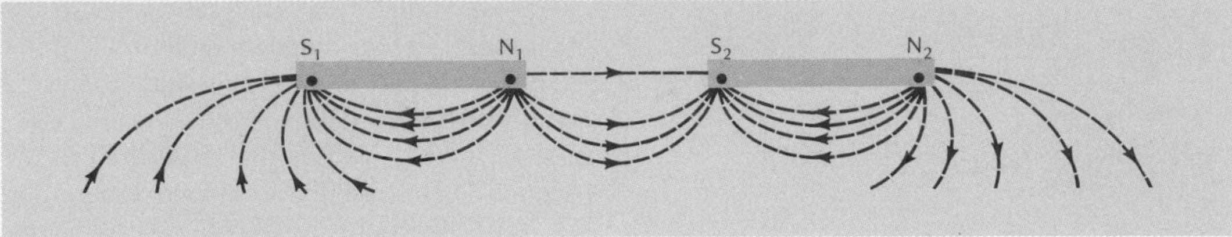

Figure 10.17
Magnetic field between unlike poles. Only the lower half is shown.

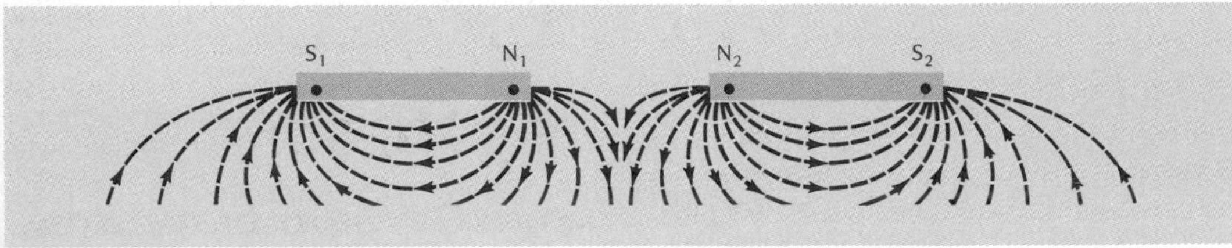

Figure 10.18
Magnetic field between like poles. Only the lower half is shown.

soft iron is brought near the south pole of the compass needle, attraction again takes place. This is spoken of as *induced magnetism*. Evidently the soft iron, although not itself magnetized, does become temporarily magnetized in the presence of a strong magnet, its near end always being of opposite polarity to the inducing pole. From this it follows that the only sure test of kind of polarity is the repulsion test. If one end *X* of a bar of iron is repelled by the north pole of a compass, it must be a north pole; but if end *X* attracts a south pole, it may be a north pole, but it also may not be magnetized at all, except temporarily, owing to the presence of a strong magnet.

The molecular theory is used to explain induced magnetism. When the south pole of a strong magnet is brought close to the right end of the unmagnetized bar of Fig. 10.19 (top), according to the law of attraction and repulsion, the north poles of the individual magnets tend to turn toward the strong south pole of the nearby magnet. This makes the soft iron temporarily a magnet because the individual molecular magnets are lined up as in Fig. 10.19 (bottom). When the strong magnet is removed, however, the random vibratory motion of the molecules of the soft iron soon cause a return to the unmagnetized condition.

Although the molecules of soft iron soon vibrate enough to cause the bar to become unmagnetized, once a steel bar is magnetized, it will remain so for a long time.

HOW TO MAGNETIZE

A steel needle can be magnetized by stroking it repeatedly in the same direction with the same pole of a strong magnet. To make the point end a north pole, stroke the needle several times from eye to point with the south pole of a strong magnet, as in Fig. 10.20.

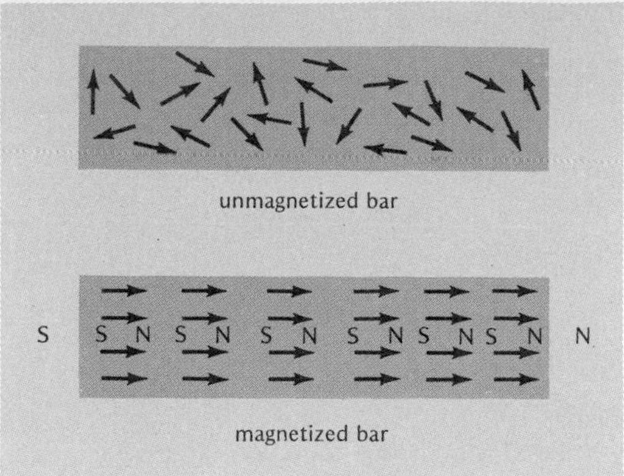

Figure 10.19
Unmagnetized bar of iron, at top, with haphazard arrangement of molecular magnets (arrows). Magnetized bar, at bottom, with all individual magnets aligned in the same direction.

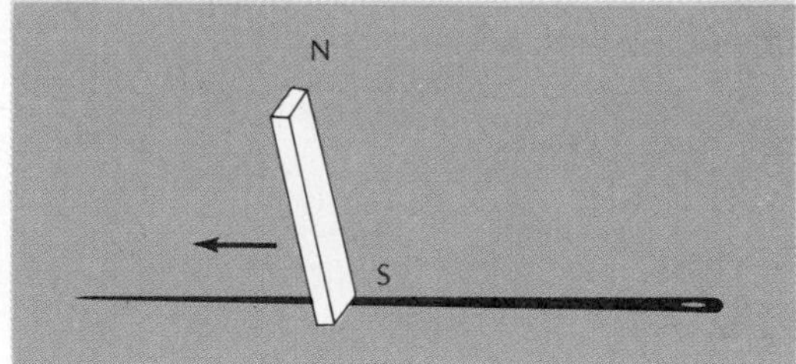

Figure 10.20
Magnetizing a steel needle so that the point will be a north pole.

METHODS OF DEMAGNETIZING

There are three ways to demagnetize a steel magnet: (1) Heat the magnet (the effect of heat is to cause the molecules to vibrate faster, so that they are soon out of magnetic alignment); (2) pound the magnet (the individual molecules are jarred to such an extent that they are soon out of alignment); (3) place the magnet in the magnetic field of a 60-cycle alternating current (the electric current causes the individual molecular magnets to point first in one direction, 1/120 sec later in the opposite direction, etc., soon resulting in a haphazard arrangement).

If you do wear a wrist watch while handling a strong magnet, the steel hairspring may become magnetized and the first two methods of demagnetizing are not recommended. A jeweler, however, can damagnetize your watch by the alternating-current method.

MAGNETIC FIELDS ASSOCIATED WITH ELECTRON CURRENTS

If a wire is held over a magnetic compass needle and the circuit containing the wire and a battery is closed by pressing a key, the north pole N of the compass needle below the wire will swing toward the east, as shown in Fig. 10.21 (top) if the electron current is south to north. The north pole N′ of the same compass needle held above the same wire will deflect toward the west.

If an electron current flows up along the vertical wire in Fig. 10.21 (bottom), which goes through a hole drilled through a horizontal glass plate, iron filings scattered on the plate will become arranged in circles. If instead of using iron filings, one moves a small magnetic compass needle along a circle around a vertical wire carrying current, it can be observed that the needle adjusts continuously and aligns itself tangent to a circular path around the wire.

These two experiments lead to the conclusion that a wire carrying an electron current is surrounded by circular magnetic lines of induction—that is, by a magnetic field. The direction, clockwise or counterclockwise, of the circular lines of induction is arrived at from Fig. 10.21 (bottom) (remembering that the north pole of a compass needle tends to point in the direction of the magnetic lines).

The first discovery of any connection between electricity and magnetism was made by Oersted in 1819. He observed the effect of an electric current on a magnetic compass needle. The compass needle was parallel to the wire carrying the current, but the needle rotated until it came to rest at right angles to the wire. When he reversed the direction of the current in the wire, the compass needle reversed its direction also and again came to rest at right angles to the wire. The observation occurred during one of his demonstra-

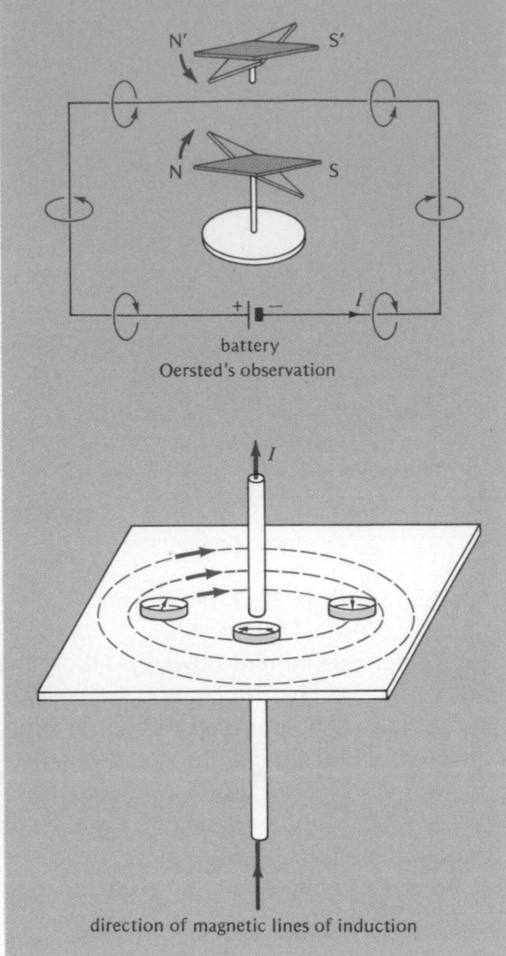

Figure 10.21
Top, the turning of a compass needle placed in a magnetic field due to electron current *I*; bottom, the direction of circular magnetic lines of induction due to an electron current flowing up through a wire.

tion lectures at the University of Copenhagen, Denmark.

Oersted's discovery, published in the form of a pamphlet dated July 21, 1820, was reported to the French Academy of Science. Capable experimenters immediately set to work on one aspect or another of the field thus opened up. The first thing that Ampère did was to refine and generalize Oersted's results by formulating in a graphic way a relation between the direction of the current and that of the motion of the needle.

Ampère's observation can be expressed

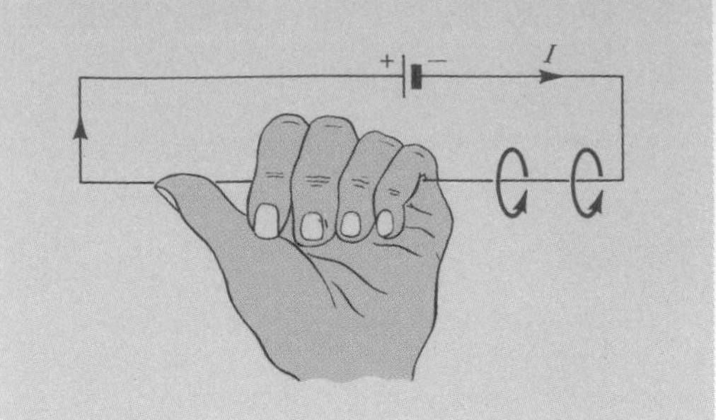

Figure 10.22
Left-hand rule for the direction of circular magnetic lines due to electron current *I*.

by the left-hand rule. The *left-hand rule* for determining the direction of the circular magnetic lines around a wire carrying a current is as follows:

If a wire is grasped with the thumb of the left hand pointing in the direction of the electron current, the fingers of the left hand give the direction of the magnetic lines around the wire.

(See Fig. 10.22; also apply the rule to Fig. 10.21.)

SOLENOIDS AND ELECTROMAGNETS

If a straight wire is made into a coil by being wrapped around a pencil, it is found that with an electron current passing through the coil one end of the coil acts like the north pole of a bar magnet and the other end like a south pole. This can be understood by applying the regular left-hand rule to any part of the coil; it will be found that on the part of the wire toward the inside of the coil magnetic lines of induction will be traveling through all parts of the inside of the coil in the same direction.

Applying the left-hand rule to the section of coil *ab* in Fig. 10.23, magnetic lines are seen to extend inside the coil from left to right. Likewise, considering a rear section *cd*, magnetic lines again are left to right inside the coil. Thus magnetic lines come out of the right end of the coil, which acts as the north-pole end of a magnet, and magnetic lines go into the left end, and this acts as a south pole.

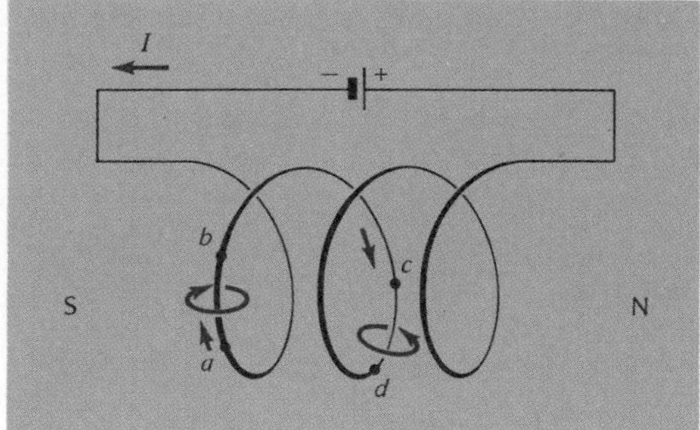

Figure 10.23
Solenoid. A current-carrying coil, or solenoid, acts as a bar magnet, with a north pole at one end and a south pole at the other.

A current-carrying coil is called a *solenoid*. If the right-hand end of the solenoid in Fig. 10.23 is brought near the north pole of a compass needle, it will repel it weakly. If a soft iron bar is placed inside the coil, the repulsion effect will be greatly increased. In other words, the use of an iron core within a solenoid greatly increases the magnetic effect. A solenoid having a soft iron core is called an *electromagnet*. Electromagnets are used as lifting magnets, in electric doorbells, in electric relays, in solenoid shutoff valves for gas furnaces, in telephone receivers, in electromagnetic circuit breakers (used in place of electric circuit fuses), etc.

The electromagnet has a great advantage over a permanent magnet, for the strength of the electromagnet depends on the magnitude of the current in the wire of the electromagnet. This current can, of course, be varied readily by changing the voltage applied to the electromagnet. In addition, a simple electric switch allows one to turn the electromagnet on and off.

THE ATOM AS A MAGNET

In the previous discussion it was pointed out that moving charges (that is, an electric current) produce a magnetic field. At the beginning of this chapter the Bohr theory of the atom was described. In Fig. 10.3 the Bohr model of a hydrogen atom is pictured. The electron is pictured as moving in a circular orbit. This moving electron will produce a magnetic field and results in the atom appearing to be a magnet.

In addition to the atom producing its own magnetic field it has been found that the electron (and also the proton) spins about an axis through its center much the way the earth spins daily on an axis through its center. This spinning of the electron also produces a magnetic field because the spinning of the electron represents a movement of charge. This results in each electron (and proton) having its own magnetic field. The properties intrinsic to the electron are its mass, its discrete and smallest unit of electric charge, and its magnetic field.

Although this discussion is based on a simple theory or model of the atom, the above results are retained by more modern and refined theories. A further discussion of the atom, the electron, and their properties is given in Chaps. 12 and 13.

Electromagnetic induction

Today we owe the large-scale production of electrical energy to Michael Faraday and Joseph Henry who discovered independently, but about the same time, how to induce an electromotive force (emf) or voltage and convert mechanical energy directly into electrical energy. If a wire is moved through a magnetic field or a permanent magnet is moved past a coil of wire, a current will be induced in the wire. The relative motion of a wire and a magnetic field gives rise to an induced emf or voltage. This principle is employed in our present electric generators in power plants or in an automobile. In a generator, normally a loop of wire or several loops of wire are rotated in a magnetic field produced by a permanent magnet or an electromagnet. The rotation of the wire loops is

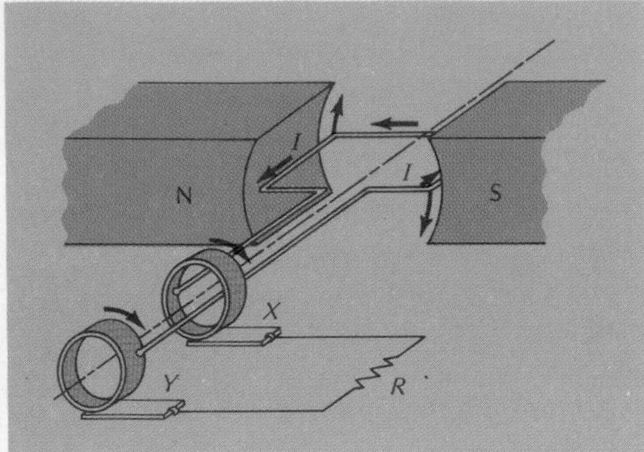

Figure 10.24
A single-loop alternating current generator. The loop is connected to a commutator consisting of two slip rings in contact with brushes *X* and *Y* and is connected to a load *R*. Note that the currents in the two halves of the loop reinforce each other and leaves from brush *X*. After a half rotation the two halves of the loop interchange places and the current leaves from brush *Y*.

produced by moving water (hydroelectric power) or from steam under pressure.

It should be noted that generators naturally produce an alternating voltage; for as the wire passes a magnet down past the south magnetic pole in Fig. 10.24, the induced current in the wire will be in one direction but when that same wire moves up past the north pole, the induced current will be in the opposite direction. This means that one end of the wire alternates from positive to negative during each half cycle (one-half rotation).

The discovery of electromagnetic induction and the invention of electric generators were major factors contributing to our modern life and the abundant use of electrical energy. We derive very little electrical energy from chemical cells and other such sources but rely on electromagnetic induction as our principal source of electrical power. In Chap. 19 we shall discuss the energy crisis and some of the problems created by the large demand for electrical power.

SUMMARY

There are two, and only two, kinds of electric charges. The kind of charge on glass that has been rubbed with silk is positive; on rubber rubbed with fur, negative.

Unlike electric charges attract each other, and like charges repel each other. The amount of the force of attraction or repulsion is in accord with Coulomb's law for electrostatics. The electrostatic force between two charged objects is proportional to the magnitude of the charge on each object and inversely proportional to the square of the distance between the two objects. The Bohr theory of the hydrogen atom is an aid to the understanding of electrostatics.

Solid conductors are substances along which negative charges, or electrons, move freely. Some electrons move from atom to atom in a wire conducting electricity.

A neutral body becomes polarized when a charged object is placed near it, and the neutral body is attracted to the charged object. The only sure test of kind of charge is repulsion. Charges tend to accumulate on the more pointed portions of a conductor. If there is sufficient buildup of opposite charges on two clouds or on a cloud and the ground, a discharge of lightning will occur. The idea of charges leaking off points may be utilized in protecting buildings from lightning.

Current electricity in a wire is a flow of electrons and is called *electron current*. The relation

$$\text{current} = \frac{\text{voltage}}{\text{resistance}}$$

is Ohm's law. Ohm's law applies to circuits with direct current and needs to be modified only slightly in order to apply to circuits with alternating currents.

Electrical power (in watts) = voltage (in volts) × current (in amperes). We pay the electrical companies for electrical energy in units of kilowatt-hours.

The use of electricity depends to a great

extent on the magnetic effect of electron currents.

Unlike magnetic poles attract each other, and like poles repel each other. The amount of the force of attraction or repulsion is in accord with Coulomb's law of magnetism.

The space around a magnet that is affected by the magnet is called the *magnetic field*. The magnetic field around a bar magnet is portrayed by magnetic lines of induction coming out of the north pole and entering the south pole.

The molecular theory of magnetism is used to explain induced magnetism.

Oersted observed the turning of a compass needle placed in a magnetic field due to an electric current. A wire carrying an electron current is surrounded by a magnetic field. The left-hand rule is used for determining the direction of the magnetic lines around a wire carrying a current.

A current-carrying coil is called a *solenoid*. A solenoid having a soft iron core is called an *electromagnet*. Electromagnets are used in common instruments, such as doorbells, telephone receivers, and lifting magnets. Magnetic effects in materials can be traced to the magnetic fields produced by moving charges and ultimately to the electrons and protons within the atoms.

If a coil of wire is moved relative to a magnetic field, a current will be induced. This process is called *electromagnetic induction* and accounts for most of our electrical energy.

Important words and terms

static electricity	electrical power
Coulomb's law	electrical energy
electron	magnetic poles
proton	magnetic field
Bohr theory of the atom	how to magnetize
conductor	how to demagnetize
insulator	electromagnet
charging by conduction	atomic magnet
charging by induction	electromagnetic induction
electroscope	ampere
lightning rod	volt
ion	ohm
resistance	watt
Ohm's law	solenoid

Questions and exercises

1. An electroscope has a positive charge. (a) Describe the condition of the electroscope leaves. Describe what happens to the electroscope (b) if a positive charge is brought near the electroscope; (c) if a negative charge is brought near the electroscope.
2. In terms of electrons and their motion describe what happens when a negatively charged rod is brought near an uncharged small piece of paper.
3. In terms of electrons and their motion describe what happens when a positively charged rod is (a) brought near an uncharged electroscope and then (b) touched to the electroscope.
4. When you walk across a rug in a dry room with leather-sole shoes you may become charged negatively. In terms of the motion of electrons describe what happened to you and the rug. What happens in terms of the motion of electrons when you touch a metal door knob?
5. It is not uncommon in dry weather for a long-haired person to comb his or her *clean* hair and find that the hair becomes electrically charged. Explain what takes place. What is the state of the comb?
6. You are given a charged object. How can you tell the sign of the charge on this object?
7. People inside a steel-framed building are safe from lightning. Why?
8. Explain how a lightning rod protects a building.
9. In terms of the movement of electrons describe what happens to a piece of silk and a glass rod when they are rubbed together.
10. List the similarities and differences among the electrostatic force law, the gravitational force law, and the magnetic force law.
11. Outline the similarities between the Bohr model of the hydrogen atom and Newton's description of the heliocentric solar system. What keeps the electron in its orbit?
12. Although the drift of the electrons in a wire is rather slow, an electric light bulb turns on al-

most immediately when the switch is closed. Explain. (Hint: The analogy with the flow of water through a pipe may be helpful.)

13. How can electric charges account for all magnetic effects?
14. What is meant by electromagnetic induction?
15. Explain in terms of atoms how a steel magnet can be demagnetized by heating it.
16. All atoms contain moving electrons. How is this fact connected with the observation that all atoms exhibit magnetic properties?
17. If the earth's magnetism is assumed to be due to a large circular loop of current in the interior of the earth, in what plane is the loop current?

Problems

1. A positive and a negative charge are initially 4 cm apart. When they are moved closer together so that they are only 1 cm apart, how will the force between them change?

Answer: Increased attraction by a factor of 16.

2. When two unlike charges are placed 6 cm apart, they exert a force of 600 dynes on each other. (a) Is this force attractive or repulsive? (b) If the charges are moved so that they are 3 cm apart, how large a force will they exert on each other?
3. Suppose we have two small charged spheres placed a distance D apart. If the charge on each is $+Q$, the force between them will be F. What will be the new force in each of the cases noted below? Only the parameters mentioned are changed from the original values.
   a. The charge on one of the spheres is changed to $-Q$.
   b. The charge on one of the spheres is doubled.
   c. The charge on both of the spheres is doubled.
   d. The distance between the spheres is doubled.
   e. The distance between the spheres is reduced by one-quarter.
   f. The physical size of the spheres is reduced by one-half.
   g. One of the spheres is moved to the opposite side of the first sphere, but they are still separated by a distance D.
   h. If one of the spheres is released, what will happen to it and how will the force between the two spheres change?
4. Is the resistance of a 120-watt light bulb greater or less than that of a 60-watt light bulb? Note that the voltage applied to either bulb is 110 volts.
5. If a resistance of 6 ohms is connected across the terminals of a 120-volt source, how much current flows in it? How much heat is developed in it per second?

Answer: 20 amp, 2400 joules/sec.

6. An electric lamp carries a current of 2.5 amp when a voltage of 110 volts is placed across it.
   a. What is the resistance of the lamp?
   b. If the current is on for 60 sec, how much energy is delivered to the lamp?
   c. What happens to this electrical energy?
7. Normally in a household each circuit (110 volts), which may contain several outlets or lighting fixtures, is fused to prevent more than 15 amp being drawn from that single circuit.
   a. Why is it unsafe to by-pass the fuse so that we can draw, say, 30 or 60 amp?
   b. Could you plug a 1000-watt toaster and a 400-watt coffee percolator into this circuit without blowing a fuse?
   c. Could you plug a 150-watt electric blanket, a 40-watt radio, four 100-watt lamps, and a 2-watt electric clock into this circuit without blowing the fuse?
8. When the north pole of a bar magnet is 2 cm from the south pole of a second bar magnet, the force of attraction between the poles is 36 dynes. If the distance is increased to 6 cm, what would the force be? Assume $k = 1$.

Answer: 4 dynes.

9. If at a distance of 12 cm two unlike magnetic poles attract each other with a force of 900 dynes, compute the force of attraction when the poles are 4 cm apart.

Answer: 8100 dynes.

10. A glass-blowing shop has an electric furnace for annealing glass. Glass is left in the furnace for 4 hours while the temperature rises to about 1050°F; then the electricity is turned off, and the furnace cools gradually for 16 hours before the glass is removed. This electric furnace uses 35 amp on a 220-volt circuit when in use only 4 hours a day on only 2 days a week. (a) Compute the rate at which

electrical energy is used by this furnace. (b) Compute the resistance of the heating elements (coils of wire) in this furnace. (c) At a rate of 3 cents/kWh, compute the cost of operating this electric furnace for an 8-week billing period. Round off all numbers up to the final answer. The latter is given to the nearest dollar and cents. Do not figure tax and discount.

Answer: (a) 7700 watts; (b) 6.3 ohms; (c) $14.78.

11. A 1440-watt room air conditioner was used on a 120-volt house circuit for a total of 300 hours during the month of August. (a) Compute the current through the air conditioner. (b) Compute the resistance of the air conditioner. (c) Compute the cost of using this air conditioner during the month of August. Use the rate of 3.5 cents/kWh. Do not figure tax and discount.

Projects

1. Electric meters and bills

a. Draw diagrams to show position of hands on the dials of an electric meter in your own home or the home of a friend. Below the diagram write the reading in kilowatt-hours.

b. Examine a recent electricity bill, and calculate from the data of this bill the average net cost per kilowatt-hour of electricity to nearest 0.1 cent.

2. Cost of electricity

a. Compute, separately, the cost of one lamp or group of lights and three electric appliances used in your home or the home of a friend for a 2-month period, using the average cost per kilowatt-hour from Project 1.

b. The form below is suggested.

Supplementary readings

Books

Bitter, Francis, *Magnets: The Education of a Physicist,* Anchor Books, Doubleday, Garden City, N. Y. (1959). (Paperback.) [A lively autobiography of a physicist's delight in probing the field of magnetism.]

Bondi, Hermann, *The Universe at Large,* Anchor Books, Doubleday, Garden City, N. Y. (1960). (Paperback.) [The earth's radiation belts and northern lights are considered in Chap. 8. In Chap. 12 the source of the earth's magnetism is discussed.]

Lemon, Harvey B., *From Galileo to the Nuclear Age,* University of Chicago Press, Chicago (1946). [Electricity is discussed clearly in Chaps. 25–27.]

Magie, W. F., *Source Book in Physics,* McGraw-Hill, New York (1935). Excerpts from original papers:

Gilbert, W., "On Magnetism and Electricity," pp. 387–393.

Franklin, B., "The One Fluid Theory of Electricity," pp. 400–403.

Coulomb, C., "Law of Electric Force—Law of Magnetic Force," pp. 408–420.

Galvani, L., "On Animal Electricity," pp. 420–427.

Volta, A., "On Electricity," pp. 427–431.

Oersted, H. C., "Actions of Currents on Magnets," pp. 436–441.

Ampère, A. M., "Actions between Currents," pp. 446–460.

Faraday, M., "Electrical Experiments," pp. 472–511.

Moore, A. D., *Electrostatics,* Anchor Books, Doubleday, Garden City, N. Y. (1968). (Paperback.)

Pierce, J. R., *Waves and Messages,* Anchor Books,

location	electric appliance or lamp	volts, V	amperes, I	watts, $P = VI$, or rated watts	hours used for 2 months: hour/day × 60	watt-hours, Wh	kilo-watt-hours, kWh	average cost per kWh from bill	cost of operating each item for 2 months
		120							
		120							

Doubleday, Garden City, N. Y. (1967). (Paperback.)

Rogers, Eric M., *Physics for the Inquiring Mind,* Princeton University Press, Princeton, N. J. (1960). [Chapter 28 deals with electrostatics; Chap. 34, with magnetism.]

Roller, Duane E., and Duane H. D. Rolle, *The Development of the Concept of Electric Charge,* Harvard University Press, Cambridge, Mass. (1954). (Paperback.) [A case history: developments that led to the establishment of the electric charge as a physical quantity and to the introduction of quantitative methods into electrical science.]

Taylor, Lloyd William, *Physics: The Pioneer Science,* Houghton Mifflin, Boston (1941); or, a paperback, L. W. Taylor, and F. G. Tucker, *Physics: The Pioneer Science,* Vols. 1 and 2, Dover Publications, New York (1962). [Facts, principles, and historical aspects of static electricity (Chap. 40); of current electricity (Chaps. 41, 44); and of the magnetic effect of a current (Chaps. 42, 45).]

The Project Physics Course, Holt, Rinehart and Winston, New York (1970).

Articles

Barschell, H. H., "Electrostatic Accelerators," *The Physics Teacher,* **8,** no. 6, 317 (September, 1970).

Becker, J. J., "Permanent Magnets," *Scientific American,* **223,** no. 6, 92 (December, 1970).

Dyal, P., and C. W. Parkin, "The Magnetism of the Moon," *Scientific American,* **225,** no. 2, 62 (August, 1971). [Recent magnetic measurements yield unique data on the moon's nature and history.]

Long, D., "Electrical Conduction in Solids," *The Physics Teacher,* **7,** no. 5, 264 (May, 1969). [A discussion of the mechanism for electrical conductivity in metals, semiconductors, and superconductors.]

Moore, A. D., "Electrostatics," *Scientific American,* **226,** no. 3, 46 (March, 1972). [A description of modern uses of electrostatics in Xerox copying and particle precipitators.]

Shiers, G., "The First Electron Tube," *Scientific American,* **220,** no. 3, 104 (March, 1969).

Shiers, G., "The Induction Coil," *Scientific American,* **224,** no. 5, 80 (May, 1971). [A historical review of the induction coil, the forerunner to the present transformer, and how it played a role in three major discoveries in physics.]

Walker, D. K., and O. Jefimenko, "Electrostatic Motors," *The Physics Teacher,* **9,** no. 3, 121 (March, 1971).

Weaver, K. F., "Magnetic Clues Help Date the Past," *National Geographic,* **131,** no. 5, 696 (May, 1967).

Reprint from *Scientific American*

825 Elsasser, Walter M., "The Earth As a Dynamo," **198,** no. 5, 44–45 (May, 1958).

STELLAR ASTRONOMY, AGE AND ORIGIN OF THE UNIVERSE

The astronomer's problem is not a lack of information but an embarrassing excess of it. His is often a problem of disentanglement rather than one of synthesis.

FRED HOYLE, 1955

The sun is a typical star and a member of a large system called the Milky Way, which is estimated to contain as many as 100 billion stars similar to the sun. The Milky Way, like all other galaxies, contains huge quantities of interstellar gas, mainly hydrogen, and interstellar dust. It is estimated that the total mass of interstellar gas in the universe is about equal to the mass of the stars. Probably the mass of the interstellar dust constitutes about 1 percent of the total mass of the universe. The universe is made up of hundreds of millions, perhaps 10 billions, of galaxies similar to the Milky Way. These external galaxies are scattered in space to at least about 5 billion light-years in all directions, and probably beyond. One light-year is the distance light travels in 1 year at its speed of 186,324 mi/sec. One light-year = 63,000 AU, roughly 6×10^{12} miles, or 6 million million miles.

Several rival theories of cosmology have received considerable discussion since mid-century. Basically, the models for these theories are either evolutionary or steady state. As we consider them, some aspects of our present knowledge of the age and origin of the universe, the earth, and the planets will evolve.

Figure 11.1
Spiral galaxy NGC 628. (Photograph courtesy of the Hale Observatories.)

Galaxies and stars

TYPES OF GALAXIES

In terms of their general appearance, galaxies have been divided into three general groups: spiral, which make up about 80 percent of the total; elliptical, comprising about 17 percent; and irregular, 3 percent.

The first catalogue of galaxies and star clusters was made by Messier in 1781. Orion, for example, is Messier 42, or M42. This catalogue is now superseded by the *New General Catalogue* (NGC), published by Dreyer in 1888, and two supplementary lists. Thus, the Great Andromeda Galaxy (Fig. 11.3) is referred to as M31 or as NGC 224.

Figure 11.1 is a photograph of the spiral galaxy NGC 628, with its plane at about right angles to the line of sight. In Coma Berenices NGC 4565 is a spiral galaxy oriented so that its plane includes the line of sight (Fig. 11.2). The nearest and best-known of the spiral galaxies is the Great Andromeda Galaxy (M31 or NGC 224), shown in Fig. 11.3. The plane of this huge galaxy makes an angle of about 15° with the line of sight. Although 870,000 light-years away, it is visible to the unaided eye under good conditions of visibility.

In Fig. 11.3, two of the elliptical types of galaxies are also visible. The one below the center of M31 is NGC 221 (M32); the one above and to the right of the center of M31 is NGC 205, which is resolved into thou-

Figure 11.2
Spiral galaxy NGC 4565 in Coma Berenices, seen edge on. (Photograph courtesy of the Hale Observatories.)

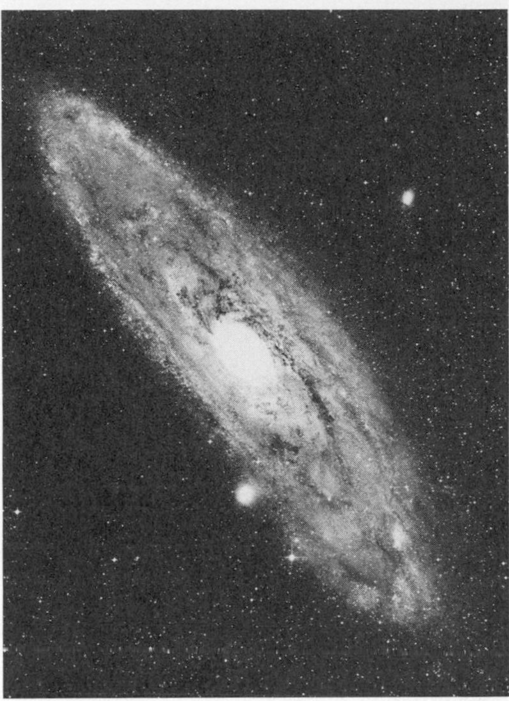

Figure 11.3
Great Andromeda Galaxy (M31) near Cassiopeia. (Lick Observatory photograph.)

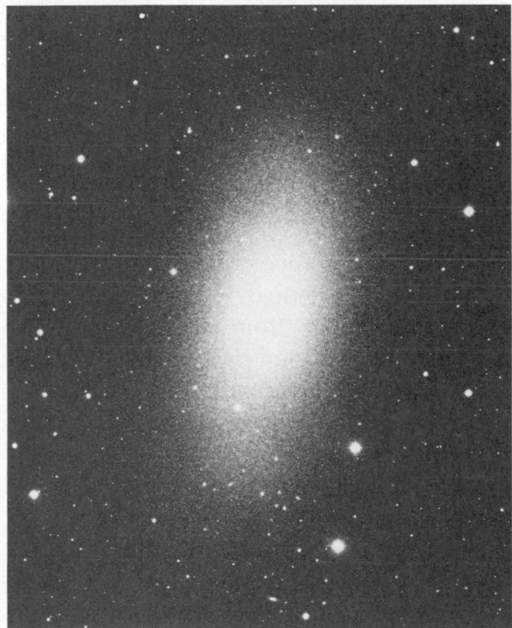

Figure 11.4
Elliptical galaxy NGC 205, photographed with the 200-in. telescope. (Photograph courtesy of Hale Observatories.)

sands of individual stars in Fig. 11.4, taken with a 200-in. telescope.

The nearest of the irregular galaxies is the Large Magellanic Cloud (Fig. 11.5), located near the south celestial pole in the constellation Dorado.

THE MILKY WAY

Our own galaxy is of the spiral type. It is thought to be in the shape of a disk that is about 100,000 light-years in diameter (*x*-axis; Figs. 11.6 and 11.7) and about 15,000 light-years in thickness at the center (*y*-axis). The sun is located about two-thirds of the way from the center of the Milky Way disk to the outer edge, in the direction of the constellation Taurus (Fig. 1.3), as shown in Fig. 11.6. The center of the Milky Way is in the direction of the constellation Sagittarius.

Figure 11.5
Large Magellanic Cloud. (Lick Observatory photograph.)

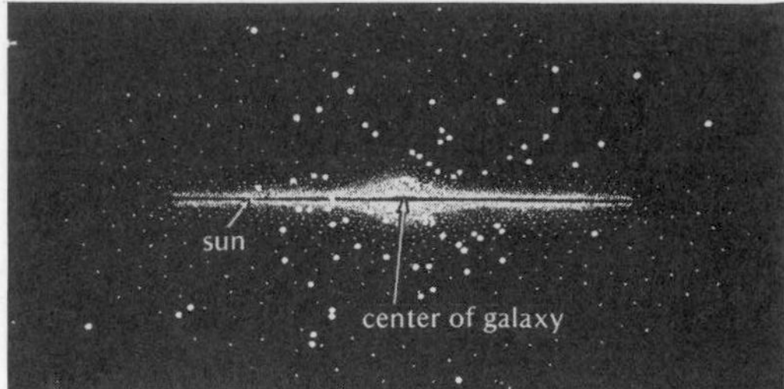

Figure 11.6
Diagram of the Milky Way, edge-on. The solar system, with the sun, is at the left, the large dots are globular star clusters. Most of the stars (small dots) lie near the plane of the Milky Way. (Yerkes Observatory photograph.)

The entire Milky Way is rotating about its center counterclockwise around an axis joining the galactic poles (north galactic pole above; view from above). The period of the Milky Way is about 200 million years, which means a speed of about 175 mi/sec for the sun.

The Milky Way contains, in addition to about 100 billion stars, much interstellar gas and dust, which give rise to bright and dark nebulae. Also, groups of stars called *galactic clusters* and *globular clusters* occur in considerable numbers.

INTERSTELLAR GAS AND DUST; BRIGHT AND DARK NEBULAE

From the shape of our galaxy and the location in it of the sun, one would expect the hazy bright band called the Milky Way to be much brighter in the direction of Sagittarius than in other directions, but it is not. This is because of the very great amount of interstellar gas and dust located between us and the center. Because shorter-wavelength energy is scattered more by the interstellar gas and dust than are longer wavelengths, the longer-wavelength infrared energy penetrates through the interstellar material better, so that infrared photographs are helpful in locating the center of the Milky Way. More recently, microwaves a few centimeters in length have enabled us to locate the center much more accurately.

Most of the obscuration is probably caused by the fine interstellar dust rather than the interstellar gas. Figure 11.7 shows the Milky Way in the direction of the galactic center, with bright star clouds obscured irregularly by interstellar dust. This dust is not distributed uniformly in space. The galactic center lies in the direction of the constellation Sagittarius.

There are instances of interstellar dust or gas that is near enough to stars to be illuminated by the light from those stars. The resulting glowing cloud brought out by a long time exposure is called a *nebula*. Diffuse nebulae result if the interstellar dust or gas is irregular in distribution near the illuminating star or stars.

Figure 11.8 shows a large bright nebula—the North American Nebula. The nebula is near Deneb (large star, upper right) in the constellation Cygnus, the swan (Fig. 1.2). The dark "bay" that forms the outline of the "Gulf of Mexico" is dark because there is little matter to reflect the light from illuminating stars. The aptly named Pelican Nebula is at the right.

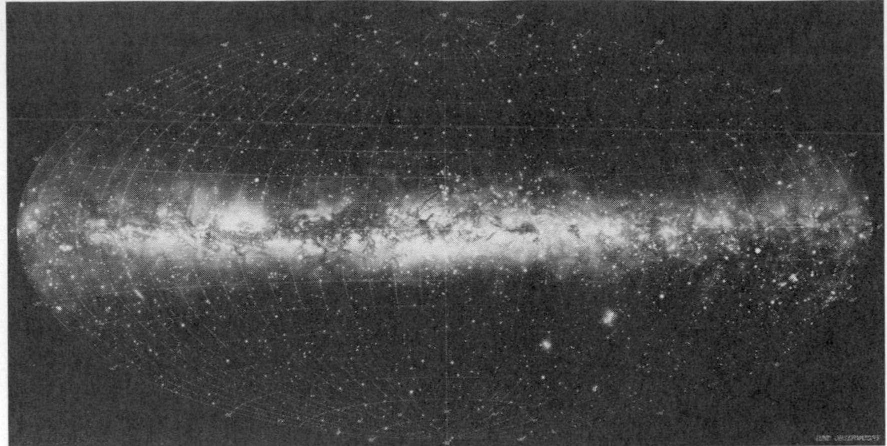

Figure 11.7
The Milky Way, drawn by Martin Keskūla and his wife under the direction of Knut Lundmark. The two bright spots in the lower right portion are the Magellanic Clouds. (Courtesy of the Lund Observatory, Sweden.)

Figure 11.8
North American Nebula. (Lick Observatory photograph.)

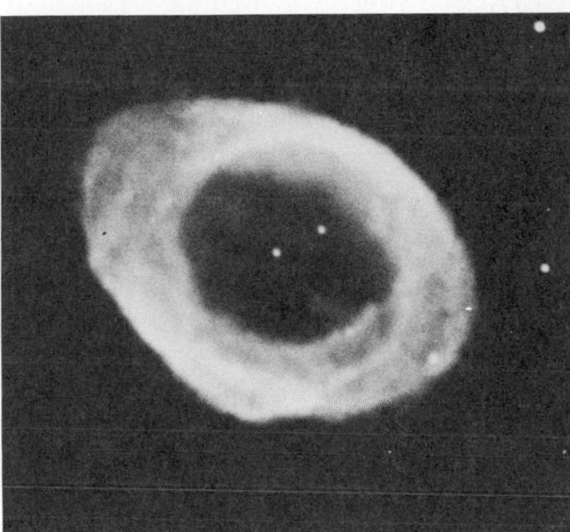

Figure 11.9
Ring Nebula in Lyra. (Photograph courtesy of the Hale Observatories.)

In some cases light is reflected from dust particles forming concentric shells around a central bright star, giving rise to planetary nebulae. One of the best known of these is the Ring Nebula in Lyra (Fig. 11.9).

Most of the light from the bright nebulae is light from the illuminating star reflected by neighboring interstellar dust and hence gives the same absorption spectrum as the nearby star. If the nearby star is one of the hotter stars, some of the short-wavelength radiation is absorbed by interstellar gas, which then reradiates energy of longer wavelength, thus giving a bright-line spectrum. If the enveloped star is of the particular type classified as B1 (temperature about 20,000°K), both types of spectra occur.

Clouds of interstellar dust too far from a star to be sufficiently illuminated to appear as bright nebulae will, if between us and star clouds or bright nebulous material, appear as dark as in Figs. 11.7 to 11.9. The presence of cold gas clouds, mainly hydrogen, although transparent to light, has recently been detected with radio telescopes.

GALACTIC AND GLOBULAR STAR CLUSTERS

The stars of our galaxy are likewise not distributed uniformly. Near the galactic plane are a few hundred groups called *galactic clusters* or *open clusters*. Typical of these condensations of stars are the Pleiades (Figs. 11.10 and 1.3) and the double cluster in Perseus.

Star clusters of a different type, the globular clusters, are located more or less uniformly in all directions from the center of the Milky Way out to a distance of about 160,000 light-years. These clusters are spherical, and each contains an enormous number of stars, probably often exceeding 100,000. The best known of these globular star clusters, of which over 100 have been photographed in our galaxy, is the globular cluster in the constellation Hercules (Fig. 11.11).

TYPES OF STARS

Spectra of stars—absorption spectra—are found to differ considerably, and stars have been classified according to their spectra. All but a few hundred stars fall into the groups O, B, A, F, G, K, and M. Class B stars contain neutral helium lines, lines of ionized oxygen, carbon, and silicon, and strong hydrogen lines; they are the hottest stars. In class A stars, the helium lines disappear, the silicon and oxygen lines are weaker, and the hydrogen lines become stronger, dominating the spectrum. Type F stars show a decrease in strength of hydrogen lines and the appearance of absorption lines of many metals, such as iron, calcium, sodium. Through G, H, and K, the hydrogen lines continue to become less prominent and the metallic lines stronger, with molecular absorption bands beginning to appear in class K. In type M the molecular bands strengthen, particularly those of titanium oxide (TiO_2).

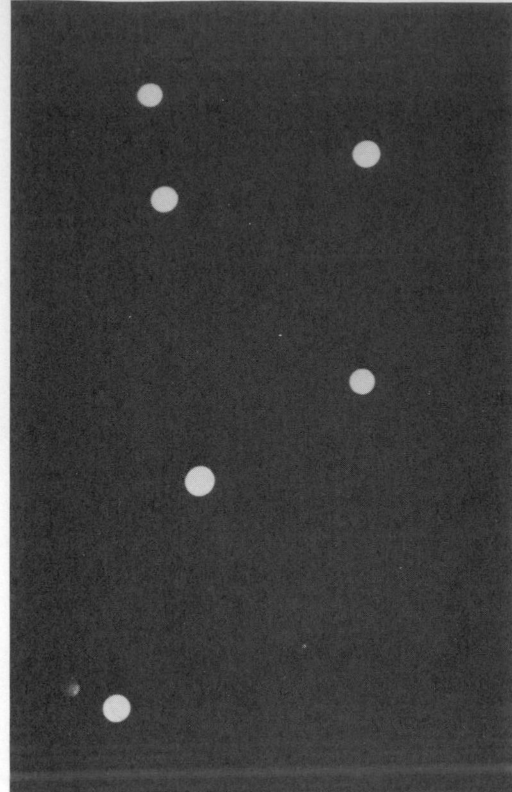

Figure 11.10
The Pleiades. (Yerkes Observatory photograph.)

The spectral types are undoubtedly more a matter of temperature than of intrinsic differences in chemical composition, the temperature being highest for the O stars and declining to lowest values for M stars. Because hotter stars give more light at the blue end of the spectrum, this classification also forms a color sequence progressing from blue-white stars of class O to the red stars of class M.

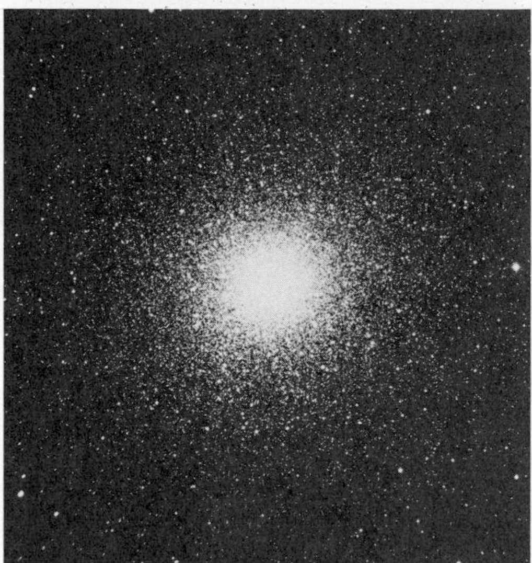

Figure 11.11
Globular star cluster in the constellation Hercules. (Photograph courtesy of the Hale Observatories.)

When stars are plotted with brightness along one axis and spectral type along the other, almost all of them fall along a sloping line called the *main sequence*. The sun lies in the main sequence and acts as a reference for us in Fig. 11.12. The sun is a yellow star with a surface temperature of about 6000°C (10,000°F) and a diameter of about 8.6×10^5 miles.

A few stars fall above the main sequence. They are the red giants (few in number) whose diameters range from about 6×10^7 miles to 7×10^6 miles and whose surface temperatures range from about 5500°F to 9000°F, making these very large but relatively cool stars. In addition to the red giants there are a few stars called *supergiants*. These stars also fall above the main sequence. The temperatures of these stars range across the diagram (Fig. 11.12) from 5000°F to 45,000°F and these stars have diameters ranging from 2×10^9 miles to about 3×10^7 miles, respectively. Note that the larger the supergiant, the lower is its temperature.

A few of the stars fall below the main sequence of stars and have been named *white dwarfs* due to their very high density, small size, and their color. These stars have diameters in the range of 5000 miles which is smaller than the diameter of the earth.

LIFE OF A STAR

The life of a star is so long that we have, of course, not been able to record the changes that occur in a single star during its lifetime. However, if one makes a single, basic assumption that most of the different stars we see in the sky are essentially the same objects but that their difference in appearance is due to their difference in age, then a rather consistent model of the development of stars can be constructed. That is, we shall assume that the multitude of different stars and gas clouds we observe are just "stars" in different stages of their lives. A typical star would have a "life" as described below.

At its birth a cloud of dust and gas collects in outer space due to gravitational attraction. It begins to contract perhaps forming several gravitational centers. Such a collection of gas and dust may result in a single star, a multiple star, or stars and associated planets. As this collection of gas contracts, the gas and dust are heated as the gravitational energy changes to kinetic energy of the random motion of the molecules and atoms. Eventually this mass of material may become so hot that a nuclear fusion process begins. Nuclear fusion is described in greater detail in Chap. 19. The most prevalent atom in the gas clouds in space and in stars is hydrogen, and it is the nuclear fusion of hydrogen—the process responsible for the hydrogen bomb—that takes place. In the fusion process the hydrogen atoms are fused together to form helium nuclei, and an immense amount of energy is given off. The fusion process acts to blow apart the condensed gas cloud, and no doubt in many cases the gas cloud explodes and spreads out into space. However, if the mass of the collected dust and gas is of an appro-

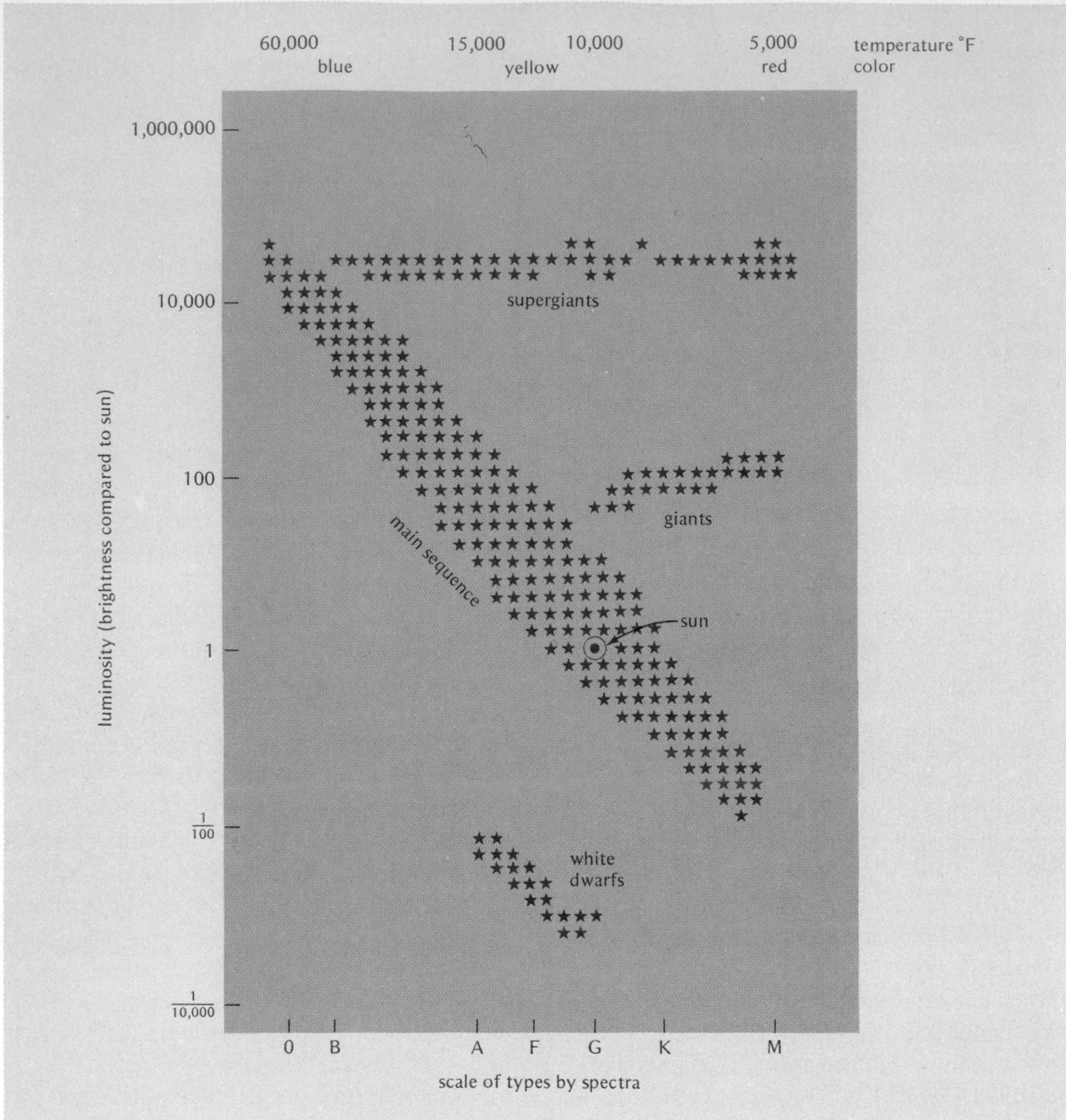

Figure 11.12
Classification of stars: Hertzsprung-Russell diagram. Points on this diagram are determined by measurement of the size and temperature (color) of stars.

priate value when the fusion occurs, a balance between the gravitational attraction that holds the star together and the exploding, nuclear fusion of hydrogen may be established. When such a balance exists, the star will reside somewhere within the main sequence of stars shown in Fig. 11.12. Apparently this equilibrium balance is rather critical, existing only for the regions of star size and temperature in which the stars of the main sequence lie.

As the hydrogen within the star is transformed to helium via the fusion process, the star moves down the main sequence becoming smaller and cooler. Eventually the star consumes about 10 percent of its original hydrogen and then begins to grow much brighter and abnormal. This change is due to the helium "ash" that resides in the core of the star. This helium contracts due to gravitational attraction, producing additional heating which results in an increased temperature of the outer regions of the star in which the hydrogen fusion is still taking place. This increased temperature causes carbon to undergo fusion, a process that rather quickly increases the output of energy from the star. The star increases in size and gets brighter. However, because of the increase in size, the outer layers of the star are pushed farther from the inner regions in which the nuclear fusion is occurring. These outer gas layers of the star cool. The star looks red and cool. A red giant has been formed.

The sun, which lies in the main sequence of stars, is estimated to be close to the point where 10 percent of its original hydrogen will be used up. However, although an exact date on which this change will occur is indefinite, it is expected that the expansion will not begin for another 3 or 4 billion years; this is actually a short time with respect to the lifetime of a star.

When a star is in the red giant stage of its life, the hydrogen is used up at an increased rate. As the helium piles up in the core, there is an increased gravitational contraction of the core. This results in further increase in temperature and an increased rate of hydrogen fusion. The star will become more blue. Eventually the core may become so hot—say, as hot as 200,000,000°F—the helium will begin to undergo nuclear fusion and result in an explosion of the core. At this stage it appears that either this helium explosion of the core results in a blowup of the star producing a supernova or the star may again contract due to gravity and form a very small but intensely bright white dwarf. Both present theory and experimental observations indicate that a red giant which has exploded its core will move quickly toward becoming a white dwarf.

Although this brief description of the life of a star may appear to be only an interesting hypothesis, it is based on a great deal of study by a large number of scientists of the nuclear processes taking place in stars. The theory is far from complete and is being altered each year by new discoveries.

VARIABLE STARS

Many thousands of stars vary in brightness, some at repeating short intervals, some with long periods, and some irregularly.

Many stars are multiples, a large number being binary, or double; a lesser number are triple, several quadruple, and a few even quintuple or sextuple. In our galaxy there are about 2000 stars of the binary type, with their planes of motion so near the line of sight that, as the two components move around their common center of mass, one star completely or partly cuts off light from the other. These are called *eclipsing variables*. Probably the best known is Algol.

Single stars with variable brightness fall into two general classes. Pulsating variables, as the name indicates, are thought to expand and contract, thus causing a variation in temperature and brightness. The principal kinds of pulsating variables are the short-period variables (less than 1 day), the Cepheids (1 to 50 days), many semiregular variables (40 to 150 days), and the long-period variables (100 to 1000 days).

The second main class is the exploding variables—novae of various types. A *nova* is not a "new star," as the name suggests, but an existing star that, because of some kind of explosion, suddenly becomes much brighter. Figure 11.13 shows the nova Herculis before and after its explosion. This star changed from a very faint star to a star brighter than Polaris, about a 170,000-fold increase, in a few days and then gradually subsided to a very faint star.

OTHER GALAXIES

As large as the Milky Way is, it occupies only about a trillionth (1/1,000,000,000,000) of known space. As mentioned at the beginning

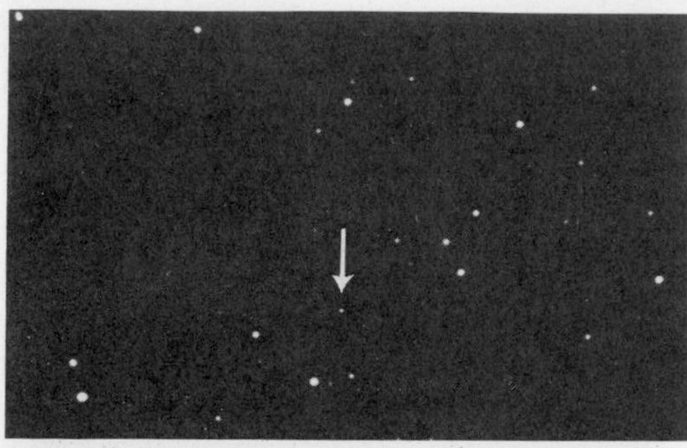

Figure 11.13
Nova Herculis before (top) and after (bottom) its outburst in December, 1934. (Yerkes Observatory photograph.)

of this chapter, there are probably several hundred million galaxies, mostly of the spiral type, but some of the elliptical type, and relatively few of the type classified as irregular. Galaxies are observed in greatest numbers near the galactic poles, 90° from the plane of the Milky Way, and diminish as the galactic plane is approached, with very few near the galactic plane. This is probably due to the obscuring effects of the interstellar dust and gas of our own galaxy. It seems likely that actually the distant galaxies are more or less uniformly distributed in space as far as we can photograph (out to 5 billion light-years) and presumably beyond.

In many of the distant galaxies, most of the features of the Milky Way can be recognized: stars of all types including giants and dwarfs; all kinds of variable stars, including novae; diffuse nebulae, indicating the presence of interstellar dust and gas; and galactic and globular star clusters.

When we contemplate the enormous extent of space and the tremendous amount of material present, we are greatly awed. Even though we have theories regarding the development of galaxies and individual stars from gas and dust, how are we to account for all the starting material, whatever its original form? It is easy to understand why most scientists, particularly astronomers, believe in a creator.

Before considering the origin of the universe, the earth, and the planets, it is appropriate that we ask, "What is the age of the earth?"

Age of the earth and universe

AGE OF THE EARTH BY SEDIMENTATION

Geoscientists accept the concept that the present is the key to the past and that, given sufficient time, the processes now at work could have produced all the geologic features of the earth, both past and present. This philosophy, usually known as the *doctrine of uniformitarianism,* demands an immensity of geologic time. It is difficult for any one of us to comprehend the significance of such staggering time periods. We quickly realize that the "everlasting hills" are very temporary. We may gain an idea of the amount of time involved by estimating the rate of deposition of sediments, the rate of erosion (p. 515), or the accumulation of sodium in the seas.

The Greek historian Herodotus observed about 450 B.C. that the Nile River overflowed its banks every year, spreading a layer of sediment over the lowlands. He concluded that the Nile Delta had grown by this annual deposition of river muds and that its building must have required many thousands of years. In modern times the statue of Ramses II at Memphis, Egypt, was discovered beneath 9 ft of river-deposited sediments. Since this statue is known to be about 3200 years old, the rate of deposition in this part of the Nile Delta is approximately 3.5 in. each century.

Some of the sedimentary rocks appear to have seasonal layering, so that their rates of deposition can be estimated with reasonable accuracy. For example, the 2600 ft of thinly bedded shales in the lake deposits of the Green River formation in Wyoming represent a period of accumulation of about 6,500,000 years. For most sedimentary rocks, however, a precise rate of accumulation cannot be estimated. Furthermore, we cannot be sure that the rate of deposition of sediments has always been anywhere near the same as today. Even if we do use some rate of accumulation for the total thickness of all sedimentary rock in the crust, we still are not close to the total age of the earth. We will never know how many times some sediments have been deposited, compacted into rock, and then eroded and returned to the seas to start the cycle all over again for a second, third, tenth, or perhaps even fiftieth time.

AGE OF THE EARTH BY RADIOACTIVE CLOCK

In Chap. 23 (p. 500) it is shown that radioactivity can be used to measure the age of the oldest rocks, of meteorites, and of the earth itself. As discussed there in detail, determinations of an estimated age for the earth now seem to be converging to a value of 4.5 billion years, with general agreement that the earth could not have existed in its present form at any time earlier than that.

AGE OF THE UNIVERSE

Can the red shift (p. 209) be used to estimate the age of the universe? Recessional velocities of galaxies may be deduced from the red shift. Dividing the distance to a galaxy by the speed of its recessional motion gives the number of years that have elapsed since the galaxy left the starting position ($d = \overline{v}t$). This time, about 4.5 billion years, is the same for all galaxies and has been referred to as the age of the universe.

Using the recession velocities of various galaxies and the distances of certain galaxies from us, as measured by Hubble and Humason, together with later determinations by Baade of distances between galaxies—the red shift implies that distances between galaxies increase with time—Baade and others computed the age of the universe as about 4.5 billion years. On the basis of recent evidence, however, Hoyle[1] thinks the best estimate for the age of our galaxy is from 10 to 15 billion years. This means that the age of the universe must be at least that much. Fowler's[2] estimate of the time of the birth of our galaxy is 12 billion years ago.

Theories of cosmology

MEANING OF COSMOLOGY

It is appropriate now to consider theories of the origin of the universe as a whole, of course including the earth; let us consider a selection from different views of modern cosmology. *Cosmology* may be defined as the theory of the origin, structure, and development of the universe as a whole.

Remember that in cosmology, as in other branches of science, theories are not necessarily useless because they are eventually discarded. Theories evoke discussion and criticism while we await advance in observational techniques and discovery that may cause revision of our ideas of the universe.

WHAT IS MEANT BY THE UNIVERSE?

A difficulty peculiar to cosmology is the uniqueness of the object of its study, the universe. In order to examine the extent of the

[1] Fred Hoyle, *Galaxies, Nuclei, and Quasars*, Harper & Row, New York, 1965, p. 91.

[2] William A. Fowler, *Nuclear Astrophysics*, American Philosophical Society, Philadelphia, 1967, p. 23.

universe we must first discuss the manner in which we obtain information about objects in the universe, or how we "see" objects in the universe.

Most scientists accept the postulate of Albert Einstein that forms the basis for his theory of relativity; that is, the speed of light in vacuum is the maximum speed attainable. This means that the fastest way to transmit information is by sending a message with an electromagnetic wave, all types of which travel at the speed of light. All of the information we have obtained from stars and galaxies has come from electromagnetic waves including γ rays, X rays, ultraviolet radiation, visible light, infrared radiation, and radio waves that emanate from these objects. When we look at light from a distant galaxy which is, for example, 100,000 light-years away, we are observing light that was emitted 100,000 years ago and has spent the last 100,000 years moving from the galaxy to us. In fact, we look backward in time each time we look at stars and galaxies. A star we saw last night which is 100,000 light-years away (a rather short distance astronomically speaking) may not even exist now; but because the light from that star takes 100,000 years to reach us, we have no way of learning this fact or some other change that has occurred on that star until 100,000 years later.

Some galaxies could exist at such great distances that light and radio waves from them have not yet reached the earth. The entire physical universe as far as we are concerned is contained in a sphere in which the galaxies are receding at speeds less than that of light. The actually possible observations using the most powerful optical and radio telescopes we could devise could concern only a very limited part of a more extensive universe. Bondi has said: "Our universe is what it is, because it was what it was." Perhaps some of the sources of light and radio waves coming to our telescopes no longer exist; they must have changed. It is also possible that similar energy from other sources has not arrived at earth.

AN EXPANDING UNIVERSE

There is some evidence, based on the Doppler effect, that we live in an expanding universe. The solar system is in the galactic system of stars known as the Milky Way (Fig. 11.7). At various distances from the Milky Way are other galaxies of star systems; the nearest of these "island universes" is the Great Spiral Nebula in the constellation Andromeda, often referred to as the Andromeda Galaxy (Fig. 11.3). These

Other galaxies are all apparently moving away from us at velocities proportional to their distances from us (Hubble's law).

This recessional motion is revealed by the increasing shift of certain strong spectrum lines toward the red, or lower, frequencies—hence the name *red shift*—as attention passes from the nearer galaxies to more distant galaxies. The most widely accepted hypothesis interprets the displacement of the spectral lines toward the red end of the spectrum—the red shift—as a Doppler effect due to the motion of galaxies away from us. This is evidence to most scientists that we live in an expanding universe.

Nearby galaxies are moving outward at several million miles per hour, whereas the most distant ones that can be seen with our biggest telescopes are receding at more than 200 million miles per hour. One of the most distant galaxies measured is 3C 295 (Source No. 295 of the 3rd Cambridge survey) in constellation Bootes. It is believed to be at a distance of about 5 billion light-years. The 200-in. Mount Palomar telescope of the Hale Observatories was used in the measurement. The spectrum of this object shows a large shift of the lines toward the red. The object is a source of radio noise, a topic to be discussed briefly later in this chapter.

Observations indicate that the clusters of galaxies are constantly moving apart from each other. Hoyle[3] illustrates this with a "raisin cake analogy":

[3] Fred Hoyle, "When Time Began," *The Saturday Evening Post,* Feb. 21, 1959, p. 96.

Think of a raisin cake baking in an oven. Suppose the cake swells uniformly as it cooks, but the raisins themselves remain the same size. Let each raisin represent a cluster of galaxies and imagine yourself inside of one of them. As the cake swells, you will observe that all the other raisins move away from you. Moreover, the farther away the faster it will seem to move.

RELATION BETWEEN RED SHIFT (VELOCITY) AND DISTANCE FOR EXTRAGALACTIC NEBULAE

Large Doppler shifts were first detected in 1912 by E. C. Slipher, who analyzed the spectral lines in the Andromeda Nebula and showed that this galaxy seemed to be approaching us. He later investigated the spectra of about 50 galaxies and discovered that in all but two of the galaxies the Doppler shifts indicated speeds of recession and the speed of recession increased with distance (Fig. 11.14).

In 1929 Humason investigated with the 100-in. Mount Wilson telescope the spectra of very distant nebulae; he obtained velocities for distant galaxies which confirmed Slipher's results. Because the relationship between distances of nebulae and their speeds of recession can be established only if the distances of nebulae are known, Hubble, also in 1929, undertook determination of these distances and discovered that

The velocity of recession of nebulae increases linearly with (strictly proportionally to) the distance (Hubble's law).

A mathematical equation was obtained for Hubble's law that was best suited for comparison with the results of observations. The constant H in the equation is called the *Hubble constant.* Hubble's law appeared not to be accurate for the very distant galaxies and clusters of nebulae. Within the last decade astronomers have revised the value of the Hubble constant; at the present time there is a 25 percent uncertainty in the value. As of now, the universe is at least twice as large as had been thought. Someone has said that this is perhaps the "biggest mistake" of all time. Will Hubble's law break down if we go out far enough in space? If the law disagrees with evidence obtained by observation, it is important to determine the nature of the discrepancy because this will help tell us which of the various theoretical models of the universe is correct.

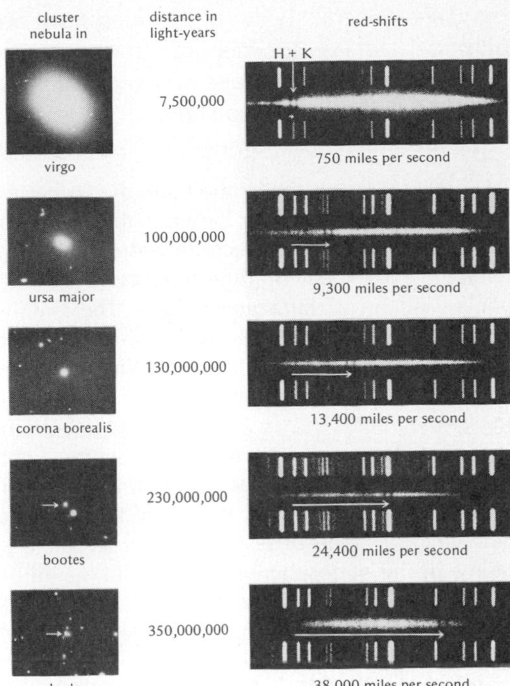

Figure 11.14
Relation between red shift (velocity) and distance for extragalactic nebulae. In Virgo the H and K lines of the element calcium, seen as two dark lines in the center band of each photograph, appear shifted only slightly toward the red side (right) of the spectrum. In the other spectra, the two lines are shifted more and more to the right (as seen above arrowhead); that is, the lines exhibit greater and greater red shifts. The distances given should be multiplied by approximately five to agree with the presently accepted value of the Hubble constant. One light-year equals about 6 million million miles. (Photograph courtesy of the Hale Observatories.)

The estimated age of the universe is derived from observations of the red shift of light from distant galaxies. In Fig. 11.14 (top right) we see the spectrum received from an elliptical galaxy (or nebula) in the Virgo cluster of galaxies. Spectra from galaxies in other clusters are also shown.

As described by Fowler,[4]

In Fig. [11.14] the redder the light, the farther to the right it falls; the bluer the light, the farther to the left. . . . The Virgo spectrum was obtained by letting the light entering the telescope from the galaxy fall on a diffraction grating. The grating reflects light of different wave length or color at different angles and thus the reflecting light can be made to fall in different locations on a recording photographic plate. . . . The Virgo spectrum is seen to consist of a continuum of "white" light broken by two closely spaced, dark lines which are known as the H and K lines of the element calcium. . . . The lines occur because calcium atoms in the atmosphere of the stars in the elliptical galaxy absorb light of this wave length. . . . The lines exhibit greater and greater red shifts [Fig. 11.14]. . . .

It will be noted [Fig. 11.14] that the size of the images of the galaxies chosen from the . . . clusters decreases from top to bottom, that is, as the red shift increases. It is assumed that all these galaxies have the same actual size and that the apparent size is a distance indicator—the more distant the galaxy the smaller the image. In practice the information obtained from the images on the photographic plates is translated into apparent luminosity. In general the smaller the image the smaller the apparent luminosity. It is assumed that this latter quantity varies as the inverse square of the distance to the galaxy. Thus if one galaxy has one-fourth the apparent luminosity of another galaxy, assumed to have the same absolute luminosity, then it is twice as far away as this other galaxy. . . . The greater the velocity of the emitting galaxy the greater is the red shift expected to be. Thus the correlation of increasing red shift with decreasing apparent size or luminosity in Fig. [11.14] is interpreted to mean that the more distant galaxies are moving with the greater velocity. . . . In fact, the observations indicate a linear relation between red shift or velocity and the luminosity distance, as it is properly called.

The simplest explanation of this relationship is that of an expanding universe in which all of the matter was at one time ejected with a spread of velocities from a common region; the galaxies whose matter received the greatest velocities relative to that of our own are now the most distant from us. Note that this will be the conclusion of observers in other galaxies, too. The general expansion means that all galaxies are in relative recessional motion. [See "raisin cake analogy" of Hoyle, p. 254.]

If the distance can be established on an absolute scale and if it is divided by the velocity, then the time back to the origin of the expansion—the age of the universe, no less—can be calculated. This is just the reciprocal of Hubble's constant and in round numbers according to current observations is about 10 billion years. . . . There are numerous reasons for uncertainty as to the age of the universe. . . . For the moment we must accept even the range from 7 to 15 billion years as the age of the universe with considerable reservation.

THE EXPLOSION (BIG BANG) THEORY

The main feature of the big bang theory is that a cosmical event is postulated with the whole of the material of the universe instantaneously created in a stupendous explosion which started the material everywhere expanding outward with the necessary velocities. The essential concept is that universal matter was originally in a state of high density and is therefore assumed to be explosive. The whole universe expanded rapidly, its initial state of very high density lasting only a few minutes. After almost a billion years of expansion and decreasing density, the clusters of galaxies formed. They have since continued to move apart.

It looked as if complex elements might be relics of the earliest period in this history of the universe, but the argument contained hypotheses that ran counter to our current knowledge of nuclear physics. This led to a modification of the explosion theory to the expansion–contraction theory. Some people, however, are especially attracted to the explosion theory because it requires a definite moment of creation for the whole uni-

[4] Fowler, *op. cit.*, pp. 71–75.

verse. This feature does not appeal to others.

THE EXPANSION–CONTRACTION THEORY

In the expansion–contraction theory a cyclic universe is considered to be alternately expanding and contracting. Each cycle is similar to the previous one and lasts about 30 billion years. During expansion, galaxies and stars are formed. Hydrogen supplies energy inside the stars and is gradually changed into complex elements. During contraction, the galaxies and stars are disrupted, and the complex elements are broken down by the high temperature generated at the state of greatest compression.

Physicists consider no particle of matter permanent; new particles can be created; one particle can be changed into another. The explosion theory and the expansion–contraction theory demand permanent particles, for the two theories rest on the assumption that all matter now existing also existed in the past; that is, they exclude the possibility of the continuous creation of matter.

A steady creation of matter could not occur without a steady expansion of the universe; conversely, a steady annihilation of matter could not occur without a contraction of the universe. Expansion and contraction of space could not occur at the same time.

Hoyle[5] reminds us that, in all dynamically oscillating systems that he knows of on earth, a static-equilibrium state or a steady state exists somewhere between the extremes of oscillation; he regards the continued failure to find either a static solution or a steady state one (next topic) for the universe as a strong indication that oscillations are not possible. In his opinion, the alternative to an oscillating universe (alternately expanding and contracting) is to admit that nucleons (constituents of the nucleus of an atom) must originate in some way. He finds that a theory for the origin of a particle can be constructed. In 1965 Hoyle[6] stated that "an oscillating model is in satisfactory agreement with all available data."

However, there are some objections to this theory, one being that a series of oscillations must eventually damp out and the universe will come to rest. For this reason and others it is of interest to examine a model involving creation of matter.

THE COSMOLOGICAL PRINCIPLE

The theoretical bases of modern cosmological theories all stem from the general theory of relativity. We do not know the exact distribution of energy and matter in the universe; we must proceed by making some reasonable assumption about it as Einstein (1879–1955) did. His assumption, which is now the basis for most cosmological theories, is called the *cosmological principle* or postulate; it states that,

Aside from random fluctuations that may occur locally, the universe must appear the same for all observers.

That is, regardless of whether the universe is seen from our own galaxy or from another galaxy a billion light-years away from us, the overall picture must be the same.

THE STEADY STATE THEORY

This theory was formulated by Gold and Bondi in 1948.

The cosmological principle was restated in slightly different words by Bondi and Gold (1948):

Apart from local irregularities, the universe presents the same aspect from any place at any time.

[5] Hoyle, *Galaxies, Nuclei, and Quasars, op. cit.*, p. 24.

[6] Fred Hoyle, "Recent Developments in Cosmology," *Nature*, **208,** 111–114, October 9, 1965.

This principle forms the basis of the steady state theory. Thus the fundamental assumption of the theory is that the universe presents on the large scale an unchanging aspect.

Because the universe must be expanding, new matter must be continually created in order to keep the density constant. (According to Hoyle the mean density of matter in the universe is 10^{-8} g/cm^3.)

If the distances between the galaxies are increasing all the time, it follows that the same matter now fills a larger volume; however, this contradicts the postulate that the universe is the same at all times. The only way out of this difficulty, according to Bondi, is to suppose that there is a process of continual creation going on—a process by which, in the enormous spaces between the galaxies, new matter constantly appears. The creation here considered is the formation of matter out of nothing, not out of radiation. This new matter condenses and forms new galaxies to fill the increasing spaces between the older galaxies. By this means the average age of galaxies is kept constant, and we arrive at a universe that is on the large scale uniform and unchanging.

The steady state concept, as a strict precept, is at variance with the counts of radio sources (next topic). The data show that radio sources were either systematically more frequent or more powerful, or both, in the past than they are at present. Realization that objects with masses perhaps up to 10^8 times that of the sun exist in the strong radio galaxies and in quasi-stellar sources makes the problem a practical one.

QUASI-STELLAR RADIO SOURCES: QUASARS

Cosmic radio waves were first detected by Jansky in 1932. This led to the new science of radio astronomy based on the use of radio telescopes. In 1946 Hey, Parsons, and Phillips discovered a remarkably great radio intensity coming from a small patch of sky in the constellation Cygnus. This was the first radio source known. In 1952 Baade and Minkowski identified optically the source in Cygnus, and Baade immediately identified what seemed like a pair of high-speed colliding galaxies—distant some half billion light-years.

The astonishing fact emerged that as much energy was being emitted in the form of radio waves as in the form of visible light. What process could cause a galaxy or pair of galaxies to emit such amazing quantities of radio energy? Alfvén and Herlofson had the right idea as early as 1950, which was that the energy emitted came from very high energy electrons moving in a magnetic field but they spoke of emission from stars; however, two Russians soon thought of emission from galaxies.

Quasi-stellar radio sources, often called *quasars,* are

Celestial sources that radiate energy intensely in the form of radio waves (10-cm to 3-m wavelength) *and also in the optical region of the energy spectrum.*

Geoffrey and Margaret Burbidge use the name "quasi-stellar objects" (QSO's) whether the sources emit radio energy or not (see the supplementary readings at the end of this chapter). A tentative picture of quasars is gradually emerging. They are believed to consist of an intensely bright nucleus, or core, surrounded by an envelope of hot luminous gas that is 100 times or more larger than the nucleus. The diameter of the nuclei of two quasars (3C 279 and 3C 446) is, for example, only 20 to 30 times the diameter of the solar system. Although a quasar may be 100 times brighter than a galaxy of billions of stars, its nucleus is only one-millionth of the diameter of such a galaxy. How such a small object can produce so much light is still a major mystery.

Obtaining a "picture" of the nucleus and its gaseous envelope cannot be done by photographing a quasar using only a big telescope. The objects appear optically as faint point images. A photoelectric spec-

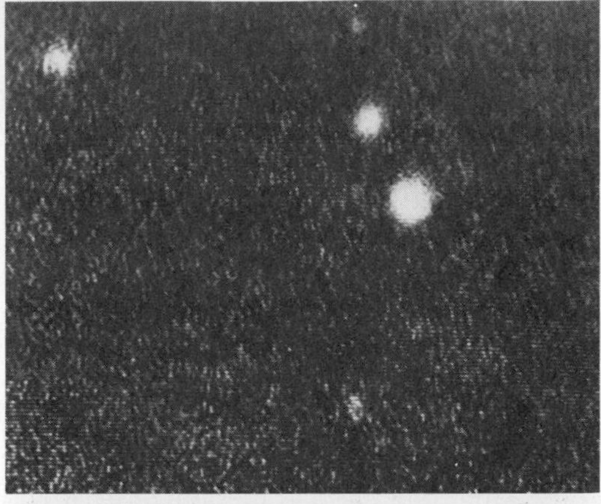

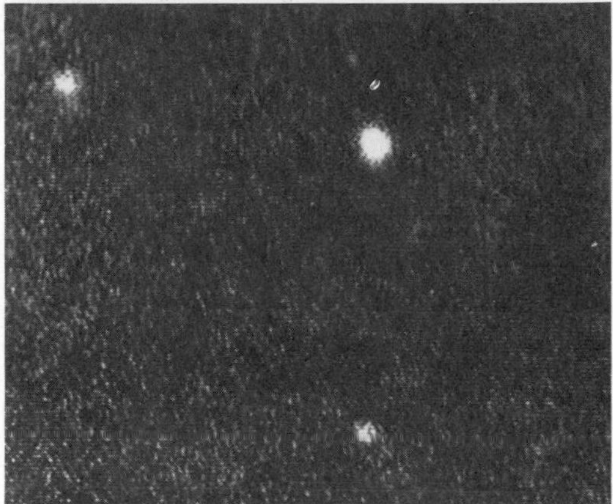

Figure 11.15
First photograph of a pulsar. Like a giant traffic light in space, this pulsar blinks on and off 15 times a second. In the top photo it is on; in the bottom photo it is off. It is believed to be composed of a superdense ball of matter that rapidly spins, thereby accounting for the pulsating effect. (Lick Observatory photograph.)

trum scanner is being used with the 200-in. Mount Palomar telescope. The spectrum of a quasar is obtained by sorting out the different wavelengths of light. The overall line pattern gives information about the chemical composition, temperature, and density of the gaseous envelope around the nucleus. The nucleus itself produces no pattern of lines and its spectrum must be analyzed by other techniques.

About 150 quasars are known. Many appear to be farther away than any other known objects. They certainly are the brightest objects in the universe.

In addition, some quasars exhibit red shifts that are much larger than normal. Such red shifts can be interpreted as ordinary Doppler shifts when corrected for relativistic effects. However, some scientists have interpreted such an unusually large red shift as being due to gravitational effects. Gravitational theory predicts that a source of an extremely large gravitational field will emit light that is red shifted because the light will lose energy as it passes through the gravitational field, which results in a change in wavelength toward the red or longer wavelength. Quasars may, in fact, be very massive bodies in a state of gravitational collapse. General relativity predicts such a process: A very massive object emits radiation as the collapse begins; this continues until the body collapsing on itself becomes so dense the resulting gravitational field is large enough to keep all radiation from escaping from the massive body. When this occurs, the object will emit no electromagnetic radiation and become a *black hole*. Such an object can be observed only by its gravitational field.

Another rather recent group of objects observed by their emitted radio waves is the so-called *pulsar*. A pulsar is a source of radio waves that varies in intensity at regular intervals. A picture of a pulsar is shown in Fig. 11.15. This pulsar blinks on and off 15 times a second. In making such photographs the radio waves are converted electronically to a visible picture somewhat in the manner electromagnetic waves are converted in a television set. At the present time the origin of pulsars is thought to be a neutron star. The neutron is a neutral particle of mass about the same as a proton, and neutrons exist in the nucleus of ordinary atoms. A neutron star is pictured as an immensely dense (10^{10} tons/in.^{3}) object made of neutrons. The pulsating radio emission presumably is produced by the rotation of such a star. The neutron star is proposed to result from the burn-out, or last stage, of an ordinary star.

Scientists continue to discover different objects in the universe—for example, the quasar-type object discovered in 1972. Five objects were observed which had rapid variations in output of radio, infrared, and visible radiation. These objects have been described as compact extragalactic nonthermal sources. These quasi quasars differ from quasars in several ways. First, the radio signals come from the core rather than from throughout the core and surrounding gas core, as is the case for quasars. The light and radio waves are polarized; that is, there is a particular orientation of the electromagnetic waves, which observation is not the usual case, though radiation from quasars does show some slight polarization. These newly observed objects have no characteristic spectra, which makes them unique among objects observed in the universe. Because no characteristic line spectra are observed, no red shift can be determined and consequently the distance to these rather dim objects is unknown.

A deeper understanding of the nature of quasars, pulsars, and quasi quasars may be helpful in determining the size and shape of the universe. We may use knowledge obtained from the study of these objects in an emerging cosmological history of the universe.

SUMMARY

The sun is a typical star. It is one of about 100 billion stars in the Milky Way, which occupies about a trillionth of known space. The universe is made up of perhaps 10 billion similar galaxies, which are scattered in space to at least about 5 billion light-years in all directions, and probably beyond.

Galaxies are grouped as spiral, elliptical, and irregular. Most galaxies are of the spiral type, as is our own galaxy. In addition to about 100 billion stars, the Milky Way contains much interstellar gas and dust. The latter may cause bright and dark nebulae.

Stars have been classified according to their spectra. The hotter stars give more light at the blue end of the spectrum. Stars vary in temperature and brightness. A star that suddenly becomes much brighter is called a nova. A "star" that appears to the unaided eye as one star may be multiple. Many stars are multiples.

The estimated age of the earth, based on techniques using radioactivity, is about 4.5 billion years. The red shift has been used in computing the age of the universe to be about 4.5 billion years. Hoyle, however, estimates that the age of our galaxy is from 10 to 15 billion years. Fowler estimates it as 12 billion years. This means that the age of the universe would be at least that much.

There are no techniques as exact as those of radioactivity to chart the course of events in the origin of the earth and the planets. It is thought by some scientists that primeval hydrogen gas condensed into stars under the influence of gravitation forces. Thus the sun could have been formed.

Cosmology may be thought of as the theory of the origin, structure, and development of the universe as a whole. The models for rival theories of cosmology are basically either evolutionary or steady state. The latter may have to be discarded because of information derived from the study of radio sources. The former include the explosion theory and the expansion–contraction theory.

There is some evidence, based on the Doppler effect, that we live in an expanding universe. This evidence led to the explosion theory, in which it is postulated that universal matter in a state of high density exploded and is expanding outward. The theory requires a definite moment of creation of the universe. As a result of the expansion and decreasing density, the galaxies were formed.

In the expansion–contraction theory the universe is considered to expand and contract with a cycle of about 30 billion years. The two theories exclude the possibility of the continuous creation of matter. A steady creation of matter cannot occur without a steady expansion of the universe, and a steady annihilation of matter cannot occur without a contraction of the universe. The fact that expansion and contraction of the universe could not occur at the same time led to the concept called the steady state theory. The fundamental assumption of the steady state theory is that the universe presents on the large scale an unchanging aspect.

The theoretical bases of modern cosmological theories all stem from the general theory of relativity. The basis for most cosmological theories is the cosmological principle:

Aside from random fluctuations that may occur locally, the universe must appear the same for all observers.

The rival theories of cosmology are likely for some time to serve as points of departure for further consideration of the origin, structure, and development of the universe as a whole.

Quasars are celestial sources which radiate energy intensely in the form of radio waves and also in the optical region of the spectrum. About 150 are known. A deeper understanding of quasars may be useful in determining the size and shape of the universe.

A pulsar is a source of radio waves that varies in intensity at regular intervals. It is believed to be a rotating, exceedingly dense neutron star.

It is suggested that the gravitational field around some dense stellar objects is so great that electromagnetic radiation from it and its vicinity cannot escape; the result is a "black hole."

Important words and terms

Milky Way
galaxy
nebulae
galactic clusters
light-year
stellar classification
main sequence
life of a star
stellar evolution
red giant
white dwarf
variable star
nova
cosmology
expanding universe
red shift
Hubble's law
big bang theory
expansion–contraction theory
steady state theory
quasars
pulsars

Questions and exercises

1. Arrange the following in correct order to indicate a complete mailing address, perhaps your own: earth, city, North American continent, universe, Northern Hemisphere, number and street, Milky Way, state, United States, solar system, zip code.
2. Give the general types into which galaxies have been grouped, and indicate the approximate percentages of each type.
3. Compute the approximate number of miles light would travel in 1 year (365.25 days). Express this distance in astronomical units.
4. (a) Draw a line diagram to indicate the shape of our own galaxy, edge-on. Place an X about where you live. (b) Give several important facts that describe our galaxy.
5. What is a nebula? Distinguish between a nebula and a galaxy.
6. Discuss, in general, interstellar gas and dust, for example, the amount, the distribution, and the effect it has in causing what astronomers observe.
7. How do astronomers distinguish, in general, between hotter stars and cooler stars on the basis of color?
8. Explain how an analysis of a sample of soil may be used to determine the age of the earth. Why can the age of the earth not be determined accurately by sedimentation?
9. Distinguish between the following: age of the universe, cosmology, the universe.
10. Explain why observations by astronomers may concern only a very limited part of the universe.
11. Briefly describe a possible course of events in the origin of the universe, the earth, and the planets.

12. Explain the meaning of the red shift, and tell how it is used by some scientists as evidence that we live in an expanding universe.
13. Which one of the rival theories—explosion, expansion–contraction, steady state—do you tend to favor? Describe that theory, and tell why you think it is the best one.
14. Define quasars. Describe them. What do astronomers hope to learn by studying them?
15. Describe the life of an ordinary star. How are stars born?
16. When we see an event such as the explosion of a star, we are actually "looking back into history." Explain this statement.
17. Describe the ways optical spectra from stars are used to obtain information on the nature of stars and their motion.
18. What is a Hertzsprung-Russell diagram?
19. When astronomers propose a model of how a star transforms during its lifetime, do they base their model on the observation of changes occurring in a single star over a period of time or on the different type of stars observed in the universe? Explain.
20. What is the difference between a red giant and a white dwarf?
21. On the basis of the latest theories of the origin of the universe, star formation, and the solar system, why is it reasonable to suppose that there are many other stars that have planetary systems circling them? How reasonable is it for us to visit such planetary systems in the lifetime of a human?
22. Can you think of reasons why we have not observed planets moving about some other star besides our sun? Assume that other planetary systems would be similar to our solar system.
23. After having studied some of the observations and theories of modern astronomy, can you write down some reasons to justify the government's supporting further expensive astronomical work such as an orbiting space observatory? Which would appear more important to you, further observations of stars, galaxies, pulsars, and similar objects *or* visiting Mars or Jupiter? Discuss.
24. Does the fact that the objects in the universe are receding from us mean that we are in the center of the universe? As a help to answering this question consider yourself to be the size of an ant and to be on a rubber inner tube (not inflated). If you were to observe some pieces of dirt placed on the rubber as the rubber is stretched during inflation, how would your "ant-sized" universe appear to you if you were at different places on the rubber?

Supplementary readings

Textbooks

Books written for use in introductory courses in astronomy are an excellent source of additional information for this chapter. See Chap. 1, "Supplementary Readings."

Other books

Alfvén, Hannes, *World-Antiworlds: Antimatter in Cosmology,* W. H. Freeman, San Francisco (1966). [Some topics discussed are: Cosmology; What does the world consist of?; Matter and antimatter; Development of the metagalaxy; How did the world begin and become what it is now?]

Asimov, Isaac, *The Universe,* Walker, New York (1971).

Asimov, Isaac, *The Universe: from Flat Earth to Quasar,* Discus Books, Avon, New York (1965).

Bondi, H., *Cosmology,* Cambridge University Press, New York (1960).

Bondi, H., *The Universe at Large,* Anchor Books, Doubleday, Garden City, N. Y. (1960). (Paperback.) [Among important topics are the expansion of the universe; different theories of cosmology; tests in cosmology—the age of galaxies, the evolution of galaxies, the origin of the elements; the stars and between the stars; the motion of celestial bodies.]

Bondi, H., et al., *Rival Theories of Cosmology,* Oxford University Press, Fair Lawn, N. J. (1960). [This fascinating little book, written in nontechnical language, is a symposium and discussion of modern theories of the structure of the universe.]

Burbidge, Geoffrey, and Margaret Burbidge, *Quasi-Stellar Objects,* W. H. Freeman, San Francisco (1967). [Summarizes and interprets the research data available about QSO's from the identification of the first quasi-stellar radio source through early 1967, with about 150 QSO's identified. The authors discuss theories of the nature of the QSO's and their energy sources.]

Editors of *Fortune, Great American Scientists,* part III, *The Astronomers* (pp. 63–68), Prentice-Hall, Englewood Cliffs, N. J. (1961). [Men and their concepts in astronomy from 1910 to the sixties.]

Fowler, William A., *Nuclear Astrophysics,* American Philosophical Society, Philadelphia (1967). [The current state of affairs in nuclear astrophysics as presented in a series of lectures for the general public. Among the topics discussed are the origin of the elements; neutrons, neutrinos, red giants, and supernovae; the age of the elements, of the universe, of the galaxy, of the solar system; quasars.]

Gamow, George, *Matter, Earth, and Sky,* Prentice-Hall, Englewood Cliffs, N. J. (1965). [General relativity and cosmology are discussed in the last two chapters.]

Hawkins, Gerald S., *Splendor in the Sky,* Harper & Row, New York (1969). [Of interest for this chapter are the chapters on the earth's beginning, a star is born, man-made moons, island universe, the red shift, a radio view, and a present-day view of the universe.]

Hoyle, Fred, *Frontiers of Astronomy,* Harper & Row, New York (1955). (Also available as a paperback from Mentor Books, New American Library of World Literature, New York.) [Here are set forth theories of how the stars and planets were formed and how the sun operates internally to provide energy. The author discusses his theory of the continuous creation of matter and the origin of the universe.]

Hoyle, Fred, *The Nature of the Universe,* Harper & Row, New York (1960). [There are intriguing chapters on the origin of the earth, the stars, and the planets. The author combines generally accepted findings and theories of modern cosmology with daring concepts of his own as he considers the expanding universe.]

Hoyle, Fred, *Galaxies, Nuclei, and Quasars,* Harper & Row, New York (1965). [In six chapters based on lectures given to university audiences, Hoyle discusses galaxies, radio sources, X rays, gamma rays, and cosmic rays, steady state cosmology and departure from that concept, the history of matter.]

Kahn, F. D., and H. P. Palmer, *Quasars,* Harvard University Press, Cambridge, Mass. (1967). [Deals with such topics as radio galaxies, the discovery and properties of quasars, relativity, cosmology, stars, and synthesis of the elements.]

Kruse, W., and W. Dieckvoss, *The Stars,* University of Michigan Press, Ann Arbor, Mich. (1957). [This well-illustrated and well-written book tells how astronomers, with starlight as their source of information, have gathered a vast store of knowledge about celestial bodies.]

Lovell, A. C. B., *The Individual and the Universe,* Mentor Books, New American Library of World Literature, New York (1961). (Paperback.) [An authority in the use of the radio telescope discusses the "new astronomy" of the past 30 years with emphasis on the origin of the solar system and of the universe.]

Lyttleton, R. A., *Man's View of the Universe,* Little, Brown, Boston (1961). [A highly readable nontechnical presentation of material to supplement Chaps. 1, 2, 3, and 11 of this textbook. The chapters that supplement Chap. 11 particularly well are "The Origin of the Solar System," "Our Galaxy," and "The Expanding Universe."]

Page, Thornton (ed.), *Stars and Galaxies: Birth, Ageing, and Death in the Universe,* Prentice-Hall, Englewood Cliffs, N. J. (1962). (Paperback.) [Writers selected by the American Astronomical Society discuss active areas of current astronomical research in simple terms. Each describes the present state of knowledge and technique, with emphasis on how it was obtained and may be extended and the basic assumptions on which it rests.]

Whitrow, G. J., *The Structure and Evolution of the Universe: An Introduction to Cosmology,* Harper Torchbooks, Harper & Row, New York (1959). [Discussion of space-time relationships and comparison of theories of the universe in a form suitable for the student and the layman are followed by current thinking on the structure of galaxies and on the evolution of the universe.]

Articles

Bok, B. J., "The Birth of Stars," *Scientific American,* **227,** no. 2, 48 (August, 1972).

Ginzberg, V. L., "The Astrophysics of Cosmic Rays," *Scientific American,* **220,** no. 2, 50 (February, 1969).

Gorenstein, P., and W. Tucker, "Supernova Remnants," *Scientific American,* **225,** no. 1, 74 (July, 1971). [Huge shells of luminous gas in our galaxy are traced back to stellar explosions.]

Iben, I., Jr., "Globular-Cluster Stars," *Scientific American,* **223,** no. 1, 26 (July, 1970).

Maran, S. P., "The Gum Nebula," *Scientific American,* **225,** no. 6, 20 (December, 1971). [A description of the largest known nebula in our galaxy.]

Ostriker, J. P., "The Nature of Pulsars," *Scientific American,* **224,** no. 1, 48 (January, 1971).

Rees, M. J., and J. Silk, "The Origin of the Galaxies," *Scientific American,* **222,** no. 6, 26 (June, 1970).

Ruderman, M. A., "Solid Stars," *Scientific American,* **224,** no. 2, 24 (February, 1971). [A discussion of white-dwarf stars and pulsars (neutron stars).]

Schmidt, M., and F. Bello, "The Evolution of Quasars," *Scientific American,* **224,** no. 5, 54 (May, 1971).

Thorne, K. S., "The Death of a Star," *The Physics Teacher,* **9,** no. 6, 326 (September, 1971).

Weymann, R. J., "Seyfert Galaxies," *Scientific American,* **220,** no. 1, 28 (January, 1969). [Galaxies which display violent changes and appear to be related to quasars.]

Wheeler, J. A., "Our Universe: The Known and the Unknown," *The Physics Teacher,* **7,** no. 1, 24 (January, 1969). [Discussion of the theories of the formation of the universe.]

Reprints from *Scientific American*

240 Sandage, Allan, "The Red-shift" (September, 1956). [The redness and presumably the speed of recession of most galaxies increases regularly with distance.]

253 Reynolds, John H., "The Age of the Elements in the Solar System" (November, 1960). [Studies of the inert gases found in meteorites have confirmed estimates that the earth is 4.6 billion years old and provide evidence that the elements in the solar system are not much older.]

ATOMS: EVIDENCE FOR THE DISCRETENESS OF MATTER

The eternal mystery of the world is its comprehensibility.

A. EINSTEIN, 1879–1955

Although Democritus and some other Greek philosophers decided about 400 B.C. that matter consisted of tiny indivisible particles (see p. 142), other famous philosophers, such as Aristotle, disagreed. From the viewpoint of the modern practicing scientist, it perhaps does not matter very much what they thought, because nothing tangible or concrete came about as a result. Support for such beliefs had to await the careful measurement of combining weights and volumes. In this chapter we describe the scientific developments that supported the idea of the discrete nature of matter and led to our present-day model of the atom and its nucleus.

Early atomic theory

We have seen how the physicists of the seventeenth and eighteenth centuries devoted their efforts to a description of the detailed behavior of macroscopic bodies. The success of their attack on many physical problems, such as the motions of the planets in the solar system and the flow of fluids, led to an overwhelming acceptance of Newtonian mechanics. However, there were many questions concerning the microscopic structure of matter that remained unanswered until the emergence of the atomic theory of matter at the beginning of the nineteenth century. In Chap. 7 the idea of the indivisibility of matter was mentioned, and this concept of the atomic nature of matter was employed in the discussion of the kinetic theory of matter. There we attempted to use a model in which atoms were in constant random motion to explain macroscopic phenomena associated with "heat." Now we shall examine the evidence for the existence of atoms, evidence that led to the beginnings of the modern science of chemistry.

TWO KINDS OF SUBSTANCES: ELEMENTS AND COMPOUNDS

Hydrogen and oxygen are examples of elements. From about the time of the French Revolution until recent years the most used definition of the term "element" was as follows: An *element* is a substance that cannot be decomposed into simpler substances by ordinary physical or chemical means, such as heating or the use of an electric current. At present 105 such elements are known, but the 13 beyond element 92, uranium, must be "man-made." (Another definition of an element, somewhat more modern and more theoretical, is given on p. 276.)

A *compound* is a substance that can be decomposed into simpler substances by ordinary chemical means. Water, sugar, and sodium chloride (the chief component of ordinary table salt) are examples of chemical compounds. Compounds are much more numerous than elements. Because new compounds are being prepared continuously, it would be impossible to tell exactly how many compounds there are. Several hundreds of thousands, however, have been prepared and described in the chemical literature.

THE LAW OF DEFINITE COMPOSITION OR PROPORTIONS

Proust (1754–1826), a professor in Spain, culminated a long period of careful experimentation by first stating in 1799 that the weight ratios of the elements present in a compound are fixed and independent of the origin of the sample. Berthollet (1748–1826), who had been an associate of Lavoisier, came to the opposite conclusion—that elements combine in variable weight ratios. A student of Berthollet, Gay-Lussac, summarized the results of his measurements of volumes of gases involved in chemical change. This statement, the *law of combining volumes,* first published in 1805, can be expressed as follows:

When measured at the same temperature and pressure, the comparative volumes of gases consumed and produced in a given chemical change can be expressed by means of small whole numbers.

After many years of controversy Proust's view became the accepted one and is generally expressed as the *law of definite composition:*

A pure compound always contains the same elements in the same ratio by weight.[1]

A *compound* may now be thought of as a substance formed by the combination of two or more elements in definite ratios or proportions by weight. For example,

	iron	+	sulfur	→	iron sulfide
(1)	56 g		32 g		88 g
(2)	77 g		44 g		121 g

In (1) the ratio of iron combining with sulfur to form iron sulfide is 7:4 by weight. In (2) the amounts of iron and sulfur combining are in the same ratio. Regardless of the amount of iron sulfide made, the ratio or proportions of iron to sulfur is always 7:4. If the word "proportions" is substituted for the word "ratio" in the law of definite composition—and it often is—the law is known as the *law of definite proportions:*

A pure compound always contains the same elements in the same proportions by weight.

Thus the laws of definite composition and of definite proportions are the same law by different names.

A NEW AND BETTER ATOMIC THEORY

About 1800 an English school teacher, John Dalton (1766–1844), became interested in meteorology. As a result of his observations he announced:

When two or more elastic fluids [we now call these "gases"] whose particles do not unite chemically upon mixture are brought together one measure of each, they occupy the space of two measures, but become uniformly diffused through each other, and remain so, whatever may be their specific gravities.

All subsequent experiences of this nature have confirmed the truth of Dalton's observation.

Dalton's experimental observation that two gases always mix spontaneously (without any outside help) seems to have been the major stimulus encouraging him to elaborate on the atomic theory of the Greeks. They had only done "thought experiments," which were scarcely more than speculation. Dalton had the new knowledge resulting from his own observations on gases, Gay-Lussac's law of combining volumes, etc. He used these facts to devise a much more comprehensive theory about the nature of atoms than had ever before existed. But so many of Dalton's assumptions about the nature of atoms have been proven untrue that they are not worth enumerating here.

DALTON'S KEY CONTRIBUTIONS

The fact that most of the assumptions constituting Dalton's atomic theory have been rejected does not mean that they were of no value in the development of science. Quite the contrary is true, for these reasons: (1) his assumptions not only appeared to be plausible but were more related to objective experience than those previously made; (2) he made the weights of atoms the keystone of his theory, thus giving it a validation denied all previous theories; (3) the rule of greatest simplicity (in general, atoms combine in simple ratios) was fruitful, for it led, for example, to the prediction of the law of multiple proportions; (4) he developed simple

[1] Whenever the term "weight" is used we mean *mass*. Most chemists use weight; the convention is followed here.

picture symbols to represent atoms and their combinations, making it possible for scientists to manipulate atoms on paper in "thought experiments."

DEFECTS IN DALTON'S ATOMIC THEORY

Dalton reasoned that the composition of a stable binary compound was ordinarily very simple. He thought that one atom of one element joined with one atom of another element to form what he called one atom of the compound. By means of picture symbols an example may be represented thus:

○ plus ● yields ○●

This expression might represent one atom of hydrogen joining one atom of oxygen to form an atom of water. But the word "atom" originally signified "indivisible"; so this term seemed somewhat inappropriate for describing a particle of water, which could be divided, or decomposed, to form an atom of hydrogen and one of oxygen.

If all oxygen atoms are indivisible and alike, it should be possible to assign a combining weight, or atomic weight, to oxygen that would be invariably true. An attempt was made to use combining ratios for determining atomic weights, but some of the results seemed inconsistent.

AVOGADRO'S HYPOTHESIS

For at least a half century prior to 1811 many scientists had noted a simplicity in the behavior of gases. This simple variable behavior of gases makes possible those generalizations which we call the "gas laws." For example, all gases have essentially the same coefficient of expansion with temperature rise. Two substances that are solids or liquids do not have this same coefficient of expansion. Likewise, it had been previously noted that relative combining volumes of gases could be expressed by means of small whole numbers. No such simple relationship exists between the combining volumes of solids or liquids.

How is the uniform behavior of combining gases to be explained? According to the atomic theory of Dalton, chemical combination takes place between atoms in whole-number ratios; according to Gay-Lussac, chemical combination of gases takes place between volumes in whole-number ratios. What is more natural than to assume that equal volumes of gases at the same temperature and pressure contain equal numbers of atoms and that a ratio of the numbers of atoms automatically becomes the same as the ratio of the volumes?

Later atomic and molecular theory

In 1811 the Italian physicist Amadeo Avogadro (1776–1856) (Fig. 12.1) proposed an important hypothesis involving the ideas discussed in the preceding paragraph, in an attempt to intermediate between the atomic theory of Dalton and the experimental results of Gay-Lussac. A *hypothesis* is a guess as to what is true, and it should be based, insofar as possible, on available facts. But it is not just a summary of facts; it explains or interrelates the facts in terms of something yet unknown. It enables the scientist to make predictions that he then can attempt to verify by experiment.

Although Avogadro's hypothesis deals with invisible particles and has not been proved directly, so much indirect evidence has been accumulated to support it that it is now often called Avogadro's law. A current statement of *Avogadro's law* is as follows:

Equal volumes of all gases contain the same number of molecules, provided that the samples are under the same conditions of temperature and pressure.

This law, like the gas laws, is a very good approximation.

Figure 12.1
Amadeo Avogadro (1776–1856). Born in Italy, Avogadro was educated for the law and practiced for some years. His interests gradually turned toward physics and chemistry, and he received an appointment as professor of physics. His famous hypothesis concerning molecules was published in 1811. It was ignored for almost 50 years. In 1858 an Italian chemist, Stanislao Cannizzaro, pointed out the extraordinary usefulness of Avogadro's generalization. (Courtesy of the Edgar Fahs Smith Memorial Collection, University of Pennsylvania.)

Figure 12.2
Stanislao Cannizzaro (1826–1910), famous Italian professor of chemistry. Cannizzaro rendered his greatest service to chemistry by suggesting changes in Dalton's particle theory of matter, by giving his interpretation and explanation of Avogadro's hypothesis, and by showing how molecular weights of gases could be determined. (Courtesy of the Edgar Fahs Smith Memorial Collection, University of Pennsylvania.)

REVISION OF DALTON'S ATOMIC THEORY: CONCEPT OF THE MOLECULE

Avogadro's proposals were almost completely ignored by the vast majority of chemists in the years between 1811 and 1858. Eventually, some chemists came to realize that something essential was still lacking in the particle theory of matter. The missing part of the puzzle was supplied in 1858, when the Italian Cannizzaro (1826–1910) (Fig. 12.2) published *Sketch of a Course in Chemical Philosophy*, a pamphlet in which a full return was made to Avogadro's hypothesis. In 1860 he addressed principal chemists of Europe in assembly and distributed the pamphlet. The impact of the content was great.

Essentially, Cannizzaro's argument was that two changes should be made in Dalton's particle theory of matter: (1) He believed that the unit particle of which most matter consists was more complex than an atom and should be designated by the new term "molecule," and (2) he showed how molecular weights could be determined.

In the chemical revolution of 1775–1789 the phlogiston theory, according to which a substance, while burning, gives off phlogiston, had been overthrown. The atomic revolution, the second major revolution in chemistry, was essentially complete by 1858 insofar as chemistry was concerned.

CHOICE OF UNIT FOR MOLECULAR WEIGHTS

The hydrogen molecule is the lightest one known. It is not surprising, therefore, that at first scientists expressed a molecular weight as a certain number of times the weight of a hydrogen molecule. The density of oxygen is about 16 times that of hydrogen, so we could say the molecular weight of hydrogen is 1 and that of oxygen is approximately 16. This standard (molecular weight of hydrogen = 1) was very soon discarded, however, because it was found that hydrogen, and also oxygen, appeared as diatomic molecules in the gas; that is, hydrogen gas is made up of molecules of hydrogen that are composed of two hydrogen atoms bound together.

By definition the molecular weight of hydrogen was made 2, and, of course, when this change was adopted, the molecular weight of oxygen also automatically became double its former value; so the new value for the molecular weight of oxygen was approximately 32. This standard was used for a half century or so until the time when a shift was made to taking the molecular weight of oxygen as exactly 32. Therefore the standard unit for expressing molecular weights was $\frac{1}{32}$ the weight of the oxygen molecule. Perhaps for simplicity we should call one of these a molecular weight unit, because originally that is exactly what it was.

In 1961 the International Union of Pure and Applied Chemistry, in conformity with similar action of the equivalent union in physics, adopted a new standard for the weight of all the building blocks of the universe, with the selection of the atomic weight of carbon as 12. Previous to 1961 there had been one scale of atomic weights used by physicists and another by chemists. Now that physicists and chemists are using the same scale, it is often referred to as the "unified" scale. Now the *atomic weight* of an element is the weight of an atom of that element with reference to an atom of a common isotope of carbon, carbon 12 (isotopes are discussed on p. 275). The unit of measurement now used by scientists in general is thus $\frac{1}{12}$ the weight (or better, mass) of the carbon-12 isotope and is given the name *atomic mass unit* (amu).

In the unified scale all atomic weights are lowered by 0.0037 percent from the chemical scale. For example, the new atomic weight of oxygen is 15.9994. The small changes in atomic weights, which result from using the reference carbon-12, instead of oxygen-16, are insignificant to *us* (though not to an atomic physicist or a radiochemist) and so we shall generally disregard this slight difference.

DEVELOPMENT OF THE PERIODIC LAW WHICH LED TO THE PERIODIC TABLE

Although specific methods of operation may vary considerably between two different sciences, such as physics and chemistry or chemistry and geology, there is always the necessity to relate, classify, and organize things according to some system.

The acceptance in about 1860 of a consistent scheme for determining atomic weights led quickly to further advances in physics and chemistry. For example, it not only furnished a ready theoretical explanation for the crude combining weights that had been of some use to chemists for a decade or two but made it possible to refine them rapidly. Perhaps more important, it facilitated the development of that extremely useful tool the periodic table.

As early as the French Revolution some chemists found it helpful to classify elements as metals or nonmetals. A decade or two later Dobereiner organized some elements into families of three (his so-called triads). Thus he emphasized the great similarities of chlorine, bromine, and iodine.

But the greatest step toward systematic organization of facts about the elements came in 1869, when the Russian chemist Dmitri Mendeleev (1834–1907) suggested that when the elements were arranged in the order of their increasing atomic weights, their physical and chemical properties tended to be repeated in cycles. During some 40 years this idea was so fruitful in predicting the existence and properties of undiscovered elements with remarkable accuracy that it became generally known as *Mendeleev's peri-*

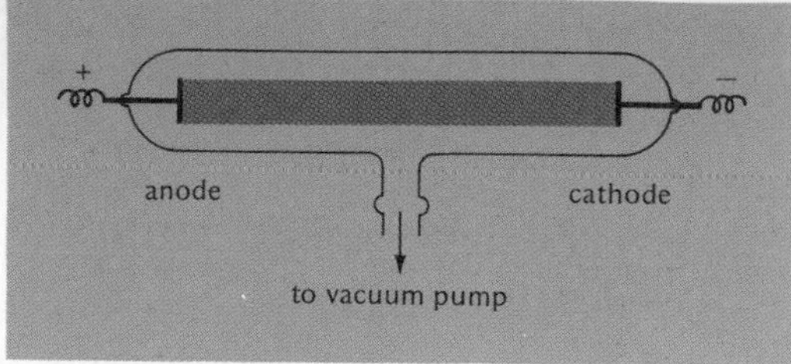

Figure 12.3
Cathode ray tube. A bluish discharge occurs as the air pressure in the tube is reduced.

odic law. A common formal statement of the periodic law is as follows:

If the elements are arranged in the order of their increasing atomic weights, there is at regular intervals a repetition or a recurrence of similar properties.

A revision of this law was made early in this century as a result of experiments by the Englishmen Rutherford and Moseley. (For this revision see p. 275 and the periodic table on the inside of the front cover.)

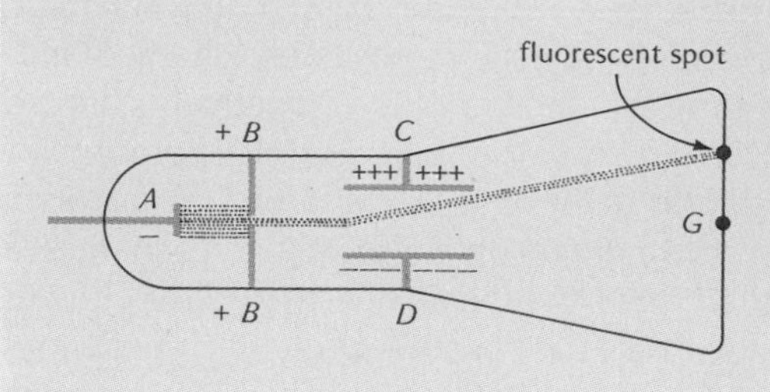

Figure 12.4
Cathode rays. The rays consist of a beam of negatively charged particles called *electrons*. If the plate *C* is charged positively, the beam of particles is deflected upward from the horizontal path. When the particles hit the glass at the end of the tube, they produce a fluorescent spot from which emanate light and X rays. Point *G* indicates the spot where the undeflected beam would terminate.

Rays and radioactivity

During the period 1860–1890, scientists "believed" that the world was composed of round, hard, indivisible atoms which quite often appeared naturally as molecules or groups of atoms. We shall now present the results of numerous experiments which produced evidence that the atom itself has structure; in fact, the concept of the indivisible atom is no longer tenable.

CATHODE RAYS

Near the end of the nineteenth century, numerous scientists were studying the discharge of electricity through rarefied gases. In such experiments a source of high voltage, such as an induction coil or an electrostatic generator, is attached to the terminals (electrodes) of a tube such as that shown in Fig. 12.3. If the tube is of the order of 1 ft long, it will probably be impossible to get any detectable conduction between the terminals when the tube is filled with dry air at ordinary atmospheric pressure. But as the air is exhausted from the tube, some easily attainable voltage, such as 5000 volts, will cause a discharge of electricity between the electrodes. By 1895 Crookes and Thomson in England and Perrin in France had carried out experiments which convinced scientists that most of the discharge between electrodes in an evacuated tube consisted of small negatively charged particles that emanated from the negative electrode (cathode) and traveled at high velocities in straight lines toward the positive electrode (anode). The stream of particles became known as "cathode rays."

The conclusion that cathode rays were, in fact, fast-moving negatively charged particles was based on the manner in which these cathode rays were influenced by the presence of a magnet or the presence of an electric field. For example, if a high voltage is placed across the terminals *A* and *B* of the tube shown in Fig. 12.4, the cathode rays would go straight through the tube producing a fluorescent spot on the end of the tube. However, if a potential of about 100 volts was placed across plates *C* and *D* as shown in Fig. 12.4, the cathode rays would be deflected upward as they passed between plates *C* and *D* (plate *C* is positively charged) indicating that the "rays" are negatively charged. Using a similar apparatus, J. J. Thomson was able to determine the ratio of the magnitude of the charge on these negative charges to their mass. Eventually these negative charges, which are the constituent of the cathode rays, were named "electrons." It might be noted that the tube shown in Fig. 12.4 is the forerunner of the cathode ray oscilloscope tube and the television picture tube.

It was not until about 1909 that the full significance of these negatively charged particles was realized. Robert A. Millikan in his famous "oil drop experiment" by measuring the charge on oil drops was able to determine a smallest unit of charge to be 1.6×10^{-19} coulomb, and he found that the charge on an object (oil drops in his case) was always an integral multiple of this unit of charge. In modern terms we say the charge is "quantized."

DISCOVERY OF X RAYS

Later in 1895 W. K. Roentgen discovered that cathode rays generated a new and strange kind of rays when they struck a metal target. This new kind of emanation was named "X rays" because its properties seemed so strange. The new rays had the astonishing capacity of penetrating cloth, leather, and even human flesh, but they did not penetrate bone or metal so well. Unlike cathode rays they were not deflected by a magnet or an electric field.

The mystery surrounding the nature of these X rays was not solved until 1912 when Max von Laue showed that X rays were electromagnetic waves of extremely high frequency. He employed an interference experiment using crystals as diffraction gratings. Such experiments indicated that X rays consisted of electromagnetic radiation quite similar to light; their extremely short wavelengths (of the order of only 1/1000 those of visible light) accounted for their astonishing ability to penetrate matter that is opaque to light.

X RAYS AND THE DISCOVERY OF RADIOACTIVITY

Scarcely had Professor Roentgen announced the discovery of X rays when a French physics professor, Becquerel, wondered if perhaps some kinds of matter might naturally give off X rays. Because it was known that X rays affected photographic plates (even carefully wrapped in black paper), he stored such plates near various chemicals. After developing plates that had been stored near a uranium compound, he found that they had indeed been blackened. The uranium compound did give off X rays. Figure 12.5 shows that a uranium compound will photograph itself under such circumstances.

As so often happens, this almost accidental discovery was not as simple as it seemed at first. A student in the school, Marie Sklodowska, devised a method for measuring the quantity of the apparent "X radiation" given off by uranium and was astonished to find that some waste eliminated during the purification of the uranium compound gave off even more radiation than the uranium itself. This led within a year or two to the discovery of two new elements, polonium and radium. A study of these elements indicated that other "radiations" besides the X radiation were given off. We now know that all elements with atomic masses greater than bismuth, which has an atomic number of 83, spontaneously emit radiations. Such substances are called "naturally radioactive." The radiation is emitted from their nuclei. Whenever an unstable nucleus decays, energy is released and radiated in all directions as photons and sometimes as material particles. *Photons* are defined as individual or discrete "packages" or quanta of electromagnetic energy. For example, light and the other elec-

Figure 12.5
Radioactivity. At top, ordinary photograph of a rock containing uranium compounds; bottom, self-photograph of the same rock, taken by means of its radioactivity. (From J. J. G. McCue, *The World of Atoms,* copyright © 1956, The Ronald Press Company.)

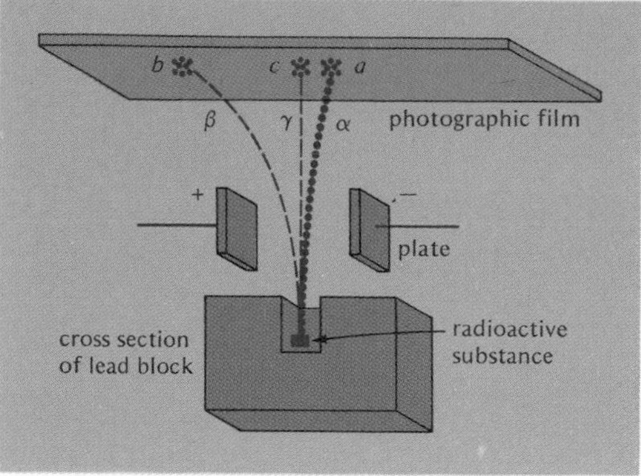

Figure 12.6
Experiment to show that rays from radioactive substances are of three kinds.

tromagnetic radiations (Fig. 9.27) consist of photons (quanta) of electromagnetic energy.

The emission from nuclei of photons or material particles is called *radioactivity*. The nuclei that radiate photons or material particles are said to undergo radioactive decay. Thus *radioactivity* is the decay of the atomic nucleus with the emission of photons or material particles. What are these particles?

If some radioactive material is placed at the bottom of the opening of a lead block, as in Fig. 12.6, a beam of emitted rays goes upward. When a strong electric field is applied between the plates charged positively and negatively as shown, the photographic film is found to be blackened in the three spots *a*, *b*, and *c*. This shows that the radiation has three components. The radiation falling at *a* has been found to consist of a stream of positively charged particles called alpha (α) particles, and the streaming of these particles is called α rays. The spot at *b* is caused by negatively charged particles called beta (β) particles. The radiation falling at *c*, called gamma (γ) radiation, is not affected by a charge on either plate. This is electromagnetic radiation of the same general nature as X rays but called γ rays to indicate the special way in which it originates.

Experiments have been performed which prove that α particles are helium nuclei He^{2+}, that is, helium atoms that have lost both outer electrons; β particles have been shown to be electrons; and γ rays are electromagnetic radiation of high frequency and short wavelength (see Fig. 9.27).

The nucleus; the periodic law and radioactivity

NEED TO REVISE THE CONCEPT OF THE ATOM

Obviously, if radioactive matter is giving off different kinds of particles and cathodes also give off particles, there is cause to question the validity of the hard, forever-enduring, indivisible atoms of Dalton and the Greeks. In an early model of atomic structure suggested by Sir J. J. Thomson, electrons were imagined to be embedded in a large "glob" or mass of positively charged material, like raisins in a pudding. There were just enough negative electrons to make the ball neutral.

RUTHERFORD AND THE NUCLEUS

In 1906 while studying the paths of α particles in magnetic fields, Lord Rutherford observed that particles were deviated or "scattered" by very thin sheets of mica. It was found later that a stream of α particles can shoot right through a thin layer of atoms, such as a sheet of gold leaf, but some of the particles are deflected from their paths by a small angle of, say, about 10°, others through large angles of, say, about 60°, and a few through much larger angles. Some α particles are even back scattered; that is, they are returned backward along the general direction from which they originated.

By 1910 such experiments gave results that could not be explained by a "pudding" atom. If the pudding were a hard lump, all α particles would bounce back; if it were a soft lump, none would bounce back.

Rutherford asked Geiger and Marsden to investigate the large-angle scattering. They had to rely on the tiny flashes of light ("scintillations") made when α particles struck a zinc sulfide screen in the dark. They reported that some of the α particles turned around in the foil and emerged from the same side they had entered. In the words of Rutherford, "It was then that I had the idea of an atom with a minute massive center carrying a charge." This was the origin of the modern conception of the nuclear atom.

Rutherford thought of the atom as consisting of a small, concentrated positively charged nucleus surrounded by electrons—a different number of electrons for the atoms of different chemical elements but the same number for each atom of a given element. He and his coworkers even calculated the approximate size of the nucleus and estimated that the charge on any particular nucleus numerically was roughly half the atomic weight.

REVISION OF THE PERIODIC LAW

In 1912 H. G. J. Moseley (1888–1915), a young Englishman, began a series of experiments the results of which verified both Rutherford's estimate of the charges on nuclei and the order of elements in the periodic table. Moseley found that different elements, when excited, by bombardment with electrons, gave off X rays of characteristic wavelengths that distinguished the atoms of one element from the atoms of all other elements. Using some three dozen elements he determined these wavelengths and then graphed the square roots of their frequencies against position in the periodic table. All values fell on a straight line whose slope suggested that the line would intercept the base at or near the position of hydrogen, which had ordinal number 1.

Mendeleev's order of the elements using the atomic weights had some outstanding difficulties. These were all removed when the elements were ordered according to their atomic numbers. Hydrogen was not only ordinal number 1 in the periodic table but the numerical value of the charge on its nucleus was also 1; the numerical value of the charge on the helium nucleus was 2, etc. The *atomic number* is the integral number representing the amount of positive charge on the nucleus of an atom. It is therefore the same as the number of protons in the nucleus of the atom and as the order number assigned to that element in the periodic table.

THE PERIODIC LAW TODAY

The foregoing leads to a *modern statement of the periodic law:*

If the elements are arranged in the order of their increasing atomic number, there is a regular recurrence of similar properties.

Isotopes and radioactivity

DISCOVERY OF ISOTOPES

A careful examination of the periodic table shows that nearly all the elements have fractional atomic weights. Yet there are no fractional protons. This posed another problem. Thomson and Aston had proved, by the differing deflection of positive ions, or atoms less one or more of their planetary electrons, passed through strong magnetic and electric fields, that nearly all elements are composed of atoms of slightly different atomic weights. The two or more kinds of atoms of the same element having the same atomic number but differing atomic weights are called *isotopes*. To illustrate the idea, let us consider neon (atomic number 10), having an atomic weight of 20.2 as it occurs naturally. By his "mass spectrograph," Aston found that natural neon consisted of two kinds of neon, $^{20}_{10}Ne$ and $^{22}_{10}Ne$. The subscript 10 to the left of the chemical symbol Ne specifies the atomic number whereas the superscript 20 to the left of the chemical symbol denotes the atomic weight of the isotope. It should be noted that because all isotopes of the element neon have an atomic number 10, the subscript 10 and the chemical symbol Ne both give the same information. In the case of the element neon it is found that the two isotopes occur in nature in the ratio of 9 parts of $^{20}_{10}Ne$ to 1 part of $^{22}_{10}Ne$. For $^{20}_{10}Ne$, $9 \times 20 = 180$; for $^{22}_{10}Ne$, $1 \times 22 = 22$. Adding the products, 10 parts $= 180 + 22 = 202$. Thus natural neon has an atomic weight of 202 divided by $10 = 20.2$.

DISCOVERY OF THE NEUTRON

In Bohr's original theory, to be discussed in the next chapter, the number of protons in the nucleus of an atom was equal to the atomic weight. Thus for helium of atomic weight 4 and atomic number 2, usually written $^{4}_{2}He$, Bohr suggested four protons in the nucleus to account for the atomic weight, with two planetary electrons revolving about the nucleus. The discrepancy in charge—four positive charges for the nucleus but only two negative charges for the planetary electrons—was handled not very satisfactorily by Bohr by assuming two "bound electrons" in the nucleus.

A much better handling of this discrepancy in charge became possible following the discovery of the *neutron,* a neutral particle having approximately the same mass as a proton, by Chadwick in 1932.

NEUTRONS IN ISOTOPES

The nucleus of $^{20}_{10}Ne$ is composed of 10 protons (p) and 10 neutrons (n), and the nucleus of $^{22}_{10}Ne$ consists of 10 protons and 12 neutrons. More precisely, the atomic weight of neon is 20.18 (not 20.2), because there are, in addition to the two main isotopes of atomic weights 20 and 22, very small amounts of the isotopes of masses 19, 21, and 23.

It has been found that isotopes are usually extremely similar to one another in their chemical properties. This is because they have the same number of protons in their nuclei—same atomic number—and the same number and arrangement of electrons outside of the nucleus. *Chemical properties of an element are related to the arrangement of extranuclear electrons.*

However, there is one property, radio-

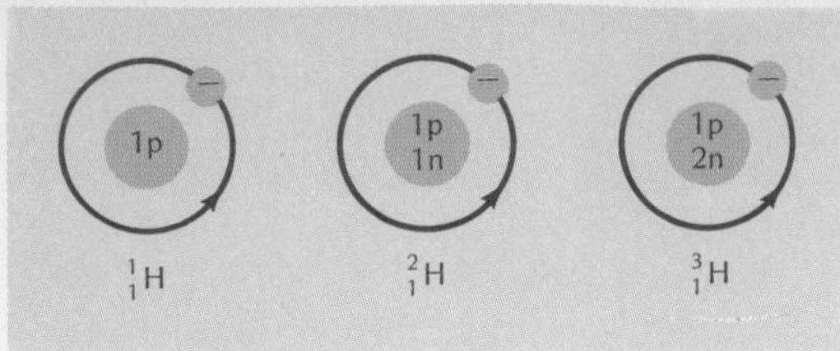

Figure 12.7
The three isotopes of hydrogen: ordinary hydrogen $^1_1$H, deuterium $^2_1$H, and tritium $^3_1$H.

activity, which is an outstanding exception to the rule that isotopes are similar. Two isotopes may differ extremely in the stabilities of their nuclei. When it becomes desirable to isolate $^{235}_{92}$U from its common abundant isotope of atomic weight 238 in order to make an atomic bomb, this separation is very expensive for the very reason that the properties of $^{235}_{92}$U and $^{238}_{92}$U are so similar.

A new and more theoretical definition may now be given for the term element. An *element* is a substance whose atoms all have the same atomic number, or the same number of protons in their nuclei.

Hydrogen has three isotopes, as shown in Fig. 12.7. Heavy water is composed of molecules formed by the combination of deuterium ($^2_1$H) with oxygen. The deuterium nucleus ($^2_1$H atom stripped of its electron) is called a *deuteron*. A very small fraction of 1 percent of natural water is heavy water. Heavy water is concentrated during the electrolysis of water, ordinary water $^1_1H_2O$ being decomposed more readily than $^2_1H_2O$. Tritium atoms $^3_1$H are used in making the hydrogen bomb. Tritium is practically nonexistent in nature.

There are three isotopes of chlorine, ^{35}Cl, ^{37}Cl, and ^{39}Cl. The periodic table shows 35.453 as the atomic weight of chlorine. The fractional atomic weight—whole number plus fraction—of an element is the average atomic weight of all the atoms, including all isotopes, of a random sample of the element. Thus the atomic weight of chlorine, 35.453, is an average atomic weight of a random sample of chlorine gas that contains chiefly atoms of atomic weight 35, some atoms of atomic weight 37, and still fewer atoms of atomic weight 39. Isotopes thus explain why the atomic weights of most elements are not whole numbers. Many of the elements of higher atomic weight have a large number of isotopes. Xenon, in which the number of protons is 54 but the number of neutrons varies, has 23 known isotopes.

MORE ABOUT RADIOACTIVITY

When an α particle $^4_2$He is emitted from the nucleus of a heavy element such as uranium, the remaining nucleus has two less positive charges and hence is the nucleus of a different element. Our original definition of an element (p. 266) stated that it is a substance which cannot be changed to a different substance by ordinary physical or chemical means. This statement is true, but it is now evident that atoms of radioactive elements do change spontaneously to atoms of different elements. This is spoken of as a natural radioactive transmutation.

Figure 12.8 indicates how an α particle is emitted from a nucleus of $^{238}_{92}$U, resulting in formation of a new element, which has atomic number 90 and atomic weight 234. The periodic table (inside front cover) tells us that the element having an atomic number of 90 is thorium, so the new element formed in this case is $^{234}_{90}$Th, one of the many isotopes of thorium.

It is easy to understand that an α particle consisting of two protons and two neutrons can be emitted from a nucleus made up of many protons and neutrons. But how can a β particle, an electron, be emitted from a nucleus consisting only of protons and neutrons? This is explained by considering that an internal explosion occurs that changes a neutron to a proton, which remains in the nucleus, and an electron, which is emitted from the nucleus. Because the emitted electron has a very small mass compared with the masses of the protons and neutrons, the remaining nucleus has approximately the same atomic weight as before. The internal change of a neutron to a proton, however, results in an increase

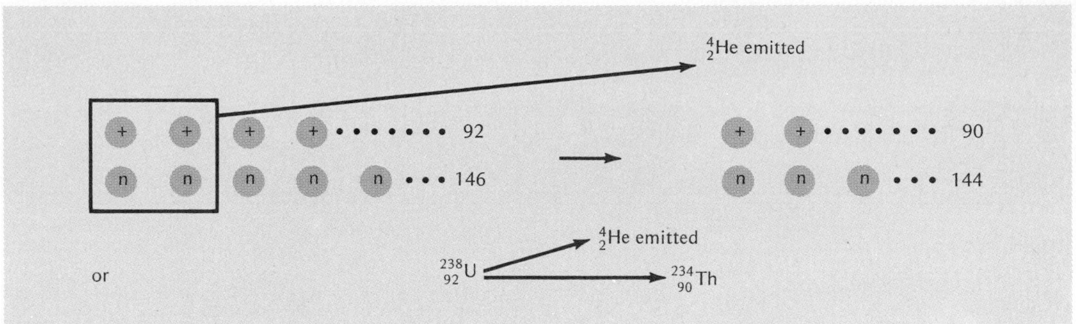

Figure 12.8
Radioactive emission of an α particle $^4_2$He from a nucleus of $^{238}_{92}$U to form $^{234}_{90}$Th.

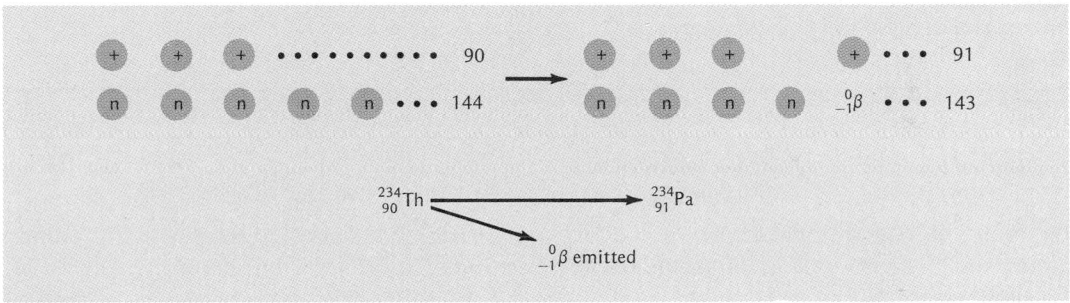

Figure 12.9
Radioactive emission of a beta particle $^{0}_{-1}\beta$ from a $^{234}_{90}$Th to form $^{234}_{91}$Pa.

of one in positive charges and therefore an increase of one in the atomic number. This kind of radioactive emission is shown in Fig. 12.9, where a β particle is emitted from a $^{234}_{90}$Th nucleus, resulting in formation of an element of atomic number 91 and atomic weight 234. Again referring to the periodic table, we find the element having an atomic number of 91 is protactinium; the new element formed in this case is $^{234}_{91}$Pa, one of the several isotopes of protactinium, because the emitted electron has negligible mass (as compared to the mass of the proton or neutron) and one negative charge, the β particle is designated as $^{0}_{-1}\beta$.

The emission of γ rays does not affect the number of protons or neutrons in the nucleus, so γ radiation will not be considered in the radioactive series to be discussed. The emission of a γ ray is apparently a mechanism by which a nucleus can lose energy and become more stable. Usually α or β emission is accompanied by γ radiation.

Relatively few of the billions of $^{238}_{92}$U nuclei present in an ordinary chunk of uranium ore undergo α disintegration. It has been computed that 4,500,000,000 years will be required for half of the atoms of $^{238}_{92}$U to make the change to the new element $^{234}_{90}$Th; hence the half-life of $^{238}_{92}$U is said to be 4.5×10^9 years. The change of $^{234}_{90}$Th to $^{234}_{91}$Pa is much more rapid, the half-life of $^{234}_{90}$Th being only 24 days.

The *half-life* of an isotope is the time re-

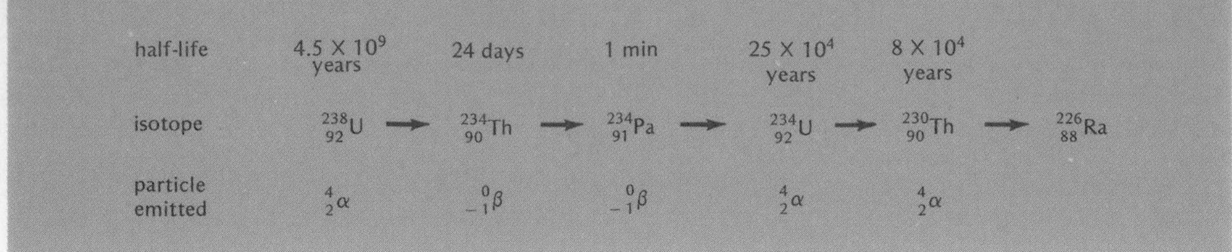

half-life
4.5×10^9 years
24 days
1 min
25×10^4 years
8×10^4 years
isotope
$^{238}_{92}U$ → $^{234}_{90}Th$ → $^{234}_{91}Pa$ → $^{234}_{92}U$ → $^{230}_{90}Th$ → $^{226}_{88}Ra$
particle emitted
$^4_2\alpha$
$^{0}_{-1}\beta$
$^{0}_{-1}\beta$
$^4_2\alpha$
$^4_2\alpha$

Figure 12.10
Portion of the uranium radioactive series from $^{238}_{92}U$ to $^{226}_{88}Ra$.

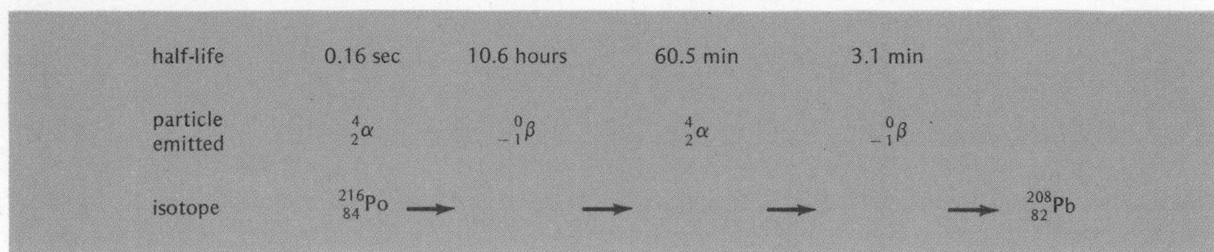

half-life
0.16 sec
10.6 hours
60.5 min
3.1 min
particle emitted
$^4_2\alpha$
$^{0}_{-1}\beta$
$^4_2\alpha$
$^{0}_{-1}\beta$
isotope
$^{216}_{84}Po$ →
→
→
→ $^{208}_{82}Pb$

Figure 12.11
Terminal portion of the thorium radioactive series.

quired for half of the nuclei in a given sample to make the change to the new isotope. For example, if the half-life of a radioactive isotope is 16 days, 2000 g of the isotope will be reduced to 1000 g in 16 days; 1000 g to 500 g in 16 additional days; 500 g to 250 g in another 16 days; etc. Thus in 48 days 2000 g of this radioactive isotope will reduce to 250 g of the original isotope, the remaining 1750 g consisting of other isotopes.

Radioactivity can now be measured with a nuclear-radiation counter over periods of time that are short compared to the half-life of the decay. The unit commonly used for such measurement is the *curie* (Ci), which is defined as 3.70×10^{10} disintegrations per second; a related unit is the millicurie (mCi) $= 3.70 \times 10^7 \text{ sec}^{-1}$.

A portion of the radioactive series, or progressive radioactive changes, which starts with $^{238}_{92}U$ and ends with the very stable lead isotope $^{206}_{82}Pb$ is given in Fig. 12.10. The formation of $^{234}_{90}Th$ and $^{234}_{91}Pa$ has already been explained. The emission of a $^{0}_{-1}\beta$ particle from a nucleus of $^{234}_{91}Pa$ gives an element with its atomic number increased by 1 to 92 and its atomic weight relatively unchanged. The periodic table tells us that this is $^{234}_{92}U$, an isotope of uranium. The emission of an α particle ^4_2He from $^{234}_{92}U$ reduces the atomic number by 2 and the atomic weight by 4, giving $^{230}_{90}Th$. Then an α particle is emitted by the $^{230}_{90}Th$ nucleus, resulting in an element of atomic number 88 and atomic weight 226. The periodic table tells us that this is one of the isotopes of radium, $^{226}_{88}Ra$.

Recalling that isotopes are different forms of the same element having the same atomic number but different atomic weights, two pairs of isotopes can be recognized in Fig. 12.10, a portion of the uranium radioactive series; $^{238}_{92}U$ and $^{234}_{92}U$ are isotopes of uranium, and $^{234}_{90}Th$ and $^{230}_{90}Th$ are isotopes of thorium.

A third natural radioactive series starts from the uranium isotope $^{235}_{92}U$ and ends with a different stable isotope of lead $^{207}_{82}Pb$.

Suggested Problem 15.1

The terminal portion of the thorium radioactive series ending in a stable isotope of lead $^{208}_{82}Pb$ is given in Fig. 12.11. The atomic numbers and atomic weights of the first substance and of the end product are given. The student should determine and write, in the space provided, the atomic numbers, atomic weights, and symbols of the three isotopes in between and indicate any pairs of isotopes of the same element.

SUMMARY

John Dalton and his revival of Greek ideas about atoms started scientists thinking about ultimate particles of matter. But it was not until approximately a half century after Dalton that Cannizzaro pointed out the lack of clarity and accuracy in these early ideas; that is, not one but two kinds of elementary particles—atoms and molecules—were needed.

Avogadro's hypothesis, or law, was the missing link that made it possible to calculate molecular weights from gas densities. Once a table of molecular weights was started, it was much easier to choose reasonable atomic weights from the so-called combining weights that had been used. Because all measurements are comparative, it should not be surprising to see that the system for expressing molecular weights has been changed from time to time.

While working with cathode rays, Thomson convinced scientists that the rays consisted of a streaming of negative particles; he chose the name electron for these particles. X rays are also given off in a cathode ray tube under proper conditions. The recognition of X rays led quickly to the discovery of radioactivity, in which α, β, and γ rays are emitted from atoms.

After the electron was discovered by Thomson in 1897, it was concluded just by rationalization that some part of the atom must be positive. The behavior of α particles shot at thin metal foils convinced scientists of the existence of a nuclear atom with almost all of the mass in the positive nucleus. The amount of positive charge on the nucleus was the basis for the new concept of atomic number and subsequent restatement of the periodic law.

The mass spectrograph is an outgrowth of the earlier simple cathode ray tube. Its use proved that most elements consisted of a number of species of atoms; all the atoms of a given element possess the same nuclear charge but they may contain different numbers of neutrons.

Important words and terms

atom
compound
law of definite proportions
law of partial pressures
Avogadro's hypothesis
atomic weight
gram-molecular weight
atomic mass unit (amu)
periodic law
periodic table
cathode rays
X ray
quantized electric charge
radioactivity
photon
quantum
α particles
β particles
γ rays
nucleus
Rutherford's model of the nucleus
isotope
symbols used to identify a nucleus
transmutation
neutron
proton
electron
heavy water
half-life

Questions and exercises

1. The Greeks theorized considerably about the atomic nature of matter. What element of modern scientific thinking was missing that made all their speculation fruitless? (Review "hypothesis" and "theory" in Chap. 1.)
2. What are the three common forms, or physical states, of matter explained earlier on the basis of the atomic theory of matter?
3. Distinguish between an element and a compound in terms of decomposition.
4. Explain the relationship between the *law of combining volumes* and the *law of definite composition*.

5. Does water always contain $11\frac{1}{9}$ percent of hydrogen by weight? Discuss.
6. Many of Dalton's ideas about the nature of atoms were in error; yet his thoughts contributed importantly to the development of science. Explain.
7. What term did Dalton use to describe the unit particle of a compound? Why was it inappropriate? What term did Cannizzaro use?
8. What advantage did the study of the behavior of gases have over that of liquids and solids in developing the theory of the atomic nature of matter?
9. In what respect did Dalton and Gay-Lussac differ in their explanation of the behavior of combining gases? Who first put forward the statement that would reconcile the two different theories?
10. State Avogadro's law. Who was principally responsible for its widespread acceptance?
11. Hydrogen, the lightest element, was originally selected as the standard for comparing the weight of other molecules and it was given the weight 1; oxygen was also designated as 16. More careful study later led to doubling these figures to 2 and 32, respectively. Why?
12. At one time the standard for the atomic weight was hydrogen; then it was oxygen. Today what fraction of what element is the standard? What is the technical designation for this unit?
13. "When you can measure what you are speaking about and express it in numbers, you know something about it; but when you cannot express it in numbers, your knowledge is of meager and unsatisfactory kind." With this remark of Lord Kelvin in mind, discuss the type of information that was needed before the elements could be arranged in meaningful form as the periodic table.
14. Look up in a dictionary the Greek roots for the word "atom." Trace the historical development from Dalton's "atom" of water, through Avogadro and Cannizzaro's revisions of Dalton's ideas, to the un-Greek atom of today. Then tie this in with the status quo of any scientific concept in the face of new evidence, as touched on in Chap. 1.
15. How was it determined that cathode rays are negatively, rather than positively, charged particles?
16. It required two men to determine the two fundamental quantities for the electron. Name them and the quantities that each determined; use the symbols you learned in Chaps. 4 and 10 to express them.
17. What two lines of evidence suggested that X rays were electromagnetic waves rather than particles like cathode rays?
18. (a) From your study of Chap. 9, indicate what kind of waves are generated when a clapper strikes a gong? (b) By analogy what material particles must strike what kind of solid to generate the electromagnetic waves known as X rays? (c) Why are X rays considered as more closely related to the rays that affect the eye than to cathode rays?
19. With regard to our knowledge of X rays, cite the contributions of Roentgen, Becquerel, and Curie (Sklodowska).
20. What are photons? How do they differ from cathode rays?
21. What is meant by radioactivity? Discuss the reasoning behind the displacement of the three spots on the photographic plate in Fig. 12.6?
22. (a) In what way does γ radiation resemble X rays? In what way does it differ? (b) Give your answer to the similar question for β rays versus cathode rays. (c) What is meant by α rays?
23. What were the major contributions to atomic theory by J. J. Thomson and Lord Rutherford?
24. Give the modern statement of the periodic law. Examine the table on the inside front cover: Does atomic weight increase with atomic number?
25. What is a neutron?
26. Define isotopes. How are they alike? How are they different? Reconsider your answer to Question 5.
27. Using the periodic table inside the front cover, write for the first 10 elements the symbol, atomic number, and atomic weight; use the subscript–superscript notation in doing this.
28. (a) Give the modern definition of an element? (b) Why do most of the elements in the periodic table have fractional atomic weights?
29. The first portion of the uranium radioactive series (Fig. 12.10) ends with $^{226}_{88}Ra$. This nucleus gives off an α particle and succeeding nuclei give off α, α, β, and β particles,

respectively. From these data construct this part of the radioactive series. Use the periodic table to get the necessary chemical symbols. List all pairs of isotopes in this part of the series.

30. If the half-life of radioactive barium $^{140}_{56}Ba$ is 12 days, how long will it be before 1000 g of radioactive barium is reduced to 250 g?

Supplementary readings

Ashford, Theodore A., *The Physical Sciences: From Atoms to Stars,* Holt, Rinehart and Winston, New York (1967). [Pages 236–286 discuss such topics as: are molecules real; Avogadro's hypothesis; atomic and molecular weights.]

Bonner, Francis T., et al., *Principles of Physical Science,* Addison-Wesley, Reading, Mass. (1971). [Chapter 7 is about the historical aspects of the atomicity of matter: the law of definite proportions (Proust, Berthollet); Dalton and the chemical atomic theory (the law of multiple proportions) and relative atomic weights; the law of combining volumes; Cannizzaro's method of atomic-weight determination; the utility of atomic and molecular weights. Chapter 9 is "Periodic Classification of the Elements."]

Cowling, T. G., *Molecules in Motion,* Harper Torchbooks, Harper & Row, New York (1960). (Paperback.) [Chapter 2 shows the triumph of the atomic theory—for example, phlogiston, Dalton's theory, Avogadro's hypothesis, Dalton's law of partial pressures.]

Dampier, William C., and Margaret Dampier, *Readings in the Literature of Science,* Harper Torchbooks, Harper & Row, New York (1959). (Paperback.) [Pages 83–111 give original writings by Lavoisier, Dalton, Gay-Lussac, and Avogadro. Lecture delivered by Mendeleev on "The Periodic Law of the Chemical Elements," pp. 112–117.]

Hughes, Donald J., *The Neutron Story,* Anchor Books, Doubleday, Garden City, N. Y. (1959). (Paperback.) [A clear account of the history, behavior, and uses of the neutron. Chapter 3 dwells on waves and particles.]

Keenan, Charles W., and Jesse H. Wood, *General College Chemistry,* 4th ed., Harper & Row, New York (1971). [Pages 20–30 give a good presentation of experimental evidence for the existence of small particles; also included are weights and sizes of atoms. Chapters 14 and 15 are about nuclear reactions and their applications.]

Leicester, Henry M., and Herbert S. Klickstein, *Source Book in Chemistry,* McGraw-Hill, New York (1952). [Excerpts from original papers: J. Dalton, "The Atomic Theory," pp. 208–220; A. Avogadro, "Hypothesis of Atoms and Molecules," pp. 231–238; S. Cannizzaro, "Sketch of a Course of Chemical Philosophy," pp. 406–417, the paper in which the method of determination of atomic and molecular weights was established.]

Nash, Leonard K., *The Atomic-Molecular Theory,* Harvard University Press, Cambridge, Mass. (1950). (Paperback.) [A case history: Dalton's atomic theory, application of the theory to chemistry, Gay-Lussac's law of combining volumes, Dalton's rejection of the law, Avogadro's contributions, Berzelius and oxygen, Cannizzaro's revival and emphasis of Avogadro's concept.]

Omer, Guy C., Jr., et al., *Physical Science: Men and Concepts,* D. C. Heath, Boston (1962). [In Part 5 are considered the nature and composition of matter; the controversy between Berthollet and Proust and the establishment of the law of definite proportions; from Dalton's *A New System of Chemical Philosophy,* his atomic theory and his views on compounds; Gay-Lussac's law of reacting volumes in his *Memoir on the Combination of Gaseous Substances with Each Other;* Avogadro's hypothesis, as in his original paper, and the calculation of molecular weights; chemistry from Avogadro to Cannizzaro, 1811 to 1860: Berzelius, Proust, Dulong and Petit, Dumas, influence of developments in organic chemistry on the atomic-molecular theory; and finally, Cannizzaro's pamphlet *Sketch of a Course of Chemical Philosophy,* with emphasis on the determination of atomic weights, which made possible the discovery in 1869 of the periodic law by Mendeleev and Lothar Meyer.]

Partington, J. R., *A Short History of Chemistry,* Harper Torchbooks, Harper & Row, New York (1960). (Paperback.) [Dalton, Berzelius's atomic-weight tables, Proust, Avogadro, and others. Summary, pp. 211–214.]

Patterson, E. C., *John Dalton and the Atomic Theory,* Anchor Books, Doubleday, Garden City, N. Y. (1970). (Paperback).

Taylor, Lloyd William, *Physics: The Pioneer Science,* Houghton Mifflin, Boston (1941); or, a paperback, L. W. Taylor and F. G. Tucker, *Physics: The Pioneer Science,* Vols. 1 and 2, Dover Publications, New York (1962). [The author does a very fine development on electric discharge in gases and cathode rays, pp. 770–779; and on X rays and radioactivity, Chap. 52.]

Wehr, M. Russell, and James A. Richards, Jr., *Physics of the Atom,* Addison-Wesley, Reading, Mass. (1967). [An appendix has a chronology of the atomic view of matter from 550 B.C. to the 1960s; dates, names, and contributions are listed.]

Whyte, Lancelot, *Essay on Atomism: From Democritus to 1960,* Wesleyan University Press, Middletown, Conn. (1961). [Contains a chronological table to serve as a record of atomism and a reminder of the most important names and dates with a brief note of each achievement.]

Wilson, Robert R., and Raphael Littauer, *Accelerators: Machines of Nuclear Physics,* Anchor Books, Doubleday, Garden City, N. Y. (1960). (Paperback.) [The preface is a vignette of Ernest O. Lawrence, whose brain produced the principle of the cyclotron. The book takes the reader on a guided tour through the stable of accelerators from the first one (X ray machine) in 1895. The book talks about accelerators—how they were developed and how they have achieved higher and higher energies—especially the cyclotron, the betatron, the cosmotron, and the synchrotron.]

ELECTRON STRUCTURE OF ATOMS

The most important trends in industry today spring from an increasing knowledge of the properties of atoms and their component parts.

DAVID SARNOFF, 1941

We have seen evidence for the belief that the world is composed of various kinds of substances, each substance having its own specific composition and its own set of properties. This is partially explained by assuming that each substance is composed of a particular kind of molecule. In turn each molecule consists of a fixed number of each kind of its constituent atoms. We have also discovered evidence that any particular atom is composed of a central region called a *nucleus* and some electrons outside this nucleus. In this chapter we shall describe the development of the modern picture of the extranuclear, or electron, structure of the atoms.

Introduction to the Bohr theory of the atom

By 1865 the atomic-molecular theory of matter was generally accepted. And by the beginning of World War I the concept of the nuclear atom was well established. Because atoms are normally neutral it was assumed that enough negative electrons were somehow grouped outside the nucleus to balance exactly, or neutralize, the amount of positive charge associated with it.

Niels Bohr (1886–1962) had just received his doctorate in physics in Denmark when he went to England in 1911. After a few months as a visiting researcher in the laboratory of J. J. Thomson at Cambridge, Bohr went to the University of Manchester and worked with Lord Rutherford's group. This group had already concluded that simple atoms, such as those of hydrogen, consisted of a positive nucleus with one or more electrons outside. But what were the electrons doing?

In 1913 Bohr proposed a theoretical structure which in a modified form was used extensively for several decades. According to his theory, a hydrogen atom (a model is shown in Fig. 10.3) is composed of a nucleus called a proton, which has a single positive charge, and an electron having a single negative charge moving around the nucleus just as a planet revolves about the sun. The mass of the proton has been found to be 1836 times the mass of the electron, which is analogous to the situation in the solar system where the mass of the sun is about 1000 times greater than the mass of the largest planet Jupiter. An atom of the next element, helium, has two planetary electrons; lithium has three, etc. Thus the number of planetary electrons in a neutral atom is equal to the atomic number of the element, which is the same as the order number denoting the place in the periodic table.

Before examining the Bohr theory of the atom in greater detail, we shall attempt to put it in perspective by discussing the work of Planck and Einstein which influenced Bohr's work.

PLANCK'S QUANTUM THEORY

In Chap. 9 we discussed light or electromagnetic radiation in the context of wave phenomena. The wave description works well when one attempts to describe phenomena such as reflection, refraction, interference, and diffraction of light. However, scientists were unable to explain the experimentally observed dependence of the amount of radiant energy and the frequency (color) of the radiation that is emitted from a heated body on the temperature of the body. This phenomenon has led to such everyday expressions as "that object is red hot" whereas a hotter object is "white hot." For an object being heated consider, for example, the heating element on an electric range. A warm heating element appears black, but as it becomes hotter, it is first deep red and then a brighter red, and perhaps if heated to a sufficiently high temperature, it will become "white hot." The amount of radiant energy of each frequency (color) was measured very carefully as a function of the temperature of the object and it was this phenomenon that could not be explained in terms of Maxwell's electromagnetic theory (a wave theory) and other classical principles. Max Planck was able to obtain a mathematical relationship that agreed beautifully with the experimental observations. To do this Planck had to postulate that energy was emitted in the radiation process in "bundles or packets of energy" which he called *quanta*. This postulate is in direct conflict with classical electromagnetic theory as developed by Maxwell in which the emission of radiation is continuous. Planck assumed that these quanta of radiation had an amount of energy proportional to the frequency of the radiation emitted. From a comparison of his mathematical equations with experimental data he found a proportionality constant $h = 6.625 \times 10^{-27}$ erg-sec, where h is called *Planck's constant*. The energy E of a quantum is given by

$$E = hf$$

where f is the frequency in cycles per second of the radiation. It should be noted that even Planck was disturbed by this "particle" theory of radiation—that is, that the quantum of energy has a discrete nature like a particle as opposed to a continuous nature associated with a wave theory. It was not until 1905 when Albert Einstein employed Planck's quantum of energy to explain experimental observations on the photoelectric effect that scientists began to "believe" in the idea of a quantum of electromagnetic energy.

PHOTOELECTRIC EFFECT

The photoelectric effect was discovered in 1888 by Heinrich Hertz while making the investigations that led to the discovery of radio (hertzian) waves. Hertz made the observation that a spark jumped more readily between two highly polished metal spheres acting as electrodes if ultraviolet light was shining on the negative electrode. This phenomenon, named the "photoelectric effect," was examined experimentally in detail. Several interesting relationships, summarized as follows, were revealed during the next few years:

1. Electrons are ejected from clean metal surfaces when light of sufficiently high frequency shines on the surface. The frequency must be higher than a certain "threshold frequency" which is different for different metals. The threshold frequency is that below which no emission of electrons occurs regardless of the light intensity.
2. The energy of the emitted photoelectrons is independent of the amount (intensity) of light but does depend on the frequency. The higher the frequency is above the threshold frequency for photoemission for a particular metal, the greater the kinetic energy of the emitted electrons.
3. The number of photoelectrons ejected is proportional to the intensity of the light striking the metal but is independent of the frequency of the light as long as the frequency is above the "threshold frequency."

These observations could not be explained on the basis of Maxwell's electromagnetic wave theory of light. For example, according to wave theory there appeared to be no way to account for dependence of the energy of the ejected photoelectron on the frequency of the incident light. In 1905 Albert Einstein proposed a simple but scientifically daring explanation. Einstein suggested that light emitted in bundles of energy, known as quanta as proposed by Planck, retains its quantum nature and continues to carry energy in bundles equal to hf. Further, when the light strikes matter such as a metal plate, the light continues to act as a bundle of energy or, in fact, acts like a particle whose energy is hf. He named these "particles of light" *photons*. Using the concept of photons striking a metal surface, Einstein was able to account for all of the experimental observations of photoelectric phenomena. When light shines on a metal surface and a photon with energy $E = hf$ strikes an electron, its energy is transferred to the electron. The electron will be ejected from the metal surface if the photon's energy absorbed by the electron is sufficient to overcome the forces holding the electron in the metal. This description explains the necessity of a threshold frequency below which the photons of light have insufficient energy to free the electron from the metal. Because the electrons in different metals would not necessarily be held with the same magnitude of force, the threshold frequency for photoemission should vary for different metals. If the frequency of the light exceeds the threshold frequency, then the excess energy will appear as kinetic energy of the ejected electron.

The number of photoelectrons emitted should increase with an increase in light intensity if one considers the intensity of the light to be just a measure of the number of photons in the light. The higher the intensity, the greater the number of photons contained in the light hitting the metal surface each second. Because each photon has the same energy (hf), the total energy per second (the intensity) in the light will be the number of photons hitting per second (N) times the energy hf or

$$\text{total energy} = N(hf)$$

This successful description of the photoelectric effect by Einstein led to a revolution in scientific thinking. It should be noted that this photon or quantum theory of light did not lead to the abandonment of the Maxwell theory and the wave picture of light. In ordinary experiments dealing with reflection, refraction, interference, diffraction, and polarization the wave description of light provides a suitable model. Apparently, when we are dealing with the interaction of electromagnetic radiation with individual atoms or electrons, the photon, or particle, model of light must be employed to explain the observed phenomena. This is called the "dual nature" of light. If this dual nature seems perplexing, it should be remembered that the wave theory of light is an invention of the human mind that enables man to describe the behavior of photons in a very satisfactory way. Waves are only a description of physical phenomena such as interference or diffraction. It is the phenomena that are the physical reality, not the waves. Hence the fact that photons (light) have properties not described by waves should not be disturbing.

THE BOHR THEORY OF THE HYDROGEN ATOM

When Bohr proposed his planetary model of the hydrogen atom, he had to provide a mechanism by which two oppositely charged objects could coexist apart from each other and maintain stability. Bohr proposed that the electrical force of attraction between the electron and the proton would provide the centripetal force (and, hence, centripetal acceleration) necessary to keep the electron moving in a circular orbit. This is analogous to the situation in the solar system where the gravitational attractive force between a planet and the sun provides the centripetal force to keep the planet in its orbit. However, there was a tremendous difference in the two systems. From the classical theory of electricity and magnetism, which was well established and accepted by scientists in 1912, an electron moving in a circle would lose energy continuously by radiating electromagnetic energy. Hence the electron would spiral into the nucleus as it lost this energy. To overcome this problem, Bohr employed the Planck concept of the quantum. It was this departure from classical ideas that enabled Bohr to develop his model of a stable hydrogen atom. Using classical electrostatics and Newton's laws of motion, Bohr derived a simple expression for the frequency of an electron moving in a circular orbit. He then introduced Planck's constant, by restricting the angular momentum to discrete values $h/2\pi$, h/π, $3h/2\pi$, $2h/\pi$, $\ell h/2\pi$, etc. (where ℓ is an integer). The angular momentum of a particle is obtained by multiplying the radius of the orbit of the electron times its momentum. This condition on the angular momentum results in restricting the electron to motion only in a discrete set of orbits (see Fig. 13.1). To obtain a stable atom, Bohr then

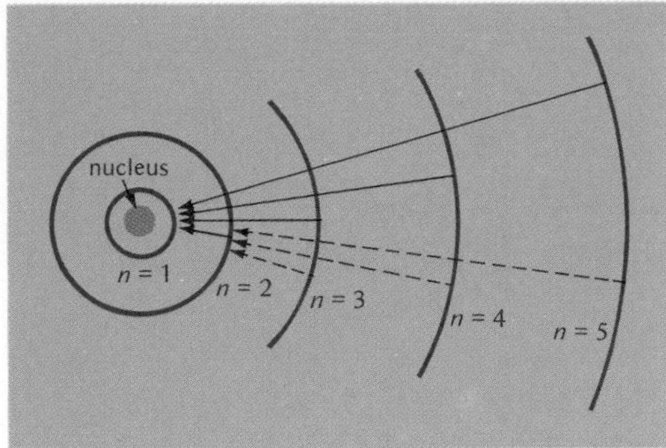

Figure 13.1
Cause of series of hydrogen-emission spectral lines. Lines due to jumps of the one electron to the innermost orbit (solid arrows) belong to the Lyman series; those due to jumps from various outer orbits to orbit 2 (broken arrows) are the Balmer series.

postulated—in contradiction to classical theory—that although the electron remains in any of these allowed orbits, it does not radiate energy. Bohr next postulated that the radiation from an atom or absorption of energy by an atom occurs only when an electron moves from one of these discrete orbits to another of the discrete orbits. Because one can assign a definite amount of energy to the electron in a particular orbit (just as can be done for each planet in the solar system), an electron in a neutral hydrogen atom would normally reside in the orbit with a smallest radius corresponding to the lowest energy or "ground state energy." The electron can be caused, by a collision with a fast-moving electron or by the absorption of electromagnetic energy, to move into one of the other discrete orbits of larger radius which corresponds to a higher energy state for the atom. The hydrogen atom with an electron in an orbit other than the inner-orbit ground state would be in an "excited" state. When an electron in an excited state moves to the ground state, a quantum of energy *hf* representing the difference between the energy of the two states is emitted. This process is responsible for discrete atomic spectra and is illustrated in Fig. 13.1.

Sometimes the energy supplied is sufficient to remove entirely the planetary electron from the hydrogen atom. The result is an ionized atom, or hydrogen ion, which must have a positive charge of one, because the balancing negative charge of the electron has been removed.

EXTENSION AND SOME LIMITATIONS OF THE BOHR THEORY

Using his theory, Bohr was able to calculate the frequencies of the spectral lines of atomic hydrogen with remarkable accuracy; the agreement was good even with the spectral lines of singly ionized helium and doubly ionized lithium. These latter ionized atoms are still one-electron systems. The Bohr theory was not applicable to neutral helium, which has two electrons, or to the more complex atoms. The greatest contribution of the theory was the idea of distinct orbits for the electrons and the assumption that unless the electron jumps from one orbit to another, its energy does not change. This represented an acceptance of the quantum hypothesis introduced by Planck and supported by Einstein.

Attempts were made to modify the Bohr model in an effort to apply it to more complex atoms. The first such modification made by Bohr and Arnold Sommerfeld was to extend the theory to incorporate elliptical orbits in place of the circular orbits. It should be noted that the mathematical problem of three interacting bodies—for example, a single helium nucleus with its positive charge interacting with the two individual negative electrons—has not yet been solved exactly. The introduction of elliptical orbits led to an additional "quantum number." In the Bohr theory one quantum number was used to specify the discrete orbit in which the electron existed. This, of course, also specified the energy level. This first, or principal, quantum number (designated *n*) must be an integer

Table 13.1
Quantum numbers

symbol	name	describes	permitted values
n	principal quantum number	energy level	1, 2, 3, 4, · · ·
ℓ	orbital quantum number[a]	angular momentum of the orbiting electron	0, 1, 2, · · · , $n-1$
m_ℓ	orbital magnetic quantum number	magnetic properties of orbiting electron pertaining to the shape of the orbit	$-\ell, -(\ell-1) \cdots 0 \cdots (\ell-1), +\ell$
m_s	electron spin quantum number	magnetic property of the "spinning" electron; (+) and (−) designate opposite spins	$-\frac{1}{2}$ or $+\frac{1}{2}$

[a]Electrons with $\ell = 0$ are designated "s" electrons. Similarly those with ℓ values of 1, 2, and 3 are designated *p*, *d*, and *f* electrons, respectively.

with $n = 1$ being the lowest energy level. The second quantum number (ℓ) necessitated by the introduction of the elliptical orbits determines the angular momentum of the electron. The permitted values of ℓ are integral and range from zero through $(n - 1)$.

A third quantum number (m_ℓ) was introduced to take account of the effect of a magnetic field on spectrum lines. An electron moving in an orbit is a moving charge and is similar to a current in a loop of wire. In Chap. 10 we saw that a current loop produces a magnetic field and, in fact, acts like a small magnet. When an atom is placed in a magnetic field, the magnetic field will influence the orbit of the electron, which results in small splitting of the energy levels originally derived from the original Bohr analysis. The splitting of the energy levels results in splitting of the spectral lines into several closely spaced lines when hydrogen or any other atom is placed in a magnetic field. Only certain values of m_ℓ are permitted as determined by the rule that values of m_ℓ range from $-\ell$ to $+\ell$ in integer steps including 0. For example, when $\ell = 0$, then $m_\ell = 0$, and when $\ell = 1$, then m_ℓ can have values −1, 0, and +1.

A fourth quantum number (s), the spin angular momentum or spin magnetic quantum number, is also necessary to account for the magnetic properties of the electron. Numerous experiments show that an electron acts as a small magnet. This is attributed to the spinning of the electron about an axis through its center. Because the electron is a quantity of negative charge, a spinning of this charge about its own axis would produce a magnetic field; that is, a spinning electron would act like a small magnet. It is found that there are only two possible values of the spin quantum number, usually denoted $m_s = \frac{1}{2}$ and $m_s = -\frac{1}{2}$. These two values represent the spinning of the electron charge clockwise and counterclockwise. These quantum numbers are summarized in Table 13.1.

All of these quantum numbers represent effects that alter the energy of the simple Bohr energy levels. There are other effects that introduce additional quantum numbers which have been introduced to describe the atomic energy levels responsible for the complex spectral lines that emanate from excited atoms. However, these four are sufficient to account for a great deal of the observed spectra in all atoms. In addition, with these quantum numbers it is possible to "build" a periodic table of the elements that agrees well with the groupings developed from the observed chemistry of the elements. Before describing this construction of a periodic table, we shall examine some of the additional modifications of our picture of the atom which have resulted in the present-day wave-mechanical or quantum-mechanical description of the atom.

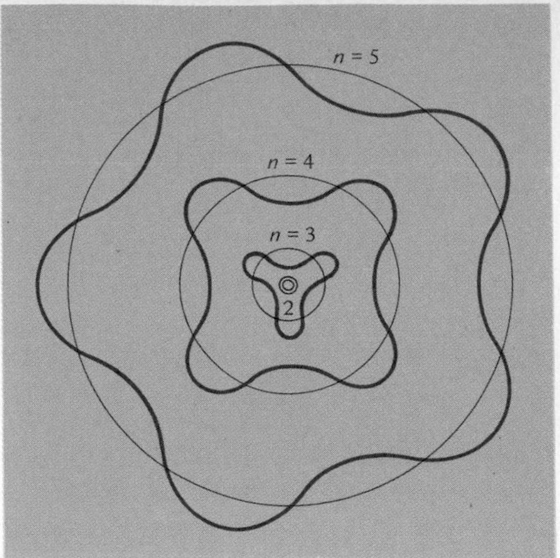

Figure 13.2
de Broglie's concept of matter waves. The electron is restricted to those orbits whose circumference equals an integer number of electron wavelengths.

de Broglie matter waves

The success of the Bohr theory in predicting frequencies of the spectral lines of hydrogen lent further support to the dual nature of light. Louis de Broglie felt that nature should be symmetrical, and if light was to have a dual nature, then why should not matter, which we have always considered as being composed of particles, have a dual nature and have wave properties? He proceeded to compare the two systems, light and matter, and arrived at the conclusion that if particles were to have a wave nature, they would have a wavelength

$$\lambda = \frac{h}{mv}$$

where h is Planck's constant and mv is the momentum of the particle whose wavelength is λ. When de Broglie proposed this hypothesis in 1924, there was no experimental evidence to support it. However, he was able to show how the idea of a moving particle having a wavelength could lead to the Bohr model of the atom. In fact he was able to derive the postulate used by Bohr that the angular momentum had to be integer multiples of the quantity $(h/2\pi)$. He could predict the size of the Bohr orbits by restricting the orbits to only those radii (R) such that the circumference $2\pi R$ of the orbit is equivalent to an even number of wavelengths of the electron in that orbit. This idea is illustrated in Fig. 13.2. Notice that the idea of a moving particle acting as a wave results in the "feeling" that the electron is no longer moving in an orbit with a strictly fixed radius but is in fact "spread" over a region centered about this radius (see Fig. 13.3).

In 1929 Davisson and Germer found experimentally that electrons, "particles," showed interference and diffraction effects when the electrons were "bounced" off a crystal. In fact the electrons produced the same diffraction patterns as one observed with X rays, which are high-energy electromagnetic waves similar to light. Their experiments verified the de Broglie relationship. Hence in this submicroscopic world of the atom a "particle" of matter, the electron, exhibits wave properties. Experiments have shown that under the proper conditions atoms exhibit wave properties.

Heisenberg's uncertainty principle

In 1927 Heisenberg proposed his *uncertainty principle* stating that for small masses, such as electrons, it is impossible to specify exactly both the position and momentum (or velocity) of the particle at the same instant of time. His principle arises from the process by

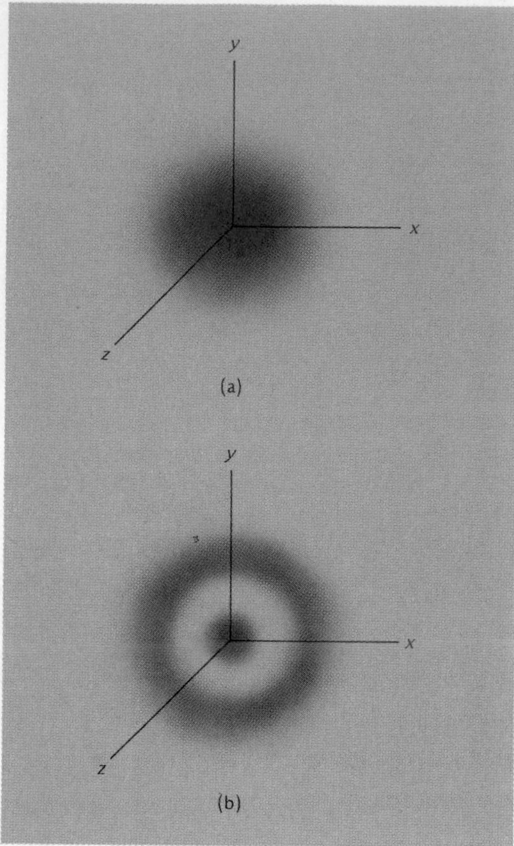

Figure 13.3
Electron cloud models of 1*s* and 2*s* orbitals. Probability representations for electrons in atoms: (a) Plot of high probability points for a 1*s* electron. An electron can be found anywhere from near the nucleus to a great distance away but it is most likely to be found approximately 10^{-8} cm from the nucleus. (b) An electron cloud model of the 2*s* orbital. The 1*s* orbital should not be thought of as being inside the 2*s* orbital. The latter has two separate regions in the same wave.

which we locate or "see" something—for example, an electron in a Bohr orbit. In order to see the electron we must turn on light, and the electron absorbs some of the light energy. The electron will instantaneously move into another energy level where it would have a different velocity—that is, a condition necessary for the observation of something causes a reaction on the thing being observed. This uncertainty principle means that Newtonian mechanics is not applicable to the world of the atom. The principle led to the development of quantum mechanics, which is a wave mechanics employing de Broglie's ideas. In the wave-mechanical treatment, the idea of definite allowable orbits of electrons (as specified in the Bohr theory) is replaced by a probability distribution—that is, that there is a region of space in the atom called an *electron cloud* where the probability of finding an electron is very high (see Fig. 13.3, electron-cloud models). This approach postulates that although the electron is not restricted to residing exactly at a fixed distance from the nucleus in an orbit, this is the place of the highest probability of finding the electron. The idea of energy levels developed by Bohr is still retained in the wave-mechanical picture of the atom and in fact it is these energy levels, or quantized states, that are important. One still speaks of electron orbits or orbitals but these are now considered as "fuzzy clouds" or probability distributions.

The periodic table

GROUPS, PERIODS, FAMILIES

As the Bohr theory was developed and refined with the introduction of quantum numbers, scientists began to attempt to explain the periods and families in the periodic table (inside front cover).

Each vertical column in the periodic table constitutes what is called a *group*. A *period* (note the arabic number in the left column) consists of a horizontal series of elements. Note that period 1 consists of only hydrogen (H) and helium (He), all other groups not being represented in this series. Period 2 consists of lithium (Li), beryllium (Be), boron (B), carbon (C), nitrogen (N), oxygen (O), fluorine (F), and neon (Ne).

There are two groups which are designated to be Group I—that is, Group IA and Group IB. That is because Group I really consists of two families of elements. The alkali metal family [lithium (Li), sodium (Na), potassium (K), rubidium (Rb), and cesium, (Cs)] is most of Group IA. Group IB also is made up of metals—namely, copper (Cu), silver (Ag), and gold (Au). Besides the fact that they are metals, as are the Group IA elements, there are other similarities. There are also differences.

Because the elements in the same vertical column tend to resemble each other more often than not and more than they resemble the elements in the other vertical column of that group, they are referred to as a family. A *family* of elements consists of those in the same subgroup (same vertical column)—for example, Group IA or Group IB.

QUANTUM NUMBERS

It is convenient and useful to think of electrons as being added outside of nuclei in "layers" or energy levels as we go up from hydrogen in atomic number. Using a basic principle of nature—that systems tend to go to a condition of lowest energy—we could predict that electrons in an atom would reside in the lowest energy level or the first Bohr orbit. However, such a situation would not explain the observed spectra of atoms that are more complicated than helium. It is necessary that electrons in complex atoms be placed in higher energy levels (larger Bohr orbits). How are the electrons distributed? An exclusion principle proposed by Wolfgang Pauli in 1925 answered this question. *The Pauli exclusion principle* states that *no two electrons in the same atom can have the same values of the four quantum numbers* n, ℓ, m_ℓ, and m_s. With this restriction, we can build up, step by step, the orbital structures of some atoms.

Consider the lowest energy state where $n = 1$. When $n = 1$, there is only one value (see Table 13.1) for ℓ, which equals 0—that is, $n - 1$. The only allowable value for m_ℓ is also 0. However, m_s can be $+\frac{1}{2}$ or $-\frac{1}{2}$. Therefore two electrons can reside in the $n = 1$ energy state with the quantum numbers $n = 1$, $\ell = 0$, $m_\ell = 0$, and $m_s = \frac{1}{2}$ or $-\frac{1}{2}$. An additional electron would have to go into a higher energy state specified by $n = 2$. Therefore only two atoms, hydrogen (one electron) and helium (two electrons), can have all of their electrons in the first or $n = 1$ energy state. For $n = 2$, the number of electrons allowed is eight and they are grouped into two subshells denoted by the two possible values of ℓ as indicated in Table 13.2. Table 13.2 could now be extended to the next energy state where $n = 3$. In this case ℓ can have three values 0, 1, and 2 with one value of m_ℓ for $\ell = 0$; three values of m_ℓ for $\ell = 1$, and five values of m_ℓ for $\ell = 2$. It is left as an exercise for the student to complete a table for $n = 3$.

Table 13.2
Filling of energy levels

quantum numbers n	ℓ	m_ℓ	m_s	number of electrons	quantum notation
1	0	0	$+\frac{1}{2}, -\frac{1}{2}$	2	*s*
2	0	0	$+\frac{1}{2}, -\frac{1}{2}$	2	*s*
	1	1	$+\frac{1}{2}, -\frac{1}{2}$	2	*p*
		0	$+\frac{1}{2}, -\frac{1}{2}$	2	*p*
		−1	$+\frac{1}{2}, -\frac{1}{2}$	2	*p*

ORBITAL STRUCTURE OF THE ATOMS: ELECTRON CLOUD MODELS AND ELECTRON CONFIGURATIONS

We shall now use these values of possible quantum numbers and a modern notation that is related to the results of quantum mechanics to build up the orbital structure of some atoms. We have seen that in the modern wave-mechanical picture the most *probable* place to find an electron replaces the exact radius of the electron orbit of the Bohr

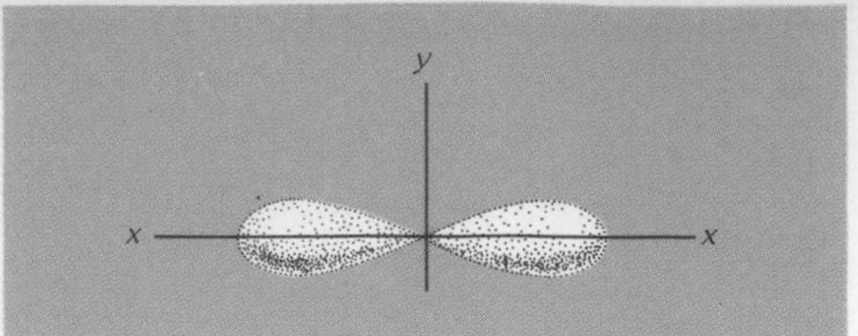

Figure 13.4
A *p* orbital. The general shape is shown for one *p* orbital with the electron density symmetrical around the *x*-axis. Compare with the spherical *s* orbitals in Fig. 13.3.

model. Electrons whose locations (that is, probability distributions about the positive nucleus) have spherical symmetry are called *s electrons* (see Fig. 13.3). These electrons, in fact, always have $\ell = 0$, as is the case for both hydrogen and helium.

A new energy level of electrons starts with lithium (Li). The new electron of higher energy is also an *s* electron, as is the fourth electron in beryllium (Be). Let us start a table of electronic structure thus:

H	$1s^1$
He	$1s^2$
Li	$1s^22s^1$
Be	$1s^22s^2$

The last notation means that a beryllium atom contains two 1*s* electrons and two 2*s* electrons.

In the electron configuration of atoms there may be energy sublevels of main energy levels or states (also referred to as subshells and shells, respectively). The number and location of electrons in atoms are summarized according to the following convention:

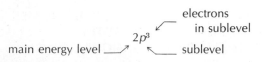

$2p^3$ — main energy level (2); sublevel (p); electrons in sublevel (3)

The notation tells us that there are three *p* electrons in the sublevel of the main energy level $n = 2$; because they are *p* electrons they have the orbital quantum number $\ell = 1$.

With boron (B) we add a higher-energy electron which has a new configuration for its orbital. Perhaps we can best picture this region as being of dumbbell shape (Fig. 13.4) along an *x*-axis with a node at the intersection of the *x*- and *y*-axes. This is called a *p* orbital.

The carbon atom (C) also has its new electron beyond the $2s^2$ structure concentrated in a dumbbell-shaped region; this new electron is therefore also a *p* electron. The new *p* orbital will be at right angles to the first *p* orbital and can be drawn either on the *y*-axis or *z*-axis, both being perpendicular to the *x*-axis. The *x*-, *y*-, and *z*-axes are each perpendicular to the other two. With nitrogen (N) we add a third *p* orbital, which can be drawn either on the *y*-axis or *z*-axis depending on which we did not use for the second *p* orbital. Let us describe the new structures and also that of oxygen as follows:

B	$1s^22s^22p^1$
C	$1s^22s^22p^2$
N	$1s^22s^22p^3$
O	$1s^22s^22p^4$

If one looks back at the structure of the helium atom, it can be seen that two electrons apparently occupy the same space (orbital) at the same time. A study of other structures indicates that this does not occur if there are still unoccupied orbitals in the same energy level. Let us represent the electron of hydrogen and the two electrons of a helium atom as occupying boxes as follows:

H [↓] $1s^1$ He [↓ ↑] $1s^2$

The boxes represent orbitals; each arrow represents an electron. The arrow pointing down represents a negative (−) spin; the one pointing up represents a positive (+) spin. Considering the helium electrons as particles, we assume that they spin in opposite directions; this opposed spin sets up a magnetic field that counteracts much of the repulsion which they otherwise would have for each other. But no orbital is ever occupied by more than two electrons; this follows from the *Pauli exclusion principle*.

Similarly, let us represent the electrons of the nitrogen atom in boxes as follows:

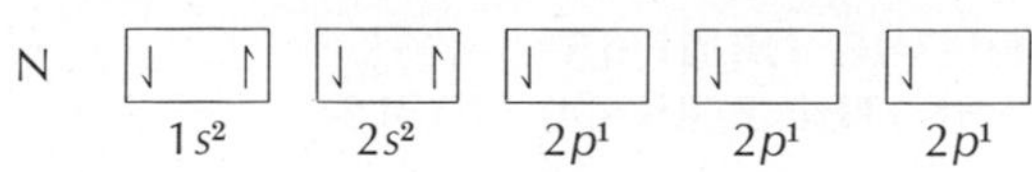

The natural repulsion of the *p* electrons for each other has forced them to occupy all three available *p* orbitals (one on the *x*-axis, one on the *y*-axis, and one on the *z*-axis). But the new electron added in the oxygen atom must go into the *p* orbital first occupied in the boron atom, leaving two unpaired electrons in the *p* energy level. Similarly, the new *p* electrons in fluorine (F) and neon (Ne) must share their orbitals with other electrons. For example, together the three pairs of electrons of opposite spin make up the three *p* orbitals in neon. Let us make brief notations of these latter structures:

F $1s^2 2s^2 2p^5$
Ne $1s^2 2s^2 2p^6$

With this much experience we can comprehend what is meant by these notations,

Na $1s^2 2s^2 2p^6 3s^1$
Mg $1s^2 2s^2 2p^6 3s^2$
Al $1s^2 2s^2 2p^6 3s^2 3p^1$
Si $1s^2 2s^2 2p^6 3s^2 3p^2$
P $1s^2 2s^2 2p^6 3s^2 3p^3$
S $1s^2 2s^2 2p^6 3s^2 3p^4$
Cl $1s^2 2s^2 2p^6 3s^2 3p^5$
Ar $1s^2 2s^2 2p^6 3s^2 3p^6$

In the argon (Ar) atom all the available orbitals "seem" to be occupied by electron pairs, so that one must start a new energy level with potassium (K),

K $1s^2 2s^2 2p^6 3s^2 3p^6 4s^1$
Ca $1s^2 2s^2 2p^6 3s^2 3p^6 4s^2$

However, when we come to the rare element scandium (Sc) a strange thing occurs; the charge on the nucleus is now sufficiently large to allow a start to be made in filling a new kind of orbital. We will not take space to draw it but only mention that there are five of this kind and they are called *d* orbitals. The 10 elements from scandium (Sc) to zinc (Zn) inclusive are illustrations of what are properly called *transition elements*. These elements are adding their new electrons in orbitals that apparently were not available when the last noble gas (Ar) seemed to fill the outermost or highest energy level.

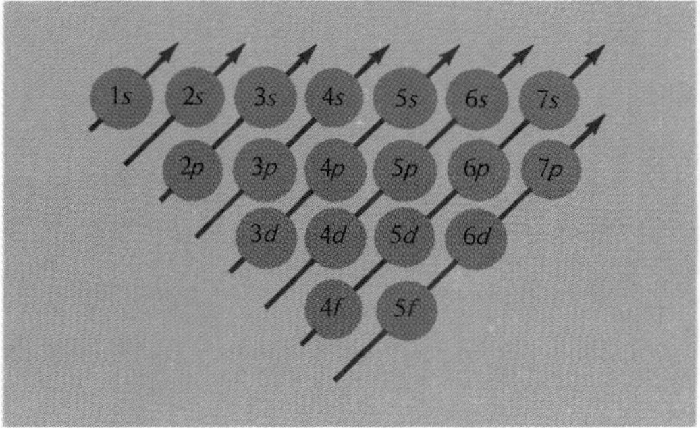

Figure 13.5
The building-up principle. The approximate order in which sublevels are filled with increasing numbers of electrons. Follow through each slanting arrow in turn, starting with the lowest and continuing with the next higher. For example, after 3*s* is filled, there follow 3*p*, 4*s*, 3*d*, 4*p*, etc. (Adapted with permission from Therald Moeller, *Inorganic Chemistry*, John Wiley & Sons, Inc., New York, 1952.)

ORDER OF FILLING ORBITALS

The boxes that represent the nitrogen atom (N) show us that the three *p* orbitals are energy sublevels of a main energy level, principal quantum number $n = 2$. As the number of electrons increases from atom to atom in the periodic table, the order in which the orbitals and sublevels, in the different energy levels, are filled usually follows the energy ranking shown in Fig. 13.5. The sublevels roughly in order of increasing energy are as follows: 1*s*, 2*s*, 2*p*, 3*s*, 3*p*, 4*s*, 3*d*, 4*p*, 5*s*, 4*d*, 5*p*, 6*s*, 4*f*, 5*d*, 6*p*, 7*s*, 5*f*, 6*d*, 7*p*. The filling of orbitals in the order of increasing energy is known as the *building-up* (*Aufbau*) principle.

Another principle, known as *Hund's rule*, also governs the order in which energy levels are filled: One electron is added sequentially to each orbital within a given

Table 13.3
Electron arrangements[a]

Main levels	1	2				3	
Sublevels	*s*	*s*	*p*			*s*	summary
H	↓						$1s^1$
He	↓↑						$1s^2$
Li	↓↑	↓					$1s^2\ 2s^1$
Be	↓↑	↓↑					$1s^2\ 2s^2$
B	↓↑	↓↑	↓	☐	☐		$1s^2\ 2s^2\ 2p^1$
C	↓↑	↓↑	↓	↓	☐		$1s^2\ 2s^2\ 2p^2$
N	↓↑	↓↑	↓	↓	↓		$1s^2\ 2s^2\ 2p^3$
O	↓↑	↓↑	↓↑	↓	↓		$1s^2\ 2s^2\ 2p^4$
F	↓↑	↓↑	↓↑	↓↑	↓		$1s^2\ 2s^2\ 2p^5$
Ne	↓↑	↓↑	↓↑	↓↑	↓↑		$1s^2\ 2s^2\ 2p^6$
Na	↓↑	↓↑	↓↑	↓↑	↓↑	↓	$1s^2\ 2s^2\ 2p^6\ 3s^1$
Mg	↓↑	↓↑	↓↑	↓↑	↓↑	↓↑	$1s^2\ 2s^2\ 2p^6\ 3s^2$

[a]Adapted with permission from Charles W. Keenan and Jesse H. Wood, *General College Chemistry*, 4th ed., Harper and Row, 1971.

sublevel before going back to add the second electron of opposite spin to each of those orbitals in the sublevel.

The foregoing principles are, with few exceptions, reliable guides for building a table to show the arrangements of electrons in atoms as the number of electrons increases from atom to atom in the periodic table. The application of these principles to 12 elements is shown in Table 13.3. Look at the table and compare oxygen (O) with nitrogen (N). Notice the application of Hund's rule: One electron must occupy each of the three *p* orbitals of nitrogen and oxygen before the new electron added in the oxygen atom occupies the one *p* orbital that was first occupied in the boron atom (B).

FILLING ORBITALS AND THE PERIODIC TABLE

In Fig. 13.6 we see a summary of how the order in which electrons are added in building up atomic structures is related to the periodic table. First, as the period increases with the first element of the period, a new main level begins to fill with the addition of an electron to an *s* sublevel. For Group IA observe the 1*s* for period 1, 2*s* for period 2, etc. Second, each period, except period 1, ends with the filling of a *p* sublevel in an atom of gas which, because the outer sublevel is filled, does not therefore react chemically with other elements, with rare exceptions. Third, a period contains a number of elements which corresponds to the filling of types of sublevels. Period 1 contains only two elements involving only an *s* sublevel; period 2 has eight elements, involving two 2*s* and six 2*p* sublevels. Period 3 has eight elements containing two 3*s* and six 3*p* sublevels, etc. Period 4 has 18 elements as does Period 5. Period 6 has 32 elements and period 7 has 19 elements through element 105, and presumably can also contain 32 elements if that many elements are ever found or synthesized.

CLASSIFICATION OF ELEMENTS BY EXAMINATION OF THE PERIODIC TABLE

On examining the periodic table on the inside front cover and also Fig. 13.6 as the need arises, we arrive at a classification of the elements into four broad categories based on the electron configuration of atoms, namely:

1. *The noble gases* (Group VIIIA), in which *s* and *p* levels are completely filled with electrons.
2. *The representative elements,* in which an *s* or *p* subshell is incompletely filled with electrons.
3. *The transition elements,* located in the left center of the periodic table, in which, in most

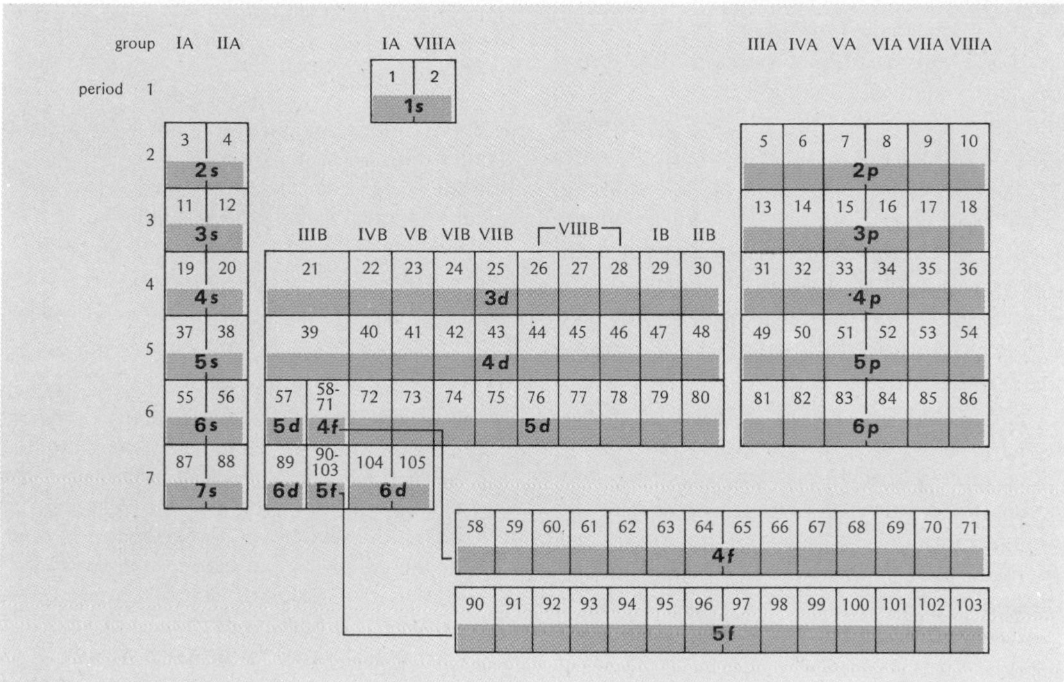

Figure 13.6
Filling orbitals and the periodic table. The order in which orbitals are filled is related to the organization of the periodic table. (Adapted with permission from Charles W. Keenan and Jesse H. Wood, *General College Chemistry*, 4th ed., Harper and Row, 1971.)

cases, the last several electrons fill the *d* subshell or energy level within the period. There is some disagreement among chemists about which elements are to be grouped as transition elements. Some would include Group IB; others, Groups IB and IIB.

4. *The inner transition* elements, comprised of the lanthanoid series in period 6 and the actinoid series in period 7, in which the last several electrons are in the *f* subshell or energy level.

The classification of some elements may also be made into another set of categories: metals, nonmetals, metalloids. These will be considered in later chapters.

At this point it should be noted that atoms of similar electronic configurations, such as lithium (Li), sodium (Na), potassium (K), rubidium (Rb), and cesium (Cs) (with a single *s* electron in their outermost orbital or highest energy level; see inside front cover, column IA below the box), constitute a family of atoms. These atoms lie one under another in the periodic table and they all have similar chemical properties. Another family arises from the atoms helium (He), neon (Ne), argon (Ar), krypton (Kr), xenon (Xe), and radon (Rn), which all have completed electron shells; these atoms also all behave similarly with regard to their chemistry. It is perhaps appropriate now to turn our attention to the "chemistry" of the elements.

SUMMARY

In 1913 Niels Bohr proposed that extranuclear electrons revolved about atoms in fixed orbits and that this did not involve the loss or gain of energy unless one of the electrons moved or jumped from one allowed orbit to another. Bohr based his theory on the idea of a quantum of electromagnetic radiation which had first been proposed by Planck and became rather well established when Einstein used the quantum to describe the photoelectric effect.

The Bohr theory had excellent verification from the spectrum of atomic hydrogen. Nevertheless the theory was eventually rejected for a number of reasons, one of the most important of which was that the spectra of most atoms and molecules (atomic hydrogen excepted) did not agree with it. However, it was a step forward because the modern wave-mechanical model retained from it the allowed energy-level states, although rejecting the idea of orbits. The wave-mechanics treatment of spectra gives the probability that an electron is located at a certain place at a certain time. The result is a kind of smear, or a shaded representation of strength of electric field, often called an *electron cloud,* rather than a particle traveling in a known direction at a known time.

Using the ideas of wave mechanics, it is possible to build up the periodic table of the elements using the quantum numbers and the Pauli exclusion principle. This process results in a prediction of the extranuclear distribution of electrons in energy levels which is useful in explaining much of the chemistry of the elements.

Important words and terms

Bohr theory of the atom
Planck's quantum of energy
photoelectric effect
quantum numbers
Bohr orbits
orbitals
energy levels
spin angular momentum
de Broglie matter waves
de Broglie wavelength
Heisenberg uncertainty principle
groups, periods, and families in the periodic table
Pauli exclusion principle
transition element
electron filling of energy levels, building-up principle
noble gases

Questions and exercises

1. (a) If the change in some phenomenon can be made in as small increments or decrements as desired, the phenomenon is said to be variable in a *continuous* way; if the change can be made only in fixed steps or units, it is variable in a *discrete* way. Which of the following vary in a continuous, and which in a discrete, fashion: (1) flow rate of water, (2) number of pennies per day in a fare box, (3) weight in daily growth of an animal, (4) number of petals on a flower selected at random, (5) speed of a car, (6) first-class postal rates. (b) In the above sense discuss the differences between the radiation theories of Maxwell and Planck. (c) Planck treated energy the way the early Greek atomic thinkers viewed matter. Elaborate, using the above concepts.
2. (a) What designation did Planck give to the discrete units of energy? (b) What was their magnitude?
3. (a) The study of what phenomenon gave the first experimental verification of the quantum nature of radiation? (b) Who made the observation and who gave the first satisfactory explanation of the basis of the new theory?
4. What are the differences between an electron and a photon?
5. Distinguish between quantum and photon.
6. What properties of X rays led to the conclusion that X rays were electromagnetic waves?
7. What is the energy of a light photon that corresponds to a wavelength of 5×10^{-7} m?
8. Which of the following types of radiation has a photon with (a) the least energy? (b) the most energy? (1) radio waves, (2) ultraviolet radiation, (3) red light, (4) X rays, (5) violet light.

9. (a) If there were no gun capable of carrying to the height of a given enemy plane, nothing would be gained by increasing the number of such guns. Compare this to the observations made of the photoelectric effect with below-threshold frequency light whose brightness was progressively increased. (b) Also consider a similar question for the case where the number of guns that did carry was doubled.
10. On a pure wave-theory basis the total energy of a beam of light depends on both the frequency f and the intensity (that is, amplitude, as we learned in Chap. 9) of the beam. (a) How would the photoelectric effect operate if radiation were a pure wave phenomenon? (b) On what two factors does the total energy depend in the quantum theory?
11. (a) In the photoelectric effect, on what two factors does the energy of ejection of an electron depend? (b) On what factor does the number of ejected electrons depend?
12. Light falling on a certain metal surface causes electrons to be emitted. (a) What happens to the photoelectric current as the intensity of the light is increased? (b) What happens as the frequency of the light is increased?
13. Einstein's idea of a quantum of light had a definite relation to the wave model. What was it?
14. Why does the photoelectron not have as much energy as the quantum of light that produced the photoelectron?
15. Discuss the dual nature of light. When do we resort to the wave explanation and when do we use the particle model of light?
16. (a) Tie together in a logical way the following ideas that worried the classical theorists: principle of conservation of energy, lower energy in orbits of smaller radius r, radiation of electromagnetic energy by orbiting electron, spiraling of electron into nucleus. (b) How did Bohr get around the above difficulty? When is electromagnetic energy radiated according to the Bohr theory?
17. In terms of the Bohr theory explain why the hydrogen spectrum contains a number of lines even though the hydrogen atom contains only one electron.
18. (a) What is an "excited" atom? (b) How is the condition brought about? (c) What happens when the atom returns to its "normal" state?
19. What is an ion?
20. Why could the Bohr theory satisfactorily account for spectral lines from singly ionized helium and doubly ionized lithium but not from neutral helium or lithium?
21. Compare the contributions of Sommerfeld and Kepler.
22. List the four quantum numbers, and their symbols, significance, and permitted values.
23. What was the main evidence that an atom could exist only in certain energy states? Why do the next heavier elements after the noble gases easily become positively charged?
24. (a) What "reasoning" led de Broglie to speculate on the existence of matter waves? (b) Would you say that this is an example of "scientific thinking" or "intuition"? (c) By making what assumptions did he arrive at his formula relating the wavelength λ for a particle of mass m moving with velocity v? (d) Who experimentally verified the existence of these matter waves?
25. In their wavelike aspect a beam of electrons can be made to yield a diffraction pattern much like that in the case of light. What difference in behavior or appearance of the pattern would you expect if the diffraction beam passed between the poles of a magnet before reaching the screen?
26. Review stationary waves in Chap. 9. Discuss the orbiting electron in Fig. 13.2 as a stationary wave.
27. A golf ball has a mass of about 50 g. Assume that it is moving at a speed of 7600 cm/sec. What would be its de Broglie wavelength?
28. (a) What is the de Broglie wavelength of an electron of mass 9×10^{-31} kg and a speed of 2×10^{8} m/sec? (b) Compare this wavelength to the electromagnetic spectrum wavelengths and pick out the type of radiation to which it corresponds.
29. What is the de Broglie wavelength of an automobile of mass 1500 kg moving at a speed of 30 m/sec (about 60 mi/hour)? Comment on the size of this wavelength; could this wavelength be observed?
30. (a) State the implications of Heisenberg's uncertainty principle in determining the posi-

tion and momentum of an electron. (b) Why is Newtonian mechanics not applicable at this level of matter?

31. Discuss where the wave-mechanical picture of matter has taken the discrete particles of the Greeks, Dalton, Thomson, etc. See also Question 1(c) to round out your discussion.
32. Distinguish between orbits and orbitals.
33. Define the terms *group, subgroup, family,* and *period.*
34. In a 20-electron atom state the principle that accounts for all these electrons not piling up in the lowest energy level ($n = 1$) according to the "basic principle of nature."
35. Match for "best answer" the most important contribution or association of each man:

a. Planck	1. wave theory of matter
b. Sommerfeld	2. exclusion principle
c. Einstein	3. wave theory of light
d. Hertz	4. discrete orbits
e. Bohr	5. electron waves
f. de Broglie	6. quanta
g. Heisenberg	7. elliptical orbits
h. Pauli	8. photons
i. Maxwell	9. photoelectric effect
j. Davisson and Germer	10. uncertainty principle

36. (a) Distinguish between shells and subshells; energy levels and sublevels. What are their corresponding quantum symbols? (b) What numerical values of which quantum symbol correspond to the *s, p, d,* and *f* orbitals?
37. Write down all the possible combinations of n, ℓ, m_ℓ, and m_s for the $n = 3$ energy levels.
38. (a) Explain the statement that for the 10 transition elements, from scandium to zinc, there are five possible *d* orbitals. (b) Refer to the periodic table for the electron configuration of the outermost levels of mercury (Hg). At the fourth level there are 14 *f* electrons. How many *f* orbitals are there at $n = 4$?
39. The electron configuration for calcium is $1s^2 2s^2 2p^6 3s^2 3p^6 4s^2$. (a) List the orbital type of electrons and the number of each at the first, second, third, and fourth energy levels. (b) Which are of spherical symmetry and which are dumbbell shaped?
40. As an exercise in extracting information from the periodic table (inside front cover), write out completely the electron configuration for uranium, beginning with $1s^2$ and ending with $7s^2$.
41. Chemical properties of an element depend on the outermost electrons. Examine the manner in which the 10 transition elements, from scandium to zinc, take on their additional electrons and explain why their chemical properties vary little from one element to the next.
42. (a) State the *Aufbau* ("building up") *principle.* (b) State Hund's rule.
43. On the basis of electron configuration, list four broad categories into which the periodic chart may be broken down. Compare Fig. 13.6 and the table inside the front cover and state in what manner the subshells build up.

Supplementary readings

Books

Arons, Arnold B., *Development of Concepts of Physics,* Addison-Wesley, Reading, Mass. (1965). [Chapters 28–36 expand on concepts discussed in this chapter and the preceding one.]

Bonner, Francis T., et al., *Principles of Physical Science,* Addison-Wesley, Reading, Mass. (1971). [Chapter 19 discusses atomic structure and the periodic table, including electron quantum numbers, the exclusion principle, and electron shells.]

Booth, Verne H., *The Structure of Atoms,* Macmillan, Collier-Macmillan, Toronto, Canada (1969).

Eddington, Sir Arthur, *The Nature of the Physical World,* Ann Arbor Paperbacks, University of Michigan Press, Ann Arbor, Mich. (1958). [Chapter 9, "Quantum Theory," considers the conflict with the wave theory of light; Chap. 10, "New Quantum Theory," considers the wave theory of matter, Schrödinger's theory, and Heisenberg's principle of indeterminacy.]

Einstein, Albert, and Leopold Infeld, *The Evolution of Physics: The Growth of Ideas from Early Concepts to Relativity and Quanta,* Simon and Schuster, New York (1967). [A simple exposition of advanced scientific thought written by the man who did the creative thinking. The authors explain the significance of the most important contributions since Newton—the inventions of the ideas of field relativity and quanta. It is a partial record of man's struggle to understand the laws governing the universe.]

Gamow, George, *The Atom and Its Nucleus,* Spectrum Books, Prentice-Hall, Englewood Cliffs, N. J. (1961). (Paperback.) [This book supplements and extends the discussion of important topics in this chapter: the electric nature of matter, radiant energy, the Bohr atom, the wave nature of particles, natural radioactivity, artificial nuclear transformations, the structure of the atomic nucleus, and large-scale nuclear reactions.]

Gamow, George, *Thirty Years that Shook Physics,* Doubleday, Garden City, N. Y. (1966). [Although much of the material may be quite difficult, the layman will find this presentation interesting and rewarding. Chap. 1, "Max Planck and Light Quanta"; Chap. 2, "Niels Bohr and Quantum Orbits"; Chap. 3, "Wolfgang Pauli and the Exclusion Principle"; Chap. 4, "Louis de Broglie and Pilot Waves"; Chap. 5, "Werner Heisenberg and the Uncertainty Principle."]

Jaffe, B., *Moseley and the Numbering of the Elements,* Anchor Books, Doubleday, Garden City, N. Y. (1971). (Paperback.)

Keenan, Charles W., and Jesse H. Wood, *General College Chemistry,* 4th ed., Harper & Row, New York (1971). [Pages 40–61 cover approximately the same range of topics as this chapter but with somewhat more emphasis on chemistry than on physics.]

March, Arthur, and Ira M. Freeman, *The New World of Physics,* Vintage Books, Random House, New York (1962). (Paperback.) [Pages 60–63 should make clear why we cannot "know" about submicroscopic "objects" such as electrons. Because we will probably never be able to observe them directly, a statistical method is used in their study. Pages 171–189 are highly recommended for those students who are interested in philosophical aspects of science. Topics included are atoms and the concept of matter; Pauli's exclusion principle; the electron as a nonmaterial structure; boundaries of our knowledge of the physical universe.]

McCue, J. J. G., *Introduction to Physical Science: The World of Atoms,* Ronald Press, New York (1963). [Chapter 27 deals with the periodic table; Chap. 45 with Bohr's hydrogen atom; Chap. 47 with quantum numbers; Chap. 48 with orbits and the periodic table; Chaps. 56–70 with transmutations, fundamental particles, nuclear reactions, nuclear fission, and cosmic physics—all well presented.]

Rogers, Eric M., *Physics for the Inquiring Mind,* Princeton University Press, Princeton, N. J. (1960). [Chap. 39, "Radioactivity and Nuclear Physics"; Chap. 40, "Atoms: Experiment and Theory"; Chap. 42, "Atom-accelerators"; Chap. 43, "Nuclear Physics"; and Chap. 44, "Physics Today," are excellent for this chapter.]

Romer, Alfred, *The Restless Atom,* Anchor Books, Doubleday, Garden City, N. Y. (1960). (Paperback.) [This book is about the experiments by which we have learned about atoms and the way they behave. It is based on the research reports of the investigators themselves and covers the period 1896–1916 as preparation for the story of atomic physics.]

Semat, Henry, and Harvey E. White, *Atomic Age Physics,* Holt, Rinehart and Winston, New York (1959). [A book used with Continental Classroom, the first nationally televised course in physics. The first 10 chapters deal with the atom and atomic structure, including a chapter on mass and energy. The last seven chapters dwell on the nucleus, for example, radioactivity, atom smashing, fusion and fission, and elementary particles. The dual character, wave and particle, of radiation is established in Chap. 9.]

Shamos, Morris H., *Great Experiments in Physics,* Holt, Rinehart and Winston, New York (1959). (Paperback.) [The original accounts of 24 experiments that created modern physics. Each chapter contains a biographical sketch of the scientist responsible for a particular discovery: Galileo, Boyle, Newton, Coulomb, Cavendish, Young, Fresnel, Oersted, Faraday, Lenz, Joule, Hertz, Roentgen, Becquerel, J. J. Thomson, Einstein, Millikan, Rutherford, Chadwick, Maxwell, Planck, Bohr, and Compton.]

Taylor, Lloyd Willian, *Physics: The Pioneer Science,* Houghton Mifflin, Boston (1941); or, a paperback, L. W. Taylor and F. G. Tucker, *Physics: The Pioneer Science,* Vols. 1 and 2, Dover Publications, New York (1962). [After developing the historical story of X rays and radioactivity in Chap. 52, the author does a very fine development in Chap. 53 of quantum theory and atomic structure. The "black body," Planck's quantum theory of radiation, Einstein's equation, Rutherford's idea of the structure of the atom, the concept of stationary states or energy levels by Bohr, X ray spectra and electron shells, Moseley's law, the Compton effect as proof of the particle nature of radiation, de Broglie's derivation of the relationship between the motion of a particle and its associated wave, experiments by Davisson and Germer, and interpretation of wave-particle duality.]

The Project Physics Course, Holt, Rinehart and Winston, New York (1970). (Book 5)

Articles

Cohen, V. W., "The Nucleus as a Spinning Top," *The Physics Teacher,* **10,** no. 1, 24 (January, 1972). [An elementary discussion of the intrinsic magnetic properties of the nucleus and atom.]

Seaborg, Glenn, T., et al., "The Synthetic Elements," *Scientific American* (April, 1950; December, 1956; April, 1963).

FUNDAMENTAL CHEMISTRY: THE NATURE OF MATTER

Knowledge comes from taking things apart.
Wisdom comes from putting them together.

ANONYMOUS

Chemistry is the investigation of matter and its transformations. In Chaps. 12 and 13, in particular, we considered matter and its transformations. We will continue to do so in this chapter and in the next ones. Thus our intensive study of chemistry began with Chap. 12 and continues. There is no clear definition of what physics is, or agreement by all on which topics belong in physics. One widely accepted definition states that *physics* is the investigation of energy and its transformations. The authors believe that many physicists would consider much of Chaps. 12 and 13 to be physics whereas many chemists would consider those chapters to be chemistry. Physics and chemistry are fields of the physical sciences with no clear-cut distinctions in many areas.

Matter: its nature and changes

RISE OF THE CONCEPTS OF MATTER AND ENERGY

Matter is defined as anything that occupies space and has mass. The term "substance" is used for any form of matter. Until relatively recently man has tended to think of matter only in terms of its gross aspects and has often said that matter may exist in three phases or physical states: solid, liquid, and gaseous. The modern understanding of matter is, however, based on the electron structure of the atoms which was discussed in the preceding chapter.

Chemistry had its early beginnings with the fermentation of wines, the production and use of metals, the glazing of pottery, and the development and use of dyes. Later, the Greek philosopher Aristotle, explained everything in terms of four basic elements: fire, water, air, and earth. Still later the pseudoscience of alchemy arose. The alchemists learned how to make and use many substances; for example, one now called "sulfuric acid." But their knowledge was empirical, unorganized, and above all lacking in understanding. They were secretive and mystical, and depended on a mixture of magic and experimentation. In short, they were not full-fledged scientists.

Two alchemists, Johann Joachim Becher (1635–1682) and Georg Ernst Stahl (1660–1734), had invented the "phlogiston theory," according to which a substance, while burning, gave off "phlogiston." There is no concept in modern science that corresponds exactly to "phlogiston." Some of the ideas expressed during the eighteenth century are acceptable to us today if the word "energy" is substituted for "phlogiston." It is difficult, however, to give rigorous definitions of many terms, one such being "matter"; but at least we recognize that "matter," or "material," is what all our familiar objects are composed of.

SCIENTIFIC REVOLUTION

As early as 1755 the Scottish student Joseph Black published a paper that exhibited several of the characteristics we expect in a modern scientific work. For example, in all cases he indicated that he had weighed his materials; and, in addition to giving his specific findings, he tried always to arrive at useful generalizations. But even more influential in bringing about the change from pseudoscience to real science were Joseph Priestley (1733–1804) and Antoine Lavoisier (1743–1794).

Priestley had somehow secured an unusually large lens and had engaged himself in the interesting task of determining what substances gave off gases—airs to him—when they were heated. In August, 1774, he was heating *mercurius calcinatus per se* (mercuric oxide) and found that it gave off a gas (Fig. 14.1). He had some 7 years earlier worked a great deal with a gas that

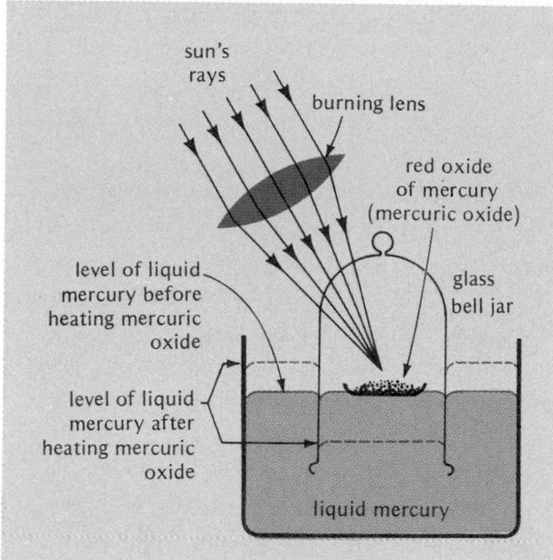

Figure 14.1
Diagram of apparatus to illustrate Priestley's method of preparing oxygen.

we call "carbon dioxide" and knew that it extinguished a flame. Although he always said it was quite by chance, it was probably the old habit developed by his previous search for unknown properties that impelled him to insert a glowing candle into a sample of this new gas. He was quite surprised and pleased with the result, for the flame instantly flashed up larger and brighter. His curiosity was aroused. He then tried various other materials, such as charcoal, and found that these materials also burned much more vigorously and for a longer time in this new "air" than they did in a similar volume of ordinary air. Next he himself breathed some of the new "air" and as a result related: "I fancied that my breath felt particularly light and easy for some time afterward."

In October, 1774, while Priestley was still enthusiastic about the results of these interesting experiments, he visited Paris. There he met the French scientist, Lavoisier, and told him the chief details about the discovery of the new "air." Lavoisier very promptly repeated Priestley's experiments and obtained essentially the same results. But Lavoisier did not stop with that; he added some new and very important experiments of his own which led to the concept of quantity in change.

CONCEPT OF QUANTITY IN CHANGE

Lavoisier found the quantitative composition of *mercurius calcinatus per se* by heating a weighed sample of mercury in a very carefully measured volume of air. He determined both the weight of *mercurius calcinatus per se* formed and the volume of air consumed. He also took a weighed sample of mercuric oxide and, by heating it to a much higher temperature than he had used in the previous experiment, caused it to decompose completely into mercury and the new air which Priestley had called "dephlogisticated air." He noted two very significant facts: (1) The weight gained in forming a given sample of the *mercurius calcinatus per se* was the same as the weight lost when the solid sample was heated to liberate the new gas, and (2) the volume of the new gas formed was equal to the volume of gas consumed when the solid *mercurius calcinatus per se* was formed.

These facts were only typical of many similar results that Lavoisier obtained while performing experiments over a period of a few years with elements and their oxides. They caused him to be more and more certain of the truth of a number of new and practically revolutionary ideas about chemistry. For one thing, the newly obtained information enabled him to unite some previous good suppositions and facts into a logical whole that gave an improved picture of the nature of the atmosphere, of combustion, and of elements and compounds.

Lavoisier pointed out that air contains two chief components and gave them two new names, *oxygen* and *azote* (nitrogen). Although Boyle had been very near our modern idea of a chemical element, it was Lavoisier who made the concept of an element more definite and clearer by pointing out that oxygen and nitrogen were elements. He soon collaborated with others in writing a chemistry textbook that attempted to name the known elements and compounds.

TWO KINDS OF CHANGE: PHYSICAL AND CHEMICAL

As previously noted, water can exist in any of the three states: (1) gaseous (vapor), (2) liquid, or (3) solid (ice). Liquid water readily evaporates into the atmosphere to become water vapor; and the liquid freezes to the solid form, ice, when the outside temperature drops below 0°C. A solid like wheat may be ground, to make flour. Such changes, which involve only change in form or appearance but not the production of a new substance, are called *physical changes*.

On the other hand, a change by which at least one substance is consumed and at least one new substance is produced is called *chemical change*. This definition of chemical change can be demonstrated as follows: A thin ribbon of grey magnesium metal disappears as the metal is being burned, and a strip of white powder of the same shape is produced. We observe that all magnesium as a metal was consumed and we also know that oxygen as a gas from the air was consumed on combining with the magnesium metal in the process of burning; we observe only one end product and this new substance produced is the white powder called "magnesium oxide." Similarly, magnesium metal burns in a photoflash bulb.

magnesium + oxygen → magnesium oxide
consumed consumed produced

PROPERTIES OF MATTER: PHYSICAL AND CHEMICAL

Every substance has characteristics or properties unique to it that distinguish it from all other substances. Refined sugar, for example, is an odorless, white, crystalline solid that will dissolve in water. Salt for table use has these same properties. Sugar has a sweet taste; salt has a salty taste. High heat will cause white sugar to melt and turn brown; salt requires more heat to melt but it will not turn brown.

There are two basic types of properties: physical and chemical. They are used to identify a substance. The distinction between them is not always easy to establish. Properties that can be appraised by the senses are called *physical properties*. Among physical properties are color, smell, taste, melting point, boiling point, heat conductivity, electrical conductivity, and hardness. Such properties are detected by using the senses, directly or indirectly. Properties that describe the chemical changes, or reactions, that a substance undergoes are called *chemical properties*. Such properties are related to the outer electrons of atoms.

Classification of matter

Matter may be classified as follows:

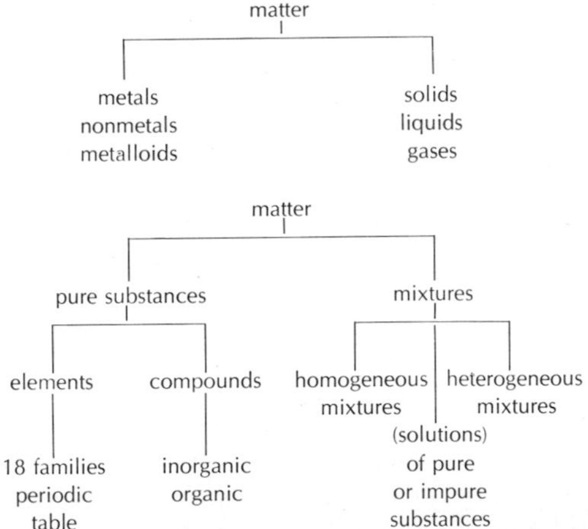

The classification chosen for use depends on the phenomenon being studied or the convenience desired. Matter may be classified as metals, nonmetals, or metalloids instead of the more common solids, liquids, gases; the former classification is also suitable for elements.

METALS, NONMETALS, AND METALLOIDS

Metals, in contrast to nonmetals, are substances that have a characteristic metallic luster, are malleable (can be hammered into thin sheets) and ductile (can be drawn into wires), and are good conductors of heat and electricity. Nonmetals are neither malleable nor ductile and are poor conductors of heat and electricity. Examples of metals are iron (Fe), sodium (Na), copper (Cu), gold (Au), and aluminum (Al); examples of nonmetals are oxygen (O), hydrogen (H), and sulfur (S).

The metals are at the left and in the center of the periodic table (inside front cover). About 80 elements are classified as metals, including some from every group except VIIA and VIIIA. The nonmetals are to the right in the periodic table, except for hydrogen. The nonmetals include about a dozen relatively common and important elements plus the noble gases (Group VIIIA). Those elements, such as boron (B), carbon (C), silicon (Si), and germanium (Ge), which exhibit to some extent both metallic and nonmetallic properties are called "metalloids." These borderline elements lie close to the zigzag line in the periodic table inside the front cover.

SOLIDS, LIQUIDS, AND GASES

Solids, liquids, and gases are known as the phases or physical states of matter. Theoretically, many substances can exist in these three states under certain conditions. The color key used with the periodic table inside the front cover indicates that at room temperature (20° to 25°C) the element is either solid, liquid, or gaseous.

PURE SUBSTANCES AND MIXTURES

A sample of matter containing only one substance is a *pure substance*. In chemistry the word "pure" does not carry exactly the same meaning as it often has outside that field. For example, when the chemist is speaking carefully as a scientist, he means by the term "pure water" a sample that consists of practically 100 percent water. A sample of water that contains something other than water is not pure, irrespective of whether the other substance is harmful.

A pure substance in nature is the exception rather than the rule. Thus if we want some pure substance, such as pure water or pure sugar, a purifying process (often called "refining") must be employed. The composition of a pure substance is constant; each part has the same composition as every other part.

Our physical world consists chiefly of mixtures. A sample of matter that contains two or more substances not chemically combined is called a *mixture*. If ground black pepper and white crystals of salt are shaken together, the result is a mixture of pepper and salt. A pair of tweezers could be used to pick all of the pepper from the salt. Thus it is usually possible to separate the components of a mixture by physical means or changes rather than by chemical changes. Air is a gaseous mixture composed chiefly of nitrogen, oxygen, argon, carbon dioxide, and water vapor.

A mixture possesses the properties of the substances of which it is composed; it has no unique set of properties—only those of each component. The various components of a mixture can be separated from one another because of differences in physical properties, such as boiling point, freezing point, and solubility. For example, by heating water, most of the dissolved gases can be driven out before the water itself boils. Then the water can be vaporized, leaving the solid impurities behind. By properly condensing the steam, a greatly improved product, from the viewpoint of purity, is secured. Water prepared in this way is called "distilled" water, and it is used in steam irons and car batteries. The process by which it is prepared is called distillation. Thus *distillation* is a process in which part of a liquid mixture is vaporized and the vapors condensed to get at least partial separation of the components from each other.

Gasoline obtained by the fractional distillation of crude oil is called straight-run gasoline. On the heating of crude oil the more volatile, lower-boiling components are vaporized and the fraction of components coming off within a certain temperature range may be condensed and collected separately as a particular grade of gasoline. Certain other (higher-boiling) fractions may be used as lubricating oils, whereas the residue may be used as asphalt for paving. In this example a physical property, the boiling point, is used to separate the components of a mixture.

Ordinary alcohol is usually made by a fermentation process in which yeast changes sugar into alcohol and carbon dioxide gas, a by-product. When the fermentation is finished, the alcohol represents only a small percentage of the total matter present. Before it is sold commercially, it is at least partially purified by distillation. Alcohol has a lower boiling point than most of the substances, such as water, present in the mixture, so that, on condensing the vapors that boil away, a distillate is obtained that contains a much higher percentage of alcohol.

CLASSES OF PURE SUBSTANCES OF MATTER: ELEMENTS AND COMPOUNDS

Pure substances are classified as either elements or compounds. In Chap. 12 elements and compounds were introduced in the discussion of early atomic theory. A shift was mentioned within the chapter from an early definition, "An element is a substance that cannot be decomposed into simpler substances by ordinary physical means or chemical means," to a modern definition that followed the discovery of neutrons and isotopes, "An *element* is a substance whose atoms all have the same atomic number." In the periodic table on the inside front cover 105 elements in 18 families are shown; they are listed in alphabetical order on the inside back cover. Among the elements that probably are familiar would be carbon (C), nitrogen (N), oxygen (O), hydrogen (H), neon (Ne), chlorine (Cl), sodium (Na), aluminum (Al), nickel (Ni), copper (Cu), gold (Au), silver (Ag), iron (Fe), iodine (I), lead (Pb), platinum (Pt), and tin (Sn).

In Chapter 12 we read that "A compound is a substance that can be decomposed into simpler substances by ordinary chemical means." For example, the compound water, or hydrogen oxide, is composed of hydrogen and oxygen and the compound can be decomposed into the simple constituent elements: hydrogen and oxygen. While the foregoing definition of a compound is still true, a modern definition, using our knowledge of the electron structure of atoms, could be as follows:

A compound is a new substance that results when two or more elements combine by sharing or transferring electrons.

The concept of the sharing or transferring of electrons by elements will be considered further in a later chapter in connection with the study of chemical bonds.

Approximately 4 million chemical compounds have been identified. Millions more are possible. Those that contain carbon (C) and hydrogen (H) are far more numerous than all compounds of the other elements. Compounds that do not contain carbon are called *inorganic compounds;* those that contain carbon are called *organic compounds*. Inorganic compounds constitute sand, clay, rocks, and other earthy materials. The study of the compounds of the elements other than carbon is called *inorganic chemistry*. A few compounds that contain carbon may be classified as inorganic rather than organic because they are rocklike or earthy. Examples of such inorganic compounds containing carbon are calcium carbonate ($CaCO_3$) and magnesium carbonate ($MgCO_3$).

Organic chemistry is that branch of science that deals with the compounds of carbon. Alcohols, carbohydrates, fats, and proteins represent familiar classes of organic compounds. Those organic compounds that occur in nature are found mostly in plants and animals, and in materials of plant or animal origin such as petroleum, natural gas, lubricating oils, and coal.

POLLUTION CONTROL AND COMPOUNDS: THE AUTOMOBILE; AIR POLLUTION CONTROL CENTERS

As an example of a compound, carbon monoxide (CO) is formed in small amounts in the gasoline automobile engine when the element carbon (C) from the gasoline combines with the element oxygen (O) from the air taken in through the carburetor; however, the compounds carbon dioxide (CO_2) and water (H_2O) are the major products from the internal combustion engine in terms of quantities. Lethal carbon monoxide is one air pollutant from automobile exhaust emissions; even though it is formed only in small amounts, the fact that small amounts of it are lethal causes the amounts to be monitored by air pollution control centers. Among other emissions pollutants are certain oxides of nitrogen: namely, the compounds nitric oxide (NO) in small amounts and nitrogen dioxide (NO_2) in much larger amounts. These two emissions are often referred to as "NOX."

Some comprehensive air pollution control centers—using computers to collect, store, and process data about air quality—monitor heavy industry in urban areas as well as the less industrialized outlying urban areas with stations equipped to record automatically wind speeds and directions along with contaminants of the atmosphere. The latter usually include suspended particulates such as dust and certain compounds such as sulfur dioxide (SO_2), carbon monoxide (CO), nitrogen dioxide (NO_2), and certain hydrocarbons and oxidants. Space-age sensors at monitoring stations are used to feed data about pollution into computers and thus to alert officials when pollutants reach a level hazardous to health.

Efforts are being made by governments, individuals, industries, factories, automobile companies, and those who service automobiles, to reduce air pollutants—undesirable compounds.

TYPES OF MIXTURES: HOMOGENEOUS AND HETEROGENEOUS

In describing mixtures of materials the chemist often finds it convenient to use the terms homogeneous and heterogeneous. Homogeneous (the same throughout) refers to material in which no differing parts can be distinguished visibly, even with a microscope; that is, any small part appears to have the same composition as every other part. Sugar dissolved in water is a solution of a solid (sugar) in a liquid (water). Such a sugar solution is a homogeneous mixture, for all parts are uniformly sweet and no differing parts can be distinguished visibly. A *homogeneous mixture* is one whose composition is uniform and it can be separated without chemical change into at least two different substances.

Heterogeneous (not the same throughout) refers to material in which there are visible differing parts. We speak of a heterogeneous mixture. Examples of heterogeneous mixtures are rocks of large mass as found in nature and whole blood. The medical technician using a microscope can count white blood cells and red blood cells—the visible differing parts. A *heterogeneous mixture* is one whose composition is not uniform and it can be separated without chemical change into its components.

FURTHER AIDS TO DIFFERENTIATION BETWEEN PURE SUBSTANCES AND MIXTURES

As a general rule the following two differences will help to avoid confusion about whether a sample of matter is a mixture or a compound.

First, physical processes are used to distinguish between pure compounds and mixtures. The parts of a mixture may be separated by physical means such as cooling, melting, evaporation, boiling, distillation, use of a centrifuge, use of a sifter, and use of a magnet. Chemical changes must occur in order to separate a pure compound into its elements, because compounds are composed of elements combined in such a way as to be inseparable by physical means.

Second, usually a compound obeys the *law of definite composition:*

A pure compound always contains the same elements combined in the same ratio by weight.

For example, the compound water (assumed to be pure) always contains the same elements, hydrogen and oxygen, in the definite fixed ratio by weight of 1:8. On the other hand there need be no definite fixed ratio of meat to potatoes in such mixtures as hash or stew.

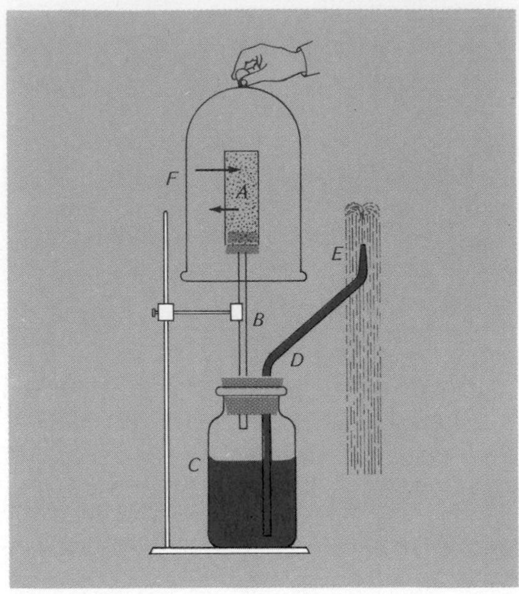

Figure 14.2
Demonstration of diffusion.

Mixing by diffusion, dispersion, osmosis

DEMONSTRATION OF DIFFUSION

The property of diffusion may be demonstrated by the use of the apparatus represented in Fig. 14.2. A small porous clay cup *A* is connected by a tight-fitting rubber stopper with a glass tube *B*, the other end of which passes just through a stopper in the bottle *C*. The bottle is half-filled with colored water and is provided with a second tube *D*, drawn to a small jet *E*, and extending to near the bottom of the bottle. A bell jar or large beaker *F* is completely filled with hydrogen and lowered over the porous cup as shown. If the apparatus is perfectly gas-tight (does not leak) and the hydrogen in the bell jar is really pure, the hydrogen will diffuse into *A* toward *C* and liquid will be forced out through the jet.

EXPLANATION OF DIFFUSION

The kinetic energy of a moving object is numerically proportional to the product of its mass times the square of its velocity ($KE = \frac{1}{2}mv^2$). This is known to be true for bodies whose masses are great enough to be determined by direct observation. It seems reasonable to assume that it is also true for very small objects, such as molecules, whose masses may be determined by some method neither simple nor obvious. If the gases are at the same temperature, their unit particles, or molecules, have the same average kinetic energies. Therefore, using the subscript O for oxygen and H for hydrogen, and after cancelling $\frac{1}{2}$:

$$KE_O = KE_H \quad \text{or} \quad m_O v_O^2 = m_H v_H^2$$

But hydrogen molecules are much lighter than oxygen molecules; therefore they must be traveling with greater velocities in order to have the same kinetic energy as the oxygen molecules. The hydrogen diffuses onto the porous cup faster than the air within the cup diffuses out, and this develops a greater pressure within the cup. This increased pressure is communicated to the surface of the water in the bottle, forcing some of the water out through the jet and forming a fountain, as indicated.

DISPERSIONS: TRUE SOLUTIONS, COARSE SUSPENSIONS, AND COLLOIDS

If one were to examine carefully a cup of water from a flowing stream, it would be found to contain particles of matter other than water. These particles are said to be dispersed or scattered throughout the water, and the whole mixture, water and other matter, is said to be a dispersion. Dispersions are usually divided into three classes: (1) true solutions, (2) coarse suspensions, and (3) colloids or colloidal solutions.

If one adds a few crystals of sugar to a glass of water, they seemingly disappear after a time. The fact that the water remains sweet is evidence that they have not been lost entirely. Such a mixture is a *true solution*. Because a solution is a mixture, it always contains at least two general types of components. One of the two components of a true solution is called the solvent. If water is one of these components, it is customarily referred to as the solvent. The component other than the solvent is called the solute. Therefore the dissolved substance in a solution is called the *solute* and the dissolving medium is called the *solvent*. If the solution contains a high solute-to-solvent ratio, then it is called a *concentrated solution*.

If one stirs some sand vigorously with water, a mixture can be obtained. Such a mixture, however, exists only temporarily, because the sand will soon settle to the bottom of the container. Such a fluid mixture which separates rather rapidly into its two components, is often called a *coarse suspension*.

When a small amount of clay or starch is stirred with water, one usually gets a mixture that seems to be extremely permanent as compared with the sand–water mixture. Some of the larger particles of clay or starch usually settle quickly, but after hours or even years such a mixture often persists. Settling either is not occurring at all or is occurring so slowly that it cannot be observed. Such mixtures as clay and water, which do not separate in any conveniently available amount of time, are said to be *colloids*, also often called *colloidal solutions* and *colloidal suspensions*.

True solutions, coarse suspensions, and colloids may be distinguished from each other by their properties.

PROPERTIES OF TRUE SOLUTIONS

Only aqueous solutions are considered. True solutions never settle; that is, the two components do not ever separate into two layers no matter how much time is allowed. True solutions may be either colored or colorless. They are always transparent if not too highly colored. A beam of light or a fine line, when viewed through parallel surfaces of a true solution, is not distorted; that is, a true solution is clear.

Also, a true solution has a freezing point lower than the freezing point of pure water. The components of a true solution cannot be separated from each other by passing the solution through an ordinary filter.

CORRESPONDING PROPERTIES OF COARSE SUSPENSIONS AND COLLOIDS

Coarse suspended matter, such as sand, is usually easy to separate from water either by allowing it to settle or by passing the water through a filter. It is very difficult to separate colloidal material from water without first destroying the colloid, because it settles so slowly that it cannot be detected in the time available. A colloid may clog the too small pores of a filter, and prevent its components from being separated by the filter.

Colloids and suspensions are often colored, though the color may be difficult to detect if there is little dispersed material present. Many colloids, such as milk, appear to be almost white in color; those such as ink may be any one of several different colors.

Colloids and coarse suspensions are probably never perfectly clear; that is, they always distort somewhat a beam of light or a fine line viewed through parallel surfaces, although again it sometimes requires very sensitive means to observe this fact (see Fig. 14.3). Often colloids and suspensions are so turbid, or "muddy," that they are practically opaque. Matter dispersed as a colloid or as a coarse suspension does not affect the freezing point of the water very much.

EXPLANATION OF THESE DIFFERENCES

Theoretically the chief differences among true solutions, colloids, and coarse suspensions are caused by great differences in the sizes of the dispersed particles. The particles in a true solution are of molecular dimensions; that is, the dimensions are about the same as those of single molecules or a cluster containing only a small number of molecules. The angstrom (Å) is a customary unit for expressing dimensions of atoms and molecules, chiefly because they are about that order of magnitude; an angstrom is 10^{-8} cm. The shortest wavelength of visible light is about 4×10^{-5} cm. The particles in true solution, being smaller than the wavelength of visible light, have little effect upon the light.

Compared to sizes of objects in our macroworld, the dispersed particles in a colloid are minute—only about 0.1 to 50 millionths of a centimeter in diameter. But this is equivalent to 10 to 5000 Å, so colloidal particles are considerably larger than the molecules of true solutions. Coarse suspended matter consists of still larger particles. Therefore colloidal particles are intermediate in size between those of true solutions and those of coarse suspensions.

Because the individual particles in a true solution defy detection by even the most powerful magnification, we usually say that a true solution is a homogeneous mixture, or has the same composition throughout. Another way of expressing the same idea is to say that a true solution consists of a single phase; that is, there are no detectable boundaries within the system. On the other hand, colloids and suspensions both show experimental evidence of being nonhomogeneous mixtures and are said to have two or more phases with boundaries between them.

EVIDENCE THAT COLLOIDAL PARTICLES EXIST

Matter in a colloidal state has several distinctive properties, including (1) Brownian motion (p. 143), (2) the Tyndall effect (next paragraph), and (3) the size of the colloidal particles.

If a beam of sunlight is admitted through a small hole into a darkened room, the minute particles of dust suspended in the air can be seen as bright flashing points in the sunbeam. Similarly, a beam of strong light passing through a colloidal solution, such as one of starch, appears as a clearly defined cone of illuminated particles because the colloid scatters the light (Fig. 14.3). The scattering of light by small suspended particles was recognized by Faraday in 1857 and was investigated extensively by the English physicist Tyndall. Hence it is called the *Tyndall effect,* and the cone of illuminated particles is called the *Tyndall cone* or *beam*.

In the absence of suspended matter in a

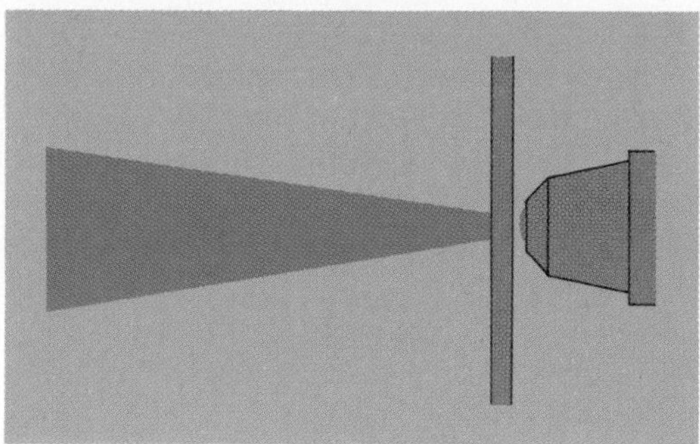

Figure 14.3
Tyndall beam. A beam of light is scattered by a colloidal solution, but not by a true solution, and produces a cone visible from the side. (Artist's reproduction from photograph by Atlas Chemical Industries, Inc.)

true solution, such as one of ordinary salt, the path of light through the true solution would not be visible from a side away from the direction of the beam. There would be no Tyndall cone or beam as in Fig. 14.3.

IMPORTANCE OF DISPERSIONS IN WATER

The abundance and universal distribution of water make dispersions in water so common that one might overlook them. One could scarcely dip a cup of water from any river without finding, if one examined it carefully, that it contained both colloidal matter and matter in true solution. The amount of such material held dispersed in any one cupful would probably be very small, though this would not be true in many flooded streams. But the aggregate attained by the summation of all moving water is very large—millions of tons per year. Fine particles of silt and clay are carried in suspension as muddy water and may be deposited on farmlands adjacent to and at the ends of rivers (deltas). Such land is usually highly productive and very valuable.

If we want to destroy insects or diseases on our flowers, shrubs, and plants, both agricultural and ornamental, we are likely to spray them with a mixture that is a dispersion in water. Even the cleansing of our bodies and our clothing is chiefly a problem of dispersing the unwanted matter, dirt in this case, and having it flow away with the water.

Although smoke, fog, and clouds, in general, cannot properly be called dispersions in water, they are closely related. The solid or liquid matter they contain is dispersed in such fine particles that they are really colloidal.

"Colloid chemistry," wrote Bancroft, "is the chemistry of grains, drops, bubbles, filaments and films." The dye industry and the paint industry, the dairy industry and agriculture, the rubber industry and the soap industry—even the contents of the family medicine cabinet—all depend on the dispersion of colloidal particles throughout a liquid medium. Dyes are particles of colored pigments. Paints also contain colored pigments, most of which no longer settle out because the pigments have been ground finely in a colloid mill. Similarly, the colored pigments in most bottles of medicine no longer settle out on standing in the medicine cabinet. Milk is homogenized by using pressure to force it through a small opening and thus break large particles of fat into smaller ones. The increased surface area results in the speeding up of chemical change in digestion. Colloidal particles of black carbon and rubber latex are used in making tires. Soaps and detergents are noted for the bubbles and thin films they can provide for cleansing.

The material within the cells of living plants and animals is itself a colloidal dispersion. Life as we know it would be impossible without true solutions and colloids. In these forms, food enters the plant or animal and is transported from place to place as needed, and waste materials are eliminated.

Table 14.1
Types of colloidal systems

dispersed phase	dispersing phase	example
liquid	gas	fog
solid	gas	smoke, dust
gas	liquid	foam, whipped cream
liquid	liquid	emulsions, such as milk, mayonnaise
solid	liquid	suspensions, such as paint, medicine
gas	solid	solid foam, such as marshmallow, cake, bread
liquid	solid	butter, oil-bearing shale
solid	solid	many alloys, selenium in glass

CLASSIFICATION OF COLLOIDAL SYSTEMS

One method of classifying colloidal systems is to group them by the physical states of the dispersed and dispersing phases. There are eight possible combinations, as shown in Table 14.1.

RECYCLING OF WATER, A COLLOIDAL PROCESS

The first step in the treatment of water by cities usually involves aerating it by spraying it into the atmosphere to get rid of bad odors and to saturate it with oxygen. Bad taste in water may be due to the presence of small amounts of substances that have objectionable taste or odor or to the lack of dissolved oxygen.

The second step usually involves coagulating the colloidal material. To accomplish this, a salt called filter alum (aluminum sulfate) is added. If necessary, some lime is also added. These substances produce a precipitate and encourage the many small colloidal particles to coalesce into a much smaller number of larger ones.

The water is then allowed to stand in large settling basins to give time for chemical and physical changes to occur. The precipitate formed chemically is gelatinous. As this gelatinous material collides with colloidal particles, such as clay, the particles increase in size, or mass. A certain minimum mass must be reached for the effect of gravity to overcome the electrical forces of repulsion between the particles. Additional mass will cause the particles to settle, because the gravitational force downward is greater than the buoyant force upward.

Next the water is filtered through large beds of sand and gravel that remove most of the remaining suspended matter. Any hardness may be removed. Finally, chlorine gas is added to kill harmful bacteria in the water before it is distributed to the consumer.

After water is used by the consumer, it may be recycled through a sewage treatment plant for reuse or for use in the next city "downstream." The recycling is completed by repeating the treatment described above at length, with the inclusion of any additional process necessary to remove unusual wastes such as those from a particular industry.

OSMOSIS

An experiment that should help to clarify the meaning of the term "osmosis" can be done in the following manner. A large carrot has a hole drilled in the top. A piece of glass tubing is inserted into a hole in a rubber stopper, which in turn is inserted into the larger hole in the carrot. The carrot is then immersed in a jar of water, and the apparatus is fixed in place by means of a proper clamp. If there are no leaky joints, water will slowly rise inside the glass tube to a level that is considerably higher than the level of water in the jar. How does this occur?

Experimental evidence indicates that water can diffuse quite freely through a carrot skin but that sugar and similar material in true solution within the carrot cannot diffuse so readily through the skin. Such a *membrane*, through which one component of a solution can pass much more readily than the other, is said to be *differentially permeable*. In this case, the water diffuses through the membrane more rapidly toward the inside of the carrot, where there is an aqueous solution of sugar and water, than outward, where there is only water. Such a case of the selective transfer of one component of a fluid mixture, such as a solution, through a differentially permeable membrane is called *osmosis*.

As the water diffuses faster toward the inside of the carrot, the pressure within also increases. This increasing pressure tends to resist the invasion of water from without and eventually causes the rate of diffusion inward to equal the rate of diffusion outward, providing, of course, that water is not removed on one side of the membrane, as it really is in a living carrot by evaporation from the leaves.

It should not, of course, be inferred that osmosis always occurs from the outside toward the inside. A carrot or a cucumber, with a skin that acts as a differentially permeable membrane, shrivels when placed in many concentrated solutions. In this case the water is again diffusing more rapidly toward the side where the solution is more concentrated.

It can be generalized from examples such as these that, *in osmosis* involving an aqueous solution and water or two aqueous solutions on opposite sides of a differentially permeable membrane,

The water diffuses more rapidly toward the side that has the greater concentration of solution.

It is just as true to say that in osmosis the water diffuses more rapidly away from the side that has the greater concentration of water.

Try now to apply this knowledge of osmosis to explain why dried prunes swell when soaked in water; why grass is killed by pouring brine, or salt solution, on it; how in stores a fine spray of water keeps lettuce crisp; why carrot strips are crisp when they are kept in water for some time before they are served. Finally, remember that osmosis occurs continuously through the membranes of the many cells in the human body.

SUMMARY

Chemistry had its beginnings very early but the real accumulation and recording of scientific knowledge in the field began in the 1600s and 1700s with the rise of concepts touching on the nature and properties of matter. Consideration of the kinds of changes occurring in matter and of the different properties of matter led to the need for the classification of matter. The classification chosen for use during the 1800s and 1900s has usually depended on the convenience desired and the phenomenon being studied. For some purposes it was only necessary to consider matter as metals or nonmetals; for other purposes as solid, liquid, or gaseous; and still others as pure substances or mixtures. The latter were thought of as being of pure or impure substances; as being either homogeneous or heterogeneous.

Refinement of the concepts of pure substances led to categorizing them into two classes: elements and compounds. Man then developed the periodic table as a means of organizing his knowledge about elements. Carbon-containing compounds are called *organic compounds*, in contrast to *inorganic compounds* formed by the other elements.

With the advent of extensive knowledge about the electron structure of the atom, chemists now consider the interactions of elements and of compounds in relation to the sharing or transferring of electrons—concepts to be explored in later chapters.

Mixing of substances may take place by the processes of diffusion, dispersion, or osmosis. Diffusion depends on the random kinetic molecular motions to effect the mixing. Dispersions may be classified as true solutions, coarse suspensions, or colloids. The size of the dispersed particles determines the type of dispersion that results and the degree of freedom from settling of the particles. In true solutions the particles are of molecular dimensions; in colloids the size of the smallest particles is of the order of the wavelength of visible light; in suspensions the particles can be so large as to be seen by the naked eye or under a microscope. The Tyndall effect is caused by the reflection of light from colloidal-size particles; Brownian motion is also a phenomenon involving colloidal-size particles. Colloidal systems in daily life can involve solids, liquids, gases with the dispersed and dispersing phases together in the varied combinations shown in Table 14.1. The recycling of water is largely a colloidal precipitation process. Osmosis is a selective diffusion process involving a semipermeable membrane between two fluids.

Important words and terms

chemistry
physics
matter
Priestley
Lavoisier
physical change
chemical change
physical properties
chemical properties
classifications of matter
distinctions between metals, nonmetals, and metalloids
distinctions between pure substance and mixture
distillation
distinction between element and compound
distinction between homogeneous mixture and heterogeneous mixture
law of definite composition
classes of dispersions
examples of true solutions, coarse suspensions, and colloids
osmosis

Questions and exercises

1. Give the conventional definitions for chemistry and physics.
2. (a) Define matter. (b) What are its gross aspects? (c) In further differentiating these gross aspects, upon what structure of matter has this added refinement been based?
3. Discuss the concept of quantity in change in the case (a) in which mercuric oxide yields pure oxygen and mercury; and (b) in which mercuric oxide is formed of mercury in the presence of oxygen.
4. To whom do we give most of the credit (a) for convincing scientists of the nature of combustion? (b) for discovering oxygen? (c) for pointing out the two chief components of the air?
5. Distinguish (a) between chemical and physical changes; (b) between chemical and physical properties. Give examples of each.
6. Distinguish among (a) metals, nonmetals, and metalloids; (b) the phases of matter.
7. Are solids, liquids, and gases always such under all conditions? Explain.
8. (a) Distinguish between a pure substance and a mixture. (b) How does the chemist's concept of "pure" differ from that of the housewife?
9. (a) Distinguish between a mixture and a compound. (b) Is air a mixture or a compound? Why?
10. What are two types of pure substances?
11. Give (a) the old and the new definition of an element; (b) the old and the new definition of a compound.
12. (a) Define and give examples of two types of mixtures. (b) Give two means of differentiating between pure substances and mixtures.
13. How would you separate (a) salt from a salt solution? (b) cream from milk? (c) oxygen from mercuric oxide?
14. On p. 308 we say, "2. Usually a compound obeys. . . ." Can you give a possible reason or reasons for the qualification "usually"?
15. On applying the physical (not chemical) process of heating, the two elements mercury and oxygen separate. Is mercuric oxide therefore a mixture? Explain.
16. Carbon tetrachloride and ethyl alcohol are both volatile liquids, but the first is heavy and its molecular weight more than three

times that of alcohol. With the two liquids at room temperature, at the same distance from you, and with no drafts, which odor would you expect to detect first on uncapping the bottles simultaneously? Explain.

17. Name three kinds of dispersions. Give examples of each.
18. (a) List the properties that enable us to distinguish among the three kinds of dispersions. (b) What is the main reason for the differences found in these dispersions?
19. (a) What is the Tyndall effect and what characteristic of the suspension is responsible for it? (b) Would you expect to see a Tyndall beam in (1) a solution of distilled water and salt? (2) water from a stream? (3) milk in a fish tank of tap water? Give your reason in each case.
20. Of what order in size are colloidal particles relative to those in other dispersions that we have studied?
21. On some nights the long beam of light in the air from a flashlight is a handy "pointer" to point out stars in the constellations. On other nights such a beam is absent. Why?
23. What are the dispersed phase and the dispersing phase in these colloidal systems: (a) paint, (b) butter, (c) smoke, (d) marshmallow, (e) whipped cream.
23. What steps are involved in recycling water?
24. Distinguish between osmosis and diffusion.
25. Fill in the following statements with the words "toward" or "away from" as may be appropriate in each case: (a) Water diffuses more rapidly __________ the side that has the greater concentration of solution; (b) __________ the side that has the greater concentration of water; and (c) __________ the side that has the greater concentration of solute.
26. Match the following:

a. Brown	____ four elements
b. Tyndall	____ phlogiston
c. Aristotle	____ first discoverer of oxygen
d. Becher and Stahl	____ concept of quantity in change
e. Priestley	____ motion of colloidal particles
f. Lavoisier	____ light scattering by colloidal particles

Supplementary readings

Books

Conant, James Bryant, *The Overthrow of the Phlogiston Theory: The Chemical Revolution of 1775–1789,* Harvard University Press, Cambridge, Mass. (1950). (Paperback.) [A case history: the discovery of oxygen as a central event in the overthrow of the phlogiston theory; Lavoisier's creation of the oxygen theory of combustion.]

Hurd, D. L., and J. J. Kipling, *The Origins and Growth of Physical Science,* Vol. 1, Penguin Books, Baltimore (1964). [Instructive introductions to original publications followed by appropriate excerpts from those publications. From p. 258 to the end of the volume the contributions of Boyle, Black, Priestley, Scheele, Cavendish, and Lavoisier are described, giving a panoramic view of the early rise of physics and chemistry as sciences.]

Jaffee, Bernard, *Crucibles: The Story of Chemistry from Ancient Alchemy to Nuclear Fission,* Fawcett World Library, New York (1960). (Paperback.) [This book, by a well-known writer, is an excellent source of information about the men discussed in this chapter and their contributions.]

Keenan, Charles W., and Jesse H. Wood, *General College Chemistry,* 4th ed., Harper & Row, New York (1971). [Pages 7–17 provide an excellent introduction to chemical science. Chapter 7 gives a somewhat fuller discussion of elements in general and hydrogen, oxygen, and water in particular than is possible in a physical science textbook, and Chaps. 10 and 11 deal extensively with solutions and the colloidal state.]

Leicester, Henry M., *The Historical Background of Chemistry,* Wiley, New York (1956). [In Chap. 15 Lavoisier and the foundations of modern chemistry are considered.]

March, Arthur, and Ira M. Freeman, *The New World of Physics,* Vintage Books, Random House, New York (1963). (Paperback.) [Pages 33–36 are devoted to the nature of explanation in science.]

Moore, F. J., *A History of Chemistry,* McGraw-Hill, New York (1931). [Chapters 4, 5, and 6, pp. 28–64, deal with material in this chapter.]

Partington, J. R., *A Short History of Chemistry,*

Harper Torchbooks, Harper & Row, New York (1960). (Paperback) [The work and experiments of Boyle, Becher and Stahl, Black, Scheele, Priestley, and Lavoisier; the controversy of Berthollet and Proust (combining proportions); Berzelius's atomic-weight tables; the contributions of Dulong, Petit, Prout, Avogadro, Dalton, and others.]

Rochow, Eugene G., *The Metalloids,* Heath, Boston (1966). [Chapter 2, "Physical Differentiation from the Metals" gives an excellent discussion of such physical properties as conductivity of heat and electricity. It also includes solid-state electronic devices such as transistors and rectifiers. The discussion of optical properties may be interesting in connection with solar batteries for radio transmitters in satellites.]

THE LANGUAGE OF CHEMISTRY

Everybody is ignorant, only on different subjects.

WILL ROGERS, 1879–1935

In the consideration of the nature of matter in the preceding chapter we were introduced to many general terms that are often used in chemistry. In Chap. 12 a few concepts central to the language of chemistry were developed somewhat, in order to tell the story of early atomic theory and of later atomic and molecular theory. In Chap. 13 the discussion of the periodic table contains several chiefly descriptive terms that are a part of the language of chemistry. Most fundamental terms and concepts central to the language of chemistry are, however, to be found in this chapter. The language will be extended with additional terms and concepts as needed at appropriate points in subsequent chapters.

Concepts necessary to understand the language of chemistry

Before we explore the meaning and use of symbols, formulas, and chemical equations, let us recall that certain concepts were considered in Chap. 12: Avogadro's law, atoms, molecules, atomic weight, atomic mass unit, and molecular weights. Now let us consider the meanings of ion; atomic weight and atomic mass unit in relation to gram-atomic weight; gram-molecular weight, the mole, gram-molecular volume; the mole and Avogadro's number in relation to the mole; the mole in relation to gram-molecular weight and gram-molecular volume; molecular weights and their relation to Avogadro's law.

THE ION

In the neutral atom the number of positively charged particles called protons is equal to the number of negatively charged particles called electrons. In Chap. 12 an ion is described as essentially an electrically charged atom or molecule; that is, an atom or a small group of atoms with a net electric charge is called an *ion*. A neutral atom that loses electrons becomes a positively charged ion; one that gains electrons becomes a negatively charged ion.

ATOMIC WEIGHTS AND GRAM-ATOMIC WEIGHT

We now express atomic weights in atomic mass units (amu). The international atomic weights (masses), approved by the International Union of Pure and Applied Chemistry in 1961 and revised in 1969, are based on the carbon isotope of mass 12 as the standard. Therefore,

One atomic mass unit (amu) is one-twelfth the mass of the isotope carbon-12 (^{12}C), or 1.66×10^{-24} g.

Similarly, the weight of the oxygen isotope 16, known to be 15.999 amu, is referred to as 16 amu, or 26.6×10^{-24} g. Thus a number of grams of an element equal numerically to the number of atomic mass units in its atomic weight is known as the gram-atomic weight. A *gram-atomic weight,* often called gram-atom, of an element is that quantity of the element whose weight in grams is numerically equal to its atomic weight. Hence for any one element, atomic mass unit and gram-atomic weight are numerically the same. Atomic weights, and also molecular weights, are relative weights only; they are not absolute weights. The list of International Atomic Weights, 1969 (see inside rear cover) does not tell the weight of any atom. It tells the weight (or mass) of an atom relative to that of carbon-12.

THE MOLE, THE AVOGADRO NUMBER

How many atoms are in one gram-atomic weight? This question was asked and an-

Table 15.1
Computation of the Avogadro number

for hydrogen: $\dfrac{1.008\text{ g}}{1\text{ g-at. wt.}} \times \dfrac{1\text{ atom}}{1.674 \times 10^{-24}\text{ g}} = 6.02 \times 10^{23}$ atoms in 1 g-at. wt. of hydrogen

for helium: $\dfrac{4.003\text{ g}}{1\text{ g-at. wt.}} \times \dfrac{1\text{ atom}}{6.65 \times 10^{-24}\text{ g}} = 6.02 \times 10^{23}$ atoms in 1 g-at. wt. of helium

for oxygen: $\dfrac{16.00\text{ g}}{1\text{ g-at. wt.}} \times \dfrac{1\text{ atom}}{26.6 \times 10^{-24}\text{ g}} = 6.02 \times 10^{23}$ atoms in 1 g-at. wt. of oxygen

swered long before the new atomic weight scale was first approved and went into effect in 1961. In order to find the answer to this question, for several elements, we divide the gram-atomic weight of the element by the measured mass of one atom of that element. The procedure and results are shown in Table 15.1. In the table, multiplication of the numerators (top) gives the gram-atomic weight; multiplication of the denominators (bottom) gives the mass of one atom.

In Table 15.1 we see that the atomic weight in grams is the weight of 6.02×10^{23} atoms, and is called a gram-mole or simply a mole of the element. However, the concept of a mole is not limited to atoms. We may apply it to matter in which the ultimate particles are atoms, molecules, ions, units of ions, electrons, other particles, or specified groups of particles.

A *mole* is now defined as

That quantity of a substance that contains the same number of ultimate particles (atoms, molecules, ions, units of ions, electrons, other particles, or specified groups of such particles) as are contained in 12 g of isotope carbon-12 (^{12}C). The number of ultimate particles (atoms, or molecules, or ions, etc.) in a mole is called the Avogadro number, 6.02×10^{23}.

X ray measurements can be used to determine Avogadro's number—to compute the number of atoms (or molecules, or ions) in a mole—as can experiments involving dispersed colloidal particles. The computations in Table 15.1 do not imply that the Avogadro number is found from experimentally determined weights of single atoms.

GRAM-MOLECULAR WEIGHT, THE MOLE, AND GRAM-MOLECULAR VOLUME

Such a small mass as one atomic mass unit (amu) is inconvenient to use in the laboratory. The gram is a much better unit for experimental purposes, so for the indirect laboratory determination of molecular weights we establish the unit called the gram-molecular weight. The gram-molecular weight always contains the same numerical quantity as the molecular weight, but it is expressed in grams instead of amu, the units for atomic weights and molecular weights. For example, 32 amu, the molecular weight, is the weight of one oxygen molecule, a particle of submicroscopic size, but 32 g is the gram-molecular weight of oxygen. One gram-molecular weight of a substance is 1 mole of that substance.

At standard conditions (0°C and 760 mm pressure) 32 g of oxygen occupies 22.4 liters of volume. One gram-molecular weight of hydrogen (2.016 g) must also occupy 22.4 liters of volume, because equal volumes of gas under the same conditions of temperature and pressure contain the same number of molecules. Therefore at standard conditions of temperature and pressure (STP), 22.4 liters is the gram-molecular volume of any gas.

Figure 15.1
Comparison of the gram-molecular volume (22.4 liters) with a standard basketball. The cube edge is almost 28.2 cm, or 11.1 in.

The *gram-molecular volume* (GMV) of a gas is the volume occupied by one gram-molecular weight of the gas, and this is 22.4 liters at standard conditions (Fig. 15.1). The *gram-molecular weight* (GMW) of a gas is the weight in grams of 22.4 liters of the gas at standard conditions.

DETERMINATION OF MOLECULAR WEIGHT OF A GAS

The problem is essentially that of determining the weight of 22.4 liters of the gas at STP. This gives the gram-molecular weight of the gas. In order to get the weight of one actual molecule, we merely write down exactly the same number of atomic mass units as we had of grams in the gram-molecular weight.

For example, we might determine experimentally that 1 liter of carbon dioxide at standard temperature and pressure weighs 1.977 g. The gram-molecular volume of carbon dioxide (22.4 liters) must weigh $1.977 \text{ g} \times 22.4 = 44.2848 \text{ g}$. This weight is the gram-molecular weight of carbon dioxide. Therefore,

$$\text{gram-molecular weight (GMW)} = \text{weight of 1 liter} \times 22.4 \quad (15.1)$$

The molecular weight and the gram-molecular weight of carbon dioxide are numerically the same quantity, but the unit for molecular weight is amu and the unit for gram-molecular weight is the gram. In other words, we have now determined the molecular weight of carbon dioxide to be 44.2848 amu.

Experiments using several different methods of determination agree that the actual number of molecules in 22.4 liters of a gas at standard conditions is about 602,300,000,000,000,000,000,000, which is written 6.023×10^{23}. This number is the Avogadro number.

The method we have used is based on the assumption that equal volumes of gas contain the same number of molecules (Avogadro's law).

Still other methods, such as observing the osmotic pressure or the freezing point of a solution, have been used for the determination of molecular weights. The most popular method during the last few decades has involved the use of the mass spectrograph. All these methods, if used properly, give results that are in good agreement. This is another illustration of how a scientist makes an assumption about something that cannot be experienced directly and later either verifies or discards the hypothesis as a result of related but somewhat indirect evidence.

RELATIVE MOLECULAR WEIGHTS: AN APPLICATION OF AVOGADRO'S LAW

Different gases have different densities. Therefore, if we accept Avogadro's assumption as being a true generalization, or law, we must immediately conclude that the molecules of different substances have different weights. We can experimentally determine the weights of equal volumes of gases and therefore calculate their relative molecular weights. This follows mathematically, for if we divide the weights of equal volumes of gases by equal numbers (of molecules), the quotients, the individual molecular weights, will be in the same ratio. It is this mathematical principle that enables us to simplify fractions. For example, the ratio of 39:52 has the same numerical value as the fraction $\frac{39}{52}$. This may seem complex and difficult to some students, but by dividing both the numerator and denominator of the fraction by 13 thus, $\frac{39}{13}/\frac{52}{13} = \frac{3}{4}$. The ratio of 3:4 probably seems simpler than that of 39:52, but both ratios have the same numerical value.

The principle might be further illustrated by assuming that some business house packaged dime and nickel coins so that one package always contained the same number of coins as any other package. Using the information given in Fig. 15.2, one could quickly determine the ratio of the weight of one dime to the weight of one nickel by weighing a package of each and dividing the weight of a package of dimes by that of a package of nickels. Thus 25 g/50 g = 2.5 g/5 g, and the ratio of the weight of one dime to the weight of one nickel is 1:2.

Figure 15.2
The ratio of the weights of two piles of equal numbers of coins is the same as the ratio of the weights of one coin from each pile.

COMPUTING THE WEIGHT OF ONE ATOM AND OF ONE MOLECULE

The weight of one atom of any particular element or of one molecule of any particular compound (except ionic compounds to be discussed in a later chapter) can be computed as follows:

$$\frac{\text{one gram-atomic weight of atoms}}{\text{the Avogadro number}} = \text{weight of one atom in grams}$$

For oxygen, as an example,

$$\frac{15.9994}{6.02 \times 10^{23}} = 26.6 \times 10^{-24}\ \text{g}$$

$$\frac{\text{one gram-molecular weight of molecules}}{\text{the Avogadro number}} = \text{weight of one molecule in grams}$$

Language of chemistry

For some time we have been using a few chemical symbols, and even fewer chemical formulas, without so naming them. We shall now consider them formally as a part of the language of chemistry in order to introduce chemical equations.

NUMBER OF ATOMS IN A MOLECULE OF SOME GASEOUS ELEMENTS

It is a fact that one volume of hydrogen is exactly sufficient to react with one volume of chlorine to form two volumes of hydrogen chloride, as shown diagrammatically in Fig. 15.3.

Let us assume that Avogadro's law is true. If so, the number of hydrogen molecules used must just equal the number of chlorine molecules consumed and must also be half as large as the number of hydrogen chloride molecules formed. For simplicity (Fig. 15.4), let us assume we have only one molecule of hydrogen reacting with one molecule of chlorine and forming two molecules of hydrogen chloride. If each hydrogen chloride molecule contains some hydrogen and some chlorine, then each hydrogen molecule and each chlorine molecule must split into two pieces before this process of combining can occur. These smaller subdivisions of the hydrogen molecule are the hydrogen atoms previously discussed. There is no evidence that a hydrogen molecule or a chlorine molecule ever breaks up to yield more than two atoms, so we always think of these molecules as each consisting of two atoms bound together by some kind of force. Thus the free molecules of the common gaseous elements

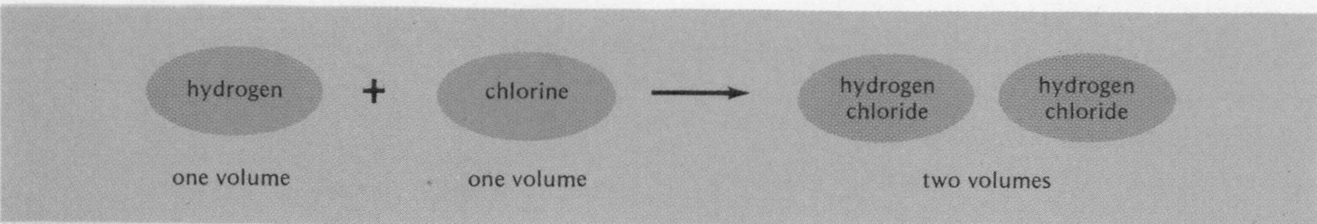

Figure 15.3
Combining volumes: formation of hydrogen chloride from hydrogen and chlorine. The number of atoms in a molecule of a gaseous element can be determined by using Avogadro's law with combining volumes.

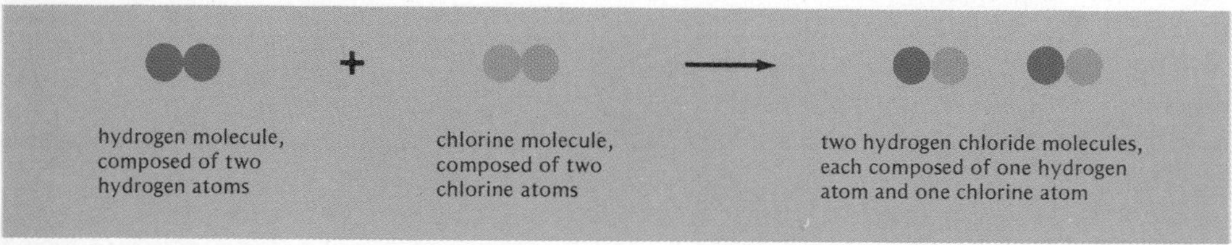

Figure 15.4
Graphic representation of chemical change: formation of two hydrogen chloride molecules from a hydrogen molecule and a chlorine molecule.

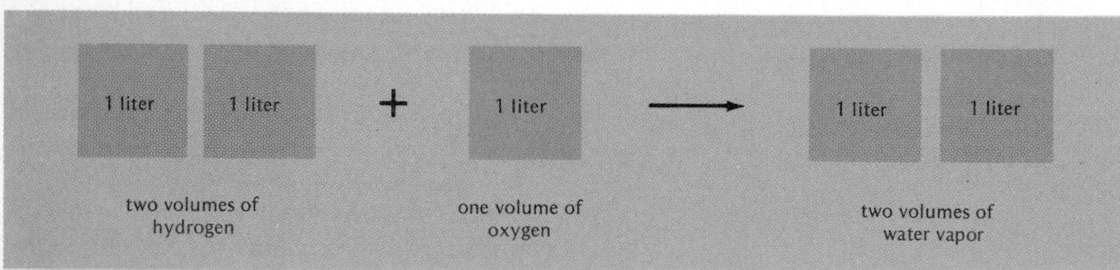

Figure 15.5
Combining volumes: formation of water vapor from hydrogen and oxygen.

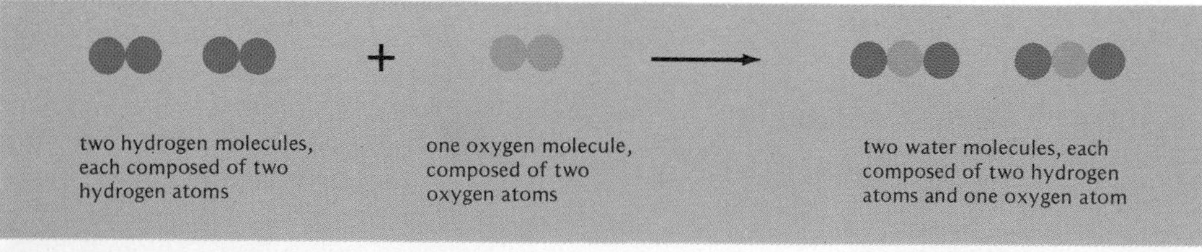

Figure 15.6
Graphic representation of chemical change: formation of two molecules of water from two molecules of hydrogen and one molecule of oxygen.

such as hydrogen (H_2), oxygen (O_2), or chlorine (Cl_2) ordinarily exist as "diatomic molecules"; that is, each consists of two atoms. There are only five gaseous diatomic molecules at room temperature: the three mentioned above and nitrogen (N_2) and fluorine (F_2). There are only six gaseous monatomic molecules at room temperature, which may also be called atoms: helium (He), neon (Ne), argon (Ar), krypton (Kr), xenon (Xe), and radon (Rn).

The less stable and hence more active form of oxygen called ozone consists of molecules each containing three oxygen atoms instead of two, as is the case with the ordinary form of oxygen. In other words, the chemical formula of ozone is O_3. Ozone, produced by an electric spark such as lightning, has a characteristic fresh odor.

GRAPHIC REPRESENTATION OF A CHEMICAL CHANGE

Such concepts as atoms and molecules are likely to seem difficult because we cannot see these particles. In order to understand them in a more definite and concrete way, let us represent them diagrammatically. The chemical change indicated by volumes in Fig. 15.3 is shown graphically in Fig. 15.4 for molecules.

A MORE COMPLICATED CHEMICAL CHANGE

It is a known fact that two volumes of hydrogen combine with one volume of oxygen to yield two volumes of water vapor (Fig. 15.5).

Consulting Fig. 15.5 and using the same kind of reasoning as before, we can conclude that the oxygen molecules must split in this process in order to have an atom of oxygen for each water molecule. The chemical change indicated by volumes in Fig. 15.5 is shown graphically in Fig. 15.6 for molecules.

SYMBOLS, FORMULAS, INTRODUCTION TO CHEMICAL EQUATIONS

Modern graphic terminology has been developed because of necessity. A small number of scientific articles published during the first half of the nineteenth century used symbolic representations very much like those we have just employed. Though they did help to show how chemical changes occurred, they were not well suited for publication because they involved setting up too much special type. Berzelius of Sweden, therefore, invented in 1814 a way of showing exactly the same ideas through the use of standard type; for example, H to represent a hydrogen atom and Cl to represent a chlorine atom.

Symbols are letters or pairs of letters used to represent atoms of elements. Pairs of letters must sometimes be used because there are 105 or more elements and only 26 letters. The element carbon, for example, is symbolized by C; chlorine by Cl. The first letter of the symbol of an element is capitalized, but any second letter is not. If we find a second letter capitalized, the expression is not a symbol of an element but a formula of a compound. For example, Co is the symbol for an atom of the element cobalt, and CO is the formula of a molecule of the compound carbon monoxide, a poisonous product of the combustion of gasoline in an automobile engine.

Symbols of several of the metallic elements are from the Latin names of the elements. These are silver, Ag (*argentum*); gold, Au (*aurum*); copper, Cu (*cuprum*); iron, Fe (*ferrum*); mercury, Hg (*hydrargyrum*); potassium, K (*kalium*); sodium, Na (*natrium*); lead, Pb (*plumbum*); antimony, Sb (*stibium*); and tin, Sn (*stannum*).

Formulas are letters and subscripts, either printed or understood, that represent the exact composition of one molecule of a substance. A formula denotes the combining ratios of atoms in the substance. For example, the formula for water is H_2O. This means that one molecule of water is composed of two atoms of hydrogen, as indicated by the printed subscript 2, and one atom of oxygen —an arabic 1 is understood to be the subscript of the symbol O.

DETERMINATION OF CHEMICAL FORMULAS

Experiments to determine the simplest formula of a substance must tell us the simplest whole number ratio of the respective atoms in the substance. For example, if we could determine the simplest ratio of moles of the respective elements in the substance, we would then know the simplest ratio of atoms. The simplest ratio of moles of the elements in a compound may be obtained from the number of grams of the respective elements in a given sample of the compound.

What is the simplest formula of silver oxide, as an example? A sample of silver oxide was found to contain 1.7271 g of silver and 0.1281 g of oxygen.

$$\text{number of moles of silver present in the sample} = 1.7271 \text{ g of silver} \times \frac{1 \text{ mole silver}}{107.9 \text{ g silver}} = 0.016 \text{ mole of silver}$$

$$\text{number of moles of oxygen present in the sample} = 0.1281 \text{ g of oxygen} \times \frac{1 \text{ mole oxygen atoms}}{16 \text{ g oxygen}} = 0.008 \text{ mole of oxygen atoms}$$

The ratio of moles of silver (Ag) atoms to moles of oxygen (O) atoms is 0.016/0.008, or 2:1. Thus the simplest formula of silver oxide is Ag_2O, a 2:1 ratio.

The molecular formula indicates the number and kinds of atoms in one molecule of a substance. It is either the simplest formula or a multiple of the simplest formula. After the simplest formula is known, the molecular formula may be determined, provided the molecular weight is known. As mentioned earlier in this chapter, the mass spectrometer is now commonly used to determine the weights of molecules as well as of atoms. Some molecular weights are multiples of the weight of the simplest formula unit. What is the molecular formula of methyl formate, for example? The simplest formula of methyl formate has been found to be CH_2O and the formula weight of CH_2O is 30 amu. The molecular weight of methyl formate has been found to be 60 amu. Because the molecular weight is twice the formula weight of CH_2O (ratio: 60 amu/30 amu = 2:1), the molecular formula of methyl formate is $C_2H_4O_2$, which is twice the simplest formula.

CHEMICAL EQUATIONS

Chemical equations are abbreviations of complete chemical sentences. The chemical sentence by which hydrogen chloride is formed (Fig. 15.4) may be read as follows: A hydrogen molecule, composed of two hydrogen atoms, reacts with a chlorine molecule, composed of two chlorine atoms, to yield two hydrogen chloride molecules, each composed of one hydrogen atom and one chlorine atom. This long chemical sentence can be abbreviated as follows:

$$H_2 + Cl_2 \rightarrow 2HCl$$

The whole preceding line is called a chemical equation. It contains the chemical formulas H_2, Cl_2, and HCl. The number 2 in front of the HCl is called a coefficient and signifies here that there are two molecules of hydrogen chloride. Similarly, H and Cl are chemical symbols. This kind of symbolic language can, of course, imply various things, but at first the beginning student is advised to employ symbols in the following way. Think of a formula as representing the exact composition of one molecule of the substance involved. For example, the formula H_2 signifies that a hydrogen molecule contains exactly two hydrogen atoms that are attached to each other. The chemical symbol H represents one atom of hydrogen, but because we very rarely encounter single hydrogen atoms, we rarely need symbols for them, except when we write formulas.

Let us consider the modern way of showing what occurs when hydrogen burns in oxygen. If one wishes to say, "Two molecules of hydrogen combine with one molecule of oxygen to form two molecules of water," he could write the symbolic equation,

$$2H_2 + O_2 \rightarrow 2H_2O$$

If one needed only to say, "Hydrogen combines with oxygen to form water," he could write only the word equation,

hydrogen + oxygen → water

BALANCED EQUATIONS

Atoms are neither created nor destroyed in chemical change (law of conservation of matter). Therefore in order to be a correctly balanced equation, both sides of the equation must have the same number of atoms of each element. In the preceding symbolic equation, there are four hydrogen atoms on the left (two molecules, each composed of two hydrogen atoms) and four hydrogen atoms on the right (two molecules of water, each containing two hydrogen atoms). Likewise, there are two oxygen atoms in the one oxygen molecule on the left, and on the right, two oxygen atoms (one oxygen atom in each of two molecules). The equation is balanced.

In order to write the chemical formula of a compound, one must know the elements composing it and the number of atoms of each element in one molecule of the compound. In order to write an equation, one must know that the substances involved will react chemically and what products will be formed.

Other examples of balanced equations follow.

$$2KClO_3 + \text{heat} \rightarrow 2KCl + 3O_2$$
$$Zn + 2HCl \rightarrow ZnCl_2 + H_2$$
$$Fe_2O_3 + 3CO \rightarrow 2Fe + 3CO_2$$

For the well-trained chemist who is compelled to use only a minimum of space, abbreviations are no doubt desirable. For college students who are nonscience majors, however, the use of abbreviations is often not so important. The use of word equations in place of those composed of symbols and formulas will probably result in sufficient understanding for their purposes. On such points, students should secure directions from their instructor. All equations should be "read"; this should help in the achievement of greater understanding. Keep in mind that formulas and equations, like chemical compositions, are exact and quantitative. If not written accurately and adequately understood, they might as well be omitted from the subject of chemistry.

Knowledge of the combining power of atoms (valence, next chapter) is useful in the balancing of equations other than the simplest ones.

An equation may be interpreted in terms of moles of reactants (left side of the equation) and of products (right side). The symbolic equation for the reaction of hydrogen and oxygen to form water may be read in terms of moles: 2 moles of hydrogen react with 1 mole of oxygen to form 2 moles of water.

$2H_2$	+ O_2	→ $2H_2O$
2 moles	1 mole	2 moles
4.03 g	32.0 g	36.03 g

The weights of a mole of hydrogen and of oxygen can be inferred from the atomic weights, and the sum of the weights of the reactants equals the weight of the product formed. The study of the quantitative aspects of chemical reactions is called *stoichiometry*.

PERCENTAGE COMPOSITION FROM THE FORMULA

Let us calculate the approximate theoretical percentage composition of ordinary water, whose formula is H_2O. The experimental atomic weight of hydrogen is now accepted to be a statistical average on account of the presence of isotopes of hydrogen. This experimental value is 1.008 amu, but the figures to the right of the decimal point are better discarded because they signify an uncalled for degree of precision. If one atom of hydrogen weighs approximately 1.0 amu, then logically two atoms of hydrogen would weigh 2.0 amu. One oxygen atom weighs 16.0 amu. The water molecule must have a weight that equals the sum of its parts; therefore 2.0 + 16.0 = 18.0 amu, which is the total weight of the water molecule. Two-eighteenths of this weight is due to hydrogen. Percent is also a fraction whose denominator is always 100, so that to get the percentage of hydrogen in water all we need do is change the fraction 2/18 into hundredths. The two fractions must have the same value; thus $2/18 = x/100$, where x represents the percentage of hydrogen in water. A statement that two fractions are equal is properly called a proportion in mathematics, and the equation can be solved to find the value of the one unknown (x). Therefore

$$18x = 2 \times 100$$

and

$$x = \tfrac{200}{18} = 11\tfrac{1}{9} \text{ percent}$$

We have just shown that, if the formula of water is H_2O, it follows that there would be $11\frac{1}{9}$ percent of hydrogen by weight in any sample of pure water. Actually, the first chemists did not know what the formula of water was. The choosing of a reasonable formula for it had to await the determination of such pertinent facts as percentage composition, gas densities, and combining volumes. Hundreds of careful analyses have revealed that water contains $11\frac{1}{9}$ percent hydrogen by weight.

The percentage composition and the molecular weight of hydrogen peroxide are such as to indicate that its formula is H_2O_2.

SUMMARY

The nature of matter and a few chemical terms closely related were introduced in earlier chapters. In this chapter we developed more concepts and terms in order to understand better those fundamentals that are basic to all chemistry—symbols, formulas, and equations.

Important words and terms

ion
international atomic weights
atomic mass unit, amu
gram-atomic weight
Avogadro number
mole
gram-molecular weight
gram-molecular volume
determination of molecular weight
relative molecular weight
computing the weight of one atom
computing the weight of one molecule
number of atoms in a molecule
symbols
formulas
determination of chemical formulas
chemical equation
balanced equation
stoichiometry
percentage composition from formula

Questions and exercises

1. (a) What is the difference between a negative ion and an electron? a positive ion and a proton? (b) What special names are reserved for the positive hydrogen ion and the positive helium ion?
2. State the relationships between atomic weight, gram-atomic weight, atomic mass unit (amu), and gram-atom.
3. (a) What is a mole of hydrogen? of oxygen? of iron? of copper? (b) How do they compare in volume? (c) Does the concept of the mole apply to electrons, molecules of compounds, or ions?
4. Look up the gas fluorine in the periodic table. (a) What is the weight of one molecule in amu? (b) How many liters would a GMV occupy at STP and what would be its weight in grams? (c) How many liters would a GMW occupy at STP? (d) How many atoms would be found in a mole of fluorine at STP?
5. (a) Which elements of the periodic table exist at room temperature as diatomic gaseous molecules? (b) Which gases are monatomic

at room temperature? (c) How does ozone differ from oxygen as we know it?

6. In Fig. 15.3 what evidence leads us to conclude that the hydrogen and chlorine gases are diatomic? In giving your answer consider whether you could explain the result if the hydrogen and chlorine were monatomic.
7. What is the significance of the 2 in Cl_2 and in 2Ag?
8. (a) When an equation is balanced, what condition obtains with regard to the number of atoms of the elements on each side of the equation? (b) Show that the three equations on p. 325 are balanced with regard to (1) the number of atoms of each element and (2) the atomic weights on each side.
9. Write balanced equations for the following reactions: (a) hydrogen and oxygen to form water; (b) hydrogen and chlorine to form hydrogen chloride; (c) hydrogen peroxide and heat to form water and oxygen. Test the last equation to see that it is balanced with regard to the number of atoms on each side.

Problems

Use the figures from the periodic table (inside front cover) in solving these problems.

1. Given the Avogadro number 6.02×10^{23}, calculate the weight in grams of a proton and an electron. (Hint: Use the atomic weight of hydrogen from the periodic table and recall that the proton is 1836 times as massive as the electron.)

Answer: 1.67×10^{-24} g; 9.11×10^{-28} g.

2. (a) What is the weight in grams of 1 mole of oxygen? (b) In a certain reaction 24 g of oxygen is produced; how many liters will this gas occupy at STP? (c) How many atoms will be in this volume?
3. (a) What is the weight of 1 mole of ammonia gas NH_3 at STP? (b) What is the weight of one ammonia molecule in amu? in grams?

Answer: (a) 17.031 g; (b) 17.031 amu, 2.83×10^{-23} g.

4. (a) What would be the weight of 1 liter of carbon dioxide gas CO_2 at STP? (b) How many molecules would be found in this volume?
5. A 10-liter volume of gas at STP weighs 25.9 g. (a) What is the GMW of this gas? (b) What is the weight in amu of one molecule of this gas?

Answer: (a) 58.02 g; (b) 58.02 amu.

6. You have a quart of water, a quart of ethyl alcohol, and a quart of carbon tetrachloride; the formulas for these compounds are H_2O, C_2H_5OH, and CCl_4, respectively. (a) From the formulas, which liquid would you expect to be heaviest? the lightest? (b) In the fashion given in Table 7.2 the specific gravity of these liquids is 1.00, 0.79, and 1.595. Compare the order given here with your results above and state your conclusion.
7. A sample of powder was found to contain 2.16 g of sodium, 3.33 g of chlorine, and 4.51 g of oxygen. (a) What is its formula? (b) How many liters would the chlorine and oxygen occupy at STP?

Answer: (a) $NaClO_3$; (b) 0.11 liter Cl_2, 3.16 liters O_2.

8. (a) What is the percentage composition by weight of potassium, chromium, and oxygen in potassium chromate, $K_2Cr_2O_7$? (b) How many grams of each element are there in a 50-g sample? (c) How many liters would the oxygen occupy under STP?
9. The formula for sugar is $C_{12}H_{22}O_{11}$. By heating the sugar, water (as represented by $11H_2O$ in $H_{22}O_{11}$) can be removed and pure carbon left behind. (a) What is the percentage of carbon in the sugar? (b) Assuming none is lost as carbon dioxide, how many grams of pure carbon can be prepared this way from 1000 g of sugar?

Answer: (a) 42.1%; (b) 421 g.

10. A sample of methyl cellosolve contains 4.737 g of carbon, 1.059 g of hydrogen, and 4.204 g of oxygen. What is the simplest formula for this compound?
11. (a) What is the percentage of iron in ferric oxide Fe_2O_3? (b) How many grams of iron would a 500-g sample yield? (c) How many liters of O_2 at STP would be produced?

Answer: (a) 69.94%; (b) 349.70 g; (c) 105.20 liters.

12. In Prob. 5 above the gas is known to consist of 82.75 percent carbon and 17.25 percent hydrogen. What is its formula?
13. In heavy water, deuterium (an isotope of hydrogen with an atomic weight of 2) is joined with oxygen. (a) What is the molecular weight of this heavy water? (b) What is the percentage of each element present in the compound? (c) Compare these figures with

ordinary water. (d) Comment on the statement; "A compound usually obeys the law of definite composition . . ." on p. 308.

Answer: (a) 20; (b) 25% H, 75% O.

Supplementary readings

Textbooks

Garrett, Alfred B., et al., *Chemistry, A Study of Matter,* Xerox College Publishing Company, Lexington, Mass. (1972). [Several chapters containing material covering approximately the same range of topics as in this chapter.]

Keenan, Charles W., and Jesse H. Wood, *General College Chemistry,* Harper & Row, New York (1971). [Chapter 5 covers much of the material contained in this chapter. Pertinent sections are on formulas and equations, calculation of formulas from data, and writing equations.]

Other Books

Rosenberg, Jerome L. (ed.), *College Chemistry* (Schaum's Outline Series), McGraw-Hill, New York (1966). (Paperback.) [There are 375 solved problems, 750 supplementary problems. A useful aid in solving problems in this chapter. Scattered throughout are also examples of problems encountered in the chemistry chapters that follow.]

CHEMICAL BONDS

We know more about the atom than we know about ourselves, and the consequences of this gap are everywhere to be seen.

CARL KAYSEN, 1972

Director of The Institute for Advanced Study, Princeton

With the increase of knowledge about the electron structure of atoms, physicists and chemists have expanded their studies of how atoms of one kind join with those of another kind; of how atoms interact by way of their outer electrons.

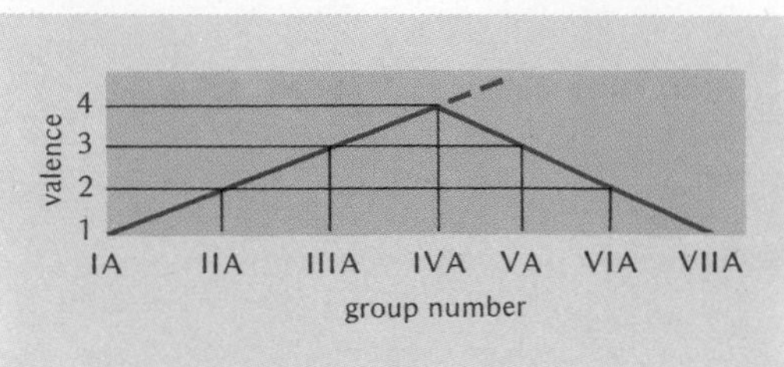

Figure 16.1
Correlation of valence with position in the periodic table.

Chemical bonds and valence

MEANING OF CHEMICAL BONDS

The forces of attraction that hold atoms together within a molecule are called *chemical bonds*. A *molecule* is the uncharged particle that results from the union of two or more atoms. The uncharged particle may be either a diatomic gas or a compound. If the atoms are two or more different kinds of atoms, we speak of the "molecules of a compound." We recall, as part of an earlier definition, that a *compound* is a new substance that results when two or more elements combine.

VALENCE

From looking at the formulas of compounds such as water (H_2O), hydrogen chloride (HCl), and carbon dioxide (CO_2), we realize that different atoms have different capacities for combining with each other. The *valence*, or combining capacity, of an element has been defined for decades as the number of hydrogen atoms that one atom of a given element will combine with or replace in forming compounds. Thus oxygen in water (H_2O) has a valence of 2 because its atom combines with two atoms of hydrogen; chlorine (Cl) in hydrogen chloride (HCl) has a valence of 1.

The early definition of valence has given way to more modern ideas about chemical reactions based on the electron structure of atoms, but valence numbers are still calculated on the basis of formulas of compounds as in the preceding paragraph.

VALENCE AND THE PERIODIC TABLE

In the main, those elements in Groups IA, IIA, IIIA, IVA, VA, VIA, and VIIA show a rather regular gradation in changes of valence (Fig. 16.1). Elements in Group IA have a valence of 1, IIA of 2, IIIA of 3, IVA of 4 usually, VA of 3 usually, VIA of 2 usually, VIIA of 1. The valence of the noble gases, Group VIIIA, is said to be zero.

USE OF VALENCE IN PREDICTING CORRECT FORMULAS

From the formula of water one readily deduces mathematically that the value for the valence of oxygen is 2. Oxygen is ordinarily listed in Group VIA in the periodic table. This is in agreement with the graph of Fig. 16.1. Lithium (symbol for the atom is Li), being in Group IA, is expected to have a valence of 1, and the formula of lithium oxide is expected to be Li_2O, which is correct. The student should check to see if formulas for other oxides of these period 2 (second row) elements would not be

BeO, B_2O_3, CO_2, N_2O_3, H_2O, F_2O

This way of calculating formulas does not invariably lead to a correct result; for example, there is another compound of hydrogen and oxygen, hydrogen peroxide H_2O_2, which is not predicted by this device.

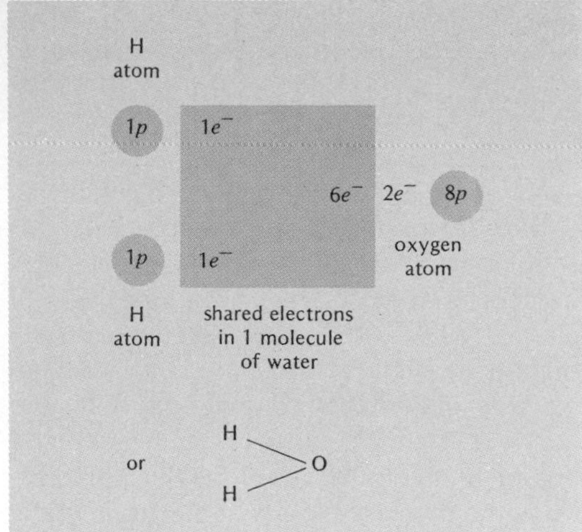

Figure 16.2
Sharing of electrons. A stable compound may be formed by the sharing of electrons. There is an overlap of orbitals.

Again, from similar reasoning one deduces that these period 2 elements form the following hydrogen compounds:

LiH, BeH_2, BH_3, CH_4, NH_3, OH_2, FH

(BH_3 does not exist. The last two formulas are usually reversed so as to be written H_2O and HF.)

ELECTRON CONFIGURATION OF NOBLE GASES

The noble gases, Group VIIIA, except for helium, have eight electrons in the highest main energy level which is the number of electrons needed to fill the *s* and *p* orbitals of that level. Helium has only two electrons, and they completely fill the first main level. Each noble gas atom, except helium, has its highest main energy level completely filled with eight electrons in the *s* and *p* orbitals as shown below,

s	*p*		
↓↑	↓↑	↓↑	↓↑

The electron configuration of the neon atom, for example, is Ne $1s^2 2s^2 2p^6$, with the outermost main energy level out from the helium core inner level $1s^2$. There is a total of eight electrons in the outermost main energy level of neon: two 2*s* electrons and six 2*p* electrons.

The electron structure of atoms and chemical reactions

VALENCE ELECTRONS

The most important structural feature of atoms in determining chemical reactions is the number of electrons in the outermost energy levels of the atoms. Electrons in the outermost energy levels of atoms are called *valence electrons*. In chemical reactions there are always some changes in the distribution of the valence electrons.

TRANSFER AND SHARING OF ELECTRONS

In the formation of compounds atoms of certain elements tend to gain electrons whereas others tend to lose electrons. In general, when a metal element combines with a nonmetal element, electrons are "lost" by atoms of the metal and "gained" by atoms of the nonmetal. Hydrogen is usually placed in Group IA of the periodic table and it acts like a metal in many chemical reactions—that is, like the metals below it in the same group. The hydrogen atom has only one electron. Two hydrogen atoms have two electrons. The oxygen atom with only eight electrons has two electrons in its inner shell and six in its outer shell. The six electrons in the outermost shell of the oxygen atom are attracted to its nucleus with greater force of attraction than is the electron in the outer shell of the hydrogen atom attracted to its nucleus, owing to the greater positive charge on the nucleus of the oxygen atom (8+) as compared to the single 1+ charge of the hydrogen nucleus. Under the proper conditions, the outer shell of the oxygen atom becomes filled with eight electrons by gaining one electron from each of the two hydrogen atoms (Fig. 16.2), and water is formed in the process. It is said that the hydrogen atom tends to "lose" electrons and the oxygen atom tends to "gain" electrons. Actually, neither loss nor gain occurs; the two kinds of atoms share the electrons in their outermost shells. Because there is no tendency for the hydrogen oxide, or water, to lose or gain additional electrons, the compound is said to be stable; it remains H_2O.

Chemical change can now be defined as

The loss, the gain, or the sharing of electrons in the outer electron shells of two or more atoms until a stable condition is reached.

In the foregoing example the chemical change occurred as a result of the sharing of electrons.[1]

Stability of the products of a chemical change is attained (1) in molecules by the sharing of electrons of two or more atoms, and (2) as ions by the actual transfer of outer-shell electrons from the atoms of one element to those of another. *Compounds* are the new substances that result when two or more elements combine by sharing or transferring electrons. The majority of compounds are formed when electrons are shared between different kinds of atoms. But what occurs when electrons are not shared by two atoms but are actually transferred from one atom to another?

When electrons are actually transferred from one atom to another, ions are formed. Some of the ions will be positive, others will be negative. Each kind of atom entering into the chemical reaction ends with a net electric charge and hence is an ion. You recall that an *ion* was defined in the preceding chapter as an atom or a small group of atoms with a net electric charge. Some oppositely charged ions may form ionic compounds such as Na^+Cl^-, discussed below.

The electrons with the highest principal quantum number—that is, those in the highest main energy level—are usually lost first in forming an ion. Atoms of elements in Group IA of the periodic table, such as sodium (Na), easily lose the one electron in the outermost shell. Chlorine (Cl) of Group VIIA, with seven electrons in the outermost shell of an atom, can gain an electron easily; it is certainly easier for the atom of chlorine to gain one electron to complete the outer shell with eight electrons than to lose the seven electrons in the outermost shell. It is generally found that elements with the same number of valence electrons are in the same vertical column of the periodic table.

When a sodium atom loses an electron to a chlorine atom, they combine to form sodium chloride (NaCl), common table salt. The reaction between the sodium and chlorine atoms can be written

$$\underset{\text{sodium atom}}{Na} + \underset{\text{chlorine atom}}{Cl} \longrightarrow \underset{\text{sodium chloride consisting of sodium ion, chloride ion}}{Na^+Cl^-}$$

The modern theory is that sodium chloride consists of positive and negative ions arranged in an ordered manner in a crystal. Each positive ion is surrounded by negative ions; each negative ion by positive ions. Further discussion of the stability of sodium chloride appears later in this chapter.

The reaction between the sodium and chlorine atoms may also be indicated as in Fig. 16.3. In this reaction when a metal element combines with a nonmetal, electrons are lost by atoms of the metal and gained by atoms of the nonmetal.

IONIC CHARGE

A chlorine atom (Cl) will combine with one atom of hydrogen (H) to form the compound hydrogen chloride (HCl). The chlorine atom will also react with one atom of sodium to form sodium chloride. Thus the valence of chlorine is 1, in terms of our earlier definition of valence. In sodium chloride the chloride ion has a valence of 1 in terms of the

[1] In an introductory textbook it is easier to present molecules in terms of the interpenetration of orbitals. No attempt is made to describe recent developments in the theories of molecular orbitals.

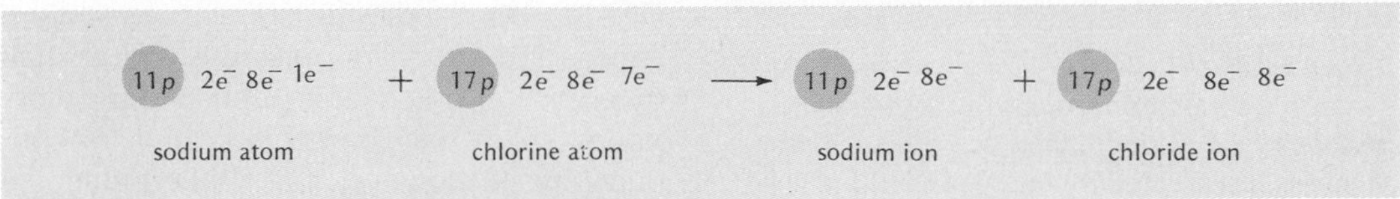

11p 2e⁻ 8e⁻ 1e⁻ + 17p 2e⁻ 8e⁻ 7e⁻ ⟶ 11p 2e⁻ 8e⁻ + 17p 2e⁻ 8e⁻ 8e⁻

sodium atom — chlorine atom — sodium ion — chloride ion

Figure 16.3
Transfer of electrons. Sodium and chlorine atoms react by the transfer of electrons to yield ions.

definition, but the ion has a net charge of −1. The ionic charge of the chloride ion is −1 and is indicated by Cl^-. The oxide ion has a net negative charge of −2. The valence of oxygen is 2; the ionic charge (sometimes designated electrovalence) of the oxide ion is −2 and may be indicated by O^{--} or $O^=$ or O^{2-}. The kind (positive or negative) and amount of net charge on an atom or group of atoms is called the *ionic charge* (sometimes designated as *electrovalence*). The ionic charge of positive ions such as those of sodium, calcium, and aluminum would be indicated, respectively, Na^+, Ca^{2+}, Al^{3+}.

Ionic (electrovalent) compounds, in which ions having opposite electric charges are attracted to each other by strong electrostatic forces, are represented by such compounds in the solid state as sodium chloride (Na^+Cl^-), lithium fluoride (Li^+F^-), magnesium oxide ($Mg^{2+}O^{2-}$), magnesium chloride [$Mg^{2+}(Cl^-)_2$], and calcium oxide ($Ca^{2+}O^{2-}$).

OXIDATION NUMBERS OR OXIDATION-STATE NUMBERS

A system of small whole numbers related to the combining ratios of elements, and called *oxidation numbers,* is useful in chemistry. The term "valence" has been defined in many ways in the past and the word has been used in several different ways such as in "electrovalence." Confusion has occurred. As a result there is an expanding tendency to forego the terms "valence," "electrovalence," etc., for a new system that consists of a set of whole numbers obtained by arbitrary rules and called "oxidation numbers" or "oxidation-state numbers." They can be used to predict the formulas of compounds after we have considered such concepts as ionization potential, electron affinity, and electronegativity. We shall defer further discussion of oxidation numbers and oxidation-state numbers to the next chapter.

ELECTRON DOT SYSTEM OF NOTATION

When it was realized that only the electrons in the outermost shells of atoms enter into chemical reactions, a simplified representation called the electron dot system was devised to show chemical reactions. The *kernel* of an atom consists of its nucleus and all inner-shell electrons. The kernel is indicated by the symbol for the element.

The *electron dot system* consists of the kernel and dots arranged singly or in pairs on the four sides of the symbol for the kernel to represent the valence electrons, those in the outermost shell. Thus dots are used only to represent electrons in the highest main energy levels. Two dots between atoms represent one pair of shared electrons; four dots represent two pairs of shared electrons; and six dots represent three pairs.

Electron dot notation of atoms in the first three periods of the periodic table is shown in Fig. 16.4. Except for helium, all members

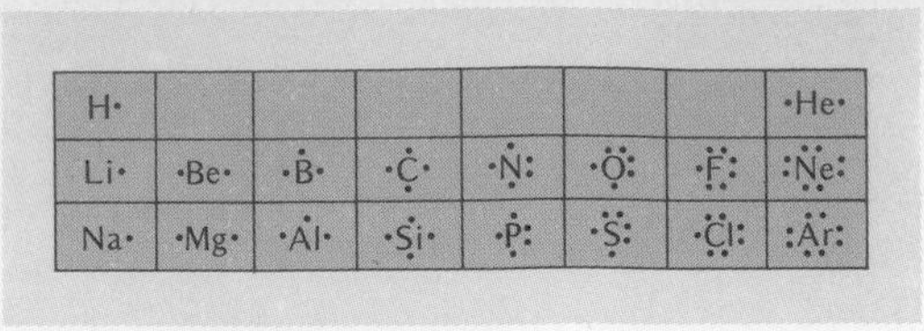

H·							·He·
Li·	·Be·	·B·	·C·	·N:	·O:	·F:	:Ne:
Na·	·Mg·	·Al·	·Si·	·P:	·S:	·Cl:	:Ar:

Figure 16.4
Electron dot notation. Atoms of the first three periods of the periodic table are shown. The kernel is represented by the symbol of the element. The dots represent the number of electrons in the outer shell.

of the same family shown in the figure have the same number of dots.

The electron dot formulas for water (H_2O), methane (CH_4), a part of natural gas, carbon tetrachloride (CCl_4), a cleaning agent, and the ionic compound sodium chloride (Na^+Cl^-) are written as follows:

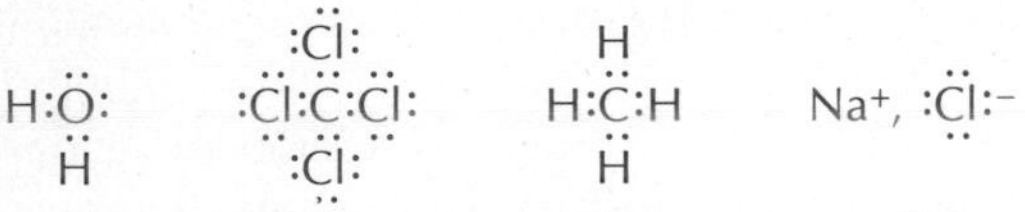

H:O: H	:Cl: :Cl:C:Cl: :Cl:	H H:C:H H	Na^+, :Cl:$^-$
water	carbon tetrachloride	methane	sodium chloride

The comma between the sodium ion and the chloride ion indicates that the ions are separate particles formed by the transfer of electrons rather than sharing.

Ionization energy, ionization potentials, electron affinity, and electronegativity

IONIZATION ENERGY AND IONIZATION POTENTIAL

Ionization energy is

The energy, expressed in electron volts, required to remove the most loosely bound electron from an atom in the gaseous phase.

If the removal is made by electrical means, the required voltage is called the *ionization potential*. For an electron the ionization energy in electron volts and the ionization potential in volts are numerically equal.

As an example, if a cathode ray tube (see Fig. 12.3) is filled with gas (such as sodium vapor) at a rather low pressure and a small difference of potential is applied to the electrodes, there may be no indication of conductivity. If, however, the applied potential is gradually increased, there will eventually come a sudden large increase in conductivity. This minimum potential drop at which there is a marked and abrupt rise in conductivity is called the *ionization potential*. The conductivity is thought to be due to the pulling or driving off of electrons from the individual sodium atoms by the electric field.

$$\underset{\text{sodium atom}}{Na} \xrightarrow{\text{electric field}} \underset{\text{sodium ion}}{Na^+} + \underset{\text{electron}}{e^-}$$

or

$$Na \cdot - e^- \rightarrow Na^+$$

Once ionization has taken place, the ions and electrons attract each other; in fact most of the sodium ions are quickly neutralized by the electrons falling back to their original positions. In the process, energy is given off, most of it appearing as the familiar yellow of sodium light (color plate facing p. 208). Like sodium, the other alkali metals (lithium, potassium, etc.) in Group IA of the periodic table have such small ionization potentials that when electrons fall back to neutralize their ions, visible light is produced.

In general the ionization potentials of the Group IIA elements (such as magnesium and calcium) are greater than those of Group IA elements. This tendency (with minor exceptions) for the ionization potentials to increase as one goes from left to right in the periodic table results in maximum ionization potentials for the noble gas family: helium, neon, etc. (see inside front cover and Fig. 16.5).

The ionization potential for the most loosely bound electron is called the *first* ionization potential; for the second most loosely bound electron, it is called the *second* ionization potential, and so on. Therefore complex atoms may have several ionization potentials.

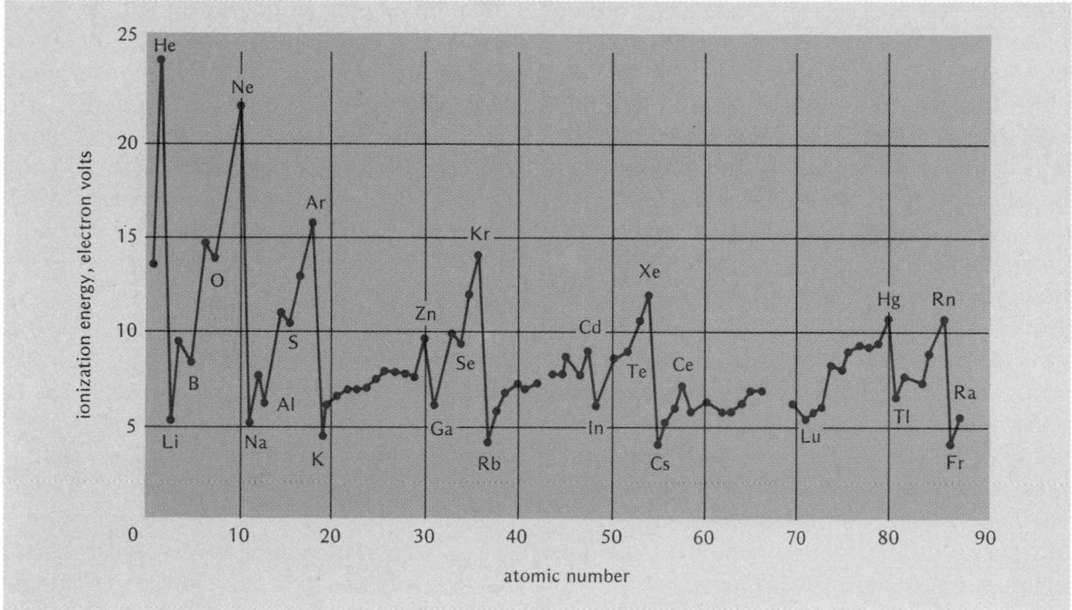

Figure 16.5
First ionization potential plotted against atomic numbers.

REASONS FOR THE MAXIMA AND MINIMA IN IONIZATION POTENTIALS

There are two chief reasons for the periodicity of ionization potentials: (1) decrease in size of atoms as one goes from left to right in the periodic table, and (2) increase in atomic number (and consequent increase in charge on the nucleus). Both increase in atomic number and decrease in size should be expected to increase ionization potential, and one could demonstrate from data that this is true.

However, in going from an inert gas such as helium to an alkali metal such as lithium there is a conflict. The increase in charge on the nucleus, in this case from two to three, should increase the ionization potential; the increase in size of the atom should lower the ionization potential. Actually the ionization potential of lithium is much lower than that of helium; atom size is a more important factor than charge on the nucleus. The most important reason for this is that in Coulomb's law, effects of size (that is, distance between charges) is raised to the second power, whereas effects of charge occurs only as its first power.

We should pause to remind ourselves that Coulomb's law describing forces of attraction and repulsion between electric charges, which can be measured in our ordinary macroworld, holds true precisely in our "common sense" world. But it would be a mistake to expect it to hold true with the same precision in the submicroscopic world of atoms. It can help us to understand and can even be used to predict, but we should think of it only as an approximation when applying it over extremely short distances.

ENERGY CHANGE DURING ION FORMATION

Obviously work is done when an electron is removed from the ordinary structure—that is, from the so-called ground state of an atom. This is because a force must be exerted through a distance to overcome the attractive force between the positive nucleus and the negative electron. If one regards the atom as a system (a region that is actually or supposedly isolated), then energy must go into the system in order to make the process take place. Because this describes an endothermic process, we can say it is a good rule that the formation of a positive ion will probably be a highly endothermic process. However, the formation of a negative ion from an electron and an atom in its ground state, which at first might appear to be the opposite kind of process, is also exothermic but only very slightly so. This should signify that these somewhat idealized "ground states," which we picture by means of models, are arrangements of nuclei and electrons in atoms representing almost minimum energies and therefore maximum stabilities. They are not, however, the ultimate in stability in the actual situation under which such ions are found.

The idealization comes about chiefly through the assumption that each atom is isolated—that is, far from any neighboring atom. Such a condition rarely exists in the real world. When another atom approaches, it is often possible for a structural rearrangement to occur that will give still lower energies and greater stabilities. The example below will clarify what is meant.

ELECTRON TRANSFER AND STABILITY

Consider a sodium atom in sodium gas (gas being the only state of matter in which comparative isolation is possible) near a chlorine atom. A possible rearrangement has already been suggested: Each of the atoms can become ions. The removal of one electron from the outer energy level of the sodium atom is decidedly endothermic, the addition of one electron to the outer energy level of the chlorine atom only slightly exothermic. So the overall process is endothermic; in other words, the products appear to contain more potential (chemical) energy than the original reactants. Common sense indicates that such a reaction should not occur, because minimum energies (the condition for greater stability) have not been attained. But sodium chloride is one of the commonest of substances and it seems to be quite stable. What we have ignored is that we stopped with gaseous sodium chloride. There is very great attraction between the positive sodium ion and the negative chloride ion. In the drawing together of the ions, the first result is the condensation to liquid sodium chloride. This is definitely exothermic; and the most exothermic process still remains—the solidification (crystallization) of the liquid. The high melting point (804°C) of solid sodium chloride is a good indication of how great the attraction is between sodium ions and chloride ions in a sodium chloride crystal and how much energy we would therefore have to supply to the system to overcome the attractive forces and melt the solid. So the overall process by which solid sodium chloride is formed from gaseous sodium and gaseous chlorine is highly exothermic; increased stability has been achieved through electron transfer aided by other effects.

ELECTRON AFFINITY

Electron affinity is usually defined as

The energy released when a neutral atom gains one electron.

It is a quantity that can be measured experimentally and is useful in predicting the properties and reactions of a given element. Electron affinity is high for nonmetals, the elements that usually gain electrons in chemical reactions. Ionization potential and electron affinity are important in determining electronegativity.

ELECTRONEGATIVITY

The capacity of an atom in a molecule to attract shared electrons is called *electronega-*

tivity. In general the higher the ionization potential of an element, the greater is its electronegativity. The higher the value of electronegativity, the greater is the atom's attraction for electrons.

A value of 4.0 on a widely used electronegativity scale has been assigned arbitrarily to the elements in Groups I–VIIA of the periodic table with the greatest tendency to gain electrons. Nitrogen and chlorine have electronegativity values of 3.0; oxygen, 3.5; and fluorine, 4.0; so fluorine would be the most electronegative element in Groups I–VIIA of the periodic table and hence the most chemically active of those elements. Cesium, with an assigned value of 0.7, is the least electronegative of the naturally occurring elements. An electronegativity scale of values is useful for predicting which bonds are ionic and which are covalent. Some common kinds of bonds formed by two or more atoms are to be discussed next.

Types of bonding

In connection with compounds we discuss the covalent bond and the ionic bond as general types of bonding.

THE COVALENT BOND

A covalent bond consists of a pair of electrons shared between two atoms. In Fig. 16.2 and also below, we observe that one short solid line, a dash, is used to represent one shared pair of electrons. Two shared pairs of electrons are called a double bond and are represented by =; three pairs are a triple bond, ≡. Compounds whose atoms are joined only by covalent bonds, by the electron-sharing process, are called *covalent compounds*. It is common practice to use a short solid line, or dash, called a bond to represent a pair of shared electrons when only the electrons that are shared are indicated.

```
            Cl               H
            |                |
H—O     Cl—C—Cl         H—C—H
  |         |                |
  H        Cl                H
water     carbon          methane
H2O    tetrachloride        gas
           CCl4             CH4
```

water H_2O; carbon tetrachloride CCl_4; methane gas CH_4

The number of covalent bonds is equal to the number of shared electron pairs.

Earlier we showed the electron dot formulas of the compounds water, carbon tetrachloride, and methane. Above we show the same compounds with a dash representing each pair of shared electrons. Formulas using dashes are often referred to as the stick, or structural, formula of a compound, in contrast to the electron dot formula.

The stick formula of ethylene gas, C_2H_4, contains a double bond and is:

```
   H  H
   |  |
H—C=C—H
```

ethylene

Ethylene gas is used by some processors to change the greenish-yellow skin of some tree-ripened oranges to an orange color.

The stick formula of acetylene gas, C_2H_2, which is used for welding, is thought to contain a triple bond and would be:

H—C≡C—H

acetylene

THE IONIC BOND

An *ionic bond* is

The electrostatic attraction that holds positive and negative ions together in pairs or groups.

This attraction that binds unlike ions together is sometimes called an electrovalent bond. As mentioned earlier, modern theory is that sodium chloride, table salt, consists of crystals composed of sodium ions and chloride ions attracted to each other, with each positive ion surrounded by negative ions and each negative ion by positive ions. Oxygen, for example, tends to form ionic compounds with metals; with nonmetals, it forms covalent compounds.

Table 16.1
Some representative sizes of ions (in angstroms)

Li^{+}		B^{3+}	O^{2-}	F^{-}
0.60		0.20	1.40	1.36
Na^{+}	Mg^{2+}	Al^{3+}	S^{2-}	Cl^{-}
0.95	0.65	0.50	1.84	1.81
K^{+}	Ca^{2+}			Br^{-}
1.33	0.99			1.95

Electron dot formulas can also be used to represent ionic compounds. Magnesium chloride, $MgCl_2$, can be written as Mg^{2+}, $:\ddot{\underset{..}{Cl}}:^-$, $:\ddot{\underset{..}{Cl}}:^-$, the commas indicating that the ions are separate particles formed by the transfer of electrons rather than sharing.

The principal structural difference between covalent and ionic compounds is that covalent compounds usually exist as discrete molecules, each containing a definite number of atoms, whereas ionic compounds occur with large numbers of individually, separately existing, positive and negative ions.

Further applications of bonding

COMPOUNDS LIKELY TO BE IONIC IN CHARACTER

The forces holding two different kinds of atoms together are forces between positive and negative ions. It is a good rule to say that negative ions are comparatively large ions (see Table 16.1, the last two columns). The net negative charge on such an ion causes an outward shift of the electrons in the various energy levels; thus the ion swells and is larger than the atom from which it was formed. By a somewhat similar line of reasoning we can conclude that positive ions are small, in fact smaller than the atoms from which they are formed. Although the exact sizes of atoms and ions are not known, it has been possible to determine accurately the distance between centers of atoms. By doing this for many substances, tables have been prepared that give close approximations of the sizes of many atoms and ions. The data of Table 16.1 illustrate the shift in size of some representative positive ions as one progresses through the periodic table. The table, arranged in an order similar to the arrangement of atomic symbols in the periodic table, shows directly under each symbolic representation of an ion its approximate radius (in angstroms, or times 10^{-8} cm) in crystals.

For a binary (two-element) compound to be saltlike and stable, its constituent ions should not differ too much in size. Examples are sodium chloride and potassium chloride. Both of these compounds dissolve in water to give solutions that are excellent conductors of electricity, but there are no signs of chemical change when they dissolve, and both have high melting points. On the other hand, when one dissolves magnesium chloride in water, a careful examination shows that the solution has become appreciably acidic, which indicates that at least some chemical change has occurred. Ions of Mg^{2+} and Cl^- are very different in size and dissociate easily.

Another class of compounds that seems to be ionic, even when pure, is the class called strong bases. Examples are sodium hydroxide, potassium hydroxide, and calcium hydroxide (see Chap. 17).

A third class of compounds, not quite as definitely ionic when pure but behaving as if ionic when in aqueous solution, is the strong acids. These are excellent conductors of electricity and undergo rapid but nonviolent, or nonexplosive, chemical changes. Examples of these are our ordinary acids of the elementary laboratory: hydrochloric, sulfuric, and nitric acids. In fact, it is probably because strong acids, strong bases, and salts are so common in elementary laboratories that students are prone to regard them as typical chemicals. More abundant substances, such as water, cellulose, starch, sugars, and hydrocarbons, are often almost ignored until advanced chemistry courses, which are not taken by most students. For further discussion of acids see Chap. 17.

MOLECULAR SUBSTANCES AND THE COVALENT BOND

A simple *covalent bond* is a pair of electrons shared between two atoms. Electrons are shared equally only if the two atoms are the same kind, such as H_2, Cl_2, and F_2. If the atoms are different, such as in HCl, the pair of electrons is attracted more toward one of the two atoms, thus making one part of a molecule partially negative and the other part partially positive. Such a pair of electrons constitutes a *polar bond*. A molecule such as the HCl molecule is called a polar molecule. It acts to some extent as though the chlorine part were negative and the hydrogen part positive.

The best examples of purely covalent substances are not compounds but elements. When two hydrogen atoms meet under proper conditions, such as a certain range of temperature, their orbitals overlap to some extent, and the two electrons arrange themselves so that their spins are opposed. The result is a linear hydrogen molecule, with interatomic distances considerably less than twice the radius of an ordinary hydrogen atom. Now the probability of finding the electrons *between* the nuclei is much greater than it was before molecule formation. The concentration of electron density between the two nuclei constitutes what is now called a covalent bond; this particular kind of covalent bond is known as a *sigma* (σ) *bond*. The center of positive charge and center of negative charge are both exactly in the center of the molecule on a line that joins the two positively charged nuclei. In this special case of bonding of two atoms of the same kind, the resulting molecule is nonpolar. Covalent bonding with a polar molecule resulting is discussed below. It may at first seem strange to consider the combination of like hydrogen atoms to form molecules as a chemical change, but why not so regard it? This reaction is exothermic; that is, energy leaves or "goes out from" the system during the reaction. Atomic hydrogen has much use in welding, etc., where a clean, nonoxidizing substance is needed for the purpose of producing a quite high temperature in a small region.

In a similar way any two halogen atoms can react with each other to form a molecule:

Cl	+	Cl	→	Cl_2
chlorine atom		chlorine atom		chlorine molecule

Such nonpolar molecules are gases unless their molecular weights are quite high; they seldom dissolve in water unless they react chemically with the water. When chlorine gas is liquefied by cooling and compression, the liquid is not a good conductor of electricity.

COMPOUNDS THAT ARE COVALENT BUT POLAR

When a hydrogen atom meets a chlorine atom, a rearrangement or reorganization of electrons can also give a new structure with lower net energy and therefore greater stability. The $1s^1$ electron of the hydrogen atom pairs its spin with the single $3p$ electron, forming what is called a hybrid orbital (hybrid because formed from two different kinds, or species). The resultant hydrogen chloride molecule (formula, HCl) is linear and has its atoms attached to each other by a σ bond. The reaction is exothermic,

H	+	Cl	→	HCl
hydrogen atom		chlorine atom		hydrogen chloride molecule

Hydrogen fluoride (HF), hydrogen bromide (HBr), and hydrogen iodide (HI) can be formed in an analogous way. This is not to say that they are necessarily prepared this way industrially (although hydrogen chloride often is), but it indicates the kind of phenomenon represented by the term "bond formation."

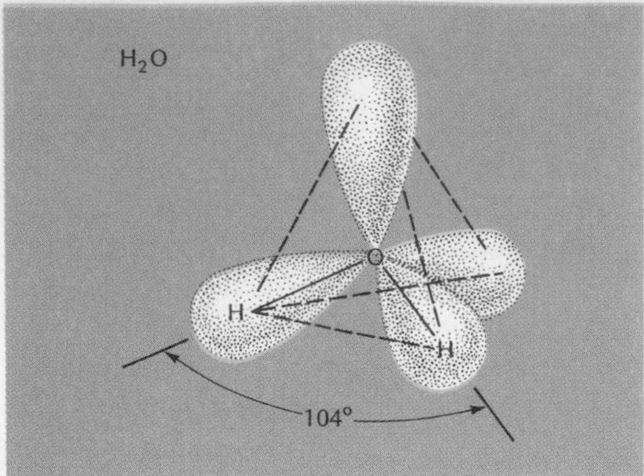

Figure 16.6
Configuration of a water molecule.

When hydrogen chloride dissolves in a substance which itself is polar and/or has a high dielectric constant, ions are formed from the molecules. This is usually termed "electrolytic dissociation." Hydrogen fluoride, hydrogen chloride, hydrogen bromide, and hydrogen iodide are all gaseous covalent compounds but are so polar that they readily form ions in appropriate solvents. See Chap. 17 for the Brønsted theory, which explains the process further.

STRUCTURES OF OXYGEN AND WATER MOLECULES

We have already seen that the oxygen molecule has the formula O_2 and water has the formula H_2O. Earlier in this chapter it was indicated that an oxygen atom in its ground state has two unpaired electrons, but these, of course, are in different orbitals. Evidence indicates that when two oxygen atoms combine to form an oxygen molecule, both of these unpaired electrons are involved in the bonding. As indicated previously, we often represent bonds by means of either a straight line (dash) or a pair of dots representing a pair of shared electrons which have their spins opposed. Examples are H—H or H:H for a hydrogen molecule and H—Cl or H:Cl for a hydrogen chloride molecule. So now let us represent an oxygen molecule in this way.

O═O or $:\dot{\underset{\cdot}{O}}::\dot{\underset{\cdot}{O}}:$

In the dot formulas the symbols represent everything in the atoms *except* electrons in the highest energy level. The four electrons between the two oxygen nuclei constitute a double bond as contrasted with the single bonds in hydrogen and hydrogen chloride. Studies of atom and molecule sizes indicate that there is an even greater overlap in the case of the double bond than in a single bond. For example, the centers of the oxygen atoms are nearer to each other in an oxygen molecule than in hydrogen peroxide, which has all single bonds as represented by the formula H—O—O—H or H_2O_2. A double bond in chemistry is a somewhat shorter and stronger bond than a single bond (though not necessarily twice as strong); two electrons form a single bond but four electrons form a double bond.

We could write the formula of water graphically thus: H—O—H, but it would perhaps be misleading because the compound is more polar than such a linear structure would indicate. Referring to the configuration of *p* orbitals one might expect that

```
O—H
|
H
```

would be a better spatial representation for the molecule, because here we are showing the hydrogen–oxygen–hydrogen atom angle as 90° rather than 180°. This is near the truth, for the angle is actually 104° (Fig. 16.6).

THE CONFIGURATION OF A NITROGEN MOLECULE

We have previously stated that the ground-state structure of a nitrogen atom is $1s^22s^22p^3$. Accordingly, with three more *p* electrons needed to complete the octet, none of the three 2*p* electrons are paired (that is, arranged so that their spins are opposed, following Hund's rule). In the nitrogen molecule they become paired, and the region of increased electron density between the nuclei is called a triple bond. The resultant molecule, sometimes represented as N≡N, is nonpolar. The theories of structure presented here in a brief and somewhat oversimplified form show how theory can be used to correlate and explain observed facts. The length of the linear nitrogen molecule, as observed, agrees with a prediction that a triple bond would represent a great deal of overlap and call for a great deal of contraction when formed. The formation of atomic nitrogen from ordinary gaseous molecular nitrogen only occurs at high temperature and is indeed endothermic, as one would expect. Perhaps this helps to explain why "free" (uncombined) nitrogen is so abundant in the atmosphere: 11.47 lb above every square inch of the earth's surface. And yet when the farmer wants a fertilizer containing a nitrogen compound, he must pay several cents per pound for it.

REVIEW OF PROPERTIES OF METALS AND NONMETALS; ELECTRON AFFINITY

It has been noted that in going from left to right through any series in the periodic table there is an increase in ionization potential. Of the elements whose properties are well known, helium has the highest first ionization potential and cesium the lowest. The large atoms of the metals do not need to have so much energy expended on them in order for an outer electron to be extracted from them; that is, they form positive ions with ease.

On the other hand the halogens, and the other active nonmetals to a lesser extent, form negative ions with ease. "Electron affinity" is the term used to express this quantitatively. Sometimes arbitrary units are used to give comparative values; other authors express this quantity in electron volts. No matter how it is expressed, fluorine and chlorine always have about the highest values. One would expect that values for electron affinity would run approximately parallel to those for ionization potential. If we ignore one very marked exception, this is true; the exception consists of the noble gas family, which has a set of extremely high ionization potentials but whose electron affinities are very low.

BORDERLINE CASES: METALLOIDS

As one goes from left to right in any period in the periodic table, the elements form smaller and smaller positive ions and the charges on the ions become larger and larger (see Table 16.1). Therefore the binding forces between ions and electrons become greater and greater. One should expect, therefore, that those properties which are characteristic of the active metals would become less and less pronounced. Evidence that this is a good generalization consists of the following: (1) Magnesium metal is harder than sodium metal and is also a poorer conductor of electricity; (2) the melting point of magnesium is higher than the melting point of sodium; (3) magnesium hydroxide is a much weaker base than sodium hydroxide.

It is impossible to draw a sharp line between metals and nonmetals. A few elements, such as carbon and silicon, have some properties like those of metals and others like those of nonmetals. The graphite form of carbon, for example, is a fairly good conductor of heat and electricity, like the metals. But carbon also forms a weak acid, as one would expect of a nonmetal. Such borderline elements are often called *metalloids*. A base or acid derived from such an element is usually very weak.

THE ELECTRICAL CONDUCTIVITY OF MATERIALS

When a sufficiently large voltage is applied, any sample of matter will conduct electricity to some extent; yet the specific resistivity of one substance may be immensely higher—perhaps 10^{30} times greater—than that of another. Materials having extremely high resistance to the flow of electricity are called *insulators*.

The best conductors of electricity are the metals. These conduct best if they are very pure and are crystalline solids. Their resistivities decrease as their temperatures decrease. Their conductivities (the reciprocals of their specific resistivities) sometimes become so large at temperatures approaching absolute zero that the materials are then said to be superconductors. A given metal becomes an appreciably poorer conductor when it is melted.

The best insulators are covalent substances. Unlike metals, their resistivities decrease with rise in temperature.

When an aqueous solution of the compound hydrogen chloride conducts an electric current, chemical change occurs at each of the two electrodes (see Fig. 10.10). Such conductivity, which is accompanied by chemical change, is called electrolytic conductance. On the other hand, when a bar or wire that is composed of metal conducts electricity, there is no resultant chemical change. As electrons enter at the negative terminal, equal numbers of electrons emerge at the positive terminal. The type of conductance involving movement only of electrons is called *metallic conductance.*

Electrolytic conductance is accompanied by chemical change and always involves the movement of ions through a fluid, usually a liquid. Although the resistivity of an electrolytic conductor is extremely small compared to that of an insulator, it is appreciably greater than that of most metals.

SUMMARY

A molecule of a compound results from the union of two or more atoms. The forces of attraction that hold atoms together within a molecule are called *chemical bonds*. The most important structural feature of atoms in determining chemical reactions is the number of electrons in the outermost energy level of the atoms.

In the formation of compounds, atoms of certain elements tend to gain electrons whereas others tend to lose electrons. *Compounds* are the new substances that result when two or more elements combine by sharing or transferring electrons. Stability of molecules is attained by the sharing of electrons, and stability of ions and ionic compounds is attained by the transfer of outer-shell electrons. When electrons are actually transferred from one atom to another, ions are formed.

When it was realized that only the electrons in the outermost shells of atoms are involved in chemical reactions, the electron dot system was devised to show the reactions. Two dots between the kernels of two atoms represent one pair of shared electrons. Formulas using dashes are often referred to as the stick, or structural, formulas of a compound in contrast to the electron dot formula.

The covalent bond and the ionic bond were discussed as general types of chemical bonds. A *covalent bond* consists of a pair of electrons shared between two atoms. An *ionic bond* is the electrostatic attraction that holds positive and negative ions together in pairs or groups. Covalent compounds usually exist as discrete molecules; ionic compounds occur as crystals with large numbers of positive and negative ions.

The applications of bonding are as numerous as the formation of the thousands of chemical compounds, the few diatomic molecules, and the ionic compounds such as sodium chloride.

Important words and terms

chemical bonds
molecule
valence
valence electrons
transfer of electrons
sharing of electrons
chemical change
compounds
ion
ionic charge
electrovalence
oxidation numbers
electron dot system
of notation or formula
kernel
ionization energy
ionization potential
electron affinity
electronegativity
stick, or structural, formula
covalent bond
covalent compound

ionic bond
polar bond
σ bond
electrolytic conductance
metallic conductance

Questions and exercises

1. (a) What is meant by chemical bonds? (b) What sort of force is involved in chemical bonding?
2. (a) What is meant by valence? (b) Compare the basis of the old definition and the new.
3. How is valence related to group number in the periodic table? Illustrate this graphically.
4. Determining valence in terms of the combining capacity of the hydrogen atom, write the formulas for the period 3 elements.
5. (a) From information in the periodic table write out the electron configuration for the krypton atom. (b) What is the designation for the electrons in the outermost level?
6. (a) What is the most important structural feature in determining the chemical properties of an element? (b) What are these units of the atom called? (c) After examining their electron configurations in the periodic table, what can you say of the chemical properties of the period 3 elements versus those in the lanthanoid series?
7. Hydrogen, which belongs to the "metal" class in the periodic table, tends to "lose" electrons; oxygen, which belongs to the "nonmetal" class, true to form, tends to gain electrons. In a reaction between the two do either really gain or lose? Explain.
8. (a) The modern definition of chemical change is given as the ________, ________, or ________ of electrons in ________. (b) In Chap. 14 a chemical change was defined as ________.
9. In a chemical change stability of the final reactant products can be attained in what two ways?
10. What is the new definition of a compound given in this chapter? Compare it with those given in Chaps. 12 and 14 (see Question 14.11).
11. (a) In a reaction between hydrogen and oxygen the chemical change comes about as a result of the ________ of electrons; in the reaction between sodium and chlorine the chemical change comes about as a result of the ________ of electrons. (b) When the chemical change involves the transfer of electrons, one atom or group of atoms becomes positive by ________; the other becomes negative by ________.
12. Distinguish between valence and ionic charge.
13. (a) What fundamental fact led to the development of the electron dot system of notation? (b) What are the two basic parts that make up the electron dot notation for an element?
14. Write the electron dot formula for elements 33 to 36.
15. For a given electron in an atom what is meant by (a) ionization energy? (b) ionization potential? (c) Having these concepts and the principle of the conservation of energy discuss the cause of the emission of light in neon signs.
16. For just the beginning and end pairs of elements from period 1 (H–He) to period 6 (Cs–Rn) draw a greatly simplified saw-toothed graph for the first ionization potentials like that in Fig. 16.5. Also show in this graph the decreasing ionization potential for the succeeding noble gases and the nearly uniform baseline for the IA elements.
17. What are the two main reasons for the periodicity of ionization potentials?
18. In the formation of sodium and chlorine ions by electron transfer it appears that the reaction to form the final products was endothermic, and minimum energy (the condition for maximum stability) was not achieved. Yet how do we account for the great stability of sodium chloride?
19. (a) What is meant by electron affinity? (b) How is it measured? (c) Which group of elements of the periodic table has high electron affinities?
20. (a) Contrast the definitions of ionization potential and electron affinity; (b) In what direction (right or left) does each increase in the periodic table? (c) Into what single term are these two concepts combined?
21. (a) What are the limits of the electronegativity scale and by which element of the periodic table are they represented? (b) What are the characteristics of elements that are highly electronegative? (c) What characteristic has a compound formed from one highly electronegative element with another only slightly electronegative?
22. What are two types of bonding connected with compounds?

23. (a) Define a covalent bond. (b) Using water as an example, draw the formula, using both the electron dot notation and the stick formula.
24. (a) Define the ionic bond. (b) Give another name for this bond. (c) Illustrate with the electron dot formula the ionic bond for Al_2O_3. (d) What is the principal structural difference between covalent and ionic compounds?
25. (a) When an atom takes on one or more electrons to become a negative ion, how does the size of the ion compare with that of the original neutral atom? (b) Answer the similar question for positive ions.
26. What three classes of compounds are ionic in character?
27. (a) What is meant by a polar bond? A σ (sigma) bond? (b) Give an example of a σ bond that is polar and another that is nonpolar, with an explanation of how the charge is distributed in each case.
28. When only two atoms combine to form a compound, we have a __________ compound; if the compound exists as discrete particles, the bonding is of the __________ type. If the positive and negative charges are located on a straight line, then we have a __________ bond. If the charges are asymmetrically located on this line, we at the same time have a __________ bond; otherwise the symmetrical distribution along this line gives us a __________ bond.
29. (a) In a column write the formulas for the oxygen molecule, water, and hydrogen peroxide. In a right column show the corresponding "stick formula" for each. (b) In terms of the strength of bonds how do you account for the ease with which hydrogen peroxide reverts to water?
30. (a) In a column write the formulas for the hydrogen, oxygen, and nitrogen molecules. In a right column show the corresponding "stick formula" for each. (b) From examination of the bonds what can you say about the size of the resulting molecules and the energy required to dissociate them?
31. Why is fertilizer nitrogen so expensive when free nitrogen is so abundant in the atmosphere?
32. The element having the highest ionization potential is __________ and the lowest is __________; from these facts we would expect __________ to be the largest atom and __________ the smallest.
33. (a) The values for both the ionization potential and electron affinity increase in what same direction in the periodic table? (b) What group of elements represents an exception and in what way does it differ?
34. (a) What do we call elements that exhibit both metallic and nonmetallic properties? (b) What are their characteristics with respect to their atomic size, binding forces, and valence? (c) In what way are these characteristics reflected in the gross behavior of these elements?
35. (a) What classes of elements are the best and the poorest conductors of electricity? (b) How does temperature affect the conductivity in each case?
36. From your knowledge of the nature of and the strength of the bond between covalent substances and the relation of heat to the motion of atomic matter, venture a guess as to why a rise in temperature is accompanied by an increase in electrical conductivity.
37. (a) If the conduction of electricity is accompanied by a chemical change in the neighborhood of the ends of the conducting medium, the phenomenon is known as __________ conduction; if no chemical change is involved, it is known as __________ conduction. (b) In one phenomenon the moving charges are __________; in the other the moving charges are __________.

Supplementary readings

Textbooks

Keenan, Charles W., and Jesse H. Wood, *General College Chemistry*, 4th ed., Harper & Row, New York (1971). [Pages 62–83 cover approximately the same range of topics as this chapter but with somewhat more emphasis on chemistry than on physics.]

Ouellette, Robert J., *Introductory Chemistry*, Harper & Row, New York (1970). [Chapter 6 discusses bonding and electronegativity.]

Quagliano, James V., and L. Vallarino, *Chemistry*, Prentice-Hall, Englewood Cliffs, N. J. (1969). [Separate chapters deal with such topics as bonding, ion and atom sizes, and valence and oxidation numbers.]

CHEMICAL REACTIONS

Balanced equations are required by nature, not by chemistry teachers.

GORDON ATKINSON AND HENRY HEIKKINEN, 1973

Matter can be changed by chemical reactions. Observations of common experiences testify to such changes. The use of food by the human body and animals and plants, the burning of fuels, the souring of milk, and the rusting of iron all involve chemical changes.

In this chapter we consider two major classes of chemical reactions in terms of atomic and molecular theory, give examples of four general types of chemical reactions, discuss factors affecting chemical reaction rates, explore the concept of the energy needed to initiate a chemical reaction, and study some of the things that can happen to a chemical system.

Two major classes of chemical reactions appropriate for study are oxidation–reduction reactions and acid–base reactions. The former class involves the transfer of one or more electrons between pairs of substances. The latter class involves the ideas about acids and bases presented by G. N. Lewis in 1923 and developed more fully during the following 15 years. Lewis suggested a concept of acid–base reactions involving unshared pairs of electrons. Chemical reactions usually involve solutions, which we shall now consider.

Solutions

Solutions are homogeneous mixtures, as mentioned in Chap. 14. The most common solutions are liquids; others are solid or gaseous. Metal alloys, such as used in coins, are solid solutions; air is a gaseous solution. Most chemical reactions probably involve liquid solutions. Before consideration of chemical reactions, we shall consider some terms and concepts dealing with solutions.

DESCRIBING TRUE SOLUTIONS

Because a solution is a mixture, it always contains at least two general types of components. The *solvent* is the substance that is in excess in a solution; it is the dissolving agent. The *solute* is the substance that is dissolved in a solvent. A water solution is called an *aqueous solution*. A *dilute solution* is one in which there is a relatively small amount of solute. Conversely, a *concentrated solution* contains a relatively large amount of solute.

If we place a large amount of sugar in a small amount of water and leave them for hours, with an occasional shaking to ensure mixing, we eventually get a saturated sugar solution. It is *saturated* if the solvent has taken up all the solute that it can dissolve at that temperature and pressure. If much less solute had been used, however, in the same volume of solvent, an unsaturated solution would probably have resulted. An *unsaturated solution* can always dissolve more solute at that temperature and pressure.

The amount of solute present in a given amount of saturated solution is called the *solubility* of the substance. The solubility of a solid solute is almost always increased by increasing the temperature. A great deal more sugar, a solid, can be dissolved in a cup of hot tea than in the same volume of iced tea.

The solubility of a gas in liquid is increased by pressure. The increase in pressure to force a gas into a liquid may result in chemical change. For example, in the production of carbonated soft drinks, carbon dioxide gas (CO_2) under high pressure is forced into a water solution with the following result:

$$H_2O + CO_2 \rightarrow H_2CO_3$$

Carbonic acid (H_2CO_3), a weak acid, is formed. The bottle is capped; the pressure remains. When the bottle is opened, the pressure decreases, and at room temperature and atmospheric pressure the carbonic acid is unstable and decomposes into water and carbon dioxide gas:

$$H_2CO_3 \rightarrow H_2O + CO_2 \uparrow$$

The arrow upward indicates that the carbon dioxide gas goes into the air as "fizz." If one leaves carbonated beverage in an uncapped bottle and returns considerably later to drink it, he may find that the "fizz" is gone and the drink tastes like ordinary water, which most of it is.

CONCENTRATIONS OF SOLUTIONS

The *concentration* of a solution refers to the weight or volume of the solute in a specified amount of the solution or the solvent. The relative amounts of solute and solvent, or concentration of solution, may be expressed in different ways: molarity, molality, percentage, or normality.[1]

The molar concentration or *molarity* (M), of a solution is the number of moles of solute per liter of solution. A 0.2 M solution of HCl contains two-tenths of a mole of HCl in every liter of solution. In general,

$$\text{molarity of solution} = \frac{\text{moles of solute in sample}}{\text{liters of solution}}$$

In order to make a 1 M solution of sodium chloride, for example, we dissolve 1 mole of NaCl (22.99 g + 35.45 g = 58.44 g of NaCl) in enough water so that the final volume of the solution is 1 liter. Molarities permit us to determine the amounts of solute by measuring a volume of solution in which they are dissolved. However, it is difficult to measure volumes very accurately because changes in temperature result in changes in volume.

Molality (m) of a solution is the number of moles of solute dissolved per kilogram of solvent. This is convenient to use and is very accurate; measurements are by weight, not volume.

The concentrations of solutions may be expressed as *percentages*—that is, percent by weight or percent by volume. A 10-percent aqueous NaCl solution contains 10 percent by weight of sodium chloride and 90 percent by weight of water. The percentage stated refers to the solute, 10 percent in this case. Unless percent by volume is stated, the term "percent" means percent by weight. A wine that is 12 percent by volume of alcohol has 12 ml of alcohol per 100 ml of wine. Volumes of liquids are not additive if one liquid disperses within another; 89 ml of water must be added to 12 ml of alcohol to make 100 ml of solution.

Concentrations, especially of acids and bases, may be expressed in terms of normality. The *normality* (N) of a solution is the number of equivalent weights of solute per liter of solution. A 1 N solution contains one equivalent weight per liter of solution. The equivalent weight of an acid is the gram-molecular weight of the acid (1 mole) divided by the number of replaceable hydrogen atoms in the formula. For example, the gram-molecular weight (GMW), or 1 mole of sulfuric acid (formula H_2SO_4) is $(2 \times 1) + (1 \times 32) + (4 \times 16) = 98$ g. There are two hydrogen atoms in the formula. Hence equivalent weight = 1 mole per number of hydrogen atoms = 98 g/2 = 49 g. Therefore a 1 M solution of H_2SO_4 (contains 98 g of H_2SO_4) would be a 2 N solution. In order to make a 1 N solution of sulfuric acid we would dissolve 49 g of acid per liter of solution.

The situation is similar for bases. The equivalent weight of a base is the gram-molecular weight (1 mole) of the base divided by the number of hydroxides (OH) in the formula. For sodium hydroxide (NaOH), $\text{GMW}/1 = (23 + 16 + 1)/1 = 40/1 = 40$, the equivalent weight. Therefore molarity equals normality for NaOH. A 1 M solution of NaOH (containing 40 g of NaOH) would be a 1 N solution. In order to make a 1 N solution of sodium hydroxide, we would dissolve 40 g of the base per liter of solution.

Before discussing oxidation–reduction reactions as a major type of chemical reactions, it is important to consider the meaning of the term "oxidation number" which was devised to indicate the state of oxidation of substances.

[1] These units are used in more advanced texts than this, so examples in the form of problems are not given here.

Oxidation numbers

As mentioned in the preceding chapter, confusion resulting from the use of the word "valence" as a part of other words has led to *a system* of small whole numbers related to the combining ratios of elements in compounds and in polyatomic ions, or radicals, and called *oxidation numbers*. The oxidation state of an ion is, for simple ions, the charge on the ion expressed as the oxidation number of the ion. Oxidation state numbers do not imply charges. For simple ions, the oxidation number is equal to the charge on the ion, the "electrovalence." However, the Arabic number and kind of charge of the ion, and the assigned plus (+) or minus (−) sign and Arabic number of the oxidation number are designated differently, as shown following rule 5 below. Certain elements have more than one oxidation number because they share or transfer their electrons in more than one way. Such elements form compounds to which different names must be given. The terms "oxidation state," "oxidation state number," and "oxidation number" are often used interchangeably.

RULES FOR ASSIGNING OXIDATION NUMBERS

Some general rules basic to the system of arbitrary assignment of oxidation numbers are as follows:

1. The oxidation number of an uncombined element is zero.
2. In a compound the more electronegative elements are assigned negative oxidation numbers and the less electronegative elements are assigned positive oxidation numbers.
3. Hydrogen has an oxidation number of +1 and oxygen has an oxidation number of −2 in most compounds and ions, when they are present. These numbers can be used in the determination of the oxidation numbers of other elements in compounds and of ions.
4. In the formula for a compound the sum of the positive oxidation numbers equals numerically the sum of the negative oxidation numbers, and the algebraic sum of the oxidation numbers of all atoms in a neutral molecule must be zero.
5. If the substance under consideration is an ion, rather than a molecule, the algebraic sum of the oxidation numbers of the atoms in the ion must equal the net charge on the ion.

In using oxidation numbers to predict the formula of a compound, one chooses the simplest ratio of atoms or ions that makes the plus and minus oxidation numbers add to zero. For example,

$$\overset{+1-1}{NaCl} \qquad \overset{+4(-2)_2}{CO_2} \qquad \overset{-2(+1)_3-1}{CH_3Cl}$$

As we see above, the oxidation number is designated by placing the number directly above the symbol, with a plus or minus sign in front of the number to indicate the electronegativity. To show the magnitude of the charge on an ion, the number is written as a superscript above and to the right of the symbol, except that the number 1 is omitted. As examples,

oxidation number: $\overset{-1}{Cl}$, $\overset{-2}{O}$, $\overset{+3}{Al}$
charge on ion: Cl^-, O^{2-}, Al^{3+}

For polyatomic ions a net oxidation number can be assigned to the entire ion; for example, the sulfate ion, SO_4^{2-}. In this case the ionic charge, or electrovalence, of 2− for the entire sulfate ion is a result of the net sum of the oxidation numbers of the sulfur (S) and oxygen (O) atoms. In polyatomic ions the term "ionic charge" or "electrovalence" is reserved for the entire radical—in this case, for the entire SO_4^{2-} radical, not for the sulfur and oxygen atoms. In a *radical* the elements behave as one unit that seldom breaks up. Sometimes radicals are ions (for example, SO_4^{2-} ions); sometimes radicals are parts of molecules (the hydroxyl radical OH in ethyl alcohol, C_2H_5OH, for example).

A parenthesis is used around radicals only when more than one radical is present in a formula; for example, one molecule of aluminum hydroxide, $Al(OH)_3$, contains three hydroxyl radicals. A subscript number used with a parenthesis multiplies everything within it.

In the SO_4^{2-} radical the charge of the sulfur ion is 6+, and of the oxygen ion, 2− (but not the 2− shown for the entire radical). In this case the oxidation number of sulfur is +6, and of oxygen, −2. The oxidation numbers of the elements in the radical are used as follows to calculate the oxidation number of the entire radical:

$1S \times +6 = +6,$	the oxidation number of S
$O_4 \times -2 = \underline{-8,}$	the total oxidation state of O_4
$-2,$	the oxidation number $\overset{-2}{SO_4}$ of the entire radical

It is often more convenient to use Arabic numbers, instead of the customary Roman numerals, to designate oxidation states.

Example 17.1

Calculate the oxidation state of sulfur in hydrogen sulfate (H_2SO_4).

Solution

The total oxidation state for H_2 is	$2 \times (+1) = +2$
The total oxidation state for O_4 is	$4 \times (-2) = -8$
The sum of +2 and −8 is −6	
The oxidation state of S in H_2SO_4 must be	$S = +6$
The algebraic sum of the oxidation states of all atoms in a neutral molecule (H_2SO_4) is	$\overline{0}$

Example 17.2

Calculate the oxidation number of chlorine in potassium chlorate ($KClO_3$).

Solution

The ionic charge of the potassium ion is 1+; hence the oxidation number for 1K is	$K = +1$
The total oxidation state for O_3 is	$O_3 = 3 \times (-2) = -6$
The sum of +1 and −6 is −5	
Therefore, the oxidation number of Cl in $KClO_3$ must be	$Cl = +5$
in order to give an algebraic sum equal to	$\overline{0}$

The formulas of *electrovalent compounds* (those in which one or more bonds are primarily due to the attraction between ions of unlike charge) can be predicted from the oxidation numbers of their elements; such formulas ordinarily indicate only the simplest ratio of ions. For an electrovalent compound, the oxidation number of an ion of that compound is the same as the charge of the ion.

In the preceding chapter it was stated that (1) covalent compounds usually exist as discrete molecules; (2) a covalent bond consists of a pair of electrons shared between two atoms; and (3) compounds whose atoms are joined by covalent bonds are called "covalent compounds." The formulas of covalent molecules show the actual number of atoms that have combined to form a molecule (for example, CO, CO_2, CCl_4); in contrast, an ionic formula as ordinarily used indicates only the simplest ratio of ions (for example, Na^+Cl^-).

It is usually not possible to use oxidation states as a basis for the prediction of the formulas of covalent compounds. In such compounds the oxidation number of an atom does not necessarily correspond to the number of covalent bonds joining the atom to other atoms. The foregoing will be seen in the examples that follow.

Carbon monoxide, for example, with the molecular formula CO contains carbon with an oxidation number +2, yet the electron dot formula for CO requires for each carbon atom a triple bond (one bond is one pair of dots between two atoms),

:C:::O:

Carbon dioxide with the molecular formula CO_2 contains carbon with an oxidation number +4, and the electron dot formula for CO_2 requires two sets of double bonds, a total of four bonds for the carbon atom,

$\ddot{\underset{..}{O}}::C::\ddot{\underset{..}{O}}$

Ethylene gas with the molecular formula C_2H_4 indicates carbon with an oxidation number −2, yet each carbon atom has four covalent bonds,

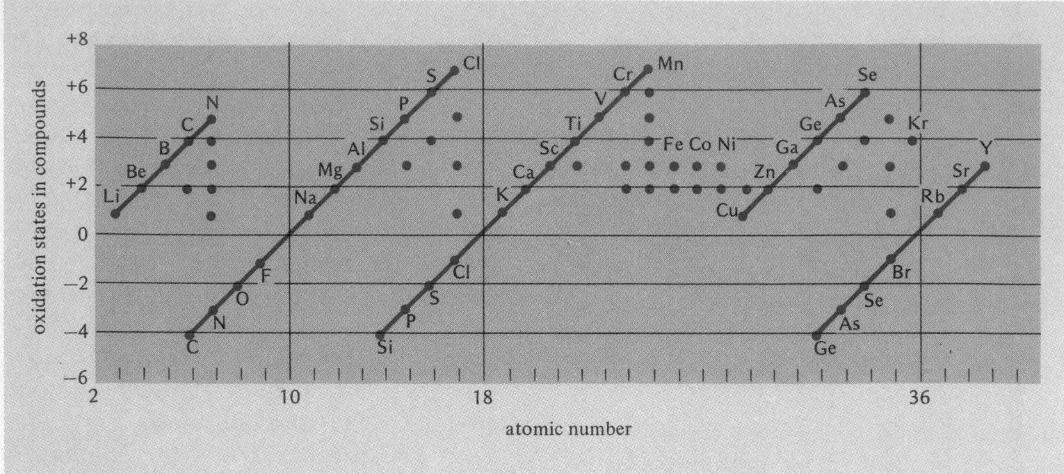

Figure 17.1
Oxidation states, a periodic function of the atomic numbers. The colored lines emphasize the periodicity. An element, such as carbon, may have more than one oxidation state, a fact not evident from the position in the periodic table. (Used with permission from Charles W. Keenan and Jesse H. Wood, *General College Chemistry*, 4th ed., Harper and Row, 1971.)

```
H—C=C—H
   |   |
   H   H
```

Acetylene gas with the molecular formula C_2H_2 appears by examination of the formula to contain carbon with an oxidation number of −1. It will soon be shown that this is impossible. The structural formula of acetylene is thought to be

H—C≡C—H

and each carbon atom has four covalent bonds.

The oxidation states of carbon in the four examples above appear to have been, respectively, +2, +4, −2, and −1; and a carbon atom has had, in these cases, three, four, four, and four bonds, respectively. Is there some way to account for an atom of one element having more than one oxidation number?

In accord with the periodic law, Fig. 17.1 shows oxidation states to be a periodic function of atomic numbers. Common oxidation states for selected elements are shown in the figure.

Oxidation numbers that we can safely predict from the Group A families in the periodic table are

group	I	II	III	IV	V	VI	VII
oxidation number	+1	+2	+3	+2, +4	+5	+6	+7
				−4	−3	−2	−1

THE MAXIMUM AND MINIMUM OXIDATION NUMBERS FOR A GIVEN ELEMENT

In Groups IA, IIA, and IIIA of the periodic table the valence numbers and oxidation numbers are usually identical, except that whereas the valence number has no sign, the oxidation number is positive (+). This enables us to write BaF_2 and BaO for the formulas of barium fluoride and barium oxide, respectively. Also in Group IVA, in which the most probable valence is 4, the oxidation number to be expected is +4 or −4. The algebraic difference between +4 and −4 is 8 (see Fig. 17.2).

Although there is much variation in oxidation number for any particular nonmetal

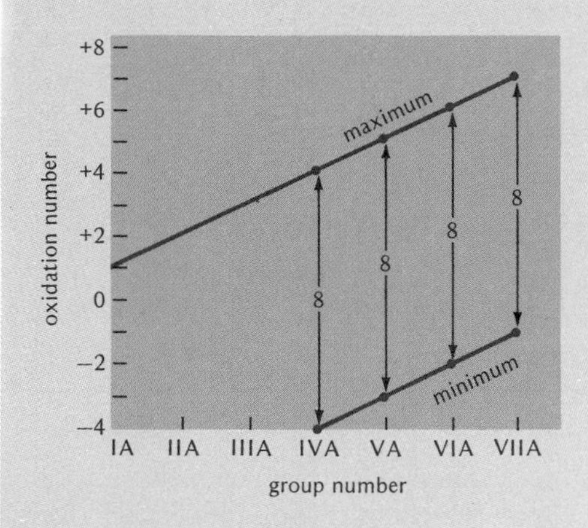

Figure 17.1
Graphic representation of maximum and minimum oxidation numbers. The difference is eight in each case.

in Groups IVA to VIIA, inclusive, one can readily see from Fig. 17.2 the limits within which the variation will take place. Moreover, one can say that whenever there is a plentiful supply of a good oxidizing agent (a substance consisting of molecules that are good electron-withdrawing agents), the reaction will occur in such a way that the nonmetal in the resulting compound will take on its maximum oxidation number. For example, free nitrogen or the nitrogen in nitrogen compounds can be oxidized under favorable conditions to nitrogen pentoxide (formula N_2O_5) in which the nitrogen has oxidation number +5. Similarly, free sulfur can be oxidized to sulfur trioxide (formula SO_3) in which the oxidation number of the sulfur atom is +6. Although chlorine heptoxide is neither very common nor very stable, the compound perchloric acid (formula $HClO_4$) exhibits the maximum oxidation number of chlorine that one might rationally expect, +7.

On the other hand, when these nonmetals react with metals or hydrogen—that is, with elements that are poor electron-withdrawing agents (low electronegativities)—the resulting compound has the nonmetal exhibiting its minimum oxidation number. For example, the very common compound ammonia (formula NH_3) exhibits the minimum oxidation number of nitrogen.

Oxidation–reduction reactions

TERMINOLOGY

Before the electron theory of chemical bonding was developed, oxidation was considered to be the reaction of oxygen with another substance to form a new compound or compounds. According to modern atomic theory, the loss of electrons by an atom or ion is called *oxidation,* and the oxidation state of the atom or ion is increased. The gain of electrons by an atom or ion is called *reduction,* and the oxidation state of the atom or ion is decreased. An oxidation reaction cannot occur without the simultaneous occurrence of a reduction reaction. When an atom or ion gives up electrons, the atom or ion is *oxidized;* when an atom or ion gains electrons, the atom or ion is *reduced.* In oxidation–reduction reactions[2] one or more electrons are transferred between pairs of substances. Oxygen need not be involved, but it may be.

A complete transfer of electrons between pairs of substances is not necessary for changes in oxidation states. A pair of electrons being shared by two different kinds of atoms may be shifted more toward one of the atoms instead of being shared equally. When sulfur burns to form sulfur dioxide (SO_2), for example, atoms of sulfur and oxygen unite by sharing pairs of electrons. Each pair of electrons is thought to be held closer to oxygen than to sulfur. Each oxygen atom is as-

[2] Oxidation–reduction reactions have been called "redox reactions" because redox reactions are generally thought of as involving a transfer of electrons, but recent studies have shown that the mechanism is often more complicated; in fact, the mechanisms of most redox reactions are unknown. Therefore the term "redox" is not used further in this book.

Table 17.1
Summary of terminology of oxidation and reduction

term	electron change	oxidation number change
oxidation	loss of electrons	increase
reduction	gain of electrons	decrease
substance oxidized	loses electrons	increase
substance reduced	gains electrons	decrease
oxidizing agent	accepts electrons	decrease
reducing agent	donates electrons	increase

signed an oxidation number of −2, resulting in an oxidation number of +4 for the sulfur atom. The sulfur is said to be oxidized because the atom gave up electrons and its oxidation state or number changed from 0 to +4. The oxygen gained electrons and was reduced, its oxidation state changing from 0 to −2.

Any atom, molecule, or ion that brings about a complete or partial transfer of an electron away from another atom, molecule, or ion in a chemical change is called an *oxidizing agent*. The substance that causes a second substance to lose one or more of its electrons is an *oxidizing agent*. A reactant that has an electron completely or partially pulled away from it in a chemical change is called a *reducing agent*. A substance that supplies electrons to a second substance in a chemical change is a *reducing agent*. The reducing agent donates electrons; the oxidizing agent accepts electrons. In the above reaction of sulfur and oxygen, oxygen was the oxidizing agent and sulfur was the reducing agent.

The terminology of oxidation and reduction is summarized in Table 17.1.

AN EXAMPLE OF AN OXIDATION–REDUCTION REACTION

When a strip of zinc (Zn) is immersed in a solution of copper sulfate ($CuSO_4$), a spontaneous reaction occurs; metallic copper plates out on the zinc strip and the zinc strip is dissolved slowly with the liberation of heat. The chemical equation for the overall reaction is

$$Zn + CuSO_4 \rightarrow ZnSO_4 + Cu$$

Each atom of zinc loses two electrons and becomes a zinc ion, as below. Each ion of copper gains two electrons and becomes a copper atom, as below; thus metallic copper plates out on the zinc strip. We have

oxidation: $Zn \rightarrow Zn^{2+} + 2e^-$
reduction: $2e^- + Cu^{2+} \rightarrow Cu$

The electrons are transferred directly from atoms to ions and no flow of electrons can be detected.

PRODUCTION OF ELECTRICITY BY CHEMICAL REACTIONS

A flow of electrons can be detected by an instrument to measure current if the above procedure is modified as follows to make one type of voltaic cell. Immerse a strip of zinc in a solution of zinc sulfate ($ZnSO_4$) as in Fig. 17.3, jar A. Immerse a strip of copper in a solution of copper sulfate ($CuSO_4$) as in jar B. Use copper wire to connect the zinc strip externally to the copper strip with an instrument placed in series in the wire circuit to measure current. The zinc sulfate solution is connected to the copper sulfate solution by a salt bridge filled with a solution of the electrolyte, potassium sulfate. The oxidation of zinc and the reduction of copper occur as

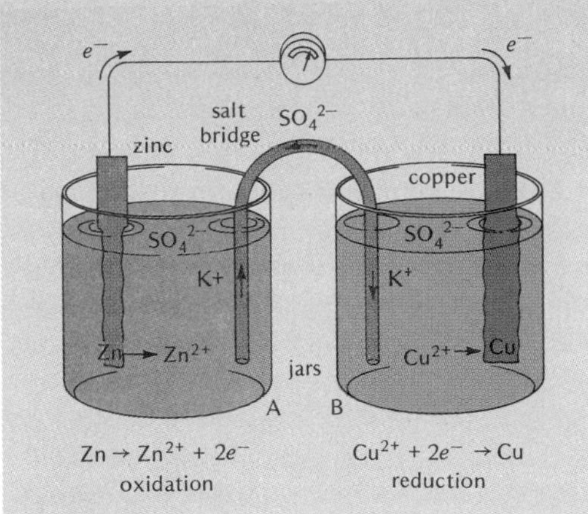

Figure 17.3
Diagram of one type of a voltaic cell. (Used with permission from Charles W. Keenan and Jesse H. Wood, *General College Chemistry*, 4th ed., Harper and Row, 1971.)

shown in the figure. For jar A, valence electrons flow from the zinc atoms of the strip into the conducting copper wire, and as the Zn^{2+} ions form they enter the solution—the zinc strip as shown is disappearing. The electrons given up by the zinc atoms (oxidation) enter the connecting wire and cause electrons at the other end of the wire to collect on the copper strip (electrode). These electrons react with copper ions to form copper atoms (reduction) that plate out on the copper electrode. The apparatus and chemicals are referred to as a "voltaic cell." It is recalled that two or more cells connected in series constitute a battery.

The flow of electrons through the external circuit constitutes an electric current. Similar oxidation–reduction reactions occur in flashlight batteries and automobile batteries to produce an electric current—a transfer of electrons through an external circuit.

ACTIVITY SERIES OF METALS

How can we find out whether one metal is more active or less active than another? Let us prepare hydrogen by using different metals to displace hydrogen from water or from an acid at room temperature.

Potassium reacts with cold water very rapidly:

potassium + water →
hydrogen ↑ + potassium hydroxide

This reaction is dangerous and hard to control. Seldom is an attempt made to collect the hydrogen being liberated when potassium reacts with water. The vertical arrow is used to indicate that hydrogen is given off as a gas.

An interesting but not very practical way to prepare hydrogen is by adding small pieces of sodium to water in a large glass beaker at room temperature:

sodium + water →
hydrogen ↑ + sodium hydroxide

The reaction of sodium with water is less rapid than that of potassium with water.

Calcium also liberates hydrogen from cold water, but this chemical change is fairly slow at room temperature:

calcium + water →
hydrogen ↑ + calcium hydroxide

Small pieces of calcium metal turnings placed in a test tube half filled with water may be passed around in class rapidly for close observation of the bubbles of gas given off.

Probably the best laboratory method of preparing hydrogen involves the reaction between an acid and one of the moderately active metals. Zinc reacts with hydrochloric acid to yield hydrogen and zinc chloride:

zinc + hydrochloric acid →
hydrogen ↑ + zinc chloride

The hydrogen gas thus generated may be collected in a bottle.

It has been noted that potassium reacts very rapidly with water, sodium somewhat less rapidly, and calcium still less rapidly. An activity series of the metals can be set up in which the metals are listed in the approximate order of their decreasing activity. Although hydrogen itself is not a metal, it is usually placed in the series because it reacts chemically like a metal and also because those metals below hydrogen in the activity

Table 17.2
Activity series of some common metals and hydrogen

element	symbol
potassium	K
sodium	Na
calcium	Ca
magnesium	Mg
aluminum	Al
zinc	Zn
chromium	Cr
iron	Fe
cadmium	Cd
nickel	Ni
tin	Sn
lead	Pb
hydrogen	H
copper	Cu
mercury	Hg
silver	Ag
platinum	Pt
gold	Au

series will ordinarily not liberate hydrogen from any hydrogen compound.

The order in this chemical activity series of metals, not exactly the same as an electromotive force series, is not absolutely fixed. A change of conditions may reverse the order of two metals adjacent to each other in the series. A partial activity series of metals in the approximate order of decreasing chemical activity is given in Table 17.2.

From evidence obtained by experimentation, certain general rules can be deduced and stated for the activity series of metals.

1. *Metals above hydrogen in the activity series will displace hydrogen from water and acids.*

For example, sodium will rapidly displace hydrogen from water, and zinc will displace hydrogen from hydrochloric acid fairly rapidly.

2. *Metals below hydrogen in the activity series will ordinarily not liberate hydrogen from any hydrogen compound.*

There is very little or no displacement of hydrogen from water by copper, which is below hydrogen in the series. For this reason, copper water pipes are used in many homes.

3. *The farther a metal is above hydrogen in the activity series, the more rapidly the metal will displace hydrogen from water and acids.*

For example, sodium can be observed to displace hydrogen from water much more rapidly than does calcium. Note that lead is close to hydrogen in the series. The reaction between lead and water or acids, both of which compounds contain hydrogen, is so negligible that lead traps are used in the drainage lines under sinks in chemistry laboratories, where acids are frequently washed away with water.

Most of the metals below hydrogen in the activity series are either found free in nature or are very easily liberated from naturally occurring compounds. It should not be surprising to learn therefore that they were among the first metals to be used in making ornaments and tools. They were prized so highly in early civilizations that they came to be referred to as "noble metals," and the term is still used to some extent. Conversely, the metals above hydrogen have sometimes been referred to as "base" metals.

In answer to the question asked in introducing the topic of activity series, if metal A replaces metal B when A is placed in an ionic solution of a compound of metal B, then A is more active than B.

THE ACTIVITY SERIES AND CORROSION

Those metals that are low in the activity series generally do not corrode as rapidly as those that are higher. Gold scarcely corrodes at all. Copper, which is below hydrogen in the series and will not ordinarily liberate hydrogen from hydrogen compounds, lasts for many years as water pipes, roofs, gutters, and screens. Also the green-colored copper compound that is formed on copper roofs seems to cover them so completely that negligible or no chemical action occurs.

Iron usually corrodes very badly unless kept covered so as to keep oxygen and moisture completely away from it. Zinc really corrodes in the atmosphere, but the zinc oxide that is formed covers the free metal so completely and tenaciously that the action soon ceases. Therefore, water pipes, pails, metal roofing, gutters, nails, and culverts are often constructed of galvanized (zinc-coated) iron.

Aluminum, which is adjacent to zinc in the activity series, has been used on the exteriors of buildings for roofing, wall panels, and window frames. Considerable corrosion occurs on such aluminum unless it has been "anodized." The anodizing process (coating with an oxide) takes place naturally on exposure of aluminum to air and forms an oxide that tenaciously and completely covers the surface of the aluminum.

EXPLANATION OF DIFFERENCES IN CHEMICAL ACTIVITY

No effort has been made to suggest why potassium is more active than sodium, etc. It is now easy to explain the reason for the differences in chemical activity. Both sodium and potassium hold their valence electrons with small attractive forces, but potassium holds its electrons with even a smaller force than does sodium. This is in accord with Coulomb's law of electrostatics. Therefore any other substance is more likely to be able to rob potassium and do it more rapidly than it can sodium.

Oxidation–reduction reactions constitute a very common and important type of chemical change insofar as the reactions of inorganic substances are concerned. Oxidation and reduction always occur simultaneously in a chemical change, in such a manner that the number of electrons lost and the number gained are equal.

An index of the metallic activity of elements is the energy necessary to cause a single electron to be detached from an atom. Theoretically the activity series is essentially an order of increasing electron attractions, from top to bottom.

When a substance such as methane is oxidized by chlorine, a halogen, the valence electrons are only removed a little farther away from their original positions. In this case no ions are formed, though the oxidizing agent did partially pull electrons away, perhaps like stretching the sweater of an opposing football player without actually tearing it off of his body.

Although the halogens—fluorine, chlorine, bromine, iodine, and astatine—are not in the so-called activity series because they are nonmetals, they could be listed in another similar series. All the halogens have great attractive forces for electrons. Fluorine has the greatest ability to attract electrons and is therefore most active. Continuing down the halogens in Group VII, the common elements chlorine, bromine, and iodine become less active nonmetals.

It should be apparent that somewhere near the middle of the periodic table there should be some elements with intermediate attractions for electrons. Carbon and silicon are examples of such elements. Although we admittedly are oversimplifying somewhat by omitting some other important factors, undoubtedly this intermediate electron affinity of carbon and silicon is to a very great extent a cause for the marked chemical inactivity of these free elements at ordinary temperature. This same intermediate electron affinity helps to explain why carbon compounds are so likely to be molecular; that is, the difference in electron affinity between carbon and any other element is too small for electron stealing. Compound formation for them must involve electron sharing, producing molecules, rather than electron stealing, which produces ions.

Now that we have considered the reasons for chemical activity, we wonder if the thousands of inorganic chemical reactions can be grouped into a few general types.

General types of chemical reactions

Chemists developed the concept of oxidation numbers to deal with the balancing of equations in complex reactions. However, as nonchemists we will be content to balance simple equations by inspection, as shown in the chemical reactions below. We recall that the number of each kind of atom of the reactants (to the left of the arrow) must equal the number of each kind in the products (to the right of the arrow), as below. We will also find it convenient in our thinking to use oxidation numbers.

Chemical reactions (inorganic) are conveniently grouped into four general types.

1. *Direct Combination (Synthesis)*
   Substances are put together to form more complex molecules.

   Examples:
   $S + O_2 \rightarrow SO_2$
   $2H_2 + O_2 \rightarrow 2H_2O$
   $4Fe + 3O_2 \rightarrow 2Fe_2O_3$
   $CaO + CO_2 \rightarrow CaCO_3$
   $SiO_2 + CaO \rightarrow CaSiO_3$
   $2Na + Cl_2 \rightarrow 2Na^+Cl^-$

   In the last example the ionic nature of common salt is emphasized.

2. *Decomposition (Analysis)*
   Complex molecules are broken into simpler substances.

   Examples:
   $2KClO_3 \rightarrow 2KCl + 3O_2 \uparrow$
   $CaCO_3 \rightarrow CaO + CO_2 \uparrow$

3. *Single Displacement*
   A metal replaces hydrogen or another metal in a compound.

   Examples:
   $Zn + 2HCl \rightarrow ZnCl_2 + H_2 \uparrow$
   $3Zn + Fe_2O_3 \rightarrow 3ZnO + 2Fe$
   $Ca + 2HCl \rightarrow CaCl_2 + H_2 \uparrow$
   $Zn + CuSO_4 \rightarrow ZnSO_4 + Cu$
   $Fe + H_2SO_4 \rightarrow FeSO_4 + H_2 \uparrow$

4. *Double Replacement*
   Two compounds react to form two new compounds.

   Examples:
   $Na_2CO_3 + Ca(OH)_2 \rightarrow CaCO_3 \downarrow + 2NaOH$
   or
   $2Na^+ + CO_3^{2-} + Ca^{2+} + 2OH^- \rightarrow CaCO_3 \downarrow + 2Na^+ + 2OH^-$
   $AgNO_3 + NaCl \rightarrow AgCl \downarrow + NaNO_3$
   $2KOH + H_2SO_4 \rightarrow K_2SO_4 + 2H_2O$

The first three types above involve oxidation–reduction reactions. The fourth involves a class known as acid–base reactions, to be discussed.

Acids and bases

Earlier in this chapter we considered a major class of chemical reactions known as oxidation–reduction reactions, in which either one or more electrons are transferred between pairs of substances, or a pair of electrons being shared by two different kinds of atoms are shifted more toward one of the atoms instead of being equally shared. We now examine a second important class of chemical reactions, acid–base reactions.

The concepts of acids and bases have evolved from attempts to classify chemical compounds. In order to facilitate the learning process, chemists have devised theories and operational definitions to classify information.

Perhaps the most widely accepted and useful theories and operational definitions to provide insights into the behavior of acids and bases have been, in chronological order: (1) the Arrhenius acid–base concept (1887); (2) the Brønsted and Lowry theory of acid–base reactions (1923); (3) the Lewis concept of acids and bases (presented in 1923 and fully developed by 1938). Each concept has advantages under appropriate circumstances and serves useful purposes. The main difference between the different theories and definitions is one of scope; those more recently developed include more compounds. No one of these theories denies the other; the chemist uses the one best suited for a particular purpose. The theory of G. N. Lewis is the broadest, for it includes the acids and bases of the other two theories and more.

ARRHENIUS ACID–BASE CONCEPT: A THEORY OF ELECTROLYTIC DISSOCIATION

Acids and bases have long been thought of in terms of the general properties of their water solutions. For example, an acid in solution has a sour taste, turns the acid–base indicator blue litmus to red, and neutralizes bases, whereas a base has a bitter, but not sour, taste, turns red litmus to blue, feels slippery to the touch, and neutralizes acids. We know that acid–base reactions do not require a solvent; a reaction between hydrogen chloride and ammonia occurs on mixing the two gases. Because water is the medium in which many chemical reactions take place, we emphasize acid–base reactions in water solutions.

If a solution containing water and sodium hydroxide, a base, has hydrochloric acid added to it very slowly, there is a point when the solution will show no properties of either acid or base. Such a solution is then said to be "neutral." The process by which acids and bases destroy each other is called *neutralization*. This kind of change is one of the most important in chemistry.

From experimental data Svante Arrhenius (Sweden, 1859–1927) concluded that when a sample of an acid, base, or salt (soon to be discussed) was dissolved in water, a fraction of its molecules dissociated into electrically charged particles which he called "ions." For example, he thought that gaseous hydrogen chloride was composed of virtually 100 percent hydrogen chloride molecules but that when it dissolved in water most of those molecules dissociated into ions.

$$\underset{\text{hydrogen chloride molecule}}{HCl} \xrightarrow[\text{of water}]{\text{in presence}} \underset{\text{hydrogen ion}}{H^+} + \underset{\text{chloride ion}}{Cl^-}$$

He noted that acids such as acetic acid, which is usually much less active than hydrochloric acid, were also poorer conductors of electricity. To explain these facts he assumed that the less active (weaker) acids had a lower percent of dissociation into ions. But because the total amount of conductance furnished per gram-molecule of acetic acid kept increasing as the acid was diluted with water, he assumed that at infinite dilution all acids, bases, and salts were 100 percent dissociated into ions.

Solutions of substances in water that conduct electricity are called *electrolytes*. For example, a solution of sodium chloride in water conducts electricity, but pure water is a very poor conductor of electricity and is usually said to be a "nonconductor." Substances whose water solutions are nonconductors of electricity are called nonelectrolytes. According to the Arrhenius theory, the ability of solutions of electrolytes to conduct electricity is accounted for by the fact that the electrolytes separate (dissociate) into positively and negatively charged particles called ions.

An acid, according to Arrhenius, is a substance that dissociates in water to yield hydrogen ions (protons, H^+). When a substance like hydrogen chloride (HCl) is added to water, some HCl molecules dissociate (separate) into H^+ ions and Cl^- ions. The acidic properties of aqueous solutions of acids are due to the H^+ ions. A base is a substance that dissociates in water to yield hydroxide, or hydroxyl, ions (OH^-). When sodium hydroxide (NaOH) is added to water, it dissociates to Na^+ ions and OH^- ions. The basic properties of aqueous solutions of bases are due to the OH^- ions.

It is common to refer to acids, bases, and

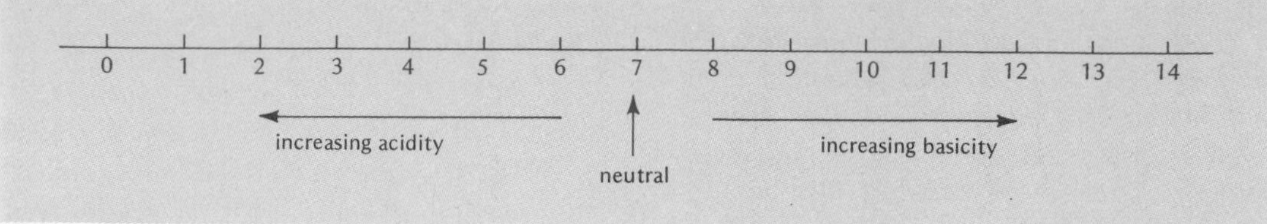

Figure 17.4
The pH scale.

salts. A *salt* is an ionic compound consisting, usually, of a positive ion of a metal and a negative ion of a nonmetal. A salt will have an ionic crystal lattice in the solid state, and its ions will move farther apart in an aqueous solution. An acid mixed with a base reacts chemically to form water and a salt, as below:

$$\underset{\text{acid}}{H^+Cl^-} + \underset{\text{base}}{Na^+OH^-} \longrightarrow \underset{\text{water}}{HOH} + \underset{\text{salt}}{Na^+Cl^-}$$

The Arrhenius theory is restricted to water solutions and has been too narrow in scope to be useful in many situations. Furthermore the theory suggests that neutralization of acids requires hydroxide ions, whereas it is known that substances other than hydroxyl ions can also react with and neutralize acids.

BRØNSTED AND LOWRY THEORY OF ACID–BASE REACTIONS

In 1923 J. N. Brønsted of Denmark and T. M. Lowry of England developed independently a theory of acid–base reactions that was not restricted to water solutions containing H^+ ions and OH^- ions. According to the Brønsted-Lowry theory, an *acid* is a substance that can donate protons; a *base* is a substance that can accept protons. Chemical reactions involving the transfer of proton(s) are *acid–base reactions.*

The chemical equation showing the dissociation of a hydrogen chloride molecule, according to the Arrhenius concept, is now revised in accord with the Brønsted-Lowry theory to show that a proton (H^+ ion) from the HCl molecule and a water molecule combine to form a hydronium ion (H_3O^+ ion) and a chloride ion (Cl^-):

$$HCl + H_2O \longrightarrow H_3O^+ + Cl^-$$

Similarly, gaseous ammonia (NH_3), a base, accepts a proton (H^+) from gaseous hydrogen chloride (HCl), an acid, to produce the ammonium ion (NH_4^+), an acid, and the chloride ion (Cl^-), a base:

$$\underset{\substack{\text{molecule,}\\ \text{base}}}{NH_3} + \underset{\substack{\text{molecule,}\\ \text{acid}}}{HCl} \rightleftharpoons \underset{\substack{\text{ion,}\\ \text{acid}}}{NH_4^+} + \underset{\substack{\text{ion,}\\ \text{base}}}{Cl^-}$$

According to the Brønsted-Lowry theory, acids and bases can be either molecules or ions, as above. The lower horizontal arrow indicates the reverse reaction in which the NH_4^+ ion functions as an acid by donating a proton to the Cl^- ion, a base.

In the Brønsted-Lowry theory there is a base that corresponds to every acid. An acid donates a proton and becomes a base; a base that accepts a proton becomes an acid. The base produced from an acid is called the *conjugate base;* an acid produced from a base is called the *conjugate acid.*

In conformity with the Brønsted-Lowry theory, acids and bases are classified according to their ability to donate or accept protons. Strong acids—such as hydrochloric, nitric, and sulfuric acids—give up protons very easily. Conversely, weak acids—such as acetic acid, in vinegar, and carbonic acid, in carbonated drinks—do not give up their protons easily. Similarly, bases are classified as strong or weak, relatively, depending on the ease with which protons are accepted.

LEWIS CONCEPT OF ACIDS AND BASES

A concept of acid–base reactions broader than those by Arrhenius, or Brønsted and Lowry was stated in 1923 by G. N. Lewis (United States, 1875–1946). The concept was fully developed by 1938. In 1916 Lewis had first proposed that a covalent bond consists of a shared pair of electrons. A *Lewis acid* is defined as a substance that acts as an electron-pair acceptor in chemical reactions; a *Lewis base* is a substance that acts as an electron-pair donor in chemical reactions.

The Lewis definitions of acid and base and the Brønsted-Lowry definitions of acid and base are not in conflict. For example, below we show that protons (H^+) act as acids when reacting with hydroxide ions, water, or ammonia, in that they are donated to the reaction and they also act as electron-pair acceptors.

$$H{:}\underset{\cdot\cdot}{\overset{\cdot\cdot}{O}}{:}^- + H^+ \longrightarrow H{:}\underset{\underset{\displaystyle H}{\cdot\cdot}}{\overset{\cdot\cdot}{O}}{:}$$

$$H{:}\underset{\underset{\displaystyle H}{\cdot\cdot}}{\overset{\cdot\cdot}{O}}{:} + H^+ \longrightarrow \left[H{:}\underset{\underset{\displaystyle H}{\cdot\cdot}}{\overset{\cdot\cdot}{O}}{:}H\right]^+$$

$$H{:}\underset{\underset{\displaystyle H}{\cdot\cdot}}{\overset{\overset{\displaystyle H}{\cdot\cdot}}{N}}{:} + H^+ \longrightarrow \left[H{:}\underset{\underset{\displaystyle H}{\cdot\cdot}}{\overset{\overset{\displaystyle H}{\cdot\cdot}}{N}}{:}H\right]^+$$

The Lewis definitions extend the concepts of acids and bases to a number of reactions that do not involve proton transfer. One example, boric trichloride is a Lewis acid but it is not an acid under the Brønsted-Lowry definition:

$$\underset{\text{Lewis base}}{H{:}\underset{\underset{\displaystyle H}{\cdot\cdot}}{\overset{\overset{\displaystyle H}{\cdot\cdot}}{N}}{:}} + \underset{\text{Lewis acid}}{\underset{\underset{\displaystyle Cl}{\cdot\cdot}}{\overset{\overset{\displaystyle Cl}{\cdot\cdot}}{B}}{:}Cl} \longrightarrow H{:}\underset{\underset{\displaystyle H}{\cdot\cdot}}{\overset{\overset{\displaystyle H}{\cdot\cdot}}{N}}{:}\underset{\underset{\displaystyle Cl}{\cdot\cdot}}{\overset{\overset{\displaystyle Cl}{\cdot\cdot}}{B}}{:}Cl$$

The BCl_3 molecule does not have a complete octet of electrons about the boron atom and can function as a Lewis acid. The ammonia (NH_3) acts as a Lewis base (an electron-pair donor) and the boron trichloride (BCl_3) acts as a Lewis acid (an electron-pair acceptor). The reaction is fundamentally the same as the reaction given earlier of hydrogen chloride with ammonia to form the NH_4 ion.

The Lewis definitions also identify as acids some nonhydrogen-containing substances that have the same function in chemical reactions as common acids have. Silicon tetrafluoride (SiF_4) is identified as a Lewis acid because it can react, for example, with fluoride ions ($2F^-$) to yield SiF_6^{2-}. The SiF_4 is an electron-pair acceptor and hence a Lewis acid.

EXPRESSING THE ACIDITY OR BASICITY OF A SOLUTION

The term pH is a brief method of expressing the hydrogen ion concentration of any acidic, neutral, or basic solution. The pH is used to indicate how acidic or basic an aqueous solution is. The pH may be defined as the exponent of the negative power of 10 used to express the hydrogen ion concentration of an aqueous solution. A neutral solution, for example, contains H^+ ions with a concentration of 0.0000001 (or, 1×10^{-7}) mole/liter and has a pH of 7. A solution with a hydrogen ion concentration of 1×10^{-5} mole/liter has a pH of 5 and is acidic; one of 1×10^{-9} mole/liter has a pH of 9 and is basic.

A pH scale is shown in Fig. 17.4. The pH of a neutral solution is 7 on a scale of 0 to 14; that is, the solution is neither acidic nor basic. Solutions with pH less than 7 are acidic, and the acidity increases as the pH number decreases from 7 to 0. Solutions with pH greater than 7 exhibit properties of a base, and the basicity increases as the pH number increases from 7 to 14. Fractional pH values also are possible. For example, normal human blood in the body has a pH of 7.40.

Note that we use the term "hydrogen" ion rather than "hydronium" ion. At present it is universally agreed that there are no hydrogen ions, as such, in an aqueous solution. But the hydration of H^+ does not cease with the addition of one molecule of water, as one might suppose by a liberal use of the term "hydronium" ion. So it seems better merely to remember that hydrogen ions in aqueous solution are hydrated but not to specify them as H_3O^+, $H_5O_2^+$, etc.

The control of pH is very useful in many processes. The growth of bacteria, such as those which consume organic wastes in sewage disposal plants, proceeds best when the pH of the sewage is well regulated. Many solutions used in the electroplating of metals function best at closely controlled pH. A great variety of plants are sensitive to and thrive best in soils that have a certain pH. For example, azaleas and blackberries thrive best in soils that are on the acid side of neutrality. Most legumes, such as clovers and alfalfa, however, would not thrive at all in such a soil. The application of ground limestone (calcium carbonate) to long-cultivated and leached acidic soils not only supplies calcium, which is used in large amounts by growing legumes, but also raises the pH to a level more appropriate for such plants.

Solutions that contain both a weak acid and a salt of the same weak acid and can maintain a nearly constant pH are known as *buffers*. A buffer solution such as one of acetic acid and sodium acetate is capable of maintaining a nearly constant pH. The acetic acid will react with any strong base that is added; the acetate ions react with any strong acid. Buffers may be prepared from any ratio of concentrations of weak acids and the salt of the weak acid (conjugate base). A ratio of 1:1 of acid to salt is the most efficient in handling the addition of either acid or base. Buffers are important in the field of medicine and in industrial processes, such as the electroplating of metals and the manufacture of dyes. Human blood, for example, is buffered within the healthy body to maintain the pH at 7.4, slightly basic. Death usually results when the pH either increases or decreases by 0.2 pH unit.

COORDINATION COVALENCE

When a positive ion, such as a hydrogen ion or a calcium ion, becomes attached to a water molecule, the linkage is through one of the unshared electron pairs on the oxygen atom of the water molecule,

$$\underset{\text{proton}}{H^+} + \underset{\text{water}}{:\overset{\substack{H\\ \cdot\cdot}}{\underset{\cdot\cdot}{O}}:H} \rightarrow \underset{\text{hydronium ion}}{\left[H:\overset{\substack{H\\ \cdot\cdot}}{\underset{\cdot\cdot}{O}}:H\right]^+}$$

Note that the water molecule supplies both of the bonding electrons; the oxygen atom is called the electron donor; and the hydrogen ion is said to be the electron acceptor. Things other than water and oxygen can play a role.

A bond for which one atom supplies both of the bonding electrons is a coordinate covalent bond and the linkage is termed *coordination covalence*. Complex compounds often contain a number of bonds properly classified as coordination covalences; those that do should be called *coordination compounds*.

Factors affecting chemical reaction rates

LIST OF FACTORS

There are five major factors that may influence the rate at which a chemical reaction takes place: (1) the nature of the reactants, (2) the temperature of the reactants, (3) the concentration of the reactants, (4) the amount of surface exposed, and (5) the presence or absence of a catalyst.

NATURE OF THE REACTANTS

In a chemical reaction bonds are broken and bonds are formed. The rate should depend on the specific bonds involved—that is, on the specific substances brought together in reaction. The nature of the chemical sub-

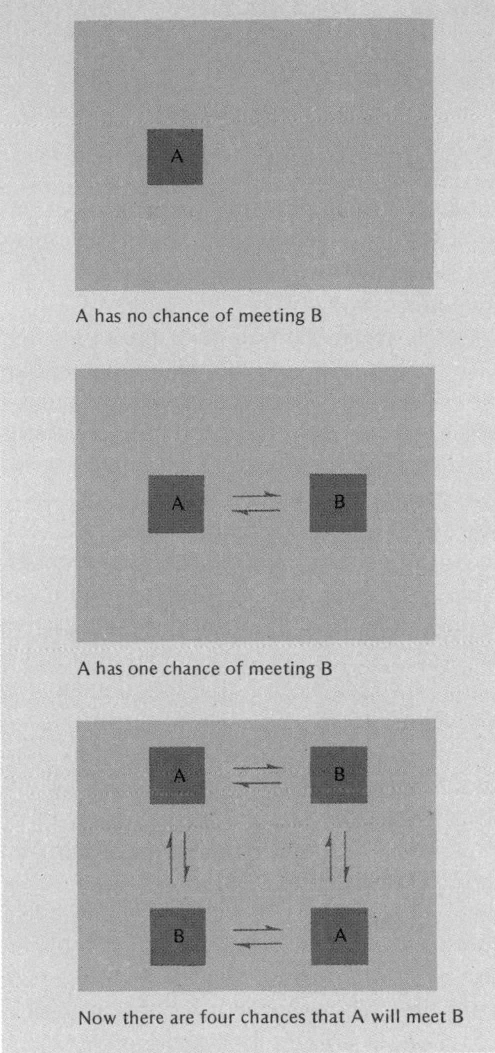

A
A has no chance of meeting B
A
B
A has one chance of meeting B
A
B
B
A
Now there are four chances that A will meet B

Figure 17.5
Effect of increased concentration in a fluid mixture.

stances is the most important variable controlling a reaction. A useful rule is that reactions involving ionic substances are rapid, and reactions of most covalent substances are slow.

TEMPERATURE OF THE REACTANTS

A rise in temperature almost always increases the rate of any chemical reaction. The reactions that occur with the absorption of energy (endothermic reactions) will be favored by a rise in temperature. Much of the heat energy that is absorbed by the reactants goes into breaking bonds. Also, the heat content of the products is higher than the heat content of the initial materials. The majority of chemical reactions are, however, exothermic (energy is released). A decrease in temperature decreases the rate of any chemical reaction independently of whether the reaction is endothermic or exothermic.

CONCENTRATION OF THE REACTANTS

Suppose that a given chemical change in a fluid mixture consumes two different reactants such that

$$A + B \rightarrow C$$

A study of Fig. 17.5 indicates that increasing the concentration of either reactant increases the reaction rate because it increases the number of contacts between molecules or ions per unit of time. It should be noted that a higher temperature also increases the number of contacts per second; this is an additional reason why reactions are faster at higher temperatures.

The rate of a chemical reaction should increase as the reacting molecules crowd more closely together—become more concentrated—because the frequency of collision increases. This is usually the case, hence the general rule that the reaction rate increases with increase in the concentration of reactants. The effect of concentration on the rate of chemical reaction is explained by the fact that a reaction between two molecules, such as hydrogen and iodine, for example, can only occur when they are in contact, or at least close together, so that new bonds may form between certain atoms while existing bonds between other atoms are broken.

$$\begin{matrix} H & & I \\ | & + & | \\ H & & I \end{matrix} \rightarrow \begin{matrix} H\text{—}I \\ \\ H\text{—}I \end{matrix}$$

the bonds are broken — new bonds are formed

For reactions in liquid solutions, the concentrations are usually expressed in moles per liter. For reactions in gaseous mixtures, the concentrations are often expressed in terms of the pressure of the individual gases.

AMOUNT OF EXPOSED SURFACE

When the amount of exposed surface of the reactants is increased, the number of reacting particles in contact with one another at any given time is also increased, resulting in a faster reaction. A pile of wood shavings burns more rapidly than does a log of equal weight. Food is chewed into smaller pieces in order to increase the surface area so the chemical changes of digestion can occur faster. In the carburetor of an automobile liquid gasoline comes out of a small opening as a fine spray or vapor with increased surface area for mixture with oxygen from air prior to combustion in the engine.

CATALYSTS

A fifth major factor of importance in influencing the speed of chemical reaction is the presence or absence of a catalyst. A *catalyst* is a substance that influences the speed of a chemical reaction but is not consumed and can be recovered completely after the reaction is over. Any consumption of a catalyst at a given step in a chemical reaction is always balanced by regeneration of the catalyst at a later point. The total amount of catalyst among the final products is the same as that among the original reactants.

When potassium chlorate ($KClO_3$) is heated, it slowly decomposes into oxygen and potassium chloride (KCl). If a small amount of the catalyst manganese dioxide (MnO_2) is mixed with the $KClO_3$ before heating is begun, the decomposition of the potassium chlorate is accelerated on heating. At the conclusion of the reaction, the $KClO_3$ is completely consumed but all of the MnO_2 remains. The catalyst is usually written over the arrow in chemical equations,

$$2KClO_3 \xrightarrow[\text{heat}]{MnO_2} 2KCl + 3O_2$$

Catalysts that speed up chemical reactions are called "positive catalysts." Substances that slow down chemical reactions are called "negative catalysts" or "inhibitors."

Catalysts are common in the world and have much to do with our well-being. In biochemical processes in the human body catalysts called "enzymes" act to control various physiological reactions. Components of saliva, gastric juice, and other secretions catalyze the digestion of food. Liquid fats, such as cottonseed oil, are made to react chemically with hydrogen to yield the more desirable solid fats—such as shortening—that are used in cooking. Finely divided nickel is used as a catalyst for this reaction. In the chemical industry catalysts are widely used to facilitate the economical conversion of reactants into a desired product. Another important use of a catalyst is in the manufacture of ammonia by the combination of nitrogen and hydrogen. Spectacular increases in the yields of agricultural products have resulted from the use of ammonia as a fertilizer. Among the inhibitors (negative catalysts) are the "antirust" compounds that are placed in the cooling systems of automobiles. Many antifreezes contain an inhibitor. In order to reduce undesirable emissions from automobile engines catalytic converters are used in research laboratories and on the road. These pollution-control catalysts contain elements from the platinum–palladium group and are used to convert harmful carbon monoxide into harmless carbon dioxide.

Activation energy

A certain amount of energy is needed to initiate a chemical reaction. In lighting a match, energy is contributed by the person striking it to start the chemical reaction of the burning match. Chemical reaction occurs only if the kinetic energy of the reactants is greater than a certain minimum value. The energy that

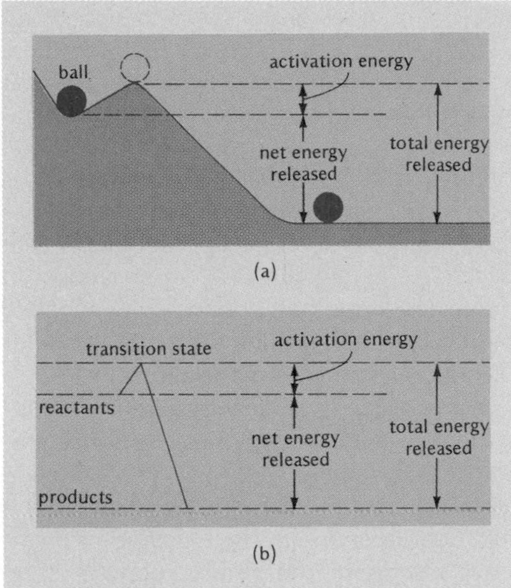

Figure 17.6
Activation energy. (a) A ball in a depression must be given additional energy to raise it to a higher point from which it can roll downhill and release energy. (b) Many reactants must be given additional energy (activation energy) in order to start the chemical reaction that releases energy as heat and light and forms products with less chemical energy. (Used with permission from Charles W. Keenan and Jesse H. Wood, *General College Chemistry*, 4th ed., Harper and Row, 1971.)

must be added in order to start a chemical reaction is called *activation energy*. The activation energy for a specific chemical reaction depends primarily on the nature of the reactants, but the method most used to supply activation energy is raising the temperature. A catalyst may be used to provide a new path for a reaction, thus avoiding the need for a high activation energy.

In Fig. 17.6 the concept of activation energy for a chemical reaction in relation to total and net energy released (b) is compared in (a) with those energies when a ball is rolled from a depression up a short hill to a high point and then let roll down a long hill to a level below the depression. In comparing (a) and (b) of the figure, it is seen that activation energy is the energy that must be added to the original energy of the ball or of the reactants before any energy is released. In certain cases a necessary amount of activation energy must be added to increase the kinetic energy of the reactants so chemical reaction can occur.

Reactions involving ions do not have to break strong bonds; they involve little activation energy. In the case of covalent bonding a double bond is generally stronger than a single bond and more activation energy is usually required to break a double bond than a single bond.

Chemical equilibrium

The examples of equilibrium discussed earlier in this book have involved physical conditions, such as a book on a desk, a saturated solution, and a rate of evaporation equal to the rate of condensation. Chemical equilibrium is a factor in many chemical reactions.

GENERAL CONCEPTS OF CHEMICAL EQUILIBRIUM

Thus far we have emphasized unidirectional chemical reactions in which reactants are converted to products. Few chemical reactions actually reach completion no matter how long the elapsed time; that is, the reactants seldom are completely changed to the end products. As the reaction proceeds, the concentrations of the reactants decrease and those of the products increase until the concentrations of both the reactants and the products remain at some fixed concentration. When this condition is achieved, the reaction is said to be in "chemical equilibrium." Let us develop an operational definition of this term.

If one confines carbon dioxide gas over water, the first thing that happens is that some carbon dioxide dissolves and then reacts with the water to form carbonic acid,

$$H_2O + CO_2 \rightleftharpoons H_2CO_3$$

But as soon as some carbonic acid is formed, it can decompose to form water and carbon dioxide. A *reversible reaction* is one that can proceed in either direction. If the condition continues long enough, the rate of the reverse reaction is equal to the rate of the forward reaction. The condition in which the rates of forward and reverse chemical reactions are equal is called *chemical equilibrium*. Thus we can say that when chemical equilibrium is established, the rate of the reverse reaction is equal to the rate of the forward reaction.

In general terms, if reactants A and B are mixed in equimolar amounts and the products consist of C and D, the chemical equilibrium eventually achieved is identical to that which can be reached by mixing equimolar amounts of C and D and allowing sufficient time for equilibrium to occur. Instead of writing two unidirectional equations, the equations for the chemical reactions can be combined into a single equation in which the arrows are shown in opposite directions, as follows:

$$A + B \rightleftharpoons C + D$$

Equilibrium results from the rate of the forward reaction being equal to the rate of the reverse reaction, as indicated by opposing arrows of equal length. In summary, for any system at equilibrium, the rates of opposing reactions are equal and the concentrations of the components remain constant.

CHANGING A SYSTEM AT EQUILIBRIUM

Equilibrium in a chemical sense is a dynamic condition; chemical activity has not lessened at equilibrium. A system remains at dynamic equilibrium until some condition is changed, thus causing the rate of one reaction to increase over the rate of the opposing reaction.

The thought or principle that any system in dynamic equilibrium tends to shift or readjust in such a way as to reduce the stress imposed on it—that is, to undo any changes made on it—was first enunciated in a general form by Henri Le Châtelier (France, 1850–1936) and is generally known as *Le Châtelier's principle*. Static systems and nonequilibrium systems may not follow the principle. In addition to applications in chemistry and other sciences the principle can be applied, in many cases, to the fields of economics and human relations.

It is possible to cause the equilibrium population to shift (1) by altering the concentration of one of the components in a system at chemical equilibrium either by adding an excess of at least one of the reactants or by removing some of one of the products; (2) by changing the temperature to which the system is subject; or (3) by changing the total pressure.

EFFECT OF CHANGING THE CONCENTRATION OF A COMPONENT OF A SYSTEM AT EQUILIBRIUM

In the dynamic equilibrium system

$$A + B \rightleftharpoons C + D$$

an increase of the concentration of A or B, or A and B, by adding an excess of at least one of the reactants would shift the reaction to the right, and this forward reaction would be greater than the reverse reaction as indicated by a new set of arrows $\rightleftharpoons$. In a similar manner, in this equilibrium system a decrease of the concentration of C or D, or C and D, by removing at least some of one of the products C or D would also shift the reaction to the right.

Conversely, starting with the equilibrium system, a reverse shift to the left would be produced either by adding an excess of reactants C or D, or C and D, or by removing at least some of the products A or B. When the reverse reaction (that is, shift to the left) is greater than the forward reaction (shift to the right), the arrows to indicate this would be $\rightleftharpoons$.

The equation given earlier for the chemical equilibrium of water, carbon dioxide, and carbonic acid would be, in the general form,

$$A + B \rightleftharpoons AB$$

EFFECT OF CHANGING THE TEMPERATURE OF A SYSTEM AT EQUILIBRIUM

A change in the temperature to which a system in dynamic equilibrium is subject, caused either by addition or removal of heat energy, can lead to a shift in equilibrium. The equation

$$2CO + O_2 \rightleftharpoons 2CO_2 + \text{heat energy}$$

represents a chemical reaction at dynamic equilibrium in the gaseous state at a constant given temperature. Heat energy is released in the system (exothermic) by the forward chemical reaction and is absorbed in the system (endothermic) by the reverse reaction.

If the temperature of the system is lowered and kept at this new temperature, the system will lose heat energy. In accord with Le Châtelier's principle, the system will readjust itself by permitting more of the CO and O_2 molecules to react to form more CO_2 molecules and thus release heat energy to the system as shown by the forward reaction (to the right). A new position of equilibrium is soon established. The foregoing indicates that exothermic reactions are favored by a decrease in temperature.

If the temperature of the system is raised and kept at a new temperature, the system will readjust itself to remove some of the added heat energy by permitting more CO_2 molecules to take on heat energy and decompose to form more CO and O_2 molecules as shown by the reverse reaction (to the left). Endothermic reactions such as this one are favored by an increase in temperature. A new position of equilibrium is again established.

EFFECT OF CHANGING PRESSURE ON A SYSTEM AT EQUILIBRIUM

Because pressure has little effect on the volume of a liquid or solid, it is said to have no effect on equilibria that involve reactants and products in these physical states. The equilibrium position of certain gaseous reactions may be shifted appreciably by changes in pressure.

Suppose we raise the pressure on a gaseous nitrogen–hydrogen–ammonia system that is in dynamic equilibrium,

$$\underset{\text{1 mole}}{N_2} + \underset{\text{3 moles}}{3H_2} \rightleftharpoons \underset{\text{2 moles}}{2NH_3}$$

We disturb the equilibrium because, in this instance, increased pressure increases the forward reaction rate more than the reverse reaction rate.

Why does increased pressure encourage the forward more than the reverse reaction? If we look at the balanced equation for the above reaction, we see that 1 mole of nitrogen and 3 moles of hydrogen (total, 4 moles) are consumed for every 2 moles of ammonia formed. For every one molecule of N_2 and every three molecules of H_2, reacting according to the equation, only two molecules of NH_3 are produced. Let us assume that we have a cylindrical container fully enclosed with a movable piston and filled with a mixture of gaseous N_2 and H_2 under a pressure of 1 atm and at a constant temperature above 400°C. If the pressure of the gases is increased by decreasing the volume of the container, the system will do what it can to reduce that pressure by permitting additional molecules of NH_3 to be produced. But the total number of molecules enclosed in the container is decreased as the result of the formation of more NH_3 molecules at the expense of twice as many molecules of the reactants N_2 and H_2, thus reducing the pressure.

Therefore the effect of increasing the pressure on the system by decreasing the volume is partly offset by the equilibrium shifting forward (to the right) and thus reducing the pressure and relieving the stress in accord with the principle of Le Châtelier. On the other hand, an equilibrium such as

$$N_2 + O_2 \rightleftharpoons 2NO$$

is not affected by pressure changes.

SUMMARY

Chemical reactions often involve solutions. The relative amounts of solute and solvent, or concentration of solution, may be expressed in different ways.

A system of small whole numbers related to the combining ratios of elements in compounds and in polyatomic ions, or radicals, and called *oxidation numbers* has been devised in order to deal with the balancing of chemical equations. The oxidation states of elements in compounds and polyatomic ions are a periodic function of the atomic numbers, and some elements have more than one oxidation state number. The formulas of electrovalent compounds can be predicted from the oxidation numbers of their elements. It is usually not possible to use oxidation numbers as a basis for the prediction of formulas of covalent compounds.

One major kind of chemical change is called an *oxidation–reduction reaction*. The loss of electrons by an atom or ion is called *oxidation;* the gain of electrons by an atom or ion is called *reduction*. Oxidation and reduction always occur simultaneously in a chemical change.

Differences in chemical activity may be explained in terms of the forces of attraction between the nuclei of atoms and their valence electrons. From evidence obtained by experimentation, an activity series of metals has been formulated and arranged in order of increasing electron attractions, top to bottom.

Chemical reactions (inorganic) are grouped as four general types: direct combination, decomposition, single displacement, and double replacement.

A second major kind of chemical change is the acid–base reaction. The theory of acids and bases by G. N. Lewis is broader than the theory of Arrhenius and that of Brønsted and Lowry for it includes these two theories and more. An acid, according to Arrhenius, is a substance that dissociates in water to yield hydrogen ions (protons, H^+), whereas a base is a substance that dissociates in water to yield hydroxide, or hydroxyl, ions (OH^-). According to the Brønsted-Lowry theory, an acid is a substance that can donate protons; a base is a substance that can accept protons; and chemical reactions involving the transfer of protons are acid–base reactions. In the Lewis concept of acids and bases, a Lewis acid is a substance that acts as an electron-pair acceptor in chemical reactions, whereas a Lewis base is a substance that acts as an electron-pair donor in chemical reactions. The Lewis definitions extend the concepts of acids and bases to a number of chemical reactions that do not involve proton transfer, and they also identify as acids some non-hydrogen-containing substances that have the same function in chemical reactions as have common acids. The pH is used to express how acidic or basic an aqueous solution is.

Major factors that may influence the rate at which a chemical reaction occurs are the nature, the temperature, and the concentration of the reactants; the amount of surface exposed; and the presence or absence of a catalyst. A *catalyst* is a substance that influences the speed of a chemical reaction but is not consumed and can be recovered completely after the reaction is over.

The energy that must be added in order to start a chemical reaction is called *activation energy*. A catalyst may be used to provide a new path for a chemical reaction, thus avoiding the need for a high activation energy. Reactions involving ions do not have to break strong bonds; they involve little activation energy.

Reactants seldom are completely changed to products. A reversible reaction is one that can proceed in either direction. The condition in which the rates of forward and reverse chemical reactions are equal is called *chemical equilibrium*. For any chemical system at equilibrium, the rates of opposing reactions are equal and the concentrations of the components remain constant. It is possible to cause the equilibrium population to shift, however.

Important words and terms

solvent
solute
aqueous solution
dilute solution
concentrated solution
saturated solution
unsaturated solution
solubility
molarity concentration of a solution
molality concentration of a solution
percentage concentration of a solution
normality concentration of a solution
oxidation numbers
radical
oxidation number of a radical
oxidation number of an ion
electrovalent compounds
oxidation
reduction
oxidized
reduced
oxidizing agent
reducing agent
activity series of metals
chemical reactions, two major kinds
chemical reactions, four general types
neutralization
electrolyte
an acid, according to Arrhenius, Brønsted-Lowry, and Lewis
a base, according to Arrhenius, Brønsted-Lowry, and Lewis
a salt
acid–base reactions, according to Brønsted and Lowry
pH
buffer
coordinate covalent bond
factors affecting chemical reaction rates
catalyst
activation energy
chemical equilibrium
Le Châtelier's principle

Questions and exercises

1. Match the following columns:

a. solvent	1. A solution containing a small amount of solute
b. solute	2. Amount of solute in a saturated solution
c. concentrated solution	3. The substance dissolved to form a solution
d. saturated solution	4. A solution containing a large amount of solute
e. dilute solution	5. The substance in excess in a solution
f. aqueous solution	6. A solution containing the maximum amount of a solute
g. solubility of substance	7. A water solution of a substance

2. (a) Write the equation for the formation of soda water. (b) Is the final product an example of a true solution according to what you learned in Chap. 14? (c) Write the equation for the products resulting when you shake violently a bottle of soda water and then remove your thumb from the opening.
3. (a) What factor can usually increase the solubility of a liquid solute in a liquid, for example, sugar that has crystallized out of honey? (b) What factors can usually increase the solubility of gas in a liquid? (c) What effect does high temperature have on a gas dissolved in a liquid? Explain the difference in the taste of tap water and boiled water.
4. (a) What is meant by the concentration of a solution? (b) In what four ways can it be expressed?
5. Distinguish between (a) molarity and molality; (b) molarity and normality.
6. A 1000-cm^3 block of gold has a mass of 19,300 g and a similar size block of copper has a mass of 8900 g. When the two are melted together, the resulting mass of the alloy is naturally 28,200 g. It appears that we have all the needed information to determine the density of the alloy from the formula $D = m/V$. Explain, however, why we *cannot* do so.

7. (a) For a 1 *N* solution, how many grams of the following would be required: (1) hydrogen fluoride, HFl; (2) phosphoric acid, H_3PO_4; (3) calcium hydroxide, $Ca(OH)_2$? (b) What would be the molarity, molality, and percentage (by weight) in each case?
8. What is the oxidation number for chromium (Cr), manganese (Mn), phosphorus (P), and selenium (Se) in the following compounds: $K_2Cr_2O_7$, $KMnO_4$, $Ca_2P_2O_7$, K_2SeO_4?

Answer: +6, +7, +5, +6

9. Why can we not determine by inspection of its formula the oxidation number of carbon in a compound like acetylene (C_2H_2)?
10. (a) What is the minimum oxidation number for sulfur as represented in the formula for sodium sulfide (Na_2S)? From the rule illustrated in Fig. 17.2 calculate its maximum oxidation number. (b) What are the maximum and minimum oxidation numbers for nitrogen as represented in the formulas for nitrogen pentoxide (N_2O_5) and ammonia (NH_3)? (c) What rule determines whether an element will take on its higher or lower oxidation number in a reaction?
11. (a) Define oxidation and reduction. (b) Is oxygen necessarily involved in oxidation? (c) In the transfer of electrons in oxidation–reduction reactions, is a complete transfer necessary? Explain.
12. Encircle the appropriate term for element X in the second column:

a. Oxidation number increased	X was reduced, oxidized
b. Electrons were lost	X was reduced, oxidized
c. X is an oxidizing agent	X accepts, loses electrons
d. Oxidation number tends to increase	X is a reducing, oxidizing agent
e. X gains electrons	Oxidation, reduction process
f. X was reduced	X gained, lost electrons
g. Oxidation number decreased	X was reduced, oxidized
h. X was reduced	Oxidation number increased, decreased
i. X donated electrons	Oxidation number increased, decreased

13. (a) Indicate what happens to the size of the zinc and copper electrodes in Fig. 17.3 and why. (b) Are any additional SO_4^{2-} ions needed in one jar or the other as the reaction progresses? Explain. (c) As the copper in solution plates out on the copper electrode, what change takes place in the copper sulfate solution? (d) Which element in the cell is oxidized and which is reduced?
14. (a) Refer to the activity series in Table 17.2 to explain the direction of the reaction in the electrolytic cell in Fig. 17.3.
15. Space batteries used by the astronauts utilize silver–cadmium cells with water as an electrolyte. (a) Which electrode supplies the electrons? (b) Which electrode tends to dissolve? (Refer to Table 17.2 and Fig. 17.3 for your answer.)
16. (a) Why are precious metals like gold, silver, and platinum found free in nature whereas other metals like iron, magnesium, and tin found combined with other elements in ores? (b) Why are copper pipes acceptable for our plumbing but not sauce pans with an inner surface of copper?
17. In which of the following reactions will replacement occur: (a) nickel in a magnesium chloride solution; (b) magnesium in a water solution; (c) silver in hydrochloric acid; (d) copper in a silver nitrate solution; (e) aluminum in a lead acetate solution; (f) iron in a copper hydroxide solution.
18. (a) It has been speculated that the lead plumbing used in ancient Rome was one of the factors that may possibly have done in this vital sector controlling the Roman Empire. With the activity series in mind and the long–term cumulative effect of lead as a toxic agent, explain what this historian had in mind. (b) Why has it been suggested that the old iron frying pan be brought back as one antidote for iron–deficiency anemia?
19. (a) Distinguish between the terms "anodize" and "galvanize." (Refer to a dictionary if necessary.) (b) Name a metal often galvanized and two that usually anodize naturally on exposure to air.
20. (a) How active do you believe the elements rubidium and cesium would be, compared to potassium, if placed in a beaker of water? (b) What are the chances of finding rubidium and cesium free in nature?
21. The human body is composed largely of carbon, hydrogen, and oxygen. Why do science

theory	definition of acid	definition of base	limitation
Arrhenius			
Brønsted-Lowry			
Lewis			X

fictionists so often like to have "silicon people" inhabiting planets in the far reaches of outer space?

22. Why are compounds made from elements in the middle of the periodic table most likely to be molecular (covalent) whereas compounds made from elements at the beginning and end of the table are most likely to be ionic (electrovalent)?
23. (a) List four types of chemical reactions. (b) Under what two general headings may we further classify these four types?
24. Identify the following according to the four types of reactions:
    a. $Ca + H_2SO_4 \rightarrow CaSO_4 + H_2 \uparrow$
    b. $4Al + 3O_2 \rightarrow 2Al_2O_3$
    c. $CaCl_2 + Na_2CO_3 \rightarrow 2NaCl + CaCO_3 \downarrow$
    d. $2CO_2 \rightarrow 2CO + O_2$
25. (a) Name three theories that touch upon acid–base reactions. (b) To what extent does each later theory supersede the earlier theory?
26. (a) Discuss the Arrhenius concept of acids, bases, salts, and electrolysis. (b) What were its limitations? (c) What was meant by neutralization?
27. (a) Discuss the Brønsted-Lowry definition of acids and bases. (b) What is transferred in acid–base reactions here and in oxidation–reduction reactions, as you learned earlier?
28. Fill in the form provided above.
    (a) What is a salt? With which theory is this concept generally associated? (b) What are conjugate acids and bases? With which theory are these concepts generally associated? (c) How are acids and bases classified in strength in the Brønsted-Lowry theory?
29. (a) What is meant by the pH of a solution? (b) What is the range of values used and their meanings? (c) What are some practical applications of pH control?
30. (a) What is a buffer? (b) How is it prepared? (c) By what mode of action does it accomplish its purpose?
31. (a) What is meant by a coordinate covalent bond? (b) Of what importance is this bond in complex compounds?
32. List five factors affecting chemical reaction rates.
33. (a) How does decreasing or increasing the temperature affect the rate of reaction of (1) an exothermic reaction; (2) an endothermic reaction? (b) Are most reactions exothermic or endothermic?
34. (a) How is the concentration of reactants expressed for liquids and for gases? (b) Why should the concentration of reactants affect the rate of reaction? (c) What two factors can increase the number of contacts per second?

35. (a) To what net extent is a catalyst consumed from the beginning to the end of a reaction? (b) What two classes of catalysts exist and how do they differ? (c) Give an example of a positive catalyst and a negative one.
36. (a) How is activation energy related to the total energy released in a chemical reaction? (b) What forms may activation energy take? (c) How do activation energies compare in the cases of ionic and covalent bonding?
37. (a) What is meant by chemical equilibrium? (b) Why is chemical equilibrium considered a dynamic, rather than a static, condition? (c) What can be done to alter this condition?
38. (a) State Le Châtelier's principle. (b) Cite three forms of "stress" that according to this principle could shift equilibrium.
39. In what two ways can the concentration of a reaction be changed so to shift equilibrium?
40. Indicate with the proper length of arrows the reaction between CO and O_2 when the combination is (a) heated; (b) cooled. (c) In what way can this reaction be compared to ordinary combustion?
41. Why does pressure have an effect on the equilibrium of a nitrogen–hydrogen–ammonia system but not on a nitrogen–oxygen–nitrous oxide system?

Supplementary readings

Books

Hurd, D. L., and J. J. Kipling, *The Origins and Growth of Physical Science,* Penguin Books, Baltimore (1964). [Pages 307–311 contain a copy of an original publication regarding the law of mass action. This article illustrates the idea that laws often undergo a period of development. The law of mass action eventually grew into the concept now usually called the law of chemical equilibrium.]

Keenan, Charles W., and Jesse H. Wood, *General College Chemistry,* 4th ed., Harper & Row, New York (1971). [Chapter 6 discusses the sources of ions, acids, and bases, structure of hydroxy compounds, and neutralization. Chapter 10 is on solutions.]

Lee, Garth L., and Harris O. Van Orden, *General Chemistry,* Saunders, Philadelphia (1972). [The chapter on ions in solution, acids, bases, and salts is an excellent development of the topics covered in this chapter.]

Ouellette, Robert J., *Introductory Chemistry,* Harper & Row, New York (1970). [Chapter 3 touches upon Le Châtelier's principle, while Chap. 10 is devoted exclusively to acids and bases, and Chap. 11 to oxidation and reduction.]

Pimentel, George C., et al., *Chemistry, an Experimental Science,* Freeman, San Francisco (1963). [Pages 132–137 present the role of energy in reaction rates; the potential energy diagram should be especially helpful in explaining activation energy, the activation complex, and net energy released.]

Sienko, Michell J., and Robert A. Plane, *Chemistry: Principles and Properties,* McGraw-Hill, New York (1966). [Pages 85–94 discuss oxidation numbers and their use.]

Article

Haensel, V., and R. L. Burwell, Jr., "Catalysis," *Scientific American,* **225,** no. 6, 46 (December, 1971). [Industrial reactions are promoted by substances that themselves stay unchanged.]

ORGANIC CHEMISTRY

Science seldom proceeds in the straightforward manner imagined by outsiders. Instead, its steps forward (and backward) are often very human events in which personalities and cultural traditions play major roles.

JAMES D. WATSON, 1968

Organic chemistry is the chemistry of carbon compounds, with a few exceptions that are considered to be inorganic compounds, such as oxides of carbon (CO, CO_2), carbonic acid (H_2CO_3), and the carbonates of metals (for example, Na_2CO_3). *Inorganic chemistry* is the chemistry of all the elements and their compounds other than the compounds of carbon, except the few carbon compounds that are considered as being inorganic. It is generally accepted now that there is no fundamental difference between the physical principles involved in organic and inorganic chemistry.

Carbon and some of its compounds were mentioned in earlier chapters, mainly in connection with molecules and chemical bonding. More than 90 percent of all known compounds contain carbon. The number and variety of covalent carbon compounds is in the millions (at least 3.5 million) because carbon atoms have the unique ability to form stable covalent bonds with other carbon atoms. This results in large networks of two or more carbon atoms linked together in millions of carbon compounds.

Organic compounds are important in relation to our fuels, foods, and other materials common in everyday living. The chemistry of life is the chemistry of carbon compounds. Our lives are dependent on the science of organic chemistry in ways not always obvious.

Hydrocarbons

At one time it was thought that carbon compounds could be produced only by living organisms. Organic chemistry now involves the study of carbon-containing compounds similar in structure to those that are present in living organisms and also includes the preparation and study of carbon-containing compounds that do not occur in nature. In a historical synthesis, Friedrich Wöhler (Germany, 1800–1882) made the organic compound urea in the laboratory in 1828. This and similar laboratory investigations were the precursors of today's chemical industry from which come a multitude of synthetic fibers, fertilizers, plastics, organic drugs, and the like—all of which seem indispensable in our modern society. The simplest class of organic compounds is the hydrocarbons.

OCCURRENCE AND SOURCES OF HYDROCARBONS

A *hydrocarbon* is a compound that contains only carbon and hydrogen. Such common materials as gasoline, natural gas, lubricating oil, mineral oil, paraffin, and fuel oil are mixtures of hydrocarbons. All of these except natural gas are derived from petroleum, often called crude oil, by a refining process. A billion dollars or more a year is usually spent in the search for new deposits of petroleum and natural gas.

When we consider how important these materials are to our homes and to modern industry and transportation, we can begin to sense the immense, and seemingly increasing, significance of hydrocarbons. It is difficult to see how modern mechanized civilization could function without them, for natural gas and fuel oil automatically heat most of our homes, and petroleum products propel our automobiles, buses, trucks, tractors, diesel ships and locomotives, and some airplanes. Oils even lubricate our machinery.

Although petroleum and natural gas are the greatest natural sources of our hydrocarbons of commerce, there are other natural sources worth mentioning. When coal is heated in the manufacture of coke, great quantities of gaseous and liquid hydrocarbons are formed as by-products. The liquid rubber latex secreted by wounded rubber trees contains a high percentage of hydro-

Table 18.1
Alkanes (C_nH_{2n+2}, general formula)

CH_4	methane
C_2H_6	ethane
C_3H_8	propane
C_4H_{10}	butanes
C_5H_{12}	pentanes
C_6H_{14}	hexanes
C_7H_{16}	heptanes
C_8H_{18}	octanes
C_9H_{20}	nonanes
$C_{10}H_{22}$	decanes

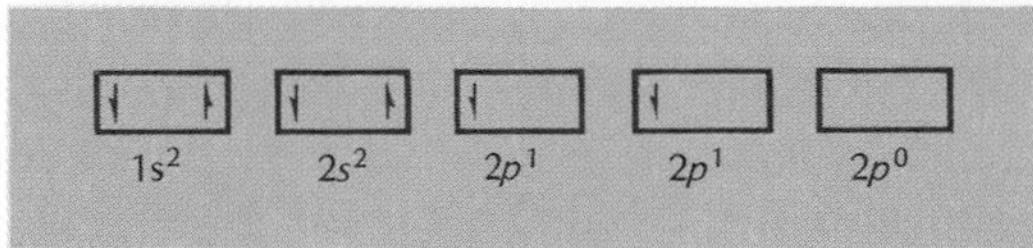

Figure 18.1
Ground state structure of the carbon atom. The $2p^0$ state is unoccupied by an electron.

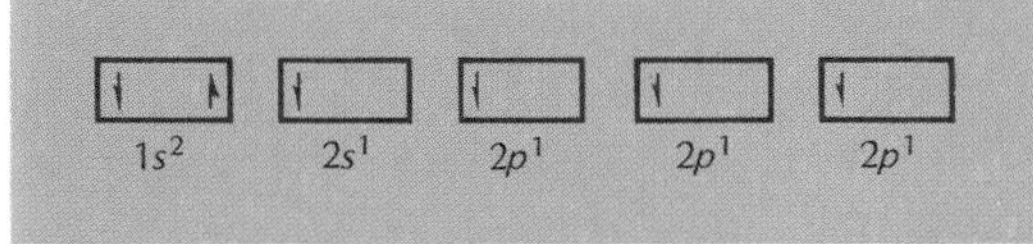

Figure 18.2
Structure of an activated (excited) carbon atom.

carbons in colloidal suspension. Certain rocks called "shales," occasionally also contain a percentage of hydrocarbons. This is only a potential source, however, as it has not yet actually been utilized extensively. The oil sands of western Canada are another large potential source of the hydrocarbons of commerce as is Alaska.

SERIES OF HYDROCARBONS

The thousands of compounds known to consist of the elements carbon and hydrogen —the hydrocarbons—can be arranged in various series in which each member of a series differs from the preceding members of the series by one carbon and two hydrogen atoms (CH_2). We will discuss briefly portions of four such series: the alkanes (the methane series), the alkenes (the ethylene series), the alkynes (the acetylene series), and the aromatic hydrocarbons (represented by the benzene series).

The general molecular formula for the alkanes is C_nH_{2n+2}, for the alkenes is C_nH_{2n}, and for the alkynes is C_nH_{2n-2}, where n is the number of carbon atoms. The most important series of aromatic hydrocarbons is derived from benzene (formula C_6H_6). The benzene series is represented by the general formula C_nH_{2n-6}.

THE ALKANES OR METHANE SERIES

The alkanes have been referred to historically as the *methane series* of hydrocarbons. The first members are listed in Table 18.1 The alkane hydrocarbons are found chiefly in natural gas and petroleum. The most abundant compound in natural gas is methane (molecular formula CH_4). There are scores of compounds that contain only carbon and hydrogen. How does it happen that hydrocarbons are so numerous and have such similar properties?

The answer involves the manner in which atoms attach themselves to each other to form molecules of compounds. In an earlier chapter we found that the electronic structure of the carbon atom in the ground state is $1s^22s^22p^2$. The first example of an atom having two $1s$ electrons was helium and we know it has a high ionization potential. When we add to this fact the further fact that in a carbon atom these $1s^2$ electrons are "buried" far from the exterior outer shell of the atom, we can assume they will remain undisturbed during chemical change. In other words, the $1s^2$ electrons in a carbon atom do not partake in bond formation. Let us represent the structure of this ground state carbon atom as shown in Fig. 18.1. Note that the $1s^2$ electrons are paired, as are also the $2s^2$ electrons, but as a result of mutual repulsion the $2p$ electrons are in separate orbitals; their spins do not need to be opposed. The $1s^2$ electrons have the lowest energies, the $2p$ electrons the highest.

Theoretically, we should be able to excite or activate a carbon atom by supplying it with energy. Suppose we succeed in giving it the right amount of energy to raise or "promote" one of the $2s^2$ electrons to the $2p$ orbital that has no electron in it—the vacant orbital (compare Figs. 18.1 and 18.2). By reacting with a hydrogen atom that is similarly excited, it can now form a bond; a new molecular bonding orbital is formed that attaches the carbon atom to the hydrogen atom by means of an ordinary σ bond. Because the new orbital was formed from electrons coming from dissimilar orbitals, it is referred to as hybrid. When four such bonds attach four hydrogen atoms to each carbon atom, the result is a molecule of methane, whose chemical formula is CH_4. In order to achieve symmetry, the bonds abandon the 90° angle in one plane (see stick formula of methane, below) and instead point to the corners of a regular tetrahedron (as in Figs. 16.6 and 21.3).

For more than a half century we have been thinking of valence bonds as containing electrons. So, using a pair of dots to represent a pair of electrons whose spins are opposed, we represent the structural electron dot formula of methane (molecular formula CH_4) as shown below. For more than a century it has been customary to express the structure of methane on a plane surface with a structural stick formula, as shown. Each dash or stick represents what has been called a "single" valence bond.

```
    H           H
    ..          |
H : C : H    H—C—H
    ..          |
    H           H
 methane     methane
```

In the structural stick formula of the methane series, and of the other series discussed in this chapter, each carbon atom has a total of four bonds.

The structural electron dot formula and stick formula of ethane (molecular formula C_2H_6) can be written as follows:

```
    H   H           H   H
    ..  ..          |   |
H : C : C : H    H—C—C—H
    ..  ..          |   |
    H   H           H   H
    ethane          ethane
```

Ethane is a component of natural gas.

Similarly, there is another hydrocarbon, propane, whose molecular formula is C_3H_8. Its structural stick formula can be written

```
   H  H  H
   |  |  |
H—C—C—C—H
   |  |  |
   H  H  H
   propane
```

Commercial liquefied petroleum gas (LPG) sold as "bottled gas" for heating homes, mobile homes, and cabins is now predominantly propane instead of butane (Table 18.2).

Methane, ethane, and propane have remarkably similar properties. When arranged in order of increasing molecular weight, each molecular formula differs from its nearest neighbor by the constant amount that can be written CH_2. Such a group of compounds in organic chemistry is called a homologous series. This particular homologous series, in which the first three compounds are methane, ethane, and propane, is usually called the *methane series* of hydrocarbons.

With this understanding of the general nature of the methane series, we can consider more detailed facts about some compounds in the series. An inspection of the values of boiling points given in Table 18.2 shows that there is a regular increase with increasing numbers of carbon atoms in the molecules. Methane has a very slight solu-

Table 18.2
Information about eleven members of the methane series

name of compound	molecular formula $C_nH_{(2n+2)}$	molecular weight	boiling point, °C[a]	physical state[b]	common name
methane	CH_4	16	−161	gas	natural gas
ethane	C_2H_6	30	− 89	gas	
propane	C_3H_8	44	− 42	gas	bottled gas
butane	C_4H_{10}	58	− 1	gas	
pentane	C_5H_{12}	72	36	liquid	naphtha
hexane	C_6H_{14}	86	69	liquid	
heptane	C_7H_{16}	100	98	liquid	gasoline
octane	C_8H_{18}	114	126	liquid	
nonane	C_9H_{20}	128	150	liquid	
decane	$C_{10}H_{22}$	142	174	liquid	kerosene
undecane	$C_{11}H_{24}$	156	195	liquid	

[a]The data given here are for the normal, or straight-chain, hydrocarbons.
[b]Room temperature and pressure of one atmosphere.

bility in water. As we go to other hydrocarbons of higher molecular weights, the solubility decreases, so that as a general rule we can state that hydrocarbons are not soluble in water.

A *homologous series* in organic chemistry is characterized by a regular gradation in physical properties, by generally similar chemical properties, and by a type formula ($C_nH_{(2n+2)}$ for the methane series).

THE ALKENES OR ETHYLENE SERIES

The alkene series of hydrocarbons is referred to also as the *ethylene series*. The first few members are listed in Table 18.3. Alkene molecules have two hydrogen atoms less than the corresponding alkane molecules, except that an alkene corresponding to methane and having the molecular formula CH_2 does not exist.

The molecular formula of ethylene is C_2H_4. In order to account for all the valence electrons furnished by two carbon atoms and four hydrogen atoms, it is assumed that a molecule of ethylene contains one double bond and is represented by the structural formula,

$$\begin{array}{c} \text{H}\quad\text{H} \\ |\quad\;| \\ \text{H—C=C—H} \end{array}, \qquad CH_2{=}CH_2$$

ethylene

Each carbon atom has a total of four bonds. Bond angles and positions of atoms in space are usually not shown in structural formulas. Compounds containing multiple bonds between carbon atoms are said to be unsaturated because they have fewer hydrogen atoms than they could have if the multiple bonds were absent. The high chemical reactivity of multiple bonds permits a wide array of chemical reactions from which to develop synthetic routes to many desired ends.

Often the mature green tomato is exposed commercially to ethylene gas to make it turn red. The gas is also used by some processors to change the greenish-yellow skin of some tree-ripened oranges to an orange color.

Table 18.3
Alkenes (C_nH_{2n}, general formula)

C_2H_4	ethylene
C_3H_6	propylene
C_4H_8	butenes
C_5H_{10}	pentenes

Table 18.4
Alkynes (C_nH_{2n-2}, general formula)

C_2H_2	acetylene
C_3H_4	propyne
C_4H_6	butynes
C_5H_8	pentynes

Similarly, one molecule of propylene (molecular formula C_3H_6) would be assumed to have one double bond and be represented by the structural formula,

```
    H
    |
H—C—C=C—H,      CH3CH=CH2
    |   |   |
    H   H   H
```

propylene

THE ALKYNES OR ACETYLENE SERIES

The alkyne series of hydrocarbons is referred to also as the acetylene series. The first few members are listed in Table 18.4. The molecular formula of acetylene is C_2H_2. In accounting for all the valence electrons furnished by two carbon atoms and two hydrogen atoms, one may assume that a molecule of acetylene contains one triple bond and is represented by the structural formula,

$$H—C\equiv C—H$$

acetylene

We note again that each carbon atom has a total of four bonds. Acetylene is produced commercially by the dehydrogenation of natural gas. It is used in cutting and welding metals, and also as a chemical starting material (for example, in the making of valuable plastics).

BENZENE: AROMATIC COMPOUNDS

The molecules of the alkane, alkene, and alkyne hydrocarbons are characterized by open chains, whereas the aromatic hydrocarbons are characterized by closed carbon rings and a bond system similar to that found in benzene (structural formula on p. 377). The open-chain hydrocarbons are referred to as *aliphatic hydrocarbons*.

The term "aliphatic" originally referred to compounds derived from fats, just as the term "aromatic" referred to natural products with a characteristic aromatic odor. These words have now been generalized in terms of the structures of molecules. Open-chain compounds not containing a benzene ring are called *aliphatic compounds*. *Aromatic compounds* are substances whose structure is related to that of benzene. The third large class of organic substances—the heterocyclic compounds, those with at least one element other than carbon as a member of the ring (for example, oxygen, nitrogen, sulfur)—will not be discussed.

Among the many homologous series of aromatic hydrocarbons, the most important ones are derived from benzene (molecular formula C_6H_6). Benzene is the simplest member of the class of aromatic compounds. Each benzene molecule contains a group of six carbon atoms linked together in the form of a hexagonal ring lying in a single plane with one hydrogen atom attached to each carbon atom. Alternate double and single bonds between the carbon atoms of the benzene molecule allow each carbon atom to have four bonds. How can this description of ring and bond structure be explained? One carbon atom with four valence electrons can help form four covalent bonds. This means that the carbon atoms are able to form chains and rings and still have valence electrons left over to form bonds with atoms of other elements.

The structural formula of benzene (molecular formula C_6H_6) is represented by the so-called benzene ring completed as follows:

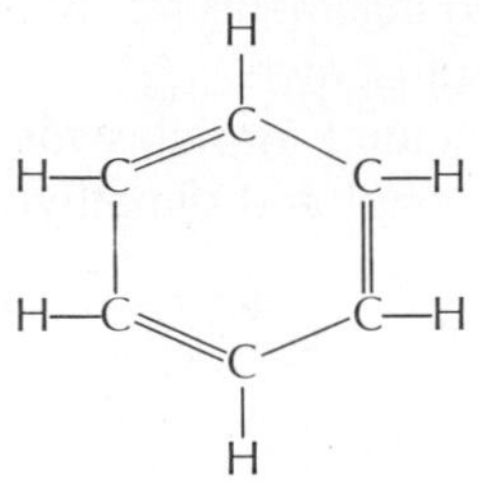

```
        H
        |
        C
     //   \
H—C        C—H
   |       ||
H—C        C—H
     \\   /
        C
        |
        H
```

benzene

Note that four valence bonds are assigned to each carbon atom.

Eventually it became evident that the bonds connecting the carbon atoms in a benzene molecule are all alike, but are different from the bonds between carbon and hydrogen atoms. Then it was assumed that the double bonds exchanged positions with the single bonds with such extreme rapidity that it was impossible to distinguish between double and single bonds. Such a condition was referred to as resonance. Resonance is not restricted to the chemistry of benzene and its derivatives; it is quite common in chemistry generally. The term has been retained and is still useful, but since the Bohr theory has been displaced by quantum mechanics, the concept of resonance has been modified accordingly. In fact, from the viewpoint of quantum mechanics, the bonds are not alternating or "resonating"; so at present the term is actually misleading. Nevertheless it has led to molecular orbital models for benzene structure that do correlate well with properties. According to these models six pairs of electrons in the plane of the benzene molecule connect the six carbon atoms to each other by means of ordinary σ bonds. Similarly, each carbon atom is attached by means of another σ bond to a hydrogen atom. But the remaining electrons are outside the plane of the benzene ring, half of them above and half of them below. The electrons that lie outside the plane of the benzene ring are called π (pi) electrons.

We have difficulty constructing adequate structural formulas for such molecules as ozone and sulfur dioxide and such ions as nitrate and formate. In these cases it is thought that there is a smear of π electrons that no one simple structural formula can represent. Fortunately, saturated members of the methane series of hydrocarbons present no such difficulty; they do not gain stability by means of resonance. The models that correspond to their structural formulas can account adequately for their properties.

Benzene and other aromatic compounds are produced from coal tar and petroleum products. Benzene and its derivatives are widely used in the production of drugs, dyes, plastics, insecticides, synthetic detergents, and nylon.

UNSATURATED AND SATURATED HYDROCARBONS

Unsaturated hydrocarbons are hydrocarbon molecules that have one or more pairs of carbon atoms linked by multiple bonds and contain less than the maximum possible number of hydrogen atoms. *Saturated hydrocarbons* are hydrocarbon molecules that have the carbon atoms linked only by single bonds and contain the maximum possible number of hydrogen atoms. Ethylene (one double bond), acetylene (one triple bond), and benzene (three double bonds) are examples of unsaturated hydrocarbons that we have studied. Molecules with multiple bonds exhibit high chemical reactivity. The multiple bonds serve as chemical reaction centers. Compounds containing carbon atoms that are attached to each other by means of either double or triple bonds will add hydrogen, chlorine, and other chemical reagents at these bonds; therefore they are called *unsaturated compounds*. As a class the alkanes—saturated hydrocarbons—are chemically very unreactive compounds, even though the designation of them in Table 18.2 as fuels would tend to indicate otherwise. Carbon–carbon and carbon–hydrogen bonds are insensitive to moderate heat; they also resist attack by acids and bases. Some examples of uses of both saturated and unsaturated hydrocarbons were given earlier in this chapter as a part of the discussion of each series of hydrocarbons.

ISOMERS

When butane-rich mixtures of hydrocarbons are very carefully separated, there seem to be two different compounds, both of which are butanes; that is, both have the same gas density and the same percentages of carbon and hydrogen. It would therefore be correct to write one molecular formula, C_4H_{10}, for both compounds. How can it be that two different substances seemingly have the same composition?

```
  H  H  H  H
  |  |  |  |
H—C—C—C—C—H
  |  |  |  |
  H  H  H  H
```

butane

The preceding formula represents the so-called straight-chain compound, but note that it is straight only when represented on a plane surface; in three-dimensional space the carbon atoms make angles similar to those of an old-fashioned rail fence. This butane is called normal butane and listed in reference books as *n*-butane. The other butane is assumed to have the following branched-chain structure:

```
  H  H  H
  |  |  |
H—C—C—C—H
  |  |  |
  H  |  H
     |
   H—C—H
     |
     H
```

isobutane

Isobutane and normal butane illustrate what is meant by the term isomers. *Isomers* are compounds that have the same molecular formula but different structural formulas, and hence different properties. The existence of isomers is called *isomerism*. Stereoisomers are compounds with the same molecular formula and the same structural formula but with different spatial arrangements of the atoms—that is, varieties of isomers.

If we look at the structural formulas for the compounds ethyl alcohol and dimethyl ether,

```
  H  H              H     H
  |  |              |     |
H—C—C—O—H         H—C—O—C—H
  |  |              |     |
  H  H              H     H
```

ethyl alcohol — dimethyl ether

we see that they are isomers. They have the same molecular formula, C_2H_6O, yet they are different compounds: One is an alcohol, and the other is an ether. The difference is emphasized by the fact that at room temperature ethyl alcohol is a liquid and dimethyl ether is a gas.

POLYMERS, POLYMERIZATION

By setting up proper conditions, molecules of unsaturated hydrocarbons can be made to react with each other. For example,

```
  H  H          H  H
  |  |          |  |
H—C=C—H   +   H—C=C—H   --catalyst-->
ethylene      ethylene

                        H  H  H  H
                        |  |  |  |
                      H—C—C—C=C—H
                        |  |
                        H  H
                       butylene
```

We see that one molecule of ethylene can be made to add itself to another molecule of ethylene to form a new substance. This new substance, butylene, contains the same elements in the same ratio by weight as ethylene, but it is not an isomer of ethylene because it has a molecular weight that is twice as great. Butylene is a dimer—two molecules that combined to form a single molecule.

A *polymer* is a compound composed of very large molecules that are formed by the repeated union of two or more, usually many, identical smaller molecules. Similarly, polymeric ions result from the repeated combination of two or more identical ions. The repeating identical molecules, or units, may be of a single kind (for example, glucose units) or they may be of several kinds (for example, consisting of some different kinds of the 24 naturally occurring amino acid units). As a rule, however, the small molecules from which the large molecules of synthetic polymers are formed are of a single kind. In this case there is a repeated union of many identical small molecules.

The process by which like molecules add to each other to form new substances with molecular weights that are multiples of the original is called *polymerization*. The high molecular weight substance formed by polymerization is called a polymer. Polymerization is important in making such modern products as plastics and synthetic rubber. The plastic polyethylene, for example, is formed by the polymerization of ethylene. Huge amounts of unsaturated hydrocarbons are required for the production of synthetic rubber.

BRANCHED-CHAIN HYDROCARBONS AND THE OCTANE NUMBER OF GASOLINES

Isobutane and butylene react to make an 8-carbon atom branched-chain product called isooctane that is used in the determination of the octane number of gasolines.

A liquid fuel determines the power of a gasoline engine primarily by the fuel's heat content and antiknock quality. Premature explosion, or detonation, in the firing chamber of an internal-combustion engine can cause vibration that we call "knock." This is unpleasant and is usually accompanied by a low efficiency in transforming the heat of combustion of gasoline (about 120,000 Btu/gal) into mechanical energy.

The knocking of an engine depends on a number of factors, two of the most important of which are (1) the amount of compression in the firing chamber and (2) the type of fuel used. A high compression of gaseous fuel (usually gasoline) and air (oxygen) in the firing chamber is desirable because as the compression increases, the ideal or theoretical efficiency of the engine also increases. At higher compressions the tendency to get improper combustion markedly increases. Efficiency would increase if it were not for improper combustion, which causes knocking.

A great amount of study and experimentation has been done on the kind of fuel required to get the proper kind of explosion. In order to get results that were to some extent quantitative, a standard for knock intensity had to be set up. This involved the use of a standard one-cylinder engine and a reference fuel. The fuel chosen was a mixture of *n*-heptane and isooctane, one of the isomers of normal octane. An arbitrary scale for knock intensity has been established with *n*-heptane given an octane number of 0 and isooctane an octane number of 100:

```
    H  H  H  H  H  H  H
    |  |  |  |  |  |  |
 H—C—C—C—C—C—C—C—H
    |  |  |  |  |  |  |
    H  H  H  H  H  H  H
```

n-heptane
0

```
        H           H
        |           |
     H—C—H      H—C—H
    H   |     H     |   H
    |   |     |     |   |
 H—C—C——C——C—C—H
    |   |     |     |   |
    H   |     H     H   H
        |
     H—C—H
        |
        H
```

isooctane
100

When *n*-heptane alone is used as the fuel in the standard engine, there is very bad knocking; when isooctane is the fuel, the engine runs much more smoothly and is practically knockless. A mixture of the two fuels performs better than *n*-heptane, but not so well as isooctane. By mixing *n*-heptane and isooctane, one can make a reference fuel that will be practically identical in knock intensity to that of any gasoline. The percentage of isooctane in a reference fuel (mixture of isooctane and *n*-heptane) that produces the same knock intensity as does the unknown gasoline being tested is designated as the *octane number* of the gasoline.

For approximately a half century, we have been rating gasolines as to their octane numbers. As methods of refining have been improved in response to the demand for better engine performance, the octane ratings have gradually increased from about 60 or less for regular gasolines to better than 90. One important factor that has brought about this improvement has been the production of increased percentages of branched-chain hydrocarbons, such as isooctane, in the gasoline. In the refineries straight-chain hydrocarbons are made to rearrange their atoms so as to increase the percentage of branched-chain molecules. The product has a higher octane rating.

A somewhat less important factor in determining octane number is the molecular weight of the hydrocarbon. In a homologous series, the lower molecular weight hydrocarbons have the higher octane ratings. For example, pure *n*-hexane would not knock as badly as pure *n*-heptane.

A spectacular way of increasing the antiknock rating, or octane number, of an otherwise finished gasoline is by adding a very small amount of tetraethyl lead (TEL) [formula $Pb(C_2H_5)_4$]. However, Federal laws against pollution of the atmosphere with poisonous lead compounds have caused manufacturers of automobiles to design engines that operate on gasolines containing no lead. The use of leaded gasolines would also gradually render ineffective the catalytic converters used with automobile engines designed to decrease emissions pollutants from automobile exhaust systems.

Oxidation products of the hydrocarbons

OXIDATION OF METHANE BY CHLORINE: THE CHLORINATED HYDROCARBONS

One should expect that chlorine, an oxidizing agent, would react with hydrocarbons, which are reducing agents. Chlorine and methane will react explosively if they are properly mixed and ignited by high temperature or sunlight. By diluting chlorine with large amounts of methane and using other precautions, one can make the reaction to occur slowly and stepwise rather than explosively:

$$\begin{array}{ccccc} \begin{array}{c} \text{H} \\ | \\ \text{H—C—H} \\ | \\ \text{H} \end{array} & + \ \text{Cl—Cl} & \rightarrow & \begin{array}{c} \text{H} \\ | \\ \text{H—C—Cl} \\ | \\ \text{H} \end{array} & + \ \text{HCl} \\ \text{methane} & \text{chlorine} & & & \text{hydrogen chloride} \end{array}$$

If properly carried out, this reaction yields methyl chloride and hydrogen chloride. Irrespective of whether it is industrially important, it is introduced here for two purposes: (1) to show a first step in the illustration of how a hydrocarbon can be converted step by step to a related substance and (2) to illustrate simple and important relationships involving the names of organic compounds.

Methyl chloride can be further oxidized to the compound whose structural formula is

$$\begin{array}{c} \text{Cl} \\ | \\ \text{H—C—Cl} \\ | \\ \text{H} \\ \text{methylene} \\ \text{chloride} \end{array}$$

Similarly, a third step in the oxidation produces chloroform, and a fourth and final step gives carbon tetrachloride:

$$\begin{array}{cc} \begin{array}{c} \text{Cl} \\ | \\ \text{H—C—Cl} \\ | \\ \text{Cl} \end{array} & \begin{array}{c} \text{Cl} \\ | \\ \text{Cl—C—Cl} \\ | \\ \text{Cl} \end{array} \\ \text{chloroform} & \text{carbon tetrachloride} \end{array}$$

Sometimes it is helpful to estimate by inspection the amount of oxidation that has theoretically produced a given hydrocarbon derivative. This can be done by counting the valence bonds on any particular carbon atom attached to elements that are oxidizing agents, remembering that the hydrogen in the hydrocarbon is a reducing agent. For example, in methane all the bonds attached to the carbon atom are attached to hydrogen; there has been no oxidation. In methyl chloride there is one single bond attaching a chlorine atom, an oxidizing agent, to the carbon atom. This carbon atom can conveniently be thought of as representing the first stage of oxidation. Similarly, in methylene chloride the carbon atom is in the second stage of oxidation. In chloroform the carbon atom is in the third stage of oxidation; and in carbon tetrachloride it is in the fourth stage because all four valence bonds attach the carbon atom to chlorine.

The halogens substitute in the methanes for hydrogen on a one-to-one basis, but substitution need not be restricted to one element of the halogens. The freons, a group of compounds used in the compressors of refrigeration units (air conditioners, refrigerators, freezers), include $C_2H_2Cl_2F_2$ and $C_2Cl_2F_4$. Freon-12, a common refrigerant, is CCl_2F_2.

INTRODUCTION TO ORGANIC NOMENCLATURE

By methods similar to those just described, one hydrogen atom of ethane can be replaced by one chlorine atom. What would be a systematic name for the compound? Probably the best system of nomenclature for the beginner to use is the one in which the name has the suffix *-ide,* as it would in a binary compound; in other words, the compound is named on the basis of the two groups making the molecule: (1) the alkyl group and (2) the halogen atom. The *alkyl group* represents everything in a hydrocarbon of the methane series except one hydrogen atom. Examples of alkyl groups are methyl (CH_3), ethyl (C_2H_5), propyl (C_3H_7), and so on. Therefore the structural formula of ethyl chloride would be as follows:

```
   H  H
   |  |
H—C—C—Cl
   |  |
   H  H
```

ethyl chloride

SOME PROPERTIES AND USES FOR HALOGEN-SUBSTITUTED PRODUCTS OF HYDROCARBONS

The physical properties of these compounds resemble those of the hydrocarbons from which they are derived. This is particularly true of solubility. Any hydrocarbon is usually completely miscible, or soluble in all proportions, in any other hydrocarbon that is at all near it in the homologous series. The mixing of two hydrocarbons is unlike the dissolving of most acids, bases, and salts in that it entails little heat effect; that is, the change is neither exothermic nor endothermic to any marked degree. If one leaves the hydrocarbons in contact, they will mix just by diffusion. This change, although involving practically no change in energy, is accompanied by an increase in entropy. The solution formed is near to what a physical chemist calls an ideal solution. Contrasted to these ideal solutions are those such as sodium hydroxide in water; heat quite evidently is evolved on mixing water and sodium hydroxide.

A halogen derivative of a hydrocarbon dissolves well in another halogen derivative and also in other hydrocarbons. Such facts explain the extensive use of chlorinated hydrocarbons as solvents.

Methyl chloride is not very important to the layman, although it has been used as a refrigerant. Likewise, methylene chloride is not a name that the general public needs to remember; this substance is used extensively in such solvents as paint and varnish removers. On the other hand, chloroform and carbon tetrachloride are names that occur so often that more attention should be given to them.

Chloroform is a liquid that evaporates rapidly because of its low boiling point. It is capable of producing complete anesthesia, or paralysis of the sensory apparatus of an animal. In the days when the family physician could not conveniently get the patient to a hospital, chloroform was the most used anesthetic, probably because it was the most convenient one to administer. With the coming of greater numbers of hospitals conveniently located and more elaborate equipment operated by specialists in the field, chloroform has decreased a great deal in relative importance. It is still used particularly for the painless killing of unwanted animals.

Many pounds of carbon tetrachloride are used each year. It is very difficult to get carbon tetrachloride to burn; its only carbon atom is in the fourth stage of oxidation. Carbon tetrachloride is a solvent. It is used in dry-cleaning plants. The very low-boiling fractions of petroleum, often called "naphthas," are excellent dry-cleaning agents and are used for cleaning those kinds of cloth which would be injured by water. These hydrocarbons, however, have one very great disadvantage; there is great danger of fire or even explosion if their vapors become ignited. Therefore naphtha is often diluted with carbon tetrachloride to make a commercial dry-cleaning agent. The use of pure carbon tetrachloride is banned because of the toxicity of its fumes.

Similarly, parts of machinery, such as typewriter keys, are often cleaned with carbon tetrachloride. It and other similar chlorinated hydrocarbons are used in large amounts in factories for degreasing metals. Prolonged breathing of the vapors of carbon tetrachloride should be avoided, for they are now known to be injurious to body tissues in the lungs and liver and can cause death.

OXIDATION OF HYDROCARBONS BY OXYGEN

When hydrocarbons are being used as fuel, they are ordinarily being oxidized by oxygen. In this case the emphasis is not so much on the products formed as on the energy released by the chemical change. In order to release the maximum amount of energy per pound of hydrocarbon consumed, the oxidation must be complete. Whenever any hydrocarbon is oxidized completely by oxygen, the products formed are always carbon dioxide and water.

Some organic compounds, derivatives of the hydrocarbons, consist of a carbon radical and a functional group, as in Table 18.5. This group often contains oxygen. The symbol R in the type formulas shown in the table represents some hydrocarbon group (usually $CH_3—$ or $C_2H_5—$). Some functional groups may be attached to two R groups, designated by R and R′. The major determinant of the chemical reactions of such compounds is the functional group. Many different oxygen derivatives, containing functional groups, exist because oxygen forms two covalent bonds, as shown in the table.

We observe under the type formula in the table that a ketone is very similar to an aldehyde, except that the $=O$ in the ketone is not bound to an end carbon as in the aldehyde. Acetone (dimethyl ketone), the most important ketone commercially, is widely used as a solvent for waxes, plastics, nail polish, and lacquers. In this chapter we mention or discuss briefly the derivatives listed in the table. The acid derivatives of the hydrocarbons are designated as *organic acids* and the amine functional group is mentioned in connection with proteins and amino acids.

METHODS OF PREPARING ALCOHOLS

Let us now consider the partial oxidation of a simple hydrocarbon, using oxygen as the oxidizing agent. This partial oxidation is difficult to control and until recently was more theoretical than practical. Under proper conditions, it is possible to cause methane to react with oxygen to yield methyl alcohol (CH_3OH):

$$\underset{\text{methane}}{2CH_4} + \underset{\text{oxygen}}{O_2} \rightarrow \underset{\text{methyl alcohol}}{2CH_3OH}$$

Table 18.5
Common derivatives of the hydrocarbons

name of derivative	functional group	name of function group	type formula
alcohol	—OH	hydroxyl	R—OH
ether	—O—		R—O—R′
aldehyde	$-\overset{\overset{\displaystyle O}{\|\|}}{C}-H$		$R-\overset{\overset{\displaystyle O}{\|\|}}{C}-H$
ketone	$-\overset{\overset{\displaystyle O}{\|\|}}{C}-$	carbonyl	$R-\overset{\overset{\displaystyle O}{\|\|}}{C}-R'$
acid	$-\overset{\overset{\displaystyle O}{\|\|}}{C}-OH$	carboxyl	$R-\overset{\overset{\displaystyle O}{\|\|}}{C}-OH$
ester	$-\overset{\overset{\displaystyle O}{\|\|}}{C}-O-$		$R-\overset{\overset{\displaystyle O}{\|\|}}{C}-O-R'$
amine	$-NH_2$	amino	$R-NH_2$

In order to show how the atoms are attached to each other in the molecule of methyl alcohol, the structural formula is necessary:

$$H-\overset{\displaystyle H}{\overset{|}{\underset{\displaystyle H}{\underset{|}{C}}}}-O-H$$

or

$$H-\overset{\displaystyle H}{\overset{|}{\underset{\displaystyle H}{\underset{|}{C}}}}-OH$$

methyl alcohol

Similarly, by the partial oxidation of ethane, ethyl alcohol can be formed as follows:

$$\underset{\text{ethane}}{2C_2H_6} + \underset{\text{oxygen}}{O_2} \rightarrow \underset{\text{ethyl alcohol}}{2C_2H_5OH}$$

The structural formula of ethyl alcohol (ethanol, grain alcohol) is

$$H-\overset{\displaystyle H}{\overset{|}{\underset{\displaystyle H}{\underset{|}{C}}}}-\overset{\displaystyle H}{\overset{|}{\underset{\displaystyle H}{\underset{|}{C}}}}-O-H$$

or

$$H-\overset{\displaystyle H}{\overset{|}{\underset{\displaystyle H}{\underset{|}{C}}}}-\overset{\displaystyle H}{\overset{|}{\underset{\displaystyle H}{\underset{|}{C}}}}-OH$$

ethyl alcohol

Such alcohols, properly called *primary* alcohols, should be thought of as consisting of an alkyl group (for example, C_2H_5) attached to the hydroxyl (OH) group, or radical. The general formula of this alkyl group is C_nH_{2n+1}. As further illustrations, propyl alcohol could have its formula written C_3H_7OH, and butyl alcohol, C_4H_9OH.

It has only been in recent years that alcohols have been prepared commercially by the partial oxidation of hydrocarbons. A half century or more ago the simplest member of this series of compounds was called "wood alcohol" more often than methyl alcohol. This was because industrially it was always manufactured by the destructive distillation of wood. Immediately following World War I a relatively inexpensive synthetic process was developed by which a purer product, methyl alcohol, could be made:

$$\underset{\text{carbon monoxide}}{CO} + \underset{\text{hydrogen}}{2H_2} \xrightarrow{\text{catalyst}} \underset{\text{methyl alcohol}}{CH_3OH}$$

A mixture of hydrogen and carbon monoxide gases that was under very high pressure was passed over a suitable catalyst and made to combine, forming methyl alcohol. Chemically, wood alcohol and methyl alcohol are identical, but wood alcohol is not so pure as synthetic methyl alcohol, commonly called *methanol*.

Until recently all commercial ethyl alcohol was prepared by the fermentation of dextrose, a sugar, that came from sources as molasses, grain, potatoes, and pineapple:

$$\underset{\text{dextrose}}{C_6H_{12}O_6} \xrightarrow{\text{fermentation}} \underset{\text{ethyl alcohol}}{2C_2H_5OH} + \underset{\text{carbon dioxide}}{2CO_2}$$

Industrial alcohol is ethyl alcohol that has been distilled from a dilute water solution of ethyl alcohol to a concentration of 95 percent alcohol and 5 percent water. About one-half of all industrial alcohol is made from petroleum by the catalytic addition of water to ethylene; most of the remainder is made by the fermentation of sugar, as above.

PROPERTIES AND USES OF ALCOHOLS

These low molecular weight alcohols (in contrast to the heavier alcohols such as glycerine, soon to be discussed) are not very soluble in hydrocarbons nor in the halogen derivatives of the hydrocarbons. They are miscible in all proportions in water. Many of their uses are related to their great solubility in water. For example, either methyl alcohol or ethyl alcohol may be added to the water in car radiators in winter in colder climates in order to get a solution that will have a freezing point far below 0°C. Methanol, synthetic methyl alcohol, is the common temporary type of antifreeze used in automobiles. A permanent type of antifreeze is discussed later in this chapter.

Ethyl alcohol is always present in intoxicating beverages and liquors. Beer usually contains approximately 3 percent of alcohol. The percentage of alcohol in wine is greater — usually from 12 to 20 percent by volume. The percentage by volume of alcohol in whiskey, gin, and brandy can often be determined by simply noting the so-called proof spirit that is advertised. Many modern whiskies are advertised to be 90 proof, meaning that they are about 45 percent alcohol by volume. Pure alcohol, often called "absolute" alcohol, is 200 proof. Such absolute alcohol, though more likely methyl alcohol than ethyl, is now sold under a trade name and added to gasoline to keep ice from forming in the fuel line and carburetor of an automobile. If properly used, it is very effective as a de-icer.

Ethyl alcohol, which may be used as a beverage and for this purpose is heavily taxed, is "denatured" for tax-free uses by the addition of an obnoxious compound to make it unfit for internal consumption. Such alcohol is not subject to the Federal internal revenue tax.

Although ethyl alcohol is widely consumed as a drink, it should be clearly understood that a person must not drink any material that contains methyl alcohol (methanol). It is exceedingly poisonous and may cause blindness if not death. Each year a rather large number of persons, who obviously thought there was only one kind of alcohol, have lost their lives because they took methyl alcohol internally.

Both methyl and ethyl alcohol are used extensively as solvents. Ethyl alcohol is much used as a solvent in preparing perfumes and medicines. A 70-percent solution of isopropyl alcohol with water is sold as rubbing alcohol.

If they were less expensive, the low molecular weight alcohols could be used extensively instead of gasoline in automobiles. They accelerate cars splendidly and have high antiknock ratings. Because some energy is lost in the exothermic reaction by which an alcohol is prepared from a hydrocarbon, the latter will always yield more heat per pound when it burns. It should not be surprising then that ordinary gasoline, a hydrocarbon, will ordinarily give more miles per gallon than alcohol. While making a short trial run, however, a race driver will often use alcohol as his motor fuel.

ALDEHYDES

A class of hydrocarbon derivatives that has terminal carbon atoms in the second stage of oxidation with oxygen is known as *aldehydes*. The common commercial method for the preparation of formaldehyde involves the partial oxidation of methyl alcohol:

$$\underset{\text{methyl alcohol}}{2CH_3OH} + O_2 \rightarrow \underset{\text{formaldehyde}}{2CH_2O} + 2H_2O$$

Similarly, we might oxidize ethyl alcohol to form its derivative, which is usually called acetaldehyde:

$$\underset{\text{ethyl alcohol}}{2C_2H_5OH} + O_2 \rightarrow \underset{\text{acetaldehyde}}{2C_2H_4O} + 2H_2O$$

The structural formulas of formaldehyde and of acetaldehyde are

```
   H            H  H
   |            |  |
H—C=O       H—C—C=O
                |
                H
formaldehyde  acetaldehyde
```

These structural formulas make clear that each compound has one carbon atom that is in the second stage of oxidation, as indicated by the double bond joining the carbon atom with an oxygen atom.

PROPERTIES AND USES OF ALDEHYDES

Formaldehyde is probably the aldehyde most widely known. It is a gas that is very soluble in water. It is often sold as a 40-percent aqueous solution called formalin. It was formerly used extensively as a germ killer and still is used in large quantity in biological laboratories for the preservation of the bodies of dead animals. The "pickling fluids" used for this purpose often contain methyl alcohol, formaldehyde, and phenol (carbolic acid). The undertaker's embalming fluid includes formaldehyde as a preservative.

By far the largest amounts of formaldehyde are used in the manufacture of plastics. Bakelite, melamine, and a number of others are made by reacting formaldehyde with some compound like phenol or urea.

Only an aldehyde can be oxidized to an acid by gaining one oxygen atom and losing no hydrogens. As a rule the aldehydes can be easily oxidized to yield their corresponding organic acids. In this respect they are all similar. In many other respects, however, they differ widely in properties. Formaldehyde has a very disagreeable odor and would be poisonous if breathed in large amounts or taken internally. Many other aldehydes, however, have pleasing odors and are considered harmless if taken in reasonable amounts. For example, benzaldehyde, sometimes called synthetic oil of bitter almonds, is commonly used for flavoring such foods as maraschino cherries.

PREPARATION AND OCCURRENCE OF ORGANIC ACIDS

When proper conditions are employed, an aldehyde can be oxidized to an organic acid. Formaldehyde, for example, can be oxidized to formic acid:

$$\underset{\text{formaldehyde}}{2CH_2O} + O_2 \rightarrow \underset{\text{formic acid}}{2CH_2O_2}$$

The structural formula of CH_2O_2 is

```
   O
   ||
H—C—O—H
```

or

```
   O
   ||
H—C—OH
formic acid
```

Formic acid is not often met by the ordinary person except when he is bitten or stung by an insect. One of the earliest sources of formic acid was the bodies of ants, and its name was derived from the Latin name for ant (*formicus*). Formic acid is used in the leather, rubber, and textile industries as an inexpensive acidifying agent.

The group of atoms characteristic of most organic acids is

$$\begin{array}{c} \text{O} \\ \| \\ \text{—C—OH} \end{array}$$

which is commonly known as the *carboxyl group*.

The second member of the homologous series that begins with formic acid is called acetic acid. It is very common. Ordinary vinegar contains from 3 to 6 percent of acetic acid, the rest being chiefly water:

$$\begin{array}{ccccccc} & & \text{H} & & \text{O} & & \\ & & | & & \| & & \\ \text{H} & — & \text{C} & — & \text{C} & — & \text{O—H} \\ & & | & & & & \\ & & \text{H} & & & & \end{array}$$

acetic acid

Acetic acid is ordinarily prepared by the oxidation of ethyl alcohol, without isolating the intermediate product, which would be acetaldehyde. Originally this was done by fermentation, starting with fruit juice. One kind of microorganism caused the sugar to change chemically to ethyl alcohol, and then another kind of microorganism caused the ethyl alcohol to react with dissolved oxygen to form vinegar. By stopping the fermentation at the desired time, one could get either the mild alcoholic beverage called "hard cider" or the flavoring and preserving material called "cider" vinegar. The modern factory method for the preparation of acetic acid is essentially the same in principle except that the ethyl alcohol is oxidized by atmospheric oxygen in the presence of a catalyst; microorganisms are not employed. "White" vinegar is colorless acetic acid and water.

Propionic acid is the third member of this homologous series of acids. It does not occur in everyday life, but one of its derivatives, lactic acid, is always present in milk that has been allowed to sour, or ferment, naturally.

$$\begin{array}{ccccccc} & & \text{H} & & \text{OH} & & \text{O} \\ & & | & & | & & \| \\ \text{H} & — & \text{C} & — & \text{C} & — & \text{C—OH} \\ & & | & & | & & \\ & & \text{H} & & \text{H} & & \end{array}$$

lactic acid

Lactic acid reacts with baking soda to cause the dough of bread, cakes, and cookies to rise:

$$\underset{\text{lactic acid}}{CH_3CHOHCOOH} + \underset{\substack{\text{sodium} \\ \text{bicarbonate}}}{NaHCO_3} \rightarrow CH_3CHOHCOONa + H_2O + \underset{\substack{\text{carbon} \\ \text{dioxide}}}{CO_2} \uparrow$$

The bubbles of gaseous carbon dioxide make the dough "rise."

Lactic acid is a product that occurs in trace amounts during muscle contraction. If it is not consumed quickly, soreness results.

Butyric acid is the fourth member of the series. It is present in large amounts in rancid butter and in some varieties of "sharp" cheese. In fact, the odors of these materials are due to butyric acid and other similar acids:

$$\text{glyceryl butyrate} + \text{water} \xrightarrow{\text{heat or catalyst}} \text{butyric acid} + \text{glycerine}$$

Other organic acids, such as citric acid and malic acid, occur very commonly in fruits and are of importance in nutrition. Vitamin C is also an organic acid.

PROPERTIES AND USES OF ORGANIC ACIDS

The physical properties of the common organic acids vary widely. Some of these acids are liquids and some are solids. Many of them have displeasing odors. Some, such as oxalic and citric acids, have very little odor. Still others, such as malic acid, present in apples, have a very pleasing odor. Their great similarity lies in the set of chemical properties that cause them to be acids. All of them will neutralize bases. Most are weak acids and will therefore increase the speed of corrosion of iron but will not rapidly dissolve iron as does hydrochloric acid.

Common paring knives frequently contain a small percentage of chromium to decrease the tendency of organic acids to corrode the metal.

As previously implied, organic acids are important in many materials consumed as food. Usually the acids are present in the foodstuff when it is harvested, as in nearly all fruits. In other cases it is supplied while the food is being prepared, as in the making of pickles. In any case there is the tendency for the acid to cause some sourness of taste, though this is sometimes at least partially masked by the presence of large amounts of sugar. The common organic acids can be assimilated easily by the body. Upon oxidation in the body they behave similarly to carbohydrate (starch, sugar). In fact, in classifying the nutrients in food such as apples, the total number of calories would probably be classified as being derived from only the nutrients (1) carbohydrate, (2) fat, and (3) protein, without any mention of those coming from organic acids. We will discuss these three nutrients shortly.

THE STRUCTURE AND SOME PROPERTIES OF GLYCERINE

In addition to the simple alcohols—methyl, ethyl, propyl, and butyl—which each contain only one —OH (hydroxyl) group per molecule and which are related to saturated hydrocarbons—methane, ethane, propane, and butane—attention should be called to alcohols containing more than one —OH group in their structures. Two widely used alcohols in this category are glycerine and ethylene glycol. The structural formula of glycerine is

```
    H
    |
H—C—OH
    |
H—C—OH
    |
H—C—OH
    |
    H
```

glycerine

Glycerine differs from those simple alcohols discussed earlier in this chapter in that there are three hydroxyl (OH) groups in each molecule of glycerine, whereas there was only one hydroxyl group in each molecule of the simple alcohols. This difference in structure causes a great difference in properties. Monohydroxyl alcohols, such as methyl and ethyl alcohol, have very low viscosities (resistances to flow) and low boiling points. On a warm day in the early spring, car radiators containing methyl alcohol, or methanol, of low boiling point may "boil over"; this antifreeze is therefore said to be temporary or nonpermanent. The presence of more than one hydroxyl group in each molecule increases both the viscosity and the boiling point. Ethylene glycol (see p. 388) and glycerine have two and three hydroxyl groups per molecule, respectively. Their boiling points are higher than that of methanol, and they would therefore be classed as a permanent type of antifreeze, because water evaporates from their solutions more rapidly than does the solute, which is the antifreeze.

Glycerine is quite soluble in water, which makes it valuable for many purposes. Glycerine is viscous (the consistency of syrup) and tastes sweet. Such properties determine its uses.

SOME USES OF GLYCERINE

Glycerine is a versatile compound with about 1600 different uses, among them the making of the explosive nitroglycerine, synthetic resins, tobacco products, cellulose films, cork products, toilet articles, and medicines.

Because of its viscosity and its sweet taste, it is used in making toothpaste. The glycerine is mixed with precipitated chalk ($CaCO_3$) or with a powdered phosphate to make the paste. Some pure, or castile, soap is included to produce a froth, and additional flavoring is added to suit the taste.

For many years glycerine has had an extensive use in medicines. It is a common component of many skin lotions. A great deal of it has been used in the preparation of tobacco products because it prevents excessive drying of the tobacco and improves the flavor.

PERMANENT TYPE OF ANTIFREEZE

The almost universally used antifreeze of the permanent type is ethylene glycol. It is useful for this purpose because it is completely water soluble and lowers the freezing point of the water; because of its high boiling point, it will not evaporate or "boil over" as will methyl alcohol, or methanol. The structural formula is

```
    H
    |
H—C—OH
    |
H—C—OH
    |
    H
```

ethylene glycol

Ethylene glycol is now made synthetically from the unsaturated hydrocarbon ethylene. It is more expensive than methanol and is the chief component of the so-called permanent type of antifreeze liquids sold under such trade names as Prestone and Zerex.

FACTORS AFFECTING BOILING POINTS

In the methane series there is a regular increase in boiling point as the molecular weight increases. This phenomenon is quite common; it also occurs among the halogens: Fluorine and chlorine are gases, bromine is a liquid, and iodine is a solid. Such nonpolar molecules as hydrocarbons and halogens have weak attractive forces between them; these are usually referred to as van der Waals forces.

If we now consider molecules that are polar, such as water or methyl alcohol, the attractive forces between molecules evidently become much greater. Only a slight inspection should convince us that this increased attraction does exist and, furthermore, that it is not primarily caused by increase in molecular weight. Water, with molecular weight of only 18, boils at 100°C, whereas the nonpolar oxygen boils at −183°C. The heavier oxygen molecule has a much lower boiling point. If we next compare ethane and ethyl alcohol, we find that ethane boils at −89°C and ethyl alcohol at about 78°C.

X ray studies of ice indicate that a hydrogen atom of one water molecule is adjacent to an oxygen atom of another water molecule.

```
H
|
O—H ··· O—H
        |
        H
```

This attractive force between a hydrogen atom of one water molecule and an oxygen atom of another water molecule is an illustration of what is meant by the term *hydrogen bonding*. It often binds hydrogen to fluorine, oxygen, nitrogen, and sometimes chlorine. It *is not a true chemical bond* because according to our present theories of atomic structure, a hydrogen atom can never be in direct association with more than one electron pair. So *these bonds must represent electrostatic attractions that are weak interactions*. Hydrogen bonds do not all have the same strength, nor do van der Waals forces, but we can approximate that, in general, a hydrogen bond is about 10 times as strong as a van der Waals force. A true chemical bond might be approximately 10 times as strong as a hydrogen bond.

As the number of OH groups in an alcohol increases, so does the boiling point increase. The reason for this is not so much an increase in molecular weight as it is an increase in the number of hydrogen bonds.

ESTERS AND RELATED MATERIALS

Methyl alcohol will react with formic acid to yield water and methyl formate:

$$\mathrm{H{-}\underset{\substack{|\\ H}}{\overset{\substack{H\\ |}}{C}}{-}O{-}H + H{-}O{-}\overset{\substack{O\\ \|}}{C}{-}H \rightarrow}$$

methyl alcohol formic acid

$$\mathrm{H_2O + H{-}\underset{\substack{|\\ H}}{\overset{\substack{H\\ |}}{C}}{-}O{-}\overset{\substack{O\\ \|}}{C}{-}H}$$

methyl formate

Methyl formate is a liquid with a low boiling point and an agreeable odor. It is not very soluble in water but dissolves well in hydrocarbons and in other esters. *Esters* may be regarded as those compounds, other than water, that are formed when alcohols react with acids. Fats such as lard, butter, tallow, olive oil, cottonseed oil, castor oil, linseed oil, and codliver oil are esters of the alcohol glycerine. *Fats* are always glyceryl esters that are formed by the reaction of fatty acids with glycerine. All fats are actually a mixture of esters, but sometimes one ester is present in large amounts.

Biological chemists often refer to fats as being *lipids*. They also use the terms "fats" and "oils," by which they merely differentiate between the ordinarily solid fats and the liquid oils; but the term "oil" tells very little about the chemical composition and chemical properties of a material.

Vegetable and animal oils are glyceryl esters; mineral oils, derived from petroleum, are hydrocarbons. In general, the glyceryl esters derived from saturated long-chain acids are solids at room temperature, and those derived from a mixture of organic acids having many unsaturated molecules are liquids. It is possible to change a vegetable oil, such as cottonseed oil, into a solid fat by a process known as "hydrogenation."

For centuries man has been extracting essential oils from naturally occurring materials, such as wood and herbs, and using them for perfumes, medicines, and flavors. Most odors and flavors are due to organic oxygen compounds belonging to the classes called alcohols, aldehydes, and esters.

CARBOHYDRATES

Most of these organic compounds, but not all, have the type formula $C_x(H_2O)_y$. Although it was this type formula that was responsible for their being called *carbohydrates,* the formula tells little or nothing about their structures or properties. The most commonly known carbohydrates are sugars and starches. Many sugars (for example, glucose, sucrose, dextrose, fructose, etc.) have an *-ose* ending.

The simpler carbohydrates are alcohols having many hydroxy groups in each molecule. In addition there is a carbonyl group (CO). If this carbonyl group is at the end of the molecule, the compound is not only a polyhydroxy alcohol but an aldehyde as well. Otherwise the presence of the carbonyl group makes the compound a ketone. Carbohydrates are usually solids that dissolve in water to give viscous, sweet solutions. The nature of these properties suggests that they are to a great extent caused by the many hydrogen bonds between molecules.

Although starch is a carbohydrate, it would at first seem not to fit the above description. It consists of giant, or macro, molecules that have resulted from what we call a "condensation" reaction (one in which the product consists of molecules that are larger than those constituting the starting material) between many molecules of simple low molecular weight sugars. Until its giant molecules are hydrolyzed, a starch is insoluble in water. Fortunately, the digestive fluids of a human being contain enzymes that ordinarily catalyze this reaction; the result is that many starches can be digested, and thus they have traditionally been one of the most important human foods. In nutritional tables starches that can be digested are listed under the heading "available carbohydrate." Other carbohydrates, like undigestible cellulose, can be assimilated by lower animals that have the necessary digestive apparatus.

PROTEINS

Among the great masses of the world population, proteins have been, and probably will continue to be, the most difficult food items to get in sufficient supply. If we ignore the water content, lean meat is chiefly protein. Upon digestion, protein yields a mixture of amino acids:

$$\text{protein} + \text{water} \xrightarrow{\text{catalyst}} \text{amino acids}$$

Like starch, the protein consists of giant molecules that are insoluble and cannot pass through the walls of the small intestine and get into the blood stream; the amino acids are soluble and can be assimilated. Amino acids always contain an amine group, which can be regarded as ammonia minus one hydrogen atom; symbolically it is represented as $\text{—}NH_2$. Glycine, also called aminoacetic acid, is an example of an amino acid that is common and simple but is not essential for human nutrition:

```
  H  H  O
  |  |  ||
H—N—C—C—OH
     |
     H
```

aminoacetic acid
(glycine)

The chemical change by which a protein is synthesized in a living plant or animal is the reverse of the one involved in digestion. Let us use aminoacetic acid to illustrate the first step in this process as follows:

```
  H  H  O           H  H  O
  |  |  ||          |  |  ||
H—N—C—C—OH + H—N—C—C—OH
     |              |
     H              H

                       H  H  O  H  H  O
                       |  |  ||  |  |  ||
 catalyst
──────→ H2O + H—N—C—C—N—C—C—OH
                          |        |
                          H        H
                        dipeptide
```

The preceding reaction is another example of the class called "condensation." The product of this reaction can further condense with a molecule of an amino acid. The result is the polymerlike macromolecule that we recognize as protein. The formation of insoluble giant molecules of starch (glycogen if in an animal body) and protein illustrates nature's way of building and storing material for living bodies.

Biochemistry

Biochemistry is that subdivision of chemistry which is concerned with the chemical study of materials that occur in living matter, and with the chemical processes that take place in plants and animals. Only during the last two centuries has knowledge about biochemistry been organized sufficiently to merit its being given the name "science." It is not difficult to find the reasons for this. Living matter is very complex and highly organized. It was easier to study and systematize inorganic chemistry first, chiefly because many of its materials were either already simpler or were more easily purified and characterized. After this and other related fields were developed, their findings were applied to the mysteries of biochemistry. In addition, during the twentieth century new techniques and new equipment, such as the electron microscope and much more efficient centrifuges, became available. The result has been such rapid progress during the last half century as to be almost revolutionary.

Although biochemistry is usually not considered a part of the physical sciences, it bears a close relationship to them. The brief introduction to biochemistry that follows is meant only to open the subject to the student's view. Some concepts may not be entirely clear, if only because in some instances the biochemists have a language of their own. But in a much greater number of instances we shall see that ideas previously learned will still be useful.

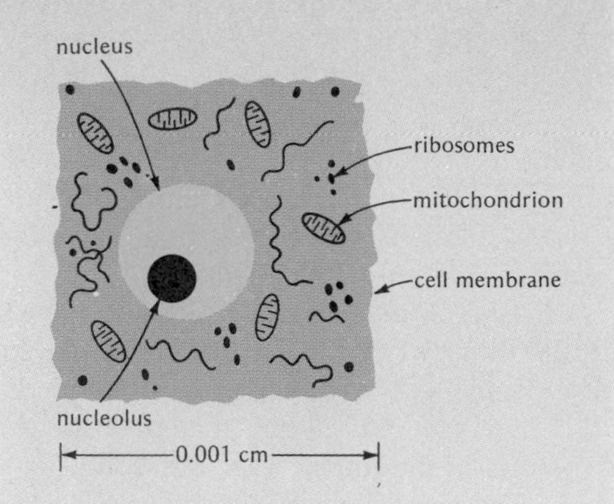

Figure 18.3
Model of an animal cell.

CELLS

Most plants and animals are composed of units or contained systems, called cells (Fig. 18.3). In the case of an animal the container usually consists of a thin wall; in the case of a plant the cell wall is often thicker and stiffer. Until rather recent times it was thought that the cell was filled with nothing more than a jellylike material called *protoplasm*. But twentieth century methods of separating and observing cell contents have revealed that instead of only a smooth, unordered jelly there are a number of discrete objects of various shapes in the cell. In fact, the term "protoplasm" has recently been falling into disuse.

The chemical changes in living cells by which energy is provided for the vital processes and activities and by which new material is assimilated for growth and repair purposes are called *metabolism*.

SOME BASIC CHEMISTRY OF CELL COMPONENTS

To the biochemist, monocarboxy acids such as acetic, propionic, and butyric acids are fatty acids. They have already been briefly discussed in this chapter in connection with organic acids and esters. Although they take part in the functioning of biological materials, we shall not stress them further in this introduction to biochemistry.

The carbohydrates that are oxidized in cells to liberate energy are chiefly the hexoses—those containing six carbon atoms per molecule. Earlier in the chapter we referred to a hexose called "dextrose." There are many isomeric forms of the hexoses; one with a terminal aldehyde group is D-glucose, and one in which the carbonyl (CO) group is not on the end is D-fructose. Another sugar which is very important in metabolism but which is not an article of commerce is the 5-carbon atom ribose, a pentose sugar, $C_5H_{10}O_5$.

When in aqueous solution, ribose molecules exist not as chains but as rings. Ribose has the following structure:

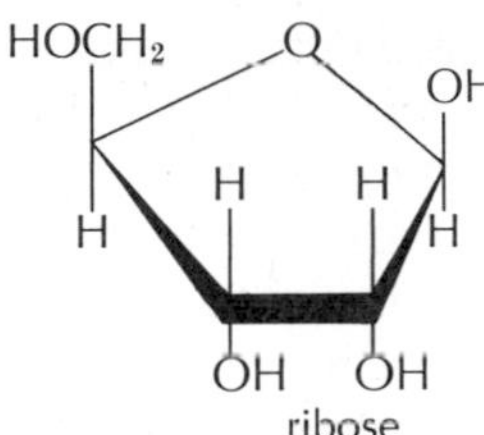

ribose

In this structure four carbon atoms (understood to be at the vertices) and the lone oxygen atom are in one plane. Half of the other groups are above the plane and half below. Another sugar that is very important as a building material in cell nuclei is deoxyribose. Its structure is surprisingly similar to that of ribose, considering the different functions of ribonucleic acid (RNA) and deoxyribonucleic acid (DNA), which will be discussed later.

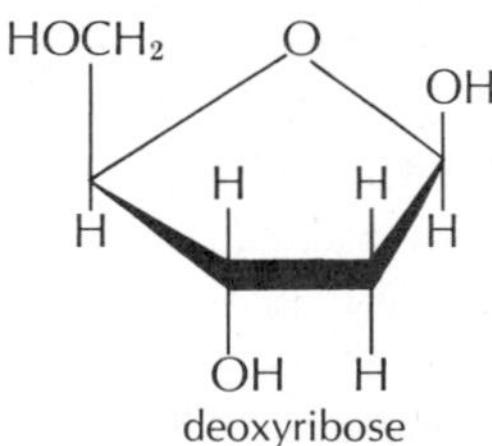

deoxyribose

The only difference is that the OH group, which in ribose was on the 2-carbon atom counting clockwise from the oxygen on the ring, has been replaced in deoxyribose by a hydrogen atom ("deoxy," alluding to this loss of oxygen by ribose).

A process in which large particles are formed by the union of small ones with the simultaneous elimination of small molecules (usually water) is called *condensation*. Phosphoric acid will undergo condensation reactions not only with itself but also with many other compounds, such as alcohols and sugars that contain OH groups. For example,

$$\underset{\text{ribose}}{C_5H_{10}O_5} + \underset{\text{phosphoric acid}}{H_3PO_4} \rightleftharpoons H_2O + \underset{\text{ribose monophosphate}}{C_5H_9O_5 \cdot H_2PO_3}$$

The H_2PO_3 group of atoms consists of the original H_3PO_4 less one of its OH groups. The biochemist has frequent need to refer to this H_2PO_3 group; he therefore usually uses the abbreviation Ⓟ instead of writing H_2PO_3 or trying to draw its structure in detail. Although we will not have occasion to use the abbreviation in this book, the student should be acquainted with it.

But why are biochemists interested in phosphoric acid and its derivatives? Essentially there are two reasons: (1) A great many of the organic derivatives of phosphoric acid react more readily and rapidly than the original organic compound would react, and (2) the organic phosphates serve as an extremely useful bank, or reservoir, of energy. For example, adenine triphosphate (ATP) will hydrolyze[1] ("break down") rapidly to give adenine diphosphate and phosphoric acid. Substances like phosphoric acid, enzymes, vitamins, and hormones are vital in that they allow essential chemical change to proceed rapidly without the necessity of raising the temperature to hasten the reaction. Unlike the physical condensation of gas to liquid, the chemical condensation of phosphoric acid to give di- and triphosphates is highly endothermic. Much of the energy released during the oxidation of sugar is absorbed as potential energy in the condensed phosphates. During the rapid hydrolysis reaction, this stored energy can be released, perhaps to do the mechanical work of muscle contraction. The energy was originally released by the oxidation of food; storing it as condensed phosphate makes it more readily available when needed.

The amino acids are of great importance in the chemistry of living matter. When we discussed them previously, we pointed out that they are the starting materials for the synthesis of the giant molecules called proteins. This synthesis is facilitated by the presence of phosphoric acid in the cell contents.

Another class of comparatively simple compounds that are vital in biochemical processes consists of organic bases. Important examples of these compounds are pyrimidine and purine:

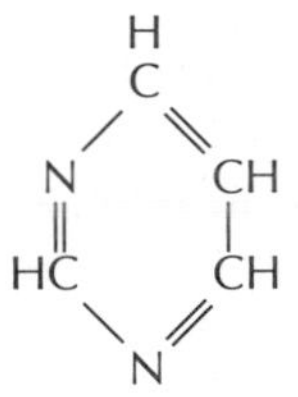

pyrimidine

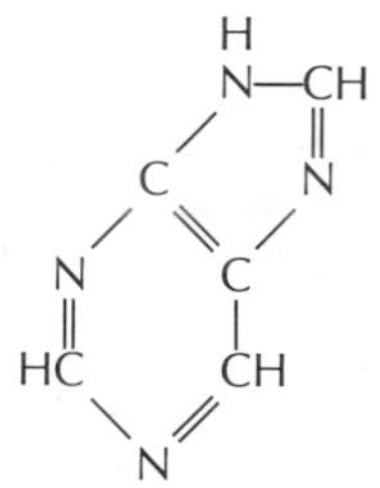

purine

Cytosine, thymine, and uracil are similar to pyrimidine. Adenine and guanine belong to the purine class of bases.

NUCLEOTIDES

Nucleotides are large molecules built up from organic bases, ribose or deoxyribose, and phosphate. They are called nucleotides because they are essential in the building up of the nucleic acids, DNA and RNA.

The condensation of three molecules of phosphoric acid with each other yields the group of atoms referred to as triphosphate. The large molecule built up from adenine, ribose, and triphosphate is properly called adenine nucleotide triphosphate but is abbreviated to ATP. When a muscle cell contracts, ATP hydrolyzes, releasing phosphoric acid and forming adenine diphosphate (ADP). These two products stimulate the oxidation of carbohydrate, which supplies the human body with a large amount of energy.

[1] Hydrolysis is a chemical reaction in which a compound reacts with the ions of water (H^+ and OH^-) to produce a weak acid, a weak base, or both.

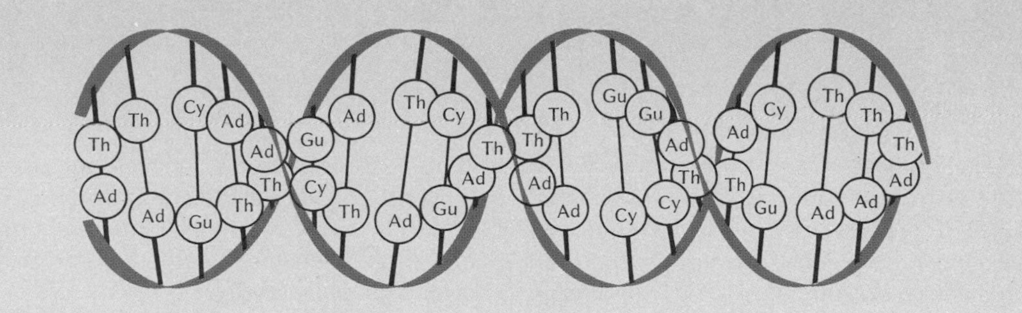

Figure 18.4
Model of DNA. Two strands of DNA in double helix are held together by hydrogen bonds between two organic bases: adenine (Ad); thymine (Th); cytosine (Cy); guanine (Gu). There can be many different sequences of organic bases besides those shown, but adenine binds only to thymine, and cytosine only to guanine.

NUCLEIC ACIDS

There are two kinds of nucleic acid, quite similar to each other: ribonucleic acid (RNA) and deoxyribonucleic acid (DNA). Most of the latter, in combination with very alkaline protein, is found in the nuclei of cells; the combination is called a nucleoprotein. When the DNA nucleoprotein is isolated it is found to be fibrous—that is, similar in appearance to asbestos or to ordinary white cotton sewing thread.

It has been demonstrated that protein macromolecules have the shape of a helix; DNA is similar except that, instead of consisting of single strands, as protein does, it has two strands twined about a common axis—a structure called a *double helix*. This is why it is fibrous. The structure of DNA may be represented as follows:

↓ one strand ↓ ↓ one strand ↓
deoxyribose—base ··· base—deoxyribose
| |
phosphate phosphate
| |
deoxyribose—base ··· base—deoxyribose
| |
phosphate phosphate
| |
deoxyribose—base ··· base—deoxyribose
↑ ↑
Complementary strands constituting DNA fiber

The two strands of DNA in the double helix are connected through organic bases by hydrogen bonding, shown above by dots (···) from one base to the other and also shown in Fig. 18.4 as the center bond of each set of three bonds—the other two (phosphate to phosphate) being ordinary covalent bonds.

FUNCTIONS OF DNA AND RNA

The chromosomes in the nucleus, the carriers of inheritance, are composed chiefly of DNA. Although all DNA is similar in general structure, there is an enormous range of variation in the order in which the purine and pyrimidine bases composing the linkages shown above and in Fig. 18.4 may appear, though always within the restriction that adenine must bind to thymine and cytosine to guanine. The specific DNA of the nucleus is the model for the building of the DNA of other nuclei. It is the DNA of the various genes, located on the chromosomes, that determines the characteristics of the individual organism—for example, whether a horse will be black or gray. Genes control the synthesis of all enzymes, as well as of the other proteins, and thus determine the basic chemical units of the cell.

The chemical structure of RNA is very similar to that of DNA except that most RNA molecules consist of a single strand;

ribose replaces deoxyribose and uracil replaces thymine as one of the pyrimidine bases. There are three classes of RNA, synthesized in the nucleus of the cell along a segment of one strand of a DNA molecule. Messenger RNA acts as a template, or mold, for protein synthesis. Carrying the genetic code formed by the sequence of the organic bases of DNA, it migrates from the nucleus to the site of protein synthesis in the cytoplasm.

If a certain protein is in excess supply, a chemical mechanism blocks the gene where the messenger coded for that specific synthesis is formed and the result is a stoppage in the building up of the particular protein. A need for a particular protein —either an enzyme or some other—produces an unblocking of a particular site on a DNA molecule and the formation on that site of the appropriate messenger RNA.

Another class of RNA—*transfer* RNA—acts as a link between messenger RNA molecules and the amino acid building blocks of protein. Ribosomal RNA, found in the ribosomes of the cell, is thought to have a structural function, but this has not yet been clearly established. It is known that the ribosomes are the site of protein synthesis.

SUMMARY

The subject of organic chemistry is introduced by way of hydrocarbons because these substances are comparatively simple and are of immense importance in everyday living. Aliphatic hydrocarbons are illustrated by the methane series of compounds, in which each carbon atom ordinarily is bonded to other atoms by means of four bonds. Aromatic hydrocarbons, illustrated by ring compounds such as benzene, have a more complicated bonding.

The promotion of a $2s$ electron to a $2p$ orbital theoretically gives rise to the opportunity to form a fourth carbon bond. In the aliphatic hydrocarbons the four valence bonds per carbon atom extend from the central carbon atoms to the four apexes of a perfect regular tetrahedron. The benzene molecule has a hexagonal structure similar to the arrangement of carbon atoms in one layer of graphite crystal.

Hydrocarbons are almost infinitely soluble in each other but not in water. Unsaturated hydrocarbons will add hydrogen and other chemical reagents. This is explained theoretically by assuming that some of the carbon atoms are attached to each other by means of double or triple bonds. Unsaturated hydrocarbons will polymerize. Such important industrial products as synthetic rubber, latex paints, and polyethylene are results of polymerization. Hydrocarbon is the basis for both the fuel and the lubricating oil of the automobile.

The halogen derivatives of the hydrocarbons have many physical properties similar to those of the hydrocarbons themselves. For example, a halogen derivative of a hydrocarbon has almost infinite solubility in a hydrocarbon.

Theoretically, an alcohol can be made from a hydrocarbon by partially oxidizing it with oxygen. A few of the higher molecular weight alcohols are actually produced industrially in this manner. Ethyl alcohol is the one that causes the intoxicating effects of alcoholic beverages. For use in beverages it is usually made by the fermentation of sugar. Much ethyl alcohol for industrial use is made by the catalytic addition of water to ethylene. The more poisonous methyl alcohol is formed by reacting carbon monoxide with hydrogen at high pressure.

Most formaldehyde is made by the partial oxidation of methyl alcohol. Large quantities of this aldehyde are consumed in the manufacture of plastics.

Organic acids are common in fruits and are an important component of our food. The physical properties of organic acids vary over quite a wide range, but the chemical properties are more nearly uniform. Organic acids taste sour and are weak acids, but they will still neutralize bases.

Low molecular weight alcohols, aldehydes,

and organic acids are soluble in water because of hydrogen bonding. Likewise, as the number of OH groups increases from ethyl alcohol to ethylene glycol to simple sugars, the viscosities of their solutions increase on account of hydrogen bonding. Having many bonds per molecule causes a compound to be a solid; sugars are an example.

Ethylene glycol is used as permanent antifreeze in car radiators. Another polyhydroxy alcohol, glycerine, is used to keep tobacco moist and to make it taste sweet; it is also used in lotions to keep human skin moist and free of cracks.

An ester is the compound other than water that is formed when an organic acid reacts with an alcohol. Low molecular weight esters have pleasing odors and are largely responsible for the fragrance of fruits. They are practically insoluble in water but quite soluble in hydrocarbons and other esters. Fats are usually esters derived from the polyhydroxy alcohol glycerine. Liquid fats are often called "oils," but this term tells little of the chemical nature of the substance.

It was once thought that there was a "vital force" in living matter which enabled it to synthesize the organic compounds necessary for plant and animal bodies. For about a century we have clearly seen that this is not strictly true. Synthetic organic products now contribute in an extremely important way toward good living.

Most chemical changes in living matter are catalyzed by enzymes. The enzymes are proteins and are made in the body. The pattern or set of instructions for making each and every enzyme is in the fibrous material DNA of the cell nucleus; RNA assembled adjacent to a strand of DNA causes the formation of needed proteins in the cytoplasm.

Important words and terms

organic chemistry
inorganic chemistry
hydrocarbon
alkanes (molecular formula)
alkenes (molecular formula)
alkynes (molecular formula)
hydrocarbon series
electron dot structural formula
stick structural formula
multiple bonds
aliphatic hydrocarbons
aliphatic compounds
aromatic compounds
benzene
unsaturated hydrocarbons
saturated hydrocarbons
unsaturated compounds
isomers
polymer
polymerization
branched-chain hydrocarbon
octane number
common derivatives of hydrocarbons and type formula
functional groups
alkyl
alcohols (examples)
aldehydes (examples)
organic acids (examples)
esters (examples)
ketones (examples)
ethers (examples)
"hydrogen bonds"
most common carbohydrates
type formula carbohydrates
nutrients (classification, examples)
esters
fats
amino acids
biochemistry
cells
nucleic acids
RNA and DNA (meaning, function)
condensation
nucleotides
double helix
hydrolysis

Questions and exercises

1. (a) Distinguish between organic and inorganic chemistry. (b) Are there any new, different physical principles involved in organic chemistry?
2. (a) Of the known compounds, what percentage contain carbon? (b) How many such compounds are known? (c) What unique property of carbon accounts for the large number of such compounds? (d) Why are carbon compounds important in life?

3. (a) For what important breakthrough about our ideas of carbon compounds is Wöhler's name important? (b) In what biological product was Wöhler's compound known to exist before it was synthesized? (Consult a dictionary if necessary.)
4. (a) What is a hydrocarbon? (b) What are some well-known hydrocarbons and the sources of each?
5. In tabular form list the names of four hydrocarbon series, a typical member of the series, the general molecular formula for each, and the permissible values of n in the formula for each series. (Hint: C_1H_2 does not exist in the alkene series, hence the permissible values are $n \geq 2$.) (See App. A if necessary.)
6. In exciting a carbon atom one of the $2s^2$ electrons fills in the $2p$ orbital. (a) What is the lowest state for the lone hydrogen electron? (b) What state does this hydrogen electron fill in when the hydrogen atom is minimally excited? (c) What do we call this orbital formed from electrons coming from the two different orbitals of these two atoms?
7. (a) What is meant by a homologous series? (b) What are the first three homologous members of the alkane series and their formulas? (c) Distinguish between natural gas, bottled gas, naphtha, gasoline, and kerosene.
8. (a) In what way do the bonds of the alkane and alkene series differ? (b) Explain why the alkene CH_2 does not exist by attempting to draw its structural stick formula.
9. In what way do the bonds of the alkane and alkyne series differ?
10. (a) What common characteristic do the alkanes, alkenes, and alkynes have that permits their classification under a common heading? (b) What is the name of this general classification of hydrocarbons? (c) Historically, to what classification of hydrocarbons did the name originally refer?
11. (a) What classification of hydrocarbons is contrasted to the aliphatic group? (b) How many carbon atoms are contained in the elementary unit of such a group and what type of bonds exist among these carbon atoms? (c) Draw the structure of the most elementary member of this fourth group.
12. (a) Distinguish between saturated and unsaturated hydrocarbons in terms of (1) multiplicity of carbon bonds and (2) number of hydrogen atoms bound. (b) How do saturated and unsaturated hydrocarbons compare in chemical reactivity?
13. (a) What is meant by the term "isomer"? (b) Illustrate in the case of butane. (c) What general designations are reserved for these two different types of isomeric chains?
14. (a) Distinguish between isomers and stereoisomers. (b) Give an example.
15. (a) What are dimers and polymers? (b) In this regard, what is the starting material for the common plastic polyethylene?
16. (a) What do the prefixes *n*- and iso- before the name of a hydrocarbon designate? (b) How do the molecular formulas of the *n*- and iso- pairs compare?
17. (a) In determining the octane number of a gasoline, what two hydrocarbons are used as reference fuels? (b) How is the octane number of an unknown gasoline determined? (c) In the past years how was the octane number of a gasoline raised? (d) What effects (beneficial and harmful) does the addition of tetraethyl lead to gasoline have?
18. Why must gasoline have some low molecular weight hydrocarbons, whereas other fuels, such as kerosene and most fuel oils, are objectionable if they contain these same low molecular weight compounds?
19. (a) In the oxidation of methane by chlorine write *balanced* equations indicating the four successive stages (beginning with methane each time) toward the complete oxidation of methane. (b) Name the final product in each case. (c) Indicate the stage of oxidation in each case. (d) Indicate a common use for each product.
20. (a) What do the refrigerants known as the freons have to do with the alkane group? (b) From inspection of the molecular formulas, which of the alkane compounds was probably oxidized to form the freon?
21. Common organic compounds of hydrocarbons are the seven derivatives known as __________. They consist mainly of an __________ radical and a __________ group. The main determinant in chemical reactions of such compounds is the __________.
22. (a) What is an alkyl? (b) Write the general formulas for the alkyls of the alkane, alkene, and alkyne series.
23. An alcohol may be considered as the organic equivalent of which of the following inorganic substances: acid? base? salt? Why?

24. (a) Look through the section on isomers and locate a compound that is an ether. (b) Show how the two isomers fit in with the two type formulas in the last column of Table 18.5. (c) Which isomer can be taken internally?
25. Earlier we considered the products of the oxidation of a hydrocarbon by chlorine. (a) What do we call the similar products of the oxidation of a hydrocarbon by oxygen? (b) What is the relation between (1) wood alcohol, methanol, and methyl alcohol; and (2) grain alcohol, ethanol, and ethyl alcohol? (c) What are the methods of preparation of alcohols?
26. (a) As a paint brush cleaner would alcohol be of any use if an oil-base paint was in question? (b) Distinguish between proof and percentage in alcoholic beverages. What is the average percentage of alcohol in beer? in wine? in whiskey? (c) What is the disadvantage of using lower molecular weight alcohols as an antifreeze?
27. (a) As an automotive fuel, how are alcohol and gasoline superior and inferior to each other? (b) Considering the law of conservation of energy and the manner in which the fuels are prepared, why would one give better mileage than the other?
28. (a) What is the stage of the oxidation of the terminal carbon atom in any aldehyde? (b) Draw the structural formula of the two simplest aldehydes. (c) Distinguish between formaldehyde, formalin, and the "pickling fluid" used in biology. (d) How are the aldehydes related to the organic acids?
29. (a) List the organic acids present in vinegar, sour milk, cheeses, oranges, and apples. (b) What properties lead us to define them as acids? (c) Would organic acids be considered as fats, carbohydrates, or proteins in your calorie column?
30. (a) Distinguish between the structural formula of the simpler and the more complex heavier alcohols. (b) What difference in physical properties is exhibited by these two classes? (c) Show the structural formula of the two simpler polyhydroxyl alcohols and list their uses.
31. (a) As the molecular weight of the hydrocarbons increases, so does the boiling point. Why? (b) In a water molecule "hydrogen bonding" refers not to the attraction between the hydrogen and oxygen atoms of that molecule, but to something else. What? (c) How does the strength of the hydrogen bond compare with those of chemical bonds and van der Waals forces?
32. (a) Show the functional structural formula of an organic acid and an ester (Table 18.5) and explain why an ester is sometimes regarded as a "salt of an organic acid." (b) Write the equation for the reaction between sodium hydroxide and hydrochloric acid and that between methyl alcohol and formic acid. Compare them in terms of their reactants and final products.
33. (a) The reaction of what two organic compounds forms fats? (b) To what hydrocarbon classification do fats belong? (c) What is the difference between fats derived from saturated fatty acids and those derived from unsaturated acids?
34. (a) Distinguish between hydrocarbons and carbohydrates. (See also content of Problem 15.9.) (b) What is the difference between starch and sugar? (c) Which would be better in the tank of an automobile, hexane or hexose?
35. (a) What chemical element, which is always a constitutent of protein, is never present in fat or carbohydrate? (b) Distinguish between protein, amine, amino acid, and peptide. (c) In what way are proteins and starches similar in regard to the body's being able to use them in the digestive process?
36. (a) What is the common "indivisible" unit of life of plants and animals called? (b) Of what is it made? How has it been further subdivided since its initial conception? (c) Compare our experience here in searching for the fundamental units of life with that in searching for the fundamental units of matter in physics.
37. (a) To what form of carbohydrate must starches, sugars, proteins, etc., be converted in order to be utilized by cells for energy? (b) What form of sugar is used in intravenous feeding? (c) Generalize: The "-ose" at the end of the name of an organic compound tells us that it belongs to what class of nutritive substances?
38. In order of importance in determining the inheritance carriers, arrange the following from largest to smallest: (a) genes, (b) nucleus, (c) DNA, (d) arrangement of purine and pyramidine bases, (e) cell, (f) chromosomes.

39. (a) What are nucleotides? (b) Draw a strand of DNA and encircle the nucleotide. (c) Why can such a strand be called a polynucleotide? (d) Distinguish between peptide and nucleotide.
40. (a) What is meant by a condensation reaction? (b) What role does condensed phosphate play in energy needed for muscle contraction? Explain in terms of endothermic and exothermic reactions when food is oxidized.
41. (a) What two general types of organic compounds bind together the strands of the double helix? (b) List examples of each. (c) Specifically, which two bases are restricted to binding together these two strands? (d) What kind of chemical bonding is involved between these two bases? (e) From what is each strand of the double helix made?
42. (a) What are ATP and ADP? (b) What role do they play in muscle physiology?
43. (a) How does the structure of RNA differ from that of DNA? (b) Name three classes of RNA and the functions of each.

Supplementary readings

Books

Asimov, I., *The World of Carbon,* Macmillan, New York (1966). (Paperback.) [Clarity and accuracy are characteristics of this introduction to the subject.]

Baxter, John F., and Luke Steiner, *Modern Chemistry,* Vol. 1, Prentice-Hall, Englewood Cliffs, N. J. (1959). (Paperback.) [Prepared for Continental Classroom, the first nationally televised course in chemistry. Organic reactions (methyl groups, alcohols, etc.) are discussed in Chap. 19; large molecules (for example, detergents) and organic polymers (for example, rubber) in Chap. 20.]

Bonner, Francis T., et al., *Principles of Physical Science,* Addison-Wesley, Reading, Mass. (1957). [The historical aspects of carbon as the key element in organic chemistry are treated. Those who wish to extend their knowledge beyond the organic chemistry presented in this textbook should read the chapter on natural and synthetic organic products.]

Hart, Harold, and Robert D. Schuetz, *A Short Course in Organic Chemistry,* Houghton Mifflin, Boston (1972). [Excellent for the student who wants to go a little deeper into the matter but not "all the way."]

Light, Robley J., *A Brief Introduction to Biochemistry,* Benjamin, New York (1968). [Excellent for the student who would like to see some of the ideas here reenunciated and elaborated further.]

McElroy, William D., *Cell Physiology and Biochemistry,* Prentice-Hall, Englewood Cliffs, N. J. (1971). (Paperback.) [This is highly recommended as a reference; it is brief, clear, and accurate up to the publication date. Chapter 1, "The Cell"; Chap. 2, "The Chemistry of Cell Contents: Proteins"; Chap. 3, "Chemistry of the Cell Contents: Enzymes"; Chap. 4, "Metabolic Energy"; Chap. 9, "Control of Protein Synthesis"; Chap. 10, "Control of Cell Metabolism."]

Partington, J. R., *A Short History of Chemistry,* Harper Torchbooks, Harper & Row, New York (1960). [The beginning of organic chemistry (Chap. 10); the theory of valency: Kolbe, Kekulé (Chap. 12); summary, pp. 294–298; the structure of the atom and the electronic theory of valency (Chap. 16).]

Rose, Steven, *The Chemistry of Life,* Penguin Books, Baltimore (1966). [This is a somewhat leisurely but elementary and quite readable presentation of biochemistry. Chapter 2, "The Small Molecules"; Chap. 3, "The Giant Molecules"; Chap. 8, "The Synthetic Pathways."]

Watson, James D., *The Double Helix: A Personal Account of the Discovery of the Structure of DNA,* Atheneum, New York (1968). [Shows how science was "done" in the discovery of the structure of DNA—the creative process in research, the excitement, the frustration, the sheer work.]

Articles

Lambert, J. B., "The Shapes of Organic Molecules," *Scientific American,* **222,** no. 1, 58 (January, 1970).

Frieden, E., "The Chemical Elements of Life," *Scientific American,* **227,** no. 1, 52 (July, 1972). [Until recently it was believed that living matter incorporated two of the natural elements. It has now been shown that four additional elements—fluorine, silicon, tin, and vanadium—play a role in life as trace elements.]

ENERGY AND ITS SOURCES

Man's expanding need for energy creates difficult economic, social, and environmental problems. The solutions call for sensible choices of technological alternatives by the market and political process.

CHAUNCEY STARR, 1971

Throughout this text numerous forms of energy have been mentioned and discussed. The conservation of energy is one of the most important laws of nature, particularly when man attempts to "control" and "manage" nature to provide for his comfort and industry. Forms of energy such as mechanical, electrical and electromagnetic, thermal, nuclear, and chemical have all been mentioned. In addition, in the remaining chapters we shall also examine the earth and its geology, from which we obtain geothermal energy. In the present chapter we shall discuss our natural sources of energy to try to provide some insight into possible sources that man can hope to tap for the tremendous quantities of energy required for the future by modern society. It should be stressed that from the principle of conservation of energy we know it is possible to transform energy of one form to energy of another form. Electrical energy is the most convenient form in which to transfer and use energy in homes and factories, so it is important to remember this as we examine the possible sources of energy that must ultimately be converted into electrical energy. In addition, it must be remembered that the efficiency with which man can economically convert one form of energy to another must also influence his choice of the most desirable energy sources. The second law of thermodynamics places restrictions on these energy conversion processes. Man must also consider what by-products or "pollution" will result from the energy transformation processes. Although we shall not examine pollution problems in detail, we shall describe some of the problems associated with obtaining useful energy from the possible sources of energy discussed.

Energy options

It is becoming increasingly apparent to many scientists, engineers, industrialists, and government officials that providing an adequate supply of energy in the forms acceptable to the public may be the most important technological challenge facing the United States and, in fact, the world in our future. As we use up our conventional sources of fuel, mainly the fossil fuels (coal, petroleum, and natural gas), new sources must be developed to produce economically feasible energy. It appears that because of increased costs in importing fuel and social restrictions in the form of pollution controls, the cost of energy will tend to rise.

There are many energy sources available to man, some of which have only been used to a very limited degree (see Table 19.1). However, to employ these sources to their full potential will require substantial scientific and engineering efforts. Such developments usually take a number of years, so new technologies and additional research are required now. To appreciate the energy problem the world faces, we need only compare the cumulative energy from now to the year 2000 with the estimates of the present reserves of economically recoverable fossil fuels. By the year 2000 the annual worldwide rate of consumption will probably be three times the present rate, and the rate of energy consumption in the United States will probably have doubled. The estimated fossil-fuel reserves are greater than the estimated demand summed up to the year 2000 by only a factor of two. It should also be realized that the sources of petroleum, coal, and natural gas are among the major raw materials used in the chemical industries. Plastics, chemicals, medicines, and hundreds of other "modern" products are derived from the resources we so lightly call "fossil fuels."

If the only energy resource were fossil fuels, the outlook would certainly be bleak, particularly when one considers the cost of obtaining some of the fossil fuel from an economic point of view. For example, some of the world's reserves are under the sea in regions where the expense of removing the fuel is considerably higher than from the

Table 19.1
Estimates of depletable energy resources in the United States[a]

resource	recoverable	total
Fossil fuels		
coal	125	1300
petroleum	5	280
natural gas	5	110
oil shale		2500
Nuclear fission		
conventional reactors	2.3	15
breeder technology	115	750
Nuclear fusion		
deuterium–deuterium		$\sim 10^9$
deuterium–tritium		$\sim 10^6$
Geothermal heat		
steam and hot water	0.2	> 60
hot rock		>600

[a]The resources are given in units of U.S. annual energy consumption (6.6×10^{19} joules); the figures in the table are equivalent to the number of years that the resource would last, if all energy came from that source, at current rates of use. Recoverable resources include those known and now available; total resource estimates include expected offshore deposits and do not necessarily represent recoverable amounts of energy. (From *Science* **177,** September 8, 1972, p. 875).

present sources. If one includes the energy available from nuclear power, then the outlook is considerably less bleak. For example, there are now 25 nuclear power plants in operation with an additional 50 under construction.

With the exception of nuclear energy, there has been very little research directed toward developing new sources of energy. There are several other promising sources now being considered, however, including solar, geothermal, and extensions of the fossil-fuel resources. The prospect for commercial breeder reactors employing nuclear fission (described later in this chapter) is very promising and the U.S. Atomic Energy Commission expects such reactors to be producing electricity in 1985. In addition to nuclear fission there are some efforts to develop a stable nuclear fusion process for power development. Nuclear fusion, the process that takes place in the sun and other stars, has a great potential as a "clean" energy source but its use is many years away from reality.

Solar energy

There are numerous nonnuclear energy sources that have tremendous potential. First, the use of solar energy in heating and for the development of electricity has almost unlimited potential if practical methods for its use can be developed. The amount of solar radiation that reaches the surface of the earth is so tremendous that if only 10 percent of the solar radiation falling on a small percentage of the land area in the United States could be harnessed, it would provide most of the energy needs for the United States in the year 2000. However, there are some major obstacles associated with large-scale use of solar radiation. For example, large areas would be required to obtain sufficient energy, and there are problems associated with conversion of the radiation to useful electricity and with the storage of this energy. The potential for solar energy is so tremendous that even small-scale use of this energy source would make an impact on the worldwide energy demand. On a small scale, experimental solar heating of homes has been employed in a few parts of the United States (Fig. 19.1). The use of solar batteries and thermoelectric devices to generate small voltages and small power outputs has provided a practical solution in remote areas in the Soviet Union for the generation of the very small quantities of power used to operate small communication relay equipment.

Several methods have been suggested to convert solar radiation into electricity on a large scale. One method involves the use of large photovoltaic cells. Such devices make use of the fact that when light strikes certain materials, the materials emit electrons that can be collected. The family of such devices includes the photocells that are used in opening doors and are employed in light-

Figure 19.1
The MIT solar heated house. (From "Solar Energy—An Option for Future Energy Production," Peter E. Glaser, *The Physics Teacher,* **10,** no. 8, 443, November, 1972.)

exposure meters used with cameras. All such devices provide low voltages and small electrical power levels. However, by employing many such cells in tandem and by perhaps developing better devices with higher power levels, practical systems could be developed.

Other techniques have been proposed for collecting the solar radiation—for example, direct absorption of the heat or infrared radiation in a selective manner much the way greenhouses or hothouses are heated by the sun. The heat would be collected and stored by heating large quantities of some liquid material. The heat could then be removed to operate a conventional steam–electrical power conversion plant. There have also been proposals to collect and concentrate with reflectors the solar radiation falling on a large area. The radiation would be concentrated or "focused" into a solar furnace where water could be turned directly into steam whose energy could then be converted into electricity. All such operations could, of course, be practical only in areas such as Arizona where there is little variation in the amount of solar radiation falling daily on the earth.

Hydropower

One source of conventional electrical power which, of course, will continue to be developed further is water power (hydropower). However, close examination of the potential of hydroelectrical power shows that it has rather limited possibilities. It has been estimated that the total hydropower capacity of the world occurring at suitable sites approximates only the total amount of energy used in industry today. At the present time less than about 10 percent of the possible, suitable water power has been developed. However, the greatest potential for development of hydropower exists in Africa, South America, and Southeast Asia where the industrial demand for power is still very low.

Besides the conventional hydropower associated with dams on rivers and inland lakes, for years there have been studies of the use of ocean water in tidal power plants. At the present time there is one operating tidal power plant in France. In the United States there has been interest from time to time in development of the tidal flow in the Bay of Fundy off the coast of Maine where the tide causes a change in water height up to 22 ft every 6 hours. In a tidal power system a scheme has to be developed to store water in a basin so that a more continuous power output can be developed. Such a project as the Passamaquoddy power project in Maine becomes a rather complex engineering adventure if maximum power is to be realized. Tidal power possibilities are, of course, limited to those areas in the world where substantial tides exist, where a large basin or estuary exists, and where the water can be regulated through the use of dams. In any case, the development of economical hydropower is far from sufficient to solve the world's power needs.

Geothermal energy

Geothermal power is obtained by extracting heat stored within the earth. To date the most practical use has been from steam that obtained its heat from volcanic sources. Such superheated steam is passed through turbines to generate electricity. Electrical power is being produced commercially from geothermal sources in seven countries including Italy, Iceland, New Zealand, Japan, and the United States. The oldest site is in Italy, where since 1904 in the Larderello region steam from wells in a geyser field has been used to produce electricity. In the United States the only site in commercial use is in The Geyser's Field near and north of San Francisco (Fig. 19.2). In Iceland hot water from shallow wells is used to heat most buildings in the capital city and for cheap heat in numerous greenhouses where plant foods, such as tomatoes and even bananas, are grown year round. Although these steam sources are in or near geologically recent volcanic areas, they are not on actual active volcanoes due to the probability of destruction of facilities during eruptions.

For useable steam to be present at shallow depths, there must be (1) a hot zone (250°C) near the surface (2000 to 4600 ft), (2) adequate circulation of surface waters to depths to be heated and thus expanded so they return toward the surface, and (3) a rock structure to trap the dry steam in pockets just as natural gas is trapped. This means layers of impervious rocks overlying porous rocks that can hold the steam. Steam is produced from drilled wells in the same way natural gas is recovered. A considerable problem exists in finding such buried steam pockets. Usually hot springs or geysers are found at the surface, which indicate that hot waters are rising from depth; but without a trap structure in the rocks, all the steam would leak to the surface. Various geological studies are needed to indicate the possible presence of a trap, but only test drilling will give a final answer. The Yellowstone National Park is an area that probably has much trapped steam. Certainly the underlying volcanic rocks are very hot (250°C at 400 ft) over nearly a thousand square mile area. Adequate rainfall supplies the circulation of dense cold water down at least 5000 ft; less dense hot water and steam move to the surface. Although the U.S. Geological Survey has recently done some research drilling to study the geothermal systems in Yellowstone, there is, of course, no thought of any commercial development in the park. A few other regions in western United States, however, appear to have more practical commercial possibilities.

There are many engineering and environmental problems associated with tapping geothermal energy. For example, the steam

Figure 19.2
A commercial geothermal plant at "The Geysers"—located in Sonoma County, California, about 75 miles north of San Francisco. In the mid-1950's, the Geysers area was redrilled, and four wells began producing at a depth of less than 1000 feet. Production was started in 1960 at a rate of 12,500 kilowatts, and the capacity has enlarged since to over 292,000 kilowatts (1972). "Hot spots" favorable for geothermal energy are related to volcanic activity in the present and the not too distant past. In the western United States, particularly along the Pacific coast, widespread and intense volcanic activity has occurred during the past 10 million years. The record of volcanism in the western states, therefore, holds promise for geothermal development. Currently, exploration for power sites is focused in California, Nevada, Oregon, and New Mexico, with some interest being displayed in the whole region from the Rocky Mountains to the Pacific Ocean. (Courtesy of U.S. Department of the Interior, Geological Survey.)

and water obtained from The Geyser's wells generally contain corrosive chemicals that attack the inside of the electric generating machines and eventually appear as a waste product, which is either put into the air or onto the earth. There is the possibility of seismic disturbances (small earthquakes) created by the pumping of the water and steam from the wells. In addition to these problems there is the question of how much steam can be extracted from a given well and for how long. The life of the Wairakie geothermal field in New Zealand has been estimated at 25 to 30 years. This field was first brought into production in 1958 from some 59 steam wells, averaging a little more than 2000 ft deep, in a valley less than 1 mile wide and about 2 miles long. The field has been developing about 25 million kilowatts weekly, enough electricity for about 300,000 people.

Most of the known steam fields are small, like the Wairakie mentioned above, and thus not very practical in heavily populated or industrialized areas. However, such sources of moderate and rather inexpensive energy may be of considerable importance if available in small, emerging nations. Exploration and development of promising geothermal energy is underway in Mexico, most Central American countries, Chile, the Philippines, and other small nations where recent volcanic activity means near-surface heat under wide areas.

At depths of 10,000 to 30,000 ft, as in wide parts of the western United States, rocks are hot, up to and well above 100°C. Consideration is now being given to finding a feasible method of extracting heat from these large sources of subterranean hot rock. Perhaps surface waters can be pumped down to be heated and returned as hot water or steam that can be used as a heat, or energy, source. The ultimate potential, in terms of heat available in deep rocks, is very great but the cost at present is prohibitive and, unfortunately, very little actual research is being done on this energy possibility.

Nuclear energy

THE EINSTEIN EQUATION

Nuclear energy must be considered under two types of processes: fission and fusion. However, the energy obtained from both of these processes is due to the conversion of mass into energy. This mass-energy equivalence was first postulated by Albert Einstein in his special theory of relativity. Einstein had been working on the problem of relative motion and developing a theory that would keep the laws of physics—that is, the physical laws of nature—independent of the reference frame in which one made the measurements. One of the outcomes of his special theory of relativity was his statement that the speed of light c in vacuum (or free space) is the highest possible; in other words, no object can travel faster than the speed of light in free space.

Einstein's special theory of relativity provides several other exciting results, but we shall not discuss them here. However, perhaps the most important practical result of Einstein's theory of special relativity is his prediction of the equivalence of mass and energy. The connection between mass and energy, oddly enough, involved that very important quantity, the speed of light or electromagnetic waves (c) in free space. The equivalence is expressed in the very famous equation,

$$E = mc^2 \qquad (19.1)$$

Here E is energy in joules or ergs and m is mass in kilograms or grams with c being the speed of light (approximately 3×10^8 m/sec or 3×10^{10} cm/sec). It is perhaps more appropriate to write the equation in the following way

$$\Delta E = \Delta mc^2 \qquad (19.2)$$

where the Δ (delta) means change. Thus Eq. 19.2 is read, "The change in energy equals the change in mass times the square of the speed of light in free space." This is a true equation which says, on the one hand, that mass may be converted into energy—say, electromagnetic energy—and, on the other hand, electromagnetic energy can be converted into mass. The latter case has been observed when high-energy γ (gamma) rays have been converted into an electron and a positron, the positron being a particle of mass equivalent to the mass of an electron but with a positive electronic charge. Note that in this conversion γ rays, which are "pure" energy, create pairs of particles—that is, two pieces of matter or mass—one positively charged and one negatively charged; and thereby no net charge is created. If this were not true and if we could convert a γ ray to a single electron, we would ultimately create more negative charge than positive and charge up our electrically neutral environment.

There are numerous examples of mass being changed into energy including the process in the sun in which approximately 5 million tons of matter of the sun is converted into energy in the form of electromagnetic energy each second.

In light of Einstein's equivalence of mass and energy let us reexamine the law of conservation of energy.

MASS-ENERGY AND THE CONSERVATION LAWS

In his textbook of 1789 Lavoisier wrote, "We must lay it down as an incontestable axiom that, in all the operations of . . . nature, nothing is created; an equal quantity of matter exists both before and after the experiment." In the nineteenth century the exactness of this principle was tested by very careful experimentation. In every case the sum of the weights of the original reactants was equal to that of the products within the small experimental error of the measurements. The principle was thus considered a generalization based on facts determined experimentally and became known as the *law of conservation of matter:*

Matter can neither be created nor destroyed.

Let us take another critical look at the law of conservation of matter and also the law of conservation of energy (Chap. 6), which is often called the *first law of thermodynamics.* Does Einstein's statement of the equivalence of mass and energy (often stated as matter and energy) mean that the conservation laws became invalid with the explosion of atom bombs in the 1940s?

Matter is the material, the "stuff," which constitutes the objects of our everyday world. Quite properly we say that the more mass we have, the more matter; and that although the weight of a sample may vary, its mass at different points on earth does not. We are, of course, ignoring now any radioactive transmutations that do take place on the earth naturally in some elements.

We usually determine the amount of mass in a given sample by comparison with standard masses on a balance. During the comparison the sample is motionless or moving only at a low velocity. In this way we get a "rest mass" of the sample. It would probably he helpful to accept as a working definition that *"rest mass" and amount of matter* are the same.

But what occurs if we give the sample of matter some extra energy? If this extra energy is appreciable, we make it impossible to "weigh" the sample. To express the thought more precisely, we cannot use a balance to determine experimentally the amount of mass in a sample if its temperature is enormously high or if it is moving at high velocity. But by use of other methods it has been observed experimentally that the ratio of charge to mass (e/m) of a high speed electron becomes smaller as the electron travels more rapidly. Also the sun bends light rays toward it as they pass near its surface, as observed during a solar eclipse. This last fact clearly indicates that light, which we have previously regarded as being energy, is affected by gravity as if it had mass. The measured inertial mass of an electron is not an invariant quantity but it varies with speed.

Look at the equation $E = mc^2$, called the Einstein equation. Suppose we increase the value of energy, E. Do we get a larger value for mass, m? It would appear therefore that a red-hot steel ingot from a furnace really represents more mass energy than it would if cold, but that we are not able to demonstrate the tiny increase in mass which accompanies that given amount of increase in energy.

A pencil in your hand represents an immense amount of potential energy ("rest mass"); but this statement is scarcely more than idle talk, because thus far there is no practical way to release this potential energy.

To avoid confusion we should now have only one *conservation law:*

The mass energy of the universe is a constant.

For example, as the sun emits energy, its mass decreases in this process at the rate of 5 million tons each second, in accordance with the Einstein equation, but the total mass energy of the universe remains constant.

To further illustrate the conservation of mass energy let us examine the results of a nuclear reaction. In previous discussions whole numbers, or integers, were used as atomic weights. Actually these atomic weights were atomic mass units (amu) as determined by the mass spectrograph with reference to the most common isotope of carbon as 12; and these atomic mass units are not exact integers. For example,

$^{1}_{1}H$ is 1.00814 amu
$^{7}_{3}Li$ is 7.01823 amu
$^{84}_{36}Kr$ is 83.9385 amu
$^{114}_{48}Cd$ is 113.940 amu
$^{208}_{82}Pb$ is 208.042 amu
$^{238}_{92}U$ is 238.126 amu

The nearest integers to these exact atomic masses, or atomic weights, are called *atomic mass numbers* and are equal to the number of *nucleons* (constituents of the nucleus—that is, protons and neutrons).

One of the earliest transmutations achieved in a laboratory was the nuclear reaction $^{1}_{1}H + ^{7}_{3}Li \rightarrow ^{4}_{2}He + ^{4}_{2}He$, in which $^{7}_{3}Li$ atoms were bombarded by protons (hydrogen nuclei) to form α particles (helium nuclei). Using accurate masses obtained by the mass spectrograph, the above relation becomes (in amu)

$$\begin{array}{ccccccc} ^{1}_{1}H & + & ^{7}_{3}Li & \rightarrow & ^{4}_{2}He & + & ^{4}_{2}He \\ 1.00814 & + & 7.01823 & \rightarrow & 4.00387 & + & 4.00387 \end{array}$$

Atomic masses will be used rather than nuclear masses for convenience; the number of electrons on the two sides of the equation is the same and their masses on the two sides of the equation cancel out. The sum of the initial masses is 8.02637 amu, and the sum of the product masses is 9.00774 amu, giving a "mass discrepancy" of 0.01863 amu. What happened to the missing mass?

The Einstein mass-energy relationship says the energy equivalence E of a mass m is mc^2 where c is the speed of light. Hence the energy equivalent to 1 amu can be obtained by expressing 1 amu in grams (1 amu = 1.6598×10^{-24} g) and multiplying by the square of the speed of light (2.99793×10^{10} cm/sec). This gives an energy of 1.4917×10^{-3} ergs. In discussing nuclear energy, however, the common unit of energy used is the electron volt (eV), which is the energy required to move an electron through a difference of potential, or voltage, of 1 volt. An MeV is 1 million electron volts. Computations show that 1 amu is equivalent to 931.2 MeV. Thus the "mass discrepancy" in the nuclear reaction under discussion is $0.01863 \text{ amu} \times 1.4917 \times 10^{-3}$ ergs/amu $= 2.768 \times 10^{-5}$ ergs, or $0.01863 \text{ amu} \times 931.2 \text{ MeV/amu} = 17.34$ MeV.

The additional energy for the end products means that the resultant α particles, or helium nuclei, will have high kinetic energies, each α particle having a kinetic energy of 17.34/2 = 8.67 MeV. This value of the kinetic energy of the resultant α particles agrees very closely with the kinetic energy computed from the measured penetration of these α particles; that is, to measure the kinetic energy of the α particles, one measures how far the α particle travels in a material substance. The higher the kinetic energy, the greater distance the α particle will penetrate into the material. This nuclear reaction is of the exothermic type; that is, energy is set free by the nuclear reaction, "seemingly" at the expense of mass.

The reverse type of nuclear reaction, the endothermic type, also occurs and is illustrated by the nuclear reaction:

$$\begin{array}{ccccccc} {}^{4}_{2}\text{He} & + & {}^{14}_{7}\text{N} & \rightarrow & {}^{1}_{1}\text{H} & + & {}^{17}_{8}\text{O} \\ 4.00387 & + & 14.00752 & \rightarrow & 1.00814 & + & 17.00453 \end{array}$$

The sum of the two masses on the left is 18.01139 amu and for the end products is 18.01267 amu.

In this case, the apparent gain in mass of the end products of 0.00128 amu means from the Einstein relation ($E = mc^2$) that the final products have 1.192 MeV more energy than the initial substances. This means that the α particle must have at least this much kinetic energy in order for the reaction to occur.

NUCLEAR SYNTHESIS, OR NUCLEAR FUSION

Under certain conditions helium is built up from hydrogen. This is a process of nuclear synthesis, or nuclear fusion. The tremendous energy radiated by the sun and other stars, according to a widely accepted theory originated by Bethe, is due to a series of six nuclear reactions called the "carbon cycle" and to another set of nuclear reactions called the "proton chain." The net result of the six reactions that make up the carbon cycle is to change four hydrogen atoms to one helium atom along with the emission of two positrons that are "annihilated" by collision with free electrons causing γ-ray energy.

Positrons are so-called antimatter and when an electron and its antiparticle, a positron, come close together, they are attracted electrically and come together "annihilating" each other producing γ-ray energy. Because there are electrons everywhere in our environment, positrons live only a very short time. In this case the mass of the helium atom (4.00387) is less than the mass of the four hydrogen atoms ($4 \times 1.00814 = 4.03256$) by 0.02869 amu, or 26.71 MeV of energy. The proton chain is a more direct synthesis of hydrogen into helium, but the net result again is a final product of one helium atom at the expense of four protons. In order for either of these hydrogen-to-helium reactions to occur, a very high temperature of several million degrees is necessary. These temperatures do, of course, exist in the sun and stars.

The development of the first atomic bomb (A-bomb) in 1945 furnished a means of obtaining on earth the very high temperatures needed for hydrogen-to-helium synthesis. This has led to the development of the hydrogen, or fusion, bomb (H-bomb), first exploded in 1952. The nuclear reactions in the explosion of an H-bomb are probably neither of the two sets just mentioned. More likely the final helium atom is arrived at by reactions involving the heavier isotopes of hydrogen, deuterium ${}^{2}_{1}\text{H}$, and tritium ${}^{3}_{1}\text{H}$:

$${}^{2}_{1}\text{H} + {}^{3}_{1}\text{H} \rightarrow {}^{4}_{2}\text{He} + {}^{1}_{0}\text{n} + \text{energy}$$

The energy released during nuclear fusion, called *thermonuclear energy,* will likely be a major source of energy on earth when our supplies of gas, oil, and coal are depleted. Thermonuclear power plants are expected to provide energy less expensively than nuclear reactors using uranium and plutonium.

It should be noted that controlled thermonuclear or fusion power presents extreme engineering problems. First, the reacting

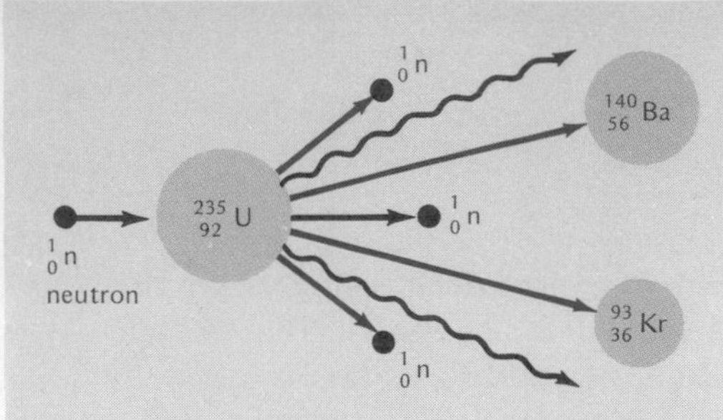

Figure 19.3
Typical fission of a uranium-235 nucleus. The wavy lines indicate the emission of energy, partly γ-ray energy from the radioactive end products but mainly kinetic energy of the end-product particles.

material must be kept at extremely high temperatures; second, the reaction must be controlled so that a runaway explosion does not occur. Much research and development will be necessary before thermonuclear power becomes a reality. It should be noted that one great advantage of thermonuclear power using the fusion process is that the "ash," or resulting end product, is not a pollutant or a radioactive hazard.

NUCLEAR FISSION

From the preceding discussion of fusion it is seen that the atomic mass per nucleon is less for helium (4.00387/4 = 1.00097) than for hydrogen (1.00814), so that when helium is made from hydrogen—nuclear fusion or synthesis—there is a "loss" of atomic mass, which is accounted for by the large amount of kinetic energy (mass) that has left the system.

Careful measurements show a gradual decrease of atomic mass per nucleon, as the atomic mass increases, for the light elements with a nearly constant value all through the center of the atomic table (periodic table) from about magnesium $^{24}_{12}Mg$ to tantalum $^{181}_{73}Ta$. From tantalum on, the elements of increasing atomic weight have an increasingly larger atomic mass per nucleon. This suggests that if by some means atoms of elements in the central part of the atomic table could be obtained from the elements at the high mass end of the atomic table, there would be a reduction in end-product mass, which would result in a large increase in the kinetic energy of the end products (from $E = mc^2$).

A series of experiments started by Fermi in 1934 culminated in 1938 in the splitting, of fission, of the uranium-235 ($^{235}_{92}U$) nucleus through bombardment by slow, or thermal, neutrons. Further study showed that many pairs of middle-table elements, such as $_{56}Ba$ and $_{36}Kr$, $_{38}Sr$ and $_{54}Xe$, are formed with nearly all end products being radioactive. Further, it was found that such fissions were accompanied by the emission of two or three fast-moving neutrons. The amount of energy released per fission, mostly in the form of kinetic energy of the end products, owing to the decrease in mass of the end products amounts to nearly 200 MeV.

The amount of energy obtained by fission will probably be better understood if presented in the following manner. In the fission of uranium-235, the sum of the masses of all the end products is about 0.999 of the mass of the initial products. Thus if 1 kg (1000 g) of fissionable material results in 999 g of end products, 1 g of mass energy is converted into other forms of energy. Converting mass units into energy units by using $E = mc^2$ for 1 g, $E = 1 \text{ g} \times (3 \times 10^{10})^2 \text{ cm}^2/\text{sec}^2 = 9 \times 10^{20} \text{ ergs} = 9 \times 10^{13} \text{ joules} = (90 \times 10^{12} \text{ joules})/(36 \times 10^5 \text{ joules/kWh}) = 2.5 \times 10^7 \text{ kWh}$, or 25,000,000 kWh. This is enough energy to supply a large city for several days.

Figure 19.3 represents a typical reaction in the fission of $^{235}_{92}U$.

The two or three neutrons that are set free at each uranium-235 fission make a continuing reaction, or chain reaction, possible. If for every neutron used two or three are set free, and if more than one, on the average, causes the fission of another uranium atom, as indicated in Fig. 19.4, a rapid spread of nuclear fissions occurs, causing a tremendous

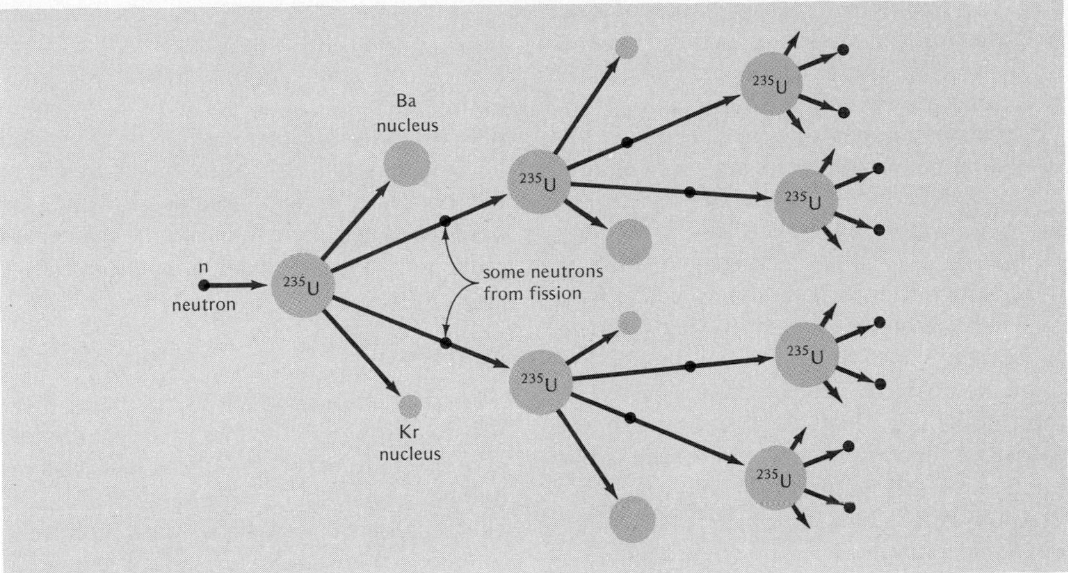

Figure 19.4
Generalized diagram of a chain reaction. The reaction is self-sustaining if at least one neutron from each fission is captured by a uranium nucleus and another fission is thereby induced. The reaction is explosive, with a tremendous release of energy if, on the average, several neutrons per fission induce additional fissions, thus causing a rapid buildup of fissions. Some of the particles emitted and waves representing energy released are not shown.

release of energy. From Fig. 19.4 it is evident that if there are too few uranium-235 atoms, the freed neutrons will go between the uranium atoms and will soon get entirely outside the uranium material, in which case the reaction will soon die out. Thus there is a minimum critical mass necessary to continue the reaction. Also, the two or three neutrons set free at each fission are very fast-moving neutrons because of the release of mass as energy. These fast neutrons are more likely to miss uranium-235 atoms and quickly leave the entire uranium chunk. To slow down these fast neutrons, some kind of a moderator is used, such as carbon in the form of graphite or the heavy hydrogen $^{2}_{1}H$ contained in heavy water, or deuterium oxide.

For an A-bomb, fast and uncontrolled fissions are desirable. The presence of more than 99 percent pure uranium-235 is required. Because in nature uranium-235 occurs with uranium-238 in the ratio of 1:140, enormous separation plants are required. At Oak Ridge, the uranium-235 is separated out by repeated diffusion of the two uranium isotopes in the form of gaseous uranium hexafluoride, the lighter isotope ($^{235}_{92}U$) diffusing more readily than the leavier one ($^{238}_{92}U$). The uranium-235 is then kept in chunks less than the critical size. Perhaps in an A-bomb two pieces of uranium-235, each less than the critical size but more than half the critical size, are so arranged that a time mechanism or an impact mechanism will force the two chunks together.

Controlled nuclear reactors are now made using natural uranium in the form of rods embedded in blocks of graphite (or placed in heavy water), which acts as a moderator, the whole unit being encased in a shield of thick concrete to protect the operators from harmful radiation. A neutron that happens to hit a uranium-235 atom will free two or three neutrons. If one of these, slowed by the moderator, hits another uranium-235 atom, the reaction is maintained. Meanwhile another of the neutrons set free may hit a uranium-238 atom, in which case a new element, plutonium, may be produced as follows:

$${}^{1}_{0}\mathrm{n} + {}^{238}_{92}\mathrm{U} \rightarrow {}^{239}_{92}\mathrm{U} + \gamma \text{ radiation}$$
$${}^{239}\mathrm{U} \rightarrow {}^{239}_{93}\mathrm{Np} + {}^{\;0}_{-1}\beta$$
$${}^{239}_{93}\mathrm{Np} \rightarrow {}^{239}_{94}\mathrm{Pu} + {}^{\;0}_{-1}\beta$$

The radioactive isotope ${}^{239}_{92}\mathrm{U}$ has a half-life of 23 min, and radioactive ${}^{239}_{93}\mathrm{Np}$ has a half-life of 2.3 days; but plutonium ${}^{239}_{94}\mathrm{Pu}$ is relatively stable, with a half-life of 24,000 years. Plutonium can be separated from uranium by chemical means and is fissionable by slow neutrons, as is uranium-235. Probably most A-bombs are now made of ${}^{239}_{94}\mathrm{Pu}$.

Large nuclear reactors are being used as power plants, and considerable research along this line is now being carried out in many countries. Although energy thus obtained is often more costly than energy obtained from natural fuels, such as coal, oil, and gas, these natural sources will some time be exhausted, and scientists must plan to be ready to use nuclear energy. One of the great difficulties to be overcome lies in the disposal of the dangerous radioactive by-products of fission. The present government-owned reactors are the sources of many of the artificially radioactive substances already being used in scientific investigations, in medicine, and in industry.

The purified ${}^{235}_{92}\mathrm{U}$ or ${}^{239}_{94}\mathrm{Pu}$ is already being used as fuel for submarines—the first of many such applications. Nuclear fuel may easily be used for ships. It is much more difficult to use nuclear fuel for airplanes and automobiles because of the great weight of the concrete or lead shields that must be used to protect humans from the dangerous radiations of the radioactive by-products.

SUMMARY

It is apparent that we are using up our inexpensive supply of fossil fuels so that the less economically obtainable fossil fuel reserves and alternative sources of energy must be developed. Solar energy shows great promise as a major energy source in areas of the world where there is little variation of solar radiation falling daily on the earth. Hydropower and geothermal energy could be developed in appropriate regions throughout the world to provide a solution to the energy demands in these local regions. It appears that the source of energy showing the greatest potential for the industrialized areas of the world is nuclear energy. Nuclear energy is obtained by fission in the present nuclear reactors, and a great deal of scientific research is now being devoted to the search for a stable nuclear fusion energy source. In the nuclear processes energy is obtained by the conversion of mass to energy, a process which obeys the famous Einstein relationship ($E = mc^2$). Energy and mass are equivalent, and one can be changed into the other. In fission processes heavy nuclei like uranium nuclei are "split" by bombardment with a neutron, producing two smaller atoms, additional neutrons and energy. Nuclear fusion is a cleaner process whereby light nuclei like hydrogen are fused to produce heavier nuclei. This is the process which takes place in stars. A great deal of additional scientific research will be required before these newer sources of energy are developed to the point where they are providing an appreciable fraction of the world's energy needs.

Important words and terms

fossil fuels
solar energy
nuclear energy
fission
fusion
hydropower
geothermal energy
Einstein mass-energy relation
significance of the speed of light
rest mass
conservation of mass energy
nucleon
positron

Questions and exercises

1. What are some of the problems the United States faces with the "energy crisis"?
2. Are all energy sources of equal value? Explain.
3. What are some of the problems associated with large-scale use of solar energy for our electrical power requirements?
4. What are some of the advantages of nuclear fission power over fossil fuels? What are some disadvantages?
5. What are the advantages and disadvantages of nuclear fusion vs. nuclear fission as processes for producing electrical energy?
6. Why do scientists not feel that geothermal energy is the answer to all of our energy needs?
7. What is so significant about the speed of light with regard to nuclear energy?
8. Distinguish between nuclear fission and nuclear fusion. Which process is responsible for the sun's maintaining a high temperature?
9. What is meant by energy and mass being equivalent?

Supplementary readings

Books

Andrade, E. N. da C., *An Approach to Modern Physics*, Anchor Books, Doubleday, Garden City, N. Y. (1956). (Paperback.) [Heat and energy, sound and vibrations, light and radiations, electricity, solids and liquids, quantum theory, structure of the atom, structure of the nucleus, applications of nuclear transformations, and uncertainty.]

Beiser, Arthur (ed.), *The World of Physics*, McGraw-Hill, New York (1960). [Selected writings: the rise of science (Bertrand Russell); conservation of energy (Einstein); entropy (Gamow); action at a distance (James Clerk Maxwell); special relativity; the discovery of radium (Marie Curie); the cyclotron; quantum physics (Condon); elementary particles; cosmic radiation; fusion power.]

Bondi, Hermann, *Relativity and Common Sense*, Doubleday, Garden City, N. Y. (1964). [An expert in the field tells the general public what relativity is by showing that it is an extension of ordinary ideas to the realm of high velocities.]

Einstein, A., and L. Infeld, *The Evolution of Physics*, A Clarion Book, Simon and Schuster, New York (1967). [A historical development leading up to relativity and quanta.]

Foreman, H. (ed.), *Nuclear Power and the Public*, Anchor Books, Doubleday, New York (1972). (Paperback.) [A series of articles on nuclear power and the effects on the environment.]

Grayson-Smith, Hugh, *The Changing Concepts of Science*, Prentice-Hall, Englewood Cliffs, N. J. (1967). [Relativity is discussed in Chap. 33.]

Holton, Gerald, and Duane H. D. Roller, *Foundations of Modern Physical Science*, Addison-Wesley, Reading, Mass. (1958). [Part 9, "The Nucleus," pp. 657–749, shows the historical development of our knowledge of radioactive isotopes as well as nuclear structure and nuclear energy, with a good presentation of the principle of conservation of mass and energy (p. 704).]

Karplus, Robert, *Physics and Man*, Benjamin, New York (1970). (Paperback.) [The theories of energy, nuclear energy, mechanics are presented.]

Articles

Barnea, J., "Geothermal Power," *Scientific American*, **226,** no. 1, 70 (January, 1972).

"Energy and Power," *Scientific American* (September, 1971). [The total issue is devoted to energy, its sources, conversion, and relationship to society.]

Glaser, P. E., "Solar Energy," *The Physics Teacher*, **10,** no. 8 (November, 1972).

Gough, W. C., and B. J. Eastland, "The Prospects of Fusion Power," *Scientific American*, **224,** no. 2, 50 (February, 1971). [Recent advances with experimental plasma containers are described as applied to practical fusion power.]

Lubin, M. J., and A. P. Fraas, "Fusion by Laser," *Scientific American,* **224,** no. 6, 21 (June, 1971). [It seems feasible to build a fusion reactor with the aid of the laser.]

Meinel, A. B., and M. P. Meinel, "Physics Looks at Solar Energy," *Physics Today,* **25,** no. 2, 44 (February, 1972).

Seaborg, G. T., and J. L. Bloom, "Fast Breeder Reactors," *Scientific American,* **223,** no. 5, 13 (November, 1970).

Stephens, W. E., "Origin of the Elements," *The Physics Teacher,* **7,** no. 8, 431 (November, 1969). [Discussion of the elementary nuclear reactions taking place in stars.]

Weaver, K. F., and E. Kristof, "The Search for Tomorrow's Power," *National Geographic,* **142,** no. 5, 650 (November, 1972).

THE EARTH AND GEOLOGY

True knowledge can only be acquired piecemeal by the patient interrogation of nature.

SIR EDMUND WHITTAKER, 1951

Planet Earth may be divided into three main parts: atmosphere, hydrosphere, and lithosphere. The atmosphere is a layer of mixed gases a few hundred miles thick. The hydrosphere is an irregular layer of salt and fresh water, snow and ice, existing as oceans, lakes, rivers, glaciers, and ground waters. The atmosphere and hydrosphere together make up only a little more than 0.024 percent of the total mass of our planet; but they not only are essential to all life, they also are responsible to a great extent for the appearance of the earth's surface. The lithosphere is the solid part of the earth on which we live.

The geologic processes acting on and within the earth may be grouped under the headings igneous activity, diastrophism, and gradation.

Geology

Geology is the study of the earth. This term was first applied by Richard de Bury in 1473 to earthly science as different from heavenly science or theology. Geology is, in effect, the physics and chemistry of the earth and its atmosphere. In recent usage all of the physical sciences dealing with the earth in a broad sense are known as the *earth sciences* and the term *geology* is now limited to a study of the rocks of the earth, the modification of the earth's surface by water, wind, and ice, and the earth's history throughout geologic time. The more physical aspects of earth science are known as *geophysics,* which includes the study of earthquakes, the earth's interior, composition, magnetism, physical oceanography, and meteorology. The latter is the study of the atmosphere and its phenomena including weather.

Geology, therefore, is the study of man's natural physical environment. An evaluation of geological factors and processes is important in understanding many of the problems of damage to, or misuse of, lakes, rivers, landscapes, and mineral deposits under the pressures of rapidly expanding populations and the increasing industrialization of today's world.

The lithosphere

The *lithosphere* is the outer solid, or rock, part of the earth. As we have previously seen, this solid sphere is slightly flattened at the poles, having the shape of an oblate spheroid approximately 7900 miles in polar diameter and 7926 miles in equatorial diameter with a circumference at the equator of 24,902 miles (p. 26). The world's highest mountain, Mount Everest of the Himalaya Mountains, rises 29,028 ft above sea level. The deepest known spot in the oceans is in the Marianas Trench in the western Pacific between Guam and Yap and is 36,198 ft below sea level. The maximum relief, or vertical distance between highest and lowest points, of the lithosphere is about $12\frac{1}{2}$ miles. Yet this distance is so insignificant when compared with the nearly 4000-mile radius of the earth that if the earth were shrunk to the size of a billiard ball, it would be as smooth. The surface irregularities of the earth would then be no more than tiny scratches in the surface coating on the billiard ball. The smoothness and curvature of the earth is apparent on photographs taken from outer space (Fig. 2.1).

CONTINENTS AND OCEAN BASINS

The major irregularities, or first-order relief features, of the lithosphere are the continents and ocean basins. The continents are characterized by relatively smooth and extensive plains, elevated and somewhat rougher plateaus, and rugged mountains. Several of the continents are typified by high mountain systems along one or more

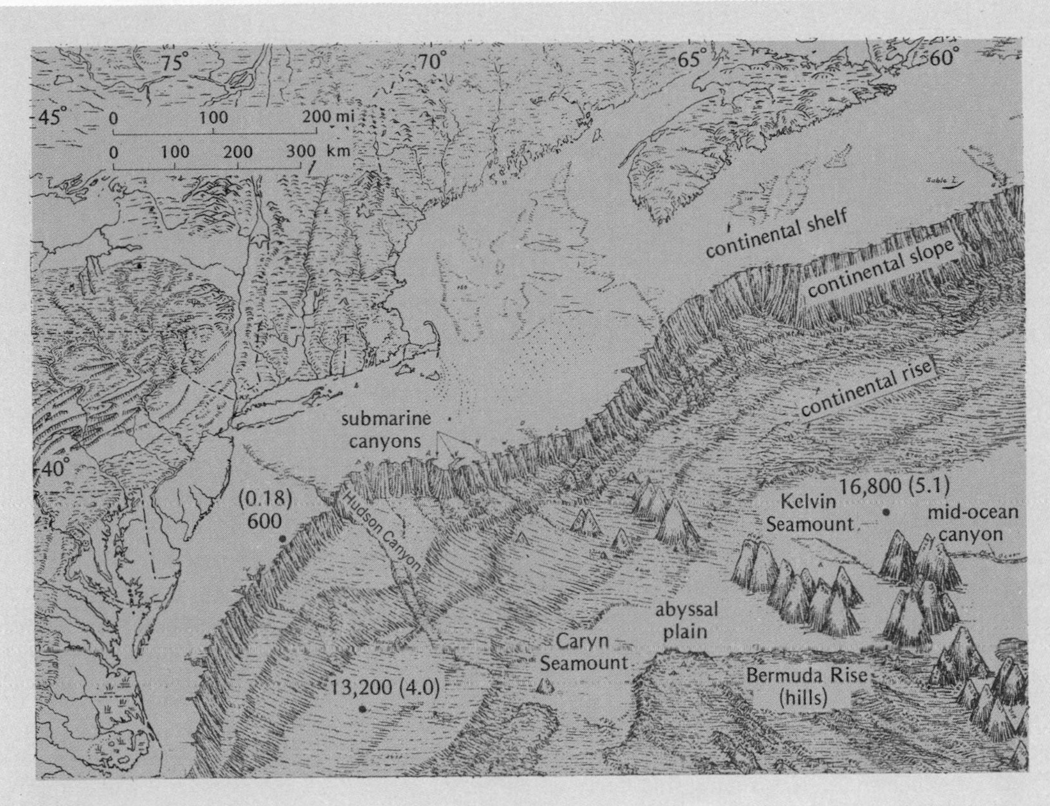

Figure 20.1
Features of the continental margin and ocean-basin floor off the coast of the northeastern United States. Depth in feet, kilometers in parentheses. (Portion of *Physiographic Diagram of the North Atlantic Ocean,* 1968, revised, by B. C. Heezen and M. Tharp, Geological Society of America.)

of their margins with wide interior plains. The irregular floors of the ocean basins are broken by long ridges and by volcanic mountains that may (the Hawaiian Islands) or may not extend above sea level, and by long narrow troughs, called *deeps,* usually located near the periphery of the ocean.

The margins of the continents are covered by ocean waters to depths of about 600 ft before they drop off abruptly to the deep-sea basins. These shallow submerged zones are known as the *continental shelves;* they vary in width from less than 1 mile to 800 miles but average 40 miles in width along the shores of all continents. These relations may be illustrated in the northwest Atlantic Ocean (Fig. 20.1) by the wide continental shelves along Nova Scotia and New England which drop off sharply to the southeast and south.

Although continental elevations range up to 29,000 ft above sea level, by far the major part (one-fifth of the entire earth's surface) of all the continental areas, including their

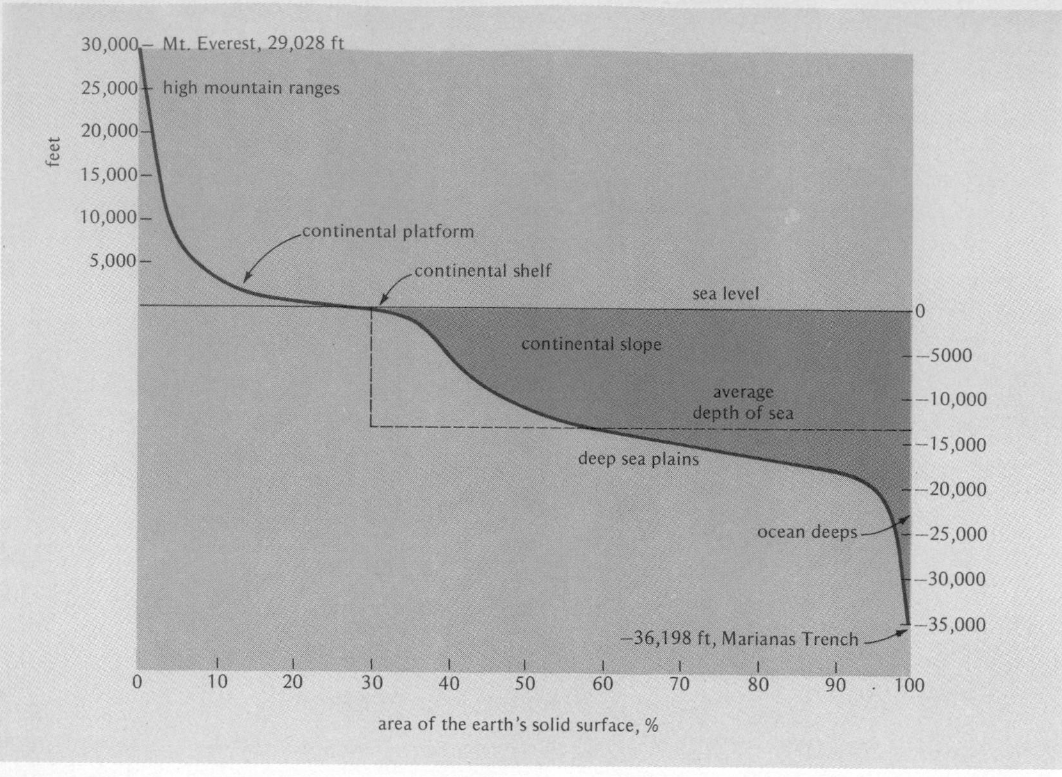

Figure 20.2
The hypsographic curve. This curve shows the percentage of the area of the earth's solid surface above and below sea level, from the top of Mt. Everest to the bottom of the Marianas Trench. It indicates that a relatively small percentage of the ocean floor is shallow and a very small percentage is below the 20,000-ft level, whereas considerable areas lie in the range of 12,000 to 18,000 ft.

shelves, lies between 600 ft below and 1600 ft above sea level. At the edge of the shelves the continental masses slope steeply into the ocean basins (Fig. 20.1). Much of the ocean bottom (over one-fourth of the earth's surface) lies between 12,000 and 18,000 ft below sea level. The percentage of area of the lithosphere surface at any level above or below sea level may be determined from Fig. 20.2, the so-called *hypsographic curve*. The lithosphere has, therefore, two dominant levels nearly 3 miles apart separated by the relatively steep continental slope. To find the explanation for these two levels separated by an abrupt slope at the edge of the continental shelves is among the great problems of geology. We will deal with some of these problems in succeeding chapters. Many earth scientists believe that the two different levels indicate different materials in the continental and oceanic parts of the earth's crust. There is evidence that the continents are composed of material of lower density than the material underlying the oceans and that the lower-density continental crust is relatively thicker than the oceanic crust. Therefore the continents stand high in comparison with the ocean basins.

Although the continents have remained relatively high, extensive parts of them have been submerged from time to time in the geologic past under shallow water. Such submerged continental areas are known as *epicontinental seas*. Present-day examples of these submergences include the North Sea, the Baltic Sea, and Hudson Bay, a small epicontinental sea on the North American Continent. Because of the presence on all continents of shallow-water deposits containing marine fossils, it is safe to say that every part of the present land areas of the world has been under seawater several times during the existence of the earth.

THE CRUST AND INTERIOR

The crust of the earth, as delimited by the geoscientists, is less than 1 percent of the total mass of the lithosphere. The outer part of the crust has been observed in mines and deep wells; the deepest penetration (through 1972) is an oil-well test hole in Oklahoma—about 30,050 ft deep. In canyons and deeply eroded mountains, we see rocks that were once many thousands of feet below the earth's surface. From these observations, geologists have a good idea of the types of rock that make up the first 5 to 10 miles of the crust under the continents.

Our knowledge of the earth's interior comes from indirect observations, largely from the study of earthquake waves. A few miles (3 to 50) below the surface of the lithosphere earthquake waves change abruptly in velocity and are also partly reflected and refracted. This change in behavior of the waves indicates the presence of a zone boundary which is called the *Mohorovičić discontinuity;* it has been used to delimit the base of the earth's crust proper. Deep within the earth, 1800 miles below its surface, another earthquake wave discontinuity indicates that the earth has a core about 4320 miles in diameter. The behavior of earthquake waves in passing through the core and the failure of some types of waves to pass at all suggest that the core is probably liquid, at least in part. The great 1800-mile zone between the crust and the core is known as the *mantle;* it has approximately two-thirds of the mass of the earth. The structure of the interior of the earth is discussed more fully in Chap. 25.

MASS AND DENSITY OF THE EARTH

The total mass of the earth can be calculated from Newton's law of universal gravitation once the gravitational constant G has been determined. Using the equation

$$f_G = G\frac{m_1 m_2}{d_e^2}$$

as discussed in Chap. 4 (p. 115), f can be calculated for a small ball m_1, the mass of the small ball is known, and d_e is the radius of the earth. Now the equation can be solved for the one remaining unknown, m_2, the mass of the earth. The mass obtained from this solution is 6×10^{21} tons (6,000,000,-000,000,000,000,000 tons). Because the volume of the earth can be calculated from its known diameter, the earth as a whole is found to be about 5.5 times as dense as water; that is, a volume of earth weighs 5.5 times as much as the same volume of water. One must not get the idea, however, that the earth is uniformly 5.5 times as dense as water. Rock material in the crust averages only about half of this, but material in the interior is much denser.

The hydrosphere

The oceans of the world make up the major part of the *hydrosphere*, an irregular and discontinuous layer of water, ice, and snow between the atmosphere and the lithosphere. Approximately 71 percent of the earth is covered by ocean water for a total volume of some 331,000,000 cubic miles of salt water.

The lakes and rivers on the continents and the ground waters in cracks and pore spaces of the earth are also considered part of the hydrosphere. The ice and snow of

Figure 20.3
Frozen hydrosphere. The Eagle Glacier system and its snow-covered ice field in the Coast Ranges of British Columbia near the Alaskan border. (Courtesy of the Geological Survey of Canada, photograph no. 99471.)

glaciers and the polar ice caps are a part of the hydrosphere which is solid during the present geologic epoch (Fig. 20.3).

The atmosphere and the hydrosphere are intermixed to a certain extent, in that water vapor occurs in the atmosphere and water particles may be suspended as clouds and fog. Furthermore, atmospheric gases are dissolved in ocean and lake waters. This two-way exchange between air and water is controlled by the vapor pressure at the air-sea boundary (p. 169). The evaporation of water from the surface of the hydrosphere to water vapor in the atmosphere, with the subsequent condensation of this vapor to form visible clouds from which precipitation may eventually fall, returning the water to the hydrosphere, is called the *hydrologic cycle* (see p. 512).

Ocean water contains about $3\frac{1}{2}$ percent by weight of dissolved matter, or 151,000,000 tons of solids per cubic mile. Over half of the known elements of the earth's crust have been found in the ocean waters, which therefore represent a tremendous reserve of many substances for man's use in the future. Already, we are extracting the elements magnesium and bromine in commercial quantities, and such metals as gold, lead, zinc, and copper are present in amounts potentially significant for the future.

The commonest ions in seawater are given in Table 20.1. The salt content varies slightly but measurably in the open ocean

Table 20.1
Commonest ions in seawater

ion	symbol	percentage of all dissolved matter
chlorine	Cl^-	55.0
sodium	Na^+	30.6
sulfate	SO_4^{2-}	7.7
magnesium	Mg^{2+}	3.7
calcium	Ca^{2+}	1.2
potassium	K^+	1.1
bicarbonate	HCO_3^-	0.4
total		99.7

from a low of 3.3 percent in Arctic and Antarctic waters, where precipitation and the melting of ice contribute more than the average amount of fresh water, to a high of 3.6 percent in the subtropics, where lack of precipitation and a high rate of evaporation are responsible for the higher salt content. Oceanographers are able to trace these slight differences of salinity over thousands of miles to determine the origin of the water at various depths in the oceans.

All atmospheric gases are found dissolved in sea water, but oxygen and carbon dioxide are the most important: the oxygen content because it maintains animal life in the water and carbon dioxide because it allows plant growth. Oxygen is dissolved in seawater in small quantities at the air–sea interface. As water moves below the surface, oxygen content decreases slowly with time owing to animal use and the decay of organic matter. In some deep waters the oxygen content has been completely used and there is no life in such "stagnant" waters. The decrease in oxygen content can thus be utilized as an indicator of the "age" of the water; that is, how many years have passed since a particular mass of water was last at the ocean surface.

However, oxygen-bearing waters do circulate from the surface to great depths; thus in some parts of the oceans animal life has been found at extremely great depths. Plant life, on the other hand, is dependent on the presence of sunlight. Because only about 7 percent of the total sunlight reaching the surface penetrates below 300 ft, plant life is restricted to the upper layers of the oceans. Therefore, carbon dioxide is used up because of plant photosynthesis only in the upper few tens of feet of the ocean. The percentage of carbon dioxide remains constant in the water that circulates at greater depths.

A large amount of carbon dioxide is dissolved in ocean water—about 60 times as much as in the atmosphere. Because the capacity of the oceans to dissolve carbon dioxide is great, the oceans act as a regulator of the carbon dioxide content of the atmosphere. This prevents the atmospheric concentration of carbon dioxide from rising too rapidly. Variations in the carbon dioxide content of the atmosphere can cause major worldwide climatic changes because carbon dioxide absorbs, or blankets, heat radiation from the earth's surface.

Temperatures of ocean surface waters vary from as high as 86°F in equatorial regions to a low in the Arctic Ocean of 28°F, the freezing temperature for oceanic salt water. The source of heat for seawater is incoming solar radiation, which, of course, reaches its greatest extent in the equatorial zones of the earth. This process of heating the ocean waters is not particularly efficient as compared with heating water from below. Heated surface waters expand, thus have decreased density, and therefore remain near the surface. As a consequence, the deeper waters of the oceans never do become heated. Below about 5000 ft, ocean water is cold everywhere, probably in the temperature range of 34° to 38°F.

The water of the Antarctic Ocean is the coldest in the world and thus the densest, or heaviest, in spite of its relatively low salinity. This heavy water sinks down in quantity and flows northward as the bottom layer in the Atlantic, Indian, and Pacific oceans. It has been recognized well north of the equator in the bottom of the Atlantic by its low temperature (33° to 35°F), low salinity, and decreased oxygen content. A similar but much smaller quantity of cold, dense water sinks off Labrador and Greenland and flows southward in the deep Atlantic just above the layer of Antarctic water.

The heat capacity of the oceans, however, is very great, because any additions or subtractions of heat are spread out, by the motions of the oceans, through large masses of water. This means that the oceans are the major climatic regulators, modifying to a large degree the temperatures of air masses moving across their surfaces. Water temperatures respond only very slowly to daily or seasonal variations in air temperature or amount of sunshine.

OCEAN CURRENTS

There is evidence of water motion at all depths in the oceans, but the major surface circulations, such as the Gulf Stream, are obviously the best known and the most important to man in ocean travel. The worldwide system of surface ocean currents is characterized by great whirls centered in the subtropical latitudes of both the Atlantic and Pacific oceans. These large whirls are known as *gyres,* and they rotate clockwise in the Northern Hemisphere and counterclockwise in the Southern Hemisphere. The major circulations or gyres are very closely related to the major wind systems of the earth. The eastward motion of the circulations in the middle latitudes, both north and south (that is, the poleward sides of the gyres), is driven by the prevailing westerlies, whereas the westward-moving currents in the tropical parts of the systems, both north equatorial and south equatorial, are driven by the trade winds. These gyres in the Northern Hemisphere carry warm waters northwestward across both the Pacific and the Atlantic and move colder waters southward along the eastern sides of both oceans, resulting in cold currents along the California coast and along the coasts of Portugal and the Canary Islands. The best known of these currents is the Gulf Stream, which moves from the Caribbean area past Florida and along the Atlantic Coast of the United States northeastward across the North Atlantic to spread warm water near Great Britain and southern Scandinavia. The Gulf Stream averages some 50 miles wide and has a volume of water equal to approximately 1000 times the average flow of the Mississippi River. The main part of the stream north of Florida is at least 1500 ft deep and flows at velocities of 2 to 6 mi/hour.

Recent oceanographic research on currents has established that return, or counter, currents flow at depths of a few thousand feet in opposed directions to those of the major surface circulations. For example, a countercurrent about 5000 ft deep beneath the Gulf Stream moves in the opposite direction.

Vertical circulations occur in the oceans in addition to the major surface horizontal-circulation systems. Such circulations result in a complete overturn of ocean waters every few thousand years; that is, surface waters descend to depth and deep waters are slowly brought back to surface. These vertical circulations are due primarily to differences in water density based on (1) temperature differences of ocean waters, (2) variations in salinity, and (3) turbidity currents.

As previously mentioned, very cold water moves downward in quantity in the Antarctic Ocean and northward along the bottom of both the Atlantic and Pacific oceans. This downward motion is due to the sinking of cold and thus dense waters produced in polar regions.

The water of the Mediterranean Sea has a higher salinity than the average for the oceans. At the western end of this sea, at the Straits of Gibraltar, a current in the

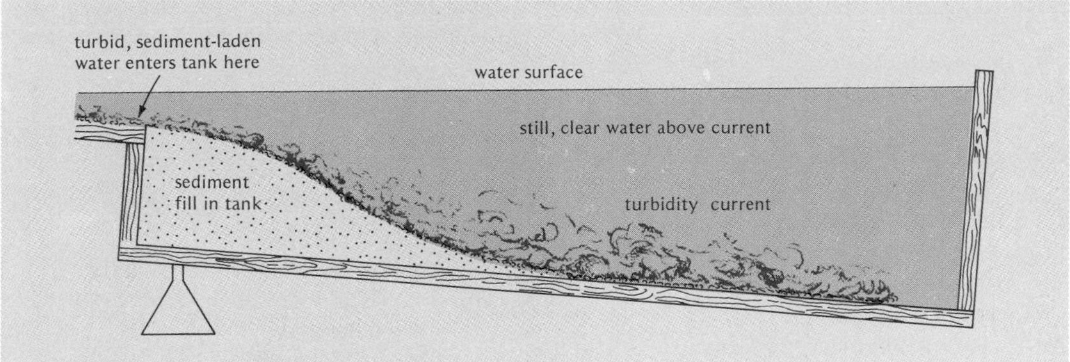

Figure 20.4
Turbidity current in a laboratory tank. (From *Principles of Geology,* 3rd ed., by James Gilluly, Aaron C. Waters, and A. O. Woodford. W. H. Freeman and Company. Copyright © 1968.)

Mediterranean moves into the Atlantic. Because this water is more saline than Atlantic water, it is denser and moves down into the depths of the eastern Atlantic; it is traceable downward and then westward for some hundreds of miles.

Water with great quantities of mud in suspension is heavier than relatively clear water. Thus muddy water moves downslope beneath clear water. Currents produced by the movement of muddy or sediment-laden water are known as *turbidity currents* (Fig. 20.4). Such currents were first observed and studied in lakes and reservoirs; for example, turbidity currents have been observed to move the entire length of Lake Mead (Nevada and Arizona). Two lines of evidence suggest that powerful turbidity currents are major features of the ocean floor. The first is the presence of sand and even gravel lenses in many places on the ocean bottom, particularly at the base of the continental slope, near the mouths of submarine canyons, and on deep-sea fans built beyond the continental slope on some parts of the ocean bottom. Apparently such coarse sediment could only have been carried seaward by massive turbidity currents. Many of these deposits of sand and gravel are far from any known ocean currents that might have been capable of their transport. The second line of evidence suggesting the presence of powerful turbidity currents comes from the breaking of submarine cables following an earthquake off the Grand Banks in 1929. Many cables were snapped successively over a period of some hours after the quake. The evidence which has been collected suggests that strong turbidity currents, or watery landslides, were set in motion by the earthquake and swept downslope, breaking cables in sequence as they moved away from the source area. Turbidity currents may be likened to underwater landslides; they develop when a large, thick mass of water-saturated sediment is suddenly thrown into an aqueous suspension. When this happens, an aqueous landslide or turbidity current is generated. Turbidity currents are capable of transporting even very coarse sediment into deep water. They are efficient scouring agents and thus can produce a great deal of erosion on the continental slope and adjacent ocean floor.

The atmosphere

The layer of mixed gases with locally included particles of dust and moisture is called the *atmosphere*, or air, and extends upward several hundred miles. Close to the surface of the earth the atmosphere is about 0.013 times as dense as water, but this gradually decreases to nearly 0 at 3000 miles up. This falling off in density is not directly proportional to elevation. About half of the total weight of the atmosphere is located within 3.5 miles of the surface of the earth, half of the remaining half is located in the next layer of 3.5 miles, and so on. Accordingly, almost 90 percent of the atmosphere is within about 10 miles of the surface of the earth. A few years ago scientists placed the boundary between the earth's atmosphere and interplanetary space at an altitude of about 600 miles, where atmospheric gases may be free from the earth's gravity and escape into space. Recently, however, using artificial satellites, scientists have discovered an envelope of charged solar particles, trapped in the earth's magnetic field, and have therefore extended the earth-space boundary to about 37,000 to 62,000 miles.

Until the advent of earth satellites and space and lunar probes, it was customary to divide the earth's atmosphere into three main zones: the *troposphere,* the *stratosphere,* and the *ionosphere* (Fig. 20.5). Because of varying thermal and electrically charged atomic conditions at different heights, as determined by satellites and probes, its additional zones were named *mesosphere, thermosphere, exosphere,* and *magnetosphere*.

The troposphere extends to about 10 miles above the earth's equator and diminishes to about 5 miles above the poles. This zone contains over 75 percent of our air and nearly all the water vapor. Hence, our weather—the movement of air currents, condensation of water vapor to form clouds, rain, snow, etc.—is almost entirely a phenomenon that occurs in the troposphere

Table 20.2
Composition of the troposphere and stratosphere

material	percentage by volume of dry air
nitrogen molecules	78
oxygen molecules	21
argon atoms	1
carbon dioxide molecules	0.03
ozone (largely in stratosphere)	traces
15 or more other gases (including pollutants[b])	0.04[a]
particulate matter (dust)	traces[a]
total, for dry air,	100
water vapor	trace to 4

[a] Locally at much higher concentrations in the troposphere during high wind storms, near volcanic eruptions, or over industrial areas.

[b] Pollutants include: carbon monoxide (nearly half of all man-made pollutants), sulfur oxides, hydrocarbons, nitrogen oxides, particulate matter.

(Fig. 20.5). Most of our airplane travel is through this layer. This part of the atmosphere is important geologically; it is wind and weather and their effects, together with the chemical changes between gases in the atmosphere and materials on the surface of the earth, that are of interest to the student of earth science.

COMPOSITION

The air is a mixture of gases of which oxygen, nitrogen, water vapor, and carbon dioxide are important to life. Dry air is about 78 percent nitrogen by weight and 21 percent oxygen (Table 20.2). Carbon dioxide, although only 0.03 percent, plays an all-important part, together with oxygen, in the processes of weathering, as we shall discuss later. Of equal importance in this respect is the presence of water vapor; its content in the atmosphere varies greatly. Over hot, dry deserts, it may be only a small fraction of 1 percent of the entire atmosphere, but in warm, humid regions, it is

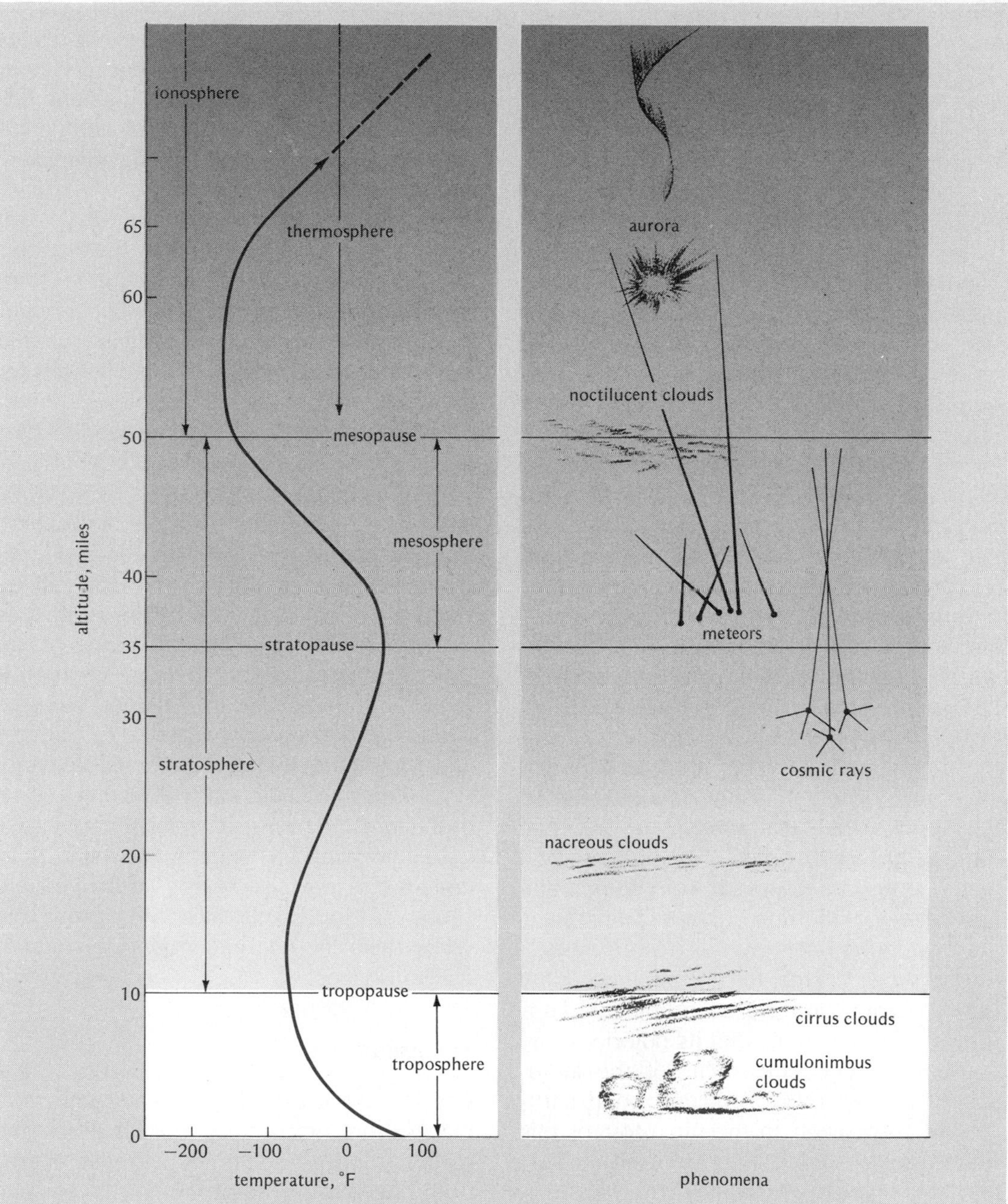

Figure 20.5
The earth's atmosphere. Temperature changes (reversals) define the various regions of the lower atmosphere. The heights shown for temperature changes and boundaries between zones (left) and for natural phenomena (right) are therefore approximations. Above the thermosphere are the exosphere and magnetosphere; these are not shown. The atmosphere may be considered to extend to about 62,000 miles at times. Almost 90 percent of the total weight of the atmosphere is within the troposphere, the gaseous envelope extending from the earth's surface to a height of about 10 miles.

sometimes as much as 4 percent. The world average is near 1 percent. The amount of water vapor that air will hold before becoming saturated increases as the temperature is raised. The water-vapor content of a given air mass is usually expressed as relative humidity, a concept that has been explained in Chap. 8.

In addition to the permanent gases of the atmosphere, many others are occasionally present because of volcanic eruptions or the activities of man. In some eruptions, large quantities of gases of sulfur compounds, hydrochloric acid, and even hydrofluoric acid are passed out into the atmosphere. These can be locally very dangerous to both life and property, but they soon become diffused into the vast quantity of permanent atmospheric gases, where their presence is no longer significant. In heavily populated and industrial areas many gases are added to the atmosphere from factory wastes and the products of combustion of all kinds of fuel. Although these gases are insignificant in amount in comparison with the permanent gases of the atmosphere, they may become locally concentrated to the point of personal annoyance or actual harm. One of the most common of such gases is carbon monoxide from automobile and other combustion engines. Some areas are plagued at times with distressing smogs: the Los Angeles area, the industrialized Pittsburgh district, and the overpopulated London, England, region, with its notorious fog.

An important constituent of the atmosphere is dust, millions of tiny, solid particles in suspension in the air. Most of this dust is fine material swept up from the surface of the land by winds and carried throughout the troposphere by the major air circulations. In addition, important quantities of dust are added not only to the troposphere but also to the stratosphere by volcanic eruptions. It is estimated that dust from the violent explosion of Krakatoa Volcano in 1883 was blown 50 miles high and entirely encircled the earth in 15 days. This dust remained suspended in the atmosphere for over a year. Salt particles are added to the atmosphere from ocean spray, especially from the surf, and may be found in air hundreds of miles inland. Man has been adding his bit with industrial smokes and more lately with radioactive particles.

Atmospheric dust is significant in several respects. Dust particles are the centers around which condensation occurs to form raindrops and snowflakes. The proper quantities of dust may result in brilliant sunsets; the entire world experienced red sunsets for several months after Krakatoa blew dust into the stratosphere in 1883. Suspended dust aids in the diffusion of sunlight, but perhaps most important of all, dust particles absorb heat passing through the atmosphere from the sun to the earth's surface. Excessive quantities of dust can so reduce the amount of heat that reaches the surface as to appreciably lower world temperatures. World records indicate that cold years followed a number of the violent volcanic eruptions of the past few centuries, notably following the outbreak at Krakatoa Volcano. Continuous volcanic activity in many parts of the world throwing excessive amounts of dust into the atmosphere has been suggested by some geologists as a possible cause of colder climates throughout the world, with the consequent development of glacial periods.

TEMPERATURE

The atmosphere itself serves as a thermal blanket or temperature regulator for the earth's surface. Some of the heat energy from the sun is absorbed in its passage toward the earth, preventing the days from being too hot. At night when the earth's surface is cooling by radiation of heat toward outer space, absorption of this radiant energy by the atmosphere decreases the amount of heat loss and so keeps the night temperatures from becoming excessively cold. Water vapor and carbon dioxide have a much greater capacity for absorbing heat energy than any other atmospheric gases. These two gases along with dust and moisture particles—clouds—are the thermal blankets of the atmosphere. The importance of water vapor in this capacity is well illustrated by the fact that deserts lose heat much more rapidly at night than do humid regions. And we all know that it can get much colder on a clear night than on a cloudy one.

Starting a short distance above the earth's surface, the temperature of the air in the troposphere falls off at a quite constant rate with elevation. This *lapse rate,* as it is called, amounts to about 19°F (Fahrenheit) per mile rise. When an elevation of about 10 miles is reached, the lapse rate quickly falls to 0; that is, the temperature remains practically constant. Above this critical elevation, called the *tropopause,* the stratosphere begins—at a height of about 10 miles.

STRATOSPHERE

The stratosphere is usually considered to extend from about 10 to about 50 miles above the earth's surface. The average temperature of the stratosphere is about −67°F. Actually there is a considerable increase in temperature followed by a greater decrease as one goes higher in the stratosphere. Airplanes may fly in the lowest region of the stratosphere, but at altitudes of above about 8000 ft their cabins are pressurized, so that the concentration of air (and oxygen necessary for life) is increased to about the equivalent of that at ground level. Probes (balloons, rockets, satellites) have been sent into various parts of the stratosphere to obtain samples of air for study.

Starting in the upper troposphere and extending well into the stratosphere, we find an abundance of ozone (O_3, the triatomic oxygen molecule), which readily absorbs ultraviolet radiation (Chap. 9). If this ozone were not present, life as we know it on earth would not be possible, because intense ultraviolet radiation is destructive of most forms of life.

UPPER ATMOSPHERE

The very thin atmosphere above the stratosphere (above 50 miles) is called the ionosphere because it includes various layers of small charged particles called *ions.* The most important of these ionized layers is the Kennelly-Heaviside layer, about 18 miles thick and starting about 50 miles above the earth's surface (Fig. 20.5). This layer is mainly responsible for reflecting radio waves back toward the earth, thus making long-distance reception possible.

The air of the ionosphere is so thin that meteors do not become incandescent, as a result of heat of friction in passing through the air, until they get within about 60 miles of the earth's surface. Northern lights (aurora borealis), which are a phenomenon of the upper atmosphere, have been measured as high as 600 miles above the earth's surface.

As a result of the new data gained in this space age, we now know that temperature changes (reversals) define various regions of the atmosphere: the *troposphere,* the *stratosphere,* the *mesosphere,* the *thermosphere,* as seen in Fig. 20.5.

Going upward from the earth through the troposphere to the stratosphere, the temperature drops to about −67°F at 10 miles, remains almost constant up to about 15.5 miles, and rises to about 18°F at about 35 miles. This last temperature value marks the beginning of a zone known as the mesosphere, where the temperature decreases again to about −112°F at about 50 miles, the top of the mesosphere (and of the stratosphere). The next zone, the thermosphere, is one of rapidly increasing temperatures to a plateau of almost 1400°F at about 450 miles. This high temperature means that the molecules of the atmosphere are moving very rapidly, but there are so few of them that a body passing through would be hit by only a very few each second and would therefore remain cold.

Above the thermosphere is the exosphere, which extends from a height of about 450 miles to 3000 miles. The outermost zone is called the *magnetosphere* because particle populations in this farthest region of the atmosphere are determined by the geomagnetic field.

The *atmosphere* may be defined in a broad sense as

The region surrounding the earth in which the presence of our planet appreciably perturbs the interplanetary medium.

Space probes and satellites have shown that the population of geomagnetically trapped particles may extend out as far as about 62,000 miles in periods of unusual activity in the sun. This distance marks the outer limit of the atmosphere—the earth-space boundary—as defined above.

The earth's magnetism

THE EARTH A HUGE MAGNET

The fact that a compass needle assumes a north-south direction leads to the conclusion that the earth must be a magnet, with the space all around it constituting a huge magnetic field. The earth's magnetic poles, however, do not coincide with the geographical poles, as shown in Fig. 20.6. We can think of the earth's magnetic field as a series of lines of force. A magnetized needle, free to move in space, will align itself parallel to one of these lines. At a place some 1800 miles north of Winnepeg, Canada, at about 100.5°W longitude and 75.5°N latitude, and just north of Prince of Wales Island, the north-pointing end of our magnetic needle will dip straight downward into the earth. This point is the north magnetic pole, or dip pole. Along the coast of Antarctica at about 67°S and 143°E, the same end of our needle will point directly skyward at the south magnetic pole, or dip pole. Between these dip poles, magnetic needles assume positions of intermediate dip and will point in a horizontal position on the magnetic equator (Fig. 20.6). The angle between the magnetic needle and the earth's surface is called the *magnetic inclination* or *dip*. This angle is 90° at the magnetic pole, or dip poles, and 0° at the magnetic equator.

A compass needle points to the magnetic pole nearest the north geographic pole, and for most places on the earth's surface this direction does not coincide with true geographic north. The exception, the meridian that passes through both the geographic and magnetic poles, and incidentally near Lansing, Michigan (*L*), is called an *agonic line*. Therefore a compass needle will, except for places on the agonic line, point west or east of true geographic north. *Variation*, or *magnetic declination*, is the angle that the compass needle makes with the direction of true, or geographic, north.

In reference to, and on maps of, land areas we indicate magnetic declination, but in nautical and aeronautical navigation the term *variation* is preferred. As shown in Fig. 20.7, the declination for places east of the agonic line, or line of zero declination, is a certain number of degrees west. For Boston, Massachusetts (*B*), the declination is 15°W. Portland, Oregon (*P*), has a variation of 21°E. Variation is very important in navigation, where it must be applied as a correction to the compass reading to give the direction of true north. For example, if an airplane over Boston is headed 70° east of north according to the magnetic compass, the true, or geographic, direction of flight is $70° - 15° = 55°$ east of geographic north.

Man has known for several centuries that magnetic declination at a given place changes slowly with time. Such so-called secular changes in the magnetic field are irregular and variable in different geographic locations but predictable within local areas.

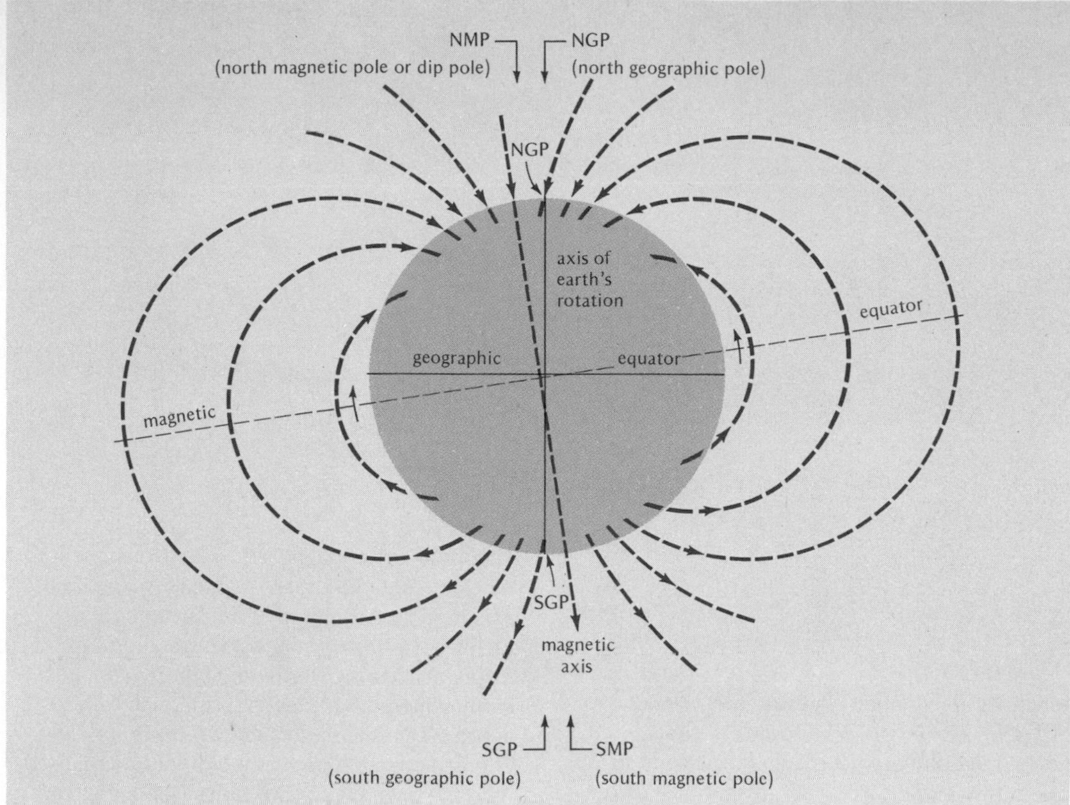

Figure 20.6
The earth's magnetic field. A diagrammatic sketch of the lines of force of the earth's magnetic field, showing magnetic poles, geographic poles, the magnetic axis, and the axis of the earth's rotation. The magnetic poles, indicated by SMP and NMP, shift positions slowly, but independently, through the years, and are not always opposite each other in the same locations on the same axis. Thus the intersection of the magnetic axis and the plane of the magnetic equator does not coincide with the intersection of the earth's axis of rotation and the plane of the geographic equator. The small arrows on broken lines indicate positions of a magnetic needle free to move in space at various locations on the earth's surface—in short, the inclination, or magnetic dip.

CAUSE OF EARTH'S MAGNETISM

The cause of the earth's magnetic field is not yet fully understood but most of it must have an internal origin. The magnetic field does have a small external component related largely to activity of charged particles from the sun. This solar component apparently explains the radiation belts and the magnetosphere that extend out from the earth for thousands of miles, changes in the ionosphere such as magnetic storms, and the aurorae (see flares, p. 55).

The earth's measured magnetic field is close to that which would result if a large bar magnet was thought to be centered

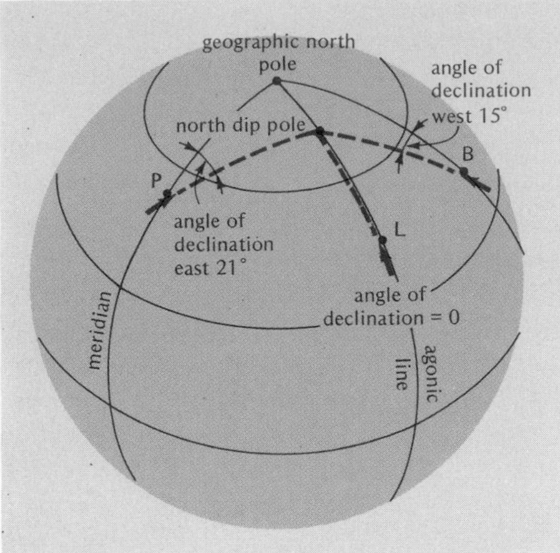

Figure 20.7
Magnetic declination. Because the magnetic or dip poles do not coincide with the geographic poles, the compass needle does not point to true north, except along the agonic line. The angle of divergence of the compass from geographic north is the declination and is measured east or west of north. B = Boston, Massachusetts; L = Lansing, Michigan; P = Portland, Oregon.

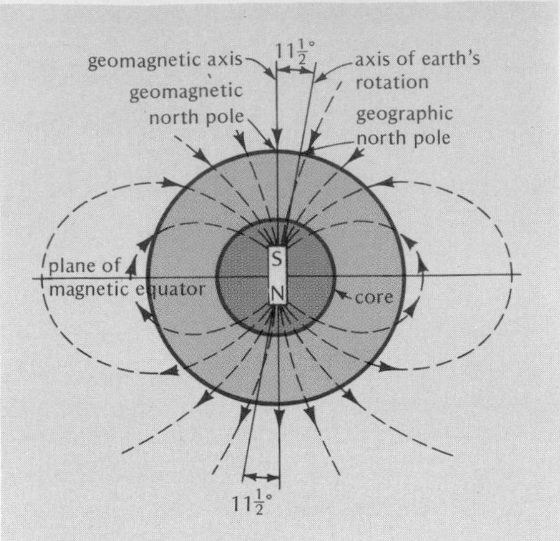

Figure 20.8
Theory of earth magnetism. The geomagnetic poles are defined by an imaginary magnetic axis passing through the earth's center and inclined 11½° from the axis of rotation. This magnetic axis is determined by a hypothetical, earth-centered bar magnet positioned to best approximate the earth's known magnetic field. The geomagnetic poles should not be confused with the magnetic or dip poles (see Fig. 20.6); the former are hypothetical, the latter are actual and observational. (From L. Don Leet and Sheldon Judson, *Physical Geology,* 4th ed., © 1971, p. 34. Reprinted by permission of Prentice-Hall, Inc., Englewood Cliffs, N. J.)

within the earth at a small angle from the earth's axis of rotation. In Fig. 20.8 this bar magnet in the earth's core is oriented with its south (S) pole northward as if, by convention, to attract the north end of compass needles (which are tiny bar magnets). This type of magnetic field is known as a *dipole field.* A similar dipole field can be produced by passing an electric current through a coil of wire. (The student should review the section on magnetism, Chap. 10, pp. 231–235.)

The core of the earth, as we shall see in Chap. 25, is believed to be molten iron-nickel alloy at very high temperature. We might suspect that this mass of iron was magnetic and thus the cause of the earth's magnetic field. But it has been demonstrated that magnetic materials lose all their magnetism above a certain temperature. This is called the *Curie temperature* for each material. The Curie temperature for iron is about 760°C and for nickel about 350°C. Because the temperature of the core of the earth must be at least 4000°C, the earth's core cannot have any permanent magnetism.

The presently accepted theory of earth magnetism is called the *dynamo theory,* which suggests that random motions in the fluid core generate electromagnetic currents. The core has a composition that would make it an excellent conductor of electric currents, and its physical state would allow such movements. These movements are probably convection-type currents caused by temperature differences in the core. The dynamo theory implies that the many irregular motions and the resulting random electromagnetic fields must become a single, united dipole magnetic field. It is believed that the rotation of the earth is the force that imposes such order on the electromagnetic fields within the core. Although the axis of this dipole electromagnetic field must, according to present theory, correspond approximately with the earth's axis of rotation, variations of a few degrees seem possible such that the magnetic axis may wobble slowly around the axis of rotation every few thousand years. This would explain the very slow wandering of the magnetic pole in the general area around the geographic pole, the secular variations in declination, and the other minor irregularities in the earth's actual magnetic field. The theory of a self-exciting dynamo as the explanation of earth magnetism is still not fully established, but it explains quite satisfactorily the puzzling facts about the earth's magnetic field.

Geologic processes

Modern geology was perhaps born in 1785 when James Hutton (1726–1797), a Scottish geologist, put into words in a convincing fashion the *doctrine of uniformitarianism* which has become the foundation of most geologic thought. This doctrine holds that earth processes in operation today have operated in the past in a similar manner. Today's processes are sufficient, allowing enough time, to explain the surface features and materials of our earth. In other words, *the present is the key to the past.* Certainly we learn about ancient volcanoes by watching modern lava flows and explosive eruptions. We discover that north-central and eastern United States were covered by an ice sheet by studying present-day glaciers and ice caps in Greenland and Antarctica. We have observed that these modern glaciers deposit piles of debris containing rock fragments of all sizes from dust to boulders, and we find this same jumble of unsorted material scattered over parts of northern North America: for example, in Michigan, the plains of Ohio, the New England hills, as well as in deep valleys in the Rocky Mountains. We have assumed, therefore, that ancient glaciers once covered all these areas and formed these deposits.

When we see tiny parallel ridges, or ripple marks, in the sand of a river bottom or beach at low tide and then find similar marks in a consolidated sandstone, we correctly infer that the sand of the sandstone was originally deposited by a shallow river or the waves along an ancient beach. And again when we observe the shell-bearing lime muds being formed today on the Bahama Banks and then find identical material as rock with fossil shells in some of the hills and mountains of the continents, we may conclude that these rocks were formed in an ancient ocean. Therefore a study of present processes tells us much about the environments and conditions in which different rock associations were formed.

With this concept of uniformitarianism geologists have been able to explain most earth features in a logical way. Past processes, presumably, have operated at the same slow pace as those of today. Therefore very long periods of time must have been available for these processes to bring about what we see on the earth's surface now. For example, let us consider the Grand Canyon of the Colorado River, a gigantic gash in the earth's surface, 1 mile deep and averaging 10 miles wide. Some early explorers wanted to say that this canyon was caused by a great catastrophe that ripped open the earth in a short time. However, by understanding the way in which weathering and the action of running water are working today, we can appreciate that such a large canyon has been formed by slow continuing processes over a long period of time, probably several million years. This concept of almost unlimited time for the development of earth features was a necessary outgrowth of the principle that the present is the key to the past. Before

Hutton's time, it was generally believed that the earth was only a few thousand years old, and major features of the landscape were all explained as being produced by catastrophic events that suddenly upheaved mountains or downdropped sea bottoms. There was a controversy for many years between the theories of catastrophism and uniformitarianism. Today we know that occasional small catastrophes—floods, earthquakes, volcanic eruptions, landslides—do occur but are relatively minor in their effect on the production of mountains, canyons, and other surface landforms. The large features of the earth's surface have been formed by a slow continuity of processes that does not seem to produce much change in the lifetime of a single man.

The major geologic processes of the earth may be considered in two broad categories which, in a sense, are acting against each other: one set whose origin is within the earth's interior and the other whose origin is on the surface of the lithosphere. The first group tends to deform the crust of the earth and elevate the land surface; they might be called *constructive forces* and are discussed under the headings "Igneous Activity" and "Diastrophism." The second set of processes is wearing away the land areas of the world and thus called *destructive processes*. This leveling activity is known collectively as *gradation* and is primarily the work of forces acting in the atmosphere and hydrosphere. The end result of gradation alone, given sufficient time, would be the wearing down of all land to sea level. The combined result of all earth processes during the history of the earth up to the present time is the landscape of the world as we know it today.

IGNEOUS ACTIVITY

The processes and phenomena associated with molten rock material, known as *magma*, and the crystallization of the latter to form the igneous rocks are all included in *igneous activity*. Magma is a hot solution of various silicates with a few dissolved oxides and gases. Until a few hundred years ago man believed that the earth's interior was molten and covered by a thin solid crust; lavas were interpreted as the molten interior breaking through to the surface. Today we know, chiefly from earthquake wave data, that the earth is solid down at least 1800 miles. Magmas, therefore, are not residual liquids from an original molten earth but are generated from time to time during the earth's existence by local melting within or below the crust. A pocket or chamber of magma, thus formed by local melting in the lower part of the crust, continues to be molten for a few million years, a relatively short time in terms of earth history.

Magma occasionally breaks out or is extruded on the surface of the lithosphere, with the resulting phenomena of volcanoes and lava flows. A *volcano*, in the strict geologic sense, is any opening in the earth's crust from which magma and its associated products are thrown out or erupted onto the surface from within the earth. Most volcanoes build up a large conical mountain around this opening composed of the erupted material. A long crack or fissure in the surface from which lava has been poured out is a true volcano even though no mountain exists.

Only very rarely does magma actually reach the surface in volcanoes; most commonly magma fills existing cracks and other spaces in the crust or actually makes room for itself by lifting overlying rocks or by removing rock materials in a combined melting and chemical-reaction process. The crustal rocks are cooler than the magma, which is chilled below its freezing temperature, gradually crystallizing to form *intrusive bodies* of igneous rock of many sizes and shapes. The crack filling, or dike (Fig. 20.9), is a very common type of intrusive igneous body. Movement of magma within the earth's crust is known as *plutonic activity*

Figure 20.9
A dike. A vertical sheetlike body of intrusive igneous rock cross cutting a series of nearly horizontal sedimentary rock layers. The dike has been formed by the intrusion of magma along a former fracture in the earth's crust. (Courtesy Hugo Mandelbaum.)

as contrasted with surface or volcanic activity.

DIASTROPHISM

Movements of solid segments of the earth with respect to one another occur as an adjustment of the crust to differential stresses and pressures within the earth. Such movements may be vertical or horizontal. Rocks that were formed by deposition of sediment in the sea, and contain the remains of marine life as proof of this origin, are now found high above sea level in some of our loftiest mountain ranges and plateaus. Study of such rocks and their relationships to one another indicates that large parts of the continental areas have been elevated far above sea level and then submerged below sea level many times during the history of the earth. Some belts have been intensely deformed as part of mountain building (Fig. 20.10). The temperature and pressures in the roots of these zones were great enough for the development of the metamorphic rocks.

GRADATION

The natural cut-and-fill action of *gradation* may be divided into the wearing-down processes of erosion and the filling-in processes of deposition. The agents that do the gradational work are air, water, and ice. Not only can these agents effectively break up the minerals and rocks of the crust by chemical action or abrasion, but they are also efficient transporting agents. Of particular importance is *weathering,* the breakup of the solid rocks of the crust. The processes of weathering include the development and enlargement of tiny cracks by the expansive forces of alternate freezing and thawing, or frost action, and by the extremes of temperature from day to night in desert areas. Water and the atmospheric gases enter these cracks, and chemical changes begin. After these actions are repeated thousands of times, the rocks finally become disintegrated and decomposed. The finely divided products of weathering are transported to lower places on the earth's surface. Weathering and the transport of the resulting debris are known as *erosion.* The transported debris or sediment is ultimately deposited in the

Figure 20.10
Deformed rocks. Intensely folded sedimentary rock layers as a result of mountain-building processes, Newfoundland. (Courtesy of the Geological Survey of Canada, photograph no. 120457.)

seas, where sedimentary rocks gradually develop.

THE ROCK CYCLE

Rocks are continually being formed and remade in a never-ending cycle. The rock cycle, as illustrated in Fig. 20.11, summarizes the relationships among those earth processes that result in rock formation or change. Molten rock material, called *magma,* solidifies by crystallization of minerals to form igneous rocks. The igneous rocks were the primary rocks of the crust from which all other rock types have subsequently been formed. The weathering of these igneous rocks and the transportation of the resulting debris to be deposited as sediments lead to sedimentary rocks. If these become deeply buried under later sediments, the lower beds may be metamorphosed to metamorphic rock or may later become so heated as to melt and form magma, completing the cycle. Anywhere along the way the cycle may be shortened by weathering back to sediments again, or igneous rocks may themselves be metamorphosed. Proper identification and interpretation of rocks leads us to our knowledge of earth history. The processes and inter-

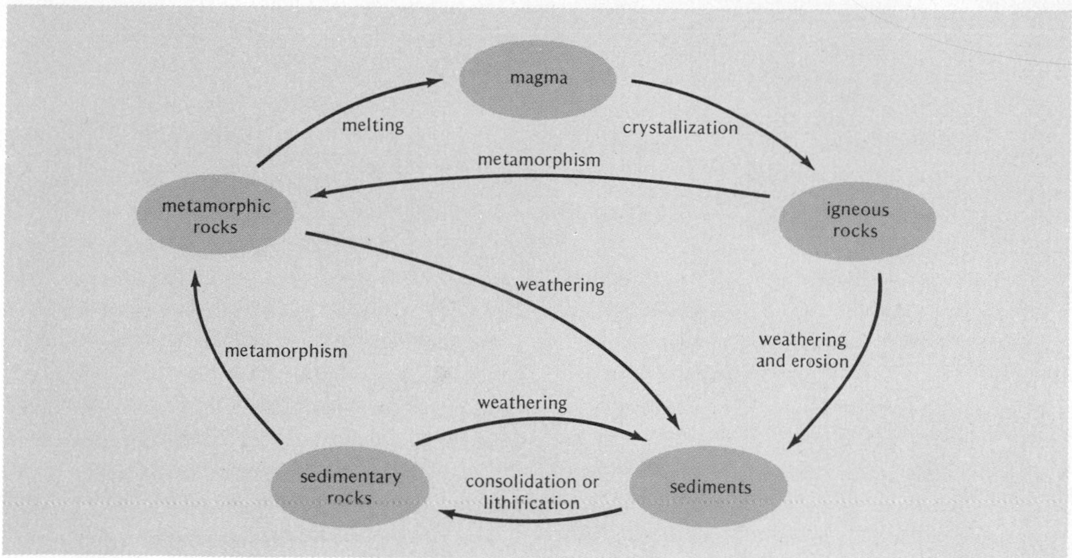

Figure 20.11
The rock cycle. Schematic diagram showing the relations among the various earth materials and the processes (arrows) by which the major rock types are formed or changed.

pretations involved in the steps of this cycle will be the basis for many of the discussions in chapters to follow.

Time, as we have seen, is an important factor in the processes outlined by the rock cycle. Weathering, deposition of sediments, consolidation of sediments, and metamorphism of sedimentary rocks are all processes involving millions of years. Metamorphic rocks and many of the deep-seated igneous rocks form at depths of 5 to 10 miles below the earth's surface. When these rocks are ultimately exposed at the earth's surface, we are forced to the conclusion that a great deal of once overlying rock material has been slowly disintegrated and eroded away. This means time—millions of years, perhaps a billion. When you see a metamorphic rock in an outcrop, remember that not only did it take millions of years for some original rock to be metamorphosed but many millions more for the processes of weathering and erosion to finally uncover the metamorphic rock. Many of the world's exposures of metamorphic and deep-seated igneous rocks occur in so-called shield areas of very old rock, as in the Canadian Shield. Shield areas are blocks of the earth's crust that have been stable areas for at least the last half-billion years. Whole chains of mountains as high as the Himalayas have been completely worn away in such areas, exposing at today's surface the roots of the former mountains where metamorphic rocks were formed. Most metamorphic rocks that have been formed in fairly recent geologic time, say the last 50 million years, are still deeply buried in the crust and entirely unknown to man.

Geology and man

Most geologists are employed in the search for mineral resources and natural fuels: metals, oil, natural gas, and ground water. In the United States we are using approximately

Figure 20.12
Landslide. The Point Firmin, California, landslip. Note slump scars and fractures extending well into the center of the photograph. (Spence Air Photos.)

30 billion dollars' worth of metals and fuels each year; therefore an equal amount of such mineral resources must be found as replacement of reserves each year. Geologists are also employed in the engineering field in foundation studies of dam sites, highways, large industrial plants, and in regional planning studies and urban development. Geology is our natural physical environment, the study of this environment, and how it may be used by man without serious damage. So far, however, man has given relatively little consideration to geological data in his use of land.

GEOLOGICAL HAZARDS

Geologists recognize areas of extreme natural hazards; for example, areas that may be affected by landslides, floods, earthquakes, or volcanic eruptions. Due to rock type and structure, many areas are especially susceptible to landsliding, especially when wet; these conditions can be recognized and such areas should not be developed or built upon. This is certainly true of many coastal areas (see Fig. 20.12), and we have all read in newspaper accounts of the destruction of expensive homes in this way. Most of these losses are unnecessary because it is known that such sites are not safe from slides.

In areas of frequent earthquakes—such as California and Alaska—some types of terrain are far more prone to severe damage than others, especially the strip of land a mile or so wide along the actual fault trace and filled or man-made land as in parts of San Francisco Bay. These areas should be drastically limited in their use; there certainly should be no schools or homes. Underocean earthquakes may generate what are erroneously called "tidal" waves, correctly known as *seismic sea waves* (p. 543). A network of stations exists along the Pacific coast and in Hawaii to give warning of such events.

Much of the loss of life from disastrous floods in recent years was really unnecessary because it should have been obvious that the threat of the flood existed. Either this threat should have been removed, perhaps by repairing an old dam, or people should have been advised against living in danger areas. Geological studies can point out areas where these and other natural or geologic hazards exist; but geologists cannot force you to heed the warnings or move out of hazardous areas. We must all work for an enlightened public awareness and for laws to prevent the development of lands with exceptionally great hazards.

ENVIRONMENTAL PROBLEMS

Many of our more severe environmental problems involve the physical landscape and therefore a disturbance of geological processes. Man must understand these processes in order to control and prevent the more serious kinds of land and water pollution. It is necessary that we mine the increasing amounts of the metals needed by today's industrialized civilization, and the coal and oil that we must have to produce energy and electrical power. We must learn how to do this mining and oil production in such a way as to minimize the damage to the environment. The scars of surface or open-pit mining can be smoothed over after mining, and precautions can be taken to prevent acid mine waters from contaminating rivers. These mined out areas can be reclaimed for other uses; but there must be an understanding of the geological factors, and this means careful studies before and during mining by sympathetic scientists.

Ground water is a great resource which is being used more and more and is becoming seriously polluted in some areas. To prevent or correct such pollution, geological data are needed on the movement of the ground waters through different types of rock and soil materials. For example, sanitary land-fill sites, where garbage and other solid refuse are being systematically buried, may or may not result in severe ground-water pollution. In gravel deposits and highly fractured rock, the subsurface water can be extensively polluted as waters move through the buried waste and leach out various chemicals. On the other hand, in some impervious clay soils it may be safe to bury refuse in this way as ground waters move very slowly and the potential pollutants are contained.

The behavior of a river is considerably changed by the building of a dam. Above the dam in the reservoir, deposition of sediment takes place gradually filling in the man-made lake. Below the dam, the water is free of sediment load and therefore can do an increased amount of erosion or downcutting. In effect, the balance of processes in the river, both above and below the dam, is changed, perhaps for the good, perhaps not. As an example of changes of this type affecting a large area, we might consider the Aswan Dam in Egypt, completed about 1964. The purpose of the dam is to regulate the flow of the Nile River so that water will be available for irrigation through the year downstream, which means it will be available to most of the populated sections of Egypt—in short, to increase crop yields in areas already under cultivation and to add new acreage to crop production. The dam holds back the annual flood of the Nile in Lake Nasser from which water can be let out gradually during the low-water months. Before the dam was built, the annual flood of the Nile spread a layer of water and mud over the flood plain lands of the Nile Valley. This mud contained a wealth of nutrients including natural phosphates and nitrates and bits and pieces of organic matter which had served for centuries as a life-giving fertilizer to these lands. This Egyptian soil had produced crop after crop for over 5000 years due to this yearly layer of enriched mud from central Africa. Now this naturally enriched mud and silt is being trapped in the reservoir. The cultivated lands downstream can be irrigated but are not getting their natural fertilization. Artificial fertilizers are needed at great expense.

Part of the Nile flood with its rich nutrients annually entered the Mediterranean Sea and provided food for a complex system of living organisms. Perhaps the sardine population was the most significant of these to man. The catch of sardines by Egyptian fishermen in 1965 was 18,000 tons; by 1968 this had dropped to 500 tons. The Aswan Dam, by trapping the nutrient-rich sediments of the Nile, has ended the Egyptian fishing industry. In addition to the disappearance of the sardines, no one knows for sure just how many other important life types will diminish in the eastern Mediterranean. The great annual flood of fresh water down the Nile into the sea reduced the salinity of the eastern Mediterranean each year. In fact, salinity varied from a low of 37.4 parts of salts per 1000 parts of water during the fresh water influx to a high of about 38.7 parts. Now the salt content of the eastern Mediterranean is near 39.8 parts of salts throughout the year. This salt content is apparently slowly increasing as new salts are leached from various soils by the irrigation waters. This increase in salinity will also have an effect on the total population of marine life in the eastern Mediterranean.

The stopping of the Nile flood is also resulting in damage to beaches along the eastern Mediterranean coasts both west of the Nile delta and to the northeast as far as Lebanon. Waves always erode sand from the beaches, but this loss has been balanced by new sand and silt brought down the Nile in its flood and carried along shore by sea currents. Thus some Egyptian beach resorts may become less attractive in the near future. The upset of the natural nutrient and sediment transport of the Nile by the Aswan Dam is becoming an economic disaster for Egypt.

It is quite obvious that man upsets many balances of nature because he does not bother to find out how geologic processes will be changed until too late. Sometimes the changes have been predicted, but man has chosen not to listen to scientists because of financial gain to a few. Geologists can get the data to determine what changes and environmental damage will be done as the consequence of certain projects, but all of us—the public—must be alert to insist that such warnings are heeded so that our natural environment can be protected for the good of the majority.

SUMMARY

The lithosphere has insignificant surface relief—that is, $12\frac{1}{2}$ miles compared with a diameter of nearly 8000 miles. Its major relief features are the continents and ocean basins. The earth's surface has two levels about 3 miles apart in elevation. The earth has a crust, 3 to 50 miles thick, surrounding a mantle and an inner core. The density of the whole earth is 5.5.

The hydrosphere consists of salt water, fresh water, snow, and ice. Ocean water contains an average of $3\frac{1}{2}$ percent of dissolved solids. The salinity and oxygen content of seawater is variable within narrow limits.

The atmosphere is a mixture of gases, principally nitrogen, oxygen, carbon dioxide, and water vapor. It contains suspended solid and liquid particles, of which dust is especially important. The zones of the upper atmosphere extend at least 600 miles above the surface, although an area affected by the earth's magnetic field may extend as far as 62,000 miles.

The earth has a magnetic field whose magnetic poles do not correspond with the geographic North and South poles. As a consequence a compass needle does not point true north, or south, over most of the earth's surface. The earth's magnetism is probably caused by movements in the liquid core; the earth behaves as a giant dynamo.

The geologic processes that modify the earth's crust may be grouped under the headings "igneous activity," "diastrophism," and "gradation." Igneous activity includes volcanic, or surface, phenomena and plutonic, or intrusive, activity and results in the crystallization of igneous rocks. Diastrophism involves vertical and mountain-making movements and helps to initiate metamorphism and the metamorphic rocks. Gradation is weathering and erosion followed by the ultimate deposition of the sedimentary rocks. The relationships between these processes and the rocks produced may be summarized in a rock cycle.

The study of geology contributes to society through the exploration and finding of natural resources—minerals and fuels—through foundation engineering and land-use recommendations, by ground water development and circulation studies, and by analysis of the behavior of streams, especially in floods and sediment deposition.

Important words and terms

geology, geophysics
lithosphere, hydrosphere, atmosphere
hypsographic curve
turbidity currents
troposphere, stratosphere
magnetic inclination and declination
dipole field
Curie temperature
doctrine of uniformitarianism
igneous activity, diastrophism, gradation
rock cycle
geological hazards

Questions

1. What kinds of solid matter are present in the atmosphere? Why are they important?
2. What dissolved compounds are most important in seawater? How do these compare with the compounds dissolved in river waters? in lakes near your home town?
3. Explain the relationships between the major earth processes and the three types of rocks.
4. Explain in your own words the significance and application of the phrase, "The present is the key to the past."
5. Why are there two relatively flat sections in the hypsographic curve of earth-surface elevations and ocean depths?
6. What interchanges take place between ocean waters and the atmosphere?
7. Point out and explain the major relationships between ocean surface circulations and local climates along the Pacific coasts of North and South America from Alaska to the southern tip of Chile.
8. What is the reasoning for and against a river-erosion origin for submarine canyons? A turbidity current origin?
9. Is the atmosphere in your region noticeably polluted? with what gases and solids? What might be done to start clearing it up?
10. Describe geologic hazards that are known to exist or might occur in your region. Have these caused any serious problems in recent years? What might be done, perhaps as a warning to uninformed persons, for protection of life and property that is not now being done adequately?
11. Is there any need for better land-use planning in your area that should involve geologic studies?
12. What is the magnetic declination at your college or university location?

Project

Select an ecological problem in your vicinity in which geological understanding may help in its solution. Collect data concerning this problem in the field and make suggestions for a possible solution as the course progresses through Chaps. 21 to 26. This might be a pollution problem, misuse of land, an unrecognized or overlooked geologic hazard, or a local mineral or rock use situation.

Supplementary readings

Adams, Frank D., *The Birth and the Development of the Geological Sciences,* Dover Publications, New York (1954). (Paperback.) [A thorough history of the earth sciences, tracing geologic thought from the earliest times to the end of the nineteenth century.]

American Geological Institute, *Dictionary of Geological Terms,* Doubleday, Garden City (1962). (Paperback.) [An abridged glossary of approximately 7500 terms for students, teachers, and others interested in earth sciences.]

Foster, Robert J., *Geology* (2nd ed.), Charles Merrill, Columbus, Ohio (1971). (Paperback.) [A short (162 pages) general text in geology that might help the student in review. Topics relate to Chaps. 20 through 27.]

Goody, Richard M., and James C. G. Walker, *Atmospheres,* Foundations of Earth Science Series, Prentice-Hall, Englewood Cliffs, N. J. (1972). (Paperback.) [Discussions of solar radiation, temperature, winds, condensation, clouds, and evolution of the atmosphere as a background for geologic studies.]

Leet, L. Don, and Florence J. Leet (eds.), *The World of Geology,* McGraw-Hill, New York (1961). [Twenty selections from the writings of well-known authors as an introduction to the various aspects of the geological sciences as well as to outstanding examples of scientific literature. The topics covered relate to Chaps. 20–26.]

Mears, Brainerd, Jr., *The Nature of Geology,* Van Nostrand Reinhold, New York (1970). (Paperback.) [An outstanding collection of 18 contemporary readings for beginning students in geology emphasizing human endeavor in earth science, geology as a way of life, and pollution problems. These readings relate to Chaps. 20 through 27.]

Shepard, Francis P., *The Earth Beneath the Sea,* Atheneum, New York (1965). (Paperback.) [An account of waves and currents that modify the sea floor, of ocean-floor topography, ocean-floor sediments, and coral reefs and atolls.]

Spar, Jerome, *Earth, Sea and Air: A Survey of the Geophysical Sciences,* Addison-Wesley, Reading, Mass. (1965). (Paperback.) [Chapters 2 and 3 discuss earth magnetism, oceans and ocean basins, currents, waves and tides.]

Turkekian, Karl K., *Oceans,* Foundations of Earth Science Series, Prentice-Hall, Englewood Cliffs, N. J. (1968). (Paperback.) [Topography of ocean basins, sediment on the sea floor, and motions, geochemistry, and history of ocean basins.]

Turner, Daniel C., *Applied Earth Science,* Wm. C. Brown, Dubuque, Iowa (1969). (Paperback.) [Natural resources, engineering, and military applications of geology, and conservation in geology are covered. Relates to Chaps. 20 through 26.]

Verduin, Jacob, *Field Guide to Lakes,* ESCP Pamphlet Series #PS-8, Houghton Mifflin, Boston, Mass. (1971). (Paperback.) [Discussion of the origin of lakes with suggestions for field projects at local lakes emphasizing ecology and pollution control. Relates also to Chap. 24.]

White, J. F. (ed.), *Study of the Earth: Readings in Geological Science,* Prentice-Hall (1962). (Paperback.) [Outstanding readings in many phases of geology emphasizing recent advances in geologic thought. Topics relate to Chaps. 20–26.]

MATERIALS OF THE EARTH

In the beginning everything was in confusion, then Mind came and reduced them to order.

ANAXAGORAS, 500?–428 B.C.

Matter is composed chiefly of the fundamental particles called *protons, neutrons,* and *electrons*—combinations of which form the 103 known elements. Only 10 of these elements make up more than 99 percent of the earth's crustal material.

Minerals are the naturally occurring, inorganic chemical compounds found in the solid part of the earth. Minerals are characterized by a definite, internal, orderly, atomic structure that is a diagnostic property of a particular mineral species. Silicates are the most abundant mineral class and may be grouped in six categories on the basis of their internal structure. This silicate structure is based on the arrangement of silicon–oxygen tetrahedrons.

Some 2000 mineral species have been recognized but only a few of these are common enough to be known as the rock-forming minerals. Aggregates of minerals are called *rocks,* and the latter comprise the great mass of the earth's crust.

Table 21.1
Composition of the earth's crust

element	percentage by weight	percentage by number of atoms
oxygen	46.7	60.4
silicon	27.7	20.4
aluminum	8.1	6.2
iron	5.0	1.9
calcium	3.6	1.9
sodium	2.8	2.5
potassium	2.6	1.4
magnesium	2.1	1.8
titanium	0.6	0.3
hydrogen	0.1+	3.0
all others	0.7	0.2
total	100.0	100.0

Elements

Protons, neutrons, and electrons combine in various ways to form structures called *atoms* which are stable units that cannot be broken down by ordinary chemical reactions. Elements are particular kinds of atoms with characteristic atomic configurations—that is, different quantities of protons and neutrons. The number of protons ranges from 1 to 103 in the known kinds of atoms; these different kinds of atoms are called *elements.* Of these 103 elements, 92 have been found in the earth's crust; but of these 92, only a very few make up most of the bulk of the earth's crust (Table 21.1). Computations have been made from the many thousands of chemical analyses of different minerals and rocks, allowing for the relative abundance of different rock types, found on the surface of the earth and in mines and deep wells. When we appreciate the amount of erosion of the surface over millions of years and realize that rocks now at the surface were formed deep within the earth, it becomes apparent that these figures give us a good approximation of the average composition of the earth's crust, or the outer part of the lithosphere. Note that 10 elements (Table 21.1) make up over 99 percent of the earth's crustal material. The well-known metal, copper, which we think of as a common metal, occurs in the earth's crust at less than 0.005 percent, just a trace compared to the 10 elements in Table 21.1. Uranium is about 0.0002 percent, gold is only 0.5×10^{-6} percent (0.000 000 5%)!

Of these 10 most abundant elements, only one, oxygen, has a negative ion. Furthermore, because the oxygen ion is a large ion and a light-weight one, it can be shown that oxygen *by volume* is almost 94 percent of the earth's crust. Actually, the earth's crust has a fairly simple composition, a close-packed array or network of oxygen ions with the nine common positive ions or metals interspersed to form a few types of crystal or atomic structures. Of these positive ions, silicon plays a very special role, as we shall see shortly.

Minerals

A *mineral* is a homogeneous, natural substance of inorganic origin having a definite chemical composition and certain characteristic physical properties. The most significant of these properties is a definite internal structure, or atomic arrangement, that finds occasional outward expression in crystals. Two fundamental properties ordinarily characterize a mineral: (1) its chemical composition, which can be expressed in a chemical formula, and (2) its distinctive internal atomic structure.

The synthetic products made in a laboratory or factory are not true minerals, despite occasional advertising to the contrary. Coal is not a mineral according to this definition because it is the remains of organic matter. Coal is best classified as a type of rock. Nevertheless, coal is usually called a "mineral fuel" and is included with other mineral resources. Ice, on the other hand, if it has formed naturally, is a perfectly good mineral, fitting the above definition in every respect. In the field of nutrition the term "mineral" is used in an entirely different sense than that used by the geologist. When you note that a food product has "vitamins and minerals added," this means that certain chemical elements have been added, usually as synthetic compounds.

ATOMIC ARRANGEMENT

Minerals are combinations of atoms of one or more of the first 92 elements. The atoms are arranged in a definite three-dimensional pattern called a *space lattice*. Minerals exhibit a great variety of these internal patterns—cubes, sheets, rings, chains, three-sided, six-sided, and others—but no two minerals have exactly the same arrangement. In most lattices the atoms have lost or gained electrons to become electrically charged ions (p. 318). The resulting electrical force, as ions of unlike charge attract one another, holds the ions in fixed geometrical arrangements. Ordinary salt (sodium chloride), the mineral halite, has a simple cubic structure that illustrates one particular space lattice (Fig. 21.1). The unit of this structure is formed by six chlorine ions equidistant from, and symmetrically placed around, one sodium ion with six sodium ions similarly placed with respect to a chloride ion. As these units interlock in all directions, the resulting space lattice consists of planes of atoms, or ions, in three directions at right angles forming cubes. Solid sodium chloride, natural or artificial, always has this structure no matter where it occurs.

The two mineral forms of carbon, graphite and diamond, illustrate in the most dramatic way possible that internal atomic structure is the most significant characteristic of minerals; it is more important than composition in determining the physical properties of the mineral.

THE GRAPHITE FORM OF CARBON

Pure free carbon is sometimes, although rarely, found in nature. When pure, it usually crystallizes in either of two allotropic forms, diamond or graphite. Either form is just as much carbon as the other. Although coal is sometimes thought of as carbon, it is really such a complex mixture that it should not be included among forms of free carbon.

Graphite (Fig. 21.2) is by far the more common form of carbon. It is a black solid and if pulverized feels slippery. It crystallizes in plates that break apart and slide over each other easily. This explains why it is

Figure 21.1
Cubic structure in salt. Atomic arrangement of sodium and chlorine atoms in the space lattice of ordinary salt (top) and part of a large natural crystal of salt, the mineral halite (bottom). (W. H. Parsons.)

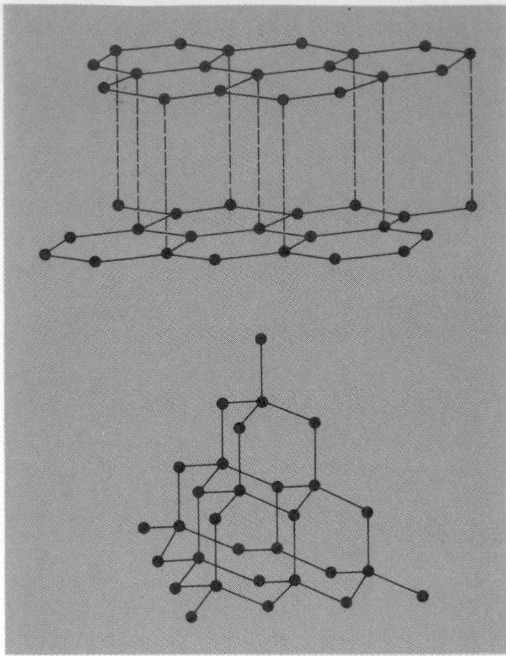

Figure 21.2
Crystal structure of graphite, top, and diamond, bottom.

so often used to decrease friction between two solid surfaces; such materials are called *lubricants*. Graphite mixed with clay makes the so-called leads in our pencils.

DIAMOND

The diamond form of carbon is much harder than graphite and is a much poorer conductor of electricity. When pure, it is usually transparent and colorless. Lavoisier once intentionally burned a diamond in oxygen, and because nothing seemed to be left except carbon dioxide, he concluded that diamond consisted of free or elemental carbon. The most spectacular use for diamonds is in jewelry, but perhaps a more important function is in making instruments and tools for marking, cutting, and grinding in industry.

In diamond, each carbon atom can be thought of as being at the center of a perfect tetrahedron (Fig. 21.2). The outer (valence) electrons are localized in four orbitals, each one directed toward a corner of the tetra-

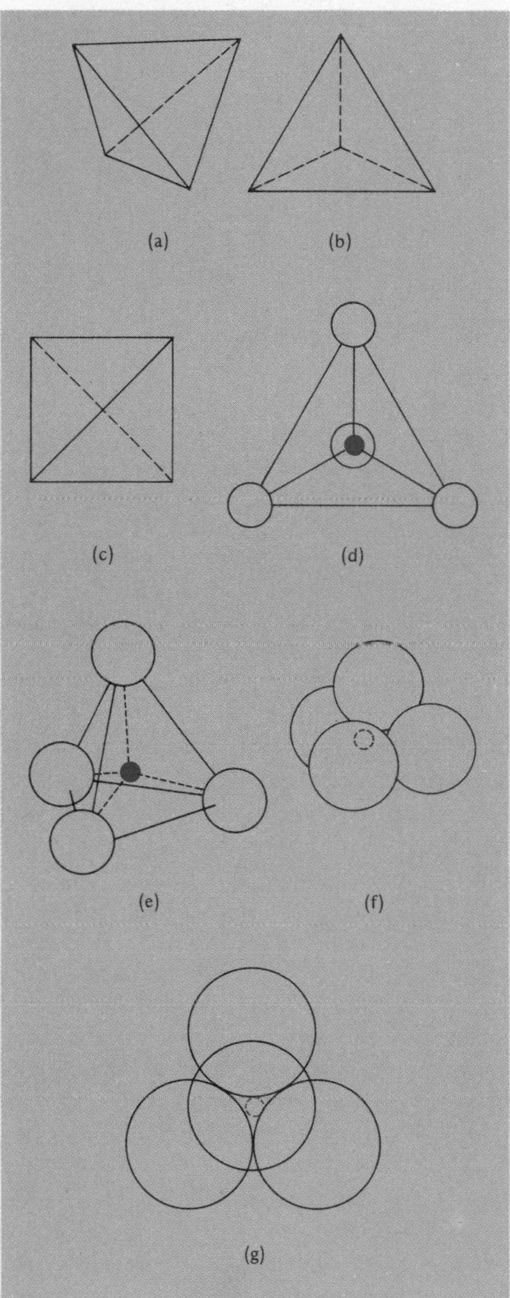

Figure 21.3
Various ways of representing the silicon-oxygen tetrahedron. (a) and (b) The geometric form, known as a tetrahedron, seen from two different angles. (c) Projection of top view of (a) used as a symbol on Table 21.2. (d) Projection of top view of (b) with oxygen ions shown at the four corners of the tetrahedron with a silicon ion (dark spot) in the center; also on Table 21.2. (e) Tetrahedral arrangement of four oxygen ions around the one silicon in an expanded view; solid lines outline the tetrahedron, dotted lines the ionic bonds. (f) Four large oxygen ions around a small and hidden (dotted) silicon ion presented more nearly as they actually occur. (g) Same as (f) but shown in view from above and in same position as (d).

hedron. These bonds are all exactly alike, each one being unusually short and strong. This structure helps to explain why diamond is extremely hard, is one of the best of materials for cutting and grinding, and is sometimes found intact in the ashes left from a burned house in which fire did not reach the high kindling temperature of diamond.

QUARTZ

The ordinary crystalline form of silicon dioxide, quartz, is quite common in nature. The chemical formula SiO_2, which has represented it for a century, is probably misleading; $(SiO_2)_x$ might perhaps be better. At any rate there are no true SiO_2 molecules. Each silicon atom is at the center of a tetrahedron and at the apexes are four oxygen atoms (Fig. 21.3). This arrangement is possible because silicon ions are quite small (ionic radius of 0.42 Å) whereas oxygen ions are much larger (ionic radius of 1.40 Å). But throughout the crystal there are oxygen "bridges" between silicon atoms: thus, —O—Si—O—Si—O—, which would go on in a specimen of quartz without any ending until one reached the outer boundary of the grain or boulder if it were a perfect single crystal. This illustrates why quartz is not a good example of a molecular substance; rather such materials consist of macromolecules or giant molecules and are

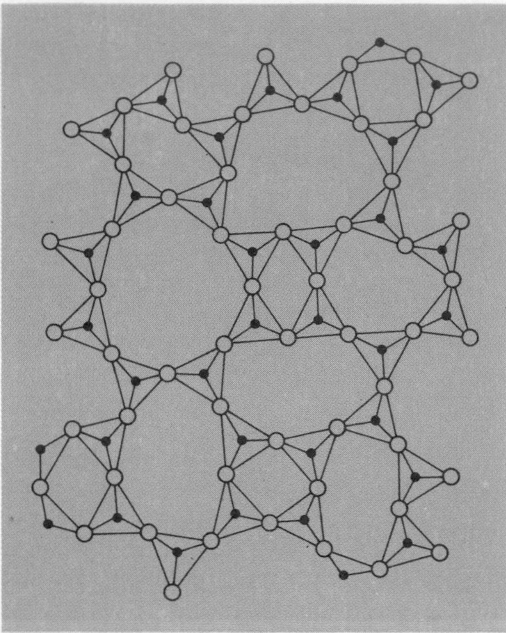

Figure 21.4
Disorder structure. A random network of silicon–oxygen tetrahedrons in a silica glass. Compare this diagram with the ordered structure of silica tetrahedrons in silicate chains and sheets in Table 21.2 (p. 451).

Figure 21.5
Crystals. Top, hexagonal prism of beryl (left) and monoclinic crystal of feldspar (right); bottom, octahedron of fluorite (left) and cubic crystal of pyrite (right). (W. H. Parsons.)

properly referred to as *polymers*. If one is interested in combining ratios, the formula SiO_2 can be quite helpful; but if silicon dioxide really contained SiO_2 molecules, one would expect it to be a liquid or a solid with low melting point. Actually, quartz not only has an extremely high melting point but is insoluble and slow in its chemical changes. All of these properties suggest that its structure is really polymeric.

GLASS

If melted, for example, by a lightning strike in quartz sand, quartz will solidify into a natural silica glass which may be thought of as a liquid that has cooled to a rigid form so rapidly that its atoms did not have time to rearrange themselves in the usual orderly pattern. The atomic arrangement of such a "rigid liquid" as a silica glass lacks continuous order throughout its mass but may exhibit some local regularity, as shown by local groupings of silicon and oxygen atoms in silica tetrahedrons; but such silica tetrahedrons are bound together in a random manner to form an overall disordered pattern (Fig. 21.4). Such a lack-of-order structure is spoken of as *amorphous* and also occurs in other natural glasses—such as obsidian, produced by the fast cooling of some lava flows. Amorphous structure is, of course, present in all man-made glass. All glass is a mixture of silicon dioxide and various silicates that has formed from a liquid of rather high viscosity which has cooled so rapidly it can no longer flow and appears to be solid. Yet it does not freeze, as water does, at a definite temperature. If a piece of glass is heated, one can observe it slowly become soft and start to sag or flow; but there is no sharp transition from solid to liquid at a particular temperature. On X ray examination, although there is some evidence of alignment of particles in regular order, as in a crystal, there is much more evidence of disorder (Fig. 21.4). It should be noted that glass is more a condition than it is a material.

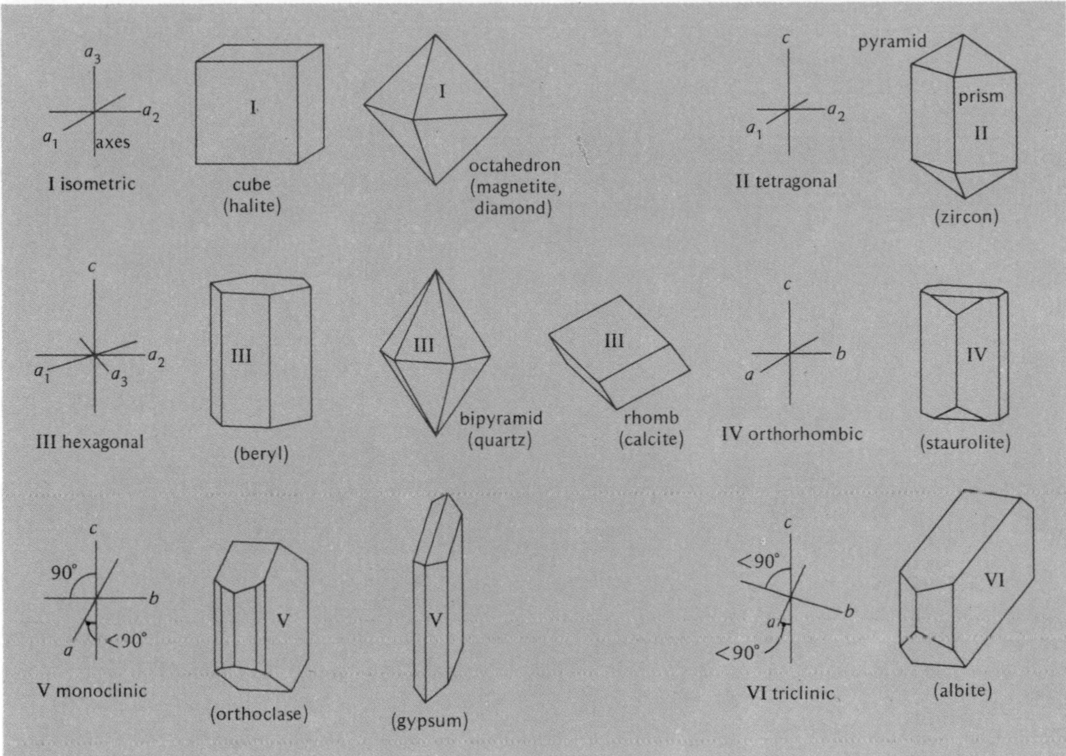

Figure 21.6
Axes and representative external crystal forms of the six crystal systems. (From *The Earth Sciences,* 2nd ed., Harper & Row, Publishers, copyright © 1963, 1971 by Arthur N. Strahler.)

CRYSTALS

All substances, whether natural or artificial, that have a definite internal atomic arrangement are said to be *crystalline*. In crystallized substances the internal planes of atoms are sometimes expressed in external planes called *crystal faces*. A solid geometric unit bounded by such faces is a *crystal*. A few of the many crystal forms of common minerals are shown in Figs. 21.5 and 21.6. In nature, geometrically perfect or nearly perfect crystals are rather rare; partly formed or distorted crystals are more common. To be perfect, a crystal must have been free to "grow" in all directions; this is rarely the case in nature. No matter how distorted the external crystal form may appear, the internal atomic arrangement, or space lattice, is essentially the same. One way to explain this external distortion is shown in Fig. 21.7.

Crystals "grow" by the addition of atoms to their outer surfaces, or faces, forming new layers of atoms that fit the space-lattice pattern of the embryo crystal. These atoms come from a surrounding liquid or, occasionally, gas. This may occur during the freezing or crystallization of a liquid or melt; for example, the freezing of water to form ice, the solidification of molten metal in a blast furnace to produce iron, and the crystallization of molten lava from a vol-

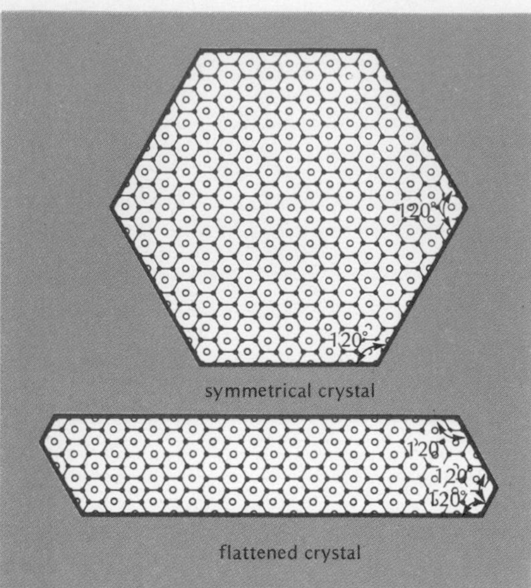

Figure 21.7
Distortion of crystal form. Diagrammatic representation of the atomic pattern, or space lattice, of a hypothetical hexagonal mineral in a geometrically symmetrical crystal (top) and in an apparently flattened or "distorted" crystal (bottom). Note that the internal atomic pattern is identical in both crystals. In the misshapen crystal there is no internal distortion. All angles between crystal faces on both crystals are 120°.

Figure 21.8
Pyrite crystals. A cluster of intergrown pyrite cubes from the wall of an ore vein in Leadville, Colorado, where they were precipitated from an ore-carrying solution. (Ward's Natural Science Establishment, Inc.)

cano to form rock. The resulting ice, iron, and rock are crystalline solids. Crystals also grow from liquid solutions in which a solid precipitate crystallizes out because of evaporation of the solution, or chemical reactions within the solution or between the solution and an adjacent solid; for example, the crust of crystals in the bottom of a dish of evaporated salt water, natural salt deposits, and almost all mineral-filled veins and ore deposits (pyrite, Fig. 21.8).

If physical conditions allow the deposition of atoms on all sides of the growing crystal, a symmetrical solid results; but if addition occurs on only one or two sides, an asymmetrical, or distorted, crystal is formed (Fig. 21.7), as for example, a crystal in a narrow crack. When many crystals grow simultaneously from numerous adjacent centers, they soon interfere and interlock and regular crystal forms cannot develop. The resulting solid is a *crystalline mass*.

IDENTIFICATION OF MINERALS

Minerals can be identified by chemical analysis, by X ray methods, by polarizing microscope examinations, or by observing their physical properties. A chemical analysis will tell how much of which elements are present. But in a few cases this analysis fails to distinguish between two or more minerals that have the same chemical composition, as we have seen for diamond and graphite. By passing an X ray beam through a crystal or powdered sample of a mineral, a pattern is obtained on a photographic plate that indicates the internal atomic arrangement. One such technique is illustrated in Fig. 21.9. This is a very precise method, because no two minerals have exactly the same internal pattern or space lattice. Most minerals in very thin slices

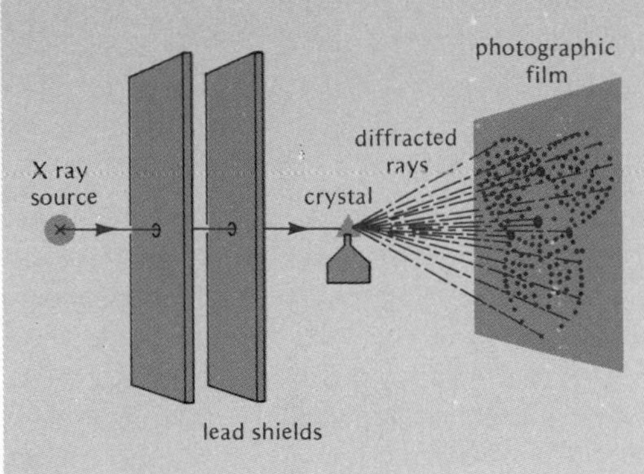

Figure 21.9
Mineral X ray method. An X ray beam is passed through a crystal to obtain a diffraction pattern. The crystal is calcite. (From *Principles of Geology*, 3rd ed., by James Gilluly, Aaron C. Waters, and A. O. Woodford. W. H. Freeman and Company. Copyright © 1968.)

or as powder are transparent to light and may be studied with a polarizing microscope. This microscope permits the determination of many optical properties, such as the index of refraction, the amount of double refraction, and variations in light absorption. This technique of mineral identification is commonly used by mineralogists and geologists. The chemical, X ray, and polarizing microscope methods all require expensive equipment and also special training.

Careful observation of certain physical properties, however, can be used for mineral identification by anyone after a little practice. We are all familiar with the properties of color, size, and shape. The nonscientist is able to identify many birds and flowers by using only such properties. Unfortunately, however, these properties have a very limited value for mineral identification. Color can be extremely variable because of the presence of slight traces of impurities. Size means nothing as a means of identification; a single crystal may be microscopic in size or as much as 20 ft long. Although crystal shapes are distinctive, well-developed crystals are all too rare.

The significant physical properties that can be observed for mineral identification include hardness, cleavage, streak, and specific gravity. These are probably new concepts to the average layman. Once these properties are understood, however, the recognition of mineral species should be as simple for the hobbyist as that of wild flowers.

Hardness is the ability to withstand scratching and is usually measured by comparing minerals to a standard set of 10 minerals, comprising a scale of hardness, in which number 1 is the softest and number 10 is the hardest:

Standard Scale of Hardness

1. Talc
2. Gypsum
3. Calcite
4. Fluorite
5. Apatite
6. Orthoclase feldspar
7. Quartz
8. Topaz
9. Corundum (sapphire)
10. Diamond

Hardness of Common Reference Items

$2\frac{1}{2}$ fingernails[1]
3 copper penny
$5\frac{1}{2}$ ordinary pocket knife
$5\frac{1}{2}$ piece of plate glass
6 steel file

This is a relative scale; the differences of absolute hardness between ranks are not uniform. If corundum were 9 in absolute hardness, then diamond would be over 40. Remember that hardness is resistance to scratching but not necessarily to breakage. Nothing will scratch a diamond except another diamond, but a diamond can be broken or split if struck a sharp blow.

Minerals vary in their strength or resistance to breakage, which is quite a different property from hardness. Some very hard minerals are rather brittle and may be shattered with a hammer blow—for example, diamond. On the other hand some very soft minerals are exceedingly tough; for example, copper in a penny has a hardness of only 3,

[1] This means that fingernails have a hardness between that of gypsum and calcite.

Figure 21.10
Cleavage in minerals. Top, perfect one-direction cleavage in mica and excellent three-direction cleavage in halite, bottom. The three cleavage directions are at right angles to one another and equally well developed; this is known as *cubic cleavage*. (Ward's Natural Science Establishment, Inc.)

but no amount of hammering will break it, although it may flatten it a little. This difference in resistance to breakage is called *tenacity*, the strength or toughness of a material. Minerals are said to be brittle, if breakable; malleable, if not breakable like the copper penny; or flexible and even elastic, as in thin pieces of mica, graphite flakes, and asbestos fibers.

A careful examination of a mineral fragment will reveal how it has broken. Many crystallized minerals split easily along the internal planes of atomic arrangement; the resulting smooth break is called *cleavage*. The separation of mica sheets is the outstanding example of cleavage, but many other minerals, such as calcite, halite, and feldspar, show this property well (Fig. 21.10). Some minerals have as many as four or even six directions of cleavage, and the angles between these cléavage planes are different in different minerals. Not all crystallized minerals exhibit cleavage, however; some break with little regard to internal structure. Such breaks are called *fractures;* they may be fairly smooth and curved, as in glass, or quite irregular and splintery.

Some minerals do have a characteristic color, but as many exhibit a wide range of colors, depending on the presence of slight impurities. The mineral quartz, for example, is colorless when pure but is also found in every color of the rainbow in addition to banded and mottled varieties. The color of the mineral in its powdered form is often more characteristic than the surface color. The color of the powder is called the *streak* of the mineral because it is easily determined by rubbing the specimen across a smooth, hard, white surface, such as a piece of unglazed porcelain. The resulting mark on the porcelain is a sample of the powdered mineral.

Relative weight or specific gravity (Chap. 7) can be a very diagnostic property. With practice a person can lift or juggle the specimen in his hand to judge whether it is slightly, moderately, or considerably heavier than an ordinary rock of the same size. More precise methods for measuring specific gravity may be used for very effective identification of pure specimens. For example, the specific gravity of cut gemstones may be accurately determined without damaging the gem in any way.

A number of other properties will also help in the study of certain minerals. The *luster*, or manner in which light is reflected from the surface of a mineral, can be quite characteristic for individual species. For example, many minerals have either metallic, glassy, silky, pearly, or earthy luster. Other properties occasionally noted include elasticity, magnetism, fluorescence in ultraviolet light, radioactivity, and even taste. In some minerals, a single property may be diagnostic enough for satisfactory identification, but more often several properties must be determined before the mineral is identified.

MINERAL SPECIES

About 2000 minerals have been identified by composition and space-lattice differences, although the majority of these are very rare compounds. About two dozen minerals make up over 99 percent of the earth's crust; one group, the feldspars, alone comprises over half the bulk of the crust. These few minerals are known as the *rock-forming minerals* (pp. 476, 482, 496). Another group of approximately 50 minerals is abundant enough to be called *economic,* or *ore, minerals*; these are useful to man as the source of metals and other elements. Iron is a constituent of several hundred minerals, but considering their relative abundance, only four or five of them can be classed as iron-ore minerals.

Minerals are grouped on the basis of their composition into a number of chemical classes. In the chemical classification a negative ion, a single atom, or a group is used to define the class to which the minerals belong. Negative ions are used because they are bigger and control properties of the mineral. The principal classes in common usage are listed with some examples of each class.

Native elements:	gold, silver, copper, platinum, sulfur, graphite, diamond
Sulfides:	galena (PbS), sphalerite (ZnS), chalcopyrite ($CuFeS_2$), pyrite (FeS_2), cinnabar (HgS), molybdenite (MoS_2)
Oxides:	ice (H_2O), hematite (Fe_2O_3), corundum (Al_2O_3), magnetite (Fe_3O_4), chromite ($FeCr_2O_4$), limonite (iron hydroxides), bauxite (aluminum hydroxides)
Halides:	halite ($NaCl$), fluorite (CaF_2)
Carbonates:	calcite ($CaCO_3$), dolomite [$CaMg(CO_3)_2$], siderite ($FeCO_3$)
Sulfates:	gypsum ($CaSO_4 \cdot 2H_2O$), barite ($BaSO_4$)
Phosphates:	apatite [$Ca_5(PO_4)_3(F,Cl)$]
Silicates:	quartz (SiO_2), feldspars, pyroxenes, amphiboles, micas, talc, clays, olivine, beryl

The silicates are by far the most abundant of any mineral group: They constitute about one-fourth of the known mineral species but in bulk they make up approximately 93 percent of the earth's crust. Most of the ore deposits of the metals are composed of sulfides or oxides (and hydroxides), although metallic ore minerals are known in all the chemical classes.

Rock-forming minerals

We have previously noted the average chemical composition of the earth's crust (Table 21.1) and the fact that oxygen and silicon are the two most abundant elements. This has led us to believe that the compounds of silicon and oxygen are the major constituents of rocks. The other metallic ions most important in these rock-forming minerals are aluminum, iron, calcium, sodium, potassium, and magnesium as can be predicted from Table 21.1. These are the elements present in the silicate minerals named above.

SILICATE STRUCTURE

The chief structural unit of the silicate minerals is the silicon–oxygen (silica) tetrahedron (Fig. 21.3) as previously described for the mineral quartz. Each of four oxygen atoms is bonded to one silicon atom by a single electron bond, leaving each oxygen with a free electron to be bonded to an oxygen in another silica tetrahedron or to a metallic ion. Thus silica tetrahedrons may be linked together in several ways and are the basic building units for all silicate minerals. The ways in which silica tetrahedrons may be linked together have become the basis for the classification or grouping of the silicate minerals. Silica tetrahedron linkage is similar in a chemical sense to the linking in organic compounds of carbon and hydrogen atoms to produce macromolecules or polymers. In the inorganic or mineral world, however, there are only six ways of linkage of these silica tetrahedrons, many fewer than the very numerous types of linkage in organic polymers. Thus, in a very real sense, the structural classification of silicates is relatively simple and the possible number of silicate minerals is limited.

The six silicate polymer structures, as illustrated in Table 21.2, include (1) independent silica tetrahedrons; (2) two tetrahedrons sharing one oxygen atom; (3) tetrahedrons linked in rings that are three, four, or six sided; (4) tetrahedrons in single and double chains; (5) tetrahedrons in sheets; and (6) tetrahedrons linked in three-dimensional frameworks. The single groups, rings, chains, and sheets are bonded together three dimensionally by the positive ions of the other common metallic elements, the latter also in rows or planes as the case may be. In the framework structure in the mineral quartz each of the four oxygen ions in one tetrahedron is linked to a silicon ion in another tetrahedron in a continuous network that is electrically neutral. Because these bonds are strong and equal in every direction, quartz is hard and has no tendency to break, or cleave, in any preferred direction.

In many minerals with framework structures, aluminum ions substitute for some of the silicon ions. Because aluminum has a valence of only three compared with four in silicon, one oxygen ion in every aluminum–

Table 21.2
Structure of silicate polymers

type and classification	oxygen shared	composition
independent (nesosilicates)	0	SiO_4^{4-}
two tetrahedra (sorosilicates)	1	$Si_2O_7^{6-}$
ring structure (only the 3- and 6-membered rings are of any importance) (cyclosilicates)	2	$Si_nO_{3n}^{2-}$
single chain (inosilicates)	2	SiO_3^{2-}
double chain (inosilicates)	alternately 2 and 3	$Si_4O_{11}^{6-}$
sheet structure (phyllosilicates)	3	$Si_2O_5^{2-}$
framework (tectosilicates)	4	SiO_2
(aluminum ions take the place of some silica ions in nepheline and feldspars		

[a]See Fig. 21.3 for an explanation of the symbols used in right column.

Si:O	example and formula	symbol[a]
1:4 (2:8)	forsterite (olivine), Mg_2SiO_4	
2:7 ($1:3\frac{1}{2}$)	akermanite $Ca_2MgSi_2O_7$	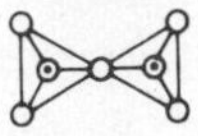
1:3	benitoite, $BaTiSi_3O_9$	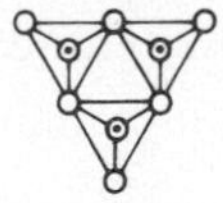
	beryl, $Be_3Al_2Si_6O_{18}$	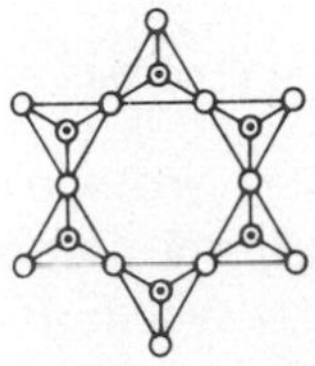
1:3	pyroxenes: enstatite, $MgSiO_3$ augite, $Ca(Mg,Fe,Al)(Al,Si)_2O_6$	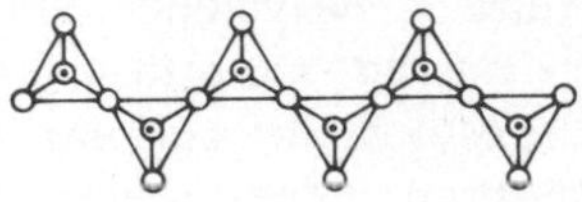
4:11 ($2:5\frac{1}{2}$) [$(1:3)(1:2\frac{1}{2})$]	amphiboles: anthophyllite, $Mg_7(Si_4O_{11})_2(OH)_2$ hornblende, $Ca_2(Mg,Fe)_5(Al,Si)_8O_{22}(OH)_2$	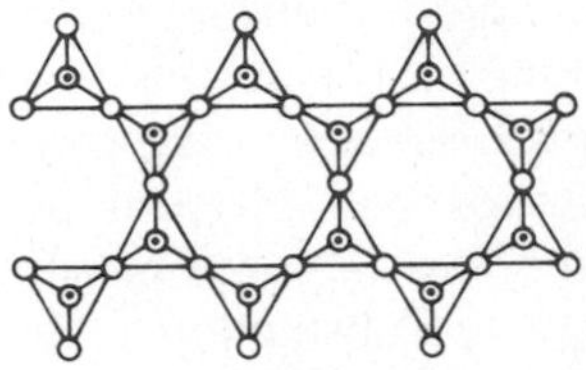
2:5 ($1:2\frac{1}{2}$)	talc, $Mg_3Si_4O_{10}(OH)_2$ kaolinite (clay), $Al_4Si_4O_{10}(OH)_8$ micas: muscovite, $KAl_2(Al,Si)_4O_{10}(OH,F)_2$	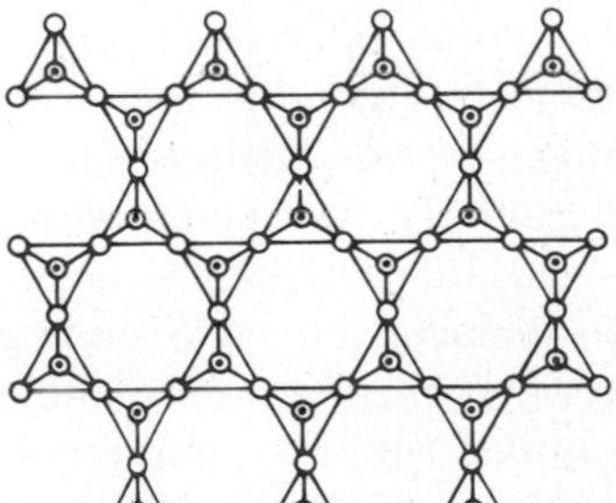
1:2	quartz, SiO_2 nepheline, $NaAlSiO_4$ feldspars: albite, $NaAlSi_3O_8$ anorthite, $CaAl_2Si_2O_8$	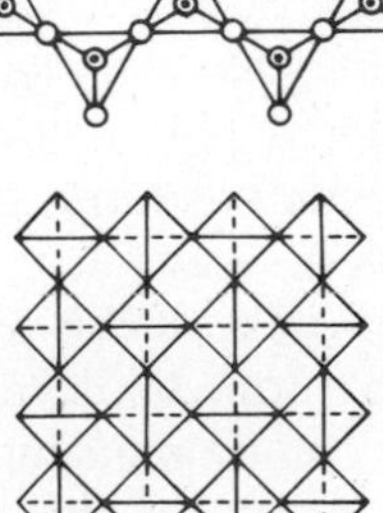

oxygen tetrahedron is left with a free valence bond. This leaves a place and need for a few positive metallic ions in the structure. The feldspars are an example of this type of framework structure; in orthoclase ($KAlSi_3O_8$) and in albite ($NaAlSi_3O_8$) one silicon ion in every four is substituted for by an aluminum ion requiring the presence of either a potassium ion or a sodium ion. In anorthite ($CaAl_2Si_2O_8$) aluminum takes half the silicon positions.

SOLID SOLUTION

The complexity of the mineral kingdom is in substitution; one ion may substitute for another if it is within 15 percent of the same size and if the valence difference between the two is not more than one. Thus sodium can substitute for calcium, ferrous iron for magnesium, ferric iron for aluminum, and aluminum for silicon; but *not* potassium for aluminum. This results in a great variety of mineral species with very similar atomic structures and physical properties. It also means that in some mineral series there is complete substitution of two elements for each other; this relationship is spoken of as *solid solution,* or isomorphism. The mineral olivine is a good example; its chemical formula is often written $(Mg, Fe)_2SiO_4$, implying that different specimens of olivine may have various ratios of magnesium to iron. The Fe^{2+} ion replaces the Mg^{2+} ion in all proportions between the two pure end-member compounds: Mg_2SiO_4 (forsterite) and Fe_2SiO_4 (fayalite) as Fe^{2+} and Mg^{2+} have nearly equal ionic radii. Nevertheless, olivine is not an aggregate or physical mixture of these two compounds but forms a homogeneous, uninterrupted system with a similar space lattice regardless of the proportion of magnesium to iron in a particular specimen. Another common example of complete solid solution exists in the plagioclase feldspar series between albite ($NaAlSi_3O_8$) and anorthite ($CaAl_2Si_2O_8$). Sodium ions replace calcium ions at progressively lower temperatures as plagioclase crystallizes from igneous melts. Because there is a difference in valence between calcium and sodium (Ca^{2+} and Na^+), silicon ions must also replace aluminum ions during the change toward albite, as you will note in the formulas above, to maintain the electrical neutrality of the minerals.

OLIVINE

Olivine is one of the first minerals to crystallize from igneous melts, and it exhibits the solid solution relationship we have just described. The iron-rich members are darker in color than the magnesium-rich forms. Olivine is a glassy, green mineral without cleavage, usually occurring in irregular, rounded grains in basic igneous rocks. It has a hardness of $6\frac{1}{2}$.

PYROXENES

The pyroxenes are a group of single-chain silicates of which the most common and the most complex in composition is *augite* [$Ca(Mg,Fe,Al)(Al,Si)_2O_6$]. Augite is dark green to black in color with a rather poor blocky cleavage and a hardness of $5\frac{1}{2}$. It is most easily confused with hornblende, but the two can be distinguished by the angle between the cleavage planes. The acute angle between these planes in augite is 87° compared with 56° in hornblende as sketched in Fig. 21.11. Augite is one of the ferromagnesium minerals in igneous rocks (p. 476). Other pyroxenes include *enstatite,* found in some mafic igneous rocks, and *diopside,* found in contact metamorphic rocks.

AMPHIBOLES

The amphiboles are the main group of double-chain silicates of which *hornblende* [$Ca_2(Mg,Fe)_5(Al,Si)_8O_{22}(OH)_2$] is the most common. This latter mineral is black, with

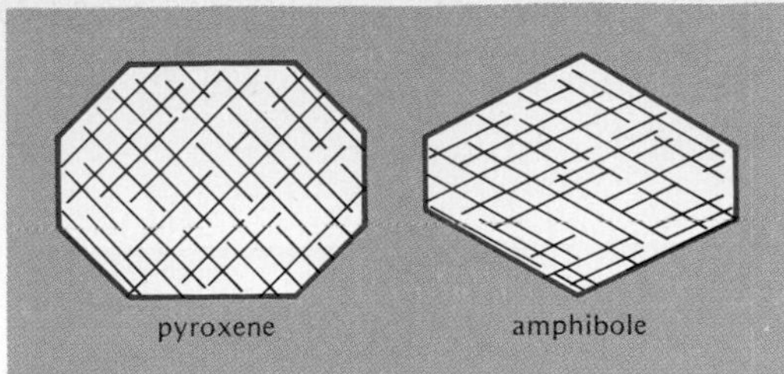

Figure 21.11
Cleavages and cross sections of crystals of pyroxene and amphibole.

two good cleavage planes at 56° and 124° (see under pyroxenes above, and Fig. 21.11); it occurs in long prismatic crystals and in small bladed or fibrous masses. Its hardness is 5½. Hornblende is common in both igneous and metamorphic rocks; it is another of the ferromagnesian minerals in the igneous rocks (p. 476). *Tremolite* and *actinolite* are common amphiboles in metamorphic rocks.

MICAS

The micas are the outstanding silicate examples of sheet structure with the resulting perfect one-direction cleavage common to all micas as their most diagnostic physical property. Micas have a brilliant or pearly luster and a hardness of only 2½. *Biotite* [$K(Mg,Fe)_3(Al,Si)_4O_{10}(OH,F)_2$] is a shiny black mica and *muscovite* [$KAl_2(Al,Si)_4O_{10}(OH,F)_2$] is the colorless or white mica, sometimes very pale greenish or yellowish. Micas may occur in large sheets up to a foot or more across, or in tiny scales or specks, still with perfect cleavage, in igneous and metamorphic rocks. The micas are the chief constituents of the foliated metamorphic rocks (p. 482).

FELDSPARS

The feldspar group includes the following minerals:

Potash feldspars

orthoclase $KAlSi_3O_8$ } slightly different crystal forms
microcline $KAlSi_3O_8$ }

Plagioclase feldspar series *(percent anorthite, An)*

end member:		
anorthite $CaAl_2Si_2O_8$		100–90 An
bytownite	↓	90–70 An
labradorite		70–50 An
andesine	to	50–30 An
oligoclase		30–10 An
end member:	↓	
albite $NaAlSi_3O_8$		10–0 An

The feldspars are characterized by two directions of rather blocky cleavage at nearly right angles to each other, hardness of 6, and thick tabular crystals. The potash feldspars are white to pale reddish or yellowish or flesh colored, occasionally green in microcline. The plagioclase series varies from white in albite and oligoclase to various shades of gray, even almost black in the calcium-rich members. Plagioclase may be distinguished from orthoclase by fine striations on cleavage planes. Albite and oligoclase are found in the felsic igneous rocks (see Table 22.3), andesine in the intermediate rocks, and labradorite and bytownite in the mafic igneous rocks. Anorthite is rather rare on earth but common in lunar igneous rocks.

QUARTZ

The very common mineral quartz, with hardness 7, is found in all rock types but in different habits. Crystalline varieties often show hexagonal prisms but no cleavage. Colors are variable from colorless to grayish or smoky, yellowish or purplish. In rocks, quartz is often shapeless, grayish but very shiny or glassy with conchoidal fracture. Very fine-grained varieties are also common where no crystal structure is apparent. Fine-grained quartz is known in all colors. *Chalcedony* is the name given to slightly translucent forms with a waxy luster. *Agate* is a banded or blotchy variety. Opaque stony-looking material is known as *chert* if light colored and *flint* if dark. *Jasper* is a red variety, the color due to microscopic specks of hematite in the quartz.

CLAY MINERALS

Many varieties of clay minerals are formed by the weathering of various silicate minerals as discussed in a later chapter (Chap. 23). The best known variety, *kaolinite* [$Al_4Si_4O_{10}(OH)_8$], is formed from the potash feldspars and is used when pure and white to make fine chinaware. Clay minerals are all submicroscopic in grain size and cannot be distinguished from one another in the hand specimen. Clay has a greasier feel than chalk, with which it may be confused. Clay will not effervesce in dilute hydrochloric acid, but chalk will. In soils, clays are mixed with other materials that discolor them.

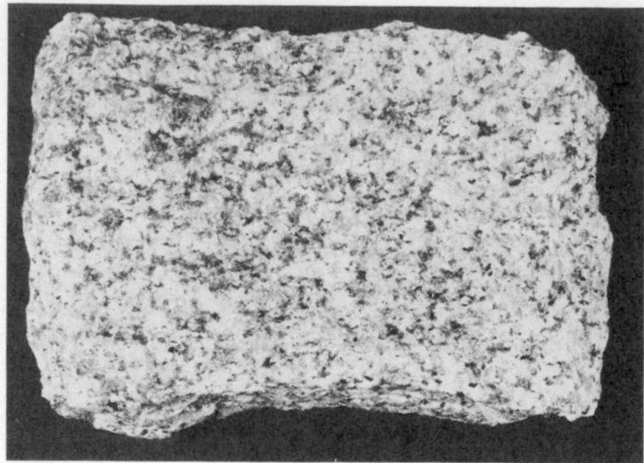

Figure 21.12
Granite. A rock composed of interlocking grains of three minerals; the whitish grains are feldspar, the gray areas are quartz, and the black particles are biotite mica. (Ward's Natural Science Establishment, Inc.)

CARBONATE MINERALS

Calcite ($CaCO_3$) and dolomite [$CaMg(CO_3)_2$] are the common carbonate minerals in sedimentary and metamorphic rocks (limestones and marbles). These two minerals look very much alike with cleavages in three directions, not at right angles, a hardness of 3, and usually of light colors. Calcite effervesces in cold dilute hydrochloric acid, whereas dolomite does not. Coarse-grained calcite or dolomite marbles may look something like feldspar-rich rocks but can be easily distinguished by their inferior hardness.

Rocks

A *rock* may be defined as any large mass making up a part of the lithosphere. Rocks are usually solid bodies, but we also include such unconsolidated masses as gravel deposits, soil layers, or a stretch of beach sand. The majority of rocks are aggregates of one or more minerals, but a few exceptions are known; for example, coal is compacted and altered organic matter and does not contain actual minerals in any quantity. The distinction between minerals and rocks might be likened to the relation between trees and forests. Just as a forest is composed of many trees, sometimes all of one species or again of many species, so a rock is a mass containing many mineral grains either of one kind or of several different kinds. A typical example of a rock in which different minerals can easily be recognized as interlocking grains in a crystalline mass is the well-known rock granite (Fig. 21.12).

Certain minerals commonly occur together, forming a great variety of characteristic mineral associations most of which are repeated many times over in various parts of the world. Each such group of minerals may be considered as a different rock type; the earth's crust therefore contains many kinds of rocks, each one different from the other. It is the goal of the geologist to find out how rocks differ and, then, why they differ. Obviously one of the major variations among rocks is in their mineral composition. They also differ, however, in the way the constituents of the rock are arranged and in the size of the individual grains. These latter variations are known as *texture*. We have seen that minerals form by crystallization from melts or by precipitation from solutions. Apparently a characteristic mineral associa-

Figure 21.13
A lava flow, igneous rock in formation. A broad flow of molten lava streaming down the side of the island of Hawaii and into the Pacific Ocean during the 1950 eruption of Mauna Loa Volcano. The incandescent flow fronts appear as light bands. Steam is being produced violently where a part of the flow enters the sea. (Air National Guard.)

tion forms in a particular environment under a special set of conditions. Therefore rocks differ because they have formed under many different conditions and in a great variety of environments.

For the present we need only recognize two major environments—those within the earth's crust and on the surface. The chief conditions within the earth are high temperature and, often, great pressure. We distinguish two main rock groups—igneous and metamorphic—as associated with these conditions. The surficial environment, on the other hand, where precipitation of material from seawater and the settling out of silt and mud in shallow seas are the most widespread processes, is characterized by low temperatures. Rocks formed under these conditions are known as sedimentary rocks. These three classes of rocks are briefly described below; their properties will be dealt with more fully in later chapters.

IGNEOUS ROCKS

Rock origin is never more dramatically illustrated than in a stream of red-hot, molten lava rushing from a volcano (Fig. 21.13). Solidified lava is an example of *igneous,* or "fire-made," *rock,* which is best defined as rock formed by the crystallization or freezing of hot, molten rock material usually deep within the crust. Actually less than 1 percent of all igneous rocks occurs as lava; most igneous rocks have crystallized below the earth's surface in cracks and other spaces in the crust. When molten rock material cools rapidly, as in a lava flow, individual crystals do not have much time to grow and consequently remain very small, giving the rock a fine-grained texture and a dense or stony appearance. On the other hand, igneous rocks forming deep within the earth's crust crystallize more slowly, and individual crystals will grow to an average diameter of perhaps $\frac{1}{2}$ in. Such rocks are coarse grained and granular in appearance; granite is a common example (Fig. 21.12). Practically all igneous rocks are crystalline masses that are composed of tightly interlocking crystals of silicate minerals.

SEDIMENTARY ROCKS

As rocks are disintegrated and decomposed at the earth's surface by weathering, their remains are carried to the lakes and seas as sand, silt, mud, and soluble salts. More than 75 percent of the land surface is underlain by beds of these same sands and muds, consolidated and cemented to form rock layers; these are *sedimentary rocks* formed by the accumulation of sediments in ancient seas but now uplifted in most plateaus and high mountains. Some of these rocks are composed of fragments of previous rocks and minerals compacted or cemented together. Others are crystalline masses of interlocking crystals formed by precipitation of the dissolved salts in ocean waters; limestone is the most important example. Sedimentary rocks

Figure 21.14
Sedimentary rocks, a canyon in southwestern United States. A uniform layering, or stratification, is the most distinctive large-scale structure in sedimentary rocks. (Union Pacific Railroad photograph.)

contain fossils, the evidence of the past life that existed at the time the rock material was being deposited. The most distinctive characteristic of sedimentary rocks is a uniform layering, or bedding, called *stratification* (Fig. 21.14).

METAMORPHIC ROCKS

Previously formed rocks, igneous or sedimentary, that are deeply buried in the crust are gradually subjected to increasing temperature and pressure as a result of deposition of overlying sediments and progressive involvement with crustal stresses. An environment of increasing temperature and pressure brings about slow changes in mineral composition and texture in the original rocks. These changes take place by atomic rearrangement in the solid state, producing new or reoriented mineral grains. The entire rock mass occupies progressively less and less space in response to the increased pressures and becomes chemically stable at the higher temperatures. The changes sometimes continue until the original rocks are completely altered in both physical properties and mineral composition; the changed rocks are called *metamorphic*. The best-known examples are slate and marble.

The relationships among the three rock types—igneous, sedimentary, and metamorphic—and the processes responsible for their formation and destruction are summarized in the previous chapter as the rock cycle (Fig. 20.11). The student should review this carefully at this time.

SUMMARY

The earth's crust is composed of 92 elements of which 10 make up more than 99 percent of its matter. Oxygen comprises 47 percent by weight but 94 percent by volume of the crust. Four oxygen ions are linked with a silicon ion to form the silica tetrahedron, which is a fundamental unit of structure.

A mineral is a naturally occurring, inorganic chemical compound with a definite internal atomic structure. The latter is illustrated by the two crystal forms of carbon, graphite and diamond, which have entirely different properties due to different internal structure. Glass is a state of matter in which a disordered structure exists in a rigid state.

Minerals may be identified by observing certain physical properties such as hardness, tenacity, cleavage, fracture, specific gravity, color, streak, and luster. More positive identification can be accomplished by X ray diffraction or chemical analysis.

Some 2000 mineral species are known, but most of these are very rare compounds. About two dozen minerals make up most of the crust of the earth and are known as the *rock-forming minerals*. These include olivine, pyroxenes, amphiboles, micas, feldspars, quartz, clay minerals, calcite, and dolomite.

Minerals may be grouped in several chemical classes of which the most important is the silicate class. The silicates may be further classified by their internal structure and the manner in which the silica tetrahedrons are linked together. There are only six such types: the independent tetrahedron, pairs of tetrahedrons, rings, chains, sheets, and three-dimensional frameworks. The complexity of the mineral kingdom is in substitution of one ion for another in an atomic structure in a manner called *solid solution*.

Rocks are aggregates of minerals that are classified by origin, mineral composition, and texture. The three groups of rocks classified according to origin, or type of environment of formation, are the igneous, sedimentary, and metamorphic rocks.

Important words and terms

elements
minerals
atomic arrangement
glass, amorphous
crystals, crystalline, crystal form
hardness, tenacity
cleavage, fracture
streak, luster
silicate structure
solid solution
rocks: igneous, sedimentary and metamorphic

Questions

1. Explain in your own words what is meant by the silica tetrahedron. Why is it so important in mineralogy?
2. What is glass in terms of atomic structure? What is the composition of glass? Is this composition always the same in all glass? Why or why not?
3. Why do we not speak of molecules in discussing the structure of minerals?
4. Mention several different ways in which minerals may be distinguished from rocks.
5. Based on the most abundant elements in the crust, what are the common minerals of the earth's crust?
6. Name some minerals of economic value that are (a) used "as is"; (b) used as the source of a nonmetallic element; (c) used as the source of a metal.
7. What mineral physical properties seem most useful to you after examining the mineral specimens supplied by your instructor? Why?
8. Do you know of any mineral pairs, other than diamond and graphite, that have exactly the same chemical composition?
9. Why are the silicates the most important mineral group?
10. Explain solid solution in minerals in your own words. How would two specimens of somewhat different chemical composition but from the same solid solution series be expected to vary from one another in physical properties?
11. Compare and contrast the environment of formation of igneous and sedimentary rocks.
12. What minerals are found in useful quantities in your vicinity, county, and state? What minerals occurring locally have mineral collector appeal?

Project

Visit a natural science museum in your town or vicinity that has a mineral display, and note the variety of minerals and their crystal forms. Note also rocks, ore specimens, and other earth materials.

Supplementary readings

Books

Ernst, W. G., *Earth Materials,* Foundations of Earth Science Series, Prentice-Hall, Englewood Cliffs, N. J. (1969). (Paperback.) [Includes several chapters on mineralogy and crystal structure as well as on rocks.]

Holden, Alan, and Phylis Singer, *Crystals and Crystal Growing,* Anchor Books, Doubleday, Garden City (1960). (Paperback.) [Explains on an introductory level the theory and practice of crystallography and the methods that can be used to grow and experiment with a number of basic types of crystals.]

Hurlbut, Cornelius S., Jr., *Minerals and How to Study Them,* 3rd ed., Wiley, New York (1963). (Paperback.) [A study of mineralogy written for the beginning student or amateur collector. Describes crystals and the common mineral species.]

Rapp, George, Jr., *Color of Minerals,* ESCP Pamphlet Series #PS-6, Houghton Mifflin, Boston, Mass. (1971). [Color development in crystals, nature of light, cause of color, luster, etc., in minerals are presented.]

Skinner, Brian, J., *Earth Resources,* Foundations of Earth Science Series, Prentice-Hall, Englewood Cliffs, N. J. (1969). (Paperback.) [Chapters on major metallic and nonmetallic mineral resources. Also included is a discussion of industrial rocks and the fossil fuels—coal and oil. This reference is thus also applicable to Chap. 23.]

Magazine

The Mineralogical Record, Editor and Publisher, John S. White, Jr. of the Mineralogy Department of the Smithsonian Institution, Washington, D. C. [A magazine affiliated with the Friends of Mineralogy which is aimed for the educated mineral collector. Has general review-type articles of college level depth on various aspects of mineralogy and on mineral groups.]

IGNEOUS ACTIVITY AND METAMORPHISM

There is nothing in all of nature that arouses more interest and more terror in mankind than a great volcanic eruption.

GORDON A. MACDONALD, 1972

Igneous activity includes all processes connected with the origin, migration, and crystallization of magma within or on the earth's crust. Extrusive igneous activity includes volcanoes and related phenomena. Intrusive activity is the action of magma below the earth's surface and its emplacement and crystallization in the crust to form igneous rocks and certain associated ore deposits. Deep-seated igneous activity and the crumpling of the crust in mountain building, or orogeny, bring about metamorphism of rocks. Intrusive igneous activity, deformation of mountains, and metamorphism of rocks are associated in time and space and presumably in origin.

Intrusive igneous bodies

Magma is a natural fluid within the crust of the earth, generally very hot, composed of a solution of silicate ions with minor amounts of dissolved oxides and gases. When thrown out on the earth's surface it is called *lava*. The actual source and origin of magma is not known, but the more logical theories suggest that local melting within or just below the crust at depths of 20 to 50 miles has produced limited pockets of molten rock material from time to time during the history of the earth. This magma does not come from a liquid interior, as was believed in medieval times; the study of earthquake waves has demonstrated that the earth is solid to a depth of 1800 miles (Chap. 25).

A pocket, or chamber, of liquid magma is under the tremendous weight of the overlying crust. In most cases the pressure resulting from perhaps 20 miles of rock is sufficient to force the liquid magma upward through any fractures or other weak zones that might exist. Some magma is squeezed all the way to the surface, sometimes many thousands of feet above sea level, to emerge from the highest volcanoes. But most magma is forced upward in stages, is cooled, and crystallizes in the upper part of the crust to form bodies of intrusive igneous rock. Such bodies are called *plutons,* in reference to the god of Greek and Roman mythology who had his domain in the underground regions. Plutons that invade space between layers of the surrounding rock and whose sides or contacts are consequently parallel to these layers are said to be *concordant*. Conversely, intrusive bodies that cut across the layering of the crust are *discordant* plutons.

A sheetlike or tabular body that is parallel to rock structure is called a *sill* (Fig. 22.1). Sills are relatively thin but laterally extensive concordant layers of igneous rock that vary from a few inches to several thousand feet in thickness and from a few feet to many miles in width. Many sills are relatively horizontal in position or attitude, but they are known at all angles, and may even be vertical, depending on the attitude of their enclosing wall rocks. A sill-like body that is thickened in the center and has bulged up the overlying rock layers is known as a *laccolith* (Fig. 22.1). This is a mushroom-shaped pluton. Magma, coming up a feeding pipe from below, has spread laterally between rock layers and bulged the overlying strata like a great blister.

A tabular body that cuts across the rock structures into which it is intruded is called a *dike* (see Figs. 20.9 and 22.1). Dikes are magma fillings of cracks and fractures. They vary from a fraction of an inch to hundreds of feet in thickness and may be many miles long and of unknown downward extension. Dikes often occur in groups—as parallel dikes or as a system of radiating dikes extending outward for miles from a central area.

A more or less vertical rod-shaped pluton is called a *pipe,* or *volcanic neck*. Such bodies represent the feeder tube, or conduit, up which magma moved to supply laccoliths or volcanoes. Ship Rock, New Mexico, a typical neck, is the igneous rock that crys-

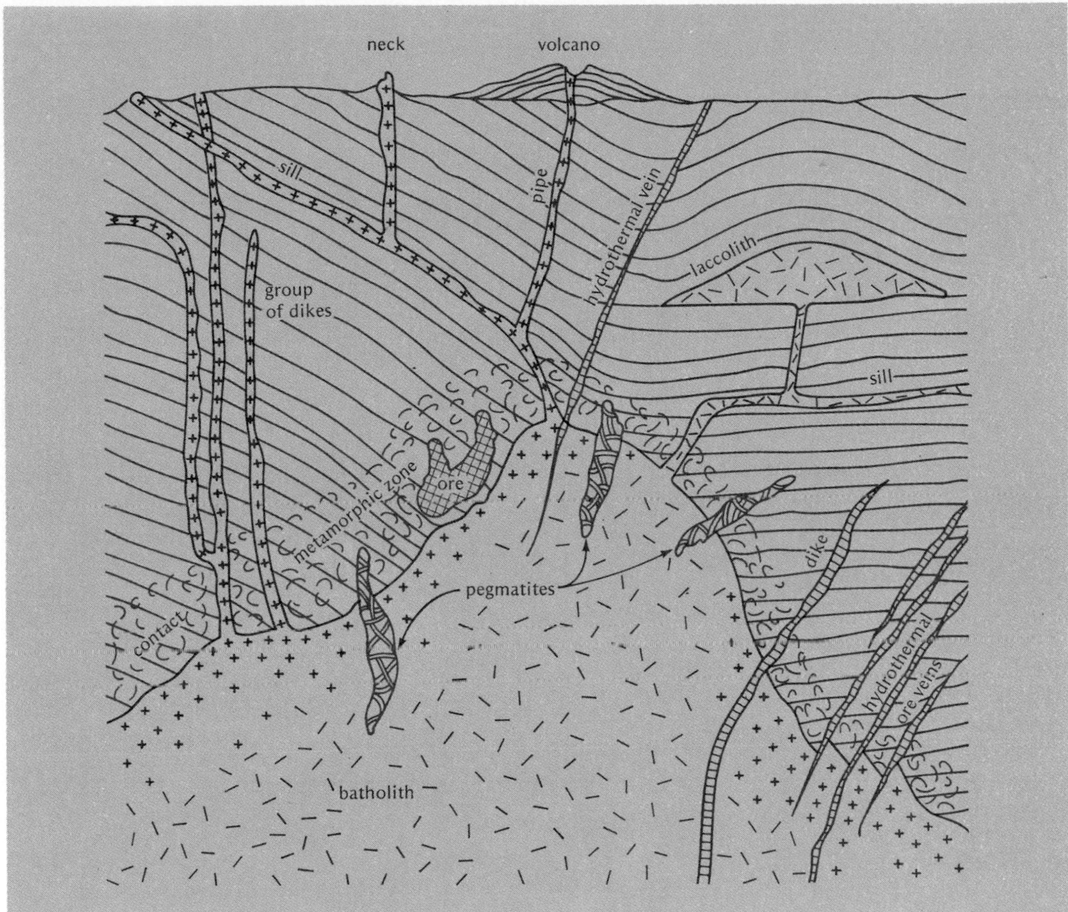

Figure 22.1
Intrusive igneous bodies, plutons. Hypothetical composite diagram showing the shapes and relationships of various types of igneous bodies and their contact metamorphic zones, pegmatites, and hydrothermal ore veins (see p. 480).

tallized in the throat of an old, now eroded volcano (Fig. 22.2). Many pipes are a few hundred to a thousand feet in diameter, but they are also known over a mile in diameter, as exemplified by the famous diamond-bearing pipes in South Africa.

An irregular pluton, with considerable extension in all dimensions, apparent increase in size downward, and no known floor, is called a *batholith* (Fig. 22.1). Some batholiths cover thousands of square miles of the earth's surface. The mechanism of emplacement of batholiths remains one of the unsolved and baffling problems of geology. The important characteristics of batholiths include their universal location in belts of mountain building, their granitic composition, their coarse, granular texture, and an appearance in some of having replaced or dissolved the invaded crustal rocks. Most batholiths are associated with metamorphic rocks and were apparently formed con-

Figure 22.2
A volcanic neck. Ship Rock, New Mexico, the eroded remnant of the throat, or pipe, of a long-extinct volcano. The long black ridges radiating from the neck are dikes, which stand up as a result of differential erosion. (From R. M. Garrels, *A Textbook of Geology,* Harper & Row, 1951; Spence Air Photos.)

temporaneously with the period of metamorphism. Their possible origin is discussed more fully later in this chapter.

Volcanoes

A *volcano* is a release of heat energy from within the earth. The characteristics of the volcano and its eruptions depend on the total amount of energy available and on how it is released—for example, in many small or few larger steps, or even in a single blast. The gas content, and thus the vapor pressure, in the magma determines the violence of an eruption. However, the character of the eruption of a volcano is greatly influenced by the viscosity of its lava. The three factors that determine the viscosity of any lava: (1) chemical composition, (2) temperature of the lava, and (3) the amount of gas in solution in the lava. In general, the lower the silica content of the lava, the less viscous it is. High temperature and, to a certain degree, high gas content also lower viscosity. But if the lava is highly viscous, owing to composition and low temperature, the enclosed gas has difficulty in escaping and must build up high pressure before its release. When it finally escapes, it does so with explosive violence. On the other hand, if the lava is less viscous, the dissolved gas is able to escape easily and continuously without explosion.

Volcanoes vary in size from small openings with cones a few feet high formed in one short eruption to some of the world's loftiest mountains. Cotopaxi, Ecuador, over 19,600 ft above sea level, is the world's highest active volcano. The Hawaiian Islands stand approximately 30,000 ft above the ocean floor and almost 14,000 ft above sea level and have been active for millions of years. Between 450 and 500 volcanoes have been active in historic time but the number of re-

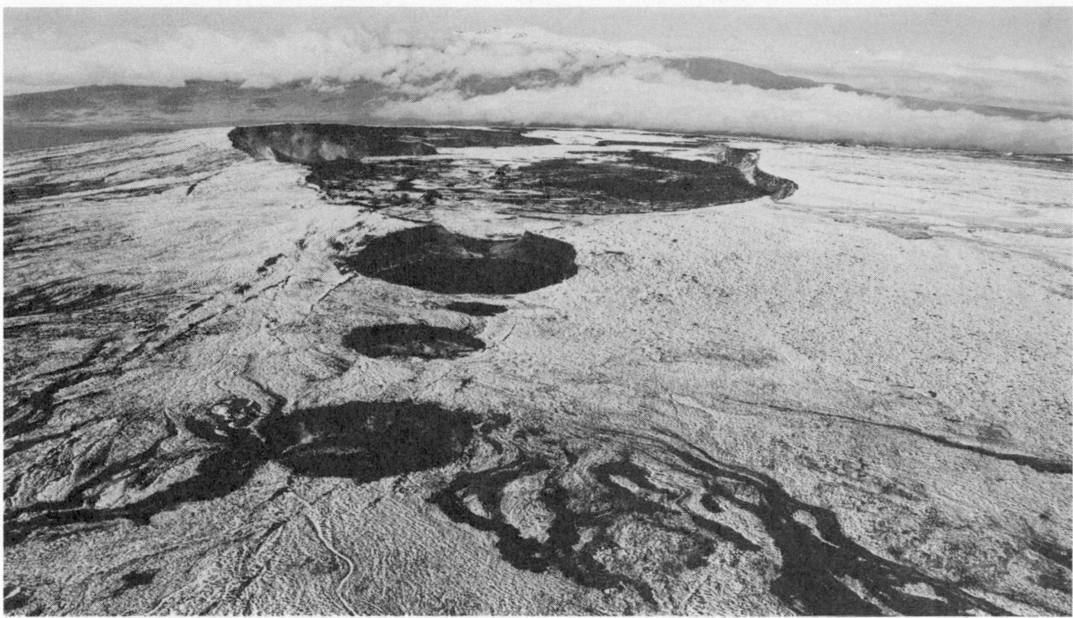

Figure 22.3
The summit of the shield volcano of Mauna Loa, Hawaii, with Mauna Kea in the background. In the foreground are three pit craters along the upper end of the southwest rift zone of the volcano, and behind them is Mokuaweoweo Caldera. The mountain is covered with a light fall of snow. (Photo by U.S. Army Air Force; courtesy of Dr. Gordon A. Macdonald.)

cently extinct volcanoes runs into several thousand, judging by the limited amount of erosion of their cones.

On the basis of characteristic activity and shape of cone, volcanoes can be divided into two major groups: (1) the quietly erupting shield volcanoes, often called the Hawaiian type, and (2) the violent, explosive composite volcanoes, such as Stromboli, Vesuvius, and Pelée. This separation, however, is more convenient than real, as there are many intermediate types. No two volcanoes behave exactly alike any more than do any two people, nor are the eruptions of any one volcano always similar. But a study of the two extremes of volcanic behavior brings out the cause and effect relationships of volcanic activity.

SHIELD VOLCANOES

Those volcanoes whose major activity is the outpouring of great quantities of highly fluid lava, usually in relatively nonexplosive eruptions, may be called *shield volcanoes*. They are characteristically associated with oceanic-type crust. They build gently sloping lava cones whose slopes are usually less than 10° (Figs. 22.3 and 22.4). Eruptions occur at central vents and also from fissures along the flanks (Fig. 22.5). Individual flows are thin but usually long because the lavas are highly fluid and stream down the flanks of the volcano easily and rapidly. Lavas of shield volcanoes are basaltic in composition, with low silica content, and are emitted at high temperatures, 1100°C or over. The consequent

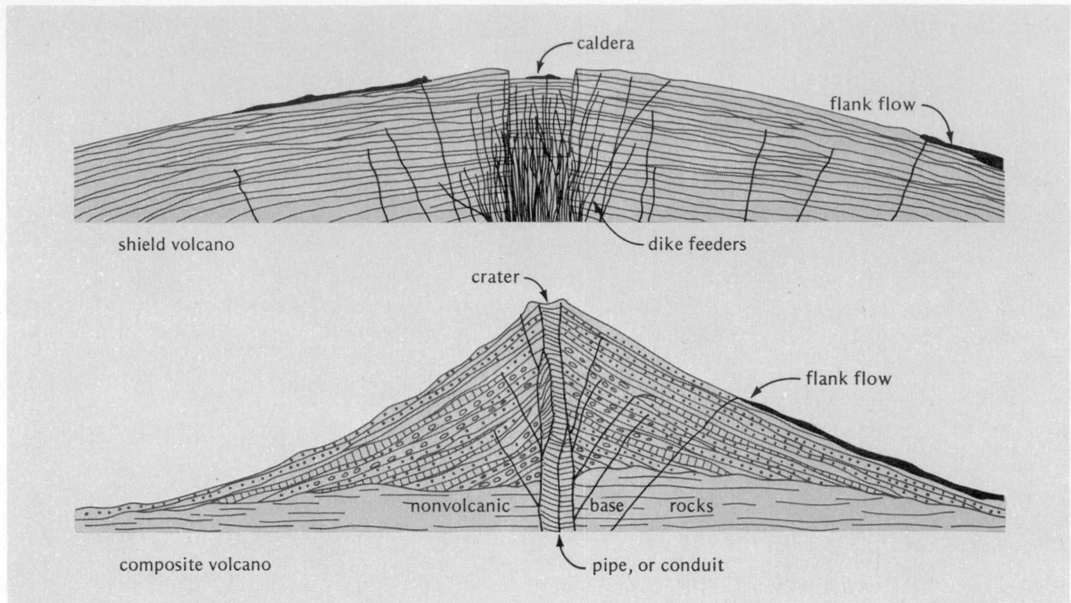

Figure 22.4
Shield and composite cones. Top, generalized diagram of the cone of a shield volcano, composed almost entirely of lava flows cut by feeding dikes and with a typical collapsed summit area, or caldera; and, bottom, a composite volcano composed of alternating pyroclastic beds and lava flows with a circular conduit, or throat, and a summit crater.

Figure 22.5
A fissure eruption. The summit of Mauna Loa Volcano during the eruption of 1949, showing an erupting fissure extending down the flank of the mountain for several miles. Eruptions from fissures or fractures, 1000 ft to 13 miles long, is characteristic of all Hawaiian eruptions. (Official photograph, U.S. Navy.)

low viscosity of these lavas permits free release of their gases. The lava is thus emitted without violence and with the formation of only minor amounts of pyroclastic (fragmental) materials. Basaltic lavas actually have a relatively low total original gas content. Many small cinder cones are formed on the flanks at the site of lava outpourings. As a basaltic shield-type volcano approaches old age and extinction, its eruptions become more explosive. Its lava is then more viscous, probably because of somewhat lower eruption temperature. Consequently large cinder cones are formed on top of the lava cone. For example, Mount Etna on Sicily started life in the geologic past as a shield volcano but has now built a large composite cone and many small cinder cones with violent eruptions in historic time.

The Hawaiian volcanoes are the outstanding examples of shield volcanoes. The island of Hawaii (Fig. 22.6) consists of five overlapping volcanoes that rest on the ocean

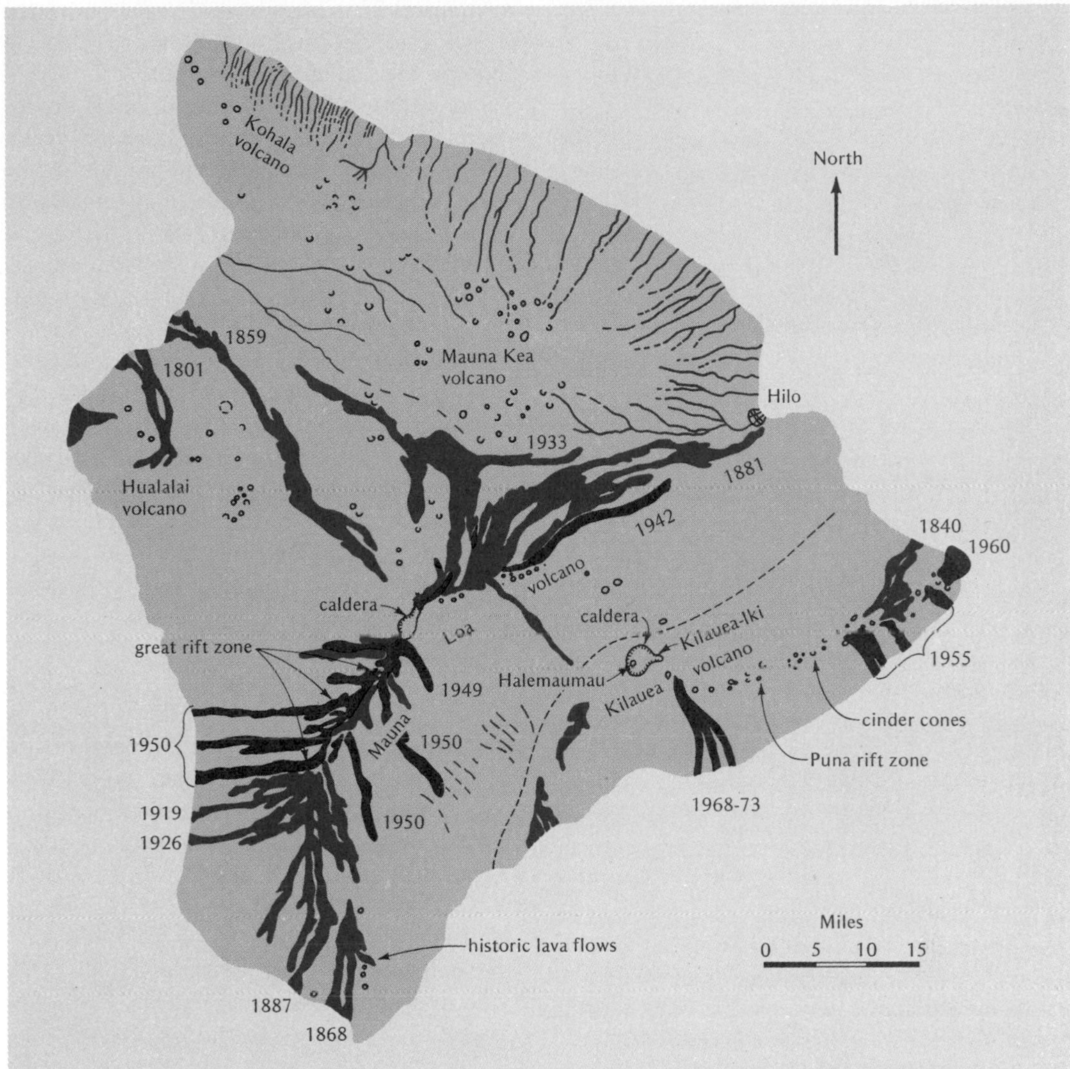

Figure 22.6

The island of Hawaii. The largest island in the Hawaiian group, composed of five volcanoes including the two great active volcanoes, Mauna Loa and Kilauea. Historic lava flows (approximately the last 170 years) are shown individually. Note the alignment of lava flows and cinder cones on the rift zones of Mauna Loa and Kilauea. (Modified after H. T. Stearns and G. A. Macdonald, Bulletin 9, Hawaii Division of Hydrography, 1946.)

floor 18,000 ft below sea level. Two of these, Mauna Loa and Kilauea, are still very active. *Mauna Loa* Volcano is one of the great lava producers and the largest single mountain of any kind in the world. Its flanks cover some 2000 square miles, about half of the island of Hawaii, but have an even greater area beneath sea level. Estimates give it a total volume of about 10,000 cubic miles, as compared with 80 cubic miles for the big

Figure 22.7
Small flank eruption. Top, viscous lava appearing at the surface in a new fissure across a paved road, Puna Rift Zone of Kilauea Volcano, March 13, 1955. This illustrates the first phase in the formation of a new volcanic vent. Bottom, spatter cone forming around a lava fountain at the site of the fissure across the road shown above, about 16 hours later. (Courtesy of G. A. Macdonald.)

cone of Mount Shasta in California. Mauna Loa has erupted at intervals of a few years throughout the past century, the last time in 1950. *Kilauea* Volcano is a relatively low, elongate dome against the flank of Mauna Loa. The summit of Kilauea is a depressed, or collapsed, area about $2\frac{1}{2}$ miles in diameter and shut in except on the south side by vertical cliffs as much as 400 ft high. Such a volcanic depression is called a *caldera*. Within this caldera is the actual crater of Kilauea, the so-called "fire-pit" of Halemaumau. For more than 100 years prior to 1924, this pit was an almost continuously active lake of liquid lava, with distinct circulation of the lava to and from unknown depths. At times the level of the lava lake sank rapidly several hundred feet; at other times it overflowed the floor of the caldera. After such a sinking in early 1924, a series of violent explosive eruptions occurred, apparently because great volumes of ground water were admitted to the hot, emptied throat of Halemaumau with the sudden and explosive formation of steam. Since then, a few short eruptions of lava have occurred in Halemaumau, as narrow fractures opened across the pit floor to permit lava to fountain upward and flood the crater floor.

Flank eruptions broke out in 1955 along a 10-mile section of the eastern, or Puna, rift zone in a forested and cultivated region of sugar cane and other tropical crops. Lava was erupted from numerous small fractures that developed intermittently over nearly a 3-month period. An excellent opportunity was afforded scientists to observe the development of fractures to be followed minutes later by the emission of lava (Fig. 22.7). This eruption was accompanied and followed by a settling and mild collapse of the summit area of Kilauea, which indicated the end of an eruptive cycle that had started 3 years earlier with the 1952 summit eruption in Halemaumau crater.

A pair of spectacular eruptions in 1959 and 1960 demonstrated again a typical eruptive cycle at Kilauea. Instruments at the Volcano Observatory indicated a slow continuing rise of magma over a 2-year period, starting at a depth of about 35 miles directly below the top of Kilauea. Late in 1959 a summit eruption started from a fracture along the side of Kilauea Iki crater, which is immediately adjacent to the caldera of Kilauea.

Figure 22.8
A lava fountain eruption. Fountains of incandescent lava about 1000 ft high from a fissure on the eastern end of the Puna Rift Zone of Kilauea Volcano near the town of Kapoho in February, 1960. Note the cloud of dark ash and fume above the jets of lava. The hill behind the buildings is part of an elongate cinder-spatter cone built to a height of about 400 ft during the first 3 weeks of the eruption. (W. H. Parsons.)

This crater was eventually filled to a depth of nearly 400 ft with new lava. Lava remained high in the volcano for a month after this eruption, as indicated instrumentally, until in early 1960 a flank eruption broke out on the Puna rift zone 28 miles east of the summit (Fig. 22.6). Again a fracture nearly 1 mile long opened up across sugar cane fields, and lava fountained (Fig. 22.8) as much as 1500 ft high for over a month, building a new elongate cone some 500 ft high. Again, the cycle closed with pronounced settling of many square miles of the summit area, which was lowered in elevation as much as 5 ft because the top of the mountain settled down into the space that had been vacated by the lava outpourings.

The eruptive cycle typical of the Hawaiian volcanoes Kilauea and Mauna Loa has four parts, as just outlined for two cycles at Kilauea: (1) the slow rise of magma, with gradual swelling of the top of the volcano, (2) a summit eruption, followed after a few months or years by (3) a flank eruption and (4) finally "collapse" of the summit area as the top of the mountain settles down into the space vacated by the lava outpourings. All parts of this cycle are not always distinguishable; sensitive instruments are necessary to detect the first and sometimes the last. Both a summit and a flank eruption may not occur, or they may be in reversed order. But this fourfold pattern is probably a characteristic norm for all shield volcanoes.

COMPOSITE VOLCANOES

The composite type of volcano is characterized by strong explosive activity and a higher ratio of gas and pyroclastics to lava than the shield volcano. The composite volcano is associated with continental margins and island arcs. The almost perfectly symmetrical cones of Fujisan in Japan, Mayon Volcano in the Philippines, and Cotopaxi in Ecuador, as well as the beautiful but partly eroded cones of Mount Rainier, Mount Shasta, and Mount Hood in our Pacific Northwest, are all composite volcanoes. High, steep cones usually composed of alternating layers of lava and pyroclastic materials characterize this type of volcano (Fig. 22.9). Compare the steep slope, often over 30°, of composite cones (Figs. 22.4 and 22.9) with the gentle slope of the shield cone (Figs. 22.3 and 22.4). All gradations occur between cinder cones and large composite cones. A cinder cone is a small, steep cone containing only pyroclastic debris, and usually represents just one eruption. The composite volcano erupts many times from the same central vent and gradually builds a small cone into a large, stately one. They erupt products that are

Figure 22.9
A composite volcano in eruption. Mayon Volcano in the Philippines in an explosive eruption. Note the steep slope of the cone and compare with Fig. 22.3. (From R. M. Garrels, *A Textbook of Geology*, Harper & Row, 1951; J. T. Stark.)

richer in silica and at lower temperatures than those of shield volcanoes.

Vesuvius is a well-known and carefully studied volcano that may serve as an example of the eruptive cycle of a composite volcano. Prior to A.D. 79 the Romans had recognized the volcanic origin of Mount Somma, the predecessor of Vesuvius, but believed it to be entirely extinct. The historic eruption of that year proved otherwise by burying Pompeii and neighboring towns in volcanic ash and other explosion debris. The volcano was dormant again during almost all of the Middle Ages but has been almost continuously active since 1631, building the present cone of Vesuvius. Continuing mild activity is climaxed by periodic violent eruptions, demonstrating a cycle of behavior 30 to 40 years in duration. The thoroughly documented major eruption of 1906 had three phases and illustrates the general character of this type of eruption. The first phase, 4 days in 1906, was a copious outpouring of lava from fractures far down the flanks of the volcano. These flows lowered the magma column in the main conduit of the volcano and lessened the confining pressure on the compressed gas below. (This is similar in principle to suddenly removing the cork from a bottle of champagne.) The gas began to expand rapidly and cleared the throat of the volcano by blowing immense quantities of incandescent lava high into the air to fall in spectacular showers on the outer slopes of the cone. The second phase followed immediately: a continuous gas blowout at tremendous speeds to heights of 8 miles above the volcano. This enormous gas blast lasted a full day and greatly enlarged the crater. The final stage, called the *dark-ash phase*, was a series of intermittent gas explosions, gradually diminishing in intensity over a 2-week period, that threw out fragmental materials from the crumbling crater walls. Avalanches from the walls widened and partially filled the crater, leaving it some 2200 ft wide and nearly 2000 ft deep. For 7 years no activity occurred, and then lava again appeared in the bottom of the crater. Mild activity in the form of small explosions and lava eruptions in the crater gradually filled the crater until, by 1926, lava overflowed the crater rim. In succeeding years other flows went down its flanks, at one time for 5 miles. During the years of variable activity, the throat of the volcano apparently becomes so clogged by solidifying lava and pyroclastics that only partial release of internal gas pressure is possible. Consequently a time finally comes when the highly compressed gas below the crater can no longer be contained, and another climactic explosive eruption occurs. This came in 1944 just after the Allied troops had occupied Naples in World War II.

Mount Pelée on the island of Martinique, in the Caribbean area, became active in 1902 after 50 years of rest. It blew out great black clouds of hot gases, charged with glowing particles of lava, which were heavy enough to sweep down the mountainside at hurricane speeds, destroying everything in their paths. The city of Saint Pierre and some 28,000 people were wiped out in a few moments. Such heavy hot-ash and gas clouds are known as *nuée ardente* eruptions, or ash flows. The mixture of solid particles and gas behaves as a fluid and flows rapidly downward and outward from the volcanic vent, sometimes for many tens of miles. In the recent geologic past, great volumes of ash have been deposited by this flow mechanism, especially in New Mexico, Nevada, Idaho, and in Yellowstone National Park.

PREDICTING ERUPTIONS

All volcanoes give warning of coming eruptions. When you read, "The volcano erupted without warning," this means that the advance notices either were not observed or not understood. Increase in fumarolic activity and in the temperature of the gases is a common warning sign of renewed activity. Small, local earthquakes are a usual premonitory phenomenon, becoming more and more frequent over a period of a couple of months prior to the eruption. These earthquakes originate at progressively shallower depths just before the final outbreak of lava. On Hawaii, instrumental measurements have shown that both Kilauea and Mauna Loa swell up prior to eruptions, owing to rise of magma. This is shown by actual increase in elevation above sea level as well as by tilting of the ground. Volcano predicting requires long, careful, and continuous study. The U.S. Geological Survey maintains the Hawaiian Volcano Observatory on the rim of Kilauea Volcano to study the Hawaiian volcanoes and volcanic problems in general. Other observatories are maintained in Italy and Japan.

MATERIALS EJECTED BY VOLCANOES

During volcanic eruptions quantities of liquid lava, pyroclastic materials, large volumes of gases, or all three are erupted onto the surface or into the atmosphere. The gases include water vapor, carbon dioxide, hydrogen sulfide, sulfur dioxide, hydrochloric acid, hydrofluoric acid, nitrogen, and many others in smaller quantities. Of these, steam predominates over all others, being 70 to 85 percent of the total gas emitted. Some of this water vapor represents ground water or seawater that has seeped into the throat of the volcano, but in the majority of eruptions the water is apparently primary; that is, it was dissolved in the magma when the latter rose through the crust. The emitted gases characteristically rise thousands of feet above the volcano, where the steam condenses to form a giant cauliflower cloud darkened by included volcanic dust.

Some of the gases escape through cracks in the volcanic cone and emerge mixed with

Figure 22.10
Section of cinder cone. Steeply dipping beds of cinders, ash, and porous bombs as exposed in a small quarry wall in a recent cinder cone. Changes in particle size from one layer to another indicate changes in violence of the eruptions. Main layer across center of picture may separate materials of two different eruptions. (W. H. Parsons.)

steam from ground water from many small openings over the surface of the crater and cone of the volcano. These so-called *fumaroles* continue more or less active between eruptions and even for hundreds of years after the volcano has become extinct. The temperature of fumaroles varies from as high as 700°C in active volcanoes down to near the boiling point of water in extinct volcanoes. These gases may heat up large quantities of ground water and develop such phenomena as hot springs and geysers.

Pyroclastic, or fragmental, materials are exploded into the air during an eruption. This material may leave the crater as globs or spray of pasty liquid lava but freezes while traveling through the air and thus falls as solid particles. Large clots that solidify on the outside during passage through the air and take on streamlined or fluted almond shapes are called *bombs*. Irregular porous fragments larger than sand-size particles and ranging up to 1 in. or so in diameter are called *cinders* (Fig. 22.10). Smaller particles are called *ash;* the finest, volcanic *dust*. Cinder and ash are fragments of porous lava and are not products of combustion; they were so named because of similarity in appearance to the combustion products. The pore spaces have originated from gases bubbling out of solution in the rapidly solidifying lava. Extremely porous and consequently very light-weight and frothy material is *pumice*. Some of the fragmental material, of all size ranges, is derived from solidified lava left in the crater from a previous eruption or was torn from the walls of the volcanic throat or conduit. Large angular fragments over 1 in. in size are termed *blocks;* some of these weigh many tons.

The coarser fragments—blocks, bombs, cinders, and some of the ash—fall near the vent and accumulate as steep-sided *cinder cones* (Fig. 22.10) or help build up the giant composite cones. Volcanic dust is carried by winds for long distances, sometimes hundreds or thousands of miles. Dust from the explosive eruption of Krakatoa Volcano in 1883 has been recognized in the ice of both Greenland and Antarctica. Heavy dust falls can smother a forest, but the dust ultimately weathers into very fertile soil. Heavy rains following eruption of considerable pyroclastic material may cause mud flows. These were a common and quite destructive phenomena on the flanks of Irazú Volcano in Costa Rica after the 1963–1965 eruptions. Heavy rains saturated the thick ash deposits on the slopes of the volcano and mobilized them in soupy landslides and floods.

Lava is the molten liquid emitted by volcanoes. Where measurements have been made, the temperature of lava leaving the crater has been as high as 1100° to 1200°C. Lava is yellow–hot when first erupted, turns to red–orange, dark red, and then soon darkens as the surface is quickly chilled. Lava is commonly of two types: pahoehoe and aa (Fig. 22.11). *Pahoehoe* is lava with a relatively smooth, ropy, or billowy surface. *Aa* has an exceedingly rough, jagged, spiny, or blocky surface and advances with a steep, clinkery front. Both pahoehoe and aa lavas may be associated with the same volcano and even in the same flow. The cause of these two different types of flow seems to be entirely physical. Pahoehoe-forming lava is highly fluid and gas charged. Aa-forming lava is relatively more viscous and has lost much of its original gas content, either while still in the volcano or during its flow downslope after eruption.

DISTRIBUTION OF VOLCANOES

Volcanoes are concentrated into a few main belts coinciding, in general, with the world's earthquake belts and with belts of recent mountain building. The most important belt is the Circum-Pacific "circle of fire." This belt includes the many high volcanoes along the Andes Mountains of South America, a very active zone in Central America, the famous but recently extinct cones of the Cascade Range in northwestern United States, and the Alaskan Aleutian chain of volcanic activity. The circle then extends southward through Kamchatka, Japan, the Philippines, Solomon Islands, and New Hebrides into New Zealand, with an important eastward branch through the East Indies. Other belts of volcanoes extend through the central Mediterranean area into Turkey, along the West Indies arc southeast of Puerto Rico, along the African rift valley, and atop the Mid-Atlantic Ridge from Iceland southward through the Azores and a number of islands in the South Atlantic. Most of the active volcanoes are located along plate boundaries as we shall see later (Chap. 27). These are on mid-ocean ridges, on island arcs or landward of deep sea trenches, or where two plates have collided in the recent geologic past.

CALDERAS

Large depressions from 2 to 20 miles in diameter are present at the summit of some volcanoes, both shield and composite, as we have already seen at Kilauea. These depressions are relatively flat bottomed and bounded by steep walls or vertical cliffs. Small volcanic cones and craters occur within them. Such depressions, known as *calderas,* are many times larger than the eruptive vent or crater of the volcano. Calderas are formed by collapse and settling of

Figure 22.11
Lava flows. Top, front of advancing aa flow showing steep, clinkery character. Flow front 10 to 15 ft high advancing about 600 ft/hour across cleared land, Puna Rift Zone, Hawaii, 1955. (Courtesy G. A. Macdonald.) Bottom, wrinkled and folded pahoehoe lava surface, Craters-of-the-Moon National Monument, Idaho. (W. H. Parsons.)

the mountain after lava has been erupted or withdrawn from below. The Krakatoa eruption of 1883 began as a violent gas and pumice eruption, partially destroying the volcanic cone. After a few weeks, what was left of the cone collapsed beneath the ocean, and as water rushed into the hot interior, a final terrifying steam blast eruption sent a dust and steam cloud 50 miles high. Volcanic dust completely surrounded the earth in a couple of weeks. Krakatoa became a depression or caldera in the ocean floor.

Crater Lake (Oregon) (Fig. 22.12) is a caldera regarded by many geologists as resulting from collapse of a former great volcano into its magma chamber. Crater

Figure 22.12
A caldera. Air view of Crater Lake, Oregon, a large circular caldera produced by collapse of a former giant composite volcano. The lake is 5 miles in diameter and its cliffs stand as much as 2000 ft above water level on the left and in the foreground. This surrounding high ground is the remains of the outer flanks of the former volcano and slopes away from the rim of the lake in all directions, especially noticeable on the right half of the foreground. In the lake at the left is Wizard Island, the top of a small new volcano that grew in the bottom of the great collapse depression before rain waters formed the lake. (Courtesy of Dr. John S. Shelton.)

Lake is about 5 miles in diameter, 2000 ft deep, and surrounded by cliffs 500 to 2000 ft high. These cliffs reveal outward-dipping beds of pyroclastic rocks and lava flows typical of a large composite volcano. By extending the angle of slope of these beds, we can picture a volcano at least 12,000 ft high, comparable perhaps to nearby Mount Hood in size and shape. Some 17 cubic miles of material is missing from this former mountain. Apparently the cone foundered after continuing eruptions had partly emptied a magma chamber beneath the volcano, as suggested in Fig. 22.13. One reason for be-

Figure 22.13
Origin of Crater Lake, Oregon. Top, a large composite volcano underlain by a magma chamber; second from top, great pumice eruptions are rapidly emptying the upper part of the magma chamber, and the cone begins to founder; third from top, collapse of the cone into the magma chamber, forming the caldera; and, bottom, minor eruptions in the caldera with crystallization of the magma in the chamber, and, eventually, accumulation of water to form the lake. (After Howel Williams, 1942.)

lieving that the caldera formed by collapse rather than explosion is the absence of sufficient quantities of pyroclastic debris around Crater Lake to represent the 17 cubic miles of cone material that has disappeared.

LAVA PLATEAUS

Extremely fluid basaltic lavas have issued from fissures in such immense quantities as to flood thousands of square miles with essentially flat-lying lava flows at intervals during the earth's history. One example of these lava floods is the Columbia Plateau of central and western Washington and Oregon and southern Idaho, where approximately 200,000 square miles is covered by basalt flows. The deep canyons cut by the Columbia and Snake rivers and their tributaries have exposed a succession of flows more than 5000 ft thick, although individual flows commonly average only 50 ft in thickness. The most recent activity occurred at Craters of the Moon, Idaho, within the last 2000 years, where fresh basalt flows and small cinder cones may be seen today almost unchanged from the day they were erupted.

Some idea of the mechanism of such eruptions can be obtained from the flank eruptions of Mauna Loa (Fig. 22.5) and from the great Laki flow in Iceland in 1783. The latter is the biggest flow to occur in historic times. Lava issued from a fracture 20 miles long and spread over more than 200 square miles in tongues 10 to 15 miles wide and up to 40 miles long.

Igneous rocks

Magmas and lavas solidify to form igneous rocks. With few exceptions this is a process of crystallization over a range of temperatures. Hundreds of varieties and subvarieties of igneous rocks have been named. These different names imply not only variations in composition but differences in appearance due entirely to grain size and the presence or absence of crystals. This is very strikingly illustrated by granite and obsidian (Figs. 21.12 and 22.14), two rocks of the same chemical composition but as different in appearance as any two rocks could be be-

Figure 22.14
Obsidian. An igneous rock composed of natural glass; obsidian has formed by the soldification of lava without crystallization. Compare with the photograph of granite, Fig. 21.12; obsidian and granite have approximately the same chemical composition but are entirely different in appearance and texture because they form in very unlike environments. (Ward's Natural Science Establishment, Inc.)

cause they have formed in very different environments.

TEXTURES OF IGNEOUS ROCKS

The manner in which the grains of a rock are put together and their relative size vary greatly. This is known as the texture of the rock. A study of the texture can tell us a great deal about where and how a rock cooled and solidified. The important textures of the igneous rocks are aphanitic, or fine-grained; granular, or coarse-grained; glassy; porphyritic; and fragmental.

Aphanitic, or *fine-grained,* texture describes a rock composed of interlocking mineral grains or crystals so small that they cannot be readily distinguished with the naked eye—the rock has a stony appearance. This texture results when crystallization takes place quickly, and none of the crystals has time to grow very large. *Granular,* or *coarse-grained,* texture implies interlocking crystals large enough to be easily distinguished by the unaided eye (Fig. 21.12). Rocks with granular texture are found in large plutons, such as laccoliths and batholiths. In general, such rocks crystallized at a considerable depth below the surface, where cooling was very slow.

Figure 22.15
Porphyritic texture. Large crystals of plagioclase feldspar in a fine-grained ground mass. (Ward's Natural Science Establishment, Inc.)

Under conditions of extremely fast cooling, lavas may solidify into homogeneous material before any minerals can crystallize. This forms a natural glass, and the resulting igneous rock has a *glassy* texture (Fig. 22.14). If the rock is dense or nonporous, it will have the smooth fracture and shiny luster of glass and will usually be black. Extremely porous rock, such as pumice, is composed of thin shells of glass around each gas bubble; this is glassy texture without a glassy appearance to the eye.

In some rocks the grains are of two conspicuously contrasting sizes: relatively large crystals set in a matrix of much smaller crystals (Fig. 22.15). This arrangement is known as *porphyritic* texture. One explanation for porphyritic texture is crystallization in two different environments. Slow cooling at depth in the crust permitted the growth of large crystals, but this process was interrupted, and the magma with its content of already formed crystals was forced upward and intruded as dikes or sills or extruded as lava

Table 22.1
Composition of igneous rocks

mineral	percentage
quartz, SiO_2	12
feldspars: orthoclase and plagioclase	60
ferromagnesian minerals:	
pyroxenes (augite) and hornblende	17
micas: biotite and muscovite	4
olivine, $(Mg, Fe)_2SiO_4$	1
oxides: magnetite, Fe_3O_4; ilmenite, $FeTiO_3$	4
all others (including apatite, $\frac{1}{2}$%)	2
total	100

flows with rapid crystallization of the remaining liquid magma.

The pyroclastic materials of volcanic eruptions may become consolidated through compaction or ground water cementation into rocks with a *fragmental* texture. The fragments—ash, cinders, and bombs—of these pyroclastic rocks are volcanic in origin but may have been acted on by wind or water during and immediately after their eruption. The resulting deposits, therefore, are often well bedded or stratified and may show excellent wind or water sorting and rounding of the fragments. It is often difficult to classify the pyroclastic rocks; on the basis of composition they are igneous rocks, but the agents and processes that acted on the materials after eruption from a volcano were sedimentary. This presents a dilemma in classification.

MINERALS OF IGNEOUS ROCKS

Table 22.1 lists the common minerals of igneous rocks; the percentages given are averages made from a large number of different rocks. Of course, no single rock contains more than a few of these, and some minerals—quartz and olivine, for instance—never or almost never occur in the same rock. The feldspar group—aluminum silicates of potassium, sodium, and calcium—is not only the most abundant mineral group but also the most important in terms of rock classification. The feldspars fall into two classes: the alkali feldspars and the plagioclase feldspars. The alkali feldspars, of which orthoclase ($KAlSi_3O_8$) is the most abundant species, are potassium feldspars usually containing a small amount of sodium in solid solution. Plagioclase feldspar is a continuous solid solution of two end members: soda feldspar ($NaAlSi_3O_8$) and lime feldspar ($CaAl_2Si_2O_8$).

The ferromagnesian minerals are dark-colored silicates of iron, magnesium, calcium, and aluminum. The commonest minerals are augite, of the pyroxene group; hornblende; and biotite, or black mica.

Rocks in which ferromagnesian minerals predominate are called *mafic* rocks (*ma* for magnesium, *f* for iron) and contain only 40 to 55 percent silica but with a concentration of calcium, magnesium, and iron. Rocks in which feldspars and quartz predominate are known as *felsic* rocks (*fel* for feldspar, *si* for silica) and consist of 65 to 75 percent silica with high sodium and potassium content.

The minor constituents of igneous rocks are quite varied and are referred to as accessory minerals. The metallic oxide minerals, magnetite and ilmenite, are normally present in small quantities, though occasionally they are concentrated enough in an igneous rock to be economically important as ores of iron or titanium, respectively. Ilmenite is the mineral in which much of the 0.6 percent of titanium in the earth's crust (Table 21.1) has crystallized. Almost all the phosphorus in the earth's crust occurs as the calcium phosphate mineral apatite.

CRYSTALLIZATION OF IGNEOUS ROCKS

Magmas are complex solutions of silicate and oxide ions, as we have seen. As a magma cools, crystallization begins and certain ions combine to form the least-soluble minerals for the temperatures involved. With gradually

Table 22.2
Order of crystallization in mafic magmas: the Bowen reaction principle

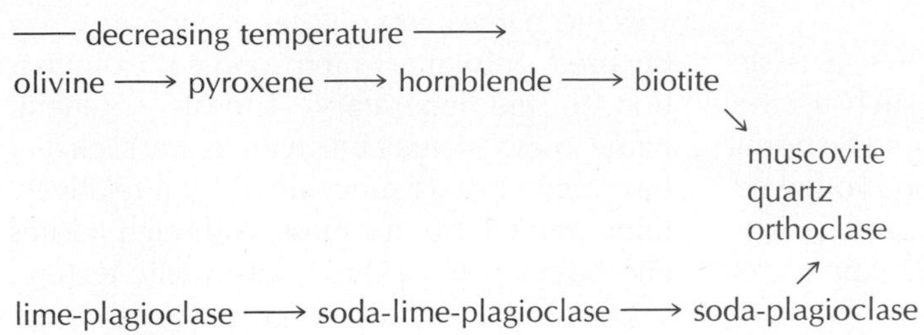

decreasing temperature, there is a definite order of crystallization—high-temperature minerals first, lower-temperature minerals next, and so on—although temperature is not the only factor involved. This order of crystallization may be generalized for primary mafic magmas as shown in Table 22.2. We note two independent mineral series: the ferromagnesian minerals and the plagioclase feldspars. Within each series the early formed crystals react with the remaining liquid magma to produce the later-formed crystals. In other words, the process is characterized by definite physical–chemical relationships that depend not only on temperature but also on the composition of the melt. This pattern of relationships is known as the *Bowen reaction principle;* it has been the subject of much experimental work on artificial melts that approach magma compositions.

In the ferromagnesian series (upper line, Table 22.2) the early formed crystals of olivine react with magmatic liquid to recrystallize as pyroxene; then later there is another reaction of the pyroxene to form hornblende; and, finally, a reaction forms biotite. These kinds of reactions are rearrangements of atoms recombining in increasingly more elaborate silicate atomic structures. The first crystals of olivine have independent silica tetrahedrons; the next crystals, pyroxene, have single chains of silica tetrahedrons; then hornblende has double chains; and, finally, biotite has sheet structures as outlined in the preceding chapter (Table 21.2). The plagioclase series is a continuous reaction in which the early plagioclase crystals remain but with sodium ions at progressively lower temperatures continuously replacing calcium ions in the crystal structures already formed. Additional plagioclase crystals grow and have a sodium-to-calcium ratio in equilibrium with the continuing reaction in the magma.

The early formed crystals may be prevented from reacting with the remaining liquid magma by either (1) the sinking of these crystals in the magma body to a bottom layer or (2) the intrusion of the remaining magmatic fluid to a new environment by some crustal pressure, leaving the crystals behind. In either case the early formed crystals are isolated from the remaining liquid and will form one type of igneous rock. The crystals might be olivine, pyroxene, and lime plagioclase forming the rock known as *gabbro*. The remaining liquid might continue to crystallize in its new location, producing hornblende and soda–lime plagioclase crystals at which time another separation of crystals and liquid might be brought about. This second set of crystals—hornblende and intermediate plagioclase—would form the rock *diorite*. The last liquid, intruded to a third location, would now finally crystallize completely to form soda plagioclase, biotite, orthoclase, and quartz—the rock called *granite*. In this example, three different igneous rock types have been produced from one original mafic magma. Such a combination of processes by which many different igneous rocks are formed from similar magmas is known as *magmatic differentiation*.

An original magma might have a felsic composition to begin with; in such a case the first minerals to crystallize could be hornblende and soda plagioclase, soon followed by biotite, muscovite, orthoclase, and quartz. Under these circumstances, no mafic rocks could be produced in any quantity. This may well be the case with magmas in deformed geosynclines (Chap. 27).

CLASSIFICATION OF IGNEOUS ROCKS

Petrologists have established precise classifications that require microscopical examination of the rocks and chemical analysis to determine their quantitative mineral and chemical compositions. Such schemes are based on the order of crystallization and the reaction principles mentioned above, so that igneous rock classification reflects the physical–chemical history of the cooling of the magma. Igneous rocks may be classified satisfactorily for visual identification, however, on the twofold basis of their mineral composition and their texture. An abbreviated classification of this type appears as Table 22.3, showing the approximate mineral composition of the rock types named. As minerals cannot be visually identified in the fine-grained and glassy rocks, these rocks are given different names from chemically similar coarse-grained rocks. Igneous rocks form a continuous series from felsic to mafic rocks. The felsic rocks, which are composed predominantly of alkali feldspar and quartz, are named on the left side of the table; the intermediate rocks in which plagioclase feldspars predominate are grouped in the center; and the mafic rocks, composed largely of ferromagnesian minerals, are to the right. The felsic rocks are usually light colored and the mafic rocks almost black, although obsidian is an exception to this color generalization.

Felsite is a general name given to all fine-grained, light-colored rocks and *basalt* to fine-grained, dark-colored rocks. The more specific names, *rhyolite* and *andesite,* may be used only after microscopical examination in the laboratory. The very general name *porphyry* may be given to any rock that has large crystals surrounded by a relatively finer-grained ground mass. Although felsites and basalts often exhibit porphyritic texture, granite porphyry, diorite porphyry, etc., are also known.

Consolidated volcanic dust and ash is named volcanic *tuff*. Larger particles—cinders, blocks, and bombs—when cemented into rock form, are known as *volcanic breccia*. These pyroclastic rocks exhibit the entire range of composition from felsic to mafic.

The various rock types are gradational both texturally and compositionally, with arbitrary dividing lines between any two rock types. For example, gabbro is a rock containing about 50 percent of ferromagnesian minerals, and peridotite contains over 80 percent. What should we call a rock with 70 percent ferromagnesians? Some geologists would call it a "mafic gabbro"; others would call it a "plagioclase-rich peridotite." Here again we have problems of classification. The purpose of rock classifications is not just to give names to rocks but to emphasize in what ways rocks differ from one another and then to try to show why they differ. This will tell us that certain rocks can be expected together, but certain other rocks will never be in the same environment. For example, basalt porphyry and gabbro may occur together, as both can form under conditions of moderately slow cooling; but obsidian and granite, though chemically alike, can never form together, as the first represents very fast cooling and the second very slow crystallization.

Two rock names, *granite* and *basalt,* stand out in the table. Granites, with granodiorites, constitute about 95 percent of all coarse-grained intrusive rocks; basalts, including the more mafic andesites, comprise nearly 65 percent of the volcanic, or extrusive, rocks of the crust. The significance of these

Table 22.3
Classification of igneous rocks, for field or hand-specimen use[a]

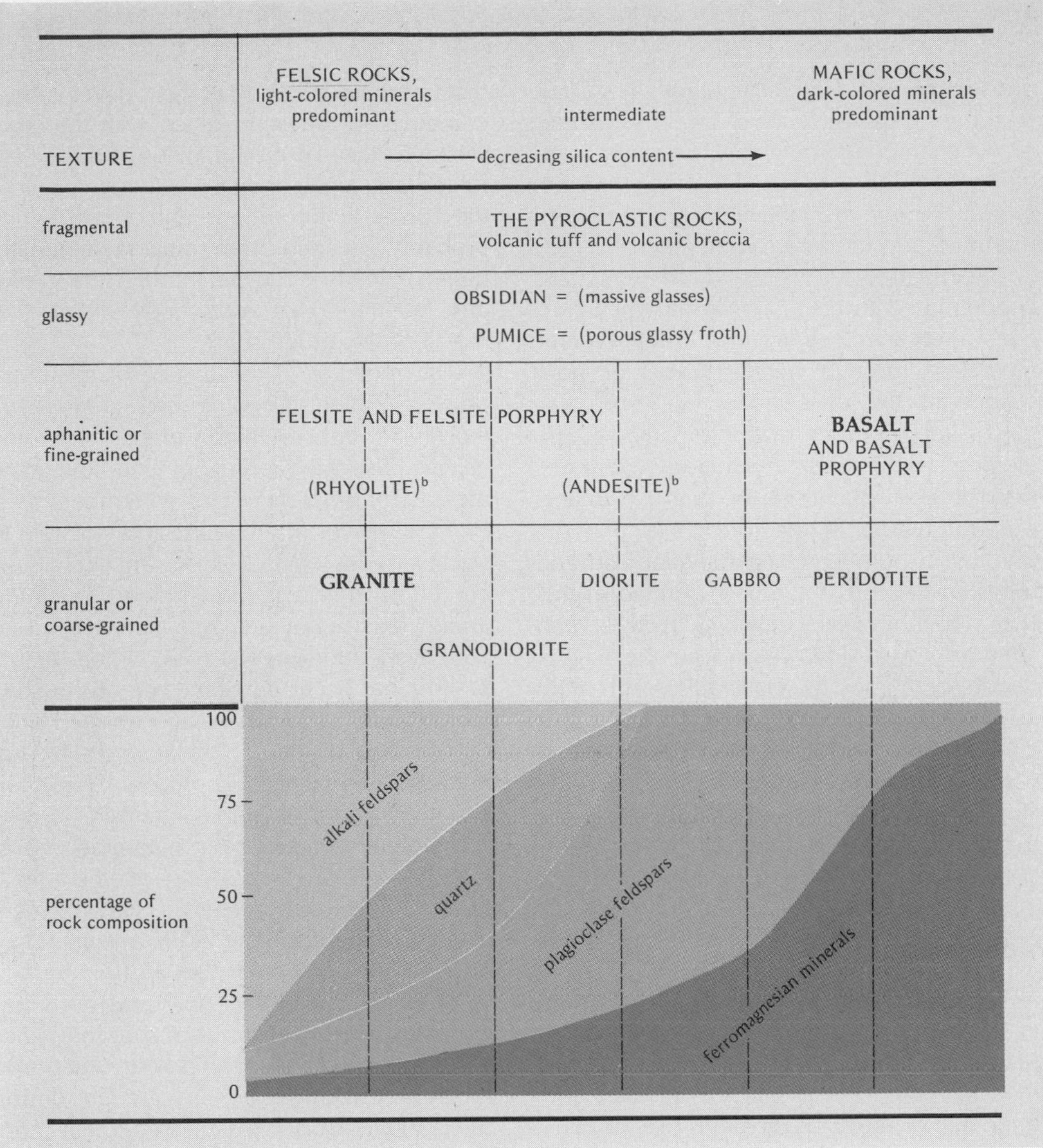

TEXTURE	FELSIC ROCKS, light-colored minerals predominant — intermediate — MAFIC ROCKS, dark-colored minerals predominant (decreasing silica content →)
fragmental	THE PYROCLASTIC ROCKS, volcanic tuff and volcanic breccia
glassy	OBSIDIAN = (massive glasses); PUMICE = (porous glassy froth)
aphanitic or fine-grained	FELSITE AND FELSITE PORPHYRY (RHYOLITE)[b] (ANDESITE)[b]; **BASALT** AND BASALT PROPHYRY
granular or coarse-grained	**GRANITE**; GRANODIORITE; DIORITE; GABBRO; PERIDOTITE

[a] Mineral composition of the rocks is indicated by the vertical line from rock name to the composition graph: granite and rhyolite contain about 50% alkali feldspars, 25% quartz, 15% plagioclase feldspars, and 10% ferromagnesian minerals. Relative abundance and importance of granite and basalt are stressed by size of lettering.

[b] Name in parentheses may be used after laboratory examinations. (Composition chart adapted from Leet and Judson, *Physical Geology*, Prentice-Hall, Inc., Englewood Cliffs, N.J., 1965, p. 67.)

abundances will be dealt with later in the chapter.

Pegmatite is an extremely coarse-grained igneous rock, usually a variety of granite, occurring in small, irregular dikes or lenses (see Fig. 22.1) but closely associated with batholiths or other large igneous bodies. Individual crystals are sometimes a few feet in diameter and have been found over 20 ft long. Pegmatites are composed predominantly of alkali feldspars, quartz, and micas but also contain an amazing array of rare and unusual minerals in small quantity but often in showy crystals. Some of the unusual elements characteristic of these rare minerals include beryllium, boron, chlorine, fluorine, lithium, niobium, tantalum, tin, tungsten, uranium, and the rare earth elements.

Dissolved water vapor and other gases are concentrated in residual magma when they cannot escape in volcanoes. They eventually seep through minute cracks or pore spaces in the surrounding wall rocks as gas or liquid phases, carrying many metallic elements with them in solution. The gases deposit vast quantities of many elements by sublimation in the wall rocks of large plutons; a number of commercial iron and tungsten deposits have been formed in this way. The hot waters known as *hydrothermal solutions* travel slowly away from the magma body, sometimes for many miles, mix with ground waters, and undergo complex chemical reactions, finally depositing a variety of minerals as vein and irregularly shaped deposits. These constitute our major metallic ore deposits.

Metamorphism

Metamorphism includes all those processes that change preexisting rocks into new, or metamorphic, rocks, with the development of new textures and structures or new minerals or both. A gradual rise in temperature, an increase in pressure, and the involvement of chemically active fluids bring about an environment in which the processes of metamorphism become dominant. The temperatures at which metamorphic processes occur range from about 150° to 800°C; at the latter temperature actual melting to form new magma may begin. The range of pressures necessary to bring about the characteristic structural changes observed in metamorphic rocks is not known, but depths of 10 to 20 miles seem probable for the site of metamorphism. It is also clear from the field occurrence of metamorphic rocks that their development coincides in time and place with the compressive stage of mountain building, where forces are great enough to intricately fold the rocks. Hot solutions and gases, which probably escaped from magma at greater depths, slowly circulate through the gradually heating rocks and effectively aid the metamorphic processes.

The measure of the degree of metamorphism is called the *grade*: high-grade metamorphism implies high temperature, and usually also high pressure, and low-grade metamorphism implies lower temperatures. In many places in the world it is possible to trace a series of rocks from an unchanged condition through low-grade, medium-grade, and high-grade metamorphic zones and even into igneous rock, where actual melting has occurred. Some igneous-looking rocks, however, may have been formed without melting by partial replacement by hot fluids in a metamorphic process. A modern hypothesis held by some scientists suggests a complete sequence from high-grade metamorphism through ionic replacement in a solid state to partial melting to generate new magmas. The resulting sialic magma may or may not have been moved from its site of origin. Some of this new magma is undoubtedly squeezed upward, but the major part may remain in its place of origin and finally recrystallize as a batholith during the gradual relaxation of the forces that cause mountain building.

Where rocks covering thousands of square miles have been involved, we speak of the changes as *regional metamorphism*. These occur in the deep zone of mountain roots, where all rocks are heated whether near magmas or not. Where changes have occurred in a narrow belt, from a few inches to perhaps a mile wide, along the contact of an igneous body, we have *contact metamorphism* (see Fig. 22.1). The only source of heat is from the magma itself, and metamorphism is limited to the narrow heated zone along the contact of the pluton.

Figure 22.16
Gneiss. An outcrop of banded gneiss in northern Canada. (Courtesy of the Geological Survey of Canada, photograph no. 86641.)

METAMORPHIC CHANGES

The structural and textural changes brought on by metamorphism are adjustments to the increasing pressure. This adjustment is a rearrangement of atoms into new crystals of the same mineral or of an altogether different mineral without actual melting. This brings about a decrease in porosity and an increase in density of the rock because the same number of atoms occupy less space. Recrystallization or reorientation of crystals results in larger crystals of equidimensional minerals, such as quartz and calcite, and in the parallel positioning of platy or elongate crystals like mica and hornblende. In both of these processes, the shape of the rock mass is changed in response to the directed pressure—a rock flowage accomplished by atomic and crystal rearrangement. Parallel positioning of mineral grains gives a rock a foliated structure, with a tendency to easy splitting in one direction. This structure is called *rock cleavage;* in fine-grained rocks it is *slaty cleavage*; in coarser grained rocks it is *schistosity*. In coarse-grained rocks without many platy minerals a general banding (Fig. 22.16) of the rock is developed rather than true cleavage. Rock or slaty cleavage develops approximately at right angles to the applied directional pressure and does not always follow original structural lines, such as bedding planes in sedimentary rocks.

Many new minerals are formed during metamorphism. Higher temperature stimulates chemical changes without any addition of outside material. Clay minerals at low metamorphic temperatures lose water and crystallize as micas and chlorite. At higher temperatures typical minerals include garnet and staurolite.

Contact metamorphism results in color changes and baking—that is, a fine-grained recrystallization that develops a hard, dense rock material, usually without foliation but with the same minerals mentioned above. In limestones a coarse-grained recrystallization to marble is typical. Calcium and calcium–magnesium silicates develop when quartz and calcite or dolomite are present together. Commonly both gases and liquids escape from the magma and add new elements to the contact zone where many new minerals are formed by a process of replacement. Some of these contain metallic elements which in thick contact zones may be of economic value as ores (see Fig. 22.1) of iron, copper, tungsten, and others.

Table 22.4
Classification of common metamorphic rocks

	metamorphic rock name	texture (and structure)	chief minerals	commonly derived from
massive	quartzite	granular	quartz	sandstones and quartz conglomerates
	marble	granular	calcite, dolomite, Ca-Mg silicates	limestones and dolomites
foliated	slate	slaty, dense, fine-grained	not visible to naked eye, micas, etc.	shales, volcanic tuffs
	schist	schistose, medium-grained	micas, talc, quartz, garnet, etc.	shales, fine-grained igneous rocks
	gneiss	banded, coarse-grained, granular	feldspars, quartz, hornblende, biotite, garnet, staurolite, etc.	coarse-grained igneous rocks and coarse-grained clastic sedimentary rocks

CLASSIFICATION OF METAMORPHIC ROCKS

The metamorphic rocks are classified not only on mineral composition and texture (grain size) but primarily on structure or arrangement of the grains in the rock, as indicated in Table 22.4. Metamorphic rocks are divided according to their structure into massive rocks (nonfoliated) and those exhibiting rock cleavage (foliated). Gradation exists between these two extremes. The massive rocks are named on composition; for example, marbles contain chiefly calcite or dolomite. The foliated rocks are named on the basis of structure and grain size. A foliated, dense, or fine-grained rock is called *slate*; a coarser-grained, well-foliated rock composed of platy minerals like mica is a *schist*; and a coarse-grained, poorly foliated or roughly banded rock is a *gneiss* (Fig. 22.16). These names are modified by the name of a prominent mineral or assemblage of minerals—for example, biotite schist, mica-garnet schist, or hornblende gneiss. When the assemblage of minerals approximates that of an igneous rock type, the rock name may be applied—for example, granite gneiss or diorite gneiss.

Origin of magma

It is easy to suggest that local increases in heat of sufficient magnitude to generate magma occasionally occur within or below the crust. But it is difficult to explain the source of such excess heat. Radioactivity has recently been suggested as a possible heat source. We know that radioactive matter is universally present, especially in felsic rocks, but we do not yet have sufficient evidence on the radioactive content of the deeper part of the crust to judge whether radioactivity can develop sufficient heat to be the main cause of magma formation.

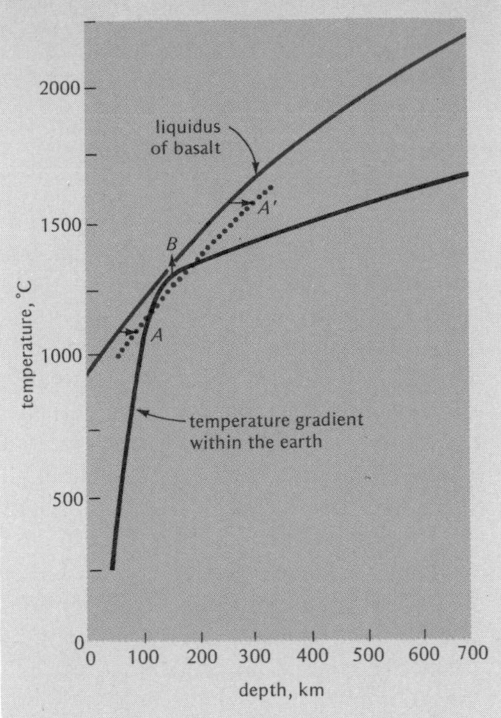

Figure 22.17
Temperature distribution in the earth. Curves showing possible temperature-melting relationships in and below the earth's crust. The red line, marked liquidus, indicates the approximate temperatures at which basalt would become completely molten. The brown line is one interpretation of the actual temperatures that may exist at various depths below the surface. Note that these lines are close together at one depth. If these lines were to intersect, melting could occur to produce magma. The liquidus curve might be shifted in the direction of the arrows *A* and *A'* to the dotted position, owing to a pressure decrease. The gradient curve might be shifted in the direction of arrow *B*, owing to a local increase in heat. (After J. Verhoogen, 1960.)

Measurements in deep mines and wells indicate that temperatures within the crust increase downward at the average rate of approximately 1°C/100 ft, or about 50°C/mile, although this gradient varies considerably in different parts of the world. The material at the base of the thicker parts of the crust, therefore, is hot, perhaps close to 1200°C, and most probably above the melting temperatures of some of the rocks concerned if the latter were at the earth's surface. But the melting temperature of material at depth is considerably increased by the confining pressures of the overlying crustal rocks. This temperature gradient probably flattens out at depth, perhaps as shown in Fig. 22.17. The temperature of complete melting, known as the *liquidus,* is progressively higher at greater depths, partly because of the weight of the overlying rocks; this curve is also plotted in Fig. 22.17. Note that the curves for temperature gradient and for melting of basalt are very close together at one place. At this point, a slight decrease in pressure might mean that the actual temperature would be higher than the liquidus for the new lower pressure, a direct function of depth, and melting of some crustal material could occur. This is like moving the liquidus curve to the right, as indicated by the arrow at *A* (Fig. 22.17), causing an intersection of the liquidus and temperature gradient curves. Note also that a slight increase of heat would move the temperature gradient curve up, as the arrow at *B* (Fig. 22.17) indicates, again resulting in an intersection of the curves and resultant melting.

Many suggestions have been made by geoscientists involving the above principle; one is worthy of comment. Major fractures that extend deep into and below the crust are known, especially in the ocean basins. A slight pressure decrease might occur along such a break at a depth where the material involved would be hot enough to melt at the reduced pressure. This theory has been proposed to explain the origin of magma beneath the Hawaiian Islands. The latter line up with a northwestward trend over 1500 miles long, indicating a fracture in the crust beneath the central Pacific Ocean. As mentioned earlier, lavas of the 1959 eruption of Kilauea Volcano were followed upward from a depth of about 35 miles, which is appreciably below the crust for the central Pacific. Magmas formed in this way have a basaltic composition, being formed by partial melting of the mantle material below the crust. It is significant that all oceanic volcanoes are basaltic in composition.

The evidence is clear that deformation of mountains, regional metamorphism, and the development of granitic batholiths are closely associated and occur in a definite sequence of time. The formation of mountain systems begins with the slow sinking of a long geosynclinal belt (more fully discussed in Chap. 26). This sinking is a downward bulging and consequent thickening of the sialic crust. Therefore sialic material, with a relatively low melting point, is pushed down into deeper and consequently hotter zones of the earth. The gradual heating of the sialic material results first in its metamorphism. If the downwarp is deep enough and continues long enough, the material may eventually become so hot that partial melting occurs to form a chamber of magma in the mountain roots. According to this view, some magmas may be formed during intense metamorphism. (Study again the rock cycle diagram, Fig. 20.11.) Magma generated in this way would be granite or granodiorite in composition. That such magmas occasionally work their way to the surface is indicated by the intimate association of present-day composite volcanoes with belts of active mountain building, especially the Circum-Pacific belt. And the products of most of these volcanoes are relatively felsic in composition.

Here, perhaps, we have the explanation for the preponderance of granite and basalt. Melting in the mantle can produce basaltic magmas that apparently move directly to the earth's surface, forming shield volcanoes in the ocean basins and the great basaltic lava plateaus. Such magmas may form outside of mountain belts. Melting of the sialic roots of orogenic belts, on the other hand, develops the granitic and granodioritic magmas that form the huge batholiths and is obviously associated with the deformational processes. Both types of magma may develop into a variety of rock types, owing to the physical–chemical processes inherent in crystallization.

Economic rocks and minerals

Rocks that are used commercially are known in the trade as *stone*. Most igneous rocks and the massive metamorphic rocks are strong enough for all building-stone purposes. Some of the coarse-grained igneous and irregularly banded gneisses are valuable as exterior finishing stone because of pleasing color or markings. Usually such material is called "granite" in the trade, although some of the stone may actually be diorite or even gabbro as well as true granite and granite gneiss. Marble is also widely used as a decorative stone both outside and inside; it is easily worked and polished, owing to its relative softness. Again, the term "marble" in the stone trade is used in a broader sense than in geology; commercial marbles include some crystalline limestones that take a polish but are not metamorphic rocks. Fine-grained igneous rocks, mostly basalts, are used as crushed stone for road foundations, railroad ballasts, or concrete aggregate. In the stone industry these basalts are called *trap rock*. Strongly foliated metamorphic rocks are not good constructional stones because they tend to split too readily. An exception is slate. Because of its excellent smooth-splitting characteristics, it is used in roofing, stair treads, and blackboards.

Many of the minerals of igneous rocks and a few metamorphic minerals are usable by man when they can be extracted at a profit. The interesting accessory minerals usually present in igneous rocks in small quantities cannot be separated except at very great cost. Only rarely is such cost justified, as in the case of diamonds. The latter occur in peridotite pipes in various parts of Africa, Siberia, and Arkansas and in stream gravels downriver from these pipes in so-called placer deposits. The concentration of diamonds is never more than one part in 1 million of the igneous rock; yet this rock is mined, crushed, and the diamonds separated, regardless of cost, because there is no alternate natural source.

Only in the *pegmatites*, with their large individual crystals, can the common igneous minerals feldspar, quartz, and mica be separated profitably during quarrying operations. Feldspars, important in the ceramic industry, occur in crystals a few feet in diameter in pegmatite bodies. Mica occurs in sheets 1 ft or so across. Many unusual minerals occur in some pegmatites, both in commercial quantities and in large showy crystals. Pegmatite minerals are the chief source of the world's lithium, beryllium, and tantalum and many of the gems, especially topaz, aquamarine, and tourmaline.

A number of metallic minerals, which occur as minor accessory components of igneous rocks, may occasionally be concentrated to become a major constituent of the rock. In this case the igneous rock is an ore deposit. Chromium is thus found associated with some peridotite plutons. Many of the world's nickel, platinum, and titanium ores, and a few iron ore deposits such as magnetite, are found in or closely associated with gabbro and related rocks. Tin and tungsten occasionally occur in economic quantities in granites. Other ore minerals occur as contact metamorphic deposits near the margin of large plutons (see Fig. 22.1). These include a few of our iron and tungsten deposits. But the major part of the world's supply of gold, silver, copper, lead, zinc, mercury, molybdenum, cobalt, arsenic, and antimony occurs in vein and replacement deposits formed from the *hydrothermal solutions* that escape at depth from magmas and move out into rocks of all types to deposit their metallic content. Here are the great mining camps of the western states: the copper and zinc at Butte, Montana; the silver and other metals at Tintic, Utah; the copper in the giant open-pit mine at Bingham, Utah; and many similar districts in Arizona; the famous Mother Lode gold belt in California, and the many well-known mining districts in the Rocky Mountains of Colorado.

A few metamorphic minerals are important to industry. Chief among these is asbestos, a fibrous form of several minerals the most important of which is serpentine. Both serpentine and its included veins of asbestos are formed by metamorphism of peridotite by chemically active fluids.

SUMMARY

Magma may be formed by melting in the mantle to produce shield volcanoes, usually associated with the ocean basins, and lava plateaus and large mafic plutons on the continents. Basalt is the most abundant rock type.

Shield volcanoes give off much fluid lava; build gently sloping cones by quiet eruptions; are characterized by high temperatures, lava fountains, and fissure eruptions, both central and flank; and develop shallow, flat calderas. Mauna Loa and Kilauea in Hawaii and various volcanoes in Iceland are examples.

Felsic magma may be formed by melting within the sial crustal layer to produce composite volcanoes associated with orogenic belts, the great batholiths of granite and granodiorite, pegmatites, and most hydrothermal solutions. High-grade regional metamorphism is associated with the same environment in the roots of orogenic belts.

Composite volcanoes give off much gas and pyroclastic material and small amounts of relatively viscous lava; build steep cones; are characterized by explosive central eruptions; and develop large, often deep calderas. Examples include Vesuvius, Stromboli, Krakatoa, Pelée, Mt. Rainier, and Fujisan.

Igneous rocks are classified on the basis of mineral composition and texture. The most important minerals are the feldspars, quartz, and the ferromagnesian minerals. Igneous rock textures are granular, aphanitic, glassy, porphyritic, and fragmental.

Metamorphic rocks are classified on structure (massive or foliated), texture, and mineral composition.

Important words and terms

magma, plutons
sill, laccolith, dike, volcanic neck, batholith

shield volcanoes, composite volcanoes
caldera, crater
nuée ardente
fumaroles
pyroclastic materials: bombs, cinders, ash, dust
cinder cones
lava, lava plateau
igneous rocks
aphanitic, porphyritic, glassy
mafic, felsic
magmatic differentiation
metamorphism: regional and contact
rock cleavage, slaty cleavage, schistosity

Questions

1. Compare the shape of shield and composite volcanoes. Explain the difference.
2. Describe in detail an eruptive cycle of Kilauea Volcano.
3. What is the evidence that most calderas are caused by collapse?
4. What textures of igneous rock might be associated with each of the following: dikes, sills, laccoliths, necks, and flows? Why?
5. How might you distinguish a sill from a buried lava flow in a section of sedimentary rock strata exposed in a canyon wall?
6. Compare the minerals of the igneous rocks and the 10 most abundant elements in the earth's crust.
7. What relationship exists between igneous and high-grade metamorphic rocks?
8. Why are granite and basalt so much more abundant than other types of igneous rock?
9. What are pegmatites, and why are they of special interest to man?
10. Discuss the various types of changes associated with metamorphism.

Project

Visit rock outcrops or quarries in your vicinity where igneous or metamorphic rocks may be seen and collected. In some regions such rocks may be found in boulder fields and gravel pits associated with glacial deposits. In larger cities, examples of polished igneous rocks and marble may be seen in the exterior and interior trim of important public buildings and offices.

Supplementary readings

Bullard, Fred, *Volcanoes: In History, in Theory, in Eruption,* University of Texas Press, Austin, Tex. (1962). [The story of several famous volcanoes and their types of eruption told in authoritative but nontechnical style.]

Eaton, J. P., and K. J. Murata, "How Volcanoes Grow," *Science,* vol. 132, (Oct. 7, 1960) 925–938. Also reproduced in J. F. White (ed.), *Study of the Earth: Readings in Geological Science,* Prentice-Hall, Englewood Cliffs, N. J. (1962). [Geology, geochemistry, and geophysics disclose the subsurface structure and eruption mechanism of Hawaiian volcanoes. The 1959–1960 eruption of Kilauea is used as a case history.]

Ernst, W. G., *Earth Materials,* Foundations of Earth Science Series, Prentice-Hall, Englewood Cliffs, N. J. (1969). (Paperback.) [A chapter on igneous rocks, another on the petrochemistry and phase equilibria of their origin, and a discussion of metamorphic rocks and processes are presented.]

Foshag, W. F., and J. R. Gonzalez, *Birth and Development of Parícutin Volcano, Mexico,* U.S. Geological Survey Bulletin 965-D, pp. 355–485 (1956). [An illustrated narrative of the first 3 years of this new volcano's eruption.]

Macdonald, G. A., and Agatin T. Abbot, *Volcanoes in the Sea, the Geology of Hawaii,* University of Hawaii Press, Honolulu (1970). [A fine review of Hawaiian-type volcanic activity and a detailed description of historic eruptions of Kilauea and Mauna Loa in Chaps. 3 and 4. Beautifully illustrated.]

Perret, Frank A., *The Vesuvius Eruption of 1906: Study of a Volcanic Cycle,* Carnegie Institution of Washington Publication No. 339 (1924). [A detailed, classic description of an eruption and the events leading up to it. The outstanding writing of its kind.]

Romey, William D., *Field Guide to Plutonic and Metamorphic Rocks,* ESCP Pamphlet Series #PS-5, Houghton Mifflin, Boston, Mass. (1971). (Paperback.) [Good discussion of metamorphic rocks and suggestions for field trips and field projects on igneous and metamorphic rocks.]

Williams, Howel, *Crater Lake: The Story of Its Origin,* University of California Press, Berkeley, Calif. (1941). [The story of Crater Lake in nontechnical language.]

WEATHERING, SEDIMENTARY ROCKS, AND GEOLOGIC TIME

All wish to know, but few the price will pay.

JUVENAL, A.D. 60?–?140

Weathering of rocks is caused by processes originating in the atmosphere and hydrosphere that result in both mechanical breakup and chemical changes. The products of weathering are both insoluble and soluble materials; the former are transported by wind, water, and ice as solid particles and the latter in solution in ground and surface waters. These transported materials are deposited as sediments, ultimately in the seas, where they are later consolidated into sedimentary rocks. The processes of weathering, transportation, and deposition serve to sort the sediments on the basis of both size and composition with the ultimate development of a wide variety of different sedimentary rock types. The superposition of these rocks and their correlation throughout the world has made possible the construction of a composite geologic column and time scale that demonstrate the relative age of rocks on all continents.

Weathering

Weathering includes all those processes whereby rocks are broken and decomposed on or close to the earth's surface by contact with air and water. During the processes of weathering, we observe a disintegration of solid rock into smaller particles and a decomposition of their minerals by chemical reactions. Although these changes go on simultaneously and are usually dependent on one another, it is convenient to discuss them separately because disintegration includes purely mechanical processes and decomposition is chemical in nature. The chief agents of mechanical weathering are temperature changes and frost action, and those of chemical weathering are water and atmospheric gases. These agents not only act at the surface of the ground but have access through pore openings and numerous large and small fractures to mineral grains even in the solid bedrock. Gravity is important in preventing the weathered debris from burying the bedrock surface on steep slopes (Fig. 23.1).

MECHANICAL WEATHERING

The most significant process of mechanical weathering is frost action. Almost all of us are familiar with the rapidity with which road surfaces are cracked or broken up by the alternate freezing and thawing conditions of a few severe winters. As water freezes to ice, an expansion of approximately 9 percent occurs. When freezing takes place in tiny cracks and irregular spaces saturated with water, the expansive force of the freezing water widens the cracks a little; this process is known as *frost action*. Effective breakup is the result of alternate freezing and thawing repeated many times. Frost action, therefore, is more effective in temperate climates than in the Arctic and Antarctic. In the latter regions, water remains permamently frozen for long periods of time. But in high mountain areas, where alternate freezing and thawing between night and day occur during a major part of the year and where surface rocks are not protected by soil, frost action is the most important of all weathering processes.

Rocks expand with heating and contract on cooling; but rock materials are such poor conductors of heat that the surface layer of a rock may become hot and expand when exposed to the sun, whereas less than 1 in. below the surface little or no heating takes place and therefore no expansion occurs. The resulting differential expansion of the surface layer of the rock when heated may cause this surface layer to split off. Daily difference in temperature of a rock surface with direct exposure to the sun may be well over 100°F, especially in desert areas. But this is not a sufficient temperature change, accord-

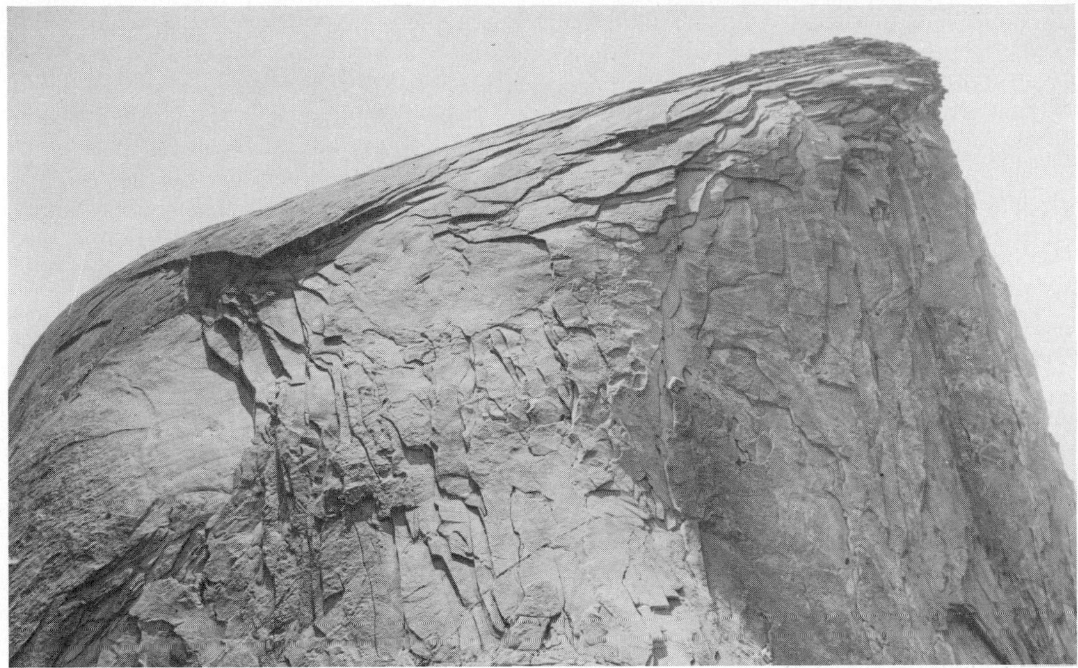

Figure 23.1
Exfoliation, Half-dome in Yosemite National Park. Part of a granite batholith in which sheets or layers of rock have split off parallel to the surface during weathering. (Courtesy of the U.S. Geological Survey.)

ing to laboratory experiments, to cause undue strain in most rocks unless repeated many thousands and thousands of times. Even then, the process of differential expansion probably is effective, at most, in slightly widening the spaces between mineral grains in the outer layer of rock to allow the entry of films of water. The final breakup of the rock, therefore, is the result not just of the alternate expansion and contraction due to temperature changes but also of chemical changes within this weakened surface layer, or zone. The spalling of layers parallel to the outside surface of a rock outcrop is common in this type of weathering and is known as *exfoliation* (Fig. 23.1).

Plant roots extend into cracks and fissures and exert a wedging action that gradually widens the cracks. This is an important process in mountainous country and on new lava flows. Digging and burrowing animals and earthworms stir up the soil and thereby promote water circulation.

CHEMICAL WEATHERING

Chemical reactions between minerals in the crust, the atmospheric gases oxygen and carbon dioxide, and water bring about the gradual formation of reaction products and the crumbling of the original rock. Oxygen is added to the minerals, usually aided by the presence of water. Iron is especially susceptible to oxidation, with the ultimate formation of hematite (Fe_2O_3) or limonite ($Fe_2O_3 \cdot H_2O$), as illustrated in this formula:

$$\underset{\text{iron}}{4Fe} + \underset{\text{oxygen}}{3O_2} + \underset{\text{water}}{nH_2O} \rightarrow \underset{\text{"rust" = iron hydroxide = limonite}}{2(Fe_2O_3) \cdot nH_2O}$$

The iron contained in such minerals as hornblende or biotite is similarly oxidized to limonite, or less commonly to hematite. This process is chiefly responsible for the discoloration of exposed rock surfaces. The resulting iron oxide and hydroxide compounds are the coloring agents in our yellowish, brownish, and reddish soils.

Hydration and carbonation involve the chemical addition of water and carbon dioxide to the minerals of a rock. Perhaps the outstanding example is the formation of clay, a hydrous aluminum silicate, from any aluminum-bearing silicate mineral, which is well illustrated in the decomposition of the feldspars:

$2(KAlSi_3O_8) + 2(H_2O) + CO_2$

orthoclase feldspars (original solid mineral) | water (liquid) | carbon dioxide (gas)

$\rightarrow Al_2Si_2O_5(OH)_4 + 4(SiO_2) + K_2CO_3$

kaolinite, a clay mineral (finely divided particles) | colloidal silica | potassium carbonate (compounds in solution)

Clay minerals are formed not only during the weathering of feldspars but also through the decomposition of micas and several of the ferromagnesian minerals, as shown in Table 23.1. It is not surprising, therefore, that clay is the major constituent of soil and the most abundant mineral in sedimentary rocks. The soluble reaction products of feldspar decay—especially the potassium carbonate from orthoclase feldspar and, similarly, sodium and calcium carbonates from plagioclase feldspar—are carried in ground and surface waters. Eventually these elements, especially the sodium, reach the ocean and have accumulated there throughout the billions of years of the earth's existence.

The weathering of limestone, composed largely of the mineral calcite, is another example of carbonation. Calcite is only slightly soluble in pure water but is readily attacked by weak carbonic acid. The latter is formed when carbon dioxide is dissolved in water. This happens as rain falls in the atmosphere, so that almost all ground water does contain a little carbonic acid. These reactions may be briefly summarized as follows:

$$\underset{\substack{\text{water}\\\text{(rain)}}}{H_2O} + \underset{\substack{\text{carbon dioxide}\\\text{(from atmosphere)}}}{CO_2} \rightarrow \underset{\substack{\text{carbonic acid}\\\text{(in ground waters)}}}{H_2CO_3}$$

$$\underset{\substack{\text{carbonic}\\\text{acid}}}{H_2CO_3} + \underset{\substack{\text{calcite}\\\text{(limestone)}}}{CaCO_3} \rightarrow \underset{\substack{\text{calcium}\\\text{bicarbonate}\\\text{(in solution)}}}{Ca(HCO_3)_2}$$

Limestones, therefore, are almost completely dissolved in humid climates and their material carried in solution to the oceans. Under certain conditions the above reactions are reversible: Carbon dioxide gas is released to the atmosphere and calcium carbonate is precipitated as a solid. This is the cause of the formation of most cave deposits.

The weathering products of the common igneous rock minerals formed by the processes we have just discussed are summarized in Table 23.1. It will be seen, furthermore, that quartz is not chemically weathered because neither oxygen, carbon dioxide, nor water have any appreciable effect on SiO_2 under the conditions that exist at the earth's surface.

CLIMATIC CONTROL OF WEATHERING

High temperatures and abundant rainfall promote chemical reaction and solution. Chemical weathering, therefore, is most rapid in wet, tropical climates but is also important in moist, temperate climates. On the other hand, chemical weathering is

Table 23.1
The weathering products of the common igneous rock minerals

igneous rock minerals	chemical composition	unaltered or residual mineral fragments	important decomposition products: solid particles	important decomposition products: materials in solution
quartz	SiO_2	quartz grains		trace of silica
feldspars:				
orthoclase	$KAlSi_3O_8$	a few feldspar grains	clay minerals	colloidal silica; potassium carbonate
plagioclase	$NaAlSi_3O_8$; $CaAl_2Si_2O_8$		clay minerals	colloidal silica; sodium and calcium carbonates
muscovite	Silicate of Al, K	a few mica flakes	clay minerals	colloidal silica; potassium carbonate
ferromagnesian minerals:				
biotite hornblende augite olivine	Silicates of Fe, Mg, Ca, Al		clay minerals, iron oxides, and hydroxides	colloidal silica; calcium, iron, and magnesium carbonates
magnetite and ilmenite	Fe_3O_4, $FeTiO_3$	"black sand" grains		

very slow in dry climates, whether they are hot or cold. Mechanical weathering becomes relatively the more important process in the Arctic regions, although all weathering is slowed to a minimum by the deep-freeze conditions. Here soil is permanently frozen below depths of 1 or 2 ft.

The response of certain rock types to variations in weathering due to climate is often quite striking. For example, crystalline limestones make up many imposing cliffs in arid regions, but similar limestones are readily dissolved in a humid climate and will therefore be expressed topographically as valleys. An ancient obelisk or monument was presented by Egypt to New York City in 1880 to stand in Central Park. This granite shaft had stood about 3500 years in the mild, dry climate of the Nile Delta region with very little weathering, so that the deep-cut inscriptions, or hieroglyphics, were still quite sharp. Set up in New York for about 90 years and exposed to frequent wetting, frost, and city smoke, this obelisk has deteriorated so rapidly that the inscriptions are completely gone on part of the monument. More weathering has occurred in 90 New York winters than in 3500 Egyptian years.

The presence of a layer of soil tends to prevent further mechanical weathering of the underlying bedrock, as the latter is protected from sudden temperature change or alternate freezing and thawing. Chemical weathering reactions, however, continue in moist climates to depths of 100 ft, and occasionally even as much as 400 ft, beneath previously formed soil material, owing to the circulation of ground water carrying dissolved atmospheric gases or organic acids.

High mountainous areas have less protective cover than lowlands, because unconsolidated material tends to fall, creep, or be washed downslope. This favors

Figure 23.2
Talus cone, Teton Range, Wyoming. Close-up view of a small cone of rock debris at the mouth of a gully on a mountainside. Note the variation in fragment size, with many blocks over 10 ft in diameter. (Courtesy Walter Nickell.)

mechanical weathering. Furthermore there is a greater tendency for rain to run off down the surface rather than sink into the ground; this also reduces chemical weathering. If the mountains are high enough to have freezing weather, frost action—depending on their latitude and height—will be the dominant weathering process. In mountains in wet, tropical climates, however, vegetation will grow profusely on very steep slopes, and chemical weathering will be the dominant process.

PRODUCTS OF WEATHERING

The immediate products of weathering are the broken pieces of rocks and minerals of all sizes that accumulate on the surface or, in mountainous areas, fall to form *talus* slopes and cones at the foot of steep mountains or cliffs (Fig. 23.2). The soil that blankets almost all our land surfaces and the material transported by wind, water, and ice and deposited as sediment are the eventual products of weathering. *Soil,* the most important product so far as man is concerned, is a thoroughly decomposed layer of unconsolidated material that will support plant growth. A century ago it was believed that the kind of rock being weathered largely determined the type of soil that was formed. It is true that bedrock composition has an important effect, but it has now been clearly demonstrated—since the pioneer work by several Russian soil scientists about the turn of the century—that climate exerts a major control on soil type. Similar soils develop on entirely different bedrocks under similar climatic conditions. Such factors as the amount of slope, the length of time the soil has been forming, and the kind of vegetation present also play a role in determining the soil type. In the United States, two main types of soil are present, pedalfers and pedocals. In the humid eastern half of the country, *pedalfer* soils—well-leached soils in which aluminum and iron compounds have accumulated—predominate. On the other hand, in the drier western half of the United States, *pedocal* soils—in which calcium compounds and other soluble materials have also accumulated—are most common.

Soils are commonly layered in structure, exhibiting zones, or soil horizons, approximately parallel to the surface, usually called the *soil profile.* Three horizons (Fig. 23.3), called the *A, B,* and *C* zones, are usually recognized, although one or more zones may be missing in some soils. Decomposition is rather complete in the *A,* or top, zone, with the exception of stable minerals like quartz, and leaching has removed not only the soluble compounds but also much of the fine particle-sized material. The soluble material has been removed in ground waters and the finely divided clay particles washed down to accumulate in the *B* zone. The *A* horizon contains varying amounts of organic matter and is often dark colored toward the top. In dry climates the *A* zone is less leached, lighter in color, or even missing. In very dry climates a surficial crust of precipitated lime and salts may form as moisture is drawn up from below, owing to capillarity, and evaporated at the surface. The *B* horizon, also known as the zone of accum-

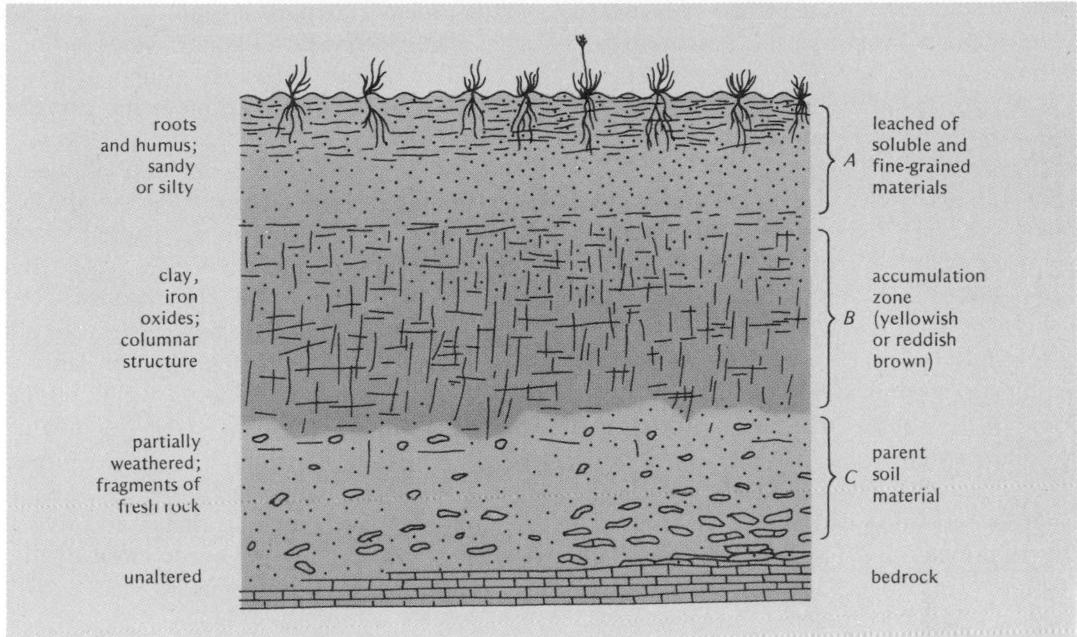

Figure 23.3
Soil profile. Diagram of the *A, B,* and *C* zones in a typical pedalfer soil in a humid temperate climate. The *A* zone is sandy or silty, with varying amounts of organic matter, and is consequently sometimes dark colored; the *B* zone is characterized by an accumulation of clay and iron oxides, is brownish in color, and somewhat columnar in structure; the *C* zone contains partially weathered material, with fragments of the underlying bedrock.

ulation, is well decomposed but only partially leached and contains the fine particles washed down from the overlying *A* horizon. In pedalfer soils the accumulation is largely clay and iron oxides; in pedocal soils it includes not only clay but also calcium compounds, especially calcite. The *C* zone, also called *parent soil material,* is composed of partially weathered minerals and broken fragments of rock and grades downward into bedrock. In this zone chemical weathering processes are slowly progressing to produce a gradually thickening soil and a continuous supply of the soluble compounds necessary for plant growth. In some 10,000 years since the ice age, the soils of the northeastern United States have developed a profile 2 or 3 ft thick. On the other hand, many soils in the southeastern states are several tens of feet thick after hundreds of thousands of years of continuous weathering.

In wet, tropical climates leaching of soluble products is extremely rapid; even the usually insoluble silica is carried away, and the soil produced is an iron- and aluminum-rich, reddish material known as *laterite.* In fact leaching of all elements except iron and aluminum hydroxides may be so complete that the resulting laterite is an ore of iron (limonite) or of aluminum (bauxite). All the aluminum ores of the world were formed as tropical, aluminum-rich soils of this type.

Soil slowly creeps or slides downslope under the impetus of gravity, rain wash, and frost action. In this way solid debris is delivered to and picked up by the transporting agents—streams, winds, glaciers, and shore currents. Solid debris in transport is commonly called *sediment,* especially when deposited at some new location. The soluble products of weathering are carried in solution first in ground waters and later by rivers. Some of the soluble material is precipitated as solid particles when the waters reach the ocean. Such material is also *sediment.*

An examination of Table 23.1 shows that the solid materials belong in two categories: unaltered fragments and decomposition products. The first includes the broken fragments or grains of original minerals, which are the products of mechanical weathering and are residual from rocks whose other components were chemically weathered. These are compounds that are stable under atmospheric conditions; by far the most abundant compound in this category is quartz. Many of the accessory minerals of igneous rocks, such as magnetite and ilmenite—the "black sand" minerals—occur here. In addition, a few grains of incompletely altered feldspar and white mica are often present. The second group of solid compounds consists of minerals formed as reaction products of the chemical weathering of less resistant minerals. Chief among these are clay minerals and iron oxides and hydroxides, the result of the decomposition of feldspars and ferromagnesian minerals.

The materials carried in solution consist of carbonates and bicarbonates of calcium, sodium, potassium, iron, and magnesium and of colloidal solutions of silica from the decomposition of various silicate minerals. Some of these dissolved compounds are precipitated relatively soon after river waters are merged with salt waters of the ocean; others remain as part of the dissolved content of the oceans for millions of years.

Sediments are deposited in many places when a transporting agent is no longer able to carry them, as along stream courses, especially toward the lower end of rivers where the gradient is low. Deposition also occurs in lakes and swamps, at the foot of mountain ranges, and in desert basins by both wind and water. Glaciers deposit sediments wherever the ice melts. These are all examples of sediments deposited on land—*continental sediments.* The vast majority of all sediments reaches the oceans or inland seas, where it is spread laterally by shore currents and finally deposited as *marine sediments,* which include deltas and many sediments laid down along or close to the seashore, such as beach deposits.

Sedimentary rocks

Sedimentary rocks are formed by compaction and cementing together of sediments. These rocks may be separated into two groups: (1) the clastic and (2) the nonclastic, or chemical and organic, sedimentary rocks. The *clastic sedimentary rocks* are composed of more than 50 percent of fragments carried as solid particles to the site of deposition. The nonclastic rocks, on the other hand, contain more than 50 percent of material carried in solution in river waters and then precipitated at the site of deposition. If this precipitation is the result of a chemical reaction or of evaporation, the rock produced is called a *chemical sedimentary rock,* but if precipitation is directly or indirectly due to organic processes, the rock is an *organic sedimentary rock.*

TEXTURES OF SEDIMENTARY ROCKS

The clastic sedimentary rocks are composed of particles or fragments of previous minerals or rocks and therefore have a *fragmental*

Figure 23.4
Fragmental textures. Sedimentary rocks composed of various types of fragments cemented or compressed together. Top, conglomerate: water-worn quartz pebbles and sand cemented together as a clastic sedimentary rock; bottom, breccia: angular chert fragments in a limestone matrix. (Bottom, Ward's Natural Science Establishment, Inc.)

texture (Fig. 23.4). All clastic sedimentary rocks show this texture. Organic sedimentary rocks, however, may also have a fragmental texture—for example, pieces of shells and coral fragments cemented to form shell limestones and the compressed plant material of peat and coal. These are classified as nonclastic rocks, meaning they are not composed of fragments of previous rocks. The calcium carbonate in the shell and coral fragments was extracted from material in solution in seawater by the original living organism in the shells or corals.

Many of the chemical and some of the organic sedimentary rocks have a *crystalline texture*. As material in solution in seawater is slowly precipitated on the ocean floor, layers of interlocking crystals of various sizes are built up. The original precipitate may be a layer of tiny unconnected crystals forming an unconsolidated mudlike sediment. At a later time additional crystals grow between the first ones to produce a solid crystalline mass.

CLASSIFICATION, COMPOSITION, AND ORIGIN OF SEDIMENTARY ROCKS

Clastic sediments are sorted according to size of particles by wind, waves, and running water. The larger fragments are carried by fast currents but deposited quickly when these currents slow down. Consequently, coarse-grained sediments, such as sand and gravel, are deposited near the shore. Very small particles are carried in suspension even by very slowly moving water and settle out gradually. Therefore, fine-grained sediments are spread far from the shore and eventually deposited in an entirely different environment from the sand grains and pebbles. Grain size in the clastic sedimentary rocks therefore reflects the environment of deposition of the sediment and as such is an important basis for classification. On the other hand, crystals precipitated from solution in lake or ocean waters may be either very small or grow to comparatively large size in essentially the same location. As a result grain size does not reflect the physical environment in which the chemical and organic sediments are precipitated and is not used as a main basis for classification of such sedimentary rocks.

Clastic sedimentary rocks, therefore, are classified on the basis of their grain size, and the names "shale," "sandstone," and "conglomerate" imply only size of frag-

Table 23.2
Classification of sedimentary rocks

	rock name	original unconsolidated materials	diameter and composition of clastic fragments
clastic sedimentary rocks, fragmental textures	conglomerate	gravel (rounded pebbles)	More than 2 mm; various
	breccia	angular fragments	mineral or rock fragments
	sandstone	sand	$\frac{1}{16}$–2 mm; various mineral grains
	shale	silt, clay, mud	Less than $\frac{1}{16}$ mm (silt and clay size), largely clay minerals

	rock name	source material	chemical (and mineral) composition
chemical and organic sedimentary rocks, crystalline and fragmental textures	limestone	chemical and organic precipitates	$CaCO_3$ (calcite)
	dolomite	calcareous precipitates (altered by solutions)	$CaMg(CO_3)_2$ (dolomite)
	chert	chemical and organic precipitates	SiO_2 and $SiO_2 \cdot nH_2O$(opal)
	evaporites:		
	gypsum	chemical precipitates due	$CaSO_4 \cdot 2H_2O$ (gypsum)
	rock salt	to evaporation	$NaCl$ (halite)
	peat and coal	plant debris (accumulated in swamps)	Carbon plus hydrogen and oxygen

ments, as indicated in Table 23.2, regardless of composition. These clastic rocks are composed of the residual minerals and solid reaction products of weathering, as noted in Table 23.1. Consequently, the most abundant minerals of the clastic rocks are the clay minerals and quartz; with minor amounts of micas and accessory minerals of the igneous rocks. Because clays are the most abundant of all the products of weathering, the shales are by far the most widespread of all the sedimentary rocks. Slow and incomplete chemical weathering of the original igneous and metamorphic rocks can mean sandstones and conglomerates, with notable amounts of feldspar or fragments of unaltered rock. The colors of these rocks are due to various iron oxides and hydroxides (reds, browns, and yellows) and to organic or carbonaceous matter (dark grays and black).

The chemical and organic rocks are named on the basis of their composition; for example, limestones are rocks containing more than 50 percent of calcite, with either fragmental or crystalline textures and exhibiting a wide range of grain sizes. The most abundant compounds of the chemical and organic rocks are the carbonate minerals—calcite and dolomite—and chert, a type of very fine-grained quartz. Many chemical and organic sedimentary rocks contain a considerable quantity of clastic particles. The precipitation of soluble matter from seawater and the settling of sand, silt, or mud particles may occur simultaneously in the same location. In this way a sediment that contains both precipitated crystals and deposited clastic grains accumulates.

Calcite may be removed from seawater and deposited in shells or supporting frameworks by the life processes of many invertebrate organisms. Broken shell and coral fragments thus accumulate on the sea floor or beach to become the main bulk of limestones of fragmental texture. Calcite crystals are also precipitated as the result of the life processes of such minute organisms as algae. These crystal deposits eventually become massive organic limestones. On the other hand, calcite may be precipitated from seawater without organic action, owing largely to the ionization of the calcium bicarbonate in solution. The bicarbonate ion, HCO_3^-, may dissociate with the separation and consequent loss of carbon dioxide as a gas phase. This will result in the precipitation of calcite, because calcium carbonate, $CaCO_3$, is much less soluble than the bicarbonate ion. In effect this is a reversal of the equations (p. 490) whereby calcite was taken into solution in ground water during weathering and completes the carbon dioxide cycle.

Silica in river water is present as a colloidal solution. When such solutions come in contact with ionized ocean waters, the colloidal particles coagulate or clump together and are precipitated, often as a gelatinous mass; the latter gradually crystallizes to a very fine-grained form of quartz known as *chert.* Siliceous materials are also precipitated by organic processes, especially by diatoms. The solid remains of these are tiny amorphous particles of hydrous silica or opal ($SiO_2 \cdot nH_2O$). Such siliceous deposits gradually become dehydrated and finally crystallize as chert.

Another group of rocks, known as the *evaporite series,* contains minerals precipitated when salt water is evaporated in landlocked bays, inland seas, and desert lakes under dry climatic conditions. The outstanding minerals of these rocks are halite ($NaCl$), gypsum ($CaSO_4 \cdot 2H_2O$), anhydrite ($CaSO_4$), and a number of rather rare potassium, magnesium, and boron salts.

Coal is a compressed and chemically altered plant material. As plants grow and die in swamps, their remains accumulate in stagnant waters, where the oxygen content is soon exhausted. This slows down or stops normal bacterial decay, which would otherwise soon destroy the plant remains. Plant fragments gradually pile up on the bottom of the swamp to become a spongy, porous mass of compressed branches, stems, leaves, and spores called *peat.* If this peat is later buried under thick layers of sediment, it is very slowly changed into coal. The latter process requires compression by sedimentary rock layers over a tremendously long period of time. A slowly continuing series of chemical changes involves the progressive loss of water and other volatile components: such as, hydrocarbon gases, carbon monoxide, and carbon dioxide. The oxygen and hydrogen content of the original plant matter is reduced to a minimum while the percentage of carbon is gradually increased in the remaining coal. These processes slowly change peat to lignite (brown coal), to bituminous (soft coal), to anthracite (hard coal). The complete progression from peat to anthracite has only occurred a few times during the history of the earth, because it requires not only tens of millions of years but also an increase in heat, such as that associated with folding of the crust. Lignite and bituminous coals are found with relatively horizontal sedimentary rocks; on the other hand, most anthracite coals occur in folded sedimentary rock series.

The lithification of shales is primarily compaction of clays, reducing pore space and squeezing out excess water. In contrast the lithification of sandstones, conglomerates, and fragmental limestones is a cementing process; the cementing compounds are precipitated from material in solution in the waters filling the spaces between the various fragments. The precipitation of cement occurs some time after

Figure 23.5
Cross bedding, photographed near Kanab, Utah. An outcrop of inclined or cross bedding in the Navajo sandstone formation of southern Utah. This particular formation has been interpreted as a deposit of wind-blown sand. (W. H. Parsons.)

deposition of the sedimentary beds. The common cementing compounds are calcite, quartz, and iron oxides. The consolidation of the chemical sedimentary rocks is the result of the interlocking of crystals, growing as they are precipitated or recrystallized some time later.

FEATURES OF SEDIMENTARY ROCKS

Sediments are deposited in successive layers, each new layer on top of a previous layer. In this way *stratification,* the outstanding characteristic of sedimentary rocks, is developed (see Figs. 21.14 and 23.9). Each individual layer is called a *stratum,* or bed; the planes that separate the layers are known as *bedding planes*. These strata vary greatly in thickness, from a fraction of an inch to several feet. Stratification is caused by differences in grain size, in composition, or in color of the successive layers of deposited material. Such differences can be very slight or sharp and contrasting. Bedding also results from breaks in deposition; for example, a layer of mud deposited under high-water conditions is later exposed to the air and its surface dried before another layer is deposited. Sedimentary rocks split easily along these planes. The changes in depositing conditions may be daily, owing to tides, or seasonal—for example, coarse sediments deposited by the swift high waters of spring floods interbedded with finer sediments deposited by the low waters of late summer or drought conditions.

Most strata are deposited in an essentially horizontal position, owing to deposition on the relatively flat or gently sloping bottoms of lakes, inland seas, and continental shelves. This is the *law of original horizontality*. The bottom of a stratum may con-

Figure 23.6
Ripple marks. A layer of sandstone with ripple marks. These were once small parallel ridges in the sand deposits of a shallow-water beach zone produced by the back-and-forth motion of the waves. (Courtesy Walter Nickell.)

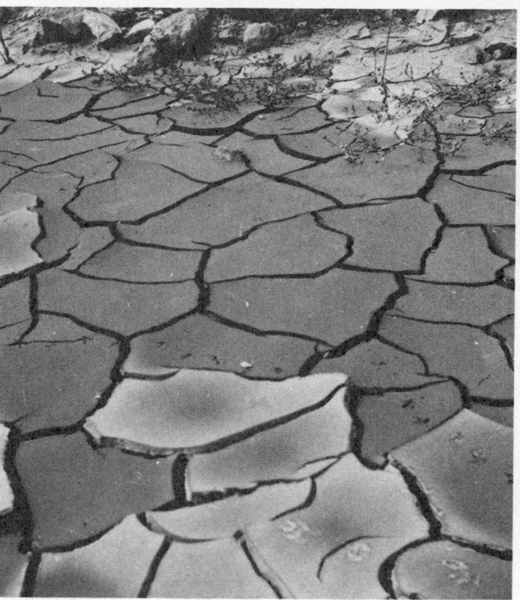

Figure 23.7
Mud cracks. Clay and silt beds in the bottom of a dried-up pond have cracked because of the shrinkage produced by dehydration. Note that the top layer of clay has curved up along the cracks in the drier (lighter-colored) spots. Note also the bird footprints made when the mud was still wet. Such mud cracks and animal footprints are often preserved in rocks. (Courtesy Walter Nickell.)

form to the irregularities of the surface on which it is deposited, but its top will be nearly horizontal. Occasionally beds are deposited locally at an angle to the horizontal by variable currents, which deposit their load irregularly. In this way *cross bedding* is developed (Fig. 23.5): short, steeply dipping beds, caused by rapid deposition, cut off at the top when the current became too swift for any deposition and then horizontal layers, created when the current again slackened. Cross bedding is characteristic in stream deposits where the current alternates from swift to gentle, in delta deposits, and in wind deposits, such as sand dunes.

Sedimentary beds contain many minor features all of which give a clue to the condition or environment of deposition of that layer of sediment. The most important of these are fossils, which tell us not only what kind of life was present but whether deposition occurred on land or in the sea and whether the climate was hot or cold, wet or dry. Other features include ripple marks (Fig. 23.6), which are a series of small, equally spaced ridges of sand or other fine sediment formed by current or wave action in relatively shallow water or on the surface of sand dunes.

Many shale beds exhibit mud cracks. When a layer of wet mud is exposed to the air, its surface dries, shrinks, and often cracks, as illustrated in Fig. 23.7. Subsequently other sediment may be deposited in these cracks to preserve the phenomena in the resulting sedimentary rock. The presence of mud cracks in a rock implies, therefore, an alternation of wet and dry conditions during deposition. Such conditions are only possible in tidal flats along a shoreline or on a delta surface or in very shallow water subject to occasional evaporation as in desert lake basins.

A variety of other minor features, due primarily to the action of ground waters, develops in sediments *after their deposition*. These include such common phenomena as geodes, concretions, and vein fillings.

USES OF SEDIMENTARY ROCKS

Thick-bedded and well-cemented sedimentary rocks are extremely satisfactory for and commonly used as building stones for both structural and decorative purposes. Sandstones and limestones are the most widely used for these purposes. Crushed limestones are widely used for all crushed-stone purposes, such as road foundations, railroad ballast, or concrete aggregate.

Limestone and shale, mixed in the proper proportions, are the raw materials of the cement industry. Sometimes shaly limestones contain just the right amounts of clay and calçite to be used directly as cement raw material.

Calcite limestones (high calcium content, very low silica or magnesium) are greatly sought after today for use as fluxes in the smelting of iron and other metals and as a source of lime (CaO) and other calcium compounds in a great variety of industries. Dolomites and dolomitic limestones also have many industrial uses, although the major share is used by the iron and steel industry as a refractory or furnace lining material.

The sedimentary processes have concentrated many elements in commercial quantities; for example, iron-rich sediments containing iron carbonate, iron oxide, or iron silicate minerals precipitated from ocean waters, under conditions that we do not yet well understand, have produced iron-bearing sedimentary rocks that are the great iron ores of the world, especially when modified by ground-water activity. Other elements concentrated by sedimentary processes to form deposits of economic value are manganese and phosphorus. All commercial phosphate is derived from sedimentary phosphate rocks.

The value and use of such sedimentary rocks as the coals, salt, gypsum, and certain shales in the clay industry are too obvious to need further comment. Some salt deposits (New Mexico and Germany) contain commercial potassium compounds as well as sodium chloride.

The sedimentary rocks are the source and reservoir rocks for petroleum and natural gas. Petroleum is believed to have originated from organic matter deposited with fine-grained sediments in shallow marine basins of quiet, warm waters. The organic source materials were the soft parts of primitive types of animals and plants, especially plankton, the floating or drifting life of the open sea. Compaction of sediment caused slow chemical changes in organic compounds, such as protoplasm, and resulted in the formation of the hydrocarbons we know as petroleum and natural gas. The continuing compaction and dehydration of the muds and silts forced tiny droplets of petroleum out of the mud, or source rock, into porous and permeable rocks of which sandstone is the commonest example. These permeable beds are often called *reservoir rocks*.

Absolute geologic time

One of the interesting and valuable contributions of the study of natural radioactivity is the determination of the approximate age in years of various minerals and rocks from the ratios of various isotopes present in them—for example, the amount of lead present in uranium-bearing minerals. Knowing that the half-life (p. 277) of uranium-238 is 4.5 billion years and noting that, compared with this long period, the half-lives of the rest of the series ending in lead are negligible, we would conclude that these uranium–lead minerals would be 4.5 billion years old if tests showed equal numbers of $^{238}_{92}U$ atoms and $^{206}_{82}Pb$ atoms to be present. Actually, fewer lead atoms than uranium atoms are found, and the ratio of lead to uranium atoms indicates ages up to about 3 billion years for such minerals. This involves the assumption that no lead was present at the time of formation of the minerals. If some lead was present at the beginning, the age of the minerals must be less than 3 billion years. On the other hand, these minerals undoubtedly formed subsequent to the beginning of the earth itself, so the earth must be well over 3 billion years old.

The study of radioactive minerals, those containing uranium or thorium, offers a precise method for measuring the time since these minerals crystallized from a magma, or hydrothermal solution. Radioactivity results in the transformation of certain elements into isotopes, or into other elements. An atom of uranium emits an α particle, the nucleus of an atom of helium gas, and some radiant energy. This decay transforms the parent atom into a new element. The atom of this new substance disintegrates, yielding another α particle and more energy and a third element. This chain reaction continues through a series of disintegrations until a stable element is formed, in this case lead-206. The rate at which these atomic disintegrations take place is slow but absolutely uniform. No action of nature or man, such as temperature changes, pressure, exposure to cosmic rays, or varied chemical environments, has any effect in altering this rate. It seems safe to assume that the rate has not varied in the slightest degree during the existence of the earth.

It is demonstrable that a certain quantity of uranium will yield in 1 year only 1/7,600,000,000 of that quantity as lead. The ratio of lead to uranium in a given mineral, may therefore be used to determine the age of the mineral originally crystallized, according to the equation,

$$\frac{\text{lead}}{\text{uranium}} \times 7{,}600{,}000{,}000 = \text{length of time in years}$$

If, for example, 10 percent of the original uranium in a mineral has altered to lead, the equation will be

$$\frac{10}{90} \times 7{,}600{,}000{,}000 = \text{about } 844{,}444{,}444 \text{ years}$$

In practice it is essential to find an unweathered occurrence of uranium minerals from which neither lead nor uranium has been leached away by ground waters. From such a deposit, a fresh sample is selected for analysis of the amounts of lead and uranium present. Many such determinations have been made of uranium minerals from various parts of the world. According to Knopf.[1]

> Lead is the stable end-product of the radioactive disintegration not only of uranium but also of thorium. Most of the radioactive minerals used in determining ages contain not only ^{238}U and ^{235}U but also thorium.
>
> Uranium-235 produces lead six times as fast as the ^{238}U. Moreover, the atomic weights of the resultant leads differ. Uranium-238 produces a lead isotope of atomic weight 206, and ^{235}U produces a lead isotope of atomic weight 207. The thorium-derived lead has an atomic weight of 208. Thus, when the chemist extracts the lead from such a radioactive mineral, the lead he obtains—the so-called "radiogenic" lead—consists of a mixture of three isotopes of lead of atomic weights 206, 207, and 208.
>
> There are four sets of data, ratios from which the age of the mineral can be calculated: $^{206}Pb/^{238}U$, $^{207}Pb/^{235}U$, $^{207}Pb/^{206}Pb$, and $^{208}Pb/^{232}Th$.

Other radioactive isotopes have been widely used in recent years, especially rubidium–strontium, $^{87}Rb/^{87}Sr$, and potassium–argon, $^{40}K/^{40}Ar$. These elements are present in such common minerals as mica and orthoclase feldspar; it is also possible to determine these isotope ratios from the analysis of an entire rock, not just a single

[1] Adolph Knopf, "Measuring Geologic Time," *The Scientific Monthly*, **85**, 229–230 (November, 1957).

mineral, the rubidium–strontium on metamorphic rocks and the potassium–argon on volcanic rocks. The potassium–argon method has been used successfully on rocks as young as 50,000 years, although the other methods are only accurate on rocks over 10 million years old.

The carbon-14 method, first devised by Willard Libby in 1947, is well known for its use by archeologists. This method is based on the presence of radioactive carbon, ^{14}C, in organic matter. Carbon-14 is produced in the upper atmosphere when cosmic rays bombard nitrogen atoms. When hit, nitrogen emits a proton and becomes carbon-14. The latter combines with oxygen to form carbon dioxide, which in turn is used by living matter. There is a constant ratio of carbon-14 to normal carbon, carbon-12, in the atmosphere and therefore the same ratio of $^{14}C/^{12}C$ in living matter. When a plant or animal dies, it stops absorbing carbon-14 and the latter begins to change back to nitrogen again. The half-life of carbon-14 is about 5730 years. In order to date the remains of an organism—wood, charcoal, shell, bone, or wool—the amount of carbon-14 is determined and compared with the carbon-14 content of living matter. The ratio of the amount of carbon-14 in the dead and living matter is a measure of the time during which the carbon-14 in the dead matter has been going through a radioactive change to nitrogen; therefore this ratio gives the age of the dead matter. Due to its short half-life, carbon-14 can only be used to date materials that are less than 50,000 years old. It has been very useful in geology in giving the approximate ages of glacial advances which have incorporated pieces of tree trunks in glacial deposits.

Age determinations using radioactive isotopes have produced tens of thousands of dates for events in earth history. Rocks from a number of localities on several continents have produced dates approximately 3.3 billion years old. Field relations show that other, still older, rocks exist. The exact age of the earth itself, however, is still undetermined; but several lines of evidence and reasoning converge to suggest an age of about 4.5 billion years, nearly the same as the half-life of uranium-238.

Relative geologic time

Geologic time may conveniently be expressed as the order in which different events have taken place throughout the history of the earth. This is relative time and usually can best be determined from the position of sedimentary rocks.

THE LAW OF SUPERPOSITION

As layers of sediment are deposited in a lake, in an alluvial fan, or in the ocean, each newly deposited layer settles on top of the previous layer. In a series of sedimentary rock strata, therefore, the top layer is the youngest and the bottom bed is the oldest, assuming that the layers have not been overturned or deformed since their deposition. This very obvious relationship is known as the *law of superposition* and is of the utmost importance in determining the relative ages of rock series. In the mile of rocks exposed in the Grand Canyon, for example, the top layers are clearly the youngest rocks present; successively deeper in the canyon older and older rocks are found.

UNCONFORMITIES

Any group of rocks that have been elevated above sea level, with or without folding, will eventually be subject to erosion. Given sufficient time, such land will be worn down to a nearly level surface near sea level (a peneplain, p. 520), and the original rocks will have been partly destroyed by the erosion. This now low-lying land surface may sink and be covered by an inland sea, with the gradual accumulation of a new and younger series of sedimentary rocks. The

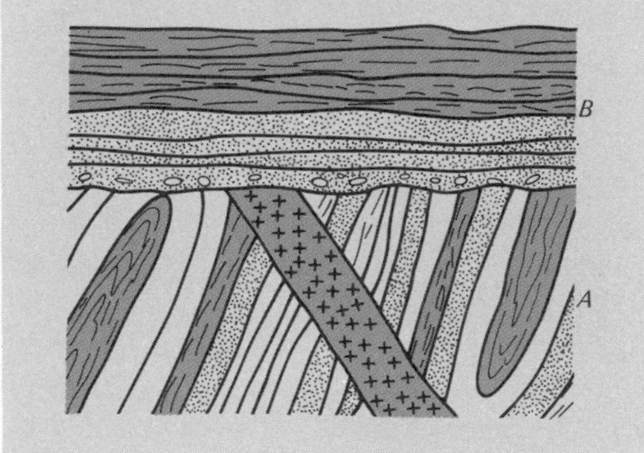

Figure 23.8
An angular unconformity. A series of older and highly folded sedimentary rocks separated by a surface of erosion, known as an *unconformity,* from the younger, overlying sedimentary strata. The major geologic events portrayed in this diagram have occurred in the following order, starting with the first, or oldest event: (1) deposition of the *A* series of sedimentary rocks, originally horizontal, (2) folding of the *A* series rocks, (3) intrusion of a dike of igneous rock, (4) a long period of erosion removing part of the *A* series rocks and part of the dike, and (5) deposition in a later sea of the *B* series of sedimentary rocks.

line of contact between the old and partly eroded rocks and the new rocks is called an *unconformity;* the latter is in reality an old erosion surface and indicates, so far as geologic history is concerned, the passage of a period of time for which no rock record is preserved. If a series of rocks was folded by mountain-building movements and then worn down to a nearly level surface before the deposition of younger rocks, an angular unconformity results, as illustrated diagrammatically in (Fig. 23.8) and in the photograph of such an unconformity in the Grand Canyon in (Fig. 23.9). On the other hand, a period of erosion may occur without deformation of the crust; in such a case the rocks below the surface of unconformity and the younger rocks above will show essentially parallel stratification (Fig. 23.10).

THE LAW OF INTRUSION

Igneous rocks are intruded into previously existing rocks of the crust. Obviously the igneous bodies are younger than the rocks they crosscut. This simple relationship is known as the *law of intrusion* and makes possible the determination of the relative age of most igneous rocks. For example, in Fig. 23.8 a dike crosscuts folded, older sediments (labeled *A*). The dike, therefore, is younger than the *A* beds. Erosion, represented by the unconformity, has removed part of the dike; this dike must be older than the period of erosion and, of course, also older than the time of deposition of the *B* beds.

THE LAW OF FAUNAL SUCCESSION

A *fossil* may be defined as any evidence of past life, such as the actual remains, a replacement, a cast, or an imprint of a former organism. We exclude arbitrarily the remains of organisms that have lived since the beginning of recorded human history. A fossil, therefore, may be the bone or shell of some ancient animal or shellfish that was buried in sand or mud and is found today in sandstone or shale strata. On the other hand, the original bone or shell may have been completely replaced by mineral material to form a petrifaction; agatized wood is a well-known example. Impressions of the original organism made in unconsolidated sediment, such as the imprint of a leaf or the footprint of an animal, are also fossils when preserved in rock. Many different species of animals probably lived at any given time and in any particular place. Such a group of organisms living in an environment at a given geologic time is known as a *fossil fauna* (Fig. 23.11).

In a thick section of sedimentary rocks, as exposed in the Grand Canyon, various fossil faunas are found at different positions in the sequence. According to the law of superposition, we know that the fauna in the

Figure 23.9
An angular unconformity. Flat-lying sedimentary rocks (with the Tapeats sandstone at their base) lying unconformably on nearly vertical metamorphic rocks (the Vishnu schist) in the Inner Gorge of the Grand Canyon. (From R. M. Garrels, *A Textbook of Geology*, Harper & Row, 1951; Santa Fe Railway.)

Figure 23.10
A parallel unconformity. An unconformity in the walls of the Grand Canyon. The upper beds lie on the eroded surface of the older layers below, showing that a period of erosion separated the deposition of the two rock series. The Red Wall limestone below with the shales and sandstones of the Supai formation on top (see Fig. 23.13). (Courtesy of the U.S. Geological Survey.)

uppermost layers is a record of more recent life. This superposition of ancient through recent faunas is called the *law of faunal succession*. Throughout geologic history this succession shows in general a gradual development of life types from primitive organisms to the complex plants and animals of today. There are some gaps in this evolutionary succession due in large part to the complete erosion of many layers of rock, with the resultant loss of part of the geologic record. Such gaps are represented by unconformities as previously discussed.

PRINCIPLES OF CORRELATION

Rocks may be correlated from one locality to another over distances of a few miles or even a few hundred miles by simply tracing the outcrop of the strata. For example, the limestones present at the brink of Niagara

Figure 23.11
A fossil fauna near Silica, Ohio. A slab of limestone covered with the fossil remains of the life that lived in the Devonian sea. Brachiopods are the most abundant fossil species represented, but fragments of crinoids, corals, bryozoans, and trilobites are present. (W. H. Parsons.)

Falls can be traced westward across Ontario through Manitoulin Island in Lake Huron, through northern Michigan to the Green Bay Peninsula in Wisconsin. This long expanse of limestones, called the *Niagaran formation,* obviously represents material deposited in an inland sea at one period in geologic time. The rock types, or lithology, are quite similar all the way from New York to Wisconsin, and identical fossil faunas occur in the various localities where the Niagaran formation outcrops.

A limited amount of correlation is possible by comparing similar unconformities. For example, nearly flat-lying sandstones and other sedimentary rocks overlie folded and metamorphosed rocks in northern New York, northern Michigan, parts of the Rocky Mountains, and in the Grand Canyon. It has been suggested that the great erosion surface represented by these various unconformities may have formed during one long period of geologic time. An unconformity, however, indicates a missing record, and we must be cautious in such correlations.

When we find the same fossil fauna in rocks in widely separated areas, perhaps even in different continents, we may safely conclude that these separate rocks are similar in age. Here we have the outstanding criteria for correlating rocks in one part of the world with those in another. Such relationships have been demonstrated many times in all parts of the world. With the use of fossil faunas, it has been possible to determine the relative age of most of the sedimentary rocks of the earth.

THE GEOLOGIC COLUMN

When sedimentary rocks in various parts of the country were correlated by their fossil faunas and the presence of unconformities was carefully noted, it became apparent that rock sequences in one locality interfinger in time with those in another section. In this way a composite geologic column of the sedimentary rocks and lavas has been gradually assembled for the entire world, in which all known rock units are arranged in their chronological order from oldest, at the bottom, to youngest; this is the *geologic column.* Various names have been applied to the different rock units in this column. When these names are thought of as representing the periods of time during which the various rocks were formed, we have the *geologic time scale,* a simplified version of which is presented in Table 23.3.

Table 23.3
The geologic time scale

eras	periods	epochs	life comments	age of
Cenozoic (began approximately 65,000,000 years ago)	Quaternary	Recent (last 10,000 years)		man
		Pleistocene: the Ice Age (last 2 million years)		
	Tertiary	Pliocene		mammals
		Miocene	grasses abundant	
		Oligocene		
		Eocene	first horses	
		Paleocene	rise of flowering plants	
Mesozoic (began approximately 230,000,000 years ago)	Cretaceous		extinction of dinosaurs	reptiles
	Jurassic		birds first appear	
	Triassic		dinosaurs appear	
Paleozoic (began approximately 600,000,000 years ago)	Permian		extinction of trilobites	amphibians
	Pennsylvanian (the great coal age)		rise of cone-bearing land plants	
	Mississippian		first amphibians	fishes
	Devonian		development of land plants	
	Silurian		first spore-bearing land plants	invertebrates
	Ordovician		first fishes and corals	
	Cambrian		rise of trilobites	
Proterozoic	together these constitute Precambrian time and represent more than 80% of the age of the earth (began at least 4 billion years ago)		primitive invertebrates	scanty fossil record
			algae	
			one-celled organisms	
Archeozoic				

The worldwide correlation of great unconformities or erosion periods has been used to separate the time scale into five eras. For example, the occurrence of a great angular unconformity in which sandstones rest on metamorphic rocks in New York, Michigan, the Rocky Mountains, and in the bottom of the Grand Canyon, as mentioned previously, represents the break between Precambrian and Paleozoic time.

The Grand Canyon may be used as an outstanding example of the application of the various laws and principles of geologic time determination (Figs. 23.12 and 23.13). In the Inner Gorge are exposed the oldest rocks, the so-called Vishnu schist. This is an Archeozoic sediment that was deformed and metamorphosed and later intruded by granite pegmatites. A long period of erosion followed; much later, Proterozoic sedimentary rocks (the Bass limestone, Hakatai shale, and Shinumo quartzite of Fig. 23.13) were deposited. These were later eroded in turn to such an extent that only small down-faulted segments of the Proterozoic rocks are preserved. This great erosion surface was finally covered by the Paleozoic deposition of about 4000 ft of sedimentary rocks. During the Paleozoic era there were two major periods of erosion — after Cambrian time and after Mississippian time, as indicated in Figs. 23.10 and

Figure 23.12
The Grand Canyon. The Archeozoic and Paleozoic rocks are visible in this view as labeled. Note the unconformity between the steep-walled black Vishna schist in the bottom of the canyon and the overlying flat beds of the Tapeats sandstone. (Courtesy Walter Nickell.)

23.13. The lower of these marks the absence of rocks of Ordovician, Silurian, and most of Devonian time, and the upper erosion surface marks the absence of rocks of Pennsylvanian age.

The Niagaran limestones in New York, Ontario, Michigan, and Wisconsin, mentioned earlier, have been shown to be Silurian in age. During Silurian time, therefore, we may postulate an extensive inland sea in what is now north-central United States in which these rocks were slowly deposited. At this same time, what is now northern Arizona was apparently a land mass suffering erosion, as indicated by the great erosion surface between the Muav and Red Wall limestones in the Grand Canyon. No fossil fauna of Silurian age has ever been found in any Grand Canyon rocks.

The relative age of intrusive igneous rocks in terms of time-scale names is determined by the law of intrusion and by consideration of unconformities. For example, the granite pegmatite dikes that cut the Vishnu schist (Fig. 23.13) were partially eroded before deposition of the Bass limestone in Proterozoic time. The dikes, therefore, must have been intruded late in the Archeozoic era.

SUMMARY

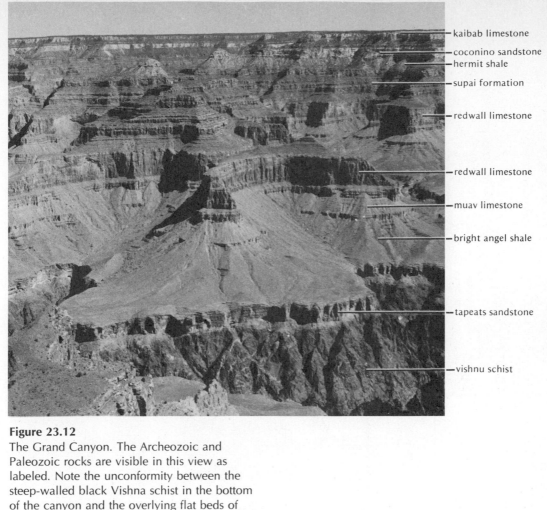

Weathering is both mechanical and chemical. The first is caused by frost action,

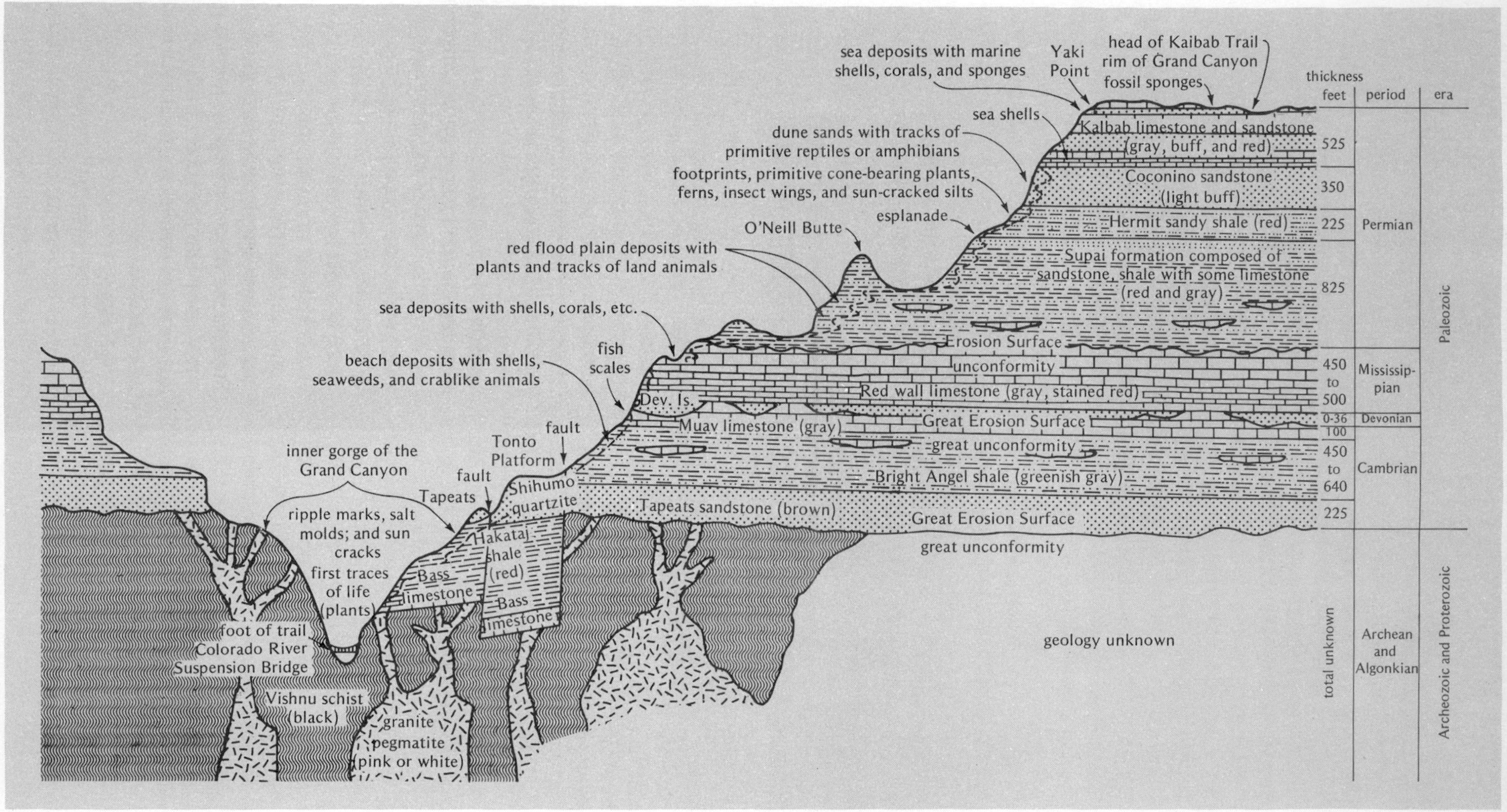

head of Kaibab Trail
rim of Grand Canyon
Yaki Point
sea deposits with marine shells, corals, and sponges
fossil sponges
sea shells
dune sands with tracks of primitive reptiles or amphibians
footprints, primitive cone-bearing plants, ferns, insect wings, and sun-cracked silts
esplanade
O'Neill Butte
red flood plain deposits with plants and tracks of land animals
sea deposits with shells, corals, etc.
fish scales
beach deposits with shells, seaweeds, and crablike animals
fault
Tonto Platform
fault
Tapeats
inner gorge of the Grand Canyon
ripple marks, salt molds; and sun cracks
first traces of life (plants)
foot of trail
Colorado River
Suspension Bridge
Vishnu schist (black)
granite
pegmatite (pink or white)
Bass limestone
Hakataj shale (red)
Bass limestone
Shihumo quartzite
Kaibab limestone and sandstone (gray, buff, and red)
Coconino sandstone (light buff)
Hermit sandy shale (red)
Supai formation composed of sandstone, shale with some limestone (red and gray)
Erosion Surface
unconformity
Red wall limestone (gray, stained red)
Dev. ls.
Great Erosion Surface
Muav limestone (gray)
great unconformity
Bright Angel shale (greenish gray)
Tapeats sandstone (brown)
Great Erosion Surface
great unconformity
geology unknown
thickness feet
period
era
525
350
225
Permian
825
Paleozoic
450 to 500
Mississippian
0-36
Devonian
100
450 to 640
Cambrian
225
total unknown
Archean and Algonkian
Archeozoic and Proterozoic

Figure 23.13
Cross section of the Grand Canyon: diagram of principle formations exposed in the Grand Canyon. Note the erosion surfaces, or unconformities. Compare with Fig. 23.12. (J. W. Stovall and H. E. Brown, *The Principles of Historical Geology*, Ginn and Company, 1955.)

sudden temperature changes, and the action of plants and animals. The latter includes oxidation, hydration, and carbonation.

The products of weathering include soils, sediments, and the sedimentary rocks.

The sedimentary rocks are divided into the clastic and the chemical and organic groups. Clastic sedimentary rocks form from material transported as solids by wind, water, or ice. Examples are shales, sandstones, and conglomerates. Chemical and organic sedimentary rocks, on the other hand, form from material carried in solution in ground and surface waters. They include limestone, dolomite, chert, rock salt and gypsum, peat and coal, iron ore, and phosphate rock.

The geologic column and time scale present earth events in relative order. These time relationships are interpreted by the law of superposition, the presence of unconformities, the law of intrusion, the law of faunal succession, and the principles of correlation.

Important words and terms

mechanical and chemical weathering
frost action, exfoliation
talus, soil, sediment
soil horizons
clastic and chemical sedimentary rocks
stratification, cross bedding
law of original horizontality
fossils
absolute and relative geologic time
laws of superposition, intrusion, faunal succession
unconformities, correlation
geologic column and time scale

Questions and exercises

1. Write a formula for the weathering of plagioclase feldspar; of a ferromagnesian silicate.
2. What happens to the chemical weathering products of plagioclase feldspar and ferromagnesian silicates in pedalfer soils? in pedocal soils?
3. Compare weathering in the Great Lakes area with that in the Colorado Plateau area.
4. What factors are used in the classification of sedimentary rocks?
5. Discuss the features that may indicate a shallow marine origin for strata of sedimentary rock.
6. What is the geologic age of the rock groups within 10 miles of your college?
7. How would you distinguish a black limestone from a basalt; a shale from a slate?
8. What holds the sand grains together in a sandstone? What holds the mineral grains together in a granite?
9. Draw a cross section to show the geologic relationships in a region in which the following events have occurred (a being the oldest); (a) deposition and folding of a series of sedimentary rocks, (b) intrusion of a sill and dike, (c) a long period of erosion, (d) deposition of a later series of sedimentary rocks, (e) emplacement of a dike and lava flow.
10. How has the atmosphere changed the earth's surface appearance?

Projects

1. Look for examples of weathering around your home or neighborhood—in road surfaces, metal objects, and buildings. What processes and factors have been most responsible for these changes?
2. Examine the rocks or building stones in old buildings, monuments, or grave stones to see the effects of weathering. What type of rock has weathered most rapidly in your climate?
3. Look for a soil profile in a road cut, stream bank, or building excavation, and examine and measure the various soil horizons.
4. Visit sedimentary rock outcrops or quarries. What local uses are being made of these rocks? Look for stratification, cross bedding, ripple marks, fossils, and other features in these rocks.

Supplementary readings

Books

Bloom, Arthur L., *The Surface of the Earth,* Foundations of Earth Science Series, Prentice-Hall, Englewood Cliffs, N. J. (1969). (Paperback.) [Energy factors at the earth's surface, rock weathering, and mass wasting appear in Chaps. 1, 2, and 3.]

Eicher, Don L., *Geologic Time,* Foundations of Earth Science Series, Prentice-Hall, Englewood Cliffs, N. J. (1968). (Paperback.) [Historical development of the concept of geologic time, the present rock and fossil record, the problems of correlation, the laws of relative geologic time, and the methods of radiometric (absolute) dating are all well discussed.]

Laporte, L. F., *Ancient Environments,* Foundations of Earth Science Series, Prentice-Hall, Englewood Cliffs, N. J. (1968). (Paperback.) [Ecology of the past as interpreted from sedimentary rocks and fossils is given.]

McAlester, A. Lee, *The History of Life,* Foundations of Earth Science Series, Prentice-Hall, Englewood Cliffs, N. J. (1968). (Paperback.) [A chronological development of life through the study of fossils. Emphasizes the evolutionary development of life forms. A very readable account of past life and its evolution.]

Matthews, William H., III, *Fossils: An Introduction to Prehistoric Life,* Barnes & Noble, New York (1962). (Paperback.) [An authoritative handbook for the amateur fossil collector with chapters on how to collect, identify, and exhibit fossils. A summary of the life of the various geologic ages and a description of the main forms in the organic world.]

Simpson, George G., *Life of the Past,* Yale University Press, New Haven, Conn. (1961). (Paperback.) [An excellent introduction to the study of fossils.]

Simpson, George G., *The Meaning of Evolution,* Mentor Books, New American Library of World Literature, New York (1951). (Paperback.) (Also available as a paperback from Yale University Press, New Haven, Conn., 1960.) [An account of the evolutionary process and an inquiry into evolution's ethical implications.]

ESCP pamphlet series

Beerbower, James R., *Field Guide to Fossils,* #PS-4.

Boyer, Robert E., *Field Guide to Rock Weathering,* #PS–1.

Foth, Henry and Hyde S. Jacobs, *Field Guide to Soils,* #PS–2.

Freeman, Tom, *Field Guide to Layered Rocks,* #PS–3. Houghton Mifflin, Boston, Mass. (1971). [Although these field guides were written for secondary school programs in earth science as part of ESCP (Earth Science Curriculum Project), they are equally applicable as very useable field and project guides on the college level. Includes many very relevant and significant field-trip activities for groups as well as individual projects. The weathering and soils projects may be conducted anywhere, in the inner city or the country. Layered rocks means sedimentary rocks.]

LANDSCAPE PROCESSES AND FORMS

I approve of reasoning if it takes observed fact as its point of departure and methodically draws its conclusions from the phenomena.

HIPPOCRATES, 460?–?377 B.C.

Landscape is produced, with the exception of the cones of active volcanoes, by the gradational agents acting on those parts of the earth's crust that have been elevated above sea level. The rocks at and near the earth's surface are disintegrated and decomposed by weathering, and the resulting debris is transported and redeposited. The agents that accomplish these tasks are streams, wind, glaciers, ground waters, and waves and shore currents. All these agents first obtain a load of weathered rock material, then transport this load, and ultimately deposit it. As a consequence of the removal of a load, erosional landscape forms, such as valleys, ridges, caves, and wave-cut cliffs, are produced. The deposition of the load produces depositional landscape features, such as flood plains, deltas, sand dunes, and beaches.

The hydrologic cycle

The *hydrologic cycle* is the circuit taken by the waters of the earth. First, they are evaporated from the hydrosphere to become water vapor in the atmosphere under the impetus of solar energy. Later, condensation to form clouds and precipitation returns the waters to the earth's surface. Finally, under the influence of gravity, the waters are returned to the oceans and the cycle is completed (Fig. 24.1).

The total annual world precipitation has been estimated at about 27,000 cubic miles of water. Precipitation, as rain, snow, sleet, and hail, falls on the earth's surface. This water is dispersed in three ways: some of it is quickly returned to the atmosphere by evaporation, including transpiration from living organisms; another part soaks into the ground to become ground water; and finally the third part of this water is carried by streams and glaciers and constitutes a process called *runoff*.

This runoff is the most important gradational agent on the earth's surface and is consequently the chief factor in erosion. The percent of rainfall that becomes runoff varies greatly in different parts of the earth, depending on climatic conditions, the total amount of annual rainfall, presence and type of vegetation, the degree of slope of the land surface, and the porosity of the rocks. For the earth as a whole it is estimated that about one-fourth of the total rainfall becomes runoff. A hot, dry climate means a much larger percent of evaporation. Lack of vegetation results in rapid and immediate runoff, but a forest delays the runoff and permits a larger percent of the rainfall to soak into the ground. Obviously, water quickly runs down steep slopes as in mountainous country, giving a larger percent of runoff. Porous ground allows much water to enter the ground quickly; for example, the lavas that compose the active volcanoes in Hawaii are so fractured and porous that even with over 100 in. of annual rainfall, there is very little runoff in some areas of the island.

Obtaining the load

PICKUP OF LOOSE MATERIAL

Most of the gradational agents have sufficient impact force when in motion to pick up and carry loose particles. Everyone has seen the wind pick up dust and papers. Running water sweeps along movable material. The size of particles that can be picked up depends on the velocity of the wind, streams, or shore currents. The size of waves or the amount of water in a river are also factors in the size of materials moved. This is forcefully demonstrated by rivers in flood and by storm waves on shorelines. Moving ice in glaciers, on the other hand, picks up and carries loose materials of all sizes – sand or boulders – regardless of the velocity of the ice. Such material becomes frozen into the ice and is then car-

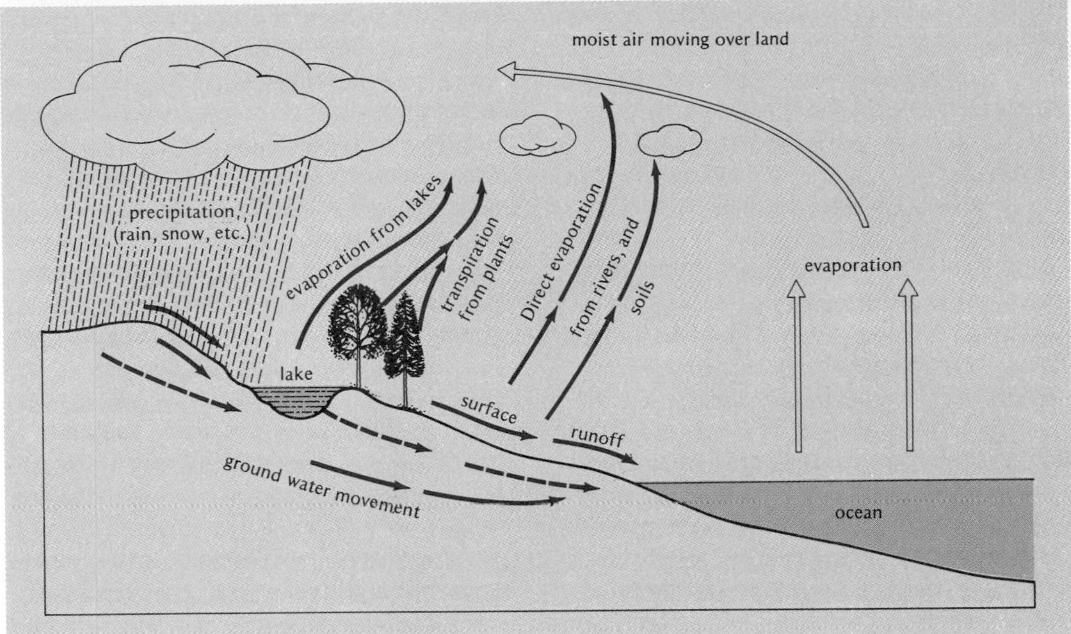

Figure 24.1
Hydrologic cycle. Water is evaporated from the oceans, and moist air blown by the prevailing winds moves over land, then falls as precipitation. This water which is now on the land surface is either (1) evaporated directly from the earth's surface, (2) transpired through vegetation to the atmosphere, (3) returned in rivers as runoff to the oceans, or (4) infiltrated into the ground to move slowly back to the oceans. A small amount of new water is given off occasionally by volcanic eruptions; some water is buried with sediments and becomes trapped for millions of years.

ried along as part of the glacier. The chief limiting factor on how much load a glacier can pick up is the total amount of ice or thickness of the glacier.

ABRASION

All the gradational agents can scrape or scour the surface over or against which they move. This abrasive action is accomplished by solid particles carried as the cutting tools of the wind, water, or ice. Sand grains carried by winds cut away at exposed rock surfaces; the sediment carried in a river scrapes the river bottom; fragments, large and small, frozen in the bottom of a glacier act as the teeth of a giant file to rasp away the underlying bedrock, and sand-laden waves hurled at cliffs along a coastline scour the rocks. In addition to wearing away or eroding the bedrock, these actions add new particles to the erosional agents.

Figure 24.2
A landslide, slump of a section of land at Pacific Palisades, California, in May, 1958. Poorly consolidated sedimentary rocks made up the land mass at this point; in addition, the rock material had been weakened by undercutting by the ocean waves. (Courtesy California Department of Transportation.)

PLUCKING

Moving ice can pick up blocks of rock in a manner different from the other agents. The ice in the bottom of a glacier freezes to the underlying surface with ice in cracks. As the glacier moves slowly forward, a pull is exerted on this underlying surface and blocks that are partly loosened by joints or other cracks are pulled or plucked away to become part of the glacier's load.

SOLUTION

Part of the load of rivers and essentially all the load of ground waters is carried in solution. This includes the soluble products of chemical weathering present in soils. In addition, certain solid unweathered rocks are dissolved as surface waters flow over them and as underground waters move slowly through them. This is true not only of such readily soluble materials as rock salt and gypsum but also of limestones, marbles, and other carbonate-bearing rocks. The presence of dissolved gases, especially CO_2, or organic acids from plant decay makes ground water a much more effective dissolving agent than chemically pure water, as discussed in the previous chapter.

MASS MOVEMENTS

The debris formed by weathering—broken rock and soil—slowly or rapidly slumps downslope under the impetus of gravity.

Eventually the loose material reaches the valley bottoms or sea coast and is delivered to rivers, glaciers, or waves and shore currents to become part of their load. The most spectacular of these gravity actions are the mass movements, or *landslides* (Fig. 24.2), in which great quantities of alluvium and also bedrock suddenly break loose and rush down the mountainside. In high mountains, individual rock fragments and small rock falls occur almost continuously and build up piles of loose rock at the foot of steep slopes and cliffs. These deposits are known as *talus* slopes (Fig. 23.2) and are characteristic features of all high mountain ranges.

The slow slump or *creep of soil* downslope, though not spectacular and at times not even noticeable, except to students of earth science, eventually delivers a tremendous load to the gradational agents. This creep of loose stuff eventually removes far more material from valley slopes than all other types of mass movements combined. Soil creep is slowed by vegetation, the roots of which tend to anchor the soil. The tilted angle of tree trunks on many hillsides, however, indicates that soil creep continues even on forested slopes. Soil creep is speeded by rain wash and by alternate freezing and thawing of the soil in temperate and arctic climates.

The Grand Canyon of the Colorado River in Arizona averages 1 mile deep and 10 miles wide. Only a very small percentage of the total rock that has been eroded was removed by abrasion by the river itself. The vast majority of the eroded rock has weathered from the canyon walls and then has crept, slumped, and fallen step by step, helped along by rain wash, until it finally reached the canyon bottom and could be picked up by the rushing waters of the Colorado River.

Transporting the load

The load of solid particles in rivers, shore currents, and winds is either carried in suspension or is rolled and pushed along. Dust may remain suspended in the air for long periods of time, and fine muds are carried in suspension in river waters. Coarser particles are rolled or dragged along river or sea bottoms. Individual grains or pebbles move by leaps and bounds as they are picked up or pushed by a surge of current, in air or water, only to drop gradually in a long curving path until bounced again by the current. Another part of the load in streams and almost all the load in underground waters is, of course, carried in solution.

The Mississippi River illustrates both the amount of material involved and certain time relationships. The Mississippi discharges into the Gulf of Mexico approximately 500,000,000 tons of rock waste every year, or about 1,500,000 tons every day, as mud, silt, sand, and various compounds in solution in its waters. This is an average of 400 tons/year from every square mile drained by the Mississippi River and all its tributaries, an area of 1,265,000 square miles. At this rate, it takes 9000 years for the total drainage area to be lowered an average of 1 ft in elevation, perhaps not a noticeable change. But if the process continues at this rate for 100 times as long, or nearly 1 million years, the average lowering is 100 ft. The reduction of the surface, however, will not be uniformly distributed but will be concentrated in valleys or on steep slopes; the lowering toward the headwaters of the river system will be considerably more than the average, and toward the sea much less than the average. We can expect, therefore, that some rivers will cut canyons a few thousand feet deep in 1 million years.

VELOCITY RELATIONSHIPS

The velocity of a stream varies widely, depending on the gradient, or slope, of its bed; the shape of the channel; the volume of water the stream is carrying; and the amount of solid load. Because gravity is the motivating force for stream flow, it is obvious that the velocity will decrease with a decrease in gradient. Water will flow much faster through a narrow, deep channel than in a wide, shallow bed, gradient and volume being equal (see Bernoulli effect, p. 157). An irregular, rough-bottomed channel will increase frictional drag and will thus result in a velocity decrease. But perhaps most important of all, as the volume of water in a stream, known as *discharge* (the amount of water passing a given point in a given time), increases after a rainy spell, the velocity is increased.

Most major rivers in humid climates have a gradual gradient decrease downstream; yet, in spite of this, measurements indicate a continuing velocity increase down these rivers. Such a velocity increase must depend largely on an increase in discharge as tributaries bring in more and more water to the main river. The valley shape—width and depth—of the lower part of the main river produces a channel much narrower and deeper than the sum total of the channels of the many tributaries that have combined to form this main stream. Thus there is much less frictional drag on the total amount of water involved in the one main channel than in the many smaller channels.

An increase in stream velocity increases the ability of a stream to erode and transport by the square of the velocity; that is, if the velocity is doubled, the transporting power is increased four times; if the velocity is tripled, the transporting power is increased ninefold; and so on. The *transporting power,* in this case, is usually measured as the diameter of the largest particle that the stream can move or roll along. Because the velocity of streams in flood stage may increase 10 to 20 times their normal velocities, it is easy to understand that boulders several feet in diameter and weighing hundreds of tons are moved downstream.

The total amount of solid load of all sizes that a stream can carry is called its *capacity.* Capacity is also increased greatly by velocity increase but not quite to the same extent as is the transporting power—the mathematical relationships involved are quite complicated. Also, a river carrying a large sediment load will have a lower velocity than a similar river of clear water with no load, other factors being equal.

SORTING ACTION

As a consequence of the velocity-particle-size relationship, running water, wind, waves, and shore currents have sorting power; that is, they can separate fine and coarse particles.

A current of a certain velocity will leave pebble-sized particles but carry sand, silt, and clay particles. A decrease in velocity will result in deposition of the sand-sized grains whereas the silt and clay are carried elsewhere. In this way, rather complete separation of different sediment sizes, important in producing different sedimentary rocks (Chap. 23), is effected.

All the gradational and transporting agents perform this sorting action on solid particles with the outstanding exception of ice. The load of a glacier is carried frozen in the ice and a large boulder is carried just as far and as fast as a clay particle. All this is deposited together when the ice finally melts. Ice has no sorting power.

Depositing the load

The load carried by running water, wind, waves, and shore currents is deposited because of a decrease in velocity of the air or water. Much deposition occurs, therefore, as flood stages in rivers subside. Such deposition takes place in the stream channel or out on the flood plain. Deposition also takes place where the gradient of a stream lessens, as where the stream enters a lake or ocean and builds a delta.

Wave deposition is common along beaches during the waning of high-wave stages. Shore currents deposit when they are slowed by entering deeper water. Simi-

Figure 24.3
A recent gully, Providence Canyon near Lumpkin, Georgia. This small canyon has been formed since the Civil War by running water that originally followed a gully in a cotton field. Even today the canyon is being extended by headward erosion into the farmland in the background. (Courtesy Walter Nickell.)

larly, winds deposit where their force slackens. Condensation of moisture and ultimate precipitation is also an important factor in bringing down dust carried in suspension in the air. Reddish rains and snows have been reported in many parts of the world.

Deposition by glaciers occurs when and where the ice melts, without regard to velocity or slope changes. Consequently, there is no sorting action, unless the melt waters pick up and transport the load beyond the ice.

Deposition by ground waters in pore spaces, cracks, small cavities, or actual caves is the result of minor chemical changes, the loss of carbon dioxide, or evaporation of water. Cave deposits are formed very slowly where water drips from cracks or other openings in the ceiling of caves. The air in caves, however, is usually very humid, often saturated with moisture, so that evaporation cannot occur. This water is normally saturated with $CaCO_3$ or, occasionally, other compounds. It seems likely that as the water droplet hangs on the ceiling a small amount of the dissolved carbon dioxide escapes from solution. The capacity of the water to hold $CaCO_3$ in solution is lessened, and the water is now super-saturated (this is a reversal of the equations given on p. 490); as a result, precipitation of a bit of $CaCO_3$ takes place.

Development of stream valleys

Stream erosion has done more to develop our present world landscape than any other group of processes. Almost everywhere we see gullies, ravines, small valleys, or large canyons as evidence of this work of running water. At times the action is dramatic, as when great floods sweep away bridges and buildings or rock masses fall from the rim of Niagara Falls. But most of the gradational work of running water is done slowly and almost unnoticed in continually repeated activity over long periods of time.

Streams are developed by rain wash. As a sheet of water runs off a surface, it is separated by original irregularities into many small channels, each of which forms a little gully. By successive rains these gullies become deeper. A few of the gullies get more water than others and deepen more rapidly to become small streams. In regions with thick soil cover or poorly consolidated rock, such erosion may proceed with amazing rapidity (Fig. 24.3). In a section of the southeastern United States, small canyons, as much as 200 ft deep, have been formed from gullies in cultivated fields during the past century.

As small gullies flow into larger ones, a pattern of master streams with many tributaries gradually develops to drain the precipitation from a land surface. This stream pattern looks very much like the branches of a tree (Fig. 24.4). Stream pat-

Figure 24.4
Stream pattern. Aerial photograph of gullies and streams on a peninsula of Queen Elizabeth Island, Canadian Arctic. Snow drifts in the gullies and along the stream banks help highlight the branching stream pattern. (Courtesy of the Geological Survey of Canada, photograph no. 110358.)

Figure 24.5
Headward erosion, upper edge of the Badlands, South Dakota. A number of branching gullies developing as small tributaries to a stream system to the left. The gullies are being cut back or elongated at their upper ends into the flat plain, now grass covered, at the right. (Photograph by David A. Rahm; © by McGraw-Hill, Inc. Used with permission.)

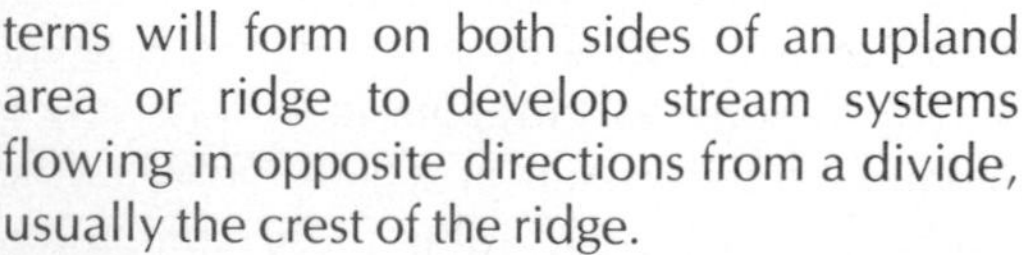

terns will form on both sides of an upland area or ridge to develop stream systems flowing in opposite directions from a divide, usually the crest of the ridge.

HEADWARD EROSION

As water rushes into the upper end, or head, of a gully or stream, its velocity is high, owing to the relatively steep slope, and has great erosive power. This rapid wash, or erosion, at the head of the valley extends the steep banks at the start of the valley further upstream (Fig. 24.5). This is headward erosion and continues until two opposite flowing streams abut against each other at a divide. Even then, one stream may advance headward more rapidly than the other and push or shift the divide in the direction of the weaker stream.

DOWNWARD EROSION

Stream valleys are deepened by abrasion and solution of the bottom of the stream channel. As long as the stream is flowing swiftly and above grade, it has an erosive power sufficient to keep its channel relatively free of deposits, so that abrasion of the rock in the stream bed may continue.

LATERAL EROSION

Stream valleys are widened by undercutting of their banks. As a bank is steepened, slumping occurs, delivering more material to the stream. Stream courses are never straight but are variously curved. The major current of a river tends to continue straight and thus impinges against the outside of a curve. The deepest part of the channel and

Figure 24.6
Meanders. A strongly meandering stream on a river flood plain in Alaska. A long history of lateral swinging is shown by the large number of cutoff meanders and abandoned channels. The concentric ridges record the lateral migration of the stream channel. (U.S. Army.)

the fastest water will be close to shore on the outside of these curves. Here, then, is where undercutting is concentrated. Conversely, the inside of the curve has shallow water and slower current and is usually the site of stream deposition—for example, the sand and gravel bars in most streams.

The cutting on the outside of the curve and deposition on the inside result in a lateral shifting of the stream channel that accentuates the curves of the river and results in a meandering pattern (Fig. 24.6). The predominance of lateral erosion and meandering occurs when a stream is no longer downcutting and is therefore characteristic of stream maturity. The curves become so accentuated that the stream channel finally doubles back on itself, and a curve or meander is cut off and by-passed. The cutoff meander may be rapidly filled or left as a narrow curved lake, called an *oxbow* lake. As a stream meanders, it develops a flood plain, covered by alluvium, several times wider than its actual channel.

Cycles of erosion

BASE LEVEL AND GRADE

Streams are limited in their downcutting by the level of the sea, or of a lake into which they flow. The water level of a river flowing into the ocean obviously cannot be lower than sea level. The bottom of the stream channel may be cut slightly below sea level, but its downcutting is distinctly limited by sea level. This lowest level to which rivers may erode by mechanical wear is called *base level*. The concept of base level envisions an imaginary plain sloping gently upstream under a river or river system and representing the limit to which the stream system can erode downward for a given set of discharge and velocity relationships. For many rivers base level is controlled by sea level, but other streams are limited by the lakes or major rivers into which they flow. Such lake levels are local or temporary base levels. Many rivers in the north central and eastern states and Canada flow into the Great Lakes. The levels of these five lakes are base levels for these rivers.

If the base level remains unchanged, a stream tends to stabilize itself by developing a gradient on which it can just carry its load to the sea, or lake, without further degrading or aggrading its channel. Such a stream is said to be flowing *at grade,* or to have a graded profile. The available energy of the stream, except at time of flood, is just enough to enable the stream to carry its load without further downcutting. If the gradient of the stream were any lower, it would be forced to deposit and thus fill up its channel.

As a land mass is elevated above sea level, weathering and the other gradational processes begin wearing it down. These processes are continually eroding the high places and depositing in the lower places on the earth's surface. Gradation is a cut-and-fill action, on a gigantic scale, toward the development of a monotonous graded surface and the removal of irregularities of the land surface. Geologists speak of a *cycle of erosion,* referring to the successive stages in the wearing down of a land mass to a graded surface or base level.

YOUTHFUL STAGE

The early stage of the erosional processes on a new land area is known as the youthful stage in the cycle of erosion. Such a land area might be a new volcano, a recently uplifted mountain range or plateau, or newly exposed marine or glacial sediments. In the youthful stage, the work of wearing that land area down to base level is just starting; only a small part of the work to be done has so far been accomplished. Streams are not graded but are actively downcutting. Steep-sided valleys or canyons with V-shaped cross section are common, but many interstream areas are still unchanged; for example, in a plateau region, many flat-topped areas still remain. That part of the Colorado River flowing in the Grand Canyon is an outstanding example of a youthful river in a youthful region. Erosion of the Colorado Plateau is in a very early stage; so far erosion has removed only a small percentage of the total rock that must be eroded before the area is reduced to base level (Fig. 24.7).

MATURE STAGE

When the work of eroding a region is well under way, the area is said to be in the mature stage of the cycle of erosion, or to have reached maturity. All traces of any original surfaces have been removed, and a maximum amount of relief exists. Early in the mature stage the flood plain is narrow and the valley walls steep (Fig. 24.8). As time progresses, valleys are widened, and streams meander strongly on extensive flood plains. Deposition becomes more dominant than erosion.

OLD AGE

When the work of eroding a region is largely done, the old-age stage in the cycle of erosion has been reached. Surfaces are graded and near base level; prominent hills or ridges are scarce. A low-graded erosion surface sloping gently upstream from sea level but near base level, formed as the end product of a cycle of erosion, is called a *peneplain.* Many rivers are in the old-age stage today—for example, the lower Mississippi River valley—but no extensive present-day peneplains are known.

The classifications youthful, mature, and old age are applied to regions and to sections of a valley. Major river valleys are

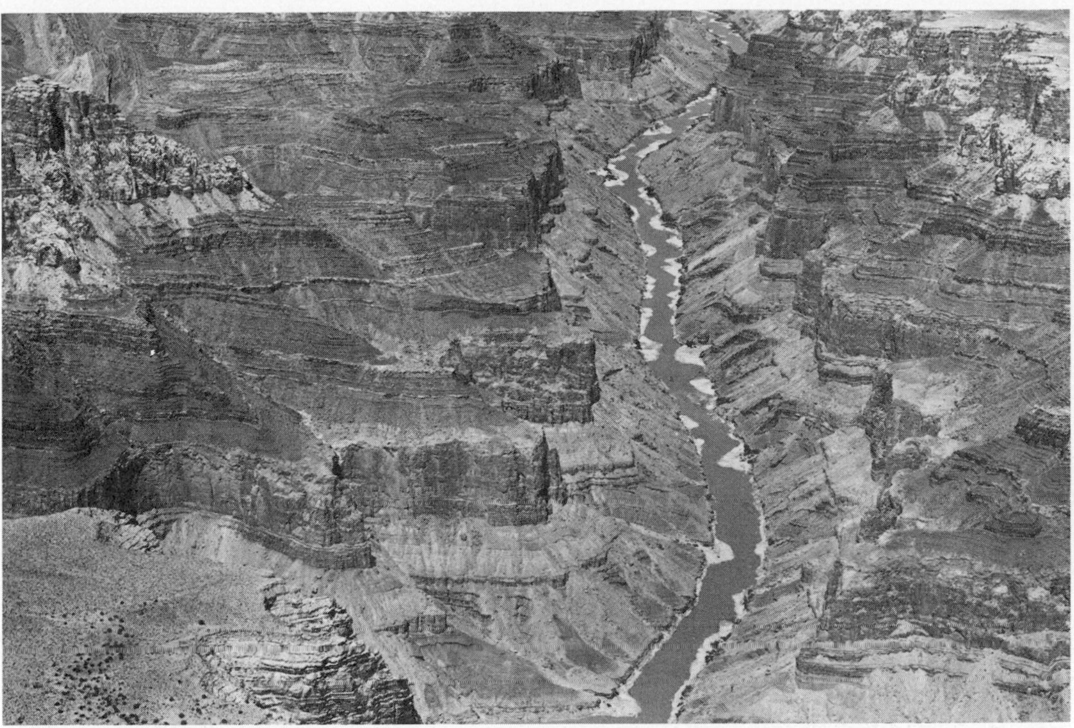

Figure 24.7

Youthful valleys. Top, a canyon of the Colorado River, showing the narrow, steep-walled, and relatively straight valley that is typical of the youthful stage of river erosion. Bottom, the steep, narrow, V-shaped valley of a tributary to the Colorado that is in an early youthful stage. (Spence Air Photos.)

Figure 24.8
Mature valleys. Top, early mature stage in Afton Canyon, near Yermo, California. A stream that has begun to widen its valley and has developed a narrow flood plain. The terrain on either side of the valley shows a maximum amount of relief. As this valley is in an arid region, the landscape has almost no vegetation, the river is intermittent, and its flood plain is choked with deposits. Bottom, late mature stage, with Needles, California in the foreground. The Colorado River flows in a wide valley, with an extensive flood plain that extends a considerable distance to the right of the photograph. Note the meandering course, the abandoned meander patterns on the right, and the amount of deposition in the present river channel. The latter is braided and choked with sandbars, some wet, some dry. (Spence Air Photos.)

not in the same stage throughout their entire length; commonly, their headwaters are youthful (the Missouri headwaters) or mature (the Mississippi between Iowa and Illinois), while the lower stretches of the river have reached the old-age stage. The cycle of erosion implies no absolute time period. The length of time necessary for completion of a cycle depends on the size of the region involved, the amount of original uplift, the character of the rocks, and climatic conditions.

The cycle-of-erosion concept can be applied to the work of all gradational agents, although with many variations. Erosional cycles by ground water are limited to areas of thick carbonate rocks. Wind erosion is not limited by a base level, as wind can scour in a dry area below sea level; for example, there are large, broad wind-scoured areas below sea level in the northern Sahara Desert.

DISTURBED CYCLES

Vertical movements of a land surface may occur before an erosion cycle is completed. An elevation of the land will mean gradually steeper gradients, increasing stream velocities, and renewed downcutting. A stream that has developed a broad, mature valley with an extreme meandering pattern now has the vigorous erosive power of youth. Through downward erosion such a stream begins to develop a narrow V-shaped valley within the old flood plain. The new canyon, however, maintains the tortuous meanders that the stream had assumed; these are known as *entrenched meanders*. The region still has many features of maturity—wide plains and rolling but gently sloping divides. In contrast to this mature pattern are new narrow V-shaped canyons occupied by swiftly flowing rivers. Such a situation represents the beginning of a second erosion cycle superimposed on the first, partially completed cycle. The region is said to be rejuvenated. There are many examples of such entrenched meanders in the southwestern United States, especially in Utah, Arizona, and New Mexico as well as in the Appalachian plateau lands in West Virginia and nearby states.

On the other hand, downward movements of the region may result in partial encroachment of the sea, flooding the lower part of valleys. In this way a pattern of drowned valleys is produced, well exemplified by Chesapeake Bay and its many branches in Maryland and Virginia.

Glaciers

In high mountains and polar regions where more snow falls in the winters than melts or evaporates during the summers, permanent snow fields occur. The *snow line* is that elevation above which this permanent snow exists and where glaciers may be born. The snow line reaches sea level in the northern Arctic and in the Antarctic, but it gradually rises to about 15,000 ft on the equator. Permanent snow and glaciers are found above this level in the Andes Mountains in Ecuador and on Mount Kilimanjaro (19,565 ft) in equatorial Africa. Some melting occurs on the surface on a sunny day, and this water trickles downward to soon refreeze. In time the light powdery snow has become a coarse granular snow known as *névé*. Eventually, these icy grains are pressed and frozen together into solid ice.

The two most important conditions under which snow and ice accumulate to form glaciers are a cool climate and heavy precipitation as snowfall. The first is obvious; winters must be long and summers cool enough so that all snow does not melt before the next winter. But low temperatures alone will not induce the formation of glaciers; rather heavy snowfall is essential. Even with temperatures well below freezing, some snow will evaporate directly without melting. Some parts of Arctic

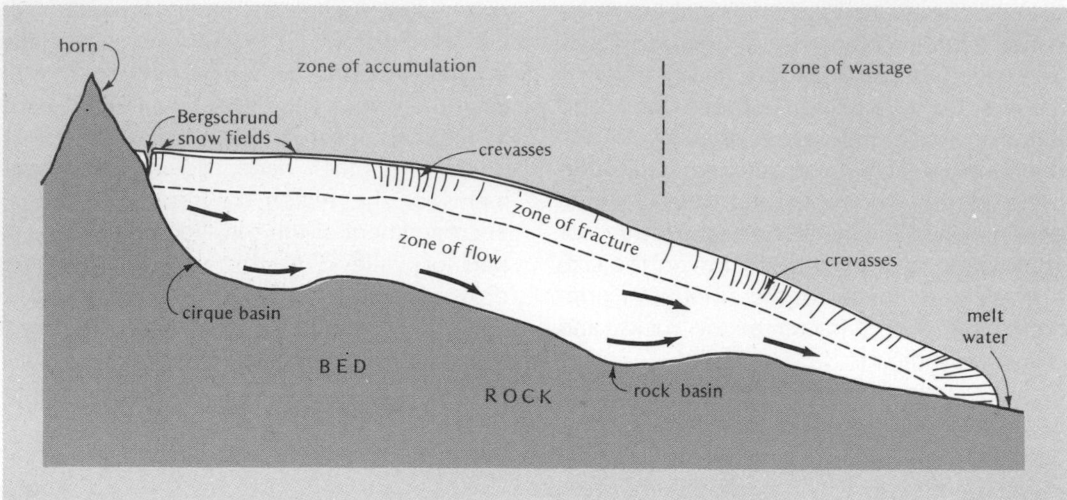

Figure 24.9
Longitudinal profile of a valley glacier. A section down the length of a glacier through the zone of accumulation in the snow fields in a cirque basin at the base of a horn to the lower part of the glacier, down valley, where melting occurs in the summer—the zone of wastage. The zone of fracture is the upper layer of the glacier where ice behaves as a brittle material and is usually about 200 ft thick. Crevasses occur in this zone and are most abundant where the ice surface or the rock mass under the ice changes slope. The zone of flow is at depth where no crevassing can take place because of confining pressure and the ice shows slow solid or plastic flow. The bergschrund is a major crevasse at the head of the glacier formed as the entire glacier pulled away from the rock wall of the horn to move down slope.

Alaska were never covered by glaciers even when glaciers extended as far south as the Ohio River in the central United States. Although this part of the Arctic must have been permanently frozen during the ice age, only a small amount of snow fell. This area was, and still is, a cold, dry region, the Arctic desert.

TYPES OF GLACIERS

Present-day glaciers fall into two major groups: the continental ice sheets of Greenland and Antarctica and glaciers in mountain ranges. A few small ice caps also exist on Iceland, Spitzbergen, and the islands north of Canada.

Ice covers about 10 percent of the present land surface of the earth—about 670,000 square miles of Greenland and nearly 5,000,000 square miles of Antarctica to an average depth well over 1 mile. The Antarctic glaciers spread out beyond the land over the oceans to merge with shelf ice. Both of these great glaciers give birth to icebergs as blocks break off the seaward wall of the glaciers. The Greenland glaciers' icebergs sometimes drift down into the North Atlantic Ocean to harass ships.

The glaciers in the high mountain ranges of the world are called *valley glaciers,* as they are streams of ice that flow down steep-walled valleys or canyons (Fig. 20.3). They reach their greatest developments in Alaska and the Himalayas, where many glaciers are 30 to 70 miles long and over 3000 ft thick. The Canadian Rockies and the Alps contain many famous glaciers

10 to 12 miles long. The Rocky Mountains, the Sierra Nevada, and the Cascade Range in the United States contain hundreds of small blunt ice masses hanging at the heads of steep valleys and hugging the shaded side of a mountain. These are *cliff glaciers,* the remnants of once-larger, typical valley glaciers. Some of these cliff glaciers are hardly more than small permanent snow fields.

Valley glaciers may flow out on a plain at the foot of a mountain range and merge to form a great apron of ice spread many miles from the mountains. Such a glacier is the Malaspina Glacier of Alaska, at the foot of the Saint Elias Range, which covers approximately 1200 square miles. A glacier of this type is called a *piedmont glacier.* During the recent ice age, many of these existed along the foot of the Rocky Mountains both in Canada and the United States.

Valley glaciers are nourished by snow accumulation in basins and flat areas high in the mountain ranges above the local snow line. Here the weight of snow and ice tends to gouge out basins at the heads of valleys and these basins become important snow accumulation areas. Farther down a glacier below the snow line, melting takes place during the summer months and the lower part of the glacier slowly wastes away. The upper part of a glacier may be called a *zone of accumulation* whereas its lower part, below the snow line, is known as the *zone of wastage,* as sketched in Fig. 24.9. If wastage is essentially equal to nourishment, the end of the glacier tends to remain in the same place over a period of time. But if accumulation is faster than wastage, the termination of the glacier moves down the valley and the glacier is said to advance. Piedmont glaciers form because there is an excess of accumulation in the high snow fields. On the other hand, if wastage and melting take place faster than snow can accumulate over several years, the termination of the glacier will be successively farther and farther up the valley—such a glacier is said to retreat; ice movement is *always down* the valley, because it is due to gravity, even though the end of the glacier may be melting away faster than its down-valley movement. Motion may cease in small cliff glaciers and the ice then becomes stagnant.

ICE MOVEMENT

When the amount of snow becomes great enough, the mass will start to move downslope under the pull of gravity, and a glacier has started. Snow fields accumulate on the upper slopes of high mountain ranges, and gradually ice tongues flow down valleys away from these source areas.

On a large, relatively flat land area, such as Greenland or Antarctica, the snow and ice become piled so high that they finally begin to spread laterally even though the rock surface has no regional slope.

Ice is a brittle material. It shatters like glass when struck with a hammer. Yet it obviously flows downhill in glaciers. It is a material with low flow strength; that is, under a moderate weight, it behaves as a plastic material. The surface of glaciers is indeed crevassed and broken (Fig. 24.10), but below a couple of hundred feet, the ice is slowly deformed by the overlying weight and actually "flows" (Fig. 24.9). This flowage is the gliding or slipping of planes of atoms by the formation of microscopic fracture or shear planes across the ice crystals and by the rearrangement of atoms into new crystal lattices.

Glaciers move at widely variable speeds. An average rate is 1 to 2 ft/day, although movements up to 150 ft/day have been recorded. The rate varies with the seasons. One way to prove that a glacier is moving is to drive a straight row of stakes across its surface. In a few weeks, this row becomes a curve, bent downstream in the center where movement is greatest. A classic experiment of this kind was carried out by

Figure 24.10
Crevasses. Highly fractured or crevassed upper part of the Bossons Glacier on the Chamonix side of Mt. Blanc in the French Alps. (W. H. Parsons.)

A. Heim on the Rhone Glacier in 1874. After 6 years, the stakes in the center of the glacier had moved down the valley about 0.5 mile, although stakes near the edge of the glacier had moved only a few hundred feet. The motion of most glaciers carries them well below the snow line.

Ground water

The interstices between particles of soil and the pores and tiny cracks in bedrocks contain water, which is important to plant life as well as to man. Almost all this ground water seeped into the ground from rainfall and may be called *meteoric* water. When sediments are deposited, seawater fills the spaces between grains. Some of this water becomes trapped when sedimentary rocks are formed and may remain for millions of years. This is *connate* water; it is the salt water so commonly encountered as wells penetrate deep into sedimentary strata. In some volcanic regions a small amount of water has escaped and risen from cooling magmas; this is known as *magmatic* or *juvenile* water.

All rocks near the earth's surface have some openings. Many sandstones have a porosity of 10 to 20 percent of their volume; lavas may also be very porous. Most ground water exists in these pores; underground streams and lakes are a rarity. For ground water to flow, these pore spaces must be interconnected; that is, the rock must have permeability.

Porosity determines the storage capacity of a rock, and permeability is the transmission factor. Openings must not only be interconnected but large enough to permit the movement of water. Water will not move in rocks with high porosity if the pores are entirely separate, as in pumice, or if the pores are too small, as in clays and shales.

WATER TABLE

Ground water moves down toward a surface below which all openings are saturated with water. This is the ground *water table*. Above the water table, openings are partly filled with air; this is called the *zone of aeration*. It is in contrast to the zone of saturation, below the water table. Water trickles downward through the zone of aeration, but below the water table it tends to move horizontally though also downward, as gravity is the motivating force for most ground water movements. If no rainfall occurred, the water table would eventually approximate a plane surface in this response to gravity.

Water moves downslope even underground. In this way, the water level can continue to rise in lakes and valleys even though wells go dry in neighboring hills where the water table is falling. Water is moving downward from under the hills to a place under the lowlands. But the addition of new water from rains and melting snow

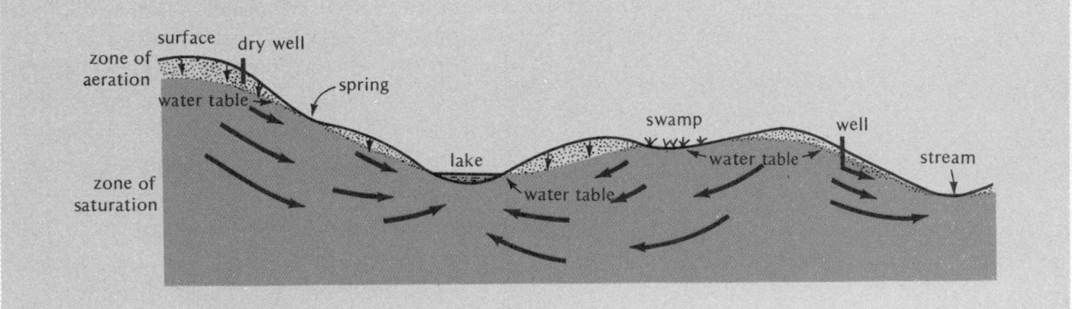

Figure 24.11
Water-table relationships. The water table rises gently under hills, joins with the water surface of lakes, swamps, and streams, and separates the zone of aeration from the zone of saturation. Water moves slowly down through the zone of aeration as indicated by tiny arrows. Water moves horizontally but gradually downward toward lakes and streams in the zone of saturation, as indicated by the long sweeping arrows. Some ground water moves into the lakes and streams to keep their levels high even in dry weather. A spring occurs where the water table intersects the surface. A well sunk below the water table will fill with water to the water-table level. A well that does not reach the water table is a dry well.

causes a water table rise under hills. The water table usually merges with the surface of lakes, rivers, and swamps. These relationships are summarized in Fig. 24.11.

Wind

The geologic work of wind, the atmosphere in motion, is known as *wind erosion*. Wind has the power to pick up and carry large quantities of finely divided materials, such as sand and dust, for hundreds or even thousands of miles. We have all seen the wind blow up a cloud of dust on a street corner; we have also read about dust storms in our southwestern states when the sky has become hazy and murky a thousand miles to the east. In arid areas dust storms can completely darken the sky at noontime. Desert dust storms are a real danger to travelers; even in our own Southwest, the windshield of a car can be quickly "frosted" (pitted) by sand in such a storm. Occasional winds of extreme velocity, as in hurricanes and tornadoes, can move very large and heavy objects. But it is the winds of moderate velocity, common in all parts of the world at all times, that are important as the agents of wind erosion.

In arid or desert areas the loose surficial materials are blown away, and in time the land surface is appreciably reduced in elevation. More than three-fourths of all desert surfaces are rock- or boulder-covered, all soil and sand having been removed. This aspect of wind erosion is known as *deflation*. The process is not limited to desert areas but

Figure 24.12
Wind-blown dust on an abandoned farm in Oklahoma in 1937. Material carried by the wind has been deposited on the fields, forming a loose, moving cover almost impossible to cultivate. (From R. M. Garrels, *A Textbook of Geology,* Harper & Row, 1951; U.S. Department of Agriculture.)

may occur wherever winds have access to loose, dry materials with no protective cover of vegetation. In humid regions, on the other hand, the moisture in the soil effectively binds fine particles against the wind.

The extent to which desert areas can be deflated is best appreciated by noting the amount of material deposited by wind elsewhere. After one dust storm in the farm lands of the Southwest, it was estimated that 125 tons of dust had fallen on each square mile at a distance of 500 miles from the apparent source, as graphically illustrated in Fig. 24.12. Dust storms originating in the Sahara Desert of North Africa carry much dust across the Mediterranean Sea to southern and central Europe. Careful study of a 4-day storm in March, 1901, originating over North Africa, indicates that dust fell, some brought down in rain, in a sheet 0.25 mm thick over central and southern Europe for a total of approximately 2 million tons. It is obvious that wind is an effective agent in transporting earth materials from one part of the world to another.

Another aspect of wind erosion is *abrasion,* or scour, a natural sandblast action. The wind employs as its tools sand grains and silt particles that are blown against exposed bedrock, or pebbles and boulders on the ground. As a result of the impact of the sand grains, additional particles are chipped off the bedrock or boulder. In the arid regions of the Southwest, wooden telephone poles are sometimes cut through near the base by this natural sandblast in a surprisingly short time. Rock surfaces become scoured, grooved, and even roughly polished. Many odd-shaped land forms are developed, including undercut hills, mushroom or pedestal rocks, and deep hollows in loose or easily eroded materials.

Most of the dust carried by the winds finally settles out of the atmosphere or is brought down by precipitation to become a part of the soil in humid regions or to settle to the bottom of the oceans. The coarser particles, sand, accumulate on the lee side of the desert regions, or inland from coastal areas, as sand dunes.

Waves and shore currents

Most of the surface waves of the ocean are the result of wind action churning up the surface of the water. Such waves vary greatly in size and velocity. The size of surface waves depends on the wind velocity, on the duration of time during which the wind continues to blow, and on the *fetch,* or the open distance over which the wind has a sweep. Wave velocity is the wavelength divided by the period of the wave. The wavelength (see Chap. 9) is the distance between two consecutive wave crests, and the wave period is the time necessary for two consecutive crests to pass a given point.

When a great many waves are formed in a storm center, the interaction of waves of many sizes and velocities produces what is known as a rough "sea." The longer, faster waves outrun the smaller waves and move far beyond the storm center, perhaps for thousands of miles across an ocean, to become regularly spaced, uniform waves called "swells." These waves move into a coastline in good weather far from any storm center and long after the storm has dissipated.

As waves move toward shore, where ocean depth decreases, wave velocity is decreased and this in turn causes a decrease in wavelength. Finally, the wave steepens, overturns, and breaks with a rush up the beach. The lower part of the wave in effect drags on the bottom to hasten the steepening and overturning of the crest and produce the breakers so characteristic of ocean coastlines.

Breakers are characterized by an uprush of water onto the beach followed by a backwash as the water runs back into the sea. The return motion of the water has sometimes been called "undertow," a term that is quite misleading, as there is really no down drag by such water. Along some coastlines at points often related to channels of deeper water, a concentration of this return water may produce a seaward current, called a *rip current*.

The fall of breakers on a beach or their pounding on a sea cliff produces great erosion by actual impact force and by abrasion on the rocky cliffs. Loose sediment is carried out into deeper water by the backwash and rip currents. When a series of breakers approaches a shoreline at an angle, the return of the water—backwash and rip currents—produces a longshore current moving laterally parallel to the shoreline. These currents are very effective as sediment-transporting agents and are responsible for the lateral movement of beach sands and other sediments along the coast.

Erosional landscape forms

Land forms are developed as a result of erosion and deposition by water, wind, ice, and waves. Many of these are characteristic of the particular agent by which they are produced; for example, deltas are features produced by deposition at the mouths of rivers. Many other land forms, however, depend not only on the erosional or depositional process involved but also on the stage in the erosion cycle to which the process has progressed and on the rock structure of the area being eroded. The shape of a valley depends not only on whether it was cut by a river or a glacier but also on whether the valley is youthful or mature—that is, on how far the erosional processes have progressed. The ridge or divide between two valleys has a shape that reflects not only the erosional agent and the stage in the erosion cycle but also the rock structure; for example, the divide may be flat topped because it is composed of flat-lying rocks, or it may be a sharp ridge of steeply tilted rocks.

STREAM EROSION

Youthful stream valleys have a general V-shaped cross section, or profile, whether they are small gullies or great canyons (Figs. 24.3 and 24.7). Such valleys are narrow and steep sided, with steep gradients characterized by rapids and waterfalls. If such valleys are cut in horizontal rocks, their slopes may be steplike, as illustrated in the Grand Canyon, owing to the differential resistance to erosion of hard and soft layers.

Mature stream valleys are wide with gently sloping sides (Fig. 24.8, bottom). Lateral erosion by the stream cuts a flood plain several times wider than the actual stream itself. The river meanders on this flood plain. The Ohio River and the Mississippi north of Cairo, Illinois, are examples of large mature valleys of this type. South of Cairo, the valley of the Mississippi gradually widens and assumes more and more the features of a late mature or old-age valley. In the lower part of this valley are extensive swamps back of natural levees. The landscape forms here, however, are mostly of depositional origin.

Figure 24.13
Glaciated valley, Yosemite Valley and Bridal Veil Falls, California. A typical U-shaped valley, with its nearly flat floor and very steep sides. The tributary valley coming in from the right and ending in a waterfall is a hanging valley. (Courtesy Teledyne Geotronics.)

ICE EROSION

Ice tongues move down valleys in high mountainous areas and gouge out steep-walled but relatively flat-bottomed, straight valleys with a *U-shaped profile* (Fig. 24.13). These land forms are distinguished from youthful stream valleys by their flat bottoms but from mature valleys by their extremely steep, precipitous sides. *Fiords* are over-deepened ocean-filled U-shaped valleys; scenic examples of this land form in Norway and Alaska are well known to all. Tributary valleys enter the main stream as *hanging valleys* (Fig. 24.13), often with impressive waterfalls. Preglacial streams enter a master valley with a steep but relatively uniform longitudinal profile, essentially at grade. But after glaciation has widened and deepened the main valley, the tributary streams enter this master valley with an abrupt steepening of their longitudinal profiles. This relationship is called a "hanging valley."

The head of a stream valley is typically a small narrow ravine down the side of a mountain. By the plucking action of glacial erosion this narrow valley is enlarged into an amphitheaterlike form bounded on three sides by steep and often vertical walls; such a basin is known as a *cirque*. Here snow accumulates, and the glacier starts its downstream movement (Fig. 24.14).

The divides or ridges between ice-carved valleys are narrowed by the widening of both valleys and left as thin, steep-walled, and very jagged ridges called *arêtes*, often

Figure 24.14
Glaciated mountains, Mt. Assiniboine, Canadian Rockies. So-called alpine scenery, as developed by intense glacial erosion. The sharp peak in the center is a horn, and the sharp ridge running from there around to the right and back to the foreground is a typical glaciated ridge or arête. The small, steep valley in the center foreground is a cirque; two other cirques are present, with small snow fields, just below the horn. (Courtesy of the Geological Survey of Canada, photograph no. CA114-23.)

with steep isolated peaks called *horns* (Fig. 24.14). The latter have been carved by ice on at least three sides, and each of these sides is the headwall of a cirque. Mountain ranges that are modified by ice erosion are generally more scenic than mountains produced by stream erosion only. The glaciated landscape, so-called alpine scenery, is characterized by long straight, flat valleys with very steep ridges and jagged peaks. This is typical of the Swiss Alps, the Canadian Rockies, the high Sierras of California, and the mountains of Glacier National Park, Montana.

On land areas that have been completely covered by ice, the hills and ridge tops are scoured to an even and sometimes streamlined surface (Fig. 24.15). The scouring is done by rock debris frozen rigidly in the ice and commonly produces a striated or grooved surface, as exemplified by the hills and rounded mountains of New York, New England, and most of eastern Canada, which were covered by thick ice of continental glaciers in recent geologic time.

WAVE EROSION

Waves erode the shorelines of lakes and oceans and gradually help to wear away the land surface. The most conspicuous land

Figure 24.15
Glaciated surface, a typical New England scene. A smoothed and striated rock surface that is the result of ice scour. (From R. M. Garrels, *A Textbook of Geology,* Harper & Row, 1951; U.S. Geological Survey.)

Figure 24.16
Wave-modified coastline, Emerald Bay, Laguna, California. Wave-cut cliffs on the projecting headlands and a few offshore rocks as a result of wave erosion. Small, typically curved beaches between headlands as a result of deposition by waves and longshore currents. (Spence Air Photos.)

form is the *wave-cut cliff,* or sea cliff. In loosely consolidated rock this is a steeply sloping bank, but in well-indurated rock it may be a vertical cliff hundreds of feet high (Fig. 24.16). The wave-modified shoreline is very irregular at first, but as erosion progresses toward maturity, the shoreline is straightened. As the cliff is gradually pushed shoreward by continuing erosion, a flat terrace sloping slightly seaward is produced just below water level at a depth where wave action becomes effective. Subsequent rise of land or withdrawal of the sea may eventually expose this *wave-cut terrace* as a nearly level plain. Sediment is often deposited on the terrace by wave action to produce beach deposits. The main beach is usually above the high-tide level deposited by major storm waves. There is also deposition of sediment in the zone between high- and low-tide levels and in the zone offshore. If the slope to the sea is gradual, deposition may form an offshore bar running parallel to the shoreline. Occasionally this bar is built above high-tide level and is known as a barrier bar. The relationships of a representative shoreline are shown in cross section in Fig 24.17, which illustrates both erosional and depositional features.

Wave erosion is one of the major geologic hazards along lake and ocean coastlines. The danger is especially great where the rock material is unconsolidated or only poorly consolidated and is therefore easily washed away by wave action. Most of the damage or loss of material takes place during major storms when waves are exceptionally large and powerful. Waves can severely undercut a steep coastline and allow large landslides to occur as we have seen earlier along the southern California coast (Fig. 24.2). At the present time (1972–1973) the Great Lakes are at a high-water stage and wave erosion has been greatly accelerated and caused mil-

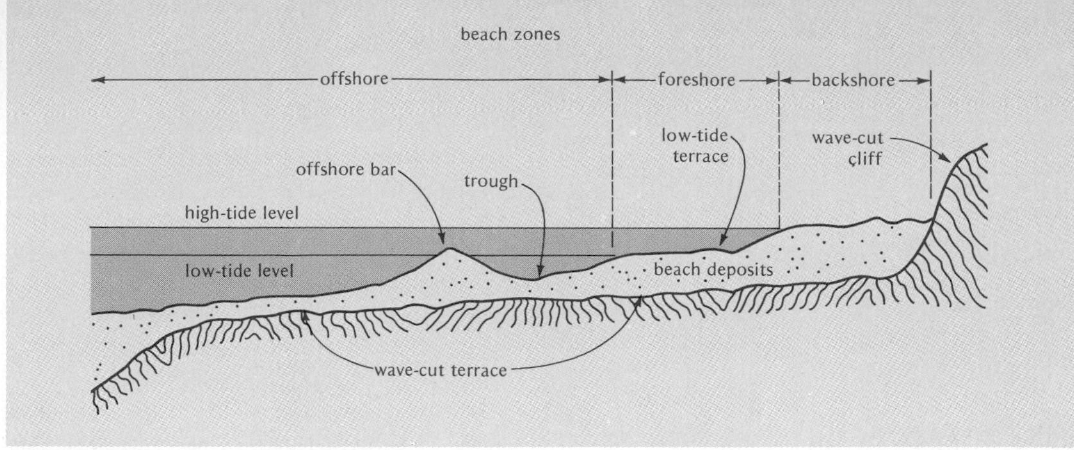

Figure 24.17
Profile of beach zone, or shoreline. Note relationships of wave-erosion features (wave-cut cliff and terrace) and beach deposits, with terminology commonly used for the principle divisions of the beach zone.

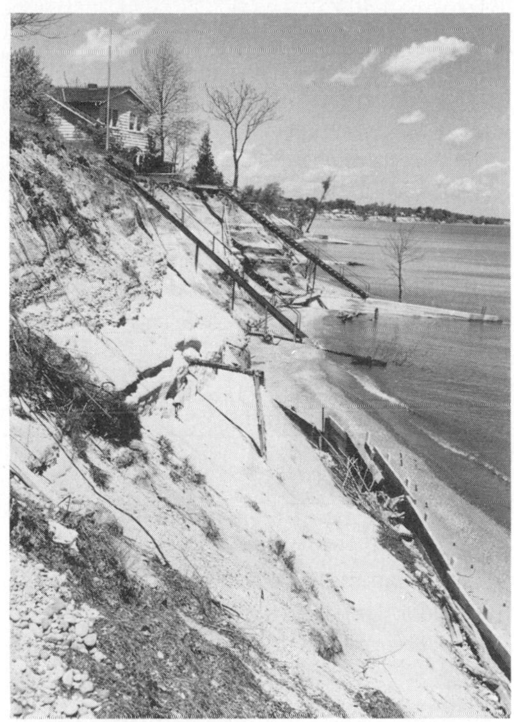

Figure 24.18
Wave erosion. A so-called wave-cut cliff cut by storm waves in unconsolidated glacio-fluvial material along the shore of Lake Michigan. Note the stratification in the glacial sand and gravel deposits. (Photograph by the Detroit News.)

lions of dollars of damage to water-front homes. The unconsolidated glacial sediments and dune sands that comprise much of the land along Lakes Michigan, Huron, and Erie are easily washed away by large waves (Fig. 24.18). On some of these shorelines, waves have eroded back into the land more than 5 ft in one storm. Some homes in the most exposed locations that were built as much as 200 ft from the sea cliff now stand on the brink ready to topple into the lake. The cost of adequate sea walls is prohibitive when hundreds of miles of shoreline are involved.

Depositional landscape features

Many landscape forms produced by the deposition of sand and other sediment are so well known as to need no special mention here. Sand dunes are common features on the sandy shorelines of large lakes and the oceans and in parts of most deserts. Their development depends only on the abundance of a supply of sand. Large, flat, swampy deltas are typically present at the mouths of the world's rivers. And we are all familiar with sand and gravel beaches

Figure 24.19
Alluvial fans, Caliente Range with Cuyama Valley in foreground, California. Streams emerging from the mountains have deposited their load of coarse sediment to build several overlapping alluvial fans. Note that the alluvial deposits have buried the lower ridges of the mountains. Note also that fan deposition has forced the course of the river in the foreground out from the mountains. (Spence Air Photos.)

at certain places on shorelines due to deposition by wave action.

STREAM DEPOSITION

Deposition of sediment by streams occurs when the velocity of the running water decreases. This occurs as volume lessens during low-water seasons or because of extreme evaporation conditions in arid areas (Fig. 24.8). A velocity drop may also occur because of change in gradient at the mouth of a stream or at the foot of a mountain range. Sand and gravel bars form in stream bottoms and along their banks in low-water periods (Fig. 24.8, bottom). These are especially common on the inside of curves or meanders. The main current is thrown against the outside of the curve. On the inside of the curve, on the other hand, velocity is low, and deposition takes place.

The entire flood plain becomes covered with deposits as flood waters spread over this area and then recede. Deposition is most pronounced near the stream channel. Therefore the highest ground on the flood plain will be close to the stream bed. These high banks are known as *natural levees* and

Figure 24.20
Moraines, Ellesmere Island, Canadian Arctic. Ice deposits surrounding the margins of a glacier. Terminal moraine at end of glacier and lateral moraines along each side. Note river of meltwater crossing terminal moraine deposits. This glacier occupies a U-shaped valley and originates from several cirque basins in the background. (Royal Canadian Air Force.)

are especially common in late-mature or old-age stages of the erosion cycle. Along the Mississippi River in Louisiana, the only really "high and dry" ground is the narrow strip, perhaps $\frac{1}{2}$ mile wide, along either side of the river.

The growth of deltas is perhaps the most obvious example of river deposition. The banks of the chief distributaries at the mouth of the Mississippi River are being pushed into the Gulf of Mexico at the rate of about 1 mile in 16 years. The Po and other rivers have filled in the northern end of the Adriatic Sea in the last 1800 years to build a strip of land 20 miles wide; the city of Adria, once a seaport, is now 14 miles inland.

As streams emerge from mountain ranges onto plains or interior basins, especially in the arid southwestern part of the United States, an abrupt velocity drop means deposition of sediment—coarse gravel and sand—that builds a semicircular feature called an *alluvial fan* (Fig. 24.19). Although these features occur in all parts of the world, they are especially well developed in semiarid mountainous regions. The many wide basins in Nevada and western Utah are filled to considerable depth with stream-deposited sands and gravels and are bordered by large, well-developed alluvial fans, often several miles across. Following a heavy rain, waters run out into the center of basins to collect as temporary lakes. As these evaporate, deposits of silt and mud accumulate to help fill the basins.

GLACIAL DEPOSITION

As ice melts, its load is deposited; some is modified by the meltwater. All these deposits are known as *glacial drift*. Some of the drift deposits are concentrated at the margins of the ice sheet or glacier (Fig. 24.20), and others are scattered over wide areas, following stagnation and final melting of glaciers. These deposits are formed of unsorted debris of all sizes, a rock material best known as "boulder clay" and named *glacial till* by geologists (Fig. 24.21). On the other hand, some of the drift deposits are carried beyond the ice margin by the meltwaters and become partially water-sorted and stratified. Such glacio-fluvial materials are mostly sands and gravels. In many localities these have important economic value as the source of commercial sand and gravel.

The most important till deposits formed along the edge of glaciers; these are known as *moraines*. In association with valley glaciers, moraines may be long ridges, formed at the sides of the ice tongue, parallel with the valley (Fig. 24.20). Such lateral moraines vary from a few feet to a few hundred feet high; they are narrow but sometimes miles long. The terminal moraine, at the end of the glacier, is more irregular in shape—a series of low hills often acting to block the valley partially and form a dam to produce long lakes in the U-shaped glaciated valleys after the ice has gone.

At the margins of the former continental glaciers in the northeastern United States, moraines that are significant landscape

Figure 24.21
Glacial till, Beartooth Mountains, Montana. Road cut in moraine showing typical boulder clay material known as *till*. (W. H. Parsons.)

features were developed. These moraines today are very irregular hilly belts up to a few miles wide and traceable for many miles, even a few hundred miles; the relief of such hilly topography is characteristically a few hundred feet or less.

The most important land forms built up with meltwater deposits are *outwash plains*. These occur downstream beyond end moraines in valleys or spread outward from hilly morainal belts often for a few miles. The outwash plain is a relatively low, flat feature compared with the neighboring moraine. These plains are usually characterized by scattered, shallow, undrained depressions called *kettle holes*. The latter may be partially filled with water to form small lakes or ponds. These depressions are small—a few hundred feet to a mile wide by only a few tens of feet deep. Kettle holes are formed where large isolated masses of stagnant ice remained at the time of deposition around them. Subsequent melting of such ice after the outwash or moraine had been deposited resulted in the hollow or depression.

GLACIAL LAKES

The Great Lakes are a result of the continental glaciation over North America in the Pleistocene period. Broad, late-mature stream valleys existed before the glacial times. These lowlands were greatly overdeepened by ice scour; some basins were dammed by glacial deposits in the former valleys, and the land surface was perhaps partially differentially downwarped as a result of stresses associated with the weight of ice on the crust. As the last glacial advances gradually began to recede about 15,000 years ago, water was dammed against the ice margins in the vicinity of Chicago and in northern Ohio to form glacial lakes. Ice blocked the normal outlet of waters to the east through the present Lake Erie basin and the St. Lawrence valley. The water accumulated, therefore, until it overflowed natural divides to the south across Indiana near Fort Wayne and across Illinois south of Chicago into the Mississippi River system (Fig. 24.22a). These early ancestral great lakes had levels more than 200 ft above the present lakes. The ice continued to melt and retreat, and lake levels were lowered as lower outlets were successively uncovered by the melting of the glaciers. These later lake stages were larger and larger in surface area as the lakes expanded northward against the retreating ice margins.

The outlet of early Lake Maumee in the Fort Wayne area was left dry as water found a lower outlet route across Michigan to Lake Chicago; thus Lake Whittlesley was formed (Fig. 24.22b) perhaps 11,000 years ago. Later, ice retreated sufficiently in New York to expose the Niagara Escarpment and the famous falls were "born." A lower outlet for the glacial meltwaters was established

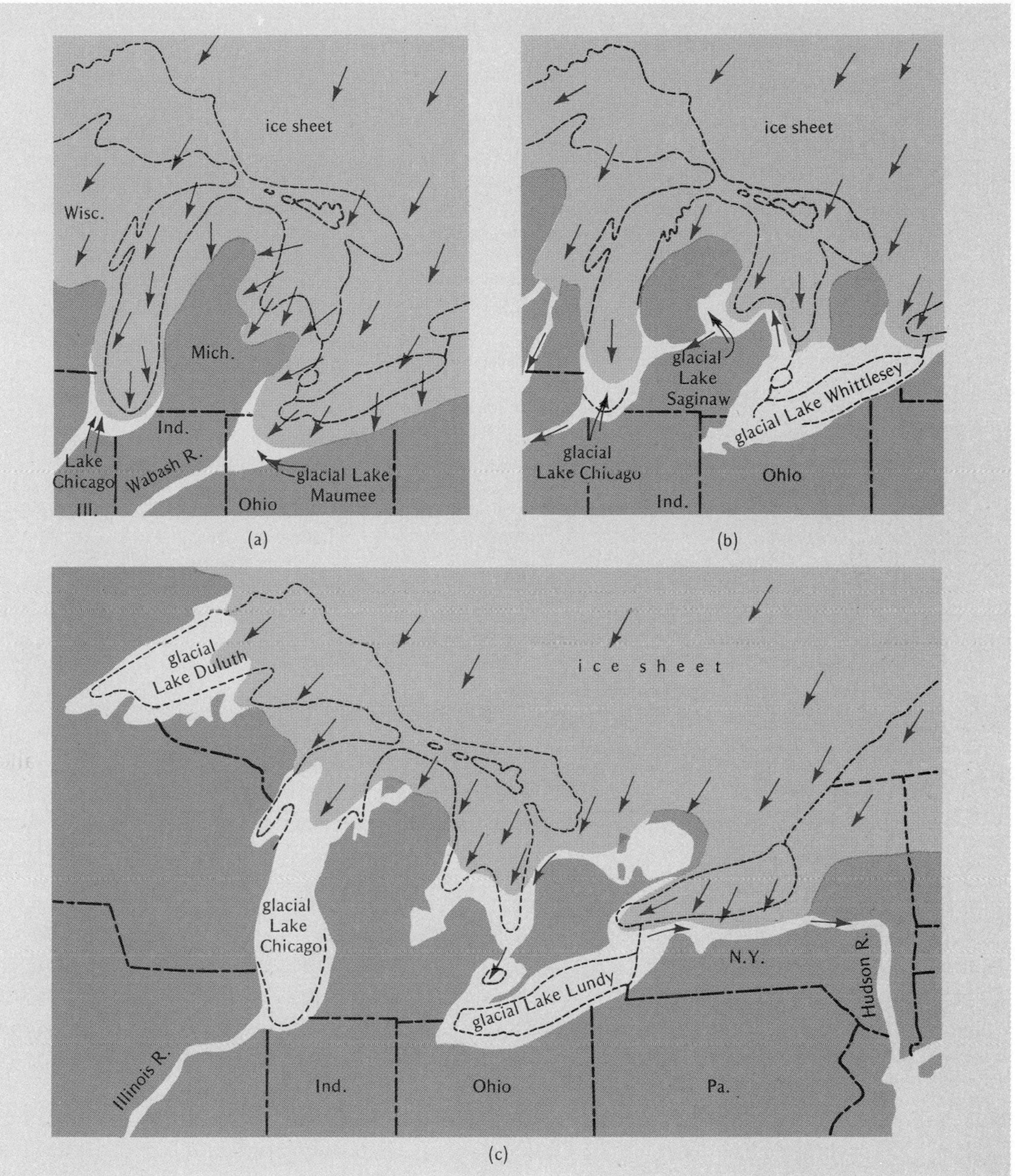

Figure 24.22
Three stages in the development of the Great Lakes; only three of the many stages in the development of lakes against the slowly retreating ice margin during the last 15,000 years. The lakes overflowed at various places at different times. At first (a) drainage was into the Mississippi River via the Illinois and Wabash outlets of glacial Lakes Chicago and Maumee; later (b) all drainage occurred via the Illinois River as glacial Lake Whittlesley drained across Michigan to expanding glacial Lake Chicago; and finally (c) the eastern lakes drained across New York State into the Hudson River. The lakes are given different names from the present Great Lakes because they stood at higher levels, as determined from their old beach deposits. (After Leverett and Taylor, U.S. Geological Survey.)

Table 24.1
Work done by gradational agents and the land forms produced

agents	load, how obtained	load, how transported	load, where and why deposited	erosional land forms	depositional land forms
rivers	impact force: pickup of loose material; abrasion, scour; dissolving action; mass movements: soil creep and landslides deliver load to stream bed	in suspension; bed load: rolled, dragged, pushed, or bounced along bottom of stream; in solution	decrease in velocity when high water subsides, slope changes, stream enters lake or ocean; loss of water by evaporation; loss of water to ground supply	V-shaped valleys, canyons, gullies; wide valleys with flood plains; peneplains; divides: flat-topped (buttes) or sharp ridges	flood-plain cover; terrace deposits; bars in streams; natural levees; deltas; alluvial fans and basin fills
wind	deflation; abrasion, scour	in suspension; rolled, dragged, or bounced along	decrease in wind velocity; brought out of atmosphere by precipitation	deflated stony areas and basins in deserts; small hollows in dune areas	sand dunes
glaciers	impact force; abrasion, scour; plucking action; mass movements: landslides of debris onto top of ice	carried frozen in ice and on surface; all sizes carried at same velocity; no sorting action	when and where ice melts; some material carried beyond ice margin by meltwaters	U-shaped valleys, fiords; hanging valleys; cirques; jagged ridges, horns; with continental ice, rounded hills	glacial drift; till: moraines, terminal and lateral (hilly); glacio-fluviatile: outwash plains
ground water	chemical action and dissolving of soluble compounds	in solution	precipitation due to chemical changes in solutions; evaporation	caves; sink holes	cave deposits: dripstone, stalactites, and stalagmites; vein deposits
waves and shore currents	impact force; abrasion, scour; dissolving action; mass movements from the shore	in suspension; rolled, pushed, bounced, tossed; in solution	decrease of wave height; decrease in velocity of shore currents	wave-cut cliffs; wave-cut terrace; offshore rocks (sea stacks) and sea caves	beaches; sand bars

along the foot of this escarpment across New York State through the Mohawk valley to the Hudson River and thence to the Atlantic Ocean (Fig. 24.22c); this new lake has been called glacial Lake Lundy and came into existence about 9000 years ago.

Finally, most of the ice had melted from the immediate area of the Great Lakes some 7000 to 8000 years ago and an outlet through the present St. Lawrence River was established for Lakes Erie and Ontario. The upper lakes, however, had an outlet through an expansion of the present Ottawa River valley. Continuing adjustments of land, and thus lake levels, to the loss of weight of ice were responsible for minor changes in lake levels, so that the present lake shorelines were established as we know them today only about 3000 years ago. This sequence of lakes against the ice margin has been worked out from a study of the old shorelines of these lakes, which are now exposed from New York to Illinois. Each successive major lake stage lasted for a few thousand years, so there was time for the development of recognizable shorelines. Several wide valleys, now dry or containing only small rivers, indicate the location of the once mighty outlet rivers of the various lake stages to the south and east.

The largest of all the glacial lakes is named Lake Agassiz after the founder of the glacial theory. It started developing in northeastern South Dakota while ice still covered central Canada. As the ice margin retreated, Lake Agassiz expanded through various stages and outlets—the latter at first to the Mississippi, later into Lake

Superior, and finally into Hudson Bay. At its maximum, Lake Agassiz stretched northward from South Dakota for 800 miles and covered parts of North Dakota, Minnesota, and Saskatchewan and at least half of Manitoba. It had a surface area greater than that of the present Great Lakes combined.

Colder climates meant increased rainfall during later glacial times in now semiarid areas such as Nevada and Utah. Extensive lakes existed in these areas contemporaneous with ice to the north and east, although with no physical contact. Such glacial-period lakes in now arid areas are known as *pluvial lakes*. Nevada contained many freshwater lakes, the largest being Lake Lahontan, and the state had more water than Minnesota has today. The largest of all the pluvial lakes in North America was Lake Bonneville, which covered most of the western half of Utah and extended into Nevada and Idaho. At its maximum stage, about 13,000 years ago, this lake was 1000 ft deep and had its outlet into the Snake River to the north and thence via the Columbia River to the Pacific Ocean. Present Great Salt Lake is a small remnant of this recent giant freshwater body.

SUMMARY

Stream valleys develop and pass through a cycle of erosion as rivers erode the earth's surface to first form V-shaped valleys and canyons, then wide valleys with flood plains and meandering streams, and finally old-age rivers on peneplains. Erosion at first is headward and downward; later it is mainly lateral.

Glaciers are present today over Antarctica and Greenland as continental ice sheets and in many mountain ranges as valley-filling ice tongues. During the past ice age continental glaciers covered northern North America, and valley glaciers were very extensive in most mountain ranges. Hanging valleys, U-shaped valleys, cirques, jagged arête ridges, and horns in ice-free mountains are the evidence for valley glaciers; smoothed, rocky hills and extensive deposits of glacial till and glacio-fluvial material for continental glaciers.

Extensive glacial lakes formed against the retreating ice margins in the Great Lakes region and in the north-central plains of the United States and Canada. The various stages of these lakes can be reconstructed from their shoreline features and deposits. Pluvial lakes developed in glacial times in now semiarid Utah and Nevada.

The work done and land forms produced by the various gradational agents in developing the present landscape of the world are summarized in Table 24.1.

Important words and terms

hydrologic cycle, runoff
obtaining the load; pickup, abrasion, plucking, solution
landslides, soil creep
transporting the load: discharge, gradient, velocity
transporting power, capacity
sorting action
depositing the load
headward, downward and lateral erosion
flood plains, meanders
base level, at grade
cycle of erosion: youthful, mature, old age
peneplains
disturbed cycles: entrenched meanders
glaciers: valley, piedmont, ice sheets
snow line, névé, crevasses
zones of accumulation and wastage
zones of fracture and flow
advance and retreat of glaciers
meteoric, connate, and magmatic waters
water table: zones of aeration and saturation
wind erosion, deflation
waves and shore currents
erosional landscape forms
V-shaped valleys, U-shaped valleys, fiords, hanging valleys, cirques
horns, arêtes, alpine scenery
wave-cut cliffs and terraces
depositional landscape forms
gravel bars, natural levees
deltas, alluvial fans
drift, till
moraines, outwash plains

glacial lakes
beaches, offshore bars

Questions

1. Compare the transport of sediments by wind, water, and ice.
2. How would you expect the factors of velocity, volume of water, and gradient to vary down the Mississippi River from Minnesota to the Gulf of Mexico?
3. Which erosional agent can cut valleys hundreds of feet below sea level? Explain.
4. What factors influence the amount of runoff in a particular area?
5. Where and why are sand and gravel beds deposited by running water?
6. Discuss a possible cycle of ground-water erosion.
7. Where, in general, are the most likely places to locate successful water wells in your vicinity? Why?
8. How does ice modify a mountain landscape?
9. What kind of erosional and depositional features would you expect to find as evidence of former glacial-lake shorelines?
10. The St. Lawrence River carries a great volume of water, and yet it has built no delta. Can you give any reasons for this?
11. Elaborate in detail on your understanding of what is meant by the hydrologic cycle.
12. Have you seen any effects of wave action on lake shores or along the ocean? If so, describe these in detail from your personal observations.

Project

Get the topographic maps of your home area or any other region with which you are familiar. Study these maps, noting scale, contour interval, map symbols, and the land forms with which you are familiar to see how their shapes are portrayed by contour lines.

Note: Topographic maps can often be purchased in large cities at some local store. They may be ordered by mail from the Director, U. S. Geological Survey, Washington, D. C., 20402, or, for maps west of the Mississippi River, from the U. S. Geological Survey, Federal Center, Denver, Colorado 80225. These maps, are issued in *quadrangles,* rectangular areas bordered by latitude and longitude lines. The names of completed quadrangles are given on a map "Index to Topographic Mapping," available for each state, free of charge, from the Geological Survey.

Supplementary readings

Bloom, Arthur L., *The Surface of the Land,* Foundations of Earth Science Series, Prentice-Hall, Englewood Cliffs, N. J. (1969). (Paperback) [Discussions of stream erosion and landscape development on the land surface are presented; similarly, the sea coasts and land where covered by ice. This book can add greatly to the rather short discussions in the present chapter.]

Davis, William Morris, *Geographical Essays,* Dover Publications, New York (1954). (Paperback) [Modern geography and geomorphology rest on the fundamental work of this earth scientist. Read especially Part 2, "Physiographic Essays." These are classic essays, first published in 1909 and presenting important theories on geographical (erosion) cycles, the peneplain, rivers, and the sculpture of mountains by glaciers.]

Dyson, James L., *The World of Ice,* Alfred A. Knopf, New York (1962). [This book discusses ice in all its forms, especially in glaciers. The past ice ages, changes in climate, sea-level fluctuations, and the record of life on the ice are all covered in stimulating layman's language.]

Hoyt, John H., *Field Guide to Beaches,* ESCP Pamphlet Series #PS-7, Houghton Mifflin, Boston, Mass. (1971). [Beach and wave erosion summary and useable suggestions for class or individual studies in the field with laboratory follow up.]

Mathews, William H., *A Guide to the National Parks: Their Landscape and Geology,* Vol. I, the Western Parks; Vol. II, the Eastern Parks, Natural History Press, Garden City, N. Y. (1968). [A park by park explanation of the geological forces which were behind the creation of each park's scenery.]

Meinzer, Oscar E. (ed.), *Hydrology,* Dover Publications, New York (1955). (Paperback) [Originally prepared in 1942 by 24 experts for the National Research Council. A complete reference library on precipitation, evaporation, snow, glaciers, lakes, infiltration, soil moisture, ground water, runoff, drought, physical changes produced by water, hydrology of limestone terrains, etc.]

EARTHQUAKES AND THE EARTH'S INTERIOR

There are secrets as yet utterly unknown to mankind and reserved for the industry of future ages.

EDMUND HALLEY, 1656–1742

Earthquakes have always been phenomena terrifying to man, both primitive and civilized. They are vibrations set in motion by movements within the earth's crust and can be recorded and evaluated. The study of earthquake records provides us with information about the structure of and conditions within the interior of the earth, which consists of a crust, 5 to 50 miles thick, an intermediate zone, or mantle, down to 1800 miles, and a metallic core. Temperatures within the earth are high, perhaps 4000°C. Quite probably the earth went through a hot, molten stage after its birth, with the sinking of molten metal to the core similar to the sinking of iron in a blast furnace.

Earthquakes

Earthquakes are vibrations of the ground caused by sudden movements or displacements along faults (see pp. 565–566) within and below the crust. Actually the earth is in an almost constant state of vibration. It is estimated that at least 1 million earthquakes capable of being recorded on standard recording instruments occur each year. Of these, 150,000 or more are strong enough to be felt by persons nearby, and about 100 each year, based on a 30-year record, are forceful enough over the source area to be destructive if they occur near inhabited regions. In addition, tiny local surface vibrations are caused by heavy waves pounding coastlines, by high wind storms, by heavy traffic in our cities, and by mine and quarry blasts.

EFFECTS OF EARTHQUAKES

We have all read of the damage caused by severe earthquakes, ranging from the loosening of plaster on walls and the shaking down or overthrow of small movable objects, such as dishes on shelves, to serious structural damage or complete destruction of substantial buildings (Fig. 25.1). When an earthquake occurs in a large modern city, fire is a far greater danger than the actual shaking; both gas and water pipes are usually ruptured. In the terrible San Francisco earthquake in 1906, and again in Tokyo and Yokohama in 1923, fire caused about 65 percent of the total loss. Earthquakes and their associated phenomena have been very destructive of human life.

Surface displacements have accompanied a number of historic earthquakes, usually producing a long, continuous scarp (a small cliff or steep bank) a few feet high. This scarp is often at the foot of a straight mountain range and probably represents renewed movement on an old break. Movement associated with the 1959 Montana earthquake produced several vertical scarps (Fig. 25.2). In the 1906 San Francisco earthquake a horizontal shift along the San Andreas fault caused an offset of as much as 21 ft, displacing roads, fences, and other surface features. After an earthquake near Disenchantment Bay, Alaska, in 1899, beaches were found to be nearly 50 ft above mean tide level, and a considerable strip of land was newly exposed above sea level.

The vibrations of an earthquake are considerably intensified in unconsolidated rock. This is analogous to shaking a bowlful of jelly; the surface of the jelly will vibrate much more than the amount of motion given to the bowl itself. In San Francisco in 1906, most buildings on the marshy and filled ground along the bay front were wrecked, but many with foundations in solid bedrock were only slightly damaged by the earthquake vibrations themselves.

Landslides often accompany earthquakes in hilly regions. Large masses of unconsolidated matter, even bedrock, are shaken

Figure 25.1
Earthquake damage. Total pancake collapse of the Mijagual high-rise apartment building in Caracas, Venezuela, during July 29, 1967 earthquake. (From the collection of Karl Steinbrugge.)

and cracked loose by the earth vibrations, with attendant fracturing and slumping of the ground. In a few major earthquakes, the landslides have caused heavy loss of life and property damage.

Seismic sea waves, or *tsunamis,* commonly but erroneously called "tidal waves," are caused by sudden changes of level or landslides on the ocean bottom as the result of a severe earthquake. These waves travel across the oceans at speeds between 300 and 500 mi/hour. In the open sea they are unnoticeable, being perhaps 2 or 3 ft high with a wavelength of many miles. But along a shoreline, they pile up many feet above normal high-tide level. The waters may withdraw considerably below the low-tide level, with a surging return in a few minutes of high water that can be very destructive to life and property. Submarine earthquakes in the deep-sea trench just south of the Aleutian Islands caused seismic sea waves in 1946, and again in 1952 and 1957, that reached the Hawaiian Islands 4 to

Figure 25.2
Recent fault scarp. Scarp developed during the Hebgen, Montana earthquake in 1959 by movement along a known fault. Note fracturing in the foreground caused by uneven settling of alluvium on downthrow side of fault. (I. J. Witkind, U.S. Geological Survey.)

5 hours later with high-water marks from 5 to 55 ft above normal high water. The disastrous Chilean earthquakes of 1960 caused a tsunami that reached Hawaii 12 hours later. At a few places along low shores, the water swept as much as $\frac{1}{2}$ mile inland. Much damage was done to bridges, roads, ships, and buildings near the shore. The 1946 and 1960 waves were also noted along the Japanese and California shores.

The Lisbon earthquake of November 1, 1755, demonstrated nearly all the damaging effects of severe earthquakes. With a terrifying, thunderous roar the ground began to shake, and jarring shocks quickly tumbled most of the buildings in the city, including great stone churches filled with people. Within 6 minutes thousands had perished; landslides in the mountains behind the city sent up dense clouds of dust, adding to the terror of the remaining population. A few minutes after the earthquake, a 50-ft seismic sea wave smashed the harbor and drowned many hundreds of refugees huddled on the open wharfs. Fire completed the destruction of the city; about 60,000 of the city's population of 235,000 were dead and less than one house in six was left standing. This earthquake, perhaps the most intense in historic times, caused widespread damage over much of Spain and Morocco. It was felt over most of Europe and North Africa, about one-thirteenth of the earth's surface. Loch Lomond which is in Scotland,

1200 miles away, was thrown into 2-ft waves.

Rumbling sounds that become a roar, accompanied by the clatter and crashing of falling objects, are heard during severe earthquakes. The ground and the floors of buildings may pitch like a ship at sea. Men standing or walking on the ground are thrown down. Some eye witnesses have seen actual waves pass across the earth's surface, like waves across water, especially in unconsolidated matter or soil. The duration of earthquakes ranges from only a few seconds to 3 or 4 min.

Major earthquakes are followed by aftershocks within a few hours, sometimes continuing for several months, some of which are almost as severe as the original tremor. This adds to the panic and occasionally to the destruction.

EARTHQUAKE WAVES AND THEIR RECORDING

The study of earthquakes is the science of *seismology*. Over 400 stations throughout the world now operate instruments to record and study earthquake phenomena. Such recording instruments are called *seismographs*. The basic principle of a seismograph is the establishment of a mass that will move as independently of the ground as possible. This has been achieved in several ways; two of the most common are a weight suspended by a spring to measure vertical movements and a heavy mass suspended by a wire from a supporting column (Fig. 25.3) to record horizontal movements. In either type, the mass tends to remain at rest while the support vibrates during an earthquake. There are many ways in which this difference of motion can be recorded and measured. A simple device is a delicate stylus or pen attached to the mass that traces a continuous line on a revolving drum attached to the framework of the instrument. On modern instruments the recording is done electrically or on magnetic tape for easy computer analysis. Time is marked automatically on the record, usually by an accurate clock, so that the arrival time of a wave can be determined to the fraction of a second if so desired.

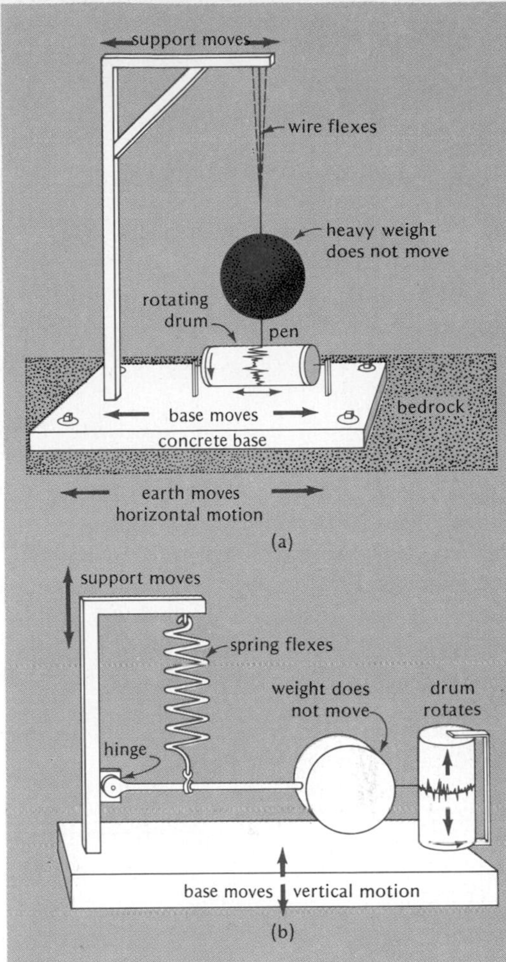

Figure 25.3
Seismographs: diagrams to illustrate the principle of earthquake recording. The seismograph on the top (a) records the horizontal component of earthquake vibration; the instrument on the bottom (b) the vertical component. When tremors vibrate the supports, including the drums, relatively rapidly, the weights maintain their position and the rotating drums move against the styli, leaving records of the vibration. (From *The Earth Sciences*, 2nd ed., Harper & Row, Publishers, copyright © 1963, 1971 by Arthur N. Strahler.)

Shock waves are generated when the crust

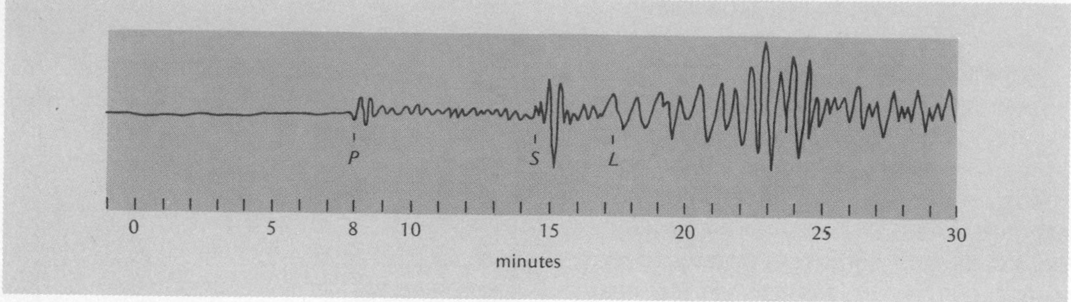

Figure 25.4
A seismogram. The record, or seismogram, of an earthquake as recorded on the rotating drum of a seismograph. The P indicates arrival of first primary waves; S indicates arrival of first secondary waves; L indicates arrival of first surface of long waves. Timing is arbitrarily diagramed for an earthquake whose focus is 3000 miles from the seismograph. This distance would be determined from the fact that the S-P interval is 6 min 27 sec.

is sharply displaced, and an energy transmission spreads in all directions from the point of original movement. Two types of waves are propagated in a homogeneous, elastic solid: *body waves,* of which sound waves are a well-known example, and *surface* or *long waves*. Body waves travel in all directions from their point of origin directly through the interior of the earth. The long waves are slower and travel or spread along the surface of the earth only. When a pebble is thrown into a pool of water, concentric surface waves radiate outward in all directions. Simultaneously, however, the noise of the pebble striking the water may be heard by a person or listening device located under the water. Such sound is transmitted through the water as body waves not visible at the surface.

A wave is defined or characterized by describing the motion of a particle in its path; two kinds of wave motion are thus defined, longitudinal and transverse (p. 195). Both types are present in the body waves of earthquakes, and they are known as the P and S waves, respectively. The *P wave,* or *primary wave,* is a push-pull or compressional wave in which particles vibrate back and forth in the path of the wave's progress—longitudinal wave motion. The *S wave,* or *secondary wave,* is a shake or shear wave in which the particles vibrate at right angles to the direction of wave progress—transverse wave motion. The P waves are transmitted in solids, liquids, or gases, but S waves can travel in solids only, as there is no shear resistance in a liquid or a gas.

Long waves are not so easily explained, but they may be looked upon as special kinds of transverse waves. They spread along the surface of the earth and cause much of the destruction associated with earthquakes because the particle movements associated with them are relatively great. Several types of surface waves have been identified in earthquake records.

The P waves travel faster than S waves and thus the former arrive first at any given station. Both types of waves travel with characteristic velocities in different types of

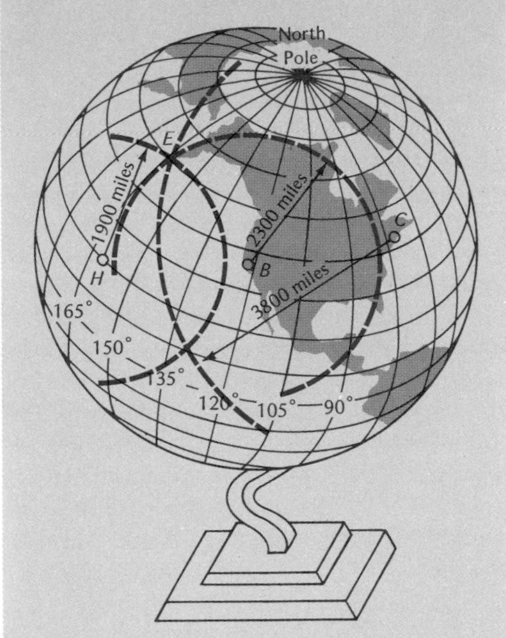

Figure 25.5
Location of epicenter. Globe showing North America to illustrate the method of locating the epicenter *E* of an earthquake off the Aleutian Islands that was recorded at Honolulu, Hawaii; Berkeley, California; and Cambridge, Massachusetts. The distance to the epicenter from each recording station is first determined by the S-P method; this distance is used as a radius to draw parts of circles around each station to find the point of intersection of all three.

rocks but with a constant ratio between their velocities. Both waves travel faster in rocks of higher densities and therefore travel at a steadily increasing rate as they move deeper into the earth. Consequently their wave fronts are elliptical rather than spherical, and their ray paths are curved, as the latter are at right angles to the wave fronts at any given point. Seismic waves are reflected and refracted when passing boundaries between layers of materials of different densities or by other inhomogeneities in the earth.

The record of an earthquake, called a *seismogram,* usually shows the arrival of the P, S, and long waves, as illustrated in Fig. 25.4. Because these three waves travel at different speeds, they arrive at a recording station at different times. The time between the arrival of the P wave and the arrival of the S wave is known as the *S minus P interval.* This interval is different for each distance and as a result can be used to compute the distance from the recording station to the source of the wave. From this data, the actual time at which the earthquake occurred can also be determined. These relations between time and distance are plotted as travel-time curves and are used for quick computation of the focus of an earthquake.

FOCUS AND EPICENTER

The point within the earth at which movement begins is called the *focus* of the earthquake. The point on the surface directly above the focus is called the *epicenter.* The majority of earthquakes have their foci at depths between 5 and 35 miles. When the depth of focus is less than 5 miles, the earthquake is not felt at any great distance, although it may be quite strong at its epicenter. A few earthquakes have foci at considerable depths, down to 435 miles below the surface. These are known as deep-focus earthquakes and are generated well below the earth's crust. Their interpretation presents many challenging problems to the student of earth structure.

The epicenter of an earthquake that occurs in an uninhabited region of the world may be determined quickly from the records of the earthquake from three different stations. Because the distance from any station to the focus of an earthquake is known, or calculable, from the S minus P interval, circles with radii equal to such distances can be drawn on a globe of the earth around the three stations. Their intersection is the approximate epicenter. (Fig. 25.5).

EARTHQUAKE MAGNITUDE

How to measure the size or intensity of an earthquake is a real problem. For many years intensity scales were used that depended on the noticeable effects of an earthquake ranging from nothing more than a slight tremor felt by a few people to complete destruction of all buildings. For example, one of the degrees of intensity was described as "felt by nearly all; many awakened; some fragile objects broken and unstable objects overturned; a little plaster cracked; trees and poles notably disturbed; pendulum clocks stopped." Such scales have definite limitations, even in inhabited regions, owing to the psychology of people and the methods used to question them, the character of the ground—bedrock versus soil, for example—and the quality of building construction. Lines can be drawn on a map connecting places that experienced essentially similar phenomena—that is, equal intensity; these are *isoseismal lines*. A systematic study of intensity distribution can be very helpful to city planners, engineers, and insurance underwriters.

Intensity scales are obviously useless in uninhabited regions, and they also fail to measure the amount of energy released by an earthquake. In an attempt to devise a system for estimating the force involved at the source of an earthquake, C. F. Richter, of the California Institute of Technology, in 1935 defined the magnitude of an earthquake in terms of the motion recorded by a seismograph of certain specifications at a standard distance (60 miles) from the epicenter. This has been extended to include equivalent responses of other seismographs at any distance. Magnitude numbers that have been determined instrumentally and not by personal appraisals can now be assigned to an earthquake. Furthermore, relationships have been worked out between magnitude and total energy release.

An earthquake rated 2.5 on the Richter magnitude scale can just barely be felt; one of magnitude 4.5 can cause some local damage; one of 6 to 7 is potentially destructive; and magnitude of 7 or over is a major earthquake. In 1906, the California earthquake was 8.25; the Tokyo earthquake of 1923 was 8.1; in 1959, the Yellowstone earthquake was 7.1; the Alaska earthquake of 1964 was 8.5; the Caracas earthquake of 1967 was only 6.5, although several buildings collapsed (see Fig. 25.1). The largest magnitude noted in this century has been 8.6, in India. The only earthquake in history that might have had a larger magnitude, perhaps up to 9, as judged by reported effects, was the Lisbon earthquake of 1755. During the 38 years from 1918 to 1955, there was an average of 232 earthquakes per year of magnitude 6 or over, including 25 per year of 7 or over.

DISTRIBUTION AND CAUSES OF EARTHQUAKES

Most of the moderate and major earthquakes occur in definite belts or zones. Over 80 percent of the total earthquake energy released in the world is in the belt bordering the Pacific Ocean. Earthquakes are frequent along the Pacific border of South America, Mexico, California, the Aleutian Islands, Japan, the Philippines, and Melanesia. A second belt, the Mediterranean–Trans-Asiatic zone, accounts for another 15 percent of earthquake energy. It extends from southern Europe through Turkey, Iran, northern India (the Himalaya Mountains), and into Burma.

These belts correspond to the world's youngest, and perhaps still growing, mountain systems and island arcs where deformation and faulting continue today. It is generally agreed that the immediate cause of earthquakes is sudden movement of rock masses along faults. Although fault displacements have been observed with a few earthquakes, in the majority the displacement must have occurred well below the surface.

A suggestion for the ultimate cause of earthquakes is the *elastic rebound theory*. Crustal movements take place slowly with gradual bending of crustal rocks. Rocks are elastic to a certain extent. Elastic strain

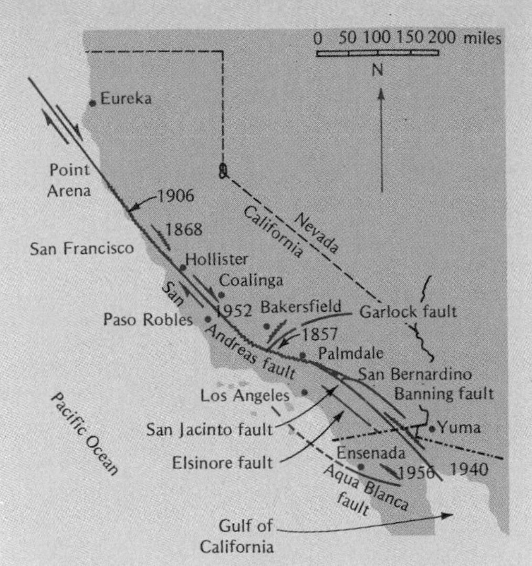

Figure 25.6
San Andreas fault trace. Map of California showing the trace of the San Andreas fault zone and related faults. The southwestern side has moved north relative to the northeastern side. Actual ground breaks during earthquakes are shown by zigzag lines and dates of movement. (Courtesy of John J. W. Rogers and John A. S. Adams, *Geology*, 1966, Harper & Row, p. 55.)

continues to build up until a critical value is reached; then a break that overcomes friction along a fault plane occurs, and the crust snaps to an unbent position, releasing the pent-up elastic strain. It is like bending the branch of a tree more and more until it finally breaks, and the end snaps back to its original place.

Careful measurements in California have lent support to this theory. A great fault, the San Andreas rift, extends diagonally across the California Coast Ranges for over 600 miles, from Point Arena on the seacoast north of San Francisco into the Colorado Desert of southeastern California and Mexico (Figs. 25.6 and 25.7). Movement and cracking of the ground occurred along the northern 270 miles of this fault in 1906 at the time of the San Francisco earthquake, with a maximum horizontal displacement of 21 ft at one place some 30 miles northwest of San Francisco. Movement of the western side was toward the north. Careful analysis of the accurate triangulation surveys by the U. S. Coast and Geodetic Survey both before and after the 1906 earthquake have demonstrated continuing relatively northerly movement of about 2 in./year for several stations on the western side of the San Andreas fault. Certainly this means a slow movement of two crustal blocks; apparently the elastic strain set up between these blocks is occasionally relieved by movement on the fault zone. Slow continuing movement on the fault is prevented by friction as the rock masses on either side press tightly against each other.

These relations can be illustrated diagrammatically. In Fig. 25.8 (top) let us assume that the line *XOY* represents a line crossing the fault *SA* at right angles at a time when there was no elastic strain in the region. Slow continuing regional drift moved point *X* northward to *X′* and point *Y* southward to *Y′*, but without any slipping or movement on the fault *SA*. The line *XOY* was then stretched to the curved line *X′OY′*. But at this time, the fault suddenly gave way, and the elastic strain was relieved, setting off an earthquake. The curved line *X′OY′* snapped into two straight segments, *X′O′* and *Y′O″* of Fig. 25.8 (bottom). Elastic rebound was apparently the mechanism by which the stored strain energy was converted to motion energy.

PRACTICAL ASPECTS OF EARTHQUAKE STUDY

The possibility of accurately predicting earthquakes in particular regions has been suggested, but so many unknown and variable factors are involved that for the present it is only a hope. The best protection against earthquakes is the construction of buildings, bridges, and water systems that will withstand the vibrations without failures. Engineers and architects have given this matter much study, especially in California and Japan. Major public buildings

Figure 25.7
San Andreas fault zone. Topographic expression of the San Andreas fault as it appears from the air in the vicinity of the Indio Hills in southern California. Estimates place the total movement along the fault at more than 150 miles. (Spence Air Photos.)

should have steel frames. Masonry walls without reinforcements are badly damaged by severe earthquakes. Above all, large buildings should be fireproof; broken electric connections and gas pipes and disturbed heating systems can start fires. Large structures should not be built on filled ground or unconsolidated rock except with special foundations. A frame dwelling stands up better than one built of brick or stone. One authority recommends iron straps and sockets to hold roof rafters to the ridgepole and floor sills in wooden buildings. It is important that the entire building move as a unit during an earthquake.

Small, man-made earth vibrations set off by dynamite blasts and recorded on properly placed seismographs have given much information about conditions below the surface. Earthquake vibrations are reflected from and refracted by layers of different elastic properties. In this manner the thickness of the soil can be determined for purposes of highway and foundation planning. The location of abrupt changes in rock composition and structure within the crust can be detected by these artificial earthquakes. Such information is extremely valuable in locating possible oil- or gas-bearing structures, and the seismic method has become one of the outstanding means of geophysical prospecting for new oil and gas fields.

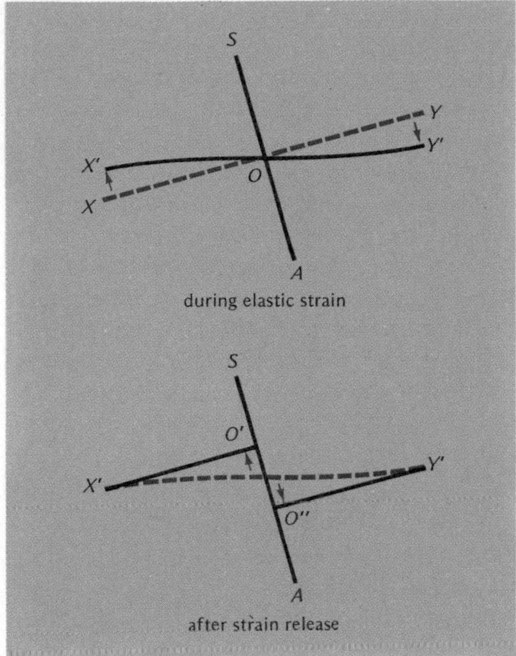

Figure 25.8
Elastic rebound theory. Slow continuing movements within the crust cause elastic yield. At top, *XOY* is a line across the old fault zone *SA* before strain has accumulated; *X'OY'* is the same line after elastic strain. At bottom, stored energy has been released by a break (sudden movement) along the fault. The curved line *X'OY'* has "rebounded" to its original shape but in two parts: *X'O'* and *O''Y'*.

Structure and composition of the earth's interior

The careful study of earthquake waves of known source from many recording stations has shown that waves travel at increasing velocities with increasing depth down to 1800 miles, where there is an abrupt reduction in their speed. In addition, reflected and refracted waves have been clearly identified. This information indicates a zoning of both the crust and the interior of the earth and gives us a basis for theories as to the probable conditions and compositions within the earth. Three major divisions are

Table 25.1
Average earthquake wave velocities in crust in miles per second

	P, or primary, wave	S, or secondary, wave
in upper, granitic layer: the Sial	3.8	2.1
in lower, basaltic layer: the Sima	4.3	2.5
in top of mantle	5.1	2.8

indicated: a relatively thin *crust,* a homogeneous *mantle* down to 1800 miles, and a metallic *core.*

THE CRUST

Variations in speed (Table 25.1) as well as reflections and refractions of earthquake waves indicate a layered structure for the crust. An upper layer, in which earthquake waves travel relatively slowly (3.8 mi/sec for P waves), underlies the continents. From the observational, geologic evidence we assume this to be a granitic, or sialic, layer (with the average composition as given on p. 440.) The lower crustal layer, probably basaltic in composition, has a remarkably uniform thickness with a sharp discontinuity at its base. Volcanic data from islands in the Pacific Ocean as well as experimental travel-time data for various rock types support the theory of a basaltic composition for this layer. The combined thickness of the crustal layers under the continents averages between 18 and 25 miles but is almost twice as thick under certain orogenetic belts, as beneath the Sierra Nevada and the Alps. Under the oceans, the sialic layer is absent or very thin and patchy, and even the basaltic layer may be thin—as little as 5 miles. The crust is approximately 0.7 percent of the total mass of the earth.

The surface or zone of sharp change in wave behavior at the base of the crust is called the *Mohorovičić discontinuity,* or the *Moho* for short, named after a Yugoslavian seismologist who first demonstrated a layered structure for the earth from an analysis of seismograms.

THE MANTLE

Below the crust is another zone, extending 1800 miles into the earth, called the *mantle;* it is approximately 67.8 percent of the earth's mass. Earthquake waves penetrate the mantle with steadily increasing velocities, indicating increasing rigidity and density. The material of the mantle may have low flow strength—that is, the force necessary to cause flow is small—but it is not liquid, because the S or shear waves pass through satisfactorily. From geologic reasoning and laboratory experiments, as well as earthquake wave data, geologists believe the mantle to be composed largely of peridotite, a rock containing magnesium-iron silicate minerals, with a preponderance of the mineral olivine.

THE CORE

The P and S waves arrive on schedule up to distances of 7000 miles around the surface from the epicenter of an earthquake. This distance is equivalent to 103° of arc around the earth. Such waves have followed a uniform curved path to a depth of about 1800 miles within the earth (line *FB* in Fig. 25.9). A short distance beyond 103°, however, P waves arrive late and are very faint, and S waves never appear. Beyond 143°, P waves are still late but somewhat stronger again; but the S waves do not reappear. A P wave passing directly through the earth (*FI* in Fig. 25.9), if it passed through the center with the same speed it has in the lower part of the mantle, should reach the opposite side of the earth in 16 min. Instead it takes 20 min; this delay can only mean a decrease in velocity within the center of the earth. It has been demonstrated that the actual velocity of the P wave below 1800 miles drops suddenly to about one-half its velocity at 1800 miles. This abrupt discontinuity in earthquake behavior indicates a core of an entirely different density from the mantle and with a radius of 2160 miles. The disappearance of the S waves implies that the core behaves as a liquid through which transverse wave motion is not transmitted. Furthermore, as P waves pass from a high-velocity medium to a lower-velocity material, they are bent, or refracted, toward the lower-velocity side. The core of the earth, therefore, acts as a huge converging lens, and the wave paths, or rays, which should emerge between 103° and 143°, are refracted to emerge in the area beyond 143°, as illustrated by *B′* and *C* of Fig. 25.9. The zone between 103° and 143° is often called the *shadow zone*—that part of the earth's surface in the shadow of the core so far as original P waves are concerned.

The P waves speed up again to some extent and are further refracted part way through the core (*FG* in Fig. 25.9), indicating a zone of greater rigidity and suggesting the existence of a solid inner core about 760 miles in radius. Furthermore, evidence from the size and character of earth tides is apparently conclusive that the whole core could hardly be in a liquid state.

ANALOGY TO METEORITES

The sharp boundary between the core and the mantle cannot be explained by physical differences alone. It seems clear for many reasons that the material of the core must be of a different composition from the material above the core boundary. But what is it? We do not know directly, of course, but, from an analogy with meteorites, the most logical material seems to be an alloy of iron and nickel. Meteorites are generally believed to be fragments of a disrupted planet that was once located between Mars and Jupiter, in the orbit of the asteroids. It is reasonable

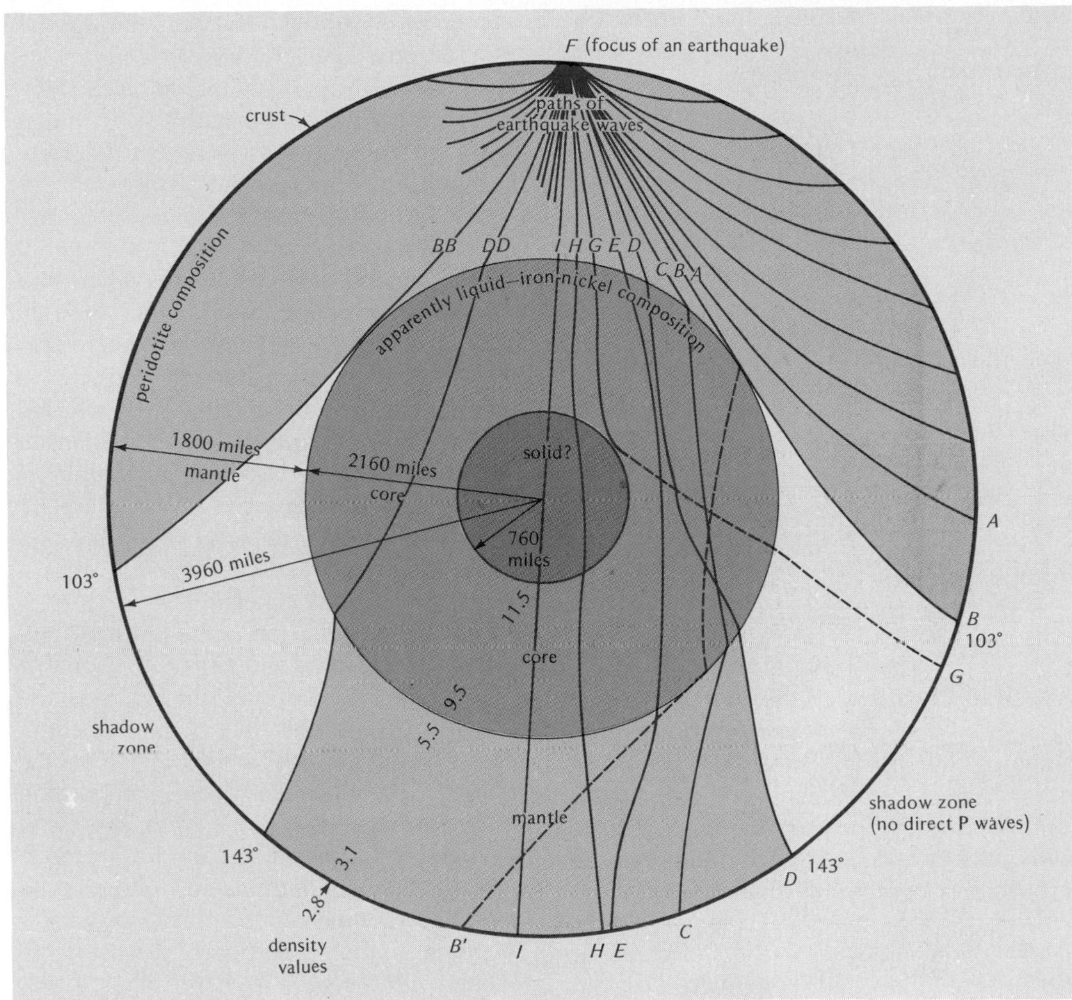

Figure 25.9
Interior of the earth. Diagram through the interior of the earth showing its main divisions: crust, mantle, and core. The paths of a few of the earthquake waves that radiate in all directions from a focus are drawn showing refraction by the core to cause the shadow zone between 103° and 143°. Numbers to the lower left of center are probable density values for various depths.

to suppose that this disrupted planet had a composition very similar to that of the earth. Meteorites are divided into three important groups: the stony, the stony–iron, and the iron–nickel meteorites (Fig. 25.10). The stony meteorites contain magnesium–iron silicates, with a preponderance of olivine, very similar to the peridotites of the earth.

Figure 25.10
Meteorite. Polished slice of an iron–nickel meteorite. The surface has been etched to show the internal crystal structure typical of this type of meteorite. (Ward's Natural Science Establishment, Inc.)

This is in accord with the suggested composition of the mantle from other lines of evidence. The metallic meteorites are alloys of iron, nickel ($8\frac{1}{2}$ percent), and cobalt ($\frac{1}{2}$ percent). If our hypothesis for the origin of meteorites is correct, then the earth's core is probably composed of a similar alloy. The stony–iron meteorites are silicate in major composition, with scattered grains of iron–nickel alloy. It is proposed that the mantle zone is composed of peridotite with an increasing number of grains of solid iron-nickel alloy in its deeper part (see Fig. 25.9).

DENSITY DISTRIBUTION

We know that the density of the earth as a whole is 5.5 but that the average density of the known rocks of the crust is not over 2.8. It is obvious that density must increase downward and that the core must be composed of an extremely dense material. Will a mantle and core of the sizes and compositions proposed above check with the earth's average density of 5.5?

The Australian geophysicist Bullen has made a careful study of density distribution within the earth. The density at the outer core boundary must be near 9.5 and rise to at least 11.5, or perhaps higher, at the center of the earth. Although this is considerably higher than iron (7.8) and nickel (8.6) at the earth's surface, the difference may reasonably be explained as due to the tremendous pressures that must exist within the earth and the resulting compressed status of its material. In the mantle, density varies from 3.1 just beneath the crust to about 5.5 at the core boundary, owing to the high pressures or, in part, to the possible presence of some iron–nickel alloy grains in the lower mantle. Density distribution is entirely consistent, therefore, with a mantle of peridotite similar to stony meteorites and an iron–nickel core 2160 miles in radius, similar to the metallic meteorites.

Here we have an excellent illustration of *scientific method* in formulating a reasonable theory. The problem was the composition and structure of the interior of the earth. Data bearing on this problem include the behavior of earthquake waves, composition of meteorites, density of the earth, and many geologic facts related to igneous rocks. A hypothesis was formulated of a solid peridotite mantle 1800 miles thick surrounding an iron–nickel core, at least partly liquid, with a radius of 2160 miles. It has been demonstrated that this hypothesis fits all known earthquake wave phenomena, experimental laboratory tests on wave travel times in various rock types, all pertinent geologic data concerning the crust, all density data, and reasoning based on physical laws of compressibility and high pressures within the earth. The hypothesis satisfies all known aspects of the problem and is acceptable as satisfactory.

Early history of the earth

Many hypotheses have been proposed for the origin of the earth and solar system. But the objections to each of these tentative explanations are stronger than any supporting data. It seems reasonable to assume, however, that the sun and planets were formed by a single grand process. At present, the most generally accepted hypothesis suggests that our sun and the rest of the solar system formed from a cold, contracting gaseous cloud, or protosun. As this cloud slowly contracted, its center heated, owing at first to compaction and later to a thermonuclear process, to produce the sun. The remaining cloud of cold dust particles continued to spiral around this new sun as a giant nebula; then, by gradual accretion of the particles, the several planets took form. The growing planets began to heat as a result of compaction and radioactivity, then 15 times more intense than at present because of the early abundance of short-lived isotopes. This caused the early melting of the still-growing earth.

The hypothesis that the earth went through a molten stage early in its history seems plausible in the light of our theories of the constitution of the earth's interior. In fact, this seems to be the only way that the zoning of the earth's interior could have been achieved. Iron and nickel gradually sank in a hot molten mass, analogous to the sinking of iron in a blast furnace, with the silicate mantle around the core analogous to the silicate slag of the blast furnace. Convection currents would keep this slaglike mantle well stirred and thus permit slow uniform cooling by heat loss at the surface. Finally, crystallization of the silicate mantle began at the base of the mantle at the core boundary and continued upward. If crystallization had started at the top of the mantle, the solid surface layer would have been such an effective insulation to further heat loss that the rest of the mantle might never have crystallized. The crust of the earth may be interpreted as the residuum from the solidification of the mantle; it is composed essentially of materials of lower crystallization temperatures than peridotite.

With the crystallization of the mantle, the core of the earth was insulated from any further heat losses. It seems logical to assume that the temperature of the iron–nickel core is about the same now as when solidification of the mantle began. A possible temperature at the core boundary may be near 4000°C; certainly it must be considerably less than 6000°C, the temperature of the sun's surface (Chap. 3).

Some scientists have suggested that the mantle is gradually cooling because of loss of heat through the crust. Other scientists, however, argue that additions of heat from within the crust and upper mantle, largely the accumulation of radioactive heat, make up for, or even exceed, the heat loss to outer space. We know that the temperature at the base of the crust must be high, perhaps near 1200°C, but whether these temperatures are increasing or decreasing with time is still an open question.

ORIGIN OF THE ATMOSPHERE AND HYDROSPHERE

A usual assumption in most of the early hypotheses of the origin of the earth held that the atmosphere and hydrosphere represented the gases that remained after crystallization of the crust. As cooling continued, steam finally condensed to form the first oceans. The primitive atmosphere and hydrosphere thus formed at the start of the earth had about their present volumes. Recent evaluation of available geochemical data, however, makes it appear more probable that the gases of the atmosphere and water of the hydrosphere have risen to the surface from the earth's interior during the long course of geologic time. According to this view the first oceans were very limited in volume but have gradually accumulated as water has been expelled by volcanoes, fumaroles, and hot springs. If we assume that fumaroles and hot springs have been about as numerous as today and that some 1 percent of their water has been magmatic water, new to the surface, throughout the 4 billion years of the earth's existence, we easily have the present volume of water in all the oceans. As the volume of water increased, the ocean basins have apparently deepened by the gradual sinking of their bottoms, in part an isostatic adjustment to the weight of water but also an adjustment to accompanying changes in the crust.

In similar manner, the large quantities of carbon dioxide gas, now locked up in the crust as carbonate sedimentary rocks, and the chief elements dissolved in seawater, such as chlorine and sulfur (in sulfate) have also gradually accumulated at the surface from volcanic and hot-spring emanations. Certainly chlorine and sulfur gases have been expelled in large quantities during historic volcanic eruptions and, less spectacularly, in present-day hot-spring waters. Certainly, also, the amount of chlorine in seawater is greatly in excess of the amount that could reasonably have been derived by the weathering of original igneous rocks.

It also seems likely that the composition of the atmosphere, especially the delicate balance of carbon dioxide, and the salinity of the oceans have been remarkably uniform through much of recorded geologic time. With the addition of new water to the oceans from within the earth's interior there were accompanying additions of chlorine, sulfur, etc., to maintain a fairly uniform seawater composition. A constant carbon dioxide content in the atmosphere would have been essential to the apparent early development of life. We can reasonably assume that carbon dioxide has been added to the atmosphere gradually from within the earth at about the same rate that it was being locked up by the precipitation of calcium carbonate as limestones. The ratio today of carbon dioxide in the atmosphere and hydrosphere, including all living matter, to the carbon dioxide buried in ancient sedimentary rocks is about 1:600. If all this carbon dioxide had been in the original primitive atmosphere, life as we know it would have been impossible for perhaps three-fourths of the earth's history, or until the excess carbon dioxide was taken out of the atmosphere by the formation of limestones. The fossil record shows that life began very early in the history of the earth; we conclude therefore that the carbon dioxide content of the early atmosphere was not very much higher than it is today (0.03 percent).

In conclusion, we picture the earth just after its birth with a crust composed entirely of igneous rocks and oceans containing much less water than today. The continued generation of magmas in the crust and their ultimate crystallization has released more water and many gases to increase slowly the volume of the oceans and maintain an approximately uniform atmospheric composition and oceanic salinity. The consequences of these assumptions are far reaching, not only in their importance for our understanding of past life, but in the explanation of a great variety of geologic phenomena.

SUMMARY

Earthquakes are vibrations caused by movements along fractures. These are sudden movements due to elastic rebound after slow deformation.

Earthquake waves can be recorded on seismographs. Surface waves and two types of body waves are recorded. The body waves consist of the P, or primary, wave, a compressional or longitudinal type of wave, and the S, or secondary, wave, a shear or transverse type of wave. The surface, or long, waves are known also as L waves.

The earth's interior consists of the crust, the mantle, and the core.

The crust is 0.7 percent of the volume of the total earth. The continental crust is 18 to 50 miles thick, with an upper layer of sial, or granitic material, and a lower layer of sima, or basaltic composition. The oceanic crust is only 5 to 10 miles thick and the upper layer is very thin and patchy or absent. The base of the crust is delimited by the Mohorovičić discontinuity.

The mantle is 1800 miles thick and is 67.8 percent of the volume of the earth. It is believed to have a peridotite composition, perhaps with grains of iron–nickel alloy in its deeper part. There is a sharp boundary at the base of the mantle, as indicated by earthquake wave behavior.

The core is 2160 miles in radius and is 31.5 percent of the earth's volume. It has been suggested that the core has an iron–nickel alloy composition, analogous to that of meteorites. Almost all the core is apparently in a dense liquid state, but an inner core, about 760 miles in radius, may be solid.

Important words and terms

earthquake
seismic sea waves, tsunamis
seismology, seismograph, seismogram
body waves: primary and secondary
long or surface waves
S minus P interval
focus, epicenter
earthquake magnitude
elastic rebound theory
San Andreas fault zone
crust, mantle, core
Mohorovičić discontinuity, Moho
shadow zones
meteorites
average density of the earth

Questions

1. Compare seismic waves as to general character and speed with some other types of waves, such as sound waves and radio waves.
2. Explain in your own words the elastic rebound theory. How would it apply to vertical faulting?
3. Explain and delimit the shadow zone for P waves; for S waves.
4. What theory for the origin of meteorites is implied when the composition of the latter is used to suggest the possible composition of the earth's core?
5. In several South American earthquakes well-constructed masonry houses were severely damaged while nearby bamboo huts were unharmed. What is a possible explanation?
6. What is the Mohorovičić discontinuity, or Moho? What is the possible nature of the materials on either side that could cause such a discontinuity?

Supplementary readings

Bascom, Willard, *Hole in the Bottom of the Sea,* Doubleday, Garden City, N. Y. (1961). [An authoritative but popularly written story of the Mohole Project to drill a hole through the earth's crust into the mantle in more than 2 miles of water.]

Bates, D. R. (ed.), *The Planet Earth,* 2nd ed., Pergamon Press, New York (1964). [Articles by experts on various aspects of the earth, its interior, origin, and atmosphere. Up-to-date reports on subjects investigated during the International Geophysical Year and since.]

Clark, Sydney P., *Structure of the Earth,* Foundations of Earth Science Series, Prentice-Hall, Englewood Cliffs, N. J. (1971). (Paperback) [Discussions of the earth's magnetic and gravity fields, seismology, the constitution of the earth's interior, and heat flow in the earth. Also good for Chaps. 26 and 27.]

Hodgson, John H., *Earthquakes and Earth Structure,* Prentice-Hall, Englewood Cliffs, N. J. (1964). (Paperback) [A very readable discussion of earthquakes, their effects, causes, and recording. Covers the information about the earth's interior which is obtained by a study of earthquake records.]

Leet, Don L. and Florence Leet, *Earthquakes, Discoveries in Seismology,* Dell Publishing, New York (1971). (Paperback) [A popular but very authoritative account of earthquakes and their study.]

Moore, Carleton B., *Meteorites,* ESCP Pamphlet Series #PS-10, Houghton Mifflin, Boston, Mass. (1971). (Paperback) [A simple discussion of meteorite falls, meteor showers, craters formed, radiometric ages, types, and mineralogy of meteorites.]

Nininger, H. H., *Out of the Sky,* Dover Publications, New York (1959). (Paperback) [A comprehensive introduction to meteorites—their composition, size, distribution, explosions, origin, and craters. A connecting link between astronomy and geology.]

San Fernando, California, Earthquake of February 9, 1971, U. S. Geological Survey Professional Paper No. 733, 1971. (Paperback.) [A preliminary but very detailed report published jointly by the U. S. Geological Survey and the National Oceanic and Atmospheric Administration discussing the important lessons affecting man's welfare learned so far from this 6.6 (Richter scale) earthquake; 254 pages of significant data evaluation designed to provide early guidance to private citizens and public officials engaged in reconstruction and future planning.]

Spar, Jerome, *Earth, Sea and Air: A Survey of the Geophysical Sciences*, Addison-Wesley, Reading, Mass. (1965). (Paperback.) [Chapter 2 is a discussion of earthquakes, geomagnetism, and the interior of the earth.]

MOUNTAINS AND MOUNTAIN BUILDING

A time to look back on the way we have come, and forward to the summit whither our way lies.

J. H. BRADLEY, 1693–1762

Mountains are land forms; their origin is interpreted by examining their rock structures and knowing the internal processes that have been active and the stage to which these processes have proceeded. All rocks show evidence of some warping, folding, or fracturing. Mountains formed where deformation has been at a minimum are chiefly the result of differential erosion. In most of the great mountain systems of the world, however, the crust has been extremely deformed by folding, faulting, and intrusion of magma. Such deformational mountains started as sinking basins of sedimentation, geosynclines, later subjected to orogenic diastrophism. The explanation of how and why geosynclines develop and are eventually deformed involves some of geology's major unsolved problems.

Vertical movements of the crust

Much evidence can be cited to show that gentle vertical movements have occurred in recent time and, in fact, are continuing today. The most striking proofs of such uplifts are beach deposits and wave-cut cliffs far above the present sea level, as seen in California, Labrador, Scandinavia, and many other coastal regions. The majority of coastal uplifts have probably taken place slowly, without sudden movements. One of the most carefully documented slow uplifts is still continuing in parts of Sweden and Finland around the northern part of the Baltic Sea. Numerous marine shells are found in farmland soils and freshwater swamps far from the present sea, which gave rise over 150 years ago to the belief that the land had risen. Markers were then placed at high-tide level, so that today we know that the land has continued to rise in a part of the region at the rate of 3 to 4 ft every 100 years. Some of the original markers are as much as a mile inland. In addition, the evidence of raised beaches and glacial deposits indicates that an uplift of at least 900 ft has occurred in parts of Scandinavia since the glacial period, perhaps 12,000 years ago.

Along nearly all the world's coastlines either elevated shoreline features or drowned valleys indicate that sea level has definitely not been a fixed level during recent earth history. Changes in sea level can be brought about in two ways: either by vertical movements of the crust or by a variation in the amount of water in the oceans. For example, if all the water locked up as ice in today's glaciers, especially over Greenland and Antarctica, were returned to the oceans, sea level would rise nearly 150 ft along all the world's coastlines. With the advance and retreat of continental glaciers over the past million years (the ice age), sea level has fluctuated both up and down many times. It was certainly a few hundred feet lower than today when vast glaciers covered parts of North America and Europe. But it has also been higher than at present, as we have reason to believe that today's polar ice may not have existed during the warmer so-called interglacial periods. Raised beaches and drowned valleys due to such fluctuations in sea level can be found on most coasts around the world at uniform and corresponding elevations above present water level.

The use of elevated beaches and other shoreline features as proof of vertical land movements requires evidence, therefore, that the features in question are local and not continuous at the same elevations along nearby coastal regions. This is true for the raised shorelines of Labrador and Scandinavia. Over a stretch of a few hundred miles these former beaches change appreciably in elevation. They could not possibly have been formed simply by a higher stand of seawater. The evidence is conclusive, therefore, that the northeastern part of Canada and the northwestern corner of Europe have been elevated several hundred feet in relation to other parts of their respective continents.

Further evidence of the recent uplift of northeastern Canada is found in former beach deposits surrounding and above the present Great Lakes in the United States and Ontario. Many of these represent a former higher water level, but some of the beach lines are not horizontal and can only be explained by differential uplift or tilting. Apparently this tilting is still continuing today. Evidence of very recent emergence can be found on the north shores of Lake Superior whereas the south shores are being flooded at the rate of 1 ft/100 years.

Many evidences of subsidence of coastlines might also be cited. Such harbors as San Francisco Bay, Puget Sound, Chesapeake Bay, and New York Harbor are all drowned river valleys; tide gauges in Denmark and Japan show that many harbors in those countries are slowly deepening. Dikes along the North Sea coast of lowland Europe have been built to between 60 and 70 ft high today as a consequence of continuously higher water levels since the Middle Ages.

The majority of sedimentary rocks were deposited under seawater; yet many, although still essentially horizontal in attitude, are high above sea level today. This is the evidence, of course, that vertical movements have been active again and again throughout geologic time. Many examples to illustrate the character and results of such processes can be cited; the Colorado Plateau is a spectacular but perhaps typical example of a complex uplift. This plateau, covering approximately a million square miles of parts of Utah, Colorado, Arizona, and New Mexico, is blanketed with thousands of feet of flat-lying sedimentary rocks, many of them of marine origin, now 5000 to 10,000 ft above sea level, and dissected by deep canyons. Although these strata are essentially horizontal, many gentle but conspicuous upfolds, wide shallow basins, and long fractures divide the plateau into many units. Here, then, is a large crustal block of sedimentary rocks that has been uplifted with only moderate warping and local fracturing.

Vertical movements of segments of the crust, either up or down, have occurred again and again during the existence of the earth without appreciably disturbing the attitude of the rocks in the affected areas. Movements of this type are known as *epeirogeny,* or epeirogenetic movements.

Orogenic movements

Observations on the deposition of sedimentary layers in water have clearly demonstrated that the strata so deposited are nearly horizontal and essentially parallel to the underlying surface. This is the *law of original horizontality,* formulated about 300 years ago by Nicolas Steno. Therefore when we note the eroded edge of sedimentary strata standing at various steep angles (Fig. 26.1), we have found the detailed evidence for deformation of the earth's crust. The tilting of the originally horizontal beds can only mean differential crustal movements. The latter have locally buckled the crust into mountain ranges.

The evidence for these mountain-making uplifts, called *orogeny,* needs little discussion. The great mountain chains of the world—the Appalachians, Rockies, Andes, Alps, and Himalayas, with their intricately folded rocks—attest to the activity of such processes. Erosional processes have entirely removed the orogenic mountains that were formed early in the earth's history, but the record of these movements is still preserved in the folded and metamorphosed strata of the roots of the former mountains.

The structures formed by this deformation of rocks fall into two categories: folds and

Figure 26.1
Folded strata near Borah Peak, central Idaho. Sedimentary rocks that have been uplifted and folded during deformation of the earth's crust. Large syncline in the foreground with an anticline and syncline in the background. (From *Geology Illustrated,* by John S. Shelton, W. H. Freeman and Company. Copyright © 1966.)

fractures. The rocks of the crust, brittle at the surface, are plastic enough at moderate depths to fold slowly in response to compressive stresses, like a blanket pushed across a tabletop or a rug over a smooth floor. Folding is a characteristic phenomenon in most mountain ranges. A fracture along which one side has moved, or been displaced, relative to the other is known as a *fault*. This displacement may amount to only a few feet or it may become as much as several miles on faults where movement has recurred repeatedly over long time intervals.

For convenience of reference and ease of plotting on maps, the geologist speaks of the *strike* and *dip* of rock strata (Fig. 26.2). Dip is the angle and direction of slope of rock strata or other surfaces in question. The dip is measured from the horizontal down to the bed, as illustrated in Fig. 26.2. When the angle of dip is 90°, the bed is vertical. Strike is the horizontal direction along the dipping, or inclined, surface. Strike may be defined geometrically as the intersection of the inclined plane with a horizontal plane. The strike direction is always at right angles to the direction of dip. By the use of strike and dip, we can refer the position and attitude of layering of any type of rock—sedimentary, igneous, or metamorphic—to the compass directions.

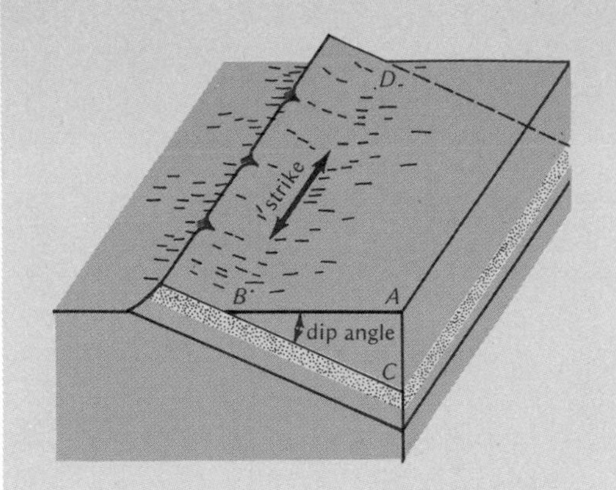

Figure 26.2
Strike and dip. Diagram illustrating the dip and strike of a tilted bed. The angle of dip is *ABC*. The horizontal direction, *BD,* is the direction of strike. The direction of dip is toward the right at right angles to the direction of strike.

Strike and dip are also used to describe the attitude of fracture planes, such as joints and faults, and some igneous bodies, such as dikes.

FOLDS

Folding is usually accepted as evidence of shortening of the earth's crust and as resulting, consequently, from a compressional stress. Three major types of folds are anticlines, synclines, and monoclines (Figs. 26.1 and 26.3). *Anticlines* are upfolds in which the sides, or limbs, of the fold dip away from the center, or axis, of the fold. *Synclines* are downfolds in which the limbs dip in toward the center. If both limbs have approximately equal angles of dip, the fold is symmetrical; if one limb has more dip than the other, the fold is asymmetrical. The latter may become overturned folds. A *monocline* is a fold that dips in one direction only or has but one limb.

A fold that dips away in all directions from a central point is a *dome;* all gradations exist between domes and the typical elongate anticline. The reverse of a dome is a *structural basin*.

Folds vary in size from small features easily visible in one outcrop (Fig. 26.4) to

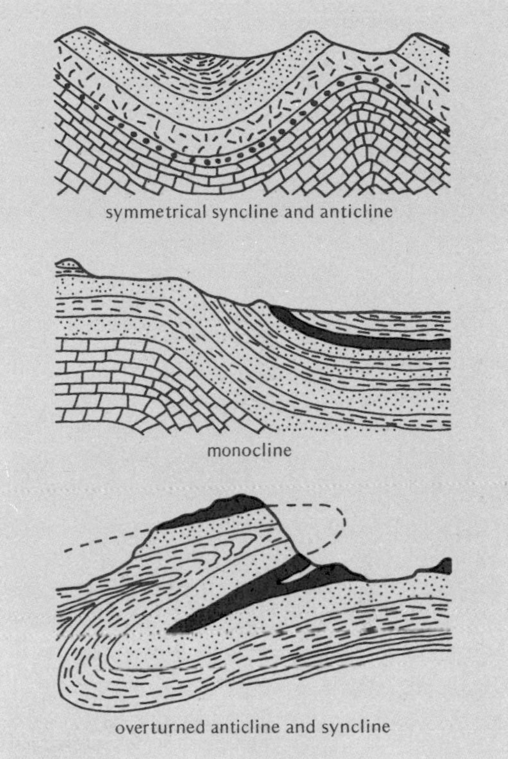

Figure 26.3
Folds. Top, symmetrical syncline and anticline (right). These folds are drawn partially eroded as they occur in the Appalachian Mountains. Ridges are found where the hard strata are at the surface. Middle, a monocline. The erosion of such a fold may result, as diagrammed, in a high plateau surface (left) adjacent to a lower plain. Examples of this relationship exist in the Grand Canyon region of Arizona. Bottom, an overturned anticline and syncline. This type of structure is common in the Swiss Alps.

features many tens or hundreds of miles in size. The San Rafael swell of central Utah is a great dome 60 miles long and almost half as wide.

The compression of rock strata within the crust suggests plastic deformation or rock flowage at considerable depth. Brittle rocks can be bent when under sufficient confining pressures. A convincing demonstration of

Figure 26.4
Anticlines. Top, a gentle anticlinal fold in thin-bedded-Miocene shales and sandstones in California. (U.S. Geological Survey.) Bottom, crest of an anticline in Clover Hollow, Virginia. The crest of a small, tight fold in limestone strata. (From *Mineral Industries Journal*, Virginia Polytechnic Institute; R. V. Dietrich.)

Figure 26.5
Normal faults. The trace of steeply dipping fault planes as exposed in a road cut. Note the offset of the beds on either side of the faults. (Courtesy H. L. Wanless.)

Figure 26.6
Joints, Beartooth Range, Montana. Closely spaced joint planes in various directions in highly metamorphosed granite. (W. H. Parsons.)

this is the complex folding often observed in glacial ice. Some rock types, especially shales and other thin-bedded sediments, are easily deformed by folding similar to this folding in ice. These incompetent beds may be folded at fairly shallow depths, with local thickening and thinning of individual strata. Massive beds, such as crystalline limestones and quartzites, on the other hand, are very competent and require burial of tens of miles before they are folded by rock flowage. Brittle and competent rocks are pushed into great folds by closely spaced fractures and not by flowage.

FRACTURES

Fractures belong to two groups: *joints,* in which no apparent displacement has taken place parallel to the break, and *faults* (Fig. 26.5), on which there is displacement, sometimes as much as 50 miles. Either type of fracture may result from tensional (pulling apart), compressional (squeezing), or torsional (twisting) stresses.

Joints form by shrinkage in drying and compaction of sedimentary rocks and by the crustal movements of diastrophism. They often occur in closely spaced, parallel, and intersecting sets, which allow the rock to break and weather in a blocklike form as shown in Fig. 26.6. The faces of vertical cliffs are often developed along joint surfaces.

Contraction occurs as the rock in a pluton crystallizes and cools. This contraction results in cracking of the igneous body in rather definite directions or patterns in response to the tensional forces set up by the cooling contraction. In tabular igneous bodies, such as dikes, sills, or thin laccoliths, these joints are commonly at right angles to the cooling surface. Occasionally they are closely and regularly spaced, so that the igneous rocks break in polygonal columns. Such columnar jointing is a characteristic feature of some lava flows and other igneous bodies (Fig. 26.7). In a horizontal flow or sill the columnar jointing is vertical, and in a vertical dike the columns are horizontal at right angles to the walls of the dike.

Figure 26.7
Columnar jointing, the Devil's Postpile National Monument, California. Joints with a regular pattern resulting from shrinkage during cooling in a basalt lava flow. The resulting columns of rock usually form at right angles to the flow surface. (From R. M. Garrels, *A Textbook of Geology*, Harper & Row, 1951; R. H. Anderson, National Park Service.)

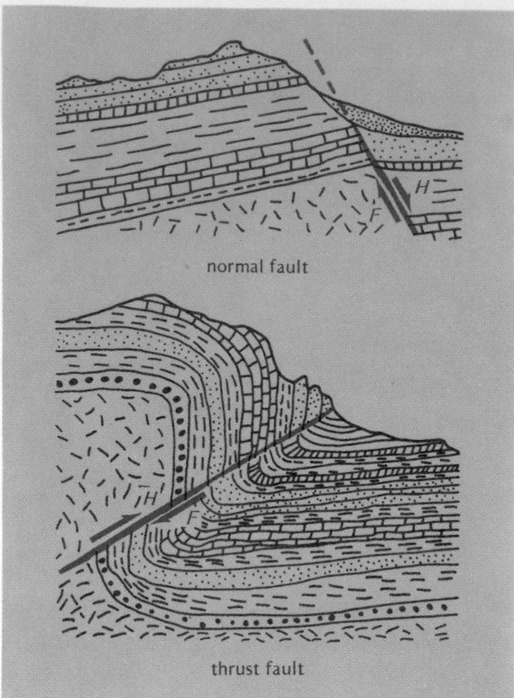

Figure 26.8
Faults. Top, a normal fault, in which the hanging wall *H* has moved down relative to the footwall *F* side of the fracture. In this way a fault-block mountain range is produced. Erosional debris from the upthrown side may accumulate as an alluvial fan deposit as shown. Bottom, a reverse or thrust fault, in which the hanging wall side has moved up and over the footwall side. This type of structure is common along the Rocky Mountain front in Wyoming and Montana.

Faults are classed as normal and reverse. *Normal faults* are generally the result of tensional stresses, and *reverse* or *thrust faults* of compression. In faults dipping other than 90°, the overlying block is called the *hanging wall,* and the underlying one the *foot wall* (Fig. 26.8). In normal faults, the hanging wall has moved down relative to the foot wall, whereas in the reverse or thrust faults, the hanging wall has moved up relatively.

Low-angle thrust faults indicate shortening of the crust and often occur with or have developed from overturned anticlinal folds (Fig. 26.3). In some faults all movement has been essentially in a horizontal direction, as in the famous San Andreas rift in California (see previous chapter, Figs. 25.6 and 25.7). Such faults are referred to as *strike-slip faults*.

Types of mountains

The land surface of the earth may be divided into plains, plateaus, and mountains. Plains are low, flat lands not very high above sea level as is characteristic of much of central United States. Plateaus are flat lands several thousand feet above sea level and often cut by spectacular, deep canyons. The Colorado Plateau in the United States is the best-known example. A mountain is a height of land rising rather prominently above the surrounding terrain. The distinction between a hill and a mountain is a local one. In flat country, such as Florida and the other Atlantic coastal states, even a small rise of land is quite prominent and may be called a mountain—for example, Stone Mountain, near Atlanta, Georgia, and the Watchung Mountains in New Jersey, a series of ridges near New York City 300 or 400 ft high. Features of the same height as this in most parts of the Rocky Mountain states or California would be considered low hills.

The elevation of a hill or mountain is its altitude above sea level, but its actual height is the difference in elevation between its summit and the surrounding country. For example, Pikes Peak in Colorado has an elevation of 14,110 ft, but the plains at its base are already nearly 6000 ft above sea level; consequently, the true height of Pikes Peak is a little more than 8000 ft. A closely spaced more or less linear group of mountains is called a *range*. A group of many ranges of essentially the same character and formed at approximately the same time is known as a *mountain system*. The Rocky Mountains and Appalachian Mountains are examples of such systems.

A classification of mountains by origin is quite arbitrary, as most mountains are the composite result of several processes. For convenience in discussion, however, mountains can be placed in three groups: depositional, erosional, and diastrophic.

DEPOSITIONAL MOUNTAINS

Hills and mountains that have resulted from the deposition or piling up of rock materials on the earth's surface may be called *depositional mountains* or *mountains of accumulation*. Most significant here are the great volcanic mountains and cones formed by the outpouring of lava and pyroclastics on the surface. The Hawaiian Islands are an outstanding example and are probably the highest mountain range on earth. The highest peaks stand more than 30,000 ft above the ocean floor.

Practically all these mountains have been modified by erosion. Although Mauna Loa is a great dome of uneroded lava, one of the neighboring inactive volcanoes is cut by 2000-ft canyons, the result of recent stream action. Mount Rainier and other famous volcanic mountains in North America are being carved by glaciers.

A few sand dunes and many glacial deposits are prominent enough to be mentioned as depositional hills, though they hardly qualify for the term "mountain." Terminal moraines that are several hundred to a thousand feet high do occur.

EROSIONAL MOUNTAINS OR PLATEAUS

Almost all mountains have been formed, at least in part, by erosion, or degradational processes. Those mountains, however, which have been carved out of flat-lying or undeformed rocks are sometimes referred to specifically as *erosional*, or *residual, mountains*, because erosion has been by far the most significant process in their origin. The majority of the erosional mountains have been cut from plateaus; for example, the Allegheny Plateau of West Virginia and the Catskill Mountains of New York are being carved from flat-lying sedimentary rocks. Both are parts of the Appalachian Plateau in a mature stage of dissection.

The Colorado Plateau is today cut by many canyons, notably the Grand Canyon. Out in the 7- to 13-mile width of this canyon are many great buttes and steep spires, separated from the canyon walls. Certainly these are mountains in every respect, but they stand within a great canyon and are thus dwarfed and overlooked in comparison with the canyon walls. They illustrate an early stage in the development of erosional mountains.

DIASTROPHIC MOUNTAINS

The majority of mountains and mountain ranges, with the exception of the deposi-

Figure 26.9
Fault scarp, MacDonald Lake, near Great Slave Lake, Northwest Territories, Canada. Trace of a fault in northern Canada with a scarp a few hundred feet high. (Royal Canadian Air Force.)

tional and erosional mountains just discussed, have been formed by the degradation of folded and faulted orogenic belts uplifted by diastrophic movements. Such mountains are commonly known as *diastrophic,* or *deformational, mountains*. The general shape and position of these mountain ranges are largely controlled by their structure. It must be emphasized, however, that not only diastrophic uplift and deformation but also erosional processes are necessary for the origin of this type.

Faulting is often expressed at the earth's surface by a fault scarp (Fig. 26.9); if the displacement is great enough, the uplifted side of the fault may be dissected into a mountain range. Fault-block mountains result in this way from normal faulting with the tilting and uplifting of long blocks of the crust. The Wasatch Range of Utah is an example of a fault-block range (Fig. 26.10). Most of the mountain ranges in Nevada are also of this origin. The Sierra Nevada Range of California is an exceptionally large fault block, 400 miles long by 75 miles wide. Its eastern side is the fault scarp, some 2 miles high.

Thrust or reverse faults also help form mountain ranges (Fig. 26.8) where resistant rocks are pushed up over softer, more quickly eroded rocks. The Lewis Range in Glacier National Park is carved from Precambrian sedimentary rocks that have been thrust at least 30 miles over Cretaceous rocks.

A great many diastrophic mountains result from folding of the crust. This may be a simple domelike uplift, such as the Black Hills of South Dakota, or a great anticlinal uplift complicated by faulting, such as the Bighorn Range of Wyoming. In either case, following erosion, old rocks are exposed in the center of the uplift or range, with the edges of progressively younger rocks lapping up on the flanks of the mountain mass.

Sedimentary rocks may be folded into a series of parallel anticlines and synclines, as in the Jura Mountains of the Alpine system and in the Appalachian Mountains. The resulting mountains are characterized by long parallel ridges where resistant strata outcrop at the surface.

Origin and history of diastrophic mountain systems

An amazing sequence of events has been outlined for almost all the great mountain systems of the world. The first and longest stage in this sequence is the deposition of sediments in a seaway or trough a few thousand miles long and a few hundred miles wide. Slowly more and more sediments accumulate as this trough, or geosyncline, sinks, until it is filled with a few tens of thousands of feet of rock. Then in the most spectacular reversal in all geologic processes, a geosyncline that has been slowly sinking for perhaps 200 million years is squeezed, deformed, and gradually uplifted, in a relatively short time, into a lofty mountain system.

Included here are such great systems as

Figure 26.10
A fault-block range, Maple Mountain, south-central Wasatch Mountains of Utah, looking east. A mountain range produced by normal faulting, the mountain being the uplifted fault block. (Compare with Fig. 26.8, top.) The actual fault plane runs from left to right across the picture at the foot of the range, and the small triangular ends of each spur are remnants of the fault plane. (Courtesy H. J. Bissell.)

the Appalachians, Rockies, Alps, Himalayas, and Andes. All these mountain systems have many features in common; they all exhibit a great thickness of sedimentary rocks, many batholiths, and other igneous bodies and all bear evidence to impressive amounts of crustal shortening. The history of the development of a typical mountain system can be summarized in three stages: the geosynclinal, the deformational, and the period of denudation and subsequent vertical uplifts.

THE GEOSYNCLINAL STAGE

The sedimentary rocks in the great mountain systems consist of interbedded sandstones, shales, and limestones, almost all of which clearly indicate deposition in shallow or only moderately deep water. In the Appalachians, these rocks are 40,000 ft thick. Even greater thicknesses are indicated in other systems. The only possible explanation is a long seaway or trough that sank gradually as the sediments accumulated. Such a structure is known as a *geosyncline,* a trough a couple of thousand miles long and tens of miles wide. Adjacent to this sinking trough must have been a rising land mass, the erosion of which supplied the sediments to the geosyncline (Fig. 26.11a). The period of sedimentation, or geosynclinal stage, has usually lasted many millions of years, perhaps as much as 200 million. The sinking was not entirely uniform, as there is evidence of minor uplift and partial erosion of some of the newly deposited sediments.

THE DEFORMATIONAL STAGE

Eventually the crust under a geosyncline becomes weakened and the entire belt is deformed in response to crustal compression; the sediments are folded and faulted (Fig.

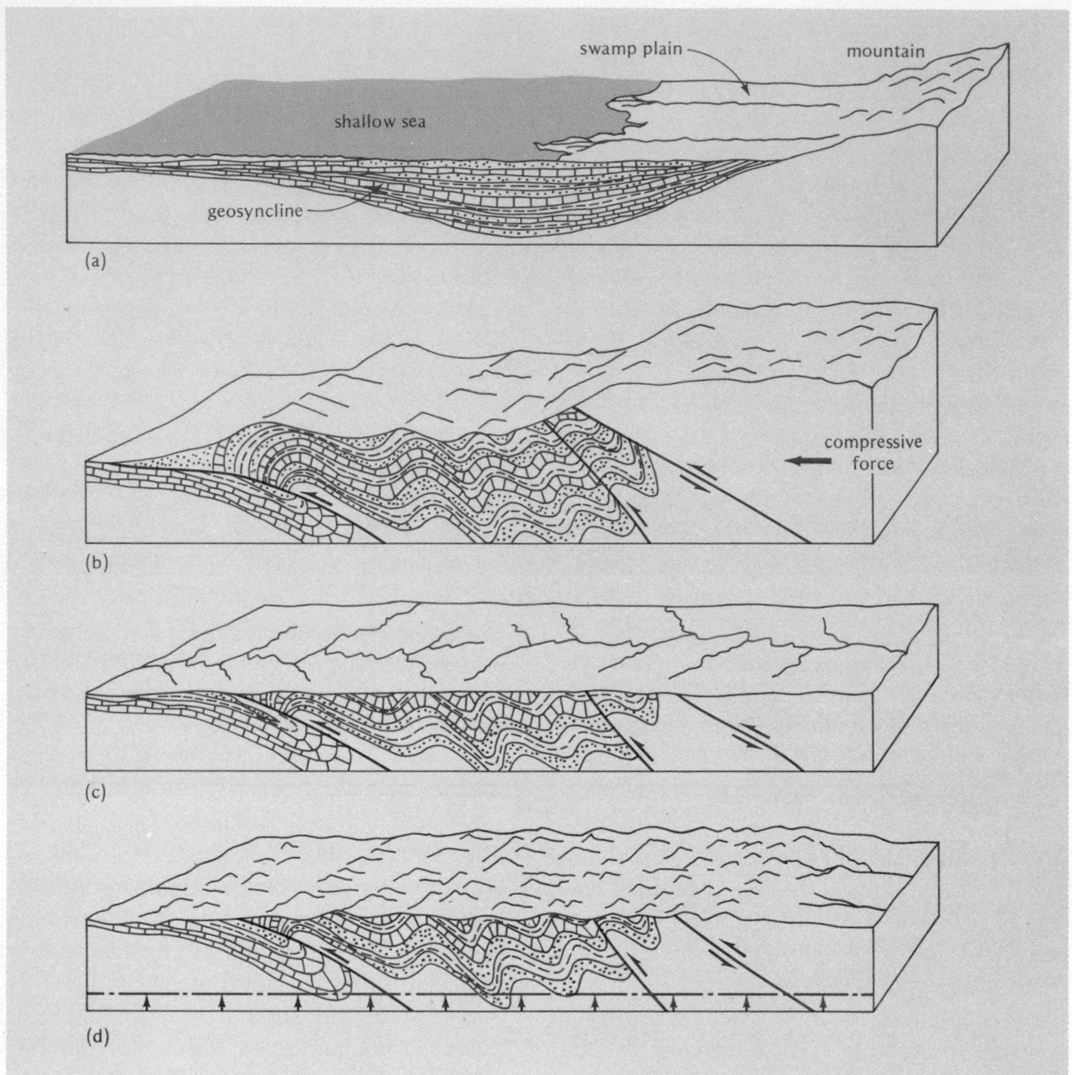

Figure 26.11
Development of geosynclinal mountains. A series of diagrams to illustrate the stages in the development of geosynclinal mountains. (a) Cross section of a geosyncline that has been filling with sediment and sinking for millions of years. (b) Compression of the crust results in folding and thrust faulting in the geosynclinal sediments, with shortening of this crustal zone. (c) Following the cessation of deformational movements, a long period of erosion has finally worn away the first generation of mountains produced by the original deformation. (d) After considerable time, vertical uplift may cause rejuvenation and renewed erosion to produce a second generation of mountains. This is the situation in the Appalachian Mountain system today. (Adapted from C. R. Longwell, et al., *Physical Geology;* copyright 1949. With permission of John Wiley and Sons.)

26.11b). Apparently the geosynclinal material is both pushed down into the crust and elevated into mountain chains. The lower parts or roots of the geosyncline become so heated that melting occurs to generate magmas, as previously outlined (p. 482), forming the batholiths and metamorphic rocks associated with most geosynclinal and orogenic belts. The period of active compression of the crust is short, compared with the period of sedimentation in the sinking geosyncline, but even so, it may last a few million years; the uplift is gradual, perhaps similar to the measured movement of 2 in./year of the crustal blocks on either side of the San Andreas fault. Igneous activity may continue long after the actual orogenic period.

PERIOD OF EROSION AND VERTICAL UPLIFTS

Erosion will begin as soon as any part of the geosynclinal belt is pushed above sea level. Thus the wearing down by erosion and the pushing up by diastrophism will go on simultaneously. When the diastrophic uplift is faster than the erosional wearing down, high mountains are produced; but perhaps these two sets of processes will balance each other, and the mountains will never attain any great height. When the compressive stresses of the orogeny are finally relaxed, erosion will become the dominant process, and in time the mountains will be worn down to a surface of low relief, or peneplain (Fig. 26.11c).

In most older mountain systems, notably in the Appalachians and the Rocky Mountains, evidence indicates several periods of simple vertical uplift or gentle arching, epeirogenic movements, long after the original deformation and after erosion of these mountains to peneplains. The chief evidence for these uplifts is the concordance or uniformity of summit levels of ridge and mountain tops and the presence of relatively flat, plateaulike areas atop many mountain ranges. These flat summit areas and the uniformity of mountain-top elevations can be the remnants of an uplifted and subsequently eroded peneplain. Such mountains are in a second cycle of erosion, and the streams have been rejuvenated to cut new canyons. Mountains have been formed again by erosion. The individual mountains and ridges are located where hard, resistant rocks outcrop at the earth's surface, and valleys where the softer rocks occur (Fig. 26.11d).

This sequence and combination of processes has been demonstrated beyond doubt, although geologists do not agree on a theory to explain the observed events. That compressive stresses have acted on the deformed geosynclinal rocks is obvious.

Causes of crustal deformation

The immediate cause of folding and faulting is response to a stress acting in the earth's crust. Some of the deformation in orogenic belts is due to lateral compressive pressures in the crust, as demonstrated by the crustal shortening evident in most mountains formed from geosynclinal sediments. But the ultimate cause of the compressive stress is due to internal forces beyond our observation.

The theory of isostasy suggests that major units of the earth's crust behave as though floating on a substratum and move vertically in response to changes in their total weight. At first glance we might assume that this would explain the sinking of the geosynclinal belts. The increasing weight of sediments will tend to deepen the geosynclinal trough but not to the extent of 30,000 or 40,000 ft. As lightweight sediments sink to lower levels where they are surrounded by denser materials, their own buoyancy will prevent further sinking. This is the same principle that prevents an iceberg from sinking indefinitely. The continued sinking of a geosyncline and the accompanying rise of an adjacent land mass, therefore, require other forces than gravitational adjustments.

CONTRACTION

One of the early hypotheses to explain crustal deformation held that the earth was cooling and consequently contracting. This meant that the crust became too long to fit the contracted interior, and lateral compression resulted in folding and thrust faulting, much as the skin of a fruit wrinkles up as it dries. Geosynclines represented weakened zones on the crust, so that they gave way to this compressive pressure. But since the discovery of radioactivity, we are quite certain that the earth's interior is not cooling off, and most geologists have abandoned the thermal contraction theory. Harold Jeffreys in 1952 proposed a modified theory suggesting that temperatures are constant down to about 100 miles, but cooling and contraction are in progress from there down to the 400-mile level.

Another modification of the contraction hypothesis assumes that the pressures within the earth are resulting in gradual increase in density and consequent decrease in volume. Thus without any temperature changes, the crust would develop lateral compressive stresses.

The chief weakness in the contraction hypothesis derives from the pattern taken by the young, active mountain systems and island arcs around the entire margin of the Pacific Ocean. As a result of internal contraction alone the pattern of recent deformation and earthquake belts should be more random; there is no mechanism inherent in the theory for localizing orogeny at the boundary between continents and ocean basins.

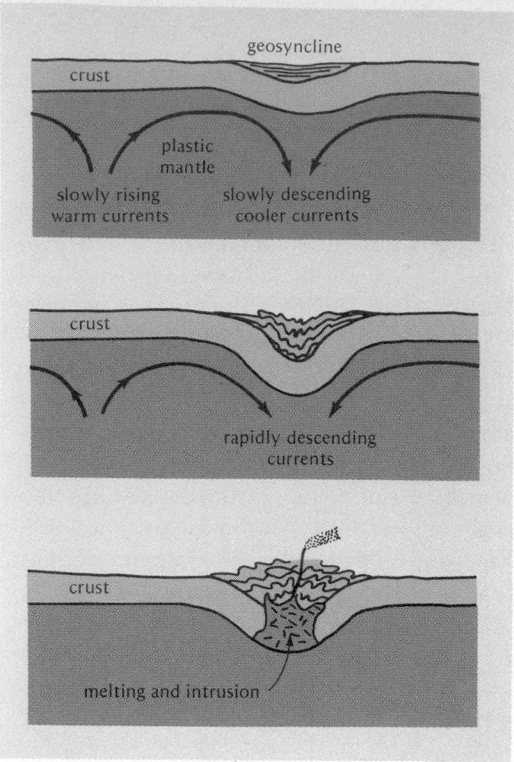

Figure 26.12
Subcrustal convection. The development of a geosyncline and its deformation on the assumption of subcrustal convection cycles. Top, the beginning of a convection cycle. Slowly rising warm currents spread beneath the crust, cool, and descend, thereby dragging down a geosynclinal trough. Middle, the convection system reaches its maximum speed; the crust beneath the geosyncline is dragged down deeper and more rapidly. This crustal downwarp is highly compressed; the sediments of the geosyncline are deformed. The lower part of the downdragged sialic crust becomes hot enough to melt and generate magma. Bottom, convection gradually ceases and the cycle ends. The new magma is intruded upward. Eventually that part of the crust which was dragged downward tends to float back up, owing to its low density, and cause the periods of vertical uplift or gentle arching that produce second-generation mountains. (After David Griggs, 1939.)

CONVECTION

Another hypothesis proposes slow discontinuous convection currents in the mantle from the crust to the core. It is proposed that relatively cool material beneath the crust sinks and warmer material from within rises in giant convective cells, thousands of miles in diameter. Where two slower sinking currents might meet, a portion of the crust could be dragged down, as suggested in Fig. 26.12.

This downdragging is just enough to maintain geosynclinal sinking at a rate equal to sedimentation for millions of years. Finally, the convective currents accelerate to the point where the geosynclinal belt becomes compressed and deformed. The downdragged sialic material becomes hot enough to melt and generate magmas. This would explain the association of orogeny, metamorphism, and vulcanism. Finally, the currents slow down and stop and the downdragged lighter sialic crust gradually rises, owing to isostatic adjustment, as expressed in the successive periods of vertical uplift and gentle arching of a mountain system during the period of erosion long after the deformational stage.

We believe, from earthquake wave behavior, that the mantle has some rigidity and is a solid. It could, however, conceivably deform over long periods by very slow plastic flow. The big objections to the convection theory have been the lack of any real evidence for it, the lack of a known inherent mechanism for localizing downward currents, and thus geosynclines, near continental margins, and the indication of a minor discontinuity or low-velocity zone in the mantle at a shallow depth. Convection currents should have destroyed any trace of zoning in the mantle. Today, however, various lines of geophysical research are suggesting new evidence that currents may be very important, indeed, in causing the earth's surface to break up into thick plates that are moving relative to each other and collide to produce mountain deformation. This is the theory of plate tectonics. So much evidence from various lines of geologic thought point to these new ideas that the entire next chapter is devoted to this theme alone.

Theory of isostasy

Is the earth's crust strong enough to support a mountain system? Before we answer this, let us imagine one of our largest steamships built on its pier or dock. Because of its tremendous weight, the ship would probably crush the dock structure and surely break its own keel. Yet this same great ship floats easily in the water. Granite has a compressive strength of about 30,000 $lb/in.^2$. If we estimate the total mass and weight of the Himalaya Mountains, we soon discover that the pressures near sea level under these mountains are greatly in excess of the strength of granitic rocks. We are led to the startling conclusion that the crust cannot possibly support such a weight. It seems the Himalayas are not resting as a dead load on a rigid crust of uniform thickness but that these mountains may be the top of a great mass projecting far into or below the crust and floating in a plastic material, just like our ship in water or a giant iceberg in the ocean.

This suggestion that mountain ranges and, in fact, whole continents stand high because of the great thickness of their mass and that ocean basins are low because their underlying rocks do not extend so deep is known as the *theory of isostasy*. Crustal blocks behave as if floating on a plastic interior. Isostasy, from the Greek meaning "equal standing," is a condition of equilibrium, or balance, of blocks of the earth's crust as a result of gravitation.

The analogy to icebergs floating in the ocean is good to a certain point. Just as only one-tenth of the iceberg is above water, with nine-tenths of its mass below water, so perhaps the mass of continents and mountains is largely below the average level of the lithosphere. The subcrustal material, however, is not a liquid; earthquake wave transmission proves it to be a solid (Chap. 25); but when we consider the tremendous pressures and high temperatures at depths of 30 to 50 miles in the earth, we must realize that even solid, brittle rocks become plastic and slowly yielding. Ice is a brittle substance and cracks to form crevasses on the surface of glaciers, but the weight of a few hundred feet of overlying ice causes plasticity of the deeper ice and slow flowage downslope. The

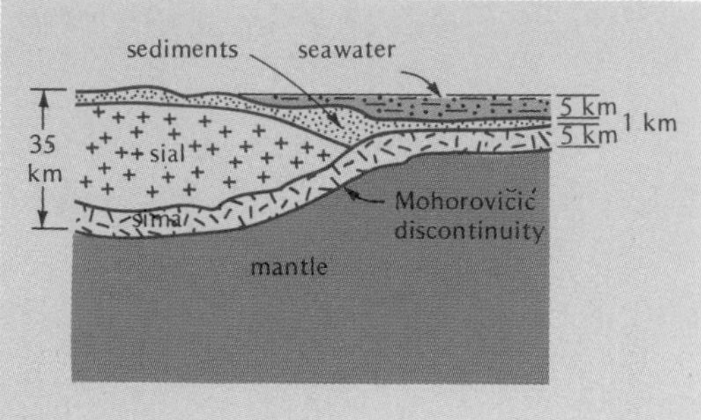

Figure 26.13
The earth's crust. An idealized cross section of a small hypothetical continent showing the structure of the crust above the Mohorovičić discontinuity. A thick layer of sial forms an upper crustal layer under the continent. A relatively thin but fairly uniform sima layer of basaltic composition underlies continents and ocean basins. The sima may be thin or absent under thickened sial.

upper couple of hundred feet of ice behaves as brittle blocks riding or floating on the plastic interior of the glacier. Hard pitch is a solid at room temperatures, and a blow with a hammer will fracture it into many pieces. Pitch has rigidity for stresses of short duration, but over long periods it behaves more like a liquid. A cube of pitch left on a flat surface will gradually flatten out into a pancakelike shape; and a hammer left on the surface of a barrel of pitch will slowly sink into it. Pitch has very little true strength (strength is continuing resistance of a body to permanent deformation or change of shape). Our subcrustal rock material behaves in the same manner as the pitch or deep glacial ice.

The theory of isostasy is supported by the behavior of the plumb bob and by gravity determinations. A plumb bob, originally a simple lead weight on a string, is pulled by gravity and points approximately to the center of the earth; at least this is true over nearly level plains away from prominent hills or valleys. If the bob is suspended over a plain at the foot of a mountain range, the bob will be deflected toward the mountains, whose mass exerts an attractive force on the bob. More than a century ago, however, while using a plumb bob near the foot of the Himalaya Mountains, a surveying crew realized that the deviation of the plumb line toward the mountains was considerably less than it should be if the mountain mass had the same average density as the rocks under the plain and the crustal materials under both mountains and plain were of approximately equal density and uniform thickness. The problem was so puzzling that it was reported to the Royal Society without a solution. Shortly thereafter, in 1855, J. H. Pratt, a clergyman, and G. B. Airy, an astronomer, proposed similar explanations that were the first formulations of the theory of isostasy. Airy suggested that the mountains had deep roots, of density comparable with that of their own material, projecting down into a subcrustal layer of higher density. This extension of lower-density material at depth under the mountains, and not under the plain, could explain the limited attraction of the plumb bob to the Himalayas.

The actual value of gravity has been measured for thousands of stations on the lands and over the oceans. After correcting these readings for latitude, altitude, and the mass of rock between the station and sea level, the corrected measured values over high land areas are usually lower (negative anomalies) than the theoretical values for the same stations. Conversely, corrected measured values for sea stations are often slightly in excess (positive anomalies) of the theoretical. These anomalies can be explained by assuming that the continents are composed of thick, relatively low-density, granitic material, or *sial* (rocks rich in silica and alumina), and that the crust under the ocean basins is composed of thin but heavier *sima* (rocks rich in silica and magnesium), somewhat as shown in Fig. 26.13. A thick, low-density, sialic mass under the continents would cause relatively low gravity readings; a thin, high-density crust of sima under the oceans explains the higher gravity values at sea stations. In addition to satisfying the gravity readings this concept is in accord with geologic data, which we cannot summarize here.

Isostasy apparently explains many vertical movements of the epeirogenic type. A great deal of data has been collected that shows a sinking of northern Canada centered near Labrador and of northern Europe centered over part of Scandinavia that corresponds in time and place with the accumulation of thick, glacial ice caps, similar to that now covering Greenland. After these glaciers finally melted, these same areas began to rise, as we have discussed earlier in the chapter. Ice accumulation placed a load on the crust, which gradually settled in a gravitational adjustment, and ice on melting caused an opposite adjustment.

Similarly, the shift of weight by gradational processes upsets the gravitational balance, or isostasy, of the crust. Long-continued erosion of a mountain range removes appreciable weight from this part of the crust. And just as the unloading of a ship causes it to rise in the water, so we might predict a rise in the eroded mountain chain. Careful study of mountainous regions has indicated that during their erosion several periodic epeirogenic uplifts characteristically do occur. The time necessary for the erosion of a mountain range to a peneplain is enormously prolongated due to these repeated isostatic uplifts.

Deposition of the eroded material places a load on the crust, usually concentrated in inland seas and deltas. Geologic studies and deep-well records have indicated that deltas are indeed sinking regions.

Finally, we are left with the belief that the major units of the crust are in or near isostasy, the ideal condition of floating equilibrium, or gravity balance. The mechanism of adjustment beneath rising and sinking blocks of the crust remains, of course, a mystery. The subcrustal material is solid, but after a certain critical stress, due to overlying weight, has been exceeded, this material must begin to deform plastically and slowly flow like pitch or ice in a glacier. We might suggest that subcrustal material could be displaced from under a sinking block to move under a rising block. But the earth is not this simple. The eroding and rising mountain chain may be separated from the sinking delta by 2000 miles of stable interior plains.

SUMMARY

Vertical movements of parts of continents are indicated by elevated and submerged coastlines and by uplifted sedimentary rock plateaus. Orogenic movements, or deformation, are indicated by folds, faults, and joints.

Three types of mountains may be recognized: depositional, erosional, and diastrophic. Most of the great mountain systems belong to the third category.

The great diastrophic mountain systems have passed through a geosynclinal stage, during which a sinking seaway or trough filled with shallow-water sediments 25,000 to 40,000 ft thick. Later, these sediments were folded and faulted during a deformational stage of mountain building. Finally, the mountain systems passed through a long period of erosion and successive vertical uplifts, or rejuvenation.

Crustal deformation may have been caused by contraction of the outer part of the earth, by slow convection currents in the mantle, by continental drift, or by a combination of two or more of these.

The major crustal blocks appear to be in gravitational balance, or isostasy, as though they were "floating" on a plastic interior.

Important words and terms

orogenic and epeirogenic movements
law of original horizontality
strike and dip
folds: anticline, syncline, monocline
fractures: faults and joints
normal, reverse or thrust, and strike-slip faults
plains, plateaus, mountains, mountain systems
depositional or accumulation mountains

erosional mountains or plateaus
diastrophic or deformational mountains
mountain building
geosynclines
geosynclinal stage
deformational stage
period of erosion or vertical uplift
isostasy
sial, sima

Questions

1. What types of folds and faults are due to tensional forces? to compressional forces?
2. A thick series of folded sedimentary rocks has risen many thousand feet above sea level. What will be the isostatic response to continuing erosion of these folded rocks?
3. The Gulf of Mexico is a modern, growing geosyncline. What relationship does the delta of the Mississippi River have in this concept? To what extent may the theory of isostasy be involved?
4. A large basin, a few hundred miles in diameter, such as the Michigan basin, continued to fill with sediment through several periods of geologic time. Can isostasy be called on as the cause for such continued sinking? Why or why not?
5. What features would you look for in the field to distinguish an angular unconformity from a low-angle thrust fault?
6. Can you relate the past great ice age and elevated shorelines in various parts of the world to the theory of isostasy?

Supplementary readings

Clark, Sydney P., Jr., *Structure of the Earth,* Foundations of Earth Science Series, Prentice-Hall, Englewood Cliffs, N. J. (1971). (Paperback.) [Chapter 2 discusses folds and faults.]

McAlester, A. L., *The History of the Earth's Crust,* Foundations of Earth Science Series, Prentice-Hall, Englewood Cliffs, N. J. (due in 1974). (Paperback.)

Omer, Guy G., et al., *Physical Science: Men and Concepts,* Part 6, "Mountain Building." Heath, Boston (1962). [The development of the theories of mountain building as advanced in the nineteenth and early twentieth centuries is reviewed, with sections of the original writings of James D. Dana, James Hall, John H. Pratt, George B. Airy, and Clarence E. Dutton. Modern explanations of mountain building are discussed and compared.]

See also readings on continental drift as listed at the end of Chap. 27.

CONTINENTAL DRIFT AND PLATE TECTONICS

Exploration is the key. Power is the hand that turns it.

E. DUKE-VINCENT, 1966

The sea floor may be spreading away from the world's mid-ocean ridge and rift system. Studies of earth magnetism and the magnetic properties of rocks give us a determination of ancient magnetic orientations and indicate that the geomagnetic pole has wandered appreciably from its earlier position. These paleomagnetic and ocean-floor spreading data strongly support a hypothesis of continental drift first proposed to explain the fit of rock types, ancient climates, geologic structures, and geography across land areas now far apart. The modern drift model proposes a mosaic of six major plates and several smaller ones which form the earth's lithosphere or outer shell.

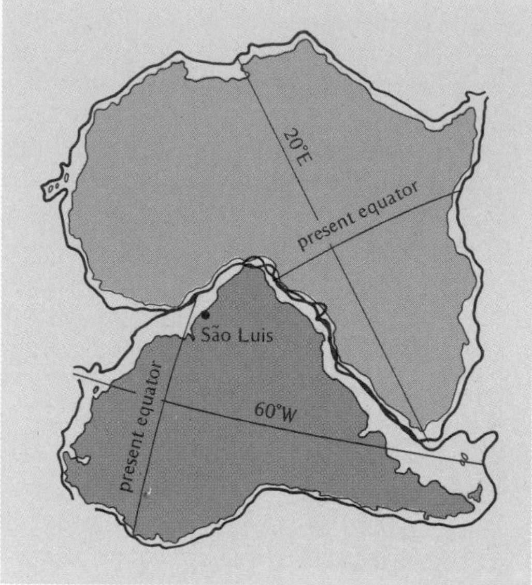

Figure 27.1
Fit of South America and Africa. South America and Africa fit very closely together, particularly when the match is made 6500 ft below sea level. (Redrawn from S. Warren Carey.)

Continental drift

Alfred Wegener, a German geophysicist, put forward in 1912 a monumental theory in which he proposed that crustal blocks of continental size may have moved laterally across the earth's surface. Similar ideas had been suggested many years earlier by others, but Wegener was the first to bring together in a scientific fashion the various lines of evidence that led him to his theory. Wegener and others had noted the apparent fit of the bulge of the east coast of South America to the west coast of Africa. Today we note that these continents do fit very well indeed, especially if the fit is made at the edge of the continental shelves as shown in Fig. 27.1. There are also good, though less spectacular, matches of other continents—for example, North America against Europe and North Africa. Wegener suggested a single super continent about 200 million years ago, which he called "Pangaea," that split up and drifted to the positions of the present continents.

These proposals seemed quite fantastic and unlikely; there was no known mechanism whereby continents could move horizontally over or through the mantle like ships plowing through the seas. As a consequence the theory of continental drift was abandoned and "buried" by most American geologists in the early 1930s, in spite of much geological and paleoclimatological supporting data. However, new discoveries were to be made; and since World War II, the idea has been revived on the basis of the data on rock magnetism, the pattern of heat flow from the continents and ocean basins, and the evidence of sea-floor spreading away from mid-ocean ridges. The idea of convection currents in the mantle has been applied as the mechanism for continental movements, and the major objections to both the convection and drift hypotheses seem to be answered by this "marriage" of the two theories. Now many geoscientists believe that continental drift must have occurred and is, in fact, still occurring. We are on the verge of what J. Tuzo Wilson calls a "major scientific revolution" in geologic thought.

GEOLOGICAL EVIDENCE

In addition to the geographic fit of South America and Africa, a considerable similarity in geology has been recognized. The rock types in Brazil, the ages of these rocks based on isotopic data, and their structural patterns are very similar to those in central Africa. At the town of São Luis on the Atlantic coast of Brazil there is a sharp contact between gneisses that are 2 billion years old and much younger rocks that are about 650 million years old. Both of these groups of rocks, with their sharp contact, can be found on the west coast of Africa, almost directly opposite São Luis when the two continents are brought together in their position of best fit as shown in Fig. 27.1. Also, the Sierras near Buenos Aires fall opposite the Cape Mountains of South Africa; both are mountain structures of the same age (Permian).

In rocks of Pennsylvanian and Permian age in several parts of South Africa and southern South America, fossil plant leaves (the Glossopteris flora) show extreme similarities. In fact, many species of fossils of these ages are found not only in South America and Africa but also in parts of India and Australia. Just recently a rare fossil reptile was found in Antarctica that had previously been known only in the earlier mentioned regions.

Another example of geologic match in terms of rocks and structures of similar ages exists between North America and western Europe; specifically, this match is between the older structures of the northern Appalachian Mountain system in New York, New England, Nova Scotia, and Labrador, and the mountainous belts in Ireland, Scotland, and Norway as shown in Fig. 27.2. In these regions there is amazing similarity of deformation of Middle Paleozoic time, the so-called Taconic revolution in New England and Nova Scotia, and the Caledonian in Scotland and Norway. Folding and metamorphism occurred on both sides of the Atlantic at approximately the same time producing similar rock types, igneous bodies, and thrust-fault structures.

Figure 27.2
Appalachian Mountain deformational zones. A trace of Taconic structures of Middle Paleozoic time in the United States, Nova Scotia, and Labrador as they line up with Caledonian structures of the same age in Ireland, Scotland, and Norway (shown in red), when the continents of North America and Europe are brought back to the position they may have had during Paleozoic time. Base map after a computer match of continents by Sir Edward Bullard and others at Cambridge. The fit was made along the continental shelf margin at the 3000-ft contour line. Regions where land masses, including the shelf areas, overlap are in dark brown; gaps in the match are light brown.

Most impressive of all, perhaps, are the Permian tillites (consolidated glacial moraine deposits or boulder clay tills) found in southeastern South America, in South Africa, India, and Australia, indicating a glacial period some 250 million years ago. If all these areas were one great continent in Permian time, the various known tillites

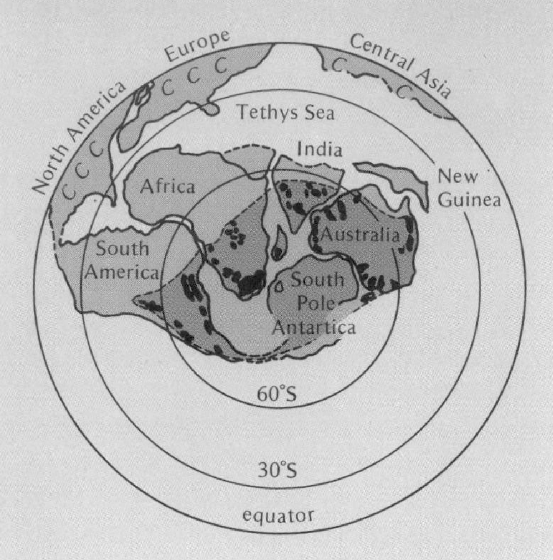

Figure 27.3
Supercontinent of Permian time. The single continent presumed to have existed from 200 to 300 million years ago approximately over the South Pole position. This continent broke up and drifted apart to form the present-day land masses as labeled. Brown spots are tillites from glacial deposits of Permian time; the dark red area is the distribution of Permian ice sheet; C indicates coal deposits from tropical forests of Pennsylvanian time. Reconstructed from paleoclimatological data and paleomagnetism.

would all have clustered around the then south pole (Fig. 27.3). When all Pennsylvanian and Permian rocks of the world are studied carefully, we find that the great coal deposits of North America, Europe, and Siberia were formed from tropical forests near the earth's equator (Fig. 27.3) and desert-type sandstone, salt, and gypsum deposits of similar ages were deposited in what would have been the trade wind desert belts of the subtropics. The "fit" of these indicated climatic zones with the postulated south pole and equator locations for Pennsylvanian–Permian time is too good to be entirely coincidental.

Paleomagnetism

Some minerals are strongly magnetic, magnetite in particular; others, such as hematite, display moderate magnetism. Rocks that contain such minerals in scattered grains are said to show *rock magnetism,* which is known as the rock's *natural remanent magnetism.* This remanent magnetism may have been developed in any of several different ways and may or may not correspond to the present magnetic field of the earth.

As magmas cool and crystallize, certain minerals acquire a permanent magnetism once they are cooling below their Curie temperatures. The magnetic polarity of these magnetic minerals takes on the orientation of the earth's magnetic field both as to declination and inclination or dip. This is called *thermal remanent magnetism* and is extremely stable for millions of years, unless the rock is later reheated above the Curie temperature (p. 428) of the magnetic minerals (580°C for magnetite, for example). This does sometimes happen during high-temperature metamorphism. The rock then loses its original magnetism but takes on a new, or secondary, thermal remanent magnetism when it again cools below the Curie temperature.

Sedimentary rocks, especially sandstones, may acquire a remanent magnetism during deposition if magnetite or hematite grains are involved. As such mineral grains are deposited, they tend to settle with their magnetic poles in the earth's magnetic field. When the sediments become consolidated, the magnetic orientation of the iron oxide minerals becomes permanent; this is known as *depositional remanent magnetism.*

The measurement and interpretation of natural remanent magnetism in rocks allow us to study the earth's magnetic field as it was in the geologic past at the time and place of rock formation. This study is called *paleomagnetism.* Analysis of the magnetic orientations in Tertiary and Pleistocene lavas tells us that the earth has had a stable dipole field, in spite of secular changes, since at least mid-Tertiary times. During Pleistocene and recent times, the geomagnetic pole has moved or wandered around the geographic north pole, but never further away than at present. This supports the dynamo theory of the earth's magnetism, which implies that the magnetic dipole axis must remain close to the axis of rotation (see Chap. 20).

The data of paleomagnetism show us the ancient directions of the geomagnetic field at given times and places. If we accept the two basic assumptions that (1) the earth has always had a dipole magnetic field and (2) the geomagnetic poles have always remained close to the geographic poles, or axis of rotation, then we find that prior to mid-Tertiary time some land areas had latitudes different from their present ones; or, in other words, the poles have wandered. These two assumptions seem logical in view of the dynamo theory of the earth's magnetism previously mentioned. As an example, the earth's paleomagnetic north poles, as plotted from magnetic data from Europe and North America for early Precambrian time, wandered from the Pacific Ocean off Lower California into the south Pacific and on westward to the central Pacific by Cambrian time, over eastern Asia and Japan by Pennsylvanian time, and on to their present positions by the mid-Tertiary, as shown in Fig. 27.4.

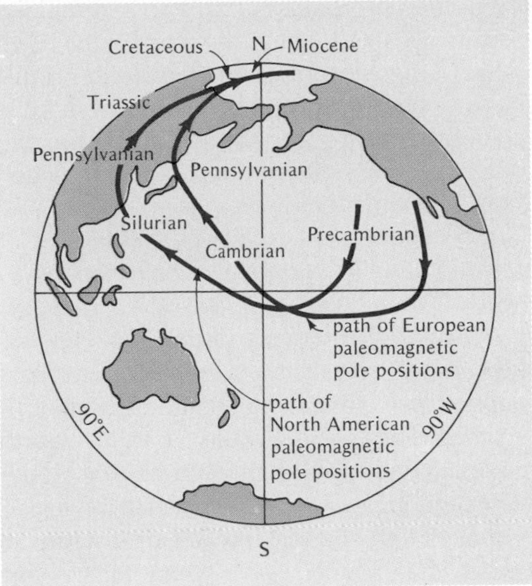

Figure 27.4
Paleomagnetic pole positions. Paleomagnetic measurements of rocks from North America and Europe show the paths followed by the magnetic poles of these two continents from Precambrian time to the present. (Redrawn from Cox and Doell, "Review of Paleomagnetism," *Geological Society of America Bulletin,* **71.**)

Because ancient pole positions before the Tertiary period were in different locations as indicated from various land areas for the same time periods, lateral shift of continents is strongly indicated. For example, the position of India (at Bombay) has shifted from latitude 40°S in Jurassic time to its present position at latitude 19°N according to the inclination of the remanent magnetism in lavas of various ages in the Deccan Plateau of India. As already mentioned, the pole-position curves (Fig. 27.4) as calculated from paleomagnetic data from Europe and North America are approximately 30° of longitude apart from Precambrian to Triassic times. This difference seems too consistent to be due to errors of measurement. The presence of two curves cannot mean that there were two geomagnetic poles, if our assumptions on the dipole magnetic field of the earth are correct. However, if we imagine North America moved 30° eastward, against Europe, then only one pole position is indicated before Triassic time. The assumption can now be made that the orientation of remanent magnetism in rocks of Precambrian to Triassic ages was set when the two continents were together and there was no Atlantic Ocean. Since Triassic time the continents have slowly drifted apart and they had approached their present positions by Tertiary time. The Tertiary paleomagnetic data from both continents point to approximately the same geomagnetic pole position, and the two curves of Fig. 27.4 thus become one.

Many geoscientists interpret these data to suggest that a slippage of the earth's crust over the underlying mantle has occurred to produce an apparent movement, or wandering, of the poles. In short, therefore, the magnetic and rotational axes have remained essentially fixed within the mass of the earth while different points on the earth's crust have moved successively over these pole positions. It also appears possible from the difference in pole positions, as calculated from paleomagnetic data from Europe and North America, that major crustal blocks, or areas—namely, continents—have shifted in relation to each other.

A startling discovery of these studies has been that reversals in polarity have occurred once or twice in each of the last few million-year periods. This means that the north geomagnetic pole has become the south geomagnetic pole and vice versa owing to some sort of sudden reversal of polarity in the earth's magnetic field. What might have caused this is very difficult to explain at present. However, a geochronology of reversals versus normals (or nonreversals) for the past 4 million years of Pleistocene and Pliocene time that holds true in all parts of the world has been worked out in basaltic lavas (Fig. 27.5). This geomagnetic time scale of polarity reversals was determined by measuring both the age and direction of magnetization in carefully selected lava flows from various parts of the world. The actual ages of the flows were determined by isotope techniques, largely potassium–argon ratios. The studies indicate a pattern of alternating polarities with a periodicity of approximately 1 million years. These major intervals are termed *polarity epochs*. The careful detail of the measurements also indicates polarity changes of much shorter intervals, about 100,000 years, which have been called *polarity events* (Fig. 27.5). The analytical errors in exact age measurements make it impossible to date these polarity events further back than about 4 million years. However, it is clear that reversals have occurred periodically throughout geologic time, and an amazingly complete history of such reversal epochs is now known extending back about 76 million years into the Cretaceous period.

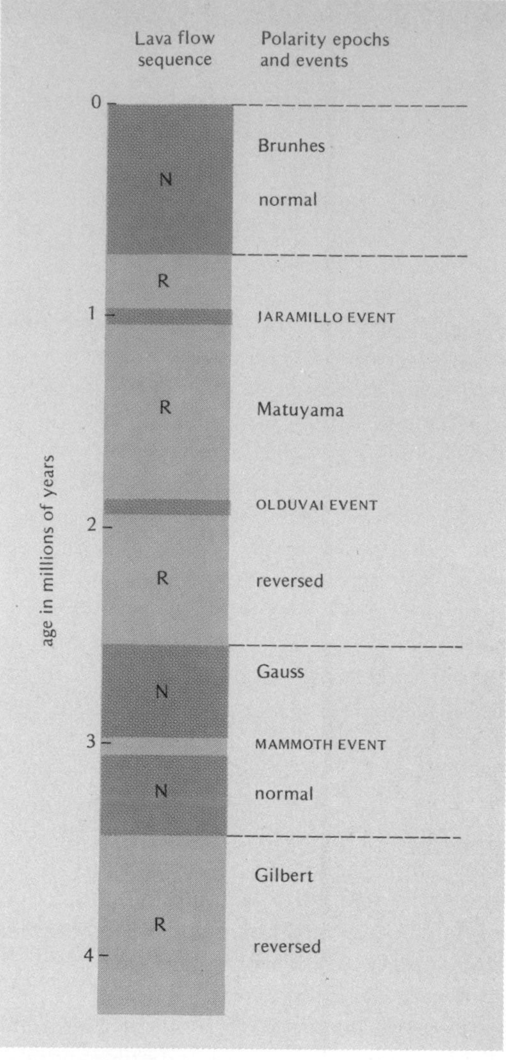

Figure 27.5
Geomagnetic time scale of polarity reversals as determined by the direction of paleomagnetic particles in lava flows dated by isotope methods.

Ocean-floor topography

The ocean floor may be divided into two major areas: (1) the continental shelves and their slopes into deeper water and (2) the deep-ocean floors. The continental shelves are under the relatively shallow water that surrounds most continents, as already mentioned, and are illustrated in the North Atlantic Ocean (Fig. 20.1). The continental slopes are relatively steep escarpments modified in many parts of the world by impressive submarine canyons. These slopes drop off rather abruptly at the edge of the shelves to the deeper parts of the oceans and there merge with the deep-ocean floor. The latter includes some very flat surfaces but is modified by a great variety of rises, ridges, scarps, seamounts, mountain ranges, and trenches. Ocean-bottom soundings by both commercial ships and oceanographic vessels have recently given a detailed picture of the ocean bottom. We know today that the ocean topography is as varied as that of the land surface.

CONTINENTAL SHELVES AND SLOPES

Continental shelves surround all continents but vary appreciably in width; their average width is about 40 miles. The water depth at the edge of the slopes averages about 500 ft, but some shelves are as deep as 1000 ft. The surfaces of the shelves are covered by sediments, most of which are similar to mud in size. However, there are many bands and filled depressions containing sand and even gravels, particularly on northern shelves in the belt of Pleistocene glaciation. It has usually been assumed that with gradual increase in depth and distance from shore, the sediment size decreased progressively, with sands near shore and muds further offshore. Recent investigations indicate, however, that this ideal distribution of sediments is rather rare and that the actual distribution is quite irregular, so that many sand lenses are formed far from shores.

At the outer edge of the continental shelves, steep slopes drop off abruptly 6000 to 8000 ft toward the deep-ocean floors. These steep scarps are known as the *continental slopes*. The exact origin of these slopes is not yet understood; some scientists have suggested that many of them may be fault scarps. The slopes do represent the seaward limit of continental-type crustal material.

The continental slopes were thought originally to be taluslike—that is, covered by thick deposits of sediment. We now know that they are covered by only a skin-deep layer of mud. It seems that all sediment has been swept away, downward and seaward, by landslides and turbidity currents, to be deposited beyond the foot of the continental scarps in deeper waters. Such deposits beyond the continental slope make up the gradual slope of the fairly smooth continental rises.

Submarine canyons are a major feature of the continental slopes. Hudson Canyon, 4000 ft deep and 180 miles long, is a typical example. It starts some miles offshore from New York Harbor at the outer edge of the continental shelf and extends through the continental slope and rise, to a depth of about 14,000 ft. It is associated near its terminus with a fanlike deposit of sediment probably brought down the canyon itself. The Congo Canyon off the west coast of Africa starts in the wide estuary of the river of the same name and extends seaward across the continental slope for 145 miles but has no known "delta" deposit.

The origin of submarine canyons has been the subject of much discussion. The similarity in shape, depth, and extent to major river canyons on land has caused some scientists to suggest river erosion when sea level was much lower. This theory does not seem plausible in the light of present knowledge. More acceptable is the suggestion that submarine canyons have been cut by turbidity currents that carried great volumes of sediments off the continental shelves with sufficient abrasive action to erode deeply. Many of the canyons do terminate within the continental rises in what appears to be a thick fanlike deposit.

DEEP-OCEAN FLOORS

The deep-ocean floors are characterized by *abyssal plains*—very flat surfaces, the flattest known on earth, in fact. These plains lie 15,000 to 18,000 ft below sea level. Rises and ridges are narrow, elongate elevations above the plains. Rises have gentle, smooth slopes, ridges more irregular and much steeper slopes. *Seamounts* are isolated conical hills or mountains rising at least 3000 ft above their surroundings. Some are flat topped and are known as *guyots*. Most of these seamounts and guyots are apparently of volcanic origin; more than 10,000 of them are known in the Pacific Ocean alone.

Basins are approximately equidimensional depressions on the sea floor. *Trenches* are steep-sided, elongated depressions, and deep trenches—those that extend below 20,000 ft—are known as *deeps*.

Deep-sea fans, cut by many shallow channels, have the shape of giant underwater alluvial fans. Such fans are closely associated with the continental rises previously mentioned. Two exceptionally extensive deep-sea fans lie on either side of India, extending southward from the continental slopes south of the deltas of the Indus River and Ganges River, respectively. The Ganges fan or cone has a surface extent greater than that of all India.

MID-OCEAN RIDGES

A worldwide system of mid-ocean ridges (Fig. 27.6) is perhaps the most amazing feature of the ocean floors. Best known of these is the Mid-Atlantic Ridge, which is traceable from Iceland southward into the South Atlantic Ocean and around between Africa and Antarctica into the Mid-Indian Ocean Ridge.

The Mid-Atlantic Ridge is an impressive mountain range rising 2 or 3 miles above the sea floor. The complete ridge with its gradual slopes is approximately 1000 miles wide, but the steeper part of the ridge is perhaps only 300 or 400 miles wide. The overall size and dimensions of this ridge are approximately comparable with those of the Cordilleran mountain system of North America, although the two great features have entirely different origins. The Mid-Atlantic Ridge and, in fact, all mid-ocean ridges are offset by many transverse fracture zones or faults as shown on the map (Fig. 27.6). These fracture zones

Figure 27.6
Worldwide pattern of sea-floor spreading is evident when magnetic and seismic data are combined. Mid-ocean ridges (*brown lines*) are offset by transverse fracture zones (*thin lines*). On the basis of spreading rates determined from magnetic data, the author and his colleagues established "isochrons" that give the age of the sea floor in millions of years (*broken thin lines*). The edges of many continental masses (*light red lines*) are rimmed by deep ocean trenches (*hatching*). When the epicenters of all earthquakes recorded from 1957 to 1967 (plotted by Muawi Barazangi and James Dorman from U.S. Coast and Geodetic Survey data) are superposed (*dark red dots*), the vast majority of them fall along midocean ridges or along the trenches, where the moving sea floor turns downward. (From "Sea-Floor Spreading," in *Continents Adrift,* by J. R. Heirtzler, W. H. Freeman and Co.

are totally unlike any on continental mountains. A sharp, deep rift zone runs down the center of the Mid-Atlantic Ridge and other mid-oceanic ridges. Active and recently extinct volcanoes are associated with the ridge. These include Iceland and the Azores in the North Atlantic and Tristan da Cunha and the Ascension Islands in the South Atlantic. The central rift zone is traceable across Iceland as a zone of down faulting and active volcanism.

Sea-floor spreading

About 1960 the late Harry H. Hess of Princeton University proposed that the ocean floor might be in motion. He suggested that the sea floor cracks open along the crest of the mid-ocean ridges, that new sea floor forms there as material rises from some depth below in the mantle, and that the sea floor then spreads apart on either side of the ridge crest (Fig. 27.7). Wegener had believed that each continent moved independently, so that it

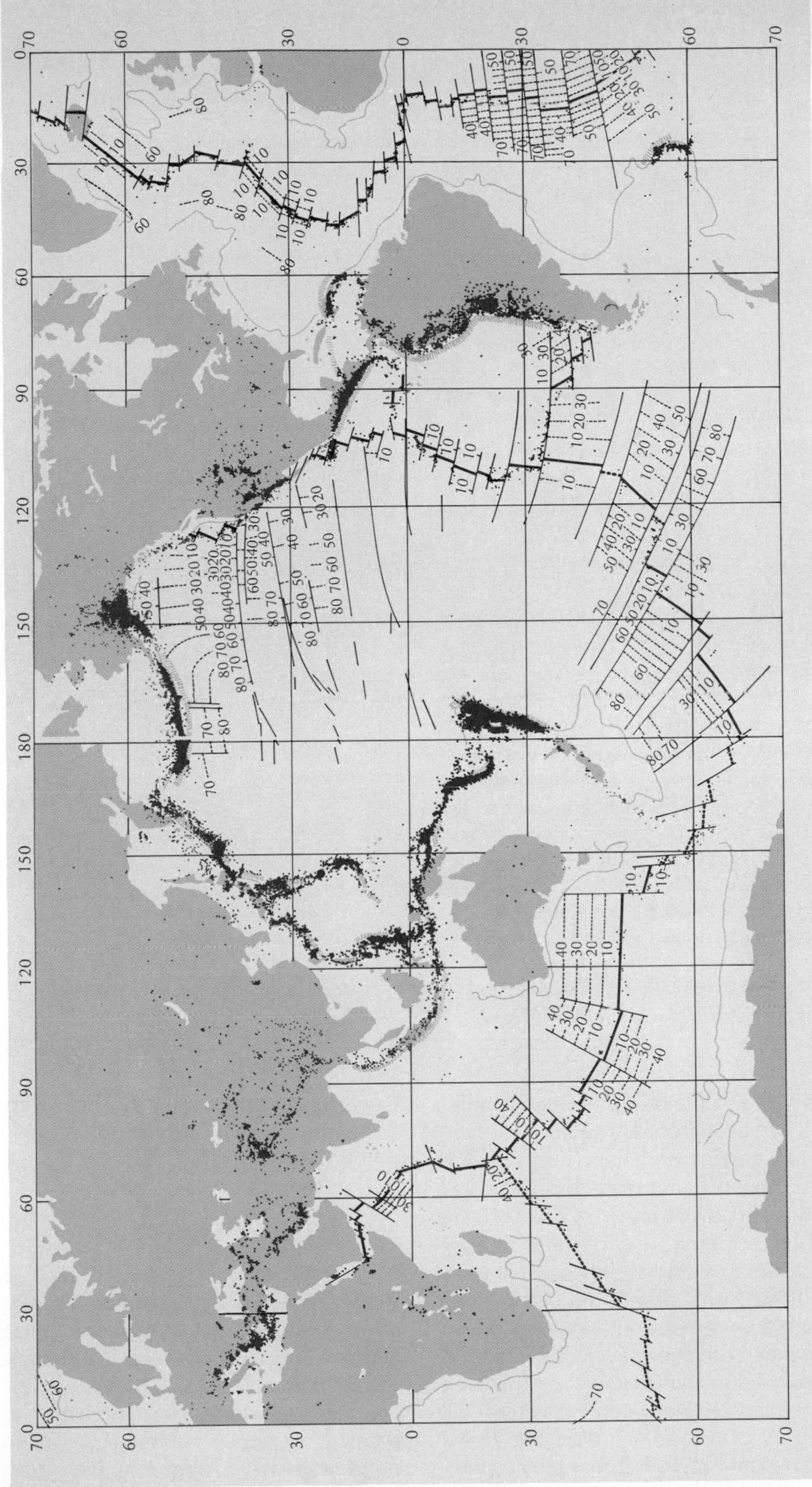

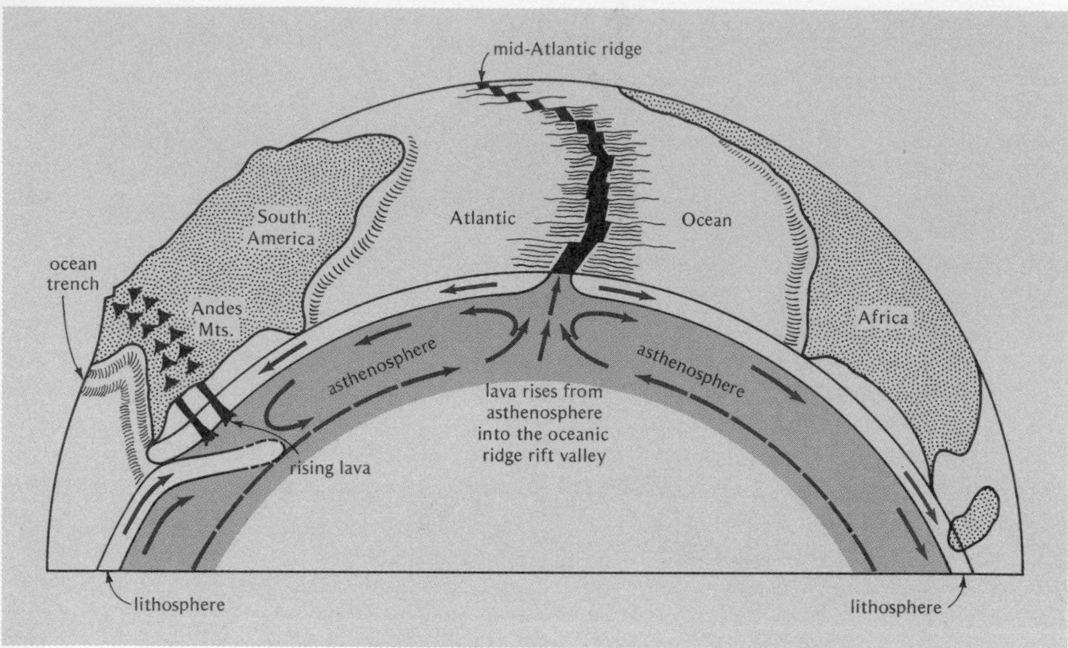

Figure 27.7
Schematic representation of sea-floor spreading and plate tectonics. Large plates of lithosphere containing the continents migrate away from the oceanic ridges as if on a conveyor belt, and the plates are carried into the earth's interior along boundaries where plates collide, called *subduction zones,* as at left. (Reprinted with permission from "Earthquakes and Continental Drift," by Peter J. Wyllie, the *University of Chicago Magazine,* January/February, 1972. Copyright, 1972, The University of Chicago.)

behaved like a ship plowing through a yielding ocean floor with new sea floor forming behind the moving continent; but no such pattern has been found. Hess proposed that the continents do not move as independent blocks but as rafts frozen into and moving with a sea floor as rigid as they are. New sea floor is formed only at the mid-ocean ridge crests, and moves away on opposite sides as if on gigantic conveyor belts. Coupled with this theory was the idea that in cases in which continents and sea floor come together, the sea floor is absorbed by plunging down under island arcs or young mountains as indicated by deep sea trenches. These were subsequently named *subduction zones* (Fig. 27.7).

MAGNETIC ANOMALIES

Geologists were finally persuaded of the reality of sea-floor spreading in the mid-1960s, mainly from the evidence of magnetic anomaly strips. A detailed magnetic survey in the northeastern Pacific Ocean in 1958 revealed a pattern of remarkably straight parallel magnetic anomalies, alternating positive and negative, that were traceable for hundreds of miles. Eventually it became apparent that such anomalies are present on either side of and parallel to all the mid-ocean ridge crests and that the pattern of such anomalies on one side of a ridge crest is the mirror image of the pattern of anomalies on the other side (Fig. 27.6). This remarkable magnetic pattern went unexplained for about 5 years until Vine and Mathews at Cambridge suggested that the linear magnetic anomalies were caused by strips of ocean-floor material magnetized in alternate normal and reversed directions. As new sea floor formed along the ridge crest from lava from below, the material became magnetized in the then prevailing magnetic field of the earth as we have seen earlier; but as time has passed, the earth's polarity has reversed itself and, therefore, belts of new sea floor sometimes have normal polarity and sometimes reverse polarity. Vine and Mathews suggested that the linear magnetic anomalies were a recording in the sea floor of the history of polarity reversals. When they compared the patterns of the sea-floor anomalies with the geomagnetic time scale of polarity reversals, they found an amazing match of all details of both features. This has been accepted as essential proof that sea-floor spreading has taken place and is indeed still taking place.

If the rocks in a certain linear zone are magnetized in the same direction as the earth's existing magnetic field, the total magnetic field is strengthened to form a positive anomaly. On the other hand, if the rocks in a linear belt are magnetized in a reversed direction to the earth's present field, the total magnetic field over that belt is somewhat reduced and a negative magnetic anomaly will exist. Figure 27.8 shows diagrammatically a sea floor of parallel bands of alternate positive and negative magnetic anomalies formed by the intrusion of new lava at the center of a ridge and its outward spreading in both directions progressively with time. For example, let us assume, quite arbitrarily, that sea-floor spreading began at a ridge about 3 million years ago during the Gauss normal polarity epoch. As lava rose, cooled, and solidified at the ridge crest, it became magnetized in the direction of the existing normal magnetic field, producing a positive magnetic anomaly. As sea-floor spreading occurred, this strip of new magnetized crust was transported away laterally, half in each direction, as if on a conveyor belt, retaining parallelism with the ridge and producing the sections *ab* and *a′b′*, on Fig. 27.8. During the next million and a half years during the Matuyama reverse epoch, additional new sea-floor crust with a negative anomaly was formed and spread laterally, as shown in bands *bc* and *b′c′*; whereas the band *cc′* was formed—and is still forming—during the Brunhes epoch of normal polarity. Even the small polarity events during the Matuyama epoch show up as narrow, although sometimes discontinuous, positive anomalies in the wide negative anomaly belt (Fig. 27.8).

Correlation of linear sea-floor anomalies and the polarity reversal time scale has not only been confirmation of the theory of sea-floor spreading but has also provided a means for calculating the rate of sea-floor movement during spreading. The average velocity of the spreading movement is given by the distance that a particular anomaly belt has moved from the ridge crest divided by the age of that polarity epoch. This is the time needed for that strip of magnetized sea floor to move from the ridge crest to its present position. Measured in this way the average rates of movement over the last few million years have varied from 1 cm/year to more than 6 cm/year. Such spreading velocities are in agreement with the rate of movement along certain fault zones in California.

TRANSVERSE FRACTURE ZONES

Major fault zones in the ocean-basin floor have long been known. The thousands-of-miles long east-west trending faults in the eastern Pacific (Fig. 27.6) were the first to be studied in detail. These faults make promi-

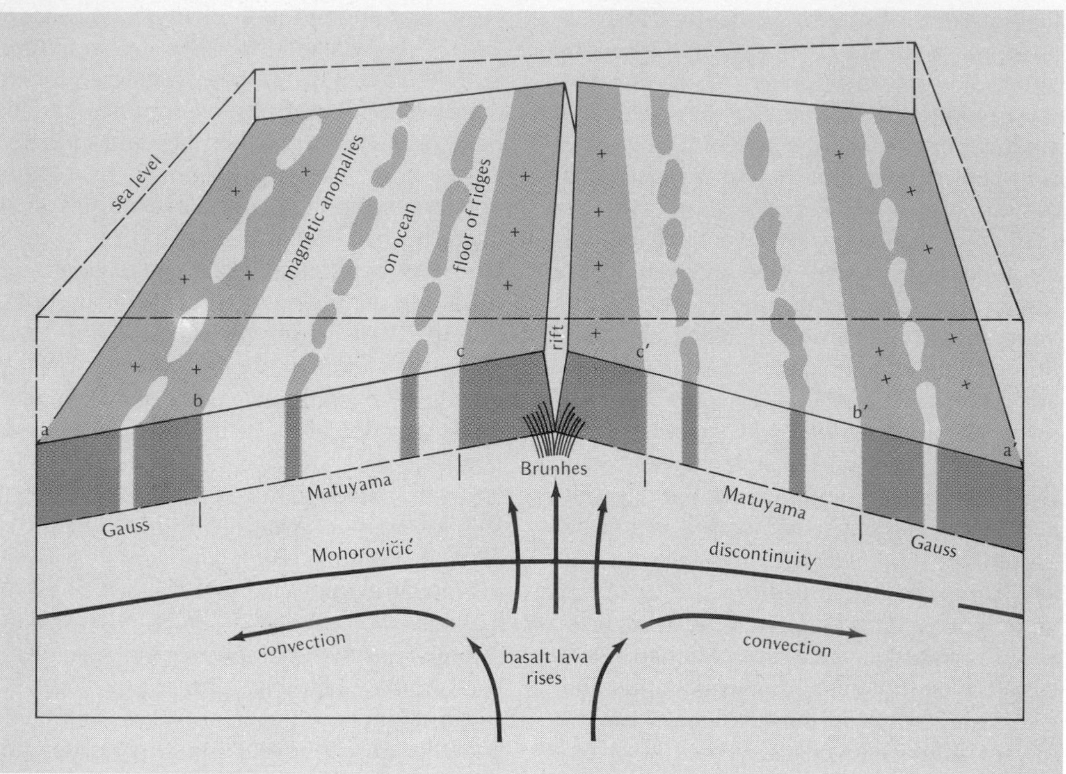

Figure 27.8
Ridge model of sea-floor spreading. Diagrammatic representation of the anomalies on the oceanic crust at a mid-ocean ridge crest, related to the polarity time scale epochs and events (see Fig. 27.5). Dark bands are normal magnetized, white bands are reverse magnetized.

nent scarps in the sea floor; some are thousands of feet high. Seismic and gravity studies have shown that these faults extend through the crust deep into the earth. When magnetic surveys were first made, it was found that the magnetic anomalies are offset. Similar faults have since been found offsetting all mid-ocean ridges (Fig. 27.6). A study of the earthquakes on those fracture zones that offset ridge crests shows that the separate pieces of ridge crest on the two sides of a fracture zone are *not moving apart,* as might seem likely at first. The two pieces of ridge crest remain fixed with respect to each other whereas on each side a plate of crust moves away as a rigid body; such a fracture is called a *transform fault.* Earthquakes occur only on the piece of the fracture between the two ridge crests; there is no relative motion on the fracture trace beyond the ridge crest (Fig. 27.9).

Plate tectonics

In modern usage the term *lithosphere* is used for the outer part of solid earth; it includes the crust and the upper part of the mantle and is in the order of magnitude of 100 km thick. This lithosphere is relatively cool and rigid, and therefore tends to fracture. It is underlain by hotter and more mobile, although still solid, material called the *asthenosphere* (Fig. 27.7), or middle mantle.

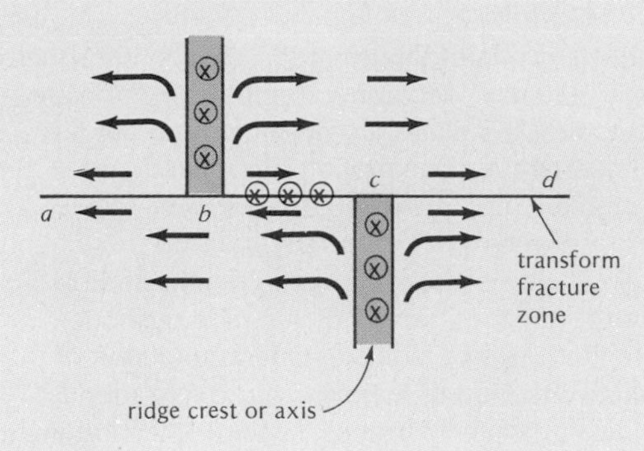

Figure 27.9
Motion on transform faults. Sketch of offset of mid-ocean ridge crest by fracture zone of transform fault type. Molten rock rises from depth along the ridge axis, solidifies, and is carried away (arrows) to allow subsequent upwelling. The spreading axis is offset by a fracture zone. Between the two offset ridge axes, material on either side of the fracture zone moves in opposite directions and the friction between the two blocks of sea-floor crust causes shallow earthquakes. Along *ab*, both sides are moving to the left; plates move in opposite directions only along *bc* and distance *bc* stays approximately the same; along *cd*, both sides move to right. Earthquake foci are shown by *X* on ridge crests and along *bc* section of fault only.

The lithosphere and asthenosphere are separated by a gradational zone 50 to 100 km thick variously called the *plastic zone*, the *low-velocity zone*, or the *upper asthenosphere*. This layering of the mantle has been demonstrated by careful analysis of earthquake records. The temperature in the plastic zone or upper part of the asthenosphere is very close to the melting range of its rocks; perhaps this zone is slightly melted and thus plastic and of lower velocity; it serves as the source zone for much magma. It is from here that material rises to the rift zone in the mid-ocean trenches according to the sea-floor model (Fig. 27.7).

The concept of plate tectonics proposes that the lithosphere is separated into a few rigid plates in motion relative to each other. These rigid plates, perhaps 100 km thick, are moving over the warmer, mobile low-velocity zone of the asthenosphere propelled by slow convection currents in the asthenosphere (Fig. 27.7). The lithospheric plates themselves are stable and almost free of earthquakes but the boundaries between plates, where there is relative motion, are marked by the world's earthquake belts. The most significant features of the earth's surface are not oceans and mountains but the plate boundaries. These boundaries are defined very exactly by the pattern of world earthquakes (Fig. 27.6). There are six major plates and several other small plates as outlined on the map of Fig. 27.10; and there are three main types of plate boundaries: tensional, compressional, and those with neither tension nor compression. The tensional boundaries are those along the mid-ocean ridge crests (red lines, Fig. 27.10) where plates are moving apart and new crust is being formed by sea-floor spreading as previously discussed. These boundaries are characterized by shallow-focus earthquakes and a few volcanoes. The compressional plate boundaries are sites where plates are colliding and, in general, where one plate is moving down into the asthenosphere beneath another plate; these are the subduction zones (heavy brown line, Fig. 27.10). These zones are marked at the earth's surface by ocean trenches that lie immediately seaward of island arcs or coastlines marked by young mountains: for example, one of the small plates (Nazca) is plunging beneath the Andes Mountains of South America. Deep-focus earthquakes occur landward of trenches and nowhere else, and appear to be caused by the movement of slabs of cool lithospheric plates into the asthenosphere at these subduction zones to depths of at least 700 km. The foci of intermediate and deep earthquakes plot out to define 40° dipping planes starting beneath trenches and dipping away from the ocean basins beneath island arcs and continental mountain chains as shown diagrammatically in Fig. 27.11. Slow downward shear movement and fracturing along the top of the rigid subducting lithospheric plate could be the cause of the deep-foci earthquakes at depths where fracturing is usually impossible in the mobile asthenosphere. These downward movements may carry to depth remnants of sediments and other crustal material which, when heated and partially melted, rises as andesitic magma to cause the volcanic activity so characteristic of most island arcs (Fig. 27.11) and young mountain chains.

The third type of plate boundary is represented by great transverse fracture zones or transform faults where plates slide past each other without actual tension or compression but with enough friction to cause severe earthquakes. The best-known plate boundary of this type is the famous San Andreas fault in California discussed in Chap. 25. Here the Pacific plate is moving northwestward past the American plate (Fig. 27.10).

SUPPORTING DATA

Perhaps the most important support for the theory of plate tectonics comes from the study of earthquakes, as we have already commented. The general pattern of earthquakes on a worldwide basis (Fig. 27.6) is a series of continuous, narrow, active belts that delineates the earth's surface into a number of stable blocks and has really led to the acceptance of the plate-tectonics theory. Zones predicted to be tensional according to the sea-floor spreading model are the sites of only shallow earthquakes where the lithosphere is thinned and being pulled apart (earthquakes cannot occur in the mobile asthenosphere). In the compressional zones—the island arcs and young mountain belts—many large-energy and deep earthquakes occur, and total earthquake activity is high (Fig. 27.6) as the cool, brittle lithosphere plunges deep into the asthenosphere ultimately to be absorbed. Deep-foci earthquakes are found only where these conditions exist.

Figure 27.10
Mosaic of plates that form the earth's lithosphere or outer shell. According to the theory of plate tectonics, the plates are not only rigid but also in constant relative motion. Boundaries are of three types: (1) ridge crests or axes, where plates diverge (shown in dark red); (2) subduction zones, where plates plunge to depth; and (3) transform faults, where plates slide past one another. Arrows indicate general direction of relative motion of major plates. (From John F. Dewey, "Plate Tectonics," May 1972. Copyright © 1972, by *Scientific American, Inc.* All rights reserved.)

The remarkable pattern of the first motions of earthquake waves offers a means of determining the type and direction of displacement on fault planes. First motion studies, also called *fault-plane solutions,* are highly mathematical analyses of the sense of direction of motion of the rock at the focus of an earthquake. This sense of motion as calculated at seismograph stations around the world can be used to determine the orientation of motion at the focus of an earthquake. This has been done for over 100 major earthquakes at various locations around the world and has given an idea of the direction of motion of the various plates that is in good agreement with the general model of plates moving away from mid-ocean ridges and subducting at island arcs. The sense of motion along transform faults, as explained in Fig. 27.9, is in agreement with the spreading pattern at ridges. In short, the general pattern of earthquakes and the direction of motion that the seismic waves indicate both support the models of sea-floor spreading and plate tectonics.

Some earthquakes do occur in areas that are neither ridge crests nor trenches but are the location of geologically recent mountains. These would seem to be the site of recent plate collisions where mountain-

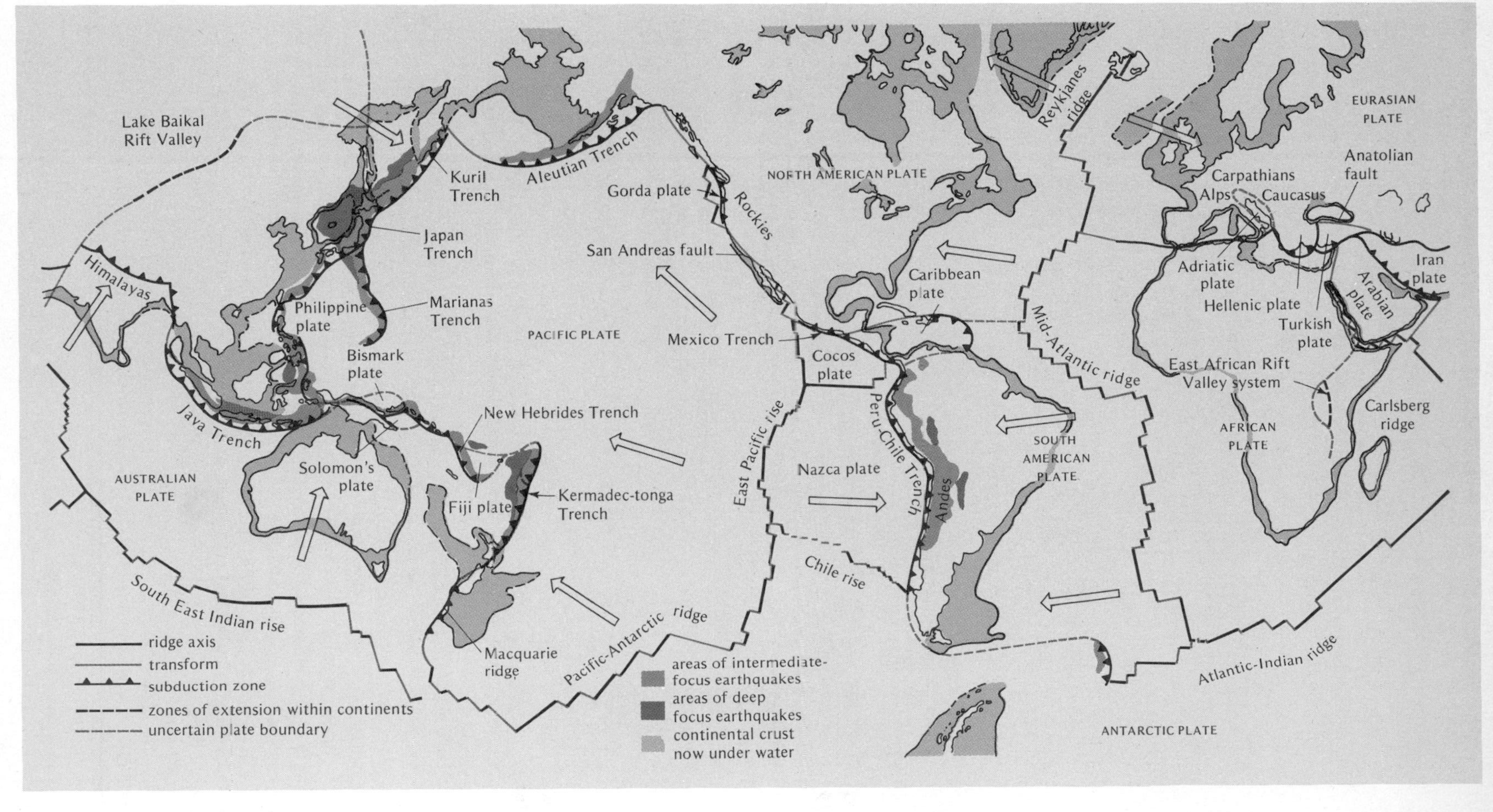
Lake Baikal
Rift Valley
Himalayas
Kuril
Trench
Aleutian Trench
Japan
Trench
Philippine
plate
Marianas
Trench
Bismark
plate
Java Trench
AUSTRALIAN
PLATE
Solomon's
plate
New Hebrides Trench
Fiji plate
Kermadec-tonga
Trench
Macquarie
ridge
South East Indian rise
Pacific-Antarctic ridge
PACIFIC PLATE
Gorda plate
Rockies
San Andreas fault
Mexico Trench
Cocos
plate
East Pacific rise
Nazca plate
Chile rise
Peru-Chile Trench
Andes
NORTH AMERICAN PLATE
Caribbean
plate
Reykjanes
ridge
Mid-Atlantic ridge
SOUTH
AMERICAN
PLATE
EURASIAN
PLATE
Carpathians
Alps
Caucasus
Anatolian
fault
Adriatic
plate
Hellenic plate
Turkish
plate
Arabian
plate
Iran
plate
East African Rift
Valley system
AFRICAN
PLATE
Carlsberg
ridge
Atlantic-Indian ridge
ANTARCTIC PLATE
ridge axis
transform
subduction zone
zones of extension within continents
uncertain plate boundary
areas of intermediate-
focus earthquakes
areas of deep
focus earthquakes
continental crust
now under water

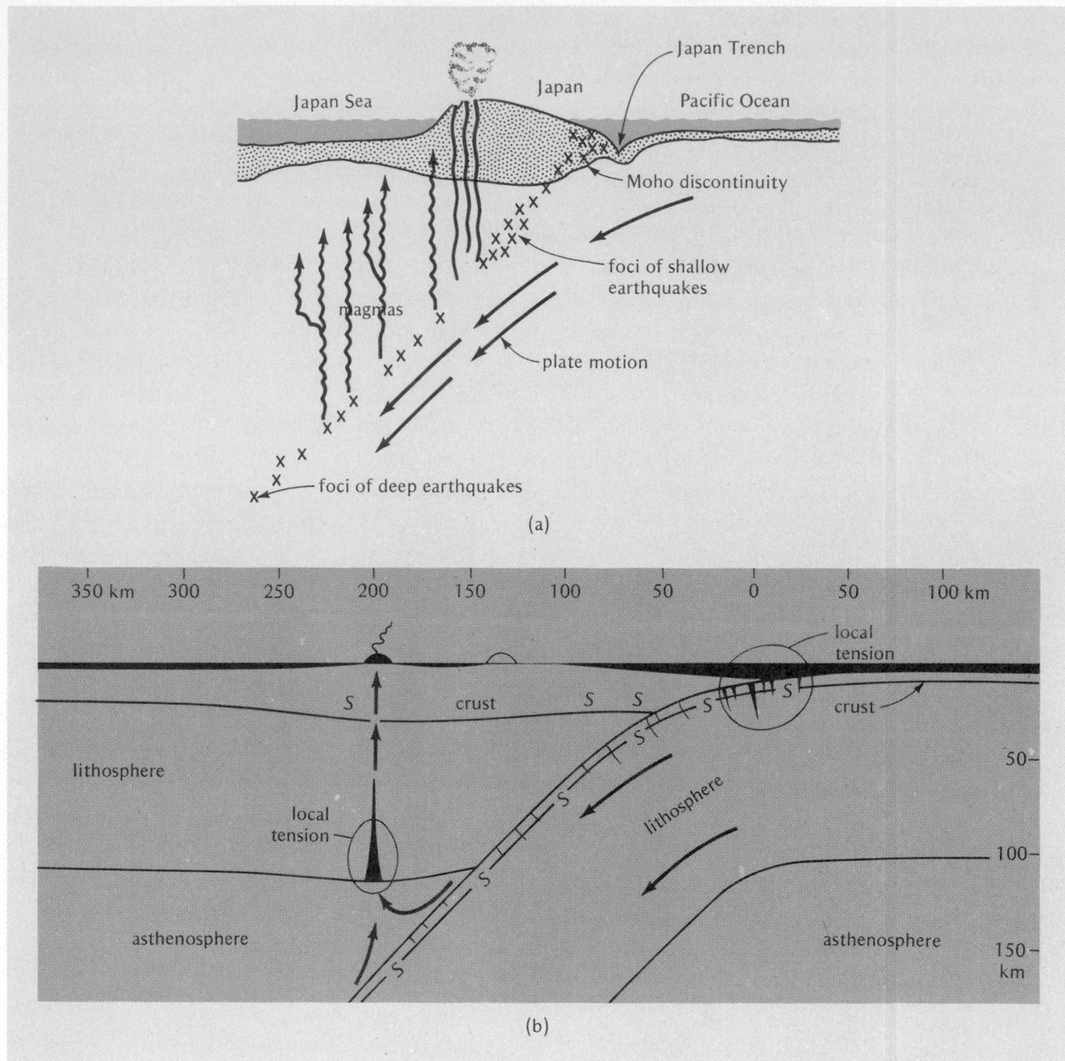

Figure 27.11

Structure beneath island arcs. (a) Schematic plot of earthquake foci beneath Japan Trench and Japan possibly caused by shear against descending plate as shown by arrows. Magmas develop along this shear zone to produce the volcanic activity of island arcs. The 40° dipping plane that contains the earthquake foci continues to depths of about 400 miles, which is below the level of this diagram. (H. Takeuchi, et al., *Debate about the Earth,* Freeman, Cooper and Co., 1967.) (b) Vertical section indicating hypothetical structures and other features. Section shows descending slab of lithosphere, seismic zones (*S*) near surface of slab and in adjacent crust, tensional features beneath ocean deep where slab bends abruptly and where overriding lithosphere in contact with the downgoing slab is bent upwards as a result of overthrusting. No vertical exaggeration. (From B. Isacks, J. Oliver and L. R. Sykes, *Journal of Geophysical Research,* **73,** 1968, 5869, copyright by American Geophysical Union.)

building compression is still active: for example, the mountainous terrain of Turkey and the many mountains to the east all the way to the Himalayas; also the Rocky Mountains system, especially in Montana.

In recent years the amount of heat escaping from within the earth to its atmosphere has been measured from hundreds of stations on the continents and the ocean floors. The actual amount of this *heat flow* is a very small amount for any one limited area but is quite uniform at most stations in all parts of the earth both on land and under the sea. As might be expected there are a few areas of very high heat flow, high heat anomalies, in areas of recent volcanic activity where the earth's crust has been heated near the surface over a period of time; for example, in Yellowstone National Park with its geysers and hot springs, heat flow is very high. Other than at volcanic areas, however, only two major types of heat-flow anomalies are important. High heat flows have been measured in the rift-zone areas of mid-ocean ridges, two to eight times the average for the earth as a whole. Such high heat flow is in agreement with a model of rising lavas from a hot asthenosphere to form new crust at mid-ocean ridge crests. Heat-flow values are unusually low on the floor of the deep-sea trenches where, according to theory, cool plates are moving to depth. If the concept of subduction zones is correct, we would predict low heat-flow anomalies at trenches. The heat-flow patterns of the earth do, indeed, support the plate tectonic concept.

When gravity measurements are made at many locations on the earth's surface and are corrected for altitude and latitude, the actual figures do not vary appreciably, except locally due to dense rock masses. This has led to the concept of *isostasy,* as discussed in Chap. 26, which indicates that most major areas, and thus plates, are in gravity balance. However, the earth does show some major gravity anomalies, both positive and negative. A low gravity reading, or negative anomaly, suggests that a part of the earth's surface is being held down against gravity. An analogy might be made to a block of ice in water that is being held down below its level of normal floating as represented in Fig. 27.12. On the other hand a positive anomaly is indicated if the block of ice is pushed up above its usual bouyancy level. In actual fact, negative anomalies have been recorded over the earth's trenches and positive gravity anomalies are present along many island arcs: for example, the islands of Java and Sumatra in Indonesia and in Puerto Rico. This suggests, by our analogy, that trenches are being pushed down and island arcs held up. This certainly supports the subduction zone model where lithospheric plates are going down, in part against gravity, and probably pushing up the edge of the upper plate where the island arcs are located.

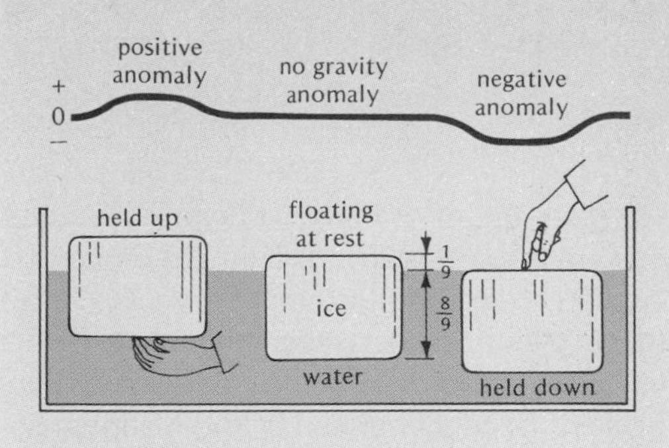

Figure 27.12
The principle of an isostatic gravity anomaly as illustrated by a floating cake of ice. (From *The Earth Sciences,* 2nd ed., Harper & Row, Publishers, copyright © 1963, 1971 by Arthur N. Strahler.)

The age, thickness, and distribution of sediment on the deep-ocean floors lend a great deal of supporting evidence to the sea-floor spreading model. If sea floor is forming at ridge crests and moving laterally away in both directions from the crest, sediment should be thin and young near ridge crests and progressively thicker and older at depth at sites farther away from ridge crests. Furthermore, if the Atlantic Ocean is less than 200 million years old, formed after the ancient continent of Pangaea began to split up and drift apart, then no sediment on its floor can be older than 200 million years. To test these theories, oceanographers developed the program known as the Deep Sea Drilling Project, largely financed by the National Science Foundation, using a remarkable ship, the *Glomar Challenger,* which has been doing research since August, 1968. This ship is 400 ft long, and carries amidships a drilling derrick that stands 200 ft above water level. Drill pipe is lowered from this derrick through an opening in the ship's bottom to allow drilling on the sea floor in very deep water. In a pamphlet released in 1970, the National Science Foundation said, after 2 years of deep-sea drilling, that the project has "Produced information of such significance as to mark it as one of the most successful scientific expeditions of all time. In this period approximately 195 holes were drilled at 125 sites in the North Atlantic and Pacific oceans. Sediment and rock cores were obtained from the earth's crust under water more than 20,000 ft deep. Several holes were drilled deeper than 3200 ft into the ocean bottom. The drilling ship has used the longest drill string ever suspended from a floating platform, 20,716 ft." Drilling has continued since 1970 in most of the other oceans of the world with several hundred more cores now on record. The data from this drilling project have completely supported the initial assumptions. No sediment more than 150 million years old has been found in the Atlantic Ocean. The sea-floor sediments are thicker away from the mid-ocean ridge crests and are progressively younger as the ridge crests are approached. The age of these sediments has been determined by a study of fossil organisms found in the sediments.

CONVECTION CURRENTS

Crustal spreading and plate tectonics must have a driving mechanism but little is known of mass movements in the mantle. Most theories of plate tectonics have suggested systems of very slow mantle circulation under the heading of *convection currents*. Several scientists, perhaps first David Griggs, whose theory was later refined by Vening Meinesz, a Dutch geophysicist, have advanced a model of convection (Fig. 26.13) involving the entire thickness of the mantle from crust to core. In terms of plate tectonics this means rising of less dense mantle material under the mid-ocean ridges and sinking of cooler and thus slightly denser material at trenches. The horizontal movements under plates would exert a drag causing the plates to move away from the mid-ocean ridges as if on a conveyor belt.

More recently a model has been suggested in which the convection currents occur only within a shallow zone of the upper mantle—all motion takes place within the asthenosphere or plastic zone. It has also been proposed that the sinking of lithosphere at subduction zones is capable of exerting a pull on the rest of the plate and response to gravity is therefore a dominant driving force. However, objections have been raised to all models advanced so far, some of which are quite bizarre.

The energy source for convection currents in the mantle according to any model may be the uneven distribution of heat from radiogenic sources. On the other hand, it may consist of kinetic energy of motions persisting from a much earlier time in the history of the earth; or the earth's rotation may somehow be involved as a driving mechanism—the axis of sea-floor spreading is near the axis of rotation of the earth.

MODERN PICTURE OF DRIFT

The modern concept of continental drift since early Mesozoic time indicates the rifting apart of the supercontinent Pangaea that existed in late Permian time in the Southern Hemisphere (Fig. 27.3) on account of the rise of convection currents beneath it. Parts of this continent drifted away to form South America, Africa, India, Australia, and Antarctica as we know them today.

Africa and the subcontinent of India drifted northward, disturbing the thick sediments in a great seaway known as the Tethys Sea. These sediments were slowly compressed against the land mass of Eurasia, as though caught in a gigantic vise, and were simultaneously deformed into the Alps, Himalayas, and other mountain systems of Europe and Asia.

Simultaneously, North and South America were drifting westward as rifting and seafloor spreading continued along the ever-growing Mid-Atlantic Ridge system. These continents were carried passively on the laterally moving or spreading convection "conveyor" system. The western, or advancing, edge of these continents became deformed into the various Cordilleran mountain systems of North and South America where the early Pacific plates sank beneath them. This resulted in compression of accumulated Mesozoic geosynclinal sediments to form some of the ranges of the Rocky Mountains. An early trench system and subduction zone probably were located just west of Montana and Wyoming and extended through Utah. Later subduction occurred farther to the west and was involved with the Coast Range development in northern California and Oregon.

The similarities of rock type and structure between Africa and South America in Precambrian and Paleozoic time do not continue into rocks of Cretaceous or younger time. Thus the extension of the Alpine Mountain system into North Africa as the Atlas Mountains has no counterpart across the Atlantic. As these mountains were formed, the Atlantic Ocean was already present and growing.

The plate model holds that convection currents originally rose beneath a continent, causing rift valleys that later widened slowly to form new oceans. It has been suggested by J. Tuzo Wilson that this process may be starting again today as evidenced by the rift valley system of East Africa and the Dead Sea in Israel. These rift valleys are connected through the Red Sea, which is a young rift valley ocean, with the rift valley at the center of the mid-ocean Indian Ocean Ridge. Perhaps a new convection current has just started to rise beneath East Africa during the past couple of million years that will produce a new narrow ocean like the Red Sea. Volcanoes are associated with the east African rift valleys indicating a rise of magma from below.

Mountain building

The cause of mountain building must certainly be related to the movement of lithospheric plates. Inasmuch as mountain deformation is the result of compression in the crust, the deformation processes must be linked with plate collisions. The latter might be divided into two types from a theoretical consideration. First, there is the island subduction-zone type (Cordilleran type) in which a plate with oceanic crust goes down into the asthenosphere under another plate, as previously discussed (Fig. 27.11). Second, there is the collision of two continental plates with or without actual subduction but with the buckling of the continental crustal materials that can thicken the sial both upward and downward. According to H. W. Menard a plate collision without subduction, in which both plates buckle, can occur if the plates are moving slowly—that is, at less than 6 cm/year. When two plates collide at a combined rate of more than 6 cm/year, neither plate can absorb the impact by buckling and one plate plunges under the other.

Plate collision, perhaps without subduction, has apparently occurred as Africa and India have moved northward across the ancient Tethys Sea (Fig. 27.3) to collide with the Eurasian plate with the gradual deformation of sediments in the Tethys during Cenozoic time. This has produced a

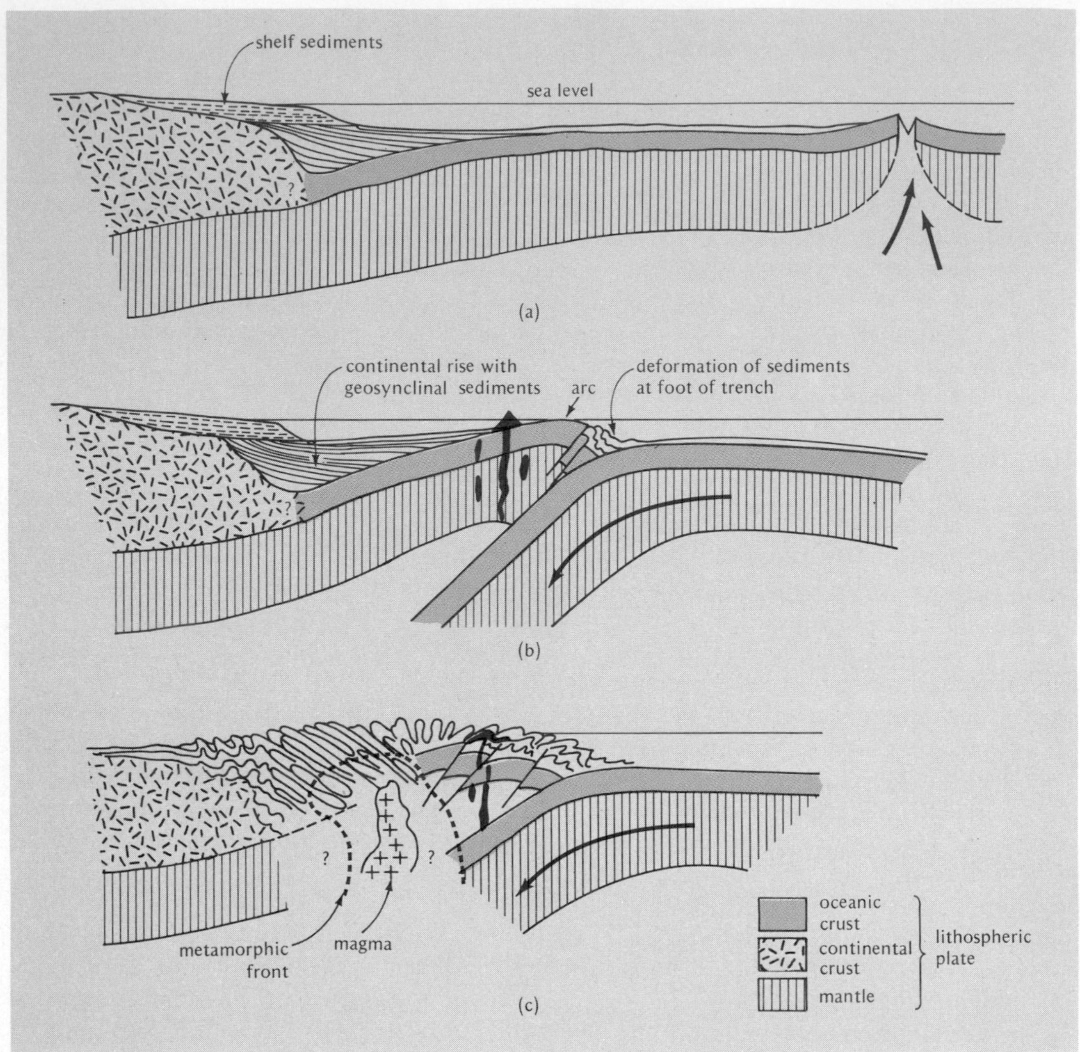

Figure 27.13
Cordilleran-type mountain building. Schematic sequence of sections illustrating (a) oceanic plate with wedge of sediments, or geosyncline on continental rise, (b) subduction with development of island arc, and (c) deformation of geosynclinal sediments as subduction continues. Compare with Fig. 27.14. (Modified after Dewey and Bird, *Journal of Geophysical Research,* **75,** 1970, copyright American Geophysical Union.)

whole series of mountain systems trending in general from west to east: the Atlas Mountains in Algeria, North Africa, and the Alps and Carpathians in southern Europe, which trend eastward through Turkey, Persia, and Afghanistan, to culminate in the Himalayas north of India. Throughout most of this mountainous belt, earthquakes are frequent today, suggesting that compression is still active. Seismic evidence indicates that under the Tibetan Plateau just north of the Himalayas, continental-type crust has

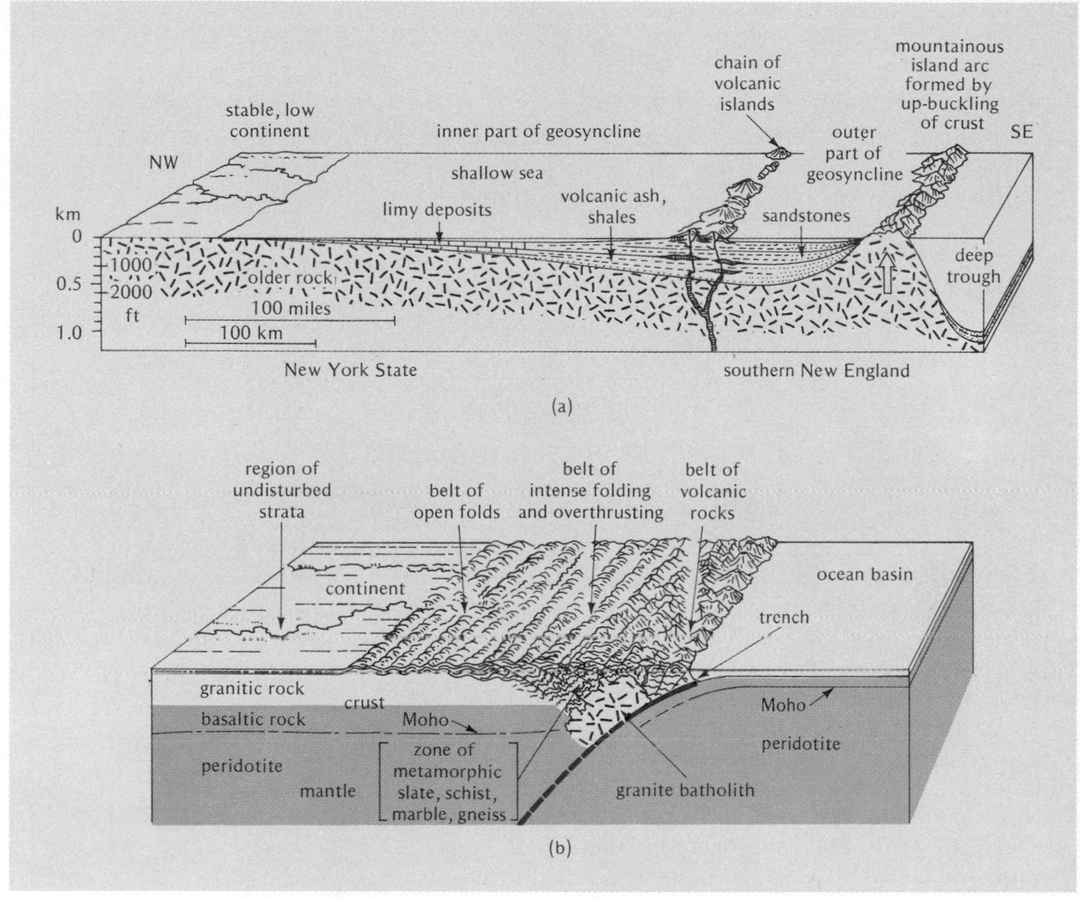

Figure 27.14
Mountain building sequence in the Appalachian region. (a) Schematic cross section of the Ordovician Geosyncline just before the Taconic Deformation showing volcanic and nonvolcanic arcs in northeastern North America. (Suggested by plate 9 of Marshall Kay, *Geological Society of America Memoir* 48, 1951.) (b) Schematic block diagram of the Appalachian orogenic belt immediately after folding and intrusion have occurred at the end of Paleozoic time. Compare with Fig. 27.13. (Both sketches from *The Earth Sciences,* 2nd ed., Harper & Row, Publishers, copyright © 1963, 1971 by Arthur N. Strahler.)

a double thickness; perhaps there has been overlap of the two plates.

The first type of plate collision with ultimate mountain-building deformation is shown diagrammatically in Fig. 27.13. Here in (a) and (b) of Fig. 27.13 a wedge of sediments of geosynclinal thickness and length has accumulated beyond the continental shelf of a continent and plate collision has formed a trench and island arc some distance offshore. As the two plates continue to move toward each other, both the sediments of the geosyncline and the rocks of the island arc become progressively more deformed as in (c) of Fig. 27.13. Perhaps this situation can be applied to the development of the Appalachian Mountains during Paleozoic time. In Fig. 27.14, two schematic

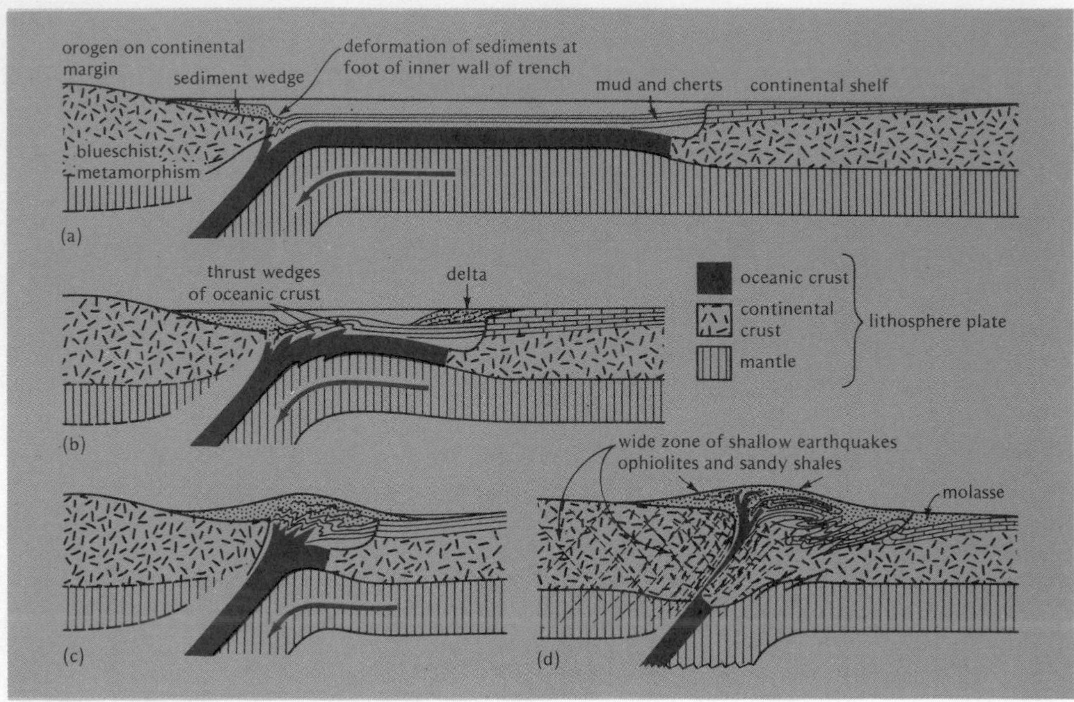

Figure 27.15
Schematic sequence of sections illustrating the collision of two continents. (From John F. Dewey and John M. Bird, *Journal of Geophysical Research,* **75,** 1970, 2642, copyright American Geophysical Union.)

diagrams suggest various stages in the development of these mountain systems. In Fig. 27.14(a) we see a geosynclinal situation back of two island arcs that is comparable to diagram (b) in Fig. 27.13. In (b) of Fig. 27.14 we have the completed Appalachian system at the end of Paleozoic time. This is comparable to (c) in Fig. 27.13. There is no trace of this old proposed subduction zone today as these two plates became welded together and have behaved as one plate for the last 250 million years. In fact a new wedge of sediments or a geosyncline of Mesozoic and Cenozoic sediments now lies off the Atlantic shelf with a maximum thickness near 30,000 ft.

The second type of plate collision, where two continental plates come together, is illustrated diagrammatically in a series of four stages in Fig. 27.15. The thin oceanic crust-covered plate is shown going down into the asthenosphere but the thick sialic continental crustal segments have buckled with the original sea-floor sediments to form the great overthrust slices so characteristic of structure in the Alps and, so far as is known, also in the Himalayas. These deformed sediments typically contain interbedded deep-sea cherts and ophiolites (altered basalt lava flows) as evidence that material from the deep-ocean floor has been mixed with the folded and faulted geosynclinal sediments from the continental-rise zone. The thick sialic crustal material and the geosynclinal and other sediments are all low-density materials as compared to the rest of the lithospheric plate material. Consequently, these low-density materials would not be expected to move downward, but would buckle up in a great pile that could spread laterally due to gravity sliding. This latter mechanism may explain some of the overturned folds and great overthrust slices shown diagrammatically in Fig. 27.15(d). Even as the upbuckling and thrusting were occurring, erosion began rapidly in the growing mountain uplift. Thick deposits of coarse-grained clastic sediments were soon deposited on the flanks of the still-growing mountain system to be partly deformed before deformation was completed. These clastic sediments are known collectively as *molasse*.

MAGMA GENERATION

As a lithospheric plate moves downward in a subduction zone, melting of the sediment and basaltic crustal material apparently takes place and develops andesitic magmas through the contamination of the basalts by sedimentary materials. These magmas rise to form the andesitic volcanoes so characteristic of island arcs and continental margins such as those found in Chile and Central America. This process has been suggested in Fig. 27.11 but has perhaps been diagrammed more dramatically by Harry Hess (Fig. 27.16). In addition, where plate subduction and collision have resulted in the deformation of thick geosynclinal sediments, these deformed sediments with segments of old sialic crustal material have been heated until partial melting has occurred to generate granite and granodioritic magmas as sketched in Figs. 27.13(c) and 27.14(b). This is the suggested origin for the great batholiths found in the core of so many mountain ranges, formed by either partial melting or by very high temperature metamorphism in the same environment.

Magmas also are generated in the asthenosphere beneath the ocean floors not only under the mid-ocean ridges, as has been discussed, but also at other scattered areas (deep fracture zones) as shown in Fig. 27.16. An outstanding example of the latter is the great group of volcanoes that make up the Hawaiian Islands. There are many other similar basaltic volcanoes on the Pacific Ocean floor, both active and extinct. Some of the oldest of these are now under several thousand feet of water; these are known as *seamounts* or, if flat topped, as *guyots*. At least 10,000 of these are charted in the Pacific basin. It is quite clear that such basaltic magmas originate in the low-velocity zone of the asthenosphere, either at ridge crests or elsewhere. At Kilauea Volcano in Hawaii, earthquake foci have been recorded at depths of at least 35 miles below the volcano, well down in the mantle part of the lithospheric plate.

Based on data that we cannot review here, a hypothesis has been advanced that the composition of the asthenosphere is about three parts peridotite and one part basalt—a hypothetical material called *pyrolite*. The hypothesis says that partial melting of this pyrolite forms a basalt magma and leaves behind olivine and pyroxene, the rock peridotite (see Table 22.3). This material has accumulated in the lithosphere as the basalt portion has been melted out to form oceanic crust and basaltic volcanoes. The basalt plateaus that are found on stable parts of the continental parts of the plates have probably formed in this same manner in the asthenosphere and then risen directly to the surface.

In summary, let us emphasize again the two-fold origin of magma: (1) andesitic and felsic magma formed relatively near surface associated with plate collision zones and (2) basaltic magmas forming deep in the mantle, by partial melting in the plastic zone, and rising through the lithosphere to the ocean floor.

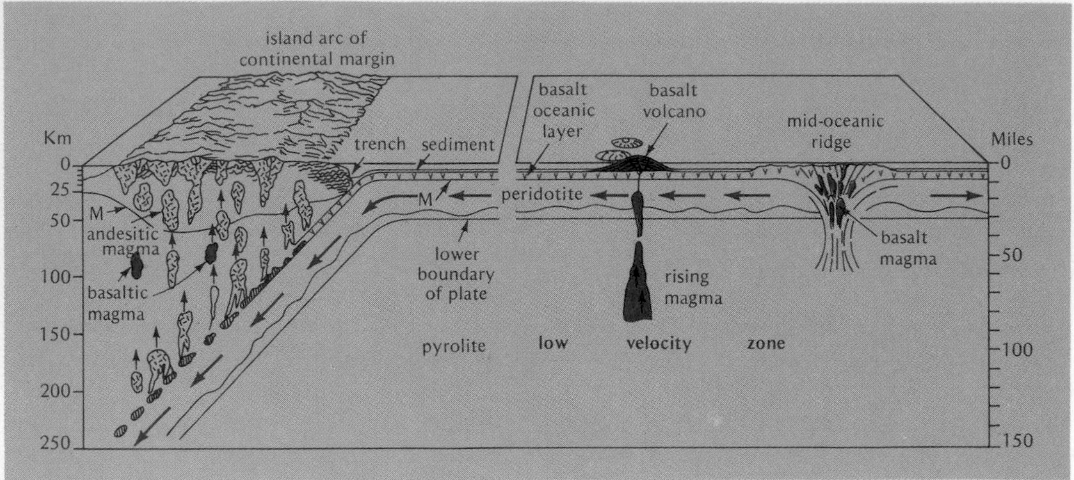

Figure 27.16
Block diagram to illustrate hypothetical generation of magma, as associated with plate tectonics. Andesitic and a little basaltic magma are formed by melting of sediment and basaltic ocean-floor material on the top of the down-going plate below the trench and island arc as shown on the left. Pure basalt magmas also rise from the low-velocity zone below plates on the ocean floor not only at the mid-ocean ridges but at other scattered locations to produce basalt volcanoes as in Hawaii. (Modified after H. H. Hess's 1962 model, as given by A. E. Ringwood in *The Earth's Crust and Upper Mantle,* Geophysical Monograph 13, 1969, American Geophysical Union, p. 12. From *The Earth Sciences,* 2nd ed., Harper & Row, Publishers, copyright © 1963, 1971 by Arthur N. Strahler.)

SUMMARY

In summary, it seems that the major mid-ocean ridges and their central rift zones of today are the surface expressions of rising mantle currents that are causing slow, continued spreading of the ocean floors. The young continental-margin mountain systems and island arcs and trenches of today represent the approximate sites of descending currents. World heat flow, gravity anomalies, and deep-earthquake data are consistent with these hypotheses. The present ocean basins, accordingly, must be fairly young features of the lithosphere.

The concept of continental drift is supported by the geographic and geologic "fit" of some continents (including the pattern of Permian glacial climates), the data from paleomagnetism, the bilateral symmetry of magnetic anomalies on mid-ocean ridges, and the presence of great transcurrent faults on the ocean floor and continental margins. The breakup of a supercontinent in early Mesozoic time is indicated by the gradual development and widening of the Atlantic Ocean as a result of sea-floor spreading. The youth of the Atlantic Ocean is supported by age and thickness distribution of sediments on its floor. The Deep Sea Drilling Project using the drilling vessel *Glomar Challenger* has obtained such data since 1968 from hundreds of holes and sediment cores from all the oceans, some holes more than 3000 ft deep and under 20,000 ft of water. Remanent magnetism in rocks and magnetic reversals are evidence for polar wandering and spreading of the ocean floors, respectively.

Ocean-floor topography near the margins of continents is characterized by continental shelves and slopes. The deep-sea floor is characterized by very flat abyssal plains modified by seamounts, guyots, deep-sea fans, trenches, and ridges. The most important of the last-named comprise the world system of mid-ocean ridges typified by the great mountainous Mid-Atlantic Ridge and its central rift zone.

The plate tectonics model suggests that the earth's outer layer is divided into six major plates whose boundaries are outlined by the pattern of earthquake foci. Three types of plate boundaries are known: (1) tensional, where plates separate—these are at mid-ocean ridges; (2) compressional, where plates collide and one goes under another in a subduction zone; and (3) boundaries, where one plate is sliding past another one on a transverse or transform fault zone as in California on the San Andreas fault.

Mountain building occurs where plates collide to cause deformation of geosynclinal sediments and local melting, at moderate depths, of sialic crustal rocks to produce felsic magmas. Basaltic magmas are generated by partial melting in the asthenosphere or low-velocity zone, usually under oceanic parts of lithospheric plates.

Important words and terms

continental drift, plate tectonics, plate boundaries
asthenosphere, lithosphere
paleomagnetism
natural, thermal, and depositional remanent magnetism
polarity epochs and events, magnetic reversals
continental shelves and slopes, abyssal plains
trenches and deeps, island arcs
mid-ocean ridges, rift zones, and valleys
transverse fracture zones, transform faults
sea-floor spreading
subduction zones, plate collision
convection currents
magnetic anomaly
heat flow anomaly
gravity anomaly

Questions and exercises

1. Without the continental drift concept, how might geologists explain the deposits of Permian glaciation in such separated areas as South America, Africa, India, and Australia?
2. If the assumption that the geomagnetic poles have always remained close to the geographic poles were to be disproved, what evidence involving magnetism in rocks would still suggest plate tectonics?
3. What is a subduction zone? Describe the kinds of evidence suggesting that they actually do exist.
4. Today we believe that lithospheric plates are moving relative to one another at 1 to 6 cm/year. Name two kinds of evidence that indicate this rate of motion.
5. Explain the difference between transform faults and ordinary strike-slip faults (refer back to p. 566).
6. In the light of the plate tectonics model or hypothesis, why are there no young, growing mountains on the Atlantic Coast of North America as there are on the Pacific side?
7. What is the significance of the continuity of the mid-ocean rift in the Indian Ocean with the rift zones in the Red Sea area and in East Africa?
8. Review shield vs. composite volcanoes (Chap. 22) in the light of the magma generation discussion in this chapter. What features of these two basic types of volcanoes fits with this magma generation discussion?
9. Study sediment data from the Deep Sea Drilling Project, which has been carried on from the *Glomar Challenger,* as supplied to you by your instructor. Compare character and ages of sediment from various parts of the different oceans. How does this support a plate tectonics hypothesis?
10. How can we reconcile the stability, for long periods of geologic time, of large sections of continents with the concept of continuing continental drift?

Supplementary readings

Bird, John M. and Bryan Isacks (eds.), *Plate Tectonics, Selected Papers from the Journal of Geophysical Research,* American Geophysical Union, Washington, D. C. (1972). (Paperback.) [A reprint compilation of about 30 technical papers over the past few years that gives a comprehensive, authoritative, up-to-date summary of research on plate tectonics and related subjects.]

Clark, Sydney P., Jr., *Structure of the Earth,* Foundations of Earth Science Series, Prentice-Hall, Englewood Cliffs, N. J. (1971). (Paperback.) [Paleomagnetic evidence for continental drift and a short general summary of the plate tectonics hypothesis (Chaps. 3 and 4).]

Takeuchi, H., S. Uyeda, and H. Kanamori, *Debate about the Earth: Approach to Geophysics through Analysis of Continental Drift* (Rev. ed.), Freeman, San Francisco (1970). [An integration of the kinds of data bearing on the concept of continental drift, including a full discussion of earth magnetism and paleomagnetism, the heat problems of the earth's interior, and hints from the ocean floor.]

Wilson, J. Tuzo (ed.), *Continents Adrift,* Freeman, San Francisco (1972). (Paperback.) [Fifteen readings from *Scientific American,* five of which consider the mobility of the earth, five discuss drift, sea-floor spreading and plate tectonics, and the last five cover some consequences and examples of drift. An outstanding group of readings in layman's science language; especially suitable for this course level.]

GEOLOGY OF THE MOON

On the surface, science is engaged in the perpetual task of turning the unknown into the known; at a deeper level, however, it succeeds equally in turning the known into the unknown.

SYDNEY J. HARRIS, 1972

Man has wondered about the moon for centuries, perhaps since he first walked this planet. What is it made of? Where did it come from? How? And now we have been there!

Lunar rocks have been brought back to earth and tested by every known chemical and physical technique. These rocks contain minerals like the igneous rock minerals of the earth: pyroxene, feldspar, olivine, and others. We now know that the moon's outer layer, at least, is composed of igneous rocks, basalts, and gabbros, including many with the textures found in lavas; and they have been radiometrically age-dated at 3 or 4 billion years.

A lot of additional data have been gathered. Some is as expected, but there have been many surprises. The moon's chemistry is somewhat different from the earth's; there is more titanium and chromium, but much less sodium and potassium. Moonquakes are less frequent than expected, whereas magnetism is a little more than predicted. And about half of the tiny particles that make up the regolith, the soillike material, is glass—a great surprise.

It has been possible to put together a general story of the geologic evolution or development of the moon, especially during its first $1\frac{1}{2}$ billion years. The moon is now surprisingly rigid and cool and apparently has been internally inactive for the past 3 billion years. The moon was active during the early history of our solar system, at a time period for which all earth records and rocks are missing, being either buried or eroded. However, we still do not know how or where the moon was formed.

Introduction

The characteristic features of the lunar surface are craters, millions of craters, of all sizes from little ones a few inches across to giant depressions 500 miles in diameter. In contrast the earth's land surface is covered by gullies or valleys from tiny channels that might form during a heavy rain to giant canyons over a mile deep. It is apparent immediately that there are processes acting on the earth's surface that are not active on the lunar surface. We have already seen in an earlier chapter (Chap. 24) that the earth's surface is constantly being modified by erosion by running water, ice, and waves. Thus the earth is completely covered by washes or valleys due to running water and rivers, and this surface must be relatively young due to the continuity of these processes in the cycle of erosion. For example, even the largest volcanoes are leveled by erosion within a few million years after activity has ceased. Geologists are convinced that none of the earth's original crustal rocks still remain unchanged. On the lunar surface, in contrast, some of the craters that were formed more than 4 billion years ago can still be seen although they have been marred by the formation of younger craters, and by a slow process that gradually rounds them off.

The planetwide processes by which the land surface is gradually built up (igneous activity and diastrophism), in opposition to erosion, result from internal sources of energy and the way in which this energy is dissipated. The amount of internal energy and its dissipation are certainly dependent on the size of a planet. On the earth this dissipation of energy has occurred (1) as a result of the explosion of volcanoes and the transport of lava and ash from the interior of the earth to its surface and (2) by the development of a low-density crust and a dense core caused by large-scale movements of the earth's interior material. These processes are responsible for such activities as plate tectonics and the building of mountains as was discussed in the last chapter. From our observations of the moon, there is no evidence of plate tectonics or mountain building, as we now understand it on the earth, even back 4 billion years. We do find evidence, however, of the transfer of material from within the moon to its surface by volcanic processes, especially by great lava flows.

Lunar exploration

Before the lunar exploration studies of the moon were started, we had already learned a great deal about it from careful observations with telescopes, radio, radar, and other techniques. The general appearance of the lunar surface was known, though not in detail, and some 300,000 craters had been identified through telescopes on the visible side of the moon. The motions of the moon around the earth, the phases of the moon, and eclipses involving the moon were well understood. Man also knew that the moon had no atmosphere or hydrosphere, and he had measured surface temperatures on the moon under both day and night conditions. This kind of information has been summarized, in part, in Chap. 3.

The major subdivisions of the moon's surface are its maria and highlands. The maria are relatively flat plains that appear dark through telescopes, or even to the naked eye, because of their low albedo (ratio of light reflected to light received). The highlands, on the other hand, are much more irregular or mountainous in topography and much lighter in color (see Fig. 3.10). Both the maria and the highlands are covered by craters, although crater density or frequency is considerably greater in the highlands.

Lunar space exploration by the United States began with the Ranger Program of missiles (first in 1963) that crash landed on the moon but sent back television pictures just before the crash. This program was followed by the Surveyor Series of soft landings of which five were successful during the period 1966–1968. These were, of course, unmanned landings but television cameras sent back pictures around the landing site, and small scoops gave some idea of the consistency of the thin outer layer of the lunar soil. These tests showed that the lunar material was strong enough to support a landing vehicle and thus man. During the time of the Surveyor Program, the unmanned Lunar Orbiter Program was also active and five of these orbiters sent back thousands of high-resolution photographs of much of the moon's surface—on both the near and back sides—even of layers of rock materials and how they overlap one another. These photos were of great value in planning exactly where man would land and what topographic problems he might encounter.

The Apollo Program, the "man-on-the-moon" program as it is popularly known, began with the *Apollo 11* landing in July, 1969, and has continued (through 1972) with a total of six successful landings, *Apollos 11, 12, 14, 15, 16,* and *17* (see locations on Fig. 28.1). Many pounds of rock samples and much other information have been returned. Instruments, notably seismographs, were installed and left on the lunar surface and will continue to send back information to earth for months or perhaps years after the landings. We now have much first-hand data about the moon, both surface and interior, which give us quite a different picture of the moon's make-up and origin than was possible only $3\frac{1}{2}$ years ago. Many more questions have been raised by our observations than old questions answered, however. Imagine creatures from space landing at six sites on the earth, collecting samples and making measurements, and then trying to answer the riddle of the earth's origin and geologic history for the last 4.5 billion years. That is what we are trying to do for the moon.

Origin of craters

Man has wondered about the origin of the lunar craters for hundreds of years—actually, since the invention of the telescope.

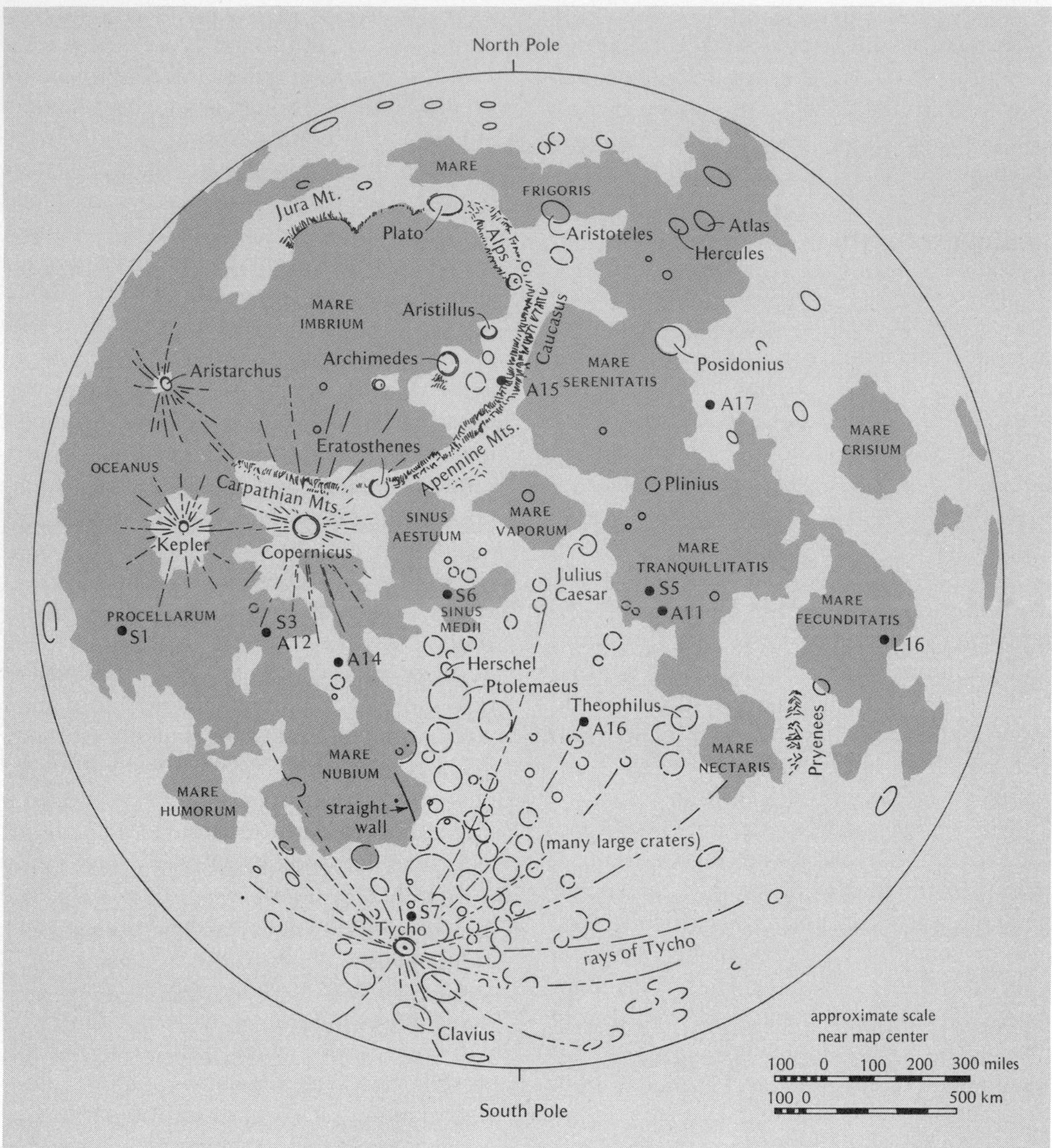

Figure 28.1
Sketch map of the major relief features of the moon. Mare surfaces are shown by light shading. Location of successful Surveyor landing sites shown by S1, S3, S5, S6, and S7; of Apollo sites by A11, A12, A14, A15, A16, and A17; and of Russian Luna unmanned site from which specimens were brought back to earth by L16. Compare with Fig. 3.10 which is a photograph of the moon in the same position.

In recent years it became apparent that there are only two reasonable, possible explanations for the craters: either an impact origin by an object from space, or a volcanic origin. For the past 25 years or more there has been a running debate over the impact vs. volcanic origin of these craters. Today it is accepted by most scientists that there are craters of both kinds, but that the great majority, both large and small, are due to impact events by meteorites, comets, or asteroids on the lunar surface. The most convincing evidence for an impact origin is the width-to-depth ratio of the craters. Regardless of the crater size, this ratio is quite constant and, in fact, is the same ratio that we can measure on a bomb crater or explosion crater here on earth. It is the same width-to-depth ratio present in the some two dozen proved meteorite impact craters on earth. The width-to-depth ratios for volcanic craters vary greatly depending on the type of volcano or its stage of growth. As seen in the high-quality pictures now available, especially from the Lunar Orbiter and Apollo missions, lunar craters have a hummocky rim of material with an outer uneven radial layer, often containing many secondary craters, all formed by throwout of material at the time of impact. All of these have the same shape, distribution, and topography as those around man-made bomb and nuclear-explosion test craters in the AEC test site in Nevada. The bottom of the crater depression or bowl is below the level of the surface in which the crater has been produced. All of these features are quite unlike the situation at most volcanic craters. The sequence of events in the impact and explosion of an object from space in the formation of a crater and its ejection products is shown diagrammatically in Fig. 28.2.

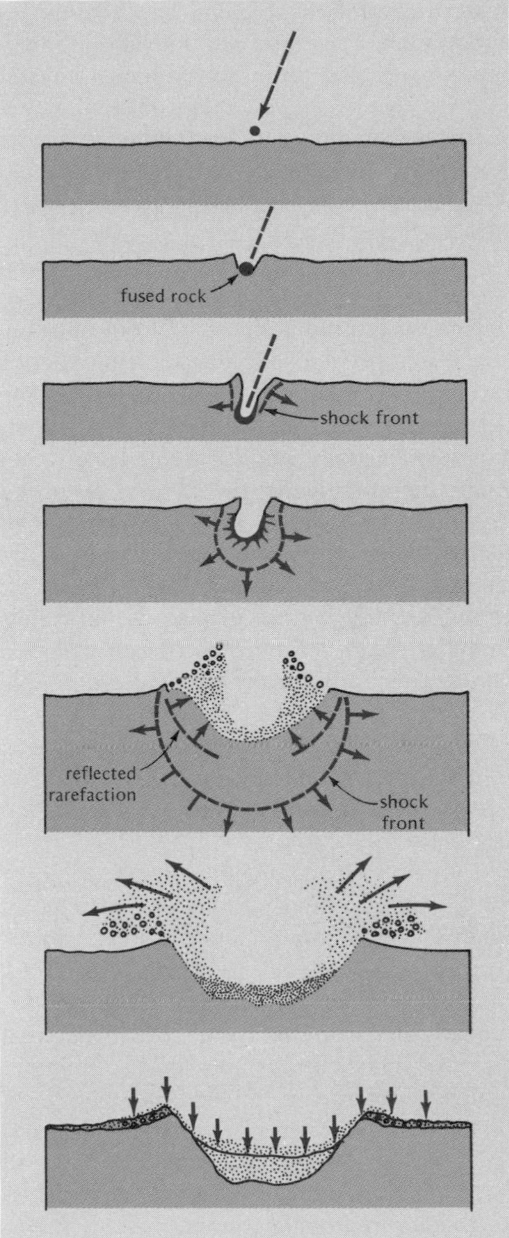

Figure 28.2
Hypothetical stages in the formation of a meteorite impact crater. (Modified from E. Anders, 1965. *Scientific American,* October 1965, p. 34. Data of E. M. Shoemaker and U.S. Geological Survey. From *The Earth Sciences,* 2nd ed., Harper & Row, Publishers, copyright © 1963, 1971, by Arthur N. Strahler.)

Craters are subdivided as primary and secondary. Those formed by the high-velocity impact of an object from space are termed *primary craters*. These are round and very regular in shape. The explosion of a large impact event throws out much material, some of it being large chunks of solid rock that land around the primary crater site to form low-velocity impact craters known as *secondary craters*. These latter are smaller, rather irregular in shape, and often clustered in groups and uneven rows roughly radial to the main primary impact crater.

It is true that the same or similar topography occurs at volcanoes due to great explosive eruptions that throw out not only ash but large blocks to form layers of debris radially away from the volcano with some secondary-type impact craters. Volcanoes, however, usually build a steep cone rather than the relatively flat apron occurring around most lunar craters. Some small features seen on the new photographs of the moon's surface do look like volcanoes: mounds with craters on their tops and craters with dark halos around them like an exploded layer of ash and cinders. It is also probable that a few of the larger lunar craters may be collapse-craters associated with volcanism, the kind of feature known as a *caldera* on earth—for example, Crater Lake, Oregon (see Figs. 22.12 and 22.13).

Other evidence for an impact origin of most craters comes from the character of fragmented soil material found at the various Apollo sites by our astronauts. This fragmented soillike material, as will be discussed later in more detail, contains a great deal of glass and partly melted mineral grains. There are numerous pieces of breccia in which such glass-rich materials have been cemented together. These all have the characteristics of material produced by impact fragmentation and melting due to the instantaneous heat produced by a major impact event. To date all the craters seen and studied by the Apollo astronauts have had the characteristics of impact events rather than volcanic eruptions.

If the moon has been impacted often enough to form millions of craters, why are there so few impact craters on the earth, which is in the same part of space as the moon? Obviously the earth must have been hit many, many times by large objects that did form craters. Only the very small meteorites are heated and vaporized in our atmosphere. The answer has to be that erosion on the earth has removed all trace of the impact scars or craters, formed throughout its 4.5 billion year history, except for those formed in the past half million years or less. Weathering and sediment formation, transportation, and deposition constitute a never-ending efficient cycle to wear away constantly and remake the earth's surface (see the rock cycle, Fig. 20.11). Again this emphasizes the importance of erosion on earth and the youth of the earth's present surface as against the lack of much erosion on the moon and the very great age of some surface features on the moon.

The lunar regolith

The surface of the moon is covered by a layer of pulverized rock material, varying in thickness from 10 to 30 ft, that is somewhat analogous to the broken rock and soil material covering so much bedrock on the earth. Such layers or deposits of broken rock material may be called *regolith*. On the earth, this broken and weathered debris has undergone chemical changes, quite extreme at the surface, to produce true soil with a profile (Fig. 23.3). On the moon, this soillike material has not been chemically weathered but only disintegrated to fine particles. It has a consistency and texture like very fine-grained soil but is not a true "soil" as we use the term on earth. The outer layer of the lunar regolith is a mixture of much dust and silt-sized particles with some larger fragments and even occasional large pieces of rock. About half of the fine-grained particles are tiny glass spheres and fragments. The rest are pieces of broken mineral grains or fine-grained rock chips. These mineral and rock fragments are highly fractured but chemically unaltered. This fractured character and the presence of much glass in the fine particles all contribute to an understanding that the lunar regolith material has been formed by impact fragmentation—a continued pounding of the rocks, and then the broken rocks, by meteorites of all sizes for several billion years. The present regolith is actually the sum total of 4 billion years or more of this impact shattering and fragmentation.

The regolith has the consistency of very fine-grained soil when walked on. The astronauts left footprints that faithfully reproduced their boot sole prints as happens on the earth's surface on wet mud or wet sand; but on the moon although this soil-like material is very dry and dusty, it still preserves every detail of the pattern of the boot sole. Why? Because the moon's surface is in a vacuum and in this situation tiny particles adhere to one another very tenaciously. They do not slide apart as fine dry dust would do in the earth's air. Thus every detail of the footprints and lunar vehicle footpads will be preserved for a very long time.

Some of the large blocks or rock fragments lying on the lunar surface have very sharp corners and edges as photographed both by Surveyor and Apollo cameras; others are quite rounded (Fig. 28.3). It is assumed that the sharp-cornered fragments have been lying on the lunar surface for a relatively short time, perhaps a few million years. On the other hand, the well-rounded rock fragments have been lying exposed to the surface for a few hundred million years. There is, therefore, a very slow type of erosion that gradually rounds off the edges of rock pieces on the moon's surface. This is apparently somewhat analogous to a sandblast action caused by the continuing bombardment of the rocks by tiny micrometeorites, particles of small dust size. This is a very slow process as compared to processes on earth, as, for example, the pounding of soil by raindrops. However, this slow lunar process does eventually round off the angular shapes of broken rocks and does slowly wear down the rims of craters and even obliterates tiny craters after a few million years. This is a gradual mixing and overturning of particles in the upper few feet of the regolith. This kind of micrometeorite churning will eventually wipe out the footprints of our astronauts, but not for nearly a million years.

The earth's atmosphere intercepts the millions of micrometeorites that enter it every day. We see some of the larger of these as shooting stars. On the lunar surface, with no atmosphere, these tiny dust-sized particles impact the surface at astronomical velocities and do their tiny bit of sandblasting. There are times when these tiny impacts create enough local heat to melt a few atoms of material. Some rocks have been brought back to earth that have microscopic-sized pits, each lined with a very thin film of glass due to this extremely concentrated, instantaneous heating.

The thin outer layer of rocks and particles at the lunar surface is being irradiated by cosmic rays (mostly protons and α particles) and noble gases (helium, xenon, and neon) in the solar wind. These produce a radiation damage layer and a trace of induced radioactivity. The measure of the latter and of the amount of noble gases absorbed in a rock fragment allows an estimate of the time a particular rock fragment has actually been within the upper couple of feet of the regolith's surface. Measured times from three Apollo sites vary from 9 to 500 million years.

Moderately small craters (up to 50 ft in diameter) usually contain no rock fragments at all, but somewhat larger craters contain many rock fragments either in the bottom of the crater or on the crater rim, or both. It is now clear that small cratering events churned up only the already-formed regolith, but larger crater-producing impacts excavated bedrock below the regolith layer depositing rock fragments on the surface. Thus the presence or absence of rock fragments in a young crater is a function of the regolith thickness. In this way the thickness

Figure 28.3
Close-up view of a large boulder in a field of boulders near the rim of Cone Crater on the *Apollo 14* mission. Most boulders are fairly angular to subangular. (NASA.)

of the regolith has been measured not only at Surveyor and Apollo sites but wherever detailed photographs have been studied. The thickness of the regolith, at least on the mare surfaces, is a function of time although this is complicated by ejected layers near major craters.

Lunar rocks and minerals

The rock samples that have been brought back by the first five Apollo teams fall into four main groups in terms of composition and texture although they are all quite closely related and all were initially igneous rocks. Almost all lunar rocks would fall in the boxes labeled gabbro or basalt in Table 22.3, the igneous rock classification used in this text. Most of the rock samples collected by the astronauts are fragments from the surface of the regolith that had been blasted out of bedrock by some cratering event. It has only once (*Apollo 15*) been possible to sample bedrock directly as a geologist does on earth at an outcrop—say in a road cut or a canyon wall. However, most of the large lunar rocks brought back for study have probably come from underlying bedrock close by the various landing sites.

Basalts were the commonest of the rock types at the *Apollo 11* and *12* sites on maria surfaces. These basalts vary in grain size from extremely fine-grained ones to some that are medium grained and might even be called *micro* gabbros. Some of the samples are vesicular (porous) and others have a porphyritic texture. However, they are very similar in appearance and mineralogy to some of the basalts on the earth, especially those of the Hawaiian volcanoes. It is safe to conclude that these lunar basalts formed in lava flows. Lunar basalts vary primarily from terrestrial basalts by being more iron- and titanium-rich and by having a very low sodium content. An overwhelming fact at the *Apollo 11* site was the high concentration of uranium and thorium in all the basalts relative to any expectations.

The second group of rocks are fine-grained noritic gabbros. These have textures characteristic of intrusive rocks on earth or of the interior of thick lava flows where cooling was slow. These rocks were particularly abundant at the *Apollo 14* site and are believed to represent igneous rocks or lavas thrown up from greater depths than the mare basalts. The noritic gabbros as compared to mare basalts contain more feldspar and slightly more potassium, rare earth elements, and phosphorus. Other than that, they might be considered as coarse-grained basalts.

The third major rock type has been described as feldspar-rich or anorthositic gabbro. This rock type seems to be the most abundant in the highlands, being important at the *Apollo 14, 15,* and *16* sites. This rock is a gabbro with a large amount of very calcium-rich plagioclase feldspar. Most of the samples average about 70 percent plagioclase, but one specimen contains 97 percent plagioclase, and the latter has a composition of almost pure anorthite, the calcium end member of the plagioclase series (Chap. 21, p. 453).

The fourth main group of rocks are breccias—that is, rocks made of fragments of other rocks fused or cemented together by crystals. The fragments are pieces of the first three rock types with many small bits of glass in among the larger fragments. These breccias are very abundant rocks at all but the mare Apollo sites; that is, they are very common materials in highland rocks.

MINERALOGY

The mineralogy of all rock types is relatively simple, the two minerals pyroxene and plagioclase feldspar being the most abundant minerals in almost every rock. The plagioclase is usually a very calcium-rich variety near anorthite in actual composition, as mentioned above.

Other minerals that usually make up more than 1 percent of the rocks and sometimes as high as 10 percent are olivine, ilmenite, cristobalite, and tridymite. The latter two are high-temperature forms of SiO_2. In addition a new mineral has been identified which does not occur on the earth, pyroxferroite, a calcium iron silicate that is related to the common pyroxenes.

Quite a few minerals occur in amounts of less than 1 percent as accessory minerals, present in some rocks, absent in others. These include globules of metallic or native iron–nickel alloy, chromite, apatite, potassium feldspar, quartz, and others. The latter include some titanium minerals at least one of which is not known on earth.

Minerals containing water—that is, the OH ion—are extremely rare. In short such common terrestrial igneous rock minerals, as hornblende and micas, do not occur in lunar rocks. No trace of organic carbon has been found.

The major difference in mineralogy between lunar basalts and gabbros and similar rocks on earth is that the lunar rocks have only calcium-rich feldspars and from 3 to 10 times more ilmenite than their terrestrial counterparts. Chemically the lunar rocks have a higher concentration of titanium, chromium, zirconium, and other so-called refractory elements, and there is a much lower concentration on the moon of sodium, potassium, chlorine, and other volatile elements. These chemical differences must be reckoned with in any theory of the origin of the moon.

SHOCK METAMORPHISM

Many of the lunar rocks of all four types show evidence of shock; that is, they are highly fractured, and contain veinlets of glass from partial melting of some of the minerals. They also contain very fine-grained secondary crystals formed by recrystallization after fracturing and melting. Such rocks are said to show *shock metamorphism*. In the true sense these are metamorphic rocks—that is, metamorphosed igneous rocks. The shock metamorphic rocks have the same mineralogy as the original true igneous rocks but with the different textures mentioned above—fractures, glass, and recrystallization. This kind of metamorphism is rare on the earth but has been found in several localities where meteorite impact craters and scars have been recognized—for example, in Arizona and Quebec. Quite obviously this type of metamorphism is the result of impact events that have partially fractured and melted the older rocks. In fact, the lunar breccias really belong here also. They are impact breccias formed by the shattering of the various igneous rocks by impact events with subsequent cohesion of the fragments partly by compaction, partly by fusion of the melted material, and in part by crystallization of some of the interstitial glass or still partially molten rock material.

REGOLITH FINES

The mineralogy and petrology of thousands of the fine particles from the lunar regolith have been carefully analyzed. About half the particles are composed of glass of various colors and shapes. The other particles are bits of minerals and rocks of the kinds already discussed, although a very few particles of other rock types are known. These latter include some so-called granites and pieces of meteoritic material. In other words, the lunar fines—other than the glassy material—are composed largely of bits and pieces of basalts, gabbros, and anorthosites, all very similar to the rock types of the large-sized rock specimens. This is especially significant because the lunar regolith fines are thought to be a statistically valid sample of the entire range of lunar surface rocks. During the continuing process of impact shattering, fine particles have been thrown for hundreds of miles and mixed with material from many other areas. Rays of material thrown out by impact can be seen to extend for more than 1000 miles from some large, younger craters. Thus the regolith anywhere on the moon should contain some fragments from almost everywhere else within a radius of more than 1000 miles. In short, the fine material of the regolith is a well-mixed sample of the entire lunar surface and we have found it to be a rather monotonous mixture of basalts, gabbros, anorthosites, and glasses formed from these rock types.

AGE OF THE LUNAR ROCKS

The various rocks from the different Apollo sites have been dated by radiometric methods, especially by rubidium–strontium ratios and by the thorium–uranium–lead methods. Age determinations have been carried out in dozens of different laboratories on the same samples. The results of all these studies are in excellent agreement. The basalts from the mare locations range in age from 3.15 billion to 3.9 billion years, although the ages of the basalts from a particular mare fall in a rather narrow age range. In short, the basalts from each of the three maria so far sampled are of different ages, although within the range mentioned above. Some of the anorthosites seem to be slightly older with a determination of 4.1 billion years on one specimen.

Some idea of the relative age of a mare

surface—that is, the age of the emplacement of the last lava flows—can be obtained by the frequency of younger craters formed on that surface. By this technique it had been determined in advance of sample collection that the mare surfaces were all very old but that individual maria were filled with lavas at different times. The radiometric data are in agreement with this. So far no rocks have been dated at less than 3.1 billion years; this is a great surprise to many scientists. It appears that all the major volcanic rock-forming processes took place during the first billion and a half years of the moon's existence and that the moon has been a relatively cold or dead planet during the last three billion years so far as surface igneous activity is concerned.

Structure and interior of the moon

Several kinds of data, in addition to the actual analysis of rock samples, have contributed to an understanding of overall lunar surface composition. For example, an X ray fluorescence experiment on the *Apollo 15* orbiter measured silicon–aluminum and silicon–magnesium ratios over a continuous strip of lunar surface. When all the various kinds of compositional data are compiled and summarized, we see that the maria are composed of rocks (basalts) rich in iron and magnesium relative to silicon whereas the highlands are rich in aluminum and calcium indicating abundant plagioclase feldspar-bearing rocks (feldspathic gabbros and anorthosites); magnesium is also present throughout the highlands, but in lesser amounts than in the maria. It is now definitely clear that the highlands and the maria do have different overall compositions.

Moonquakes recorded at three of the Apollo sites indicate that the moon has less seismic activity than the earth. Many of the moonquakes occur at the time of perigee and are apparently induced by tidal stresses. The low level of seismic activity on the moon suggests that the outer layer or crust of the moon is rigid and tectonically stable quite in contrast to the earth.

Data from seismographs at *Apollo 12, 14,* and *15* sites indicate that the moon does have a layered internal structure with a thick crust, a mantle, and perhaps a core. These zones may be somewhat more gradational than is the case on the earth. The moon apparently differentiated by igneous processes very early in its existence, about 4.5 billion years, to form an outer layer, less than 50 miles in thickness, with a concentration of calcium, aluminum, and radioactive elements. This outer layer, or crust, was plagioclase-rich or anorthositic in composition and formed the original highlands crust or surface. Partial melting of the deeper part of this layer developed magmas that were erupted on the surface as lava flows, probably before the mare basins were excavated or in the bottom of the oldest of such basins. These flows had a noritic gabbro composition.

Below the crustal layer the abundance of aluminum and calcium decreases and iron and magnesium increases, probably largely as the mineral olivine. This might be considered as the lunar mantle. Partial melting of this zone probably produced the iron- and titanium-rich basalts that fill most of the younger mare basins (3.15 to 3.7 billion years).

Several large positive gravity anomalies have been measured over some of the circular maria. These must represent a concentration of mass and have been termed *mascons*. Although the origin of the mascons and their cause is not yet fully understood, it is suggested by several scientists that the mascons are large basins filled with thick dense lavas that did not settle to achieve an isostatic balance. The mascons, therefore, also indicate a very rigid crust, at least to a depth of about 100 miles.

The moon has a very weak magnetic field but one that is more than was estimated prior to Apollo. Magnetometers on *Apollos 12* and *14* showed that the lunar magnetic field has a permanent field and a transient field. The permanent field is indicated by paleomagnetism in the lunar rocks; this indicates that in the past (3.5 billion years ago) the moon had a larger magnetic field than it has now or that it came within the influence of a larger magnetic field, perhaps the earth's when and if these two bodies were closer together than they are today. Paleomagnetism in lunar rocks indicates a magnetic field of 1000 gammas for the ancient moon, whereas the present earth's field at its equator is 30,000 gammas.

The transient magnetic fields on the moon are induced by the solar wind, a stream of electrically charged particles from the sun. As measured by *Apollos 12* and *14* this transient field varies from just less than 40 gammas to over 100 gammas, which is very small when compared to the earth's field but more than that predicted for the moon. These magnetic data provide a way of estimating the temperature of the lunar interior by measuring the electrical resistance of the moon's interior. By this technique, the suggested temperature halfway to the center of the moon, is about 1000°C which is 3000° less than the temperature in the earth at its halfway point. This indicates that the moon is cool and cannot have a molten core; it is also in keeping with a low level of moonquakes and the data from the mascons. Such information suggests that the moon has a rigid crust and cool mantle and that no magmas have been able to develop in any quantity for the last 3 billion years.

Figure 28.4
Oblique view of a portion of the moon taken after transearth injection from *Apollo 15* spacecraft. Smith's Sea is in the right foreground; Mare Marginis is at left center. Notice that many craters have been flooded by mare material and that only the tops of the crater rims are visible for some craters in the mare known as Smith's Sea. Note also the very young sharp-rimmed craters, especially two in Smith's Sea. Compare dark-colored mare surfaces with lighter-colored heavily cratered highland areas. The reader should work out more detailed time relationships among the many craters. (NASA.)

History and origin of the moon

TIME SCALE

A lunar geologic time scale has been worked out for the moon's surface, or at least part of it, using the law of superposition. Various layers of rock materials can be seen to overlap one another on the photographs now available of the moon's surface. For example, some craters have clearly been formed early, then flooded by lava-flow layers, and other craters have subsequently been formed in the lava surface as illustrated in Fig. 28.4. Many quadrangle geologic maps of the moon's surface have been published by the U.S. Geological Survey as well as a composite geologic map of the entire near side of the moon (see Supplementary Readings at the end of this chapter.) The main purpose of this mapping has been to organize data available from all earth-based studies and orbiter photographs as a guide to the Apollo Programs. Such maps have helped in the selection of landing sites and the scientific studies planning of the Apollo experiments. An abbreviated form of the lunar geologic time scale is presented in Table 28.1. This table illustrates the way in which geologists approximate the relationships of rock units from photographs. Similar photogeology on earth has been widely and successfully used in difficult terrains—such as the head waters of the Amazon River and arctic Canada—to work out geologic relationships and find

Table 28.1
A lunar geologic time scale

system	descriptions	actual ages in years
COPERNICAN	craters with visible rays, crater rim materials and ejecta blankets	Tycho-270 million Copernicus-850 million
ERATOSTHENIAN	rayless craters; crater rim materials and ejecta blankets	(none available)
IMBRIAN		
Procellarum group	basin filling mare lavas; Fe- and Ti-rich basalts	*Apollo 12:* 3.15–3.2 billion *Apollo 15:* 3.25–3.37 billion *Luna 16:* 3.41 billion *Apollo 11:* 3.59–3.71 billion
Archimedian series	crater rim materials, etc. often flooded with mare lavas	
Apenninian series	ejecta material from mare basins; *Fra Mauro Formation* at *Apollo 14* site, noritic gabbro breccias	Mare Imbrian event at 3.9 billion: some basins older, some basins younger
PRE-IMBRIAN	prebasin rocks, feldspar-rich anorthositic gabbros, noritic gabbros, original highland crustal rock	an anorthosite at 4.1 billion

mineral deposits before man has actually been on the surface in question.

GEOLOGIC EVOLUTION OF THE MOON

A model for the geologic development of the moon, whose rock units are arranged in time order in Table 28.1, is based on what we know of relative ages of lunar rocks and landforms, on absolute ages by radiometric dating, and on the petrology, chemistry, isotope ratios, and other data on Apollo rocks.

It is suggested that the moon developed 4.6 to 4.7 billion years ago by the rapid accretion of cold planetesimals building a solid moon in less than 200 million years. This growth by infall was accompanied by rapid heating to over 1000°C of at least the outer 300 miles of this new body. The heat was generated by the rapid and numerous impacts and perhaps by early short-lived radioactivity. This heating may have occurred in about 10 million years.

An early chemical, magmatic differentiation and partial melting took place in the outer 300 miles of material to develop magmas that formed the early crust of the moon, rich in aluminum and calcium (plagioclase feldspar-rich rocks). This crust still occurs today as the noritic gabbros, anorthositic gabbros, and anorthosites of the highlands. As this early crust—probably built up in lava-flow fashion—began to crystallize and become rigid, a rapid infall of numerous small to moderate-sized circumterrestrial bodies continued and formed the many overlapping craters that are still characteristic of some highland areas. The impacting bodies at this time represented the remaining material that had formed the moon in the first place. This stage of differentiation and first magma development, crustal formation, and early highland cratering took place during the interval from 4.6 to 3.95 billion years ago and is known as *pre-Imbrian time*.

There then occurred a moderately short period of time (a few hundred million years) when a dozen or more very large bodies collided with the moon to form the circular mare basins, some up to 500 miles in diameter. The best known of these is Mare Imbrian (see location on Fig. 28.1) and this time period is thus known as *early Imbrian*. The diameter of the body whose impact explosion produced Mare Imbrian has been estimated at 40 to 115 miles, depending on its impact velocity. These relatively large impacting bodies are thought by some to be protomoons (small early satellites in earth orbit) formed at the same time as the moon and finally caught or swept up by the moon in Imbrian time. These impact events not only developed the various circular mare basins but also threw out prodigious amounts of material, building great mountain-sized piles of brecciated material. Note the ring of such mountains surrounding Mare Imbrian on Fig. 28.1 and named variously as the Apennines, Caucasus, Alps, and Jura. Using a variety of radiometric techniques on the glass in local regolith layers, the age of the Imbrian impact event has been dated at 3.9 billion years.

Some of the Mare Imbrian ejecta material was examined at the *Apollo 14* site and has been named the *Fra Mauro* formation. This formation is composed almost entirely of complex breccias, in which most fragments are noritic or feldspar-rich gabbros, formed by the shattering of pre-Imbrian highland-type rocks. One specimen of anorthosite has been dated at 4.1 billion years. *Apollo 15* landed on the edge of the Mare Imbrian plain at the foot of the Apennine Mountains, which rise more than 15,000 ft above the landing site, and brought back many similar rocks.

A long period of time, a few hundred million years, separates the time of impacts to excavate the mare basins and the ultimate lava flooding of these basins. During this time interval, a number of moderately large craters of the Archimedes type (see Crater Archimedes on Fig. 28.1) were formed, some in mare basins, some in highlands. Some of the larger of these craters, as Archimedes itself, were not completely flooded during later lava flood periods. This relationship is well shown on Fig. 28.4, the lunar photograph taken from an Apollo spaceship. These crater ejecta materials are known as the *Archimedian Series* (Table 28.1).

The next major stage in the evolution of the moon was another major magmatic differentiation occurring perhaps as deep as 300 miles; basalt magmas formed by partial melting of mantle material. These magmas ascended to the surface at various times during the interval 3.7 to 3.2 billion years ago to flood first one, then another, of the mare basins with iron- and titanium-rich basalts, each probably containing thousands of separate but extensive lava flows. These lava-flow accumulations in various mare basins are known collectively as the *Procellarum Group* of upper Imbrian age. There is evidence from photographs for small-scale volcanism in the highlands at the same time or earlier during Archimedian time.

Finally, there has been the period of the last 3.1 billion years during which the surface of the moon has not changed greatly, although many new craters and their extensive rays have modified its appearance. This stage of lunar history includes the Eratosthenian system of rayless craters best recognized where they have impacted mare basalt surfaces, and the sharp-rimmed Copernican system of rayed craters formed during the past billion years or so (see Fig. 3.11). These craters have probably been caused by asteroids, meteorites, and comets from both within and without our solar system but not from objects in earth orbit as suggested for the flood of early impacts during pre-Imbrian and Imbrian time. Certainly the cratering rate was much slower during the last 3 billion years than it was during the first 1.5 billion years.

The moon's crust and mantle apparently became quite solid and cool early in this 3-billion-year stage. Volcanic activity on a large scale ceased 3.1 billion years ago but may have continued in isolated areas as small-scale activity not yet sampled by man. Young material of this type that looks volcanic has been observed on Lunar Orbiter and Apollo photographs.

Figure 28.5
Oblique view of Triesnecker Crater and Triesnecker Rilles, located at the northeastern edge of Central Bay as photographed by *Apollo 10.* Many of these rilles are fairly straight and might be fault features, but not all of them. Note the sharp rim and apron of radial ejecta around Triesnecker Crater. (NASA.)

RAYED CRATERS

Using these same radiometric techniques on glass in regolith layers, the impact producing the crater Copernicus—one of the younger rayed craters (Fig. 3.11)—has been dated at 850 million years ago. Tycho was formed 270 million years ago (Figs. 3.10 and 28.1). A relatively small crater, Cone Crater, at the Apollo 14 site was produced only 25 million years ago. Young craters, those less than 1 billion years, usually have sharp rims and light-colored rays that radiate outward for miles. The rays from Copernicus radiate for hundreds of miles. Copernicus ray material was sampled by the Apollo astronauts. These rays consist of very thin, broad bands of fine-grained material, largely dust, ejected at the time of impact. In the younger craters this ray material shows up in photographs by having a lighter color or higher albedo. With time this material darkens, perhaps because of changes brought on by the solar wind, until the rays are no longer visible at older craters.

RILLES

Of the many interesting smaller-scale features on the lunar surface, the so-called rilles are perhaps the most challenging and enigmatic (see Fig. 28.5). The rilles are valley-like features; some are straight whereas others are quite serpentine. The larger rilles are well over 2000 ft deep and some of the straight ones are a few miles wide. These great straight trenches are probably fault valleys, a narrow zone dropped down between two parallel faults. Earth analogues would be the African rift valleys in Kenya, Tanzania, and Ethiopia as well as the valley occupied by the Dead Sea in Israel. However, the sinuous rilles must have an entirely different origin. Hadley Rille, which was visited by *Apollo 15,* is an excellent example of these mysterious features. Hadley Rille, at the *Apollo 15* landing site, is 1000 ft deep and about a mile wide at the top and has a V-shaped profile. It is very snakelike in map view and is continuous with rather uniform size for nearly 100 miles (Fig. 28.6). Suggestions were once made that these rilles might be stream-produced valleys left over from an early time when the moon had water. Such rilles, however, have few branches and no evidence of a water-erosion period has been found even in rocks up to 3.95 billion years old. So the stream-valley origin must be abandoned. But then how could such rilles be formed? The rilles do occur only on the

Figure 28.6
The Hadley Rille which was visited by the *Apollo 15* mission (X shows the landing site). The rille is developed in an embayment of mare lavas into the edge of the Apennine Mountains. This mare area is called Palus Putredinus but is in the overall basin of Mare Imbrium. This sinuous rille is about 1 mile wide and 1000 ft deep for most of its length. (NASA.)

mare surfaces and are concentrated near the mare margins; that is, the rilles occur on the basalt lava-flow surfaces. On earth very similar but much smaller features are known on basalt lava surfaces. These very curvy or sinuous elongate depressions on earth are lava channels and collapsed lava tubes. They are common in the lavas of Hawaii, in several areas in Oregon, and on the Snake River Plains in Idaho. About the largest such collapsed lava tube on earth is about 20 miles long and perhaps 100 ft deep. On the moon sinuous rilles, like Hadley Rille, are a mile wide and many miles long. Initial studies of the depth-to-width ratios derived from *Apollo 15* surface and orbital photography of this amazing rille give tentative support to the theory that it may be the result of incomplete collapse of a buried lava tube of an enormous size by terrestrial standards.

ORIGIN OF THE MOON

There are three main ideas as to the possible origin of the moon itself. These are (1) that the material of the moon was torn from the earth and that the Pacific Ocean basin might be the scar from this event; (2) that the moon is a stranger to our solar system but was captured by the earth as it passed through our solar system close to the earth, and (3) that the moon and the earth originated at about the same time and place in the solar system and are, therefore, sister planets. As yet we do not have enough data to evaluate these three hypotheses; there are objections to all of them, but at present the objections to the sister planet hypothesis are least serious.

It is clear that the moon has been a separate body for more than 4 billion years. The Pacific Ocean basin on earth is a much, much younger feature and cannot be a scar from which moon material was torn out. The differences in overall chemistry of earth and moon rocks—high titanium and low sodium on the moon, for example—make it very unlikely that the moon came from earth material even at a very early time. This hypothesis seems impossible.

The earth and moon as sister planets "growing" simultaneously during the origin of our solar system by planetesimal accretion does seem reasonable. However, why the chemical differences? The great difference in size of the two bodies could mean quite a different early history. The smaller body began to melt at the surface very early and very fast and thus differentiated a crustal layer in a different way than happened on the earth. Again because of the small size of the moon, certain volatile elements such as sodium, potassium, chlorine, and atmospheric gases were lost to space. But why does the moon have no metallic iron core? The moon has an overall density of 3.3 (compared to the earth's 5.5) and thus cannot have a comparable iron-rich core. There is no information as yet with which to suggest an answer to this problem. However, by radiometric studies the moon is known to be approximately 4.6 billion years old. The same age has been indicated for the earth and most of the meteorites found on earth. Thus the moon, meteorites, and the earth were all formed (solidified) at the same time and probably in the same solar system. This is the most compelling evidence against the stranger hypothesis for the moon's origin, plus the fact that the overall chemical abundances on moon and earth are similar except for the rather minor differences we have mentioned. It may well be that the actual origin of the moon was a far more complex series of events than any of the three hypotheses mentioned above—perhaps some combination of the suggestions. New hypotheses and variations of hypotheses are appearing in print every few months by competent scholars.

Significance to man

As a result of the intensive studies related to the Apollo missions, a new highly cosmopolitan scientific community has been shaped in which thousands of scientists are recognizing the interrelationships of their own work with that of others in completely different fields. A new planetary science has emerged. Physicists, chemists, astronomers, and geologists are now cooperating as never before. Although this cooperation was started by the study of the moon, it is continuing with totally new programs aimed at a better understanding of the earth.

The rocks found on the surface of the moon are extremely old, with the oldest going back close to the birth of the moon. The youngest rocks so far found on the moon (3.1 billion) are about the same age as the oldest rocks known on earth. Thus an essentially unbroken record of the history of our solar system has been left within the materials of these two planets. The birth and early adolescence of the solar system is written in the lunar samples; the late adolescence and adulthood of the solar system is written in terrestrial rocks. The continuing attempt to read these records provides an intriguing challenge to man.

In these days when environmental science has found a much needed relevance, it is becoming increasingly obvious that nothing will contribute more to the continued life of man on earth, and to the quality of his life, than a comprehensive understanding of the vulnerable planet on which he lives. Our study of the primitive moon, in relation to the complex earth, can contribute significantly to this comprehensive understanding. The study of the moon is a great adventure in thought. An understanding of the origin and evolution of the earth's only near companion in space will be a milestone in the history of man.

SUMMARY

The moon has been studied by the Surveyor, Lunar Orbiter, and Apollo Programs since 1966. A vast volume of data and thousands of photographs are now available, some still unstudied, and 5000 to 10,000 pages of scientific literature have been published so far.

The moon is covered by millions of craters from inches to hundreds of miles in diameter. Most of these craters are due to meteorite, asteroid, or comet impact events that have shattered the surface rocks and produced, in addition to the craters, a regolith of broken rock, mineral, dust-sized particles, and impact breccias in which fragments have been compacted and recrystallized. Many fragments are composed of glass or partly melted rock and mineral fragments, the melting due to instantaneous impact-generated heat. The width-to-depth ratios of most craters correspond to those of known meteorite impact craters and bomb and atomic test craters on the earth. A few small craters are probably of volcanic origin.

Small craters have no visible rock fragments because their impact-producing events disturbed only the regolith layer. Larger craters were formed by events that excavated bedrock to throw out rock fragments. Crater size, therefore, can be used to estimate regolith thickness. Young craters have sharp rims and angular rock fragments, older craters have more subdued rims and rounded rock fragments. This erosion is due to a micrometeorite bombardment that also continually churns the regolith material.

The maria are relatively flat plains of dark color or low albedo and are composed of basalt lava flows rich in iron, titanium, and magnesium. Their major minerals are pyroxene, olivine, and plagioclase feldspar.

The highlands are lighter in color and are hilly to very mountainous. Some segments apparently represent original crust. These are more heavily cratered than younger surfaces like mare ejecta hills, which, in turn, have more craters than mare lava-covered surfaces. The highland rocks are noritic gabbros, anorthositic or feldspar-rich gabbros, and anorthosites. All these rocks contain abundant calcium-rich plagioclase feldspar. In the highlands, the elements calcium and aluminum are concentated as compared to the maria. Abundant breccia and shock metamorphosed igneous rock are present in the highlands because much of its material has been thrown out by mare basin impact events.

Rocks from the moon, especially the mare basalts, are 3.15 to 3.9 billion years old as dated by various radiometric techniques. One anorthosite was dated at 4.1 billion years.

The moon has a layered structure, according to seismic data, with a crust 25 to 50 miles thick composed of feldspar-rich rocks with a mantle rich in iron and magnesium and apparently with much olivine. Moonquakes, mascons, and interior temperature estimates all suggest a rigid crust and cool mantle as compared to the earth.

The present model of the geologic evolution of the moon indicates a pre-Imbrian period, 4.6 to 3.9 billion years ago, in which the moon formed by accretion, and differentiated rapidly to produce the feldspar-rich crust which soon became highly cratered. Some of this original cratered crust apparently still remains in highland areas. In Imbrian time, 3.9 to 3.1 billion years ago, a series of great impact events excavated the mare basins and pushed up great rims of ejecta material of mountain-range size. After a few hundred million years, these basins were filled by volcanic flooding to form the flat mare surfaces. During the period of a cooler moon, from 3.1 billion years ago to the present, some cratering has continued with the formation, more recently (the last billion years), of the striking rayed craters like Copernicus and Tycho.

Important words and terms

Surveyor, Lunar Orbiter, Apollo
primary and secondary craters
regolith and regolith fines
lunar basalts, gabbros
noritic gabbros
anorthositic gabbros, anorthosites
breccias
shock metamorphism
mascons
maria and highlands
mare material
rayed craters
rilles

Questions and exercises

1. What hazards have the Apollo astronauts successfully overcome while on the lunar surface?
2. If you have watched any of the Apollo TV broadcasts from the moon, answer these: Why was the sky black? How did the low lunar gravity field seem to affect the astronauts? What color are the lunar rocks? Did the lunar landscape remind you of any place on earth? If so, where and why?
3. Compare earth and moon erosion processes as to the agents acting, the products resulting, and the time involved.

4. Compare impact and volcanic craters on earth (refer also to Fig. 1.19 and Chap. 22) in as many ways as possible. What evidence would you consider as proof of either the impact or volcanic origin at a particular crater?
5. Would all your evidence and proofs in Question 4 remain the same on the moon? Explain.
6. Compare earth and moon basalts. Can you think of features often present in terrestrial basalts but never in lunar basalts?
7. The oldest rocks on earth are just over 3 billion years old and the youngest rocks on the moon are also just over 3 billion years old. Compare these two groups of rocks of the same age as to type of rock, textures, probably mineralogy, and thus origin. If you were given one typical specimen of each, how could you be sure which was which?
8. It was necessary to have three lunar seismograph stations (*Apollos 12, 14,* and *15*) before any data on layering within the moon could be obtained. Why? (Refer to Chap. 25, if necessary.)
9. Study geologic maps of the moon as supplied by your instructor. Note overlap of layers and descriptions of materials in the map's legend.
10. Work out a partial lunar time scale of local events from photographs supplied by your instructor, or use Figs. 28.4 and 28.5.
11. Do some thinking about the origin of lunar rilles after studying pictures of them. What do *you* think their shapes suggest? Give your reasoning.
12. Do you think the lunar exploration programs, Apollo especially, have been worth their cost to man? Why or why not? Discuss this in class.

Supplementary readings

Book

Mutch, Thomas A., *Geology of the Moon,* Princeton University Press (1970). [A lengthy book that discusses in detail the moon's shape and motions, remote sensing techniques, origin, and comparison of lunar and terrestrial craters, lunar and volcanic stratigraphy, principles of the stratigraphy of various mare basins, *Apollo 11* results and discussion of ages, relative and absolute.]

Articles

Allen, Joseph P., "Apollo 15, Scientific Journey to Hadley-Apennine," *American Scientist,* **60,** 162–174, (1972). [Summary of scientific objectives and findings of this very interesting and significant Apollo mission at the foot of the Apennine Mountains, written in layman's language.]

Hinners, N. W., "The New Moon: A Review," *Reviews of Geophysics and Space Physics,* **9,** 447–522 (1971). [A rather technical but very complete summary of what man has learned so far about the moon (through *Apollo 14*). Emphasis is on lunar rocks, regolith, chronology, surface processes, geophysical data, and their interpretation relative to lunar history and development.]

Lowman, Paul D., Jr., "The Geologic Evolution of the Moon," *Journal of Geology,* **80,** 125–166 (1972). [A review of the geologic history of the moon as we know it so far based on data through *Apollo 15.* Not too technical for college students.]

ESCP Pamphlet Series

Moore, Carleton B., *Meteorites,* #PS-10, Houghton Mifflin, Boston, Mass. (1971). [A discussion of meteorite falls, meteor showers, craters formed on earth as compared to the moon, radiometric ages of meteorites, types, and mineralogy of meteorites. Mostly data from the earth but very useful as background when considering the meteorite bombardment of the lunar surface for 4.5 billion years.]

Map

Wilhelms, D. E., and J. F. McCauley, *A Geologic Map of the Near Side of the Moon,* U. S. Geological Survey Map #I-703, 1971. [A 53 × 35 in. map with a 7-page summary text. Scale 1:5,000,000 (1 in., about 80 miles). Order from: Distribution Section, U.S. Geological Survey, 1200 So. Eads St., Arlington, Va., 22202. Prepayment of $1.00 is required.]

Articles from Scientific American

Dyal, Palmer, and Curtis Parkin, "The Magnetism of the Moon" (August, 1971).

Mason, Brian, "The Lunar Rocks" (October, 1971).

Wood, John A., "The Lunar Soil" (August, 1970).

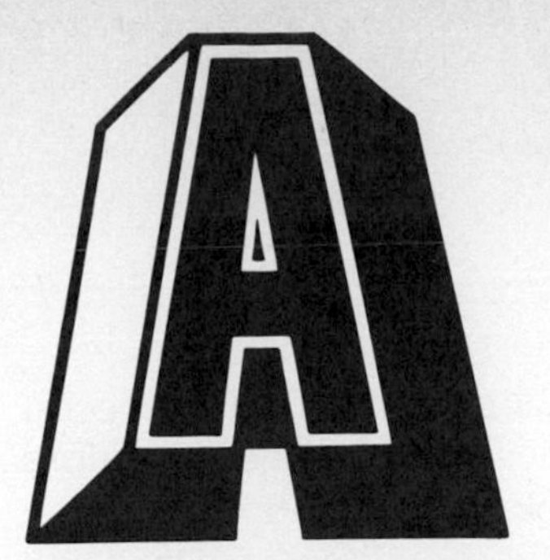

APPENDIX / MATHEMATICS

The mathematical background that is necessary for your understanding of the discussions in certain chapters of this book is summarized in this appendix. You have no doubt been exposed to the elementary mathematical concepts and operations used in this book. They are summarized here for your use as a refresher and for ready reference.

The discussions are often analytical and sometimes quantitative. However, the treatment in this book is relatively nonmathematical in the sense that only simple mathematical concepts and techniques of manipulation are used. For the most part, the reasoning employs only arithmetic and a few simple ideas from algebra and geometry. If the teacher wishes to supplement the material in this book, he may find it necessary to develop additional concepts from the fields of geometry and trigonometry in class.

The student should remember that the word "compute" as used in problems, means "show your work"; for example, show the equation, show substitution in the equation, and show the answer. The student should also remember that for increased learning and understanding he may "visualize" or write a mathematical equation in symbols but he should "read" (or say) the equations in words.

This mathematical appendix includes a brief presentation of selected concepts and operations in arithmetic, scientific notation, approximation, algebra, geometry, and proportionality.

Arithmetic

It is assumed that the student is familiar with the four fundamental operations of arithmetic: addition, subtraction, multiplication, and division; and that he understands the meaning of the common terms used therewith.

The equal sign (=) indicates an equality. For example, $2 = 2$.

The inequality sign ($\neq$) indicates an inequality and is read "is not equal to." For example, $3 \neq 2$.

The approximately equal sign is either $\cong$ or $\doteq$. For example, $2.1 \cong 2.0$ or $2.1 \doteq 2.0$ is read "2.1 is approximately equal to 2.0."

The sign $>$ is read "is greater than." For example, $3 > 2$.

The sign $<$ is read "is less than." For example, $2 < 3$.

Multiplication is represented by the multiplication sign. The multiplication sign is either $\times$ or $\cdot$. Thus 3 times $2 = 3 \times 2 = 3 \cdot 2 = 6$. In multiplication, the answer is called the "product."

When letters are used in place of numerals, the product may be indicated without the multiplication sign. For example, A multiplied by B, instead of being written $A \times B$, may be expressed as AB. Also, 4 times B may be written as $4B$.

In multiplication, 5×0 means that you take 5 zeros and add them; 0×5 means that you take none of the 5s and add them. The product of multiplying anything by zero is zero.

The division sign $\div$ or slant line / indicates a fraction.

FRACTION

A fraction is a number that may be considered as an indicated division. The *numerator* (on top) is divided by the *denominator* (on bottom). Thus we can write

$$\frac{1}{4}, \frac{3}{8}, \frac{1}{2}, \frac{3}{4}, \frac{11}{8}$$

The *dividend* is the numerator of a fraction, and the *divisor* is the denominator.

An integer, or whole number, may be considered as a fraction with 1 as the denominator. Thus

$$4 = \frac{4}{1}; \qquad 333 = \frac{333}{1}$$

In a *proper* fraction, the numerator is smaller than the denominator, and the value of the fraction is less than 1. Thus

$$\frac{3}{4} \text{ is less than 1, or } \frac{3}{4} < 1$$

In an *improper* fraction, the numerator is

larger than the denominator. Thus $\frac{11}{8}$ is larger than 1, or

$$\frac{11}{8} > 1$$

An improper fraction may be expressed as a mixed number (whole number and fraction) and vice versa. Thus,

$$\frac{11}{8} = 1\frac{3}{8}; \qquad 8\frac{1}{2} = \frac{17}{2}$$

In fractions, $\frac{0}{2}$ "asks" how many 2s fit into zero? The answer is none. Also, $\frac{2}{0}$ "asks" how many zeros fit into 2. The answer is an infinitely large number. The symbol for an infinitely large number is the one that is used for infinity (∞).

RECIPROCALS

The reciprocal of a fraction is the inverse of the fraction; that is, it is the fraction "turned upside down." The reciprocal of $\frac{3}{4}$ is $\frac{4}{3}$; the reciprocal of 8 is the inverse of $\frac{8}{1}$, which is $\frac{1}{8}$.

Division is the inverse of multiplication. Thus dividing by a fraction is the same as multiplying by the reciprocal of the fraction. For example,

$$36 \div \frac{4}{3} = 36 \times \frac{3}{4} = 27$$

Similarly,

$$36 \times \frac{1}{4} = 36 \div 4 = \frac{36}{4}$$

The value of the fraction is unchanged if both the numerator and denominator are multiplied by the same number (not including zero as a number). For example,

$$\frac{2}{3} = \frac{2}{3} \times \frac{4}{4} = \frac{8}{12}$$

Similarly, the value of a fraction is unchanged if both numerator and denominator are divided by the same number. For example,

$$\frac{8}{12} = \frac{\frac{8}{4}}{\frac{12}{4}} = \frac{2}{3}$$

The familiar "canceling" is an application of this rule.

Warning. Adding (or subtracting) the same number to the numerator and to the denominator changes the value of the fraction For example,

$$\frac{2}{3} \neq \frac{2+4}{3+4} \neq \frac{6}{7}; \text{ that is, } \frac{2}{3} \text{ is not equal to } \frac{6}{7}$$

In handling "compound fractions," in which the numerator (or the denominator, or both) is a fraction, a rule is "to divide by a fraction, invert and multiply." Thus

$$\frac{\frac{A}{B}}{\frac{C}{D}} = \frac{A}{B} \div \frac{C}{D} = \frac{A}{B} \times \frac{D}{C} = \frac{AD}{BC}$$

For example,

$$\frac{\frac{2}{3}}{\frac{4}{5}} = \frac{2}{3} \div \frac{4}{5} = \frac{2}{3} \times \frac{5}{4} = \frac{10}{12}$$

Also, in handling "compound fractions," a rule is "to divide a fraction by a fraction, divide the product of the 'extremes' by the product of the 'means.'" Thus

$$\frac{\frac{A}{B}}{\frac{C}{D}} = \frac{A \times D}{B \times C} = \frac{AD}{BC}$$

The "extremes" are *A* and *D;* the "means" are *B* and *C.*

As an example of the latter rule for handling "compound fractions" (using the same numerical fractions as in the first rule),

$$\frac{\frac{2}{3}}{\frac{4}{5}} = \frac{2 \times 5}{3 \times 4} = \frac{10}{12}$$

In working with "compound fractions," it is better not to use any slanting lines to indicate division; use only horizontal fraction lines, with the main fraction line longer or heavier. This helps one to avoid confusion.

Any quantity is not changed in value when it is multiplied or divided by one. Thus

$$\frac{\frac{A}{B}}{C}=\frac{\frac{A}{1}}{\frac{B}{C}}=\frac{AC}{1B}=\frac{AC}{B}$$

For example,

$$\frac{\frac{2}{3}}{4}=\frac{\frac{2}{1}}{\frac{3}{4}}=\frac{2\times 4}{1\times 3}=\frac{2\times 4}{3}=\frac{8}{3}$$

or

$$2\div\frac{3}{4}=\frac{2}{1}\times\frac{4}{3}=\frac{2\times 4}{1\times 3}=\frac{8}{3}$$

Also,

$$\frac{\frac{A}{B}}{C}=\frac{\frac{A}{B}}{\frac{C}{1}}=\frac{A\times 1}{B\times C}=\frac{A}{BC}$$

For example,

$$\frac{\frac{2}{3}}{4}=\frac{\frac{2}{3}}{\frac{4}{1}}=\frac{2\times 1}{3\times 4}=\frac{2}{12}=\frac{1}{6}$$

or

$$\frac{2}{3}\div 4=\frac{2}{3}\div\frac{4}{1}=\frac{2}{3}\times\frac{1}{4}=\frac{2\times 1}{3\times 4}=\frac{2}{12}=\frac{1}{6}$$

PERCENTAGE

Percentage is the fractional part of a number expressed in "parts per hundred." Thus $\frac{6}{10}$ of a group means 6 out of 10, or 60 out of 100, or 60 percent. Similarly, 40 percent of 20 means

$$\frac{40}{100}\times 20=0.40\times 20=8$$

DECIMALS

Decimals are special fractions in which the denominator is some power of 10. Thus

$$0.1=\frac{1}{10};\qquad 0.03=\frac{3}{100}$$

or

$$\frac{3}{10^2};\qquad 438.2=438\,\frac{2}{10};\qquad 35.48=35\,\frac{48}{100}$$

Multiplying a number by 10 is equivalent to moving decimal point one place to the right; by 100, two places to the right; by 1000, three places to the right; etc. For example,

$$4.538\times 10=45.38$$
$$4.538\times 100=453.8$$
$$4.538\times 1000=4538.$$

Dividing a number by 10 is equivalent to moving the decimal point one place to the left; by 100, two places to the left; by 1000, three places to the left; etc. For example,

$$4538.\div 10=453.8$$
$$4538.\div 100=45.38$$
$$4538.\div 1000=4.538$$

POWERS OR EXPONENT NOTATION

When a number appears as a factor more than one time, the multiplication can be expressed efficiently in *exponential* form or *power* form. The *base* is the number that appears as the factor, and the *exponent* (or power) is the number of times the base multiplies itself. Thus

$$A\times A=A^2$$

For example,

$$3\times 3=3^2=9$$

Conversely, A^2 ("A squared" or "A to the second power") means $A \times A$. Hence,

$$A^2=A\times A$$

For example,

$$3^2=3\times 3=9$$

Similarly, "A cubed" or A^3 (A to the third power) means "A times A times A"; thus,

$A^3 = A \times A \times A$

For example,

$3^3 = 3 \times 3 \times 3 = 27$

As another example,

$2 \times 2 \times 2 \times 2 \times 2 = 2^5 = 32$

where 2 is the base and 5 is the exponent (or power). Similarly,

$10 \times 10 \times 10 = 10^3 = 1000$

Conversely, 5^4 means

$5 \times 5 \times 5 \times 5 = 625$

2^5 is read "two to the fifth power."

The second power is often called "square." Thus 5^2 may be read as "five to the second power" but is more often read "five squared." The reason for this terminology is that the area of a square is the length of a side multiplied by itself.

The third power is often called "cube." Thus, 5^3 may be read as "five cubed" as well as "five to the third power." The reason for this terminology is that the volume of a cube is the length of a side taken three times and multiplied.

If the number appears as a factor only once, the exponent 1 is understood. Thus $10^1 = 10$.

MULTIPLICATION OF EXPONENTIALS

The definition of exponents permits the efficient multiplication of exponentials.

Because

$5^4 = 5 \times 5 \times 5 \times 5,$

and

$5^3 = 5 \times 5 \times 5$

it follows that

$5^4 \times 5^3 = (5 \times 5 \times 5 \times 5) \times (5 \times 5 \times 5) = 5^7,$

or

78,125

The same answer results by adding the exponents:

$5^4 \times 5^3 = 5^{4+3} = 5^7,$

or

78,125

This leads to the *multiplication rule:*

To multiply numbers expressed as powers of the same base, add the exponents.

In general,

$A^m \times A^n = A^{m+n}$

DIVISION OF EXPONENTIALS

Similarly, this operation may be carried out as follows:

$$\frac{5^5}{5^3} = \frac{5 \times 5 \times 5 \times 5 \times 5}{5 \times 5 \times 5} = 5 \times 5 = 5^2 = 25$$

This leads to the *division rule:*

To divide numbers expressed as powers of the same base, subtract the exponent of the denominator from the exponent of the numerator.

In general,

$$\frac{A^m}{A^n} = A^{m-n}$$

For example,

$$\frac{5^5}{5^3} = 5^{5-3} = 5^2 = 25$$

Is $5^0 = 1$? Because

$$\frac{5^5}{5^5} = 1$$

then

$$\frac{5^5}{5^5} = 5^{5-5} = 5^0 = 1$$

NEGATIVE EXPONENTS

The concept of exponents has been extended to negative powers. Applying the rule for the division of exponentials, 5^3 divided by $5^5 = 5^{-2}$; thus

$$\frac{5^3}{5^5} = 5^{3-5} = 5^{-2}$$

In terms of the original definition of exponents, 5^{-2} is meaningless. However, by carrying out the division in the extended form,

$$\frac{5^3}{5^5} = \frac{5 \times 5 \times 5}{5 \times 5 \times 5 \times 5 \times 5} = \frac{1}{5 \times 5} = \frac{1}{5^2}$$

Therefore,

$$5^{-2} = \frac{1}{5^2}$$

Similarly,

$$10^{-2} = \frac{1}{10^2} = \frac{1}{100}$$

$$\frac{1}{10^3} = 10^{-3} = \frac{1}{1000}$$

and

$$\frac{1}{10^{-6}} = 10^6$$

Any exponential may be moved from the numerator to the denominator (or vice versa) by changing the sign of the exponent. Thus a positive exponent in the denominator becomes negative in the numerator, and vice versa.

DIVISION OF BASES

In considering different bases to the same power, is

$$\frac{A^2}{B^2} = \left(\frac{A}{B}\right)^2 \quad ?$$

If $A = 20$ and $B = 10$, then

$$\frac{20^2}{10^2} = \left(\frac{20}{10}\right)^2$$

that is,

$$\frac{400}{100} = 2^2,$$

or

$$4 = 4$$

The answer to the question is "yes."

RAISING TO A POWER

To raise a number already in power form to another power, multiply the exponents. Thus,

$$(2^5)^3 = 2^5 \times 2^5 \times 2^5 = 2^{5 \times 3} = 2^{15}$$

ROOTS

The opposite of raising a quantity to a given power is called "extraction of a given root."

The root symbol is $\sqrt{\ }$ (called the "radical sign"). The desired root is indicated by a numeral as follows, $\sqrt[3]{\ }$. If no root number is used, the number 2 is understood. Thus

$\sqrt{A}$ is read "the square root of A"
$\sqrt[3]{A}$ is read "the cube root of A"

The expression $\sqrt{25}$ (read "the square root of 25") means to find the number that multiplied by itself will give 25. The number is obviously 5, because

$$5^2 = 25$$

Similarly, the expression $\sqrt[3]{64}$ (read "the cube root of 64") means to find a number that taken three times and multiplied will give 64. The number is obviously 4, because

$$4^3 = 4 \times 4 \times 4 = 64$$

If roots are not whole numbers, they can be obtained approximately by inspection. High accuracy is not required for most purposes.

For example, if $A^2 = 27$, then $A = \sqrt{27}$. What is the value of A, approximately? That is, what number multiplied by itself will give about 27? The answer is 5+ (read 5 plus, meaning a little more than 5). Therefore,

$$A = \sqrt{27} = 5+$$

Notation of numbers in science

An important application of exponents is in the handling of very large and of very small numbers in science. Such numbers are expressed easily in the efficient scientific notation that is known as "powers of 10."

POWERS OF 10

All multiples of 10 can be expressed as powers of 10. For example,

$$100 = 10 \times 10 = 10^2$$
$$1000 = 10 \times 10 \times 10 = 10^3$$

All submultiples of 10 (such as one-tenth, one one-hundredth) can be expressed as powers of 10. For example,

$$\frac{1}{10} = \qquad\qquad = 10^{-1}$$
$$\frac{1}{100} = \frac{1}{10 \times 10} = \frac{1}{10^2} = 10^{-2}$$
$$\frac{1}{1000} = \frac{1}{10 \times 10 \times 10} = \frac{1}{10^3} = 10^{-3}$$

Multiples of 10 have common names (such as hundred, thousand, million, billion). Some of the common names of submultiples of 10 are one one-hundredth, one one-thousandth, and one one-millionth. The table that follows is a partial list of multiples and submultiples of 10:

$1{,}000{,}000 = 10^6$	one million
$1000 = 10^3$	one thousand
$100 = 10^2$	one hundred
$10 = 10^1$	ten
$1 = 10^0$	one
$0.1 = 10^{-1}$	one-tenth
$0.01 = 10^{-2}$	one one-hundredth
$0.001 = 10^{-3}$	one one-thousandth
$0.000{,}001 = 10^{-6}$	one one-millionth

For numbers greater than 1, the exponent is positive and is equal to the number of zeros following the 1. For numbers smaller than 1, the exponent is negative and is equal to the number of places 1 is to the right of the decimal point.

Very large or very small numbers may be read in words by being expressed as products of appropriate powers of 10. For example, 10^{12} is "one trillion." However, if expressed as $10^6 \times 10^6$, it is read as "one million, million." Similarly, 10^{-12} expressed as $10^{-6} \times 10^{-6}$ is read as "one one-millionth of a millionth."

OTHER NUMBERS

Any number can be expressed as the product of a number and a power of 10. Thus 2973 may be expressed as

297.3×10

or

$29.73 \times 100 = 29.73 \times 10^2$

or

$2.973 \times 1000 = 2.973 \times 10^3$

Moving the decimal point one place to the left is equivalent to dividing the number by 10; two places to the left, by 100; three places to the left, by 1000. In order to restore the value, the number is multiplied by a power of 10 equal to the number of places the decimal point has been moved to the left.

In decimals smaller than 1, the decimal point is moved to the right, and the resulting number is multiplied by a negative power of 10 equal to the number of places the decimal point was moved to the right. Thus

$$0.02973 = 2.973 \times \frac{1}{100} = 2.973 \times \frac{1}{10^2}$$
$$= 2.973 \times 10^{-2}$$

Usually the decimal point is moved so that the first number is between 1 and 10. Thus the speed of light, 30,000,000,000 cm/sec, is usually expressed as 3×10^{10} cm/sec.

In "writing in powers," write digits $\times$ 10^{power}. For example,

$297.3 = 2.973 \times 10^2$

The concepts of this topic are illustrated in the following table:

$$2973. = 2.973 \times 10^3$$
$$297.3 = 2.973 \times 10^2$$
$$29.73 = 2.973 \times 10^1$$
$$2.973 = 2.973$$
$$0.2973 = 2.973 \times 10^{-1}$$
$$0.02973 = 2.973 \times 10^{-2}$$
$$0.002973 = 2.973 \times 10^{-3}$$

MANIPULATING VERY LARGE OR VERY SMALL NUMBERS

The scientific notation "power of 10" makes possible the convenient handling of multiplication or division of very large numbers and of very small numbers.

Example 1
Given the speed of light and the number of seconds in a year, compute the distance light travels in a year.

$$\begin{aligned} \text{speed of light} &= 30{,}000{,}000{,}000 \text{ cm/sec} \\ &= 3 \times 10^{10} \text{ cm/sec} \\ 1 \text{ year} &= 31{,}500{,}000 \text{ sec} \\ &= 3.15 \times 10^{7} \text{ sec} \end{aligned}$$

Therefore

$$\begin{aligned} \text{distance} &= \text{speed of light } (3 \times 10^{10}) \text{ cm/sec} \\ &\quad \times \text{ seconds in 1 year } (3.15 \times 10^{7}) \\ &= (3 \times 3.15) \times (10^{10} \times 10^{7}) \\ &= 9.45 \times 10^{17} \text{ cm} \\ &= 945 \times 10^{15} \text{ cm} \\ &= \text{"945 million, billion centimeters"} \end{aligned}$$

Example 2
Compute the wavelength of a television wave sent out at a frequency of 60 megahertz or megacycles per sec (60×10^6 cycles/sec).

$$\text{wavelength} = \frac{\text{speed}}{\text{frequency}} = \frac{3 \times 10^{10} \frac{\text{cm}}{\text{sec}}}{60 \times 10^{6} \frac{\text{cycles}}{\text{sec}}}$$

$$= \frac{30 \times 10^{9}}{6 \times 10^{7}} = 5 \times 10^{2} \frac{\text{cm}}{\text{cycle}}$$

$$= 500 \frac{\text{cm}}{\text{cycle}}$$

Notice the change of "3 to 30" and "60 to 6" in order to obtain the fraction $\frac{30}{6}$ so that the quotient will be greater than 1.

Example 3
What is the mass of 602,300,000,000,000,000,-000,000 (6.023×10^{23}) molecules of hydrogen gas if each molecule has a mass of 0.000,000,000,-000,000,000,000,003,32 (3.32×10^{-24}) g?

$$\begin{aligned} \text{mass} &= (6.023 \times 10^{23}) \times (3.32 \times 10^{-24} \text{ g}) \\ &= 19.99636 \times 10^{-1} \text{ g} \\ &\cong 20 \times 10^{-1} = 2 \text{ g} \end{aligned}$$

Approximations

Much mathematical reasoning can be carried out with approximate numerical values. The scientist often rounds off numbers, uses rough estimates of numbers, and makes estimates of complicated mathematical expressions. The ease with which the answers are obtained compensates for the loss of exactness. The use of round numbers permits placing the emphasis on the reasoning involved.

ACCURACY OF DATA

All scientific data derived from measurement have a limited accuracy. Certain errors are inherent in the method of measurement. The accuracy is indicated by the number of significant figures. Thus the statement that the equatorial diameter of the earth is 7926.68 miles indicates that it is known accurately through five significant figures (with some uncertainty in the sixth figure). The accuracy of most measurements does not exceed three or four significant figures. For most purposes, however, accuracy of even four significant figures is not necessary, and consequently the numbers are rounded off.

ROUNDING OFF

To round off a number, select the significant figure to which the accuracy is desired. Then (1) if the next digit is less than 5, replace all succeeding digits by zeros and drop the decimals; (2) if the next digit is 5 or greater, increase the significant figure by one and replace the succeeding digits by zeros.

The method is illustrated in the following successive steps in rounding off the equatorial diameter of the earth formerly used:

7926.7	to five significant figures
7927	to four significant figures
7930	to three significant figures
7900	to two significant figures
8000	to one significant figure

APPROXIMATE CALCULATIONS

The rounding off of numbers and the methods of notation of numbers in science make possible the estimation of answers quickly. The following examples will illustrate the method.

Example 1

Assume that it is desired to determine the ratio of the diameter of the earth to that of the moon. Given the diameter of the earth (equatorial) = 7926.68 miles; the diameter of the moon = 2160 miles.

The desired answer requires the division of 7926.68 by 2160. By long division the answer is 3.67.

The scientist usually proceeds as follows: The diameter of the earth is rounded off to 8000 miles; that of the moon to 2000 miles; and the ratio is quickly seen to be about 4. The answer is sufficiently accurate for most purposes and permits a meaningful comparison of the relative size of the two bodies.

Example 2

How much time is required for a microwave signal (electromagnetic wave) to go from a radiotelescope on the earth to the moon and be reflected back to the earth? Given the speed of light and other electromagnetic waves is 186,324 mi/sec. The average distance of moon from earth is 239,000 miles. Approximations frequently used are 186,000 and 240,000.

$$\text{speed of light} = 186{,}000 \text{ mi/sec} = 1.9 \times 10^5 \text{ mi/sec}$$

$$\text{distance of the moon} = 240{,}000 \text{ miles} = 2.4 \times 10^5 \text{ miles}$$

$$\text{time for round trip} = \frac{2 \times (2.4 \times 10^5)}{1.9 \times 10^5} = \frac{4.8 \times 10^5}{1.9 \times 10^5} = 2.5 \text{ sec}$$

ORDER OF MAGNITUDE

Accuracy to one significant figure is sometimes sufficient for rough estimates. Numbers indicating only order of magnitude are indicated by the symbol ~, which may be read "is of the order of." The symbol ~ is also frequently used and read "approximately."

For example, in comparing the diameter of the sun (864,000 miles) to that of the earth (~8000 miles) in order to obtain the relative size of sun and earth, a rough approximation is made; thus

$$\frac{\sim 900{,}000 \text{ miles}}{\sim 8000 \text{ miles}} \quad \text{is} \quad \sim 100$$

It may be stated that the sun is approximately 100 times as large as the earth, or that the sun is of the order of 100 times as large as the earth.

Algebra

Algebra is that branch of mathematics which treats of the relations and properties of quantities by means of letters and symbols instead of numbers. Algebra is essentially generalized arithmetic. The symbolism facilitates performing the operations and permits placing the emphasis on relations.

Algebra is concerned only with addition, subtraction, multiplication, division, powers, and the extraction of roots, in finite numbers.

SYMBOLISM

Any letter can be used as the symbol of any number, known or unknown. Known quantities are usually represented by letters from the beginning of the alphabet (usually a, b, c, d), and unknown quantities by letters from the end of the alphabet (usually x, y, z). Quantities may be positive (+) or negative (−).

In order to suggest the meaning of the symbol, the letter used is often the first letter of the word for the quantity symbolized. For example, p may stand for *p*ressure; t for *t*ime; f for *f*orce. In any discussion the meanings of the symbols should be stated clearly.

Certain symbols and arbitrary signs commonly used in arithmetic are also commonly used in algebra. For example,

$>$ is greater than
$<$ is less than
$\neq$ is not equal to

SUBSCRIPTS

The use of subscripts on symbols is acceptable practice. This helps avoid the use of many different symbols in a discussion and helps differentiate among the symbols. D_E may stand for the diameter of the planet Earth; D_S the diameter of the sun. Similarly, the velocity at the end of 1 sec may be indicated by v_1; at the end of the second second by v_2. Also, the original volume, or volume in the first case, may be indicated with V_0 or V_1; the volume in the second case by V_2.

EQUATIONS

The equation is the central part of algebra. An equation is a mathematical expression of equality. It is established by placing on each side of the "equals" sign quantities that are known or assumed to be equal. The left-hand member is equal to the right-hand member.

SOLVING EQUATIONS

Once an equation is established it may be subject to certain manipulations without destroying the equality between its left- and right-hand members. Equality is not destroyed by any operation performed on one side of an equation, provided that the same or equivalent operation is performed on the other side.

Some common manipulations that will not destroy the equation may be expressed by the axioms that follow.

1. If equals are added to equals, the sums are equal. Examples:

   If $A = B$, then $A + C = B + C$.
   If $X = Y$, then $X + 8 = Y + 8$.
   As $10 = 10$, then $10 + 8 = 10 + 8$.

2. If equals are subtracted from equals, the remainders are equal. Examples:

   If $A = B$, then $A - C = B - C$.
   If $X = Y$, then $X - 8 = Y - 8$.
   As $10 = 10$, then $10 - 8 = 10 - 8$.

3. If equals are multiplied by equals, the products are equal. Examples:

   If $A = B$, then $AC = BC$.
   If $X = Y$, then $8X = 8Y$.
   As $10 = 10$, then $8 \times 10 = 8 \times 10$.

4. If equals are divided by equals, the quotients are equal. Examples:

   If $A = B$, then $\frac{A}{C} = \frac{B}{C}$.
   If $X = Y$, then $\frac{X}{8} = \frac{Y}{8}$.
   As $10 = 10$, then $\frac{10}{2} = \frac{10}{2}$.

5. Like powers of equals are equal. Examples:

   If $A = B$, then $A^2 = B^2$.
   If $X = Y$, then $X^2 = Y^2$.
   As $4 = 4$, then $4^2 = 4^2$,
   or $16 = 16$.
   If $C = 3 \times 10^{10}$, then $C^2 = 3^2 \times (10^{10})^2$,
   or $C^2 = 9 \times 10^{20}$

6. Like roots of equals are equal. Examples (extracting the square roots of both sides):

   If $A^2 = 25 \times 10^{10}$
   $\sqrt{A^2} = \sqrt{25 \times 10^{10}}$
   $A^{2/2} = \sqrt{25} \times \sqrt{10^{10}}$
   $A = 5 \times 10^5$
   Also, $X_1^2 = 16X_2^2$
   $\sqrt{X_1^2} = \sqrt{16X_2^2}$
   $X_1 = 4X_2$

In summary, the basic rule for manipulating equations is that an equation remains an equation (an equality) if the same operation is performed on both the left-hand member and the right-hand member. In the foregoing equations, the quantities such as A, B, C may be complicated quantities that are merely symbolized by those letters.

Solving an equation means getting the unknown quantity by itself on one side of the equality sign. The most common types of equations to be solved in this book are variations of the following:

$$\frac{A}{B} = \frac{C}{D}$$

We shall now give a very simple mechanical rule for solving this equation, and will then illustrate it and explain why it works.

Rule. Any of the quantities in the equation

$$\frac{A}{B}=\frac{C}{D}$$

may be moved provided it is moved across the equality sign *and* across the fraction line; that is, the equation remains an equation under all the following manipulations:

7. Moving B, leaving 1 in its place

$$\frac{A}{B}=\frac{C}{D} \qquad \frac{A}{1}=\frac{BC}{D}$$

8. Moving A,

$$\frac{A}{B}=\frac{C}{D} \qquad \frac{1}{B}=\frac{C}{AD}$$

9. Moving C,

$$\frac{A}{B}=\frac{C}{D} \qquad \frac{A}{BC}=\frac{1}{D}$$

10. Moving D

$$\frac{A}{B}=\frac{C}{D} \qquad \frac{AD}{B}=\frac{C}{1}$$

In order to understand why the foregoing manipulations are valid, note that:

11. Axiom 7 is equivalent to multiplying both members by B:

$$\frac{A}{B}=\frac{C}{D} \qquad \frac{AB}{B}=\frac{CB}{D} \qquad \frac{A\not{B}}{\not{B}}=\frac{CB}{D} \qquad \frac{A}{1}=\frac{CB}{D}$$

Similarly, (8) is equivalent to dividing both members by A; (9) is equivalent to dividing both members by C; (10) is equivalent to multiplying both members by D. You should prove these statements by performing the indicated operations.

If one of these quantities is regarded as unknown, we can solve for it by getting it alone in the *numerator* on one side of the equation.

12. Solving for A,

$$\frac{A}{B}=\frac{C}{D} \qquad \frac{A}{1}=\frac{BC}{D} \qquad A=\frac{BC}{D}$$

13. Solving for B,

$$\frac{A}{B}=\frac{C}{D} \qquad \frac{A}{1}=\frac{BC}{D} \qquad \frac{AD}{1}=\frac{BC}{1}$$

$$\frac{AD}{C}=\frac{B}{1} \qquad B=\frac{AD}{C}$$

14. Solving for C,

$$\frac{A}{B}=\frac{C}{D} \qquad \frac{AD}{B}=\frac{C}{1} \qquad C=\frac{AD}{B}$$

15. Solving for D,

$$\frac{A}{B}=\frac{C}{D} \qquad \frac{AD}{B}=\frac{C}{1} \qquad \frac{D}{B}=\frac{C}{A}$$

$$\frac{D}{1}=\frac{BC}{A} \qquad D=\frac{BC}{A}$$

The student should learn to solve any of these types of equations (12, 13, 14, 15) very quickly "in his head" by inspection, writing down only the answer.

Some other operations commonly employed in rearranging the terms of an equation in order to facilitate solution are in reality simplified applications of some of the foregoing axioms (1, 2, 3, 4).

16. Transposing a term from one side of the equation to the other and changing its sign. Example:

If $X+A=B+C$

and we wish to obtain an expression for X in terms of A, B, and C, we may *transpose* the A to the other side of the equal sign and change its sign, obtaining

$X=B+C-A$

This is equivalent to the axiom, "If equals are subtracted from equals the remainders are equal." For, if

$X+A=B+C$

then subtracting A from each side gives

$X=B+C-A$

Likewise, transposing a negative term from one side to the other and changing its sign to positive is equivalent to "adding equals to equals."

17. *Cross-multiplying,* or multiplying each numerator by the denominator of the opposite side of the equation. Example: If

$$\frac{A}{B}=\frac{C}{D} \qquad \text{then} \qquad AD=CB$$

This is the same as multiplying each side of the original equation by BD; thus,

$$\frac{A}{\not{B}}\times \not{B}D=\frac{C}{\not{D}}\times B\not{D}, \qquad \text{or } AD=CB.$$

By cancellation, there remains $AD = CB$. Therefore, cross-multiplying is the same as "equals multiplied by equals."

Geometry

Geometry is that branch of mathematics in which the relations, properties, and measurement of lines, angles, surfaces, and solids are investigated. A list of common concepts follows (the student is advised to look them up, or to have someone explain them to him, if he does not know them):

SYMBOLS

For example: $\angle$, angle; $\triangle$, triangle; $\perp$, perpendicular; $\parallel$ parallel; $\therefore$, therefore; and the symbols used in algebra.

PLANE FIGURES

Triangle, rectangle, square, circle, ellipse (p. 12).

SOLID FIGURES

Cube, sphere, rectangular parallelepiped (such as a chalk box or a television cabinet with square corners), oblate spheroid (p. 26).

In addition, the student should know the following relations.

PLANE FIGURES

The area of a rectangle = base × height, or $A = bh$.

The area of a square is the square of the side, or $A = s^2$.

SOLID FIGURES

The volume of a rectangular parallelepiped = length × width × height, or $V = lwh$.

The volume of a cube is the length of the side cubed, or $V = s^3$.

CIRCLES

A *circle* is a plane closed figure every point on the boundary of which (called the *circumference*) is the same distance from a point within, called the *center*.

The various elements and divisions of a circle include the following:

A *radius* of a circle is a straight line from the center to any point on the circumference.

A *diameter* is a straight line that passes through the center and ends at any two points on the circumference. The diameter is twice the radius, or $D = 2r$.

A *tangent* is an external straight line that touches the circumference at one point only. A tangent is perpendicular to the radius drawn to the point of tangency.

An *angle* is formed by the intersection of two lines. A central angle is formed by the intersection of two radii at the center of a circle.

An *arc* is a portion of the circumference between two points.

A circle is divided into 360 degrees (360°). A *right angle* (square corner) measures 90 degrees (90°). A right angle is formed by the intersection of two lines that are perpendicular to each other.

Angles and arcs are measured in the same units, usually to the nearest whole number of degrees. Thus a central angle of 30° subtends an arc of 30°; that is, the two radii that form a central angle of 30° cut an arc of 30° on the circumference.

For a circle of any size, the ratio of the length of the circumference C to the diameter D is 22/7, or about 3.1416, and is represented by the symbol π (pronounced "pie" but spelled pi). Thus the *circumference* of any circle is always 22/7 times its diameter.

Because $C/D = \pi$, then $C = \pi D$. Also, $C = 2\pi r$, because $D = 2 \times$ radius r.

The *area of a circle* is equal to π times the radius squared, or $A = \pi r^2$.

An *ellipse* is the path of all points so located that the sum of their distances from two fixed points, called *foci*, remains constant.

SPHERES

A *sphere* (such as a round ball) is a portion of space bounded by a surface such that all straight lines from a point within, the center, to the surface are equal. A sphere is formed when a circle is rotated through 360° about any diameter as an axis.

The *radius* of a sphere is any straight line from the center to the surface.

A *diameter* of a sphere is a straight line that passes through the center and ends at any two points on the surface.

The *area of a sphere is*

$$A = 4\pi r^2$$

The *volume of a sphere is*

$$V = \frac{4}{3}\pi r^3$$

An oblate spheroid is formed when an ellipse is rotated through 360° about any diameter as an axis.

ANGULAR MEASURE

A circle is divided into 360 degrees (360°). One degree (1°) is divided into 60 minutes (60′), and each minute is divided into 60 seconds (60″).

A right angle measures 90°. An *acute angle* is smaller than a right angle and measures less than 90°. An *obtuse angle* is larger than a right angle and measures more than 90° but less than 180°.

Proportionality

Proportionality and the related concepts of ratio and proportion are considered in this section.

RATIO

A ratio is a fraction, one number divided by another. Thus the ratio of 6 to 4 is 6/4. The ratio of 6 to 4, for example, tells how many times 6 is as large as 4. The ratio of 6 to 4 may be written as 6:4; or 4 to 6 as 4:6.

PROPORTION

A proportion is a statement of the equality of two ratios. The equality can be stated as an equation. For example,

$$\frac{6}{4} = \frac{3}{2}$$

The proportion may also be written as 6 : 4 :: 3 : 2. It is read "six is to four as three is to two."

PROPORTIONALITY

Much quantitative thinking and experimentation in science is concerned with establishing a relation between one quantity and another quantity. Equality of ratios may be used. The concept of proportionality is essential.

Often one quantity depends on another quantity. For example, the length of the circumference of a circle is related to the length of the diameter. Simple relations can be expressed in simple mathematical forms. In many cases the relation between two quantities is not a simple one. We should attempt to determine if the relation is "direct" or "indirect" (inverse).

DIRECT PROPORTIONALITY

If one quantity increases as a second quantity increases, the two quantities are said to be "directly" proportional to each other. Also, if one quantity decreases as another quantity decreases, the two quantities are "directly" proportional to each other. (The "direct" proportion may be to the first power; to the second power—that is, squared—or to some other power.)

A good example of a direct proportionality is the relation of the length of the circumference C of a circle to the length of the diameter D. Because the circumference increases as the diameter increases (or the circumference decreases as the diameter decreases), the circumference of a circle is directly proportional to the diameter. This direct proportionality is expressed with symbols as follows:

$$C \propto D$$

which is read "C is proportional to D."

In changing a proportionality to an equation, a constant (some number, perhaps accompanied with units) must be introduced. In arithmetic, for example, $10 \propto 2$. Thus $10 = 5 \times 2$. We always multiply 2 by the constant 5 in order to obtain 10. The proportionality $C \propto D$ is transformed into an equation as follows:

$$C = k \times D$$

where k is the proportionality constant. The value of k in this example has been shown to be pi.

$$\pi = \frac{22}{7}$$

as indicated previously. Thus

$$C = \frac{22}{7} \times D$$

A quantity may be directly proportional to the square of another quantity. The fact that the area A of a circle is directly proportional to the radius squared r^2 can be read from the equation $A = \pi r^2$. Other relationships can be interpreted in a somewhat similar manner.

Two concepts of proportionality can be expressed as an equality of two ratios. As an example, if

$$A_1 \propto B_1$$

and

$$A_2 \propto B_2$$

then

$$\frac{A_1}{A_2} = \frac{B_1}{B_2}$$

Using numbers,

$$20 \propto 4$$
$$10 \propto 2$$
$$\frac{20}{10} = \frac{4}{2}$$
$$2 = 2$$

INDIRECT (INVERSE) PROPORTIONALITY

If one quantity increases as a second quantity decreases (or vice versa), the two quantities are said to be indirectly (inversely) proportional to each other.

For example, if the temperature of an enclosed volume of gas is held constant, what is the relation between the volume V of that gas and its pressure P? The results of experimentation indicate that if the volume is halved, the pressure is doubled; if the volume is decreased to one-third of the original volume, the pressure is tripled. Thus pressure is inversely proportional to the volume:

$$P \propto \frac{1}{V}$$

Mathematically, it can be seen in the foregoing proportionality that a smaller volume V would result in an increased pressure P, and a larger volume would result in a decreased pressure. A smaller number V divided into 1 gives a larger number P, and a larger number V divided into 1 gives a smaller number P.

This inverse proportionality between pressure and volume can be expressed as an equality of two ratios. As an example, if

$$P_1 \propto \frac{1}{V_1}$$

and

$$P_2 \propto \frac{1}{V_2}$$

or

$$\frac{1}{P_2} \propto \frac{V_2}{1}$$

then

$$\frac{P_1}{P_2}=\frac{V_2}{V_1}$$

The equality is verified by numbers recorded as experimental data. (When the volume is halved, the pressure is doubled—an inverse relation.) In the first case, the volume was 26 cm^3 of gas and the pressure of the gas was 76 cm of mercury. For the second case, the volume was 13 cm^3 and the pressure was 152 cm of mercury.

$$\frac{76\text{ cm}}{152\text{ cm}}=\frac{13\text{ cm}^3}{26\text{ cm}^3}$$

So,

$$\frac{1}{2}=\frac{1}{2}$$

An inverse proportion may be to the first power, to the second power, or to some other power. An excellent example is the following: The intensity of illumination I from a light source is inversely proportional to the square of the distance d from the source:

$$I \propto \frac{1}{d^2}$$

$$I=\frac{k}{d^2}$$

The intensity of illumination on a surface is only one-fourth as much if the distance from the surface to the source of light is doubled. Tripling the distance decreases the intensity to one-ninth. This type of proportionality is known as an "inverse-square" proportionality, and this generalization is called the "inverse-square law."

Cases of proportionality more complicated than the "cube" are not considered in this book.

DIRECT AND INVERSE RELATIONS

If two quantities A and B are related so that A increases when B increases and A decreases when B decreases, they are said to be in direct relation. There are many kinds of direct relation; for example, quantity A may be (1) directly proportional to B, (2) directly proportional to the square of B, (3) directly proportional to the cube of B, (4) directly proportional to the square root of B, etc. These may be written as

1. $A \propto B$ or $\frac{A_2}{A_1}=\frac{B_2}{B_1}$

2. $A \propto B^2$ or $\frac{A_2}{A_1}=\frac{B_2^2}{B_1^2}=\left(\frac{B_2}{B_1}\right)^2$

3. $A \propto B^3$ or $\frac{A_2}{A_1}=\frac{B_2^3}{B_1^3}=\left(\frac{B_2}{B_1}\right)^3$

4. $A \propto \sqrt{B}$ or $\frac{A_2}{A_1}=\frac{\sqrt{B_2}}{\sqrt{B_1}}=\sqrt{\frac{B_2}{B_1}}$

In the above, $\propto$ is the symbol meaning "is proportional to," B_1 is one value of B corresponding to the value A_1 of A, and B_2 is a second value of B corresponding to a second value A_2 of A.

Similarly, if one quantity increases as another decreases, they are said to be in inverse relation. There are many kinds of inverse relation; for example, quantity A may be (1) inversely proportional to B, (2) inversely proportional as the square of B, etc. These may be expressed as

1. $A \propto \frac{1}{B}$ or $\frac{A_2}{A_1}=\frac{B_1}{B_2}$

2. $A \propto \frac{1}{B_2}$ or $\frac{A_2}{A_1}=\frac{B_1^2}{B_2^2}=\left(\frac{B_1}{B_2}\right)^2$

etc.

In the foregoing, saying that A is proportional to $1/B$ is the same as saying that A is inversely proportional to B ($A \propto 1/B$).

For two quantities A and B some of the possible relations may be summarized as follows:

Direct relations
A directly proportional to B
A directly proportional to B^2
A directly proportional to B^3
A directly proportional to $\sqrt{B}$
A^2 directly proportional to B^3, etc.

Inverse Relations
A inversely proportional to B
A inversely proportional to B^2
A inversely proportional to B^3
A inversely proportional to $\sqrt{B}$
A^2 inversely proportional to B^3, etc.

INTERPRETING EQUATIONS

Equations express the relation between the quantities involved. The quantity on the left side of an equation is in direct relation to all quantities in the numerator on the right side of the equation and is in inverse relation to all quantities in the denominator of the right side. Consider the equation $B = k(\ell L^3/bd^3)$, where B is the bending, due to load ℓ, of a bar of length L, width b, and depth d; k is a constant depending on the material of the bar. This equation gives us the following relations:

Bending varies directly as the load: $B \propto \ell$
Bending varies directly as the cube of the length: $B \propto L^3$
Bending varies inversely as the width: $B \propto 1/b$
Bending varies inversely as the cube of the depth: $B \propto 1/d^3$

Properly interpreting the relations given by equations enables us to solve problems involving the quantities of the equation.

The idea of a constant of proportionality was used in Chap. 5 with the determination of G in the equation for Newton's law of universal gravitation. The constant permitted proceeding from a proportionality to an equation.

Chapter 4 provides another illustration of the proportionality constant. The distance d a car travels at uniform speed is proportional to the time t:

$$d \propto t$$

The ratio of the measured distance in a given direction to the measured time is the speed v; therefore,

$$\frac{d}{t} = v$$

The proportionality constant is v, the uniform speed,

$$d = vt$$

GRAPHS

Data, or quantities given or determined by experimentation, are plotted as points on a graph grid, and a curve is drawn for these points. The curve is used in the determination of values between the values of the known data. This procedure is shown in Example A.1. An understanding of the procedure used there should make it easier for one to understand and use graphs.

Example A.1

Plot the values of F and d on a graph grid, and draw a curve for the points plotted. What is the value of F in dynes when $d = 8$ cm? The corresponding values of F and d are

F	d
1200 dynes	20 cm
4750 dynes	10 cm
18,600 dynes	5 cm
29,000 dynes	4 cm

Solution

In order to plot quantities on a graph, an x-axis, which is horizontal, and a y-axis, which is vertical, are used. The intersection of these mutually perpendicular axes is the origin 0. Lines drawn parallel to these axes and equidistant from one another form a grid, as shown in Fig. A.1. One of the quantities is then plotted along the y-axis and the other along the x-axis. A proper choice of the scale to be used is important. The idea is to make each division represent a convenient number and to use nearly all of the available space.

Values for F are to be plotted along the y-axis. Because the largest value of F is 29,000 dynes and six divisions are available, a convenient scale would be one division = 5000 dynes. Do not let one division = 29,000/6 = 4833 dynes, because this would be a very inconvenient scale. A convenient scale for d along the x-axis would be one division = 4 cm or one division = 5 cm. Notice,

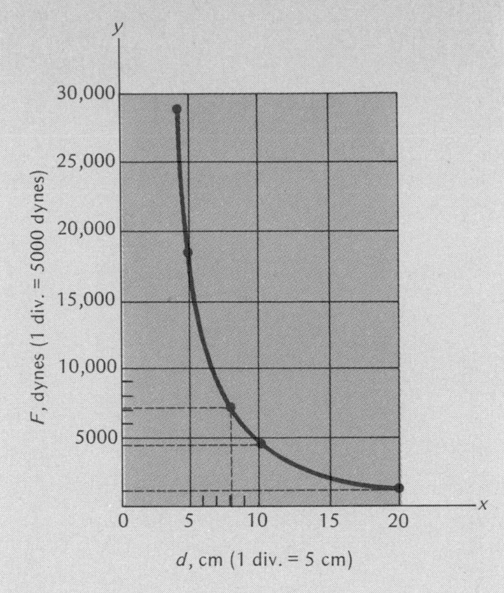

Figure A.1

however, that one division is the same length (1 cm) on each axis. To plot the point $F = 1200$ dynes, $d = 20$ cm, estimate the position of $F = 1200$ along the vertical axis; the point must be horizontally opposite this and vertically up from $d = 20$. This locates the point; a small dot has been drawn to indicate this point. Similarly, over from $F = 4750$ and up from $d = 10$ locate the second point; a dot is used to indicate this point. The other two points are located in the same manner, and a smooth curve is then drawn through the points. Do not connect adjacent points by straight lines but draw the smooth curve that the points as a whole seem to indicate.

The question was asked, "What is the value of F in dynes when $d = 8$ cm?" In Fig. A.1 the broken lines from $d = 8$ to the curve and from the curve to the vertical axis indicate that for $d = 8$ cm, $F = 7200$ dynes.

RELATIONS BETWEEN QUANTITIES

In the solution to Example A.2 we see the general procedure for determination of the mathematical relation between numerical quantities. This involves computing the relation systematically and in a logical order as we proceed from the simple toward the more complex—a good procedure in science and in other fields of human endeavor. We may use such symbols as $\neq$ (is not equal to), $>$ (is greater than), $<$ (is less than), and $\sim$ (approximately), as indicated earlier.

Example A.2

(a) Compute systematically, from the data below, the relation between the quantities P and V. *(b)* State the relation between P and V in words. *(c)* Plot points given by data of *(a)* on a graph grid with V along the horizontal x-axis and P along the vertical y-axis. Draw freehand a smooth P-V curve "through" (but not necessarily touching all) the points for values of P and V. On the graph grid show the value of P for $V = 350$.

P		V	
P_1	12	600	V_1
P_2	27	400	V_2
P_3	47	300	V_3
P_4	109	200	V_4

Solution

(a) There are two possible general relations between P and V: direct and indirect, or inverse. From observation of the corresponding values of P and V in the two columns, it is clear that the relation is indirect (inverse) because as P increases V decreases. Proceed from the simple to the more complex, using any two sets of data such as sets 4 and 1. Begin with values to the first power, then squared, etc.

Try $P \propto 1/V$.

$$\frac{P_4}{P_1} = \frac{V_1}{V_4}$$

$$\frac{109}{12} = \frac{600}{200}$$

$$\sim 9 \neq 3$$

This is read, "approximately nine is not equal to three." Thus P is not inversely proportional to V.

Try $P \propto 1/V^2$

$$\frac{P_4}{P_1} = \left(\frac{V_1}{V_4}\right)^2 \qquad \text{from } \frac{V_1^2}{V_4^2}$$

$$\frac{109}{12} = \left(\frac{600}{200}\right)^2 \qquad \text{or } 3^2$$

$$\sim 9 = 9 \qquad \text{O.K.}$$

It looks as if this is the correct relation. Let us call this "one test" of the relation. Do "one check" and stop if the one check is in agreement with the one test. In scientific laboratories many sets of experimental data are taken and many checks are made. One check is

$$\frac{P_3}{P_1} = \left(\frac{V_1}{V_3}\right)^2$$

$$\frac{47}{12} = \left(\frac{600}{300}\right)^2 \qquad \text{or } 2^2$$

$$\sim 4 = 4 \qquad \text{O.K.}$$

(b) P is inversely proportional to V squared, as determined above and as read from $P \propto 1/V^2$
(c) See Fig. A.2. Notice that one division is the same length on each axis.

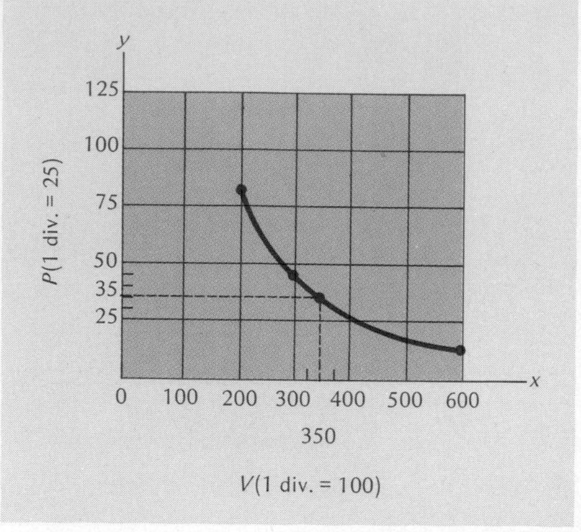

Figure A.2

APPENDIX / NUMERICAL DATA

Metric system

1 meter (m) = 39.37 inches (in.) (a little more than a yard)
1 centimeter (cm) = 0.01 m = 0.39 in. (a little less than $\frac{1}{2}$ in.)
2.54 cm = 1 in.
1 kilometer (km) = 1000 m = 0.62 mile ($\frac{5}{8}$ mile)
1 cubic centimeter (cc) (volume of a cube 1 cm on the side)
1 liter = 1000 cc (very nearly) = 1.06 quarts (qt) (a little more than 1 qt)
1 gram (g) = mass of 1 cc of water (very nearly)
454 g (approximately) = 1 pound (lb)
1 kilogram (kg) = 1000 g (weighs 2.2 lb)
1 metric ton = 1000 kg (weighs 2200 lb)

British system

1 yard (yd) = 3 feet (ft) = 36 inches (in.) = 91.4 centimeters (cm) (a little less than a meter)
1 mile = 5280 ft = 1.6 kilometers (km)
1 pound (lb) = 16 ounces (oz)
1 ton = 2000 lb

Interrelationships of some commonly used units

astronomical unit (AU) = 93,000,000 miles = 149,000,000 km = average distance from earth to sun
angstrom (Å) = 10^{-8} cm = of the order of magnitude of the diameter of atoms
light-year (LY) = 6×10^{12} miles = 9.6×10^{12} km = distance light travels in a year at speed of 186,324 mi/sec or 2.99793×10^{10} cm/sec
atmosphere (atm) = 14.7 lb/in.^{2} = 10^6 dynes/cm^2 = 76 cm of mercury = 760 mm = 1013 millibars (mb)
kilocalorie (kcal) = 1000 calories
calorie (cal) = 4.18 joules = 4.18×10^7 ergs
kilowatthour (kWh) = 3,600,000 joules
electron volt (eV) = 1.6×10^{-19} joules = 1.6×10^{-12} ergs
atomic mass unit (amu) = 931 MeV

Some important physical quantities

diameter of the earth (average) = 7912 miles = 12,740 km
distance to the moon (average) = 239,000 miles = 380,000 km
distance to the sun = 93,000,000 miles = 149,000,000 km
freezing point of water = 0°C = 273°K = 32°F
boiling point of water = 100°C = 373°K = 212°F
speed of light = 2.99793×10^{10} cm/sec or 186,324 mi/sec
Avogadro's number = 6×10^{23} (number of hydrogen atoms in 1 g of hydrogen)
mass of hydrogen atom = 1.66×10^{-24} g
mass of electron = $\frac{1}{1836}$ mass of the hydrogen atom = 9.1×10^{-28} g
Planck's constant $h = 6.625 \times 10^{-27}$ erg-sec

INDEX

74 75 76 77 9 8 7 6 5 4 3 2 1

	SYMBOL	ATOMIC NUMBER	ATOMIC WEIGHT
actinium	Ac	89	**(227)**
aluminum	Al	13	**26.9815**[a]
americium	Am	95	**(243)**
antimony	Sb	51	**121.75**
argon	Ar	18	**39.948**[b,c,d,g]
arsenic	As	33	**74.9216**[a]
astatine	At	85	**(210)**
barium	Ba	56	**137.34**
berkelium	Bk	97	**(247)**
beryllium	Be	4	**9.01218**[a]
bismuth	Bi	83	**208.9806**[a]
boron	B	5	**10.81**[c,d,e]
bromine	Br	35	**79.904**[c]
cadmium	Cd	48	**112.40**
calcium	Ca	20	**40.08**
californium	Cf	98	**(249)**
carbon	C	6	**12.011**[b,d]
cerium	Ce	58	**140.12**
cesium	Cs	55	**132.9055**[a]
chlorine	Cl	17	**35.453**[c]
chromium	Cr	24	**51.996**[c]
cobalt	Co	27	**58.9332**[a]
copper	Cu	29	**63.546**[c,d]
curium	Cm	96	**(245)**
dysprosium	Dy	66	**162.50**
einsteinium	Es	99	**(249)**
erbium	Er	68	**167.26**
europium	Eu	63	**151.96**
fermium	Fm	100	**(255)**
fluorine	F	9	**18.9984**[a]
francium	Fr	87	**(223)**
gadolinium	Gd	64	**157.25**
gallium	Ga	31	**69.72**
germanium	Ge	32	**72.59**
gold	Au	79	**196.9665**[a]
hafnium	Hf	72	**178.49**
hahnium[h]	Ha	105	**(260)**
helium	He	2	**4.00260**[b,c]
holmium	Ho	67	**164.9303**[a]
hydrogen	H	1	**1.0080**[b,d]
indium	In	49	**114.82**
iodine	I	53	**126.9045**[a]
iridium	Ir	77	**192.22**
iron	Fe	26	**55.847**
krypton	Kr	36	**83.80**
kurchatovium[h]	Ku	104	**(261)**
lanthanum	La	57	**138.9055**[b]
lawrencium	Lr	103	**(257)**
lead	Pb	82	**207.2**[d,g]
lithium	Li	3	**6.941**[c,d,e]
lutetium	Lu	71	**174.97**
magnesium	Mg	12	**24.305**[c]
manganese	Mn	25	**54.9380**[a]
mendelevium	Md	101	**(256)**
mercury	Hg	80	**200.59**
molybdenum	Mo	42	**95.94**
neodymium	Nd	60	**144.24**
neon	Ne	10	**20.179**[c]
neptunium	Np	93	**237.0482**[b]
nickel	Ni	28	**58.71**
niobium	Nb	41	**92.9064**[a]
nitrogen	N	7	**14.0067**[b,c]
nobelium	No	102	**(254)**
osmium	Os	76	**190.2**
oxygen	O	8	**15.9994**[b,c,d]
palladium	Pd	46	**106.4**
phosphorus	P	15	**30.9738**[a]
platinum	Pt	78	**195.09**
plutonium	Pu	94	**(244)**
polonium	Po	84	**(210)**
potassium	K	19	**39.102**
praseodymium	Pr	59	**140.9077**[a]